Rules and Exceptions in Biology: from Fundamental Concepts to Applications

Alfredo V. Peretti • Lucía S. Calbacho-Rosa
Paola A. Olivero • Mariela A. Oviedo-Diego
David E. Vrech

Rules and Exceptions in Biology: from Fundamental Concepts to Applications

Springer

Alfredo V. Peretti
Facultad de Ciencias Exactas Físicas y Naturales
Universidad Nacional de Córdoba. Instituto de Diversidad y Ecología Animal (CONICET – UNC)
Córdoba, Córdoba, Argentina

Lucía S. Calbacho-Rosa
Facultad de Ciencias Exactas Físicas y Naturales
Universidad Nacional de Córdoba. Instituto de Diversidad y Ecología Animal (CONICET – UNC)
Córdoba, Córdoba, Argentina

Paola A. Olivero
Facultad de Ciencias Exactas Físicas y Naturales
Universidad Nacional de Córdoba. Instituto de Diversidad y Ecología Animal (CONICET – UNC)
Córdoba, Córdoba, Argentina

Mariela A. Oviedo-Diego
Facultad de Ciencias Exactas Físicas y Naturales
Universidad Nacional de Córdoba. Instituto de Diversidad y Ecología Animal (CONICET – UNC)
Córdoba, Córdoba, Argentina

David E. Vrech
Facultad de Ciencias Exactas Físicas y Naturales
Universidad Nacional de Córdoba. Instituto de Diversidad y Ecología Animal (CONICET – UNC)
Córdoba, Córdoba, Argentina

ISBN 978-3-031-55380-6 ISBN 978-3-031-55382-0 (eBook)
https://doi.org/10.1007/978-3-031-55382-0

This Springer imprint is published by the registered company Springer Nature Switzerland AG
The registered company address is: Gewerbestrasse 11, 6330 Cham, Switzerland

Paper in this product is recyclable.

To students and professionals curious to question rules and value exceptions in biology

To our families

Preface

Science is a changing system in which many currently accepted paradigms were considered inadequate in the past. Likewise, it is accepted that well-established explanations may be greatly modified in the near future. Many fields in the life sciences, such as ecology, ethology, and evolution, are good examples of such change. Robust explanations result from the identification of patterns that reflect general and well-established rules. However, it is often the case that other patterns are exceptions or remain unclear.

It is crucial to clearly recognize the differences between what is established as a general pattern and what is an exception, as interpretations in both cases have different implications from an evolutionary perspective. In the last 20 years, some aspects of rules and exceptions have been studied in some organisms or contexts. However, they have been partially analyzed. For example, by focusing only on laws or the value of some abnormalities (teratologies), including useful biographies of important scientists in the field. Reviews or articles also evaluated these components, rules, and exceptions separately and limited to some papers on the theory of biological laws and ecogeographic rules.

From this promising current scenario, we decided to work on a book that could not only help fill a niche in the scientific education of pre- or postgraduate students, but also help generate useful and valuable discussions among scientists working in the life sciences and related disciplines. We explore rules and exceptions in biology, and the criteria that have been used to establish such definitions and boundaries. We analyze the dynamics between rules and exceptions, both due to changes in nature and in scientific thinking. We also assess how criteria and concepts have changed within disciplines, and what implicit biases may exist.

Studies of many groups of organisms are used as examples to illustrate some of this interactive relationship and to provide new information to confirm, limit, or refute concepts based on rules and exceptions in science. It is our hope that this book will be a source of knowledge about how two seemingly opposite situations, rule and exception, flow and interact from different scales of time, space, and taxa in nature. This book took many years to write, mainly because we had to learn aspects of biology not necessarily related to our specialty. As invertebrate zoologists, especially on arthropods, we can sometimes have a bias of examples, but without neglecting globality. In fact, the examples in the book are not exhaustive, but illustrative of the vast subject matter covered.

This book is intended for an international scientific audience: undergraduate students (i.e., those doing honors projects), graduate students, postgraduate students, and professional scientists. The book could be used as a reference source of ideas, perspectives, and critical views of new directions for students of life sciences. Any biologist, as well as scientists in related disciplines (e.g., biomedical researchers, psychologists) with broad evolutionary interests, would be interested in the material contained in the book.

Regarding the structure of the book, the first two chapters of the book are basic introductions to the diversity of definitions of rules and exceptions in science and other disciplines. They also relate the concepts of biological rules and exceptions to ideas in epistemology and the philosophy of science. The third chapter explores the basis of the dynamics between rules and exceptions from temporal, spatial, and group perspectives, and the main directions of change between rules and exceptions in different temporal, spatial, and organismic scenarios.

From Chap. 4–7, we explore situations of change between rules and exceptions with illustrative examples from different areas of biology. Chapter 4 analyzes the situation when a rule persists as a rule, both in biology and in scientific thinking. Chapter 5 explores the world of exceptions. Because of the common underestimation of exceptions and their value in biology, this chapter, together with Chap. 6, which examines when exceptions become rules, is probably the heart of the book. Indeed, it is where we emphasize the most. Chapter 7 looks at situations where rules become exceptions. Chapter 8 presents illustrative cases of rule-exception transitions beyond biology, including other sciences, the arts, and other areas of human activity.

Chapter 9 offers a final overview and outlook, with some interesting tips and suggestions for scientific approaches and practices.

Córdoba, Córdoba, Argentina

Alfredo V. Peretti
Lucía S. Calbacho-Rosa
Paola A. Olivero
Mariela A. Oviedo-Diego
David E. Vrech

Acknowledgments

Our first thanks go to Lars Kroener, Springer, who received and supported this proposal throughout its long development. This book began to take shape after A. Peretti was invited to give a presentation on rules and exceptions in sexual selection at a Brazilian Ethology Congress in 2010, which later resulted in a chapter on the subject. Our first thanks go to Regina Macedo and Glauco Machado, who helped with this first big step. Special thanks also go to Anita Aisenberg, who facilitated the presentation of the first drafts of this book at the Uruguayan Zoology Congress in 2012. Thanks also to Carlos Cordero for encouraging A. Peretti to give a plenary lecture on rules and exceptions in behavior at an international behavior course in Mexico in 2015. We are very grateful to Alejandro Córdoba-Aguilar and Bill Eberhard for supporting the realization and writing of this literary adventure.

We also thank our colleagues at the Laboratory of Reproductive Biology and Evolution (LABRE, IDEA, CONICET-UNC) for their input and feedback with extremely helpful comments during the writing and reading of the book, especially Franco Cargnelutti. Special thanks are due to those who provided us with original photographic material from their research, including Matías Izquierdo, Darko Cotoras, Andrea Coccuci, Gabriel De Simone, Otto Bylén Claesson, Sergio Rodríguez, and Franco Peretti-Plazas, who helped in the preparation of some of the first versions of the figures. We are also thankful to the following people for comments and suggestions on ideas and contents of this book: Franco Cargnelutti, Matias Izquierdo, Alejandra de Lourdes Ceballos, Anita Aisenberg, Eileen Hebets, and Alex Córdoba-Aguilar. We are indebted to our families for their immeasurable support throughout this work.

We gratefully acknowledge the support of the Consejo Nacional de Investigaciones Científicas y Tecnológicas (CONICET) of Argentina, the Secretaría de Ciencia y Técnica de la Universidad Nacional de Córdoba (SECYT), and the Agencia Nacional de Promoción Científica y Técnica de la Argentina (ANCYPT).

Contents

About the Authors

Alfredo Vicente Peretti is Principal Researcher of the Consejo Nacional de Investigaciones Científicas y Técnicas de la Argentina, Instituto de Diversidad y Ecología Animal (CONICET-IDEA) and Associate Professor of the Departamento de Diversidad Biológica y Ecología, Facultad de Ciencias Exactas Físicas y Naturales, Universidad Nacional de Córdoba, Argentina. His main area of interest and research is zoology, with emphasis on invertebrates. He works in behavior, ecology, and evolution, particularly in the field of reproductive biology using arthropods as models.

Lucia Calbacho-Rosa is Associate Researcher of the Consejo Nacional de Investigaciones Científicas y Técnicas de la Argentina, Instituto de Diversidad y Ecología Animal (CONICET-IDEA). Her main areas of research are behavior, reproductive biology, sexual selection, and studies of genital morphology using arthropod species as a model.

Paola Andrea Olivero is an Assistant Researcher of the Consejo Nacional de Investigaciones Científicas y Técnicas de la Argentina, Instituto de Diversidad y Ecología Animal (CONICET-IDEA) and Associate Professor in general biology at the Facultad de Ciencias Exactas, Físicas y Naturales of the Universidad Nacional de Córdoba, Argentina. Her research field is related to the study of the reproductive biology, analyzing the behavior during courtship and patterns of differential paternity in arachnids and insects.

Mariela Oviedo-Diego is Postdoctoral Fellow at the Consejo Nacional de Investigaciones Científicas y Técnicas de la Argentina, Instituto de Diversidad y Ecología Animal (CONICET-IDEA) and vocational teacher in Entomology at the Facultad de Ciencias Exactas, Físicas y Naturales, Universidad Nacional de Córdoba, Argentina. Her research has focused on sexual behavior, the study of pre- and postcopulatory characters and reproductive interference in different arachnid species.

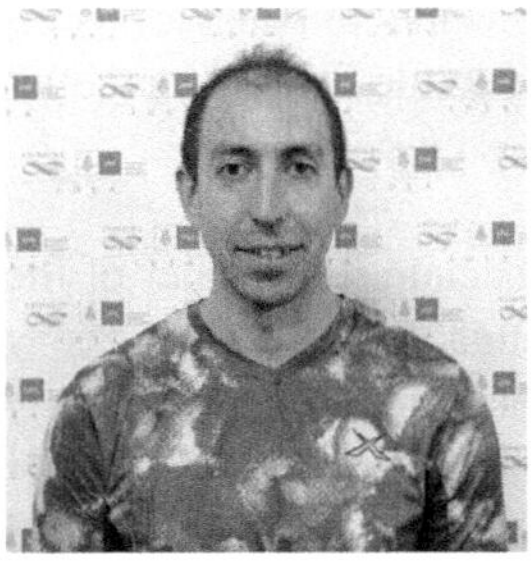

David Eduardo Vrech is Associate Researcher at the Consejo Nacional de Investigaciones Científicas y Técnicas de la Argentina, Instituto de Diversidad y Ecología Animal (CONICET-IDEA) and Assistant Professor specializing in invertebrate biology at the Facultad de Ciencias Exactas, Físicas y Naturales, Universidad Nacional de Córdoba, Argentina. His primary research areas encompass sexual selection, reproductive biology, and the investigation of sperm competition across arachnid groups.

Part I
Establishing Criteria and Theoretical Possibilities

Chapter 1
A Big World of Rules?

Everyday life weaves us, daily, a spider's web in our eyes

—Oliverio Girondo

Contents

A. V. Peretti et al., *Rules and Exceptions in Biology: from Fundamental Concepts to Applications*, https://doi.org/10.1007/978-3-031-55382-0_1

Abstract To establish criteria and theoretical possibilities for rules and exceptions in biology, it is helpful first to analyze what a rule is, as well as its many variants. This chapter presents some hierarchical levels of rules in biology, with examples of the most comprehensive and the most restrictive. In both contexts, we discuss the concepts of laws, rules, and patterns and how these boundaries are not always so clear. The topic of the search for universal laws in biology is introduced. In this context, the phenomena of emergence, biological complexity, and others are mentioned. We analyze when there are smaller rules, limited to a certain field within biology. We see as some introductory examples when there are rules of behavior, ecology, and evolution. We also discuss the links between rules and epistemology, especially regarding paradigm shifts.

1.1 Introduction

The "plasticity" of ideas, hypotheses, and theories in science is a fact. As Kuhn (1996) classically analyzed, the so-called paradigms change over time. Science is the opposite of a static system, or at least it should be, as dogmas tend to remain unchanged in time and space. There is a consensus in the scientific community that changes are bound to occur in science. However, the open-mindedness to such changes, which implicitly recognizes the flexibility of knowledge, differs among the different areas of expertise. The "Kuhnian" basis of the construction of a paradigm, its subsequent acceptance during a particular time (i.e., normal science), and finally, its crisis that will lead to its fall and replacement by new ideas (i.e., scientific revolution) (Kuhn, 1996; Bussard, 2005) are a fluid cycle. This change is understood and accepted, but with different tones, emphases, or perspectives in each discipline. The history of science in general, and biology in particular, is full of examples of information sets and concepts that became paradigms at one time in the past and are no longer valid today.

Likewise, much of the information obtained throughout humanity's history was due to the development of science. The contribution of scientific knowledge led to currently accepted ideas and concepts that might have been rejected or not even imagined in the past. This fact is evident in areas of modern biology such as genetics, molecular and evolutionary biology, nanotechnology, and quantum physics. A classic example is the field of sexual selection, which was almost "dormant" for approximately a century after the publication of the first edition of "The Origin of

Man and Selection in Relation to Sex" (Darwin, 1871). This oblivion, in large part, was due to the moral conditioning of the time about the view of "sex" and, therefore, its role as a critical factor in the evolution of living beings (Gould & Gould, 1989; Birkhead, 2000). Sexual selection was revisited in animals after several years, and its presence in plants has been almost ignored until very recently (Andersson, 1994; Moore & Pannell, 2011). The change of view on a scientific concept from being "not acceptable" to "acceptable" and vice versa, as we shall see, may be due to multiple reasons. Among them, the nature of biology itself, specifically biological evolution, may have played a preponderant role. In other situations, the change is more linked, as we said, to modifications in scientific interpretation due, for example, to advances in research techniques and technologies, differences in social biases, and schools of thought, among others.

1.2 What Is a Rule?

In this context, how can we define a rule? Definitions are inherently challenging as they frequently lack an ingredient, making them somewhat odious, but here, they become necessary for what we will analyze in this book. As we will see, allusions to the rule often link to its antonym; in other words, they are more defined by what it is not than by what it is. In this problem, it is interesting to keep in mind that for a given fact or concept, we will have, on the one hand, the rule and, on the other hand, what is an exception to it. Undoubtedly, there is no single definition of "rule" (Box 1.1). Nevertheless, for practical purposes, we can summarize a rule as "*a generalized statement that describes what is true in most or all cases*" (The American Heritage®, Dictionary of the English Language, 2003). The notion of the rule leads us almost directly to the definition of "scientific law" as "*a scientific statement that claims a constant relationship between two or more variables or factors, each of which represents (at least partially and indirectly) a property of concrete systems*" (Honderich, 1995). It is crucial to consider that, depending on the scientific discipline, there will be a plethora of terms closely associated with the term rule, such as "norm," "pattern," and "law" (Box 1.1). Indeed, they can be synonymous, although each meaning will have its own level of "applicability," as we will see below. In the natural sciences, a scientific law is "*a rule that relates events that have a joint, usually causal, occurrence.*"

BOX 1.1 Glossary and synonyms of rule

Glossary

Axiom: a statement or proposition which is regarded as being established, accepted, or self-evidently true.

Law: a statement of fact, deduced from observation, to the effect that a particular natural or scientific phenomenon always occurs if certain conditions are present. Generalization based on a fact or event perceived to be recurrent.

Norm: something that is usual, typical, or standard.

Paradigm: a typical example or pattern of something. [science paradigm: universally recognized scientific achievements that, for a time, provide model problems and solutions to a community of practitioners].

Pattern: a regular and intelligible form or sequence discernible in the way in which something happens or is done.

Precept: a general rule intended to regulate behaviour or thought.

Principle: a fundamental truth or proposition that serves as the foundation for a system of belief or behaviour or for a chain of reasoning. A general scientific theorem or law that has numerous special applications across a wide field.

Rule: one of a set of explicit or understood regulations or principles governing conduct or procedure within a particular area of activity.

Theory: a supposition or a system of ideas intended to explain something, especially one based on general principles independent of the thing to be explained. A set of principles on which the practice of an activity is based.

Synonyms

model, maxim, theorem, universal, mean

habitual, frequent, ordinary, usual, common, regularity

generalisation, typical, current, system, continuous

equilibrium, natural, guideline, measure, standard

While rules play a crucial role in constructing scientific theories, their enunciation in the biological sciences is generally less frequent compared to other scientific disciplines such as chemistry and physics (Fig. 1.1a). Dhar & Giuliani (2010) arrange biological, chemical, and physical events into trait levels and, along with them, the rules or laws of those levels (Fig. 1.1b). The enunciation and acceptance of rules apply to basic levels, such as the atomic or cellular levels. However, they can also be done for higher levels involving organisms or interactions between them (e.g., ecology and communities).

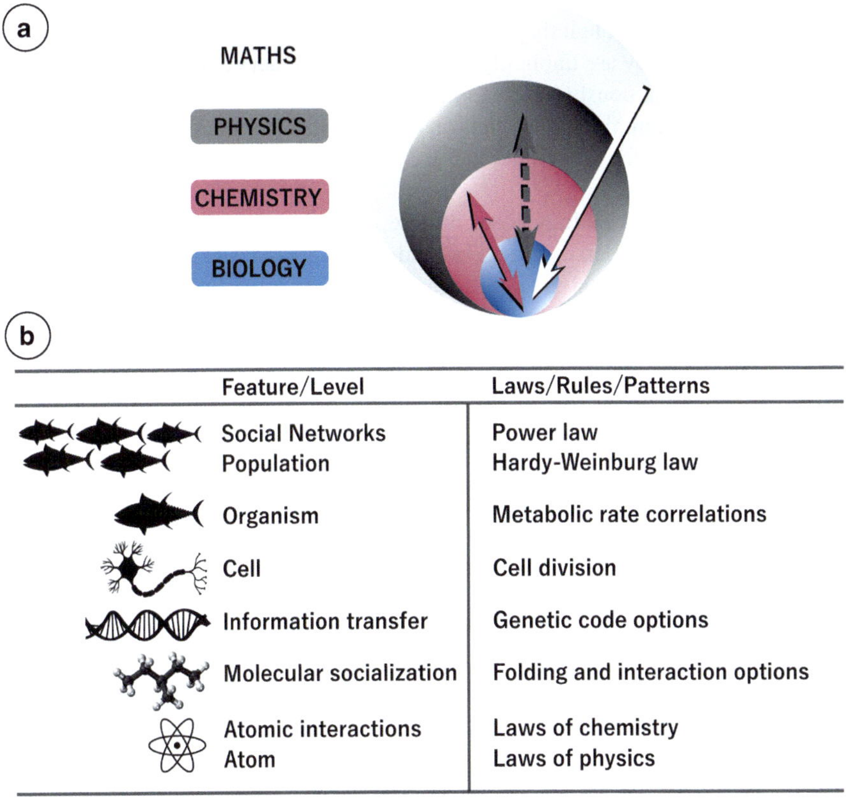

Fig. 1.1 Levels of analysis in science. (**a**) Interrelation between scientific disciplines with different degrees of inclusiveness/competence in their rules. (**b**) Level of organization associated with the level of rules and the scientific discipline that includes them. License: own creation. Free images from Pixabay. (Modified from Dhar & Giuliani, 2010)

One might question whether there are scientific disciplines devoid of rules or those that are so dynamic over time that articulating them becomes a waste of time. A narrow approach could limit the application of laws only to the so-called "hard" sciences, such as physics, chemistry, or mathematics. Indeed, it is often postulated that biology need not have any fundamental laws of its own since all the basic laws of biology are simply a subset of the fundamental laws of physics (Wayne, 2020). However, biology presents itself as a complex of disciplines and subdisciplines that have been and still are difficult to catalog, summarize, or encompass in just laws of physics. Many are bodies of hypotheses and theories that are gradually being demonstrated.

In many cases, although they do not acquire the name of "law" in common scientific language, they are implicitly taken as such in practice. In other words, basic principles are continuously sought in biology beyond physics (e.g., laws of thermodynamics). For example, the integration of biological phenomena in an evolutionary framework of living beings and the selection processes involved (natural, sexual).

On the other hand, biology poses diverse, complex, changing, and often unpredictable scenarios. Understanding how random historical events shape evolutionary processes is a critical goal of evolutionary biology (Harms & Thornton, 2014). In this context, Gould's (1989) reviews of the phenomenon of contingency in biology (i.e., what may or may not happen in nature), such as meteorite falls and volcanic eruptions, are classic. Some of these events have been key in biological evolution, for example, in great extinctions. To this must be added the intrinsic contingency, such as factors inherent to the molecules that make up the organism and the low probability of repeating the same mutations that have formed them or helped them to their current behavior and sensitivity. Biology differs from physics in this respect as to what could happen if history were rewound about what to foresee or not and whether the same chain of random events would repeat itself.

In this sense, evidence shows that the inherent complexity of proteins makes evolution deeply dependent on intrinsic random events of very low probability. This is the conclusion that Harms & Thornton (2014) arrived at after analyzing the evolution of cortisol specificity in the ancestral glucocorticoid receptor. After examining numerous potential evolutionary paths for the protein, they discovered no alternative mutations that could have given rise to the current protein, apart from the two tolerant mutations it already possesses. They concluded that the evolution of the ancestral glucocorticoid receptor was strongly dependent on improbable and nondeterministic events, and this contingency arose from the intrinsic biophysical properties of the protein. We will return to the phenomenon of contingency throughout this chapter.

As far as we know, biological concepts and predictions apply only to our planet. The same is not true of physics and perhaps chemistry, whose more "encompassing" laws allude to contexts beyond the confines of the Earth. Exobiology, whose main objective is the study of the possibilities of extraterrestrial life, tries to propose analogous situations in other worlds from the knowledge of our biological history and evolution. However, it is limited because the options could vary in different places of the universe. That is where scientific speculation can border on fantasies, with hypotheses that are not falsifiable or difficult to test. Nevertheless, exobiology exists as a discipline and is valued for its potential to attempt such extrapolation in a very *sui generis* scientific framework.

Given the abundance of exceptions in biology, it becomes pertinent to pause and ponder upon a seemingly self-evident question: Is it meaningful to formulate, establish, and apply laws, as well as their more nuanced manifestations such as rules, patterns, or regularities? This perspective would be aligned with that of an orthodox physicist who might harbor some skepticism toward biology due to the challenge of identifying general laws in this field of science directly and above all, devoid of exceptions or with exceptions that are minimal in nature.

The truth is that biology is a "flexible" science because its object of study, living beings, demands this flexibility. However, it is not because of this characteristic that it does not fail to recognize the existence of patterns and principles, which translate into different levels of rules or laws. One cannot expect, naively perhaps, the absence of exceptions. However, we will see below specific, very comprehensive, or general laws in biology where exceptions should not be expected to exist.

For any attempt to propose rules in biology, in addition to the inherent flexibility, it is important to consider that the primary driver of the evolutionary process in biology is the interpopulation variability within a species, primarily arising from the occurrence of random mutations.

In that case, we are facing a scenario more challenging than in the so-called "hard sciences" but also a frustrating one if we want to meet it with the same methodologies, epistemological approaches, and study perspectives. If we add to this "cocktail" the different "layers" or levels of study in biology (e.g., from the microscopic to the macroscopic), and not only from the cellular to the organismic but also at the level of communities and ecosystems, the challenge we face is much more significant. Therefore, we have to approach biology's study of rules, patterns, and laws with a less orthodox and more inflexible view and perspective. We must maintain scientific rigor in their accuracy, evaluation, and application, but we must be adaptable to the presence of variants of the rules that we can establish. However, it is crucial to avoid falling into a false dichotomy that either denies the existence of principles or indiscriminately accepts everything with exceptions. It is important to strike a balance by critically evaluating generalizations that may be incorrectly imposed, such as extrapolating findings from one model of a living being to another. By doing so, we can avoid overlooking critical analysis and ensure that principles are appropriately formulated and applied.

On this subject of what is a rule, it is good to remember that Charles Darwin, in The Origin of Species, already deals with the issue of laws in nature. Today, we can say that, since that work and subsequent associated articles, the theory of natural selection is a central axis in any idea about biological evolution (something that we will return to in the following paragraphs when we refer to laws of biology at more globalizing levels). Darwin recalls an ancient principle of natural history: "*Natura non facit saltum*" (nature does not proceed by leaps—, i.e., nature varies continuously and not in an abrupt manner), to reinforce that the theory of natural selection would apply both to present-day organisms and to past times, known or unknown.

Darwin emphasizes that organic beings were formed according to two great laws: "unity of type" and "conditions of existence." The first alluded to the fact that there is a general concordance in the structure observed in beings of the same kind, while the second law alludes to the conditions in which these beings inhabit; that is, it is directly connected with the adaptation of organisms to their abiotic and biotic environments. He then points out that this second law is wholly included in the "principle of natural selection." He concludes by saying, "*the law of the conditions of existence is the superior law because, through the inheritance of previous variations, it includes the law of the unity of type*." In essence, this perspective establishes a hierarchy between these two laws, expressing an opinion that one law encompasses the other, as they are mutually causative and consequential. This comprehensive viewpoint, specifically in the context of evolution, allows for a more holistic understanding. In addition to reinforcing his postulate of natural selection, the most exciting thing is that he reinvents these laws, reinterprets them, and does not discard them; in any case, he explains them better with his study and innovative perspective. He discarded dogmas that touch upon concepts such as the divine origin of organic beings, recognizing them not as laws but rather as outdated beliefs.

Let us move on a hundred years later. We have another great naturalist, postulating principles. However, these principles take the form of "problems" or questions, serving to address specific uncertainties prevalent during the study of animal behavior at that time. It is worth noting that such "problems" are inherent to any field of biology, reflecting common challenges faced in scientific inquiry. The naturalist's name was Niko Tinbergen, the great Dutch ethologist, who, in the early sixties, postulated the "four problems" for studying animal behavior (Tinbergen, 1963). A couple of years earlier, Ernst Mayr had already laid the foundations in his look at proximate and ultimate causes, extended to ecology and the potential for any study of biological phenomena (Mayr, 1961). Within this framework of seeking regularities or laws through the examination of problems, the study of animal behavior encompasses a wide range of perspectives and methodological approaches. These approaches vary depending on the specific questions being addressed, such as investigating proximate causes related to physiology and ontogeny or exploring ultimate reasons connected to adaptive function and evolution. The distinction between proximate causes (how and when) and ultimate causes (why and wherefore) guides the diverse avenues of research within the field (Bateson & Laland, 2013).

These differences in approach between the immediate and nonimmediate, although interdependent (Sherry, 2005), mean that among biologists themselves, there may be certain biases or criticisms depending on the type of questions their study addresses. Studies with questions focused on patterns and mechanisms that are more "exact" to observe and measure (e.g., proximate causes) will have a different value than those with "less exact" questions or hypotheses that are more difficult to falsify (e.g., ultimate causes). For example, a physiologist focused on how a trait is expressed during ontogeny would consider his conclusions more robust than those reached by a behavioral ecologist focused on the adaptive function of that trait. On the other hand, the latter would see the conclusions of the former as more limited and perhaps reductionist. Thus, specific rules could be considered "stronger" (would they add up to a few exceptions?) in disciplines oriented to studying proximate causes, such as molecular biology and physiology, more directly influenced by physics and chemistry. However, mistakenly, those enunciated for the organismic level and ultimate causes (e.g., certain branches of ecology, taxonomy, natural history, evolution) may be seen as "less strict" rules (can they add up to more exceptions?). Fortunately, this has been changing over time. There are already ample examples of how immediate and ultimate causes interrelate and nurture each other in approaches and information (Kirschner & Gerhart, 2006) and make the orthodoxy of a different valuation of one or the other no longer so helpful today (Laland et al., 2011, 2013).

Our planet functions as a whole, an interactive system whose key components are living beings. This perspective, as we know it today, was contemplated by Lovelock's famous "Gaia hypothesis" (1985), which states that life can transform the biosphere, fostering and maintaining environmental conditions suitable for itself, with modification of the environment. Beyond the scope, controversies, and limitations of this hypothesis, it is important to address the fact that what we know about biology would necessarily have to consider in any study the existence of interaction within an "integrated whole." Thus, wanting to dissect biology into rules and concepts that define it is challenging, and there is a significant risk of falling into reductionism.

Thus, biology invites us to incorporate other visions for a more critical understanding. In that context, we will then examine the phenomenon of "emergence," which implies that there will be properties at certain levels of organization that cannot be predicted simply by the properties exhibited at lower levels (Emmeche et al., 1997). For many, this is a valuable perspective on tracing universal patterns to all of biology, in particular, because it connects to a more holistic view of nature. For example, holism has been embedded in developing new disciplines and theoretical frameworks in recent decades, such as evo-devo and phenotypic plasticity (Pigliucci, 2014). In this book, we will go a step further, examining the nature and role of exceptions in detail. We will describe whether the presence of rules is fundamental for the biological system studied. In addition, we will review whether some instances were omitted as part of rules due to errors in our appreciation of what the rule should contemplate.

1.2.1 Some Basic Hierarchical Levels of Rules in Biology

We can only see some of the rules at the same level. There are differences in that each rule encompasses specific given biological phenomena or groups of phenomena (note: understand the term "phenomenon" as referring to a pattern, a mechanism, or a process, as the case may be). Thus, we have different levels of rules according to their degree or limit of competence. For this reason, it would not make sense to consider a rule that explains everything. In any case, it would do so from a general perspective since later, when diving in detail into the components, we would have other, more specific rules. For example, as shown in Fig. 1.1, we could start from integrative laws in biology, such as the group of laws of Mendelian inheritance, to laws or rules about specific aspects of nature, such as the almost fixed patterns of reproduction of a given clade or ways of performing instinctive-based behavior. We will see later how these rules can still vary and not be so framed.

Between one rule of a globalizing level and another more restricted one, there is a potentially broad spectrum of other rules for the intermediate levels. In other words, the jump or interval from one level to another often ends up being "operative." This criterion means that these ranges and limits are generally established by the researchers themselves, markedly influenced by the variability existing in living beings (variability embodied mainly as the "bricks" necessary as a basis for biological evolution). Thus, plotting or diagramming the interaction between levels or their degree of inclusion can be challenging, and any attempt, whether illustrative or mathematical, will be minimal. However, as seen in Fig. 1.1, one way is to plot these interactions as concentric circles from one level to another. As we will see later, this graph has its basis in one of the classic ideas of scientific thought, its construction, and change, which Imre Lakatos put forward. Later, we will see that this illustration is limited when we examine the differences and dynamics between rules and exceptions. However, it is valid, as far as this first excursion into the world of rules is concerned, where we will first focus on the most global levels (level I). It is often not easy for scientists to define which laws we could unfailingly postulate as "safe" for biology, which we can say encompass biology as a whole.

It has been mentioned that only three fundamental laws can be established for biology (Box 1.2) since these would undoubtedly cover the totality of living beings (Trevors & Saier Jr., 2010). Interestingly, the first of these laws asserts that biology is not separate from other sciences that provide the foundational knowledge for it, specifically chemistry and physics (Wayne, 2020). On its own, this first law of biology would not enable an extraterrestrial entity to comprehend the essence of what characterizes biology within our living system. However, an implicit law holds value in being explicitly stated.

BOX 1.2 The three "infallible" laws of biology

The First Law of Biology: all living organisms obey the laws of thermodynamics

First corollary ⟶ life requires the temporary creation of order, so resource utilization by living organisms increases world entropy

Second corollary ⟶ an organism at biochemical equilibrium is dead. Living organisms are in disequilibrium with their environment

The Second Law of Biology: all organisms consist of membrane-encased cells

First corollary ⟶ cell is the only structure that can grow and divide independently of another life form

Second corollary ⟶ genetic instructions are required for maintenance of life

The Third Law of Biology: all living organisms arose in an evolutionary process

First corollary ⟶ living organisms contain homologous macromolecules (DNA, RNA, proteins) that derived from a common ancestor

Second corollary ⟶ genetic code is universal

Two approaches to discover new regularities and laws

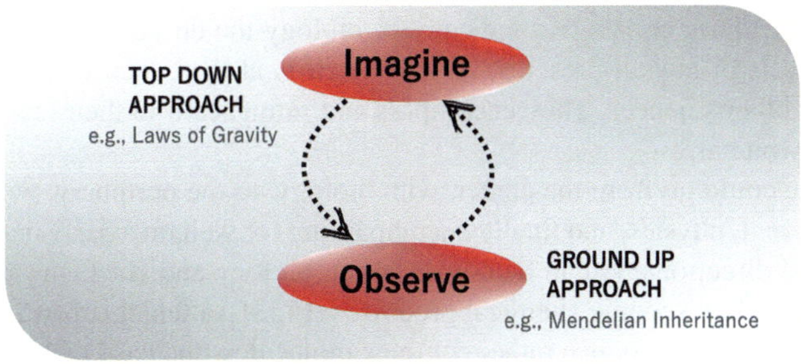

Modified from Trevors & Saier Jr. (2010) and Dhar & Giuliani (2010). License: Own Creation

The second of these universal laws of biology is that all living things are covered by membranes. Here, it already focuses on something very characteristic of biology by mentioning systems defined by their coverings, as is the case with cells and membranes. This biological characteristic does not deny, of course, that physics is excluded. Note that there is still no mention of DNA or RNA, i.e., a law that does not highlight genetics. It could well have been posed as a universal law, also one that we mention that all organisms are formed in their genetic composition by one and the other. Nevertheless, that has not been the case.

The last of the laws makes explicit that all living beings have appeared through an evolutionary process. It may be the law that reading it already reflects the existence of more than a form, a process that contains all living beings without exception, biological evolution (this, of course, also has a basis in physics and chemistry). In other words, no organism arises by spontaneous generation, the creation of a superior being, or appears from nothing by pure chance (even if this is an essential ingredient before the beginning of the process when mutations are generated by chance).

After seeing these three laws of biology, one is left with the desire to try to establish another one of equal importance, such as the genetics we mentioned. In any case, the last of the laws indirectly contemplates the genetic basis in the evolutionary process, although not explicitly in its statement. Therefore, it is a problem to limit biology to the scope of other basic sciences, such as physics and chemistry; in fact, it is much more than the action or "sum" of these. It does not deny them, but its dynamism exceeds both.

In this framework, biological theories are undoubtedly put forward, and they add up to a level almost parallel to that of laws or principles in the aforementioned "level I." For example, the theory of evolution is the most comprehensive. Even with its controversies in variants and details, as we will see later (e.g., greater or lesser emphasis on the role of chance in mutations and the importance of epigenetics), it is the paradigm that nucleates and organizes. It establishes a hierarchy among the rest of the rules explicitly about biology (Ruíz & Ayala, 1998). In this sense, just as we show concentric circles that indicate for biology the degree of inclusion of the different levels of laws it presents, biology itself is at the center of the discussion concerning other sciences. They encompass and influence it in their laws and rules (Fig. 1.1a, white arrow).

Thus, we could go from the center, with biology, to the periphery, with chemistry, and above it, physics, and finally encompassing (or well on a par?) mathematics. This relative directionality can be dual between biology and chemistry due to biological feedback in specific chemical processes (Fig. 1.1a, black arrow). For example, living organisms produce rather complex molecules, such as DNA or RNA, not outside the sphere of biology. Additionally, living organisms produce complex chemical compounds, such as proteins, fats, and oils. It can then be added to chemistry in sensu stricto (inorganic and organic) those compounds not found in the foundations of the periodic table of chemistry (biological chemistry). The biological contribution to the atmosphere is clear since living organisms emit oxygen and carbon dioxide molecules due to photosynthesis (still with incipient physiological machinery) and respiration. Methane produced by the digestion and excretion of urate and associated nitrogen-based compounds (a product of protein degradation in many animal diets) is also added. Therefore, it is evident that the concentrations of these gases in the atmosphere differ from what only strictly chemistry would provide outside of the biological (Fig. 1.1a, an arrow with dotted lines).

We could suggest a first concept regarding the feasibility of the occurrence of rules. The probability of occurrence, approach, or application (depending on the emphasis) of rules (general, globalizing) will be directly proportional to the range of coverage and influence of the latter; therefore, being more feasible in physics and less, progressively, in the other sciences that it influences as in biology. The truth is that biology, and in particular, the field of organisms (e.g., ecology, natural history, behavior), is often seen in a reductionist way. For example, biology is an area of science that hardly presents, at least in a clear way, principles that can summarize and explain its characteristics comprehensively. However, this is different for biologists. Of course, more care must be taken in "generalizations" 'because there is more inherent variability in their patterns when dealing with living beings' (e.g., limitations in stating the rules, the scope of the rules, and the possibility of exceptions). Indeed, the components of the organization of living beings, such as the existence of species, populations, communities, and ecosystems, undoubtedly add up to more complex levels. This complexity shows us that biology does not result only from the sum of the effects of other sciences, such as physics and chemistry. This implicit complexity of living systems makes the appearance of exceptions, such as oddities in behaviors and morphologies, much more probable.

This perspective is encompassed within systems biology, which aims to explore, in an interdisciplinary manner, the interactions between internal and external processes within organisms and even beyond them. Unlike the classical scientific method, which is based on the rejection of hypotheses, systems biology is based on mathematical modeling of the processes of interest, which allows us to obtain predictions of the state of that process (Tavassoly et al., 2018). This perspective opens the door to detecting the presence of modellable patterns, which can be stated as rules. In addition, this approach also favors the discovery of exceptions at certain levels.

1.3 Searching for Universal Rules in Biology

1.3.1 The Phenomenon of Emergence

Suppose we were behaving truly strictly and contemplated biology only from physics and chemistry. In that case, we could ask ourselves with much skepticism: Are there no other laws for biology than those three already mentioned and those from physics? We could reasonably consider for level I (or even a regulator of this, as a "level 0") the phenomenon of emergence. Emergence is a process where larger entities, patterns, and regularities emerge through interactions between smaller or simpler entities that do not exhibit such properties (Bunge, 2003; Sawyer, 2005; Bizzarri et al., 2020). In other words, and broadly speaking, the concept of emergence considers that a whole is more than the sum of its parts (Francescotti, 2007; Bizzarri et al., 2020). Emergence, of course, is not limited to living beings but also encompasses forms and patterns observed in abiotic components of the planet (and perhaps of the universe?). As far as biology is concerned, the phenomenon of life as studied in this science is commonly perceived as an "emergent" property of interacting molecules, something that, in general terms, is already studied in chemistry and which, in turn, reflects the particle interactions studied in physics. Thus, biology could have a "Law of Emergence" that characterizes it as an area of science as a whole, at least in terms of structural patterns (i.e., "form").

Life itself could be considered an emergent phenomenon. Life is the most significant source of complexity if we look at the progression of evolution, from the micro to the macroscopic, as well as historical events (e.g., from the initial RNA and DNA emergence to the more complex systems of communities and ecosystems). An attempt is made to frame elements inherent to all living beings that define them in their form and evolution. If we see life as a system with different levels, we need a hierarchy of analysis to accurately describe the system (Emmeche et al., 1997). It would be imprecise to look for a law that encompasses the whole system since the relationships are dynamic between the constituent elements and change with the level of systemic organization (Bizzarri et al., 2020). In other words, laws that describe relationships at lower levels may need to be revised to describe relationships at higher levels. Therefore, a law defined as a set of causal relationships between the phenomena studied may only hold at some levels (Bizzarri et al., 2020).

On the other hand, there are emerging "structures" in many groups of living beings, such as colonies and massive coordinated groupings, which can form true "superorganisms" (Wilson & Sober, 1989; Hölldobler et al., 2009). Such is the case of the conformation of something very complex, such as colonies of ants, bees, organized flocks of birds, or schools of fish (swarming patterns). Thus, the phenomenon of emergence can be at level I and at lower levels. The "theory" of emergence is something that many physicists endorse. However, it is also controversial and has a tumultuous history in science and philosophy (Pigliucci, 2014). This author highlights, "*even if emergence is only an epistemic phenomenon, there are good reasons to take it seriously, for example, because it points to theoretical or methodological deficiencies that make rigid reductionism unfeasible in practice if not in principle as such.*" We find patterns and regularities that repeat in nature, such as the ramifications of a circulatory system, of a tree or a river, as well as analogies based on the Fibonacci series present in the concentric helical structures in the arrangement of petals in a sunflower, shell of a nautilus, water whirlpool, and spiral galaxy (Lüttge & Souza, 2019).

It is interesting to note that evolutionary theory and emergence theory are not mutually exclusive visions but are based on different approaches that complement each other. The theory of emergence does not attempt to explain evolution. However, it shows the analogy between forms by arguing that there are emergent patterns in common, repeated in both abiotic and biotic contexts and perhaps extending beyond the planet.

1.3.2 On Some More Holistic Approaches

Connected with the emergence and within the framework of more general principles, attempts have been made to obtain one that best encompasses all or at least many of nature's phenomena. This goal led to a "search for the holy grail."

An example is when attempts are made in science to explain opposite patterns from the viewpoint of Eastern philosophy, even using concepts from the philosophy of the Tao. For instance, the bestselling book "The Tao of Physics," authored by Fritjof Capra in the mid-1970s, highlights the parallels between certain issues in modern physics, such as wave–particle duality, and the concepts presented in the Tao philosophy, including the notion of complementary opposites represented by Yin and Yang (Capra, 1975). It draws analogies between these domains, showcasing the interconnectedness and shared principles between Eastern philosophy and physics. As much appreciated as strongly criticized, the same author later took this vision to other books with an integral look at the emergence and holism of life (Capra, 1996). Subsequently, in collaboration with Pier Luigi Luisi (Capra & Luisi, 2014), the author has shifted their focus to biology within the framework of systems theory, including complex systems. This approach emphasizes a holistic perspective and the concept of emergence, which offers an integrative understanding of the organization of life across various aspects and levels. While the value of these contributions may be subject to debate, it is undeniable that they exemplify a nontraditional pursuit of universal patterns.

In the same sense, there are other works with holistic approaches, always with a bias toward oriental philosophy over the classical positivist scheme. Undoubtedly, this approach already places these essays at the "limit" of science, even more than what could be questioned in Capra's contributions. We can mention as examples of these approaches the study, with a more philosophical look, by Poggi & Ohsawa (1986) on "The Order of the Universe" and the work of Barry (1993). Barry proposes a holistic theory or vision of the universe to arrive at something that encompasses "almost" everything and, as he says, tries to integrate science, psychology, and religion. Both studies are just two examples in a vast bibliography of such approaches. At the level of the biological sciences, the Gaia idea of Lovelock and Margulis (1974) is not different from these holistic views. These authors suggest that living organisms interact synergistically with their inorganic environment in a complex, self-regulating system. At first, it was a very controversial idea, branded as teleological and somehow going against the principles of natural selection. Nevertheless, certain modifications have been introduced with the assistance of other disciplines, which have enhanced the robustness of the idea. However, it is important to note that despite these refinements, criticisms of the concept still persist.

To recapitulate, these are interesting approaches, but they are beyond the scope of this book since they pretend to be all encompassing. Here, we will examine, from the level of biology itself, its rules and exceptions, not above it. At the end of the book, we will come back to this and see to what extent what we analyze here is indirectly connected to complement these holistic "suprabiological" views.

1.3.3 Biological Complexity: Controversial Rules

Defining complexity is challenging and has significant differentiating biases depending on the domain under consideration (Chambers, 2015). Complexity is an attribute of systems that alludes to the interrelatedness of the system's components. "Complex" should not be confused with "complicated," which would be something intricate that is difficult to understand. Complex systems are usually simple and not complicated (Weaver, 1991; Adami, 2002; Bizzarri et al., 2020). Complex systems form a functional "whole" with interdependent and variable parts. These systems make up most of the world, from individual living beings to social systems and many natural systems. The components proper to a complex system self-organize to form potentially evolving structures exhibiting a hierarchy of emergent system properties.

Controversies over what rules might be posited for biology sometimes become very acrimonious. For example, McShea & Brandon (2010) explicitly postulated in the title of their book "Biology's First Law: The Tendency for Diversity and Complexity to Increase in Evolutionary Systems" that the first law of biology would be the increase in (taxonomic) diversity along with progressively greater complexity in organisms, seeing both components as "a whole" on an evolutionary scale. The idea postulated by McShea & Brandon (2010) is based on records and comparisons of how phylogenetic patterns and somatic complexity are associated over evolutionary time. However, it has received several criticisms, most notably one on the revision of the book by Bromham (2011), that almost tears the whole idea to shreds. This author compares some phylogenies and the expected increase in complexity that should occur with the increase in biological diversity, not finding clear support for the rule proposed by McShea & Brandon (2010).

What is certain is that everything referred to as complexity is connected and framed in a theory on which progress is being made (Chambers, 2015; Bizzarri et al., 2020). Taking into account the ongoing discussions surrounding the role of increasing complexity in biology, while also acknowledging the importance of simplification within that framework (as discussed in Chap. 2), it becomes evident that the quest to establish comprehensive laws for biology can be both enticing and passionate. However, it is important to recognize that such an endeavor is not without controversy and valid questions, making it a challenging and debatable task.

1.3.4 The General Systems Theory and Systems Biology Perspective

The issue of principles in biology has been addressed by Green (2015) in a review that focuses on design principles in the context of general systems theory and systems biology. Her detailed analysis demolishes the false premise of other observers seeing biology as a discipline lacking principles due to the variability, contingency, and complexity of living systems. Some researchers suggest that generalizations can be made but that these are related to the hard sciences or are processes that describe contingent outcomes of evolution (Beatty, 1995). This apparent shortcoming of biology does not make it weak as a scientific discipline but separates it from other fields (Green, 2015). Contingency would not be a rule in biological systems either, and in fact, there are examples where generalizations can be seen. Recall, for example, the universality of the genetic code in all living beings or the use of ATP as energy currency in physiology. We also know that there are general theories, such as the cell theory and the predator–prey theory of the Lotka–Volterra model, among others. Thus, Green suggests that generalizations can be made in biology without generating a reductionism that threatens the discipline's autonomy. She also distinguishes different types of generalizations and highlights constraint-based generality, a kind of generalization resulting from restrictions on systems to show other configurations or behaviors. Thus, a group of systems shares characteristics through formalized dependency relations, although the restrictions limit and make specific possible spaces. The restrictions make it easier for the biologist to understand the reason for the presence or absence of patterns in nature. Generalization in sciences such as biology is possible, but considering its particularities and knowing that the level of generalization may never be compared with other, more complex sciences. This divergence creates an opportunity for generalizations with certain precautions, where the presence of exceptions prompts a reassessment. Such a reassessment is not necessarily negative but rather invites the generation of new ideas and constructive relationships within and between disciplines. It encourages a fresh perspective and fosters innovation by embracing the potential for novel insights that arise from exceptions and deviations from generalizations.

1.4 Presence of Rules According to Levels and Areas of Biology

1.4.1 On "Minuscule" Rules: Certainties Even in the Narrowly Defined

The general principles in biology should be approached with a perspective of "flexibility" that is more pronounced compared to other "hard" sciences. That is, knowing that there will be logical limitations in the case of not being able to cover 100% of the variability presented by organisms. Variations and rarities will be found at more restricted levels of the norms or regularities surrounding us. Returning to Fig. 1.2, we are more likely to encounter daily changes at the levels encompassed by I (e.g., II, III, IV). In this sense, an analysis of rules and principles is sufficient for the general, as summarized in the works cited above, or whether it would be worthwhile to delve into these much more specific levels. Ultimately, whether we acknowledge it or not, the inherent nature of the natural world forces us to engage

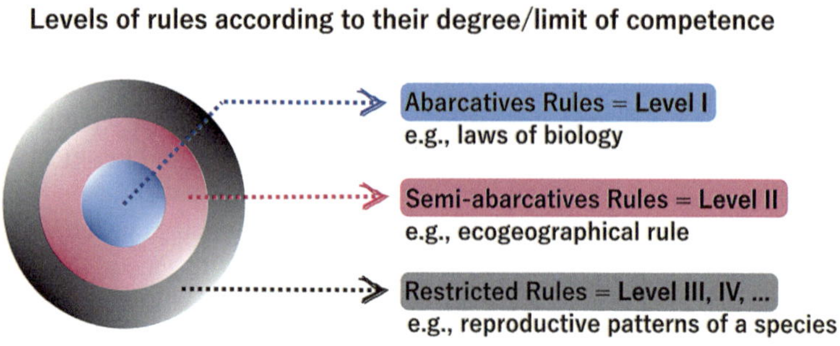

Fig. 1.2 Levels of rules taking into account their degree/limit of inclusiveness or competence

with rules at lower levels, regardless of our inclination or interest in doing so. They are not perhaps made explicit as "laws" (a term for a more globalizing level) but as "regularities" or "patterns" depending on the preferred term in the area of biology in which we find ourselves. Thus, for example, if we were to take a level II, we would zoom in on specific details of the surrounding world, such as the rules known as "eco-geographical," which will be discussed in Chap. 4.

On the other hand, at a more "modest" but no less critical level, there are affirmations from tested hypotheses, which then remain as rules of a lower level (e.g., III, IV). They are "mini rules" compared to those already mentioned, which are of greater scope. In short, within a specific discipline, the researcher usually does not question or consider the applicability of level I rules but takes them for granted. For example, we focus on an entomologist and specialist in reproductive biology. In that case, he will not consider whether insects have a cuticle and six legs since these are implicit rules (synapomorphies) to which they all conform. However, he will consider more complicated things, such as those referring to the mating success of males. For example, remember that as a "rule," it is common for sperm precedence in insects to favor the second male generally or that they tend to be a very promiscuous group. These "mini rules" on which the researcher bases arguments and new questions are vital foundations for advancing knowledge. For example, let us imagine that suddenly someone was to point out that, at the moment of sperm transfer, the male of the dragonflies under study no longer grabs the female by the dorsal part of her thorax (i.e., the "mini rule" for mating in these insects). In contrast, the male stands beside the female without physical contact and only mates with the coupling of the aedeagus, which they use to transfer sperm. This observation would be a novelty in this sublevel of their biology. This mini rule would not be taken for granted by the researcher as it has been done thus far.

Therefore, all levels will have certainties when starting an investigation or orienting oneself to discover something that fits or does not fit into all that is known for that particular level. Interestingly, the rules have mainly emerged from studies performed more than a century ago and, in many cases, in publications that were not written in English. Therefore, it is expected that there may have been some misrepresentations of the original study and statements over time because the original work needed to be read or translated correctly. This problem brings us to a central question. How much time has to pass before a rule begins to be questioned or refuted? Refuted because it is considered erroneous or because of the appearance of many exceptions to the original idea. A simple and obvious answer is that the more time passes, the more likely these changes will occur. The social and scientific context, changes in the discipline itself, or the knowledge of other associated areas, among other factors, would also be added to the moment of its postulation, as we will discuss later.

1.4.2 *Laws, Rules, and Patterns*

Box 1.1 summarizes various synonyms for rules, from laws to patterns and theories or principles. Although they are taken as synonyms, each word has its own specificity. As a scientist, it is necessary to pay attention to its meaning to express oneself adequately. Additionally, terms from other sciences, such as mathematics, are included, which are linked to those shown in the table as the main ones, for example, theorems and axioms. The most widely accepted meanings are shown, although it is clearly difficult to reach a consensus on which one or ones are more in line with each level. However, it could be considered that the laws belong to level I, rules to level II, and patterns (or "regularities") to level III. Theories would be closer to level I, although with potential components of level II concerning the presence of exceptions. The "fundamental" laws of biology, such as those exposed in previous paragraphs (Box 1.2), do not accept exceptions. Just try to imagine that living organisms that did not follow the laws of physics and chemistry were not composed of a membrane at the cellular level or had not undergone any evolutionary process during their existence or those of their ancestors. If exceptions to this level of core laws appear, we would be in serious trouble as a science.

Thus, together, these fundamental laws form a scientific paradigm (along the lines of Khun) or a central nucleus (following Lakatos), which in turn are covered by other principles that will eventually be modified sooner or later, either by advances in scientific knowledge or by biological evolution itself. Beyond these epistemological components, what is certain is that at more specific and limited levels, such as rules, we can already see the presence of exceptions. At this level, the variability inherent to biological systems has an impact on "continuous" situations or cases with qualitative "ranges" that deviate from an average value or "norm." This pattern is valid even if we limit ourselves to a rigorous statistical appreciation rather than an epistemological one. We must be clear that those exceptions are not due to misinterpretations of the rule's statements (explicit and implicit) or its range of coverage in the study system.

Finally, we will have the patterns or, if you will, "regularities" of a lower level (below III), which are not usually expressed in the papers as rules. Even if they are implicit rules, in practice, this term is not used for a more limited level of study. However, these regularities are observed as "modal" (i.e., not fixed) ways of presenting a particular event or state.

1.4.3 *Visiting Some Witness Cases*

1.4.3.1 Classification of Living Beings

When taxonomic keys are made for determining species or other taxonomic categories, the elements used indirectly trace differences between the groups included. Plesiomorphies and apomorphies summarize patterns and regularities that give entities to a taxon. The "volatility" that the history of taxonomy may have had was partly based on not having adequately contemplated characters, patterns, and regularities that reflected genuine phylogenetic relationships. Currently, the degree of taxonomic error has decreased since not only phenotypic characters are used to establish homologies but also a practical and powerful component is added, namely, molecular analyses (Yang & Rannala, 2012; Lee & Palci, 2015).

Even so, morphology aspects are essential to define (as a level I rule) a particular group. For example, "all cnidarians are diploblastic." In other words, their layers are composed of ectoderm and gastrodermis (endoderm). Indeed, the mesoderm has yet to appear on this evolutionary scale, as it will later appear in flatworms. None of the more than 11,000 species of cnidarians is known to be triploblastic (Giribet & Edgecombe, 2020; Schierwater & DeSalle, 2021). Thus far, there would be no exception to this level I rule in this phylum. Thus, this apomorphy of the phylum Cnidaria is intrinsically a rule. It has stayed the same with either the historical time of our knowledge of biology or evolutionary time. Examples such as this could be listed for all invertebrate taxa, vertebrates, plants, fungi, protists, and bacteria, among other groups (Fig. 1.3). In each clade, a set of characteristics will allow us to arrive at a general statement covering all the species contained in it. Exceptions to this high level of rules are not expected in general. However, we will see in the next chapter that they can appear, which has evolutionary and phylogenetic importance.

1.4.3.2 Behavior and Communication

Behavior is fascinating as a source for analyzing the possibility that it involves, and serves to exemplify, different levels of rules. Within animal communication, courtship in animals occupies a special place due to the richness and complexity of the signals involved during male–female interactions. It is clear that in each species, the sexual behavioral repertoire is not composed of a series of random or highly variable patterns but of "modally" established or stereotyped patterns. Indeed, courtship serves directly or indirectly as a prezygotic mechanism of specific recognition and interspecific isolation. However, in addition to having functions for specific recognition, the evolution of courtship and sexual behavior is undoubtedly under sexual selection pressures such as mate choice (Andersson, 1994; Houck & Arnold, 2003; Kallman et al., 2015). Then, despite behaviors being stereotyped, it is expected that, for example, males will show high behavioral plasticity or exhibit courtship patterns that vary in intensity, which may lead to exceptions to the stereotypes expected for their species. Curiously, the exception may be an "extreme" within a quantitative

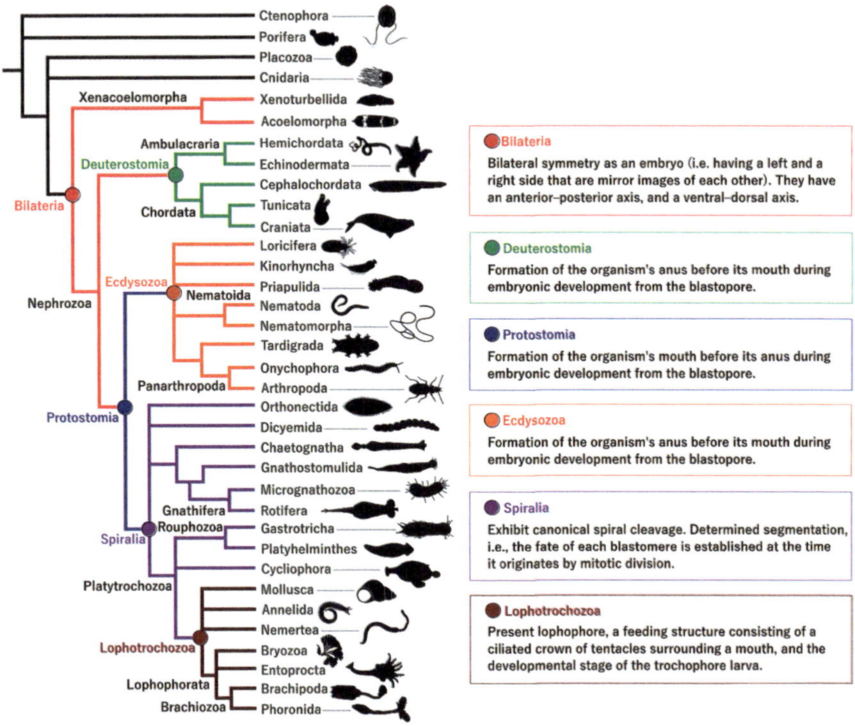

Fig. 1.3 Major rules in metazoan phylogeny. Animal phylogeny with phyla at the tips and major clades indicated at nodes. (Modified from Giribet & Edgecombe, 2020). Characteristics that group major clades are indicated in colors and can be considered "rules" for the groups considered. Free images from https://www.phylopic.org

gradient, i.e., not something "separate" from other values on a continuum. However, to our eyes, it may be perceived as something qualitatively discrete and different from the rest.

An example of this behavioral plasticity can be observed in scorpions of the species *Bothriurus bonariensis*, distributed in South America (Mattoni & Acosta, 2005; Ojanguren-Affilastro, 2005). Like most species in this group, these scorpions have an elaborate courtship consisting of ritualized behaviors (Peretti, 1993). However, there are differences in behavioral patterns between populations of the species, where males may vary the frequency and duration of some stimulatory behaviors (Olivero et al., 2015). Even when individuals from two different populations interbreed, males may modulate their behavior according to the "demands" of an exotic female from another population. This example shows us that although there is a courtship pattern that would represent the rule in these scorpions' reproductive behavior, there are exceptions that allow populations to continue interbreeding with each other.

On a more globalizing level, could a law referring to communication be added to all living beings? One could be that "*every living being communicates*," at least with its peers (Scott-Phillips, 2008). The organism may exchange information with other individuals of its species, colony, and even with other species in contexts of predation or defense. This communication could be basic, including tactics and chemistry, for example, in significantly lower organisms and then in higher evolutionary levels gradually enriched with displays of rituals, stereotyped body movements, and other much more complex ones. The question worth asking is: If a living being does not communicate, would it not be a living being? For example, do algae or plants also communicate? Recent work has shown how plants communicate, a topic we will address further in Chap. 6. We could draw rules from behavior that approach level I (Fig. 1.2) for all living beings. This option, however, requires mental flexibility to appreciate these new insights. Fortunately, it was formed in recent years (e.g., communication used to be an attribute of animals, but not of plants or even lower organisms such as bacteria—Keller & Surette, 2006; Won et al., 2020).

Suppose we focus, for example, on language in humans. In that case, it is known that all cultures handle a specific scheme of symbolism, a form of writing (whether graphic or based on a conventional alphabet in its form) derived from a language and the words that compose it. We could then say that there is a law of universality of language that governs the current human species in terms of "the existence of a language in all cultures" (Pozzo & Soloviev, 2011). This does not admit an exception as such, except for teratological cases in which an individual cannot assimilate the language patterns of his or her culture (e.g., problems in the pre and/or postnatal development of Broca's and Wernicke's areas) (Pozzo & Soloviev, 2011). At another level of analysis, within each language, we can find rules, such as grammatical and orthographic rules. Exceptions that might occur in particular contexts could be accepted. In turn, at more specific levels, we could find patterns of written expression that are more specific to certain populations, up to the formation of deviations or dialects. These more restricted patterns do not reject the previous levels of rules but add or modify certain aspects of the language that make them even more labile and changeable from one territory to another. Even if we focus on "nonverbal" communication (Duncan Jr., 1969), we also recognize constitutive elements that are repeated in different cultures and that could allow us to extract laws or rules, at least in general (e.g., facial expressions). Already at a more limited level of each village culture, specific patterns of "nonverbal" communication, such as what we have already mentioned above for "verbal" communication, could be seen as exceptions to these rules (Jena, 2020).

1.4.3.3 Rules and Basic Plant Patterns

In addition to more recent striking topics, such as the communication we will discuss later, specific rules and patterns can be found in plants, which can be assigned to certain levels. For example, a level I rule for the Plant Kingdom is that they are characterized by being autotrophic species capable of producing food from inorganic matter. They possess chloroplasts and related organs and use a pigment called chlorophyll to take light energy from the sun and use it for photosynthesis (Curtis et al., 2008). If an organism of this Kingdom does not photosynthesize, then it implies the presence of exceptions. Some plant species are not photosynthesizing but parasitic at different levels (Nickrent & Musselman, 2004; Nickrent, 2020). Examples are *Epifagus americana*, *Monotropa uniflora*, and plants of the genera *Cuscuta* and *Rafflesia* (Hidayati & Walck, 2016).

From this basis, we can examine more punctate things that will ultimately, as always, end up being the synapomorphies that characterize and give meaning to a given clade. For example, algae may be cellular or tissue-forming and photosynthesizing, but they do not have xylems such as vascular plants or the ability to tolerate environments far from water (Curtis et al., 2008). The same could be the case at the level of particular characteristics for marchantiophyta or liverworts (pores instead of stomata and elateria to disseminate spores), bryophytes or mosses (sugar and water transporting cells, gametophyte with multicellular rhizomes, capsule with operculum where the spores come out), phyllicophytes or ferns (compound leaf with leaflets on both sides, young leaves rolled in a spiral, sporangia on the underside of the leaves, antheridia formed by a superficial cell that has an apical cell as an operculum), spermatophytes (stem with secondary growth, branching from buds in the leaf axils, ovule formed by female gametophyte, nurturing substance and tegument, male gametophyte highly reduced), and in turn among spermatophytes, on gymnosperms (male and female reproductive structures in cones, ovules on the upper side of the scales of the female cone, male gametes carried by the pollen tube to the ovocell) and angiosperms (complex compound flowers, ovule protected by a carpel that will form the fruit, reduced pollen grain with three haploid cells, female gametophyte also reduced in the form of an embryonic sac, complex wood with vessels and supporting structures, phloem that helps the circulation of sugars) (Curtis et al., 2008). Therefore, a series of patterns can be recognized that conform to the level I rules in each case. Suppose we find exceptions to the rule that summarizes the characteristics of each plant form. In that case, they could be teratological or evolutionary-derived traits.

Plants also have evolutionary processes and characteristics that were only recently considered present in animals, such as sexual selection (Stephenson & Bertin, 1983; Cocucci et al., 2014) or, as we previously said, communication (Karban, 2015). Indeed, plants demonstrate the presence of such patterns and mechanisms with their particularities. It is not by chance that many of the most striking plants are those with solid interactions with animals, such as those subject to coevolution by pollination, including cases of mimicry (e.g., in orchids) (Schaefer & Ruxton, 2009). In a phylogenetic framework, many of these species are actual "taxonomic exceptions" within a clade, as we will see in Chaps. 5 and 6.

The point is that wanting to define sharp divisions at the phenotypic and behavioral level between plants and animals has been challenging (Weller, 1955; Paps, 2016; Glibert et al., 2019). One noticeable trait is that referring to photosynthesis or not. However, it is now known that what concerns the ability to move, communicate, or better yet, the "form" each group uses in each Kingdom is insufficient to make such a division. Fungi are a clear example of this difficulty when classifying due to multiple "intermediate" characteristics, something we will go deeper into in Chap. 5. Better knowledge about the origin and function of many "rare" aspects of plants has gradually made this functional difference between them and animals somewhat "fainter" (Lloyd, 2011). It should be recalled that in the past, sponges and cnidarians were even included in the Kingdom Plantae. These animals were confused with plants mainly due to their immobility (or low mobility), "branched" body appearance (with rhizomes and stems), and colorations (e.g., green in many of them due to symbiont algae). Aristotle, Pliny the Elder, and other ancient naturalists believed that sponges were plants precisely because they were sessile and had little detectable movement. It was not until 1765 that John Ellis first observed an actively provoked water current within their bodies, and they came to be considered animals (Ellis, 1765; Ruppert & Barnes, 1996).

1.4.3.4 When It Is Still Difficult to Establish Rules: The Case of "Ecological Genomics"

When talking about rules, we need to remember that many of them are "under construction," i.e., we are still trying to integrate evidence from nature that will allow us to establish them as rules per se. These situations occur particularly in relatively new areas of the biological sciences, such as evolutionary developmental biology (aka Evo-Devo) or a new field, ecological genomics (Van Straalen & Roelofs, 2011; Raghavender, 2021). The problem lies in the fact that, with a broader view, researchers detect exceptions to the "potential" rules they try to identify in their models. For example, in their study on "exceptions to the parallel genetic basis," Elmer & Meyer (2011) argue that "*the many exceptions suggest we do not yet have any rule.*" They reach this conclusion by analyzing numerous cases of parallelism and convergence in evolution, finding that there are too many exceptions. There must be many more cases, such as the one cited, perhaps some of which we still need to discuss or have yet to be made public in the scientific community. That is, rules are still "cooking" from bodies of ideas and data. Through more detailed analysis, on par with a more integrative vision, researchers identify so many exceptions that, in the end, they consider it premature to define a particular rule. Indeed, the situations where previously established rules are later revised or refuted, as discussed in the previous items, or others under construction, as well as the influence of exceptions, are closely connected with epistemological concepts. At this point, it will be helpful to review below some of the most classical epistemological frameworks linked to the general problem of rules in science.

1.5 Linking Rules with Epistemology: Changing Paradigms

The changes in scientific thinking and methodologies over time have been the object of the study of epistemology. The ideas proposed by Khun, Lakatos, and Feyerabend come from authors contemporary to each other at a time of much debate about the foundations of modern scientific thought. This book aims not to elaborate on the theory of scientific thinking in biology. Nevertheless, we will review some valuable concepts to detect, evaluate, and discuss the presence of regularities and exceptions in biology as an essential part of its nature. From this, we will contemplate its dynamics and draw valuable conclusions in basic and applied research.

The ideas about scientific thinking of these authors and others have been somewhat exposed (and learned) as "opposites" when they are not mutually exclusive, at least in their totality. As we have previously stated, the "Khunean" paradigm encompasses ideas about rules and theories. Indeed, when we speak of rules, we refer directly or indirectly to a body of ideas that are part of scientific thought for a given time (Kuhn, 1970). As we will see in Chap. 2, crises and paradigm shifts in biology are associated with modifications in what are considered to represent the rules and their exceptions. In a certain way, the "Lakatian" scientific research program was reflected when in Fig. 1.2, we plotted "levels" of rules. As he says, it would be a complication in science if essential laws of biology were demolished (Lakatos, 1978). This connects with the "moment of crisis" postulated by Khun. Likewise, the scientific system becomes more labile as "real" exceptions (i.e., not the product of misinterpretations or slight extrapolations of the author's original ideas) accumulate, resulting in principles of the central core behaving as parts of the protective belt (i.e., "variables").

In every discussion in scientific thought, the so-called construction of "paradigms" (as well as their strengthening, weakening, and renewal) takes place at any level of observation. Feyerabend was disruptive in his ideas. He postulated scientific anarchism, highlighting the limitations of a structure that wants to regulate or control processes and mechanisms, such as society or scientific thought (Feyerabend, 1993). Some consider that the anarchism proposed by Feyerabend was more a call for attention than premises to tear down any basis of science.

The idea that rules and exceptions are separate entities can be dismissed. In other words, we should always expect all laws, rules, or associated terms to have exceptions. Therefore, the point is not to uphold a naive belief that rules do not have exceptions, especially when discussing biological rules that are not at level I.

In contrast, the aim is to highlight the dynamic coexistence of rules and exceptions. This component, dynamics, is critical in a science such as biological sciences, which, by its very evolutionary basis, implies modifications of patterns. Moreover, the value of exceptions alone, even if we were sure that they would never increase in number or qualification to overthrow a preexisting rule, deserves to be highlighted to advance further studies on rarities in biology. In other words, to know "what is" an exception, one must define and show something of the vast and controversial universe of rules. In the next chapter and Chap. 5, exceptions will have their presentation.

It is important to highlight the significant influence of confirmation bias in science. Confirmation bias is the tendency of people to seek or interpret information in a way that supports their current beliefs (Nelson & McKenzie, 2009; Oswald & Grosjean, 2012; Mercier, 2016). This bias is constantly at play when evaluating whether a rule is confirmed in a study. This can occur either by confirming the rule in its entirety or by modifying it according to new evidence to make it more "modal" (i.e., not so black and white). As we will return to later, exceptions are sometimes not perceived if the researcher has confirmation bias. Indeed, the researcher might unintentionally design the experimentation methodology or interpret the results "to confirm" the rule, pattern, or hypothesis that supports his preestablished ideas (Oswald & Grosjean, 2012). It is expected that students, who still need to exercise the critical view toward ideas and information they are incorporating, tend to fall into this bias. For example, a student may focus all his/her attention on analyzing how the data from the study "fit" with what is "already known" (e.g., comparing it to current paradigms, influential papers, fashionable topics in the discipline, orthodox mentors, etc.) instead of using them to critically analyze them objectively and relativize their hypotheses. A colleague used to tell us: "*There is nothing better than overthrowing a previous traditional theory or idea with the contribution of your data.*" This opinion connects with Kuhn's classic ideas about the "crisis" moments that precede a change in a given paradigm (Kuhn, 1970; Andresen, 1999; Anand et al., 2020). We will return to the issue of confirmation bias later.

The reaction of researchers to changes in their postulates can be very variable, ranging from acceptance to a fierce defense of their previous ideas. Here, we will highlight the positive side of being receptive to change. For example, we see how recently the physicist Juan Maldacena managed to unify quantum mechanics and relativity with his studies embodied in String Theory (Maldacena & Susskind, 2013). Interestingly, this scientist claims, "*if my theory is disproved, it would still be a scientific advance.*" This comment is a precious synthesis of open-mindedness and the view that potential changes are the key to scientific progress. Some people believe that all scientific evidence is valuable, whether it refutes or supports a hypothesis, is significant or not. However, there is an implicit bias among researchers with a false premise that if the idea or working hypothesis is refuted, the data obtained have no scientific validity or are useless. This conclusion is a prejudice that must be eradicated, and the valuing of "negative data" in science should be promoted.

In the field of biology, Ruppert & Barnes (1996) end their chapter on Perspectives in Phylogeny, in their classic book on Invertebrate Zoology, with an encouraging paragraph for students and specialists: "*We encourage students interested in the history of life to critically evaluate our cladogram and any others they encounter. Let us welcome new ideas but let us analyze them carefully.*" Ultimately, these open ways of thinking promote the study of everyday phenomena in biology from different perspectives and encourage us not to see rules and exceptions as watertight compartments but as subjects to a dynamism that is worth discovering and analyzing in detail.

1.6 Evolutionary Theory as a Driving Force for Changes in Biological Rules

There is one point, at least in biology, that has had historical implications for the changes in rules, postulates, and patterns that were initially proposed: the discovery of an evolutionary background. Systems biology was a developing field before Darwin's Origin of Species provided a unifying framework. At that time, taxonomy was the most relevant discipline, and it is evident currently that the lack of an evolutionary basis has become a significant weakness. Symbolically, it would be like building a house without a foundation or building it on the sand. There are plenty of examples around us to illustrate what we have mentioned.

For example, on a walk around our city's lake, we observed a pair of mallards (*Anas platyrhynchos*). The male of this species has bright green colors on his neck with a white collar. The female, on the other hand, exhibits duller colors. These differences in the phenotype of the sexes, known today as sexual dimorphism, could only be explained by Darwin's work. As a result of sexual selection, males possess these colors as ornaments to be chosen by females. In other cases, armaments, such as horns, are used for competition between males to access females. These ideas were not embodied in a theory or paradigm at the time of Linnaeus. Linnaeus's definition of species was based on morphological similarities, so it should not be strange that he classified males and females of the same species as belonging to different species.

Something similar happened with other taxonomists and animal groups, with iconic examples of animals such as the anglerfish (*Cryptopsaras couesii*). This fish inhabits from 500 m depth in all oceans (Shimazaki & Nakaya, 2004). The male of this fish species was elusive to researchers, who later realized that it was tiny and resembled a sperm sac attached to the female's body. This pattern is known as sexual parasitism and is related to the difficulty of mating in deep environments (Vieira et al., 2013). Another example is the argonauts *Argonauta argo*, pelagic octopuses that live in warm oceans. Members of this group exhibit marked sexual dimorphism with dwarf males that are difficult to find in the wild. Females are capable of secreting a calcareous structure to care for their eggs. Conversely, males do not produce this structure and present an extremely long hectocotylus or copulatory arm (Battaglia et al., 2021). This example of sexual dimorphism in animals shows how "certainties" in specific disciplines, such as taxonomy, are relatively labile.

Currently, everything is "easier" to understand with the contribution of molecular genetics. Many changes in the valid names of species and changes to synonymy have been promoted by the lack of a solid basis within a framework of biology (e.g., evolution before Darwinian times) or on specific biology of the group (e.g., species with coloration polymorphism). Another iconic example is the so-called "birds of paradise." These birds, which are very common in Oceania, were believed to come from a heavenly place. The natives of these lands gave the explorer Fernando de Magallanes specimens without legs, as they had cut them off to make decorations. Magallanes did not understand this, as he did not speak the native language. Magallanes took the specimens to Europe, and a "logical" explanation was constructed as to why they had no legs: these birds flew continuously in the sky and only settled down to die (it was believed that they came from paradise and fed on dew) (Beehler, 1989; Kirsch, 2006). Indeed, Linnaeus classified the great bird of paradise with the specific name "*apoda*": *Paradisaea apoda* (Kirsch, 2006), a significant error considering that we are talking about the father of taxonomy. A total absence of basic knowledge about the biology of the organism and much imagination were factors that played a significant role in this misunderstanding.

In his excellent book "The Discovery of Animals" (1982), Herbert Wendt shows many classic examples of misinterpretations of various iconic animals regarding their origin and initial taxonomic position. Examples include the birds of paradise, stories of the discovery of monkeys, and the legends of unicorns and sea monsters. All of them had been initially contemplated without an evolutionary paradigm. This lack of a rigorous scientific observation method made many considered authentic, even for hundreds of years. If we pay attention to the years of those initial findings, they all date from an almost "primitive" stage of biological sciences. Currently, some aspects linked to cryptozoology can "replicate" a certain degree of imagination as it existed in those times.

However, today, it is easy to criticize these "mistakes" from our position with a vast knowledge of evolutionary biology, phylogeny, taxonomy, and molecular and developmental biology. We are still determining how many mistakes made today, while establishing a rule, will be studied and ironically criticized a century from now. Imagine we could travel back in time, without any present knowledge, and see in a Precambrian time that the unique aquatic life we find there could later move onto land or even into the air. It would be absurd. The idea of an "amphibian" would not even be in our minds, much less life in a desert or the flight of an organism called a dragonfly, pterodactyl, seagull, or bat. These pre-Darwinian historical examples help us to understand how biology changed in its fundamental bases. These changes occurred thanks to evolutionary ideas to such an extent that, as we have already seen, they are proposed as one of the three fundamental laws of biology (100% level I). Because of evolution itself, biology brings a series of often unpredictable changes, which contrasts with other "hard" sciences such as physics or chemistry.

1.7 On Mathematics and Contingency: Does It Weaken Evolutionary Predictions?

In mathematics, there are theorems, which consist of a proposition that can be logically proved from an axiom or from other theorems that were proven in advance (Manoukian, 2012). Thus, once stated, a theorem will not be proven in the short or medium term. Recently, the English mathematician Andrew J. Wiles was awarded for proving Fermat's theorem, one of the greatest mysteries of modern mathematics (Kleiner, 2000; Hellegouarch, 2001). Fermat was a great mathematician who noted in the margin of a book that he had figured out how to solve his theorem but did not have enough space to detail it. Thus, the challenge among mathematicians began to solve such a problem. It took them 350 years to solve it. Fermat's theorem is used in Internet encryption, and solving the theorem will enable the development of new, more secure forms of encryption. In physics, there are many examples of predictions that take time to demonstrate. In the 1930s, Albert Einstein predicted the existence of gravitational waves (Einstein & Rosen, 1937). However, it was only in 2016 that their existence could be demonstrated, in this case in the teamwork of great global scope at the level of research and costs, in the scientific project working with the instrument called the Laser Interferometer Gravitational-Wave Observatory (LIGO).

In the two examples mentioned above, a result is predicted and eventually proven. This raises the question of whether biology is on a different plane, in another scientific context, or whether the same can be done. That is, it could be supported by hypotheses that can be falsified in the short term and by much more speculative theoretical bodies that will be confirmed only in the medium or long term. Meanwhile, this will be provisionally accepted by the scientific community as a "rule pending verification."

For example, within the world of morphology, Olson (2012) analyzed animal teratologies as parts of a morphospace in the form of a range of "morphological possibilities." Much of the empty morphospace can come to be imagined, although not always possible (e.g., because of ontogenetic constraints), although others might come to be predicted. An example on a much smaller scale was the moth predicted by Darwin upon encountering the orchid *Angraecum sesquipedale* from Madagascar. Because of its unusual morphology and based on plant–pollinator coevolution, that portion of morphospace was imagined and predicted by Darwin; years later, the moth was found. These examples and the notion of morphospace will be further elaborated in Chap. 5.

There are an increasing number of studies in mathematical biology where models are applied to specific problems, practical to predict certain behaviors of the phenomenon under analysis. However, we are still far from what the "hard" sciences can predict in their respective fields. Paradoxically, or perhaps not, this limitation, or rather, this close and necessary link with the almost immediate factual, makes biology especially interesting and exciting. The wrong label of "less exact" science is a biased perspective from observers coming from other disciplines. Perhaps we are currently concerning the use of models and their predictions, as Linnaeus was with those species with remarkable sexual dimorphism in his time. That is, learning and waiting for new theoretical, conceptual frameworks that provide further cohesion to the tremendous existing biological diversity, at the level of patterns and mechanisms, in any of its branches.

It should be noted that at this point, what we comment on is where Gould's (1989) postulates on the phenomenon of "contingency" in biological evolution invites us to reflect on what is possible and feasible at the level of causes and consequences. For example, if we rewind the past and repeat the evolutionary process of a particular ancestral taxon from the beginning, it will likely not lead to the taxa and lineages that we know today. This situation is because contingency is a potentially robust factor in the evolutionary process, among other reasons, due to the improbability of tolerable mutations repeating themselves in the same way during their repetition.

Extrinsic contingencies have also been vital in partly shaping what we see in evolution. Evolution would have followed another course if those events had not occurred or did occur but with a slightly different impact. Although questionable, contingency and its role in biology relativize some "certainties" of what evolution was and would be if situations could be replicated on or outside the planet. Interestingly, could contingency be a concern for physics or chemistry (excluding biochemistry) if their histories for certain phenomena (compounds) could be traced back to their beginnings? Certainly not. This finding might be relevant since we might have other concentrations of chemicals on the Earth today. Nevertheless, would the components of the periodic table of elements be any different? Would there be organic compounds that are currently unknown, while others that are known would no longer appear in the "new Table"?

For physics, it is postulated that in "other" universes, there could be other laws that govern it. For example, matter would behave differently from the one we currently know. In biology, however, the impact of "rare" events would seem much more robust since evolution is a constant process and permeable to such extrinsic contingencies, in addition to the low probability of tolerable mutations as intrinsic causes. This situation marks a new difference between biology and the "hard" sciences regarding the predictability of future phenomena.

It is interesting to note that if a contingency is held to be a fundamental factor in biological evolution, then life itself and many subsequent evolutionary events have arisen more through "exception" than "norm" types of processes. Furthermore, as we will see in the next chapter, even leaving aside the "contingency" factor and its predictions, we will find that exceptions are still at the origin of many (if not all) evolutionary processes.

1.8 Possible "Principles" on the Nature of Biological Rules?

We have seen definitions, approaches, schools of thought, and general and biological examples of "rules." As we have said, they are not static but rather dynamic. Especially in biology, rules are not fixed elements. They are modified to a greater or lesser extent and can be rejected. As we will see below, changes can occur in a "time" context. For example, in the historical study of biology, something is easily understood from the epistemological approach. However, changes in the scope of rules also usually occur in a "time" context inherent to biology, such as evolutionary history.

Furthermore, this occurs in the context of "space," where the vast geography of our planet may mean that rules cannot be applied everywhere. To these two contexts, we add the "group" context, where in the following chapters, we will see how certain rules may or may not apply to some specific clades within a given taxon. In this framework, the question arises: Can we posit "laws," principles, and norms for the dynamics between rules and exceptions in biology? If so, could these also encompass other areas of knowledge, such as physics and chemistry? In this scenario of questions, if we play with the duality "rule-exception" and its fundamental dynamics, we could posit different postulates according to the dynamics of rule and exception.

1. *Every current pattern or rule began as a rarity or exception, respectively*: In evolutionary terms, this makes sense because adaptations that spread in a

population and species do not arise simultaneously and massively in all individuals. In contrast, the basis of evolution (sometimes not so explicitly stated) is that evolutionary change begins at the individual level. Indeed, this occurs in any phenotypic expression. Initially, it had a frequency of occurrence that was not the one that exists today, in the case that it has become the typical pattern or rule for the group.

Applying this principle to other sciences, such as physics, may be more complicated. For example, light did not vary in nature over the evolutionary time of the universe. The same could be said of gravity as a phenomenon, even if its values change in specific sectors of the universe. In chemistry, it could be applied to compound abundance in the history of the universe and the Earth. Additionally, the presence or absence of certain compounds underwent variations, for example, in organic compounds, such as those that in the "primordial soup" emerged as the organic precursors of life (various carbon compounds), aspects that we will see in Chaps. 6 and 8.

On the other hand, this first principle would invite us to think in any of the three contexts (time, space, or group) about "rigid" alternatives such as 2. *Every current exception always was an exception*: this may be accurate as well since it does not mean that something that is a rarity as phenomenon or process, in some time, space or group, becomes at some point a rule (or pattern if we speak of initial rarities). 3. The opposite principle is that *every current exception will become a rule*: This postulate is similar to 1. However, that one points to the past and implies rules initially arise from exceptions but does not predict every exception will be a rule (or became one at some point in the past). It is a necessary condition, but not the only one.

Another principle to take into account in a potential discussion to refute or accept is 4. *Every current pattern or rule always was a pattern/rule*: that is, nothing changed since the beginning of biology as such, since the emergence of the first living being or if the rule in question is restricted to a taxon, in the evolutionary history of the same. It sounds strange and challenging to sustain this since the spontaneous generation of an organism or a creationist principle comes to mind if it is accepted. Even if the origin of a process or phenomenon was not on our planet, its actual beginning could be traced elsewhere in the universe. This postulate could be sustained if one accepts the theory of life as a phenomenon coming from space, as exobiology might suggest, for example, brought here by a comet (we will return to this subject in Chap. 7). In this more "static" framework, option (2) exception–exception (i.e., a current exception always was in the past) may be more plausible. In contrast, the rule–rule option would be little or not at all sustainable in evolutionary terms. The caveat would be the three fundamental laws of biology, which we discussed in the previous sections since they refer to patterns and principles inherent to the existence of all living beings that occurred at the time of the origin of all life that we are analyzing today.

A prediction in this analysis could be that 5. *Every current pattern or rule will become an exception sooner or later*: either in space or in a space and group combination. Again, it is not easy to hold this prediction in general terms as a global principle or law for all of biology. However, it could apply to specific evolutionary biodiversity contexts. A particular pattern could become rare in the future without necessarily leading to its disappearance. The cases of species extinctions serve as an example, as we will see in Chap. 7. In the context of the history of science and scientific postulates, it may also sound strange that all the current rules will be maintained the same for hundreds of years. In any case, they will be adapted and molded to new data sources, even more so in biology.

In contrast, they may also be discarded and replaced by antagonistic ideas. In any case, as discussed in the epistemological aspects of the rules, it is feasible that a central "core" of the paradigm will survive. Therefore, of these postulates, the latter is the most complicated and risky to pose.

Principle 1 is the most feasible in biological evolution and even from the view of scientific ideas. It is the main contribution of what we will analyze in this book, an idea linked to classical epistemological postulates.

Summing up all that has been said up to this point: is it possible to develop a comprehensive framework, such as a "Theory of Rules and Exceptions" or a "Theory of Rule-Exception Dynamism," that illuminates the interplay between these seemingly opposing concepts and how they mutually influence each other within the context of time, space, and specific groups? This dynamic relationship, which remains largely unexplored, invites us to reframe our understanding of nature in general and biology in particular. Continuing our exploration, we observe that, similar to the examples of systems theory and emergence, our search for patterns and their contextualization in different temporal, spatial, and group contexts plays a crucial role. These proposed theories, centered on biological evolution and encompassing biodiversity and various biological aspects, reside within the epistemological realm of ideas, laws, and scientific thought. However, it is important to note that attempting to accomplish such a task is premature. The subsequent chapters of this book will provide a more comprehensive framework and pave the way for further studies on this captivating subject.

Finally, a point that we have mentioned and is worth noting is whether the probability of exceptions, i.e., the occurrence of exceptions according to the type of rule, could be predicted. We could expect each type of rule to offer a particular "tapestry," a more or less flexible scenario for the occurrence of exceptions. For example, if we focus on level I of rules, perhaps it will be more common to expect exceptions since we are dealing with something that can hardly contemplate the totality of the multiple variations of what is stated in the rule (whether explicit or implicit for science). However, if these rules are not simple generalizations but allude to essential aspects of life, then exceptions should not occur. From the above, could the probability of occurrence of exceptions be inversely proportional to the rules' level (i.e., the degree of inclusion)? In other words, for a level I rule with a significant degree of proven inclusion, the probability of exceptions is lower compared to medium- or low-level rules. In the latter cases (II, III), exceptions are more likely to occur. The intuitive thought would be the opposite, that more exceptions are expected in generalizations, which could be confused or taken as a synonym for a level I rule with a high degree of inclusion.

In contrast, at finer or more specific levels, by the mere fact of being more circumscribed, fewer exceptions are expected. However, if the comprehensive rules have arisen from an exhaustive analysis of reality, once they have been established, it may be rare for them to present exceptions. It is a controversial point that requires analyzing the intuitive and counterintuitive nature of both options in concrete facts. We leave it here and will indirectly revisit this question through examples throughout the book.

1.9 Key Concepts and Ideas in This Chapter

- Rules are general practical concepts or constructs that allow us to describe reality.
- In biology, the effects of rules seem to be more labile due to the plastic characteristics of life, emergence, and contingency effects associated with living systems.
- Comparatively, the sciences associated with biology do not seem to have the conceptual and encompassing strength of phenomena that disciplines such as physics and, to a lesser extent, chemistry show.
- It should be noted that there are basic precepts and robust central-level theories for life itself, such as that all living things follow the laws of physics and chemistry, all living organisms consist of membrane-covered cells, and all organisms have arisen from and are affected by evolution.
- The theory of evolution constitutes a globalizing rule of first degree for all living things.
- In an epistemological framework, some currents of scientific thought follow in part the logic of evolutionary theory to explain changes in scientific paradigms. Some epistemological cycles are similar to the rise and decay of species in nature.
- There is intrinsic importance in rule-exception duality. Exceptions are often seen as unimportant but have proved to be essential in comprehensively understanding biology.

References

Adami, C. (2002). What is complexity? *BioEssays, 24*(12), 1085–1094.

Anand, G., Larson, E. C., & Mahoney, J. T. (2020). Thomas Kuhn on paradigms. *Production and Operations Management, 29*(7), 1650–1657.

Andersson, M. (1994). *Sexual selection*. Princeton University Press.

Andresen, J. (1999). Crisis and Kuhn. *Isis, 90*(S2), S43–S67.

Barry, R. (1993). *A theory of almost everything: A scientific and religious quest for ultimate answers*. Oneworld Pubns Ltd.

Bateson, P., & Laland, K. N. (2013). Tinbergen's four questions: An appreciation and an update. *Trends in Ecology & Evolution, 28*(12), 712–718.

Battaglia, P., Stipa, M. G., Ammendolia, G., Pedà, C., Consoli, P., Andaloro, F., & Romeo, T. (2021). When nature continues to surprise: Observations of the hectocotylus of *Argonauta argo*, Linnaeus 1758. *The European Zoological Journal, 88*(1), 980–986.

Beatty, J. (1995). The evolutionary contingency thesis. *Concepts, Theories, and Rationality in the Biological Sciences, 45*, 81.

Beehler, B. M. (1989). The birds of paradise. *Scientific American, 261*(6), 116–123.
Birkhead, T. (2000). *Promiscuity: An evolutionary history of sperm competition*. Faber and Faber.
Bizzarri, M., Naimark, O., Nieto-Villar, J., Fedeli, V., & Giuliani, A. (2020). Complexity in biological organization: Deconstruction (and subsequent restating) of key concepts. *Entropy, 22*(8), 885.
Bromham, L. (2011). Wandering drunks and general lawlessness in biology: Does diversity and complexity tend to increase in evolutionary systems? *Biology & Philosophy, 26*, 915–933.
Bunge, M. (2003). *Emergence and convergence: Qualitative novelty and the unity of knowledge*. University of Toronto Press.
Bussard, A. E. (2005). A scientific revolution? *EMBO Reports, 6*, 691–694.
Capra, F. (1975). The Tao of physics: An exploration of the parallels between modern physics and eastern mysticism. Shambhala. distributed in the U.S. by Random House.
Capra, F. (1996). *The web of life*. Anchor Books.
Capra, F., & Luisi, P. L. (2014). *The systems view of life: A unifying vision*. Cambridge University Press.
Chambers, G. K. (2015). Understanding complexity: Are we making progress? *Biology and Philosophy, 30*, 747–756.
Cocucci, A. A., Marino, S., Baranzelli, M., Wiemer, A. P., & Sérsic, A. (2014). The buck in the milkweed: Evidence of male–male interference among pollinaria on pollinators. *New Phytologist, 203*(1), 280–286.
Curtis, H., Schnek, A., Barnes, S., & Massarini, A. (2008). *Biología de Curtis* (7th ed.). Editorial Médica Panamericana.
Darwin, C. (1871). *The descent of man, and selection in relation to sex*. John Murray.
Dhar, P. K., & Giuliani, A. L. (2010). Laws of biology: Why so few? *Systems and Synthetic Biology, 4*, 7–13.
Duncan, S., Jr. (1969). Nonverbal communication. *Psychological Bulletin, 72*(2), 118.
Einstein, A., & Rosen, N. (1937). On gravitational waves. *Journal of the Franklin Institute, 223*(1), 43–54.
Ellis, J. (1765). XXXI. On the nature and formation of sponges: In a letter from John Ellis, Esquire, FRS to Dr. Solander, FR S. *Philosophical Transactions of the Royal Society of London, 55*, 280–289.
Elmer, K. R., & Meyer, A. (2011). Adaptation in the age of ecological genomics: Insights from parallelism and convergence. *Trends in Ecology & Evolution, 26*(6), 298–306.
Emmeche, C., Koppe, S., & Stjernfelt, F. (1997). Explaining emergence: Toward an ontology of levels. *Journal for General Philosophy of Science, 28*, 83–119.
Feyerabend, P. (1993). *Against method* (3rd ed.). Verso.
Francescotti, R. M. (2007). Emergence. *Erkenntnis, 67*, 47–63.
Giribet, G., & Edgecombe, G. D. (2020). *The invertebrate tree of life*. Princeton University Press.
Glibert, P., Mitra, A., Flynn, K., Hansen, P., Jeong, H., & Stoecker, D. (2019). Plants are not animals and animals are not plants, right? Wrong! Tiny creatures in the ocean can be both at once! *Frontiers Young Minds, 7*, 48.
Gould, J. L., & Gould, C. G. (1989). *Sexual selection*. Scientific American Library.
Green, S. (2015). Revisiting generality in biology: Systems biology and the quest for design principles. *Biology & Philosophy, 30*, 629–652.
Harms, M. J., & Thornton, J. W. (2014). Historical contingency and its biophysical basis in glucocorticoid receptor evolution. *Nature, 512*(7513), 203–207.
Hellegouarch, Y. (2001). *Invitation to the mathematics of Fermat-Wiles*. Elsevier.
Hidayati, S. N., & Walck, J. L. (2016). A review of the biology of *Rafflesia*: What do we know and what's next? *Buletin Kebun Raya, 19*, 67–78.
Hölldobler, B., Wilson, E. O., & Nelson, M. C. (2009). *The superorganism: The beauty, elegance and strangeness of insect societies* (1st ed.). W.W. Norton.
Honderich, B. (1995). *Laws, natural or scientific*. Oxford Companion to Philosophy, Oxford University Press.

Houck, L. D., & Arnold, S. J. (2003). Courtship and mating behavior. *Reproductive Biology and Phylogeny of Urodela, 1*, 383–424.

Jena, A. (2020). Nonverbal communication across different cultures. *International Journal of Multidisciplinary Educational Research, 9*(5), 169.

Kallman, B. R., Kim, H., & Scott, K. (2015). Excitation and inhibition onto central courtship neurons biases *Drosophila* mate choice. *eLife, 4*, e11188.

Karban, R. (2015). *Plant sensing and communication*. University of Chicago Press.

Keller, L., & Surette, M. G. (2006). Communication in bacteria: An ecological and evolutionary perspective. *Nature Reviews Microbiology, 4*(4), 249–258.

Kirsch, S. (2006). History and the birds of paradise. Surprising connections from New Guinea. *Expedition: The Magazine of the University of Pennsylvania, 48*(1), 15–21.

Kirschner, M. W., & Gerhart, J. C. (2006). *The plausibility of life: Resolving Darwin's Dilemma*. Yale University Press.

Kleiner, I. (2000). From Fermat to Wiles: Fermat's last theorem becomes a theorem. *Elemente der Mathematik, 55*(1), 19–37.

Kuhn, T. (1970). *The structure of scientific revolutions*. University of Chicago Press.

Kuhn, T. S. (1996). *The structure of scientific revolutions* (3rd ed.). University of Chicago Press.

Lakatos, I. (1978). The methodology of scientific research programmes. In J. Worrall & G. Currie (Eds.), *Philosophical papers* (Vol. 1). Cambridge University Press.

Laland, K. N., Sterelny, K., Odling-Smee, J., Hoppitt, W., & Uller, T. (2011). Cause and effect in biology revisited: Is Mayr's proximate-ultimate dichotomy still useful? *Science, 334*(6062), 1512–1516.

Laland, K. N., Odling-Smee, J., Hoppitt, W., & Uller, T. (2013). More on how and why: Cause and effect in biology revisited. *Biology & Philosophy, 28*, 719–745.

Lee, M. S., & Palci, A. (2015). Morphological phylogenetics in the genomic age. *Current Biology, 25*(19), R922–R929.

Lloyd, F. E. (2011). *The carnivorous plants*. Lloyd Press.

Lovelock, J. E. (1985). *Gaia, una nueva visión de la vida sobre la Tierra*. Ediciones Orbis S. A.

Lovelock, J. E., & Margulis, L. (1974). Atmospheric homeostasis by and for the biosphere: The Gaia hypothesis. *Tellus, 26*(1–2), 2–10.

Lüttge, U., & Souza, G. M. (2019). The Golden Section and beauty in nature: The perfection of symmetry and the charm of asymmetry. *Progress in Biophysics and Molecular Biology, 146*, 98–103.

Maldacena, J., & Susskind, L. (2013). Cool horizons for entangled black holes. *Fortschritte der Physik, 61*, 781.

Manoukian, E. B. (2012). *Modern concepts and theorems of mathematical statistics*. Springer Science & Business Media.

Mattoni, C. I., & Acosta, L. E. (2005). A new species of *Bothriurus* from Brazil (Scorpiones, Bothriuridae). *The Journal of Arachnology, 33*(3), 735–744.

Mayr, E. (1961). Cause and effect in biology. *Science, 134*, 1501–1506.

McShea, D. W., & Brandon, R. N. (2010). *Biology's first law: The tendency for diversity and complexity to increase in evolutionary systems*. University of Chicago Press.

Mercier, H. (2016). Confirmation bias-myside bias. In R. F. Pohl (Ed.), *Cognitive illusions* (pp. 109–124). Routledge.

Moore, J. C., & Pannell, J. R. (2011). Sexual selection in plants. *Current Biology, 21*(5), R176–R182.

Nelson, J. D., & McKenzie, C. R. (2009). Confirmation bias. *Encyclopaedia of Medical Decision Making*, 167–171.

Nickrent, D. (2020). Parasitic angiosperms: How often and how many? *Taxon, 69*, 5–27.

Nickrent, D. L., & Musselman, L. J. (2004). *Introduction to parasitic flowering plants*. The Plant Health Instructor. https://doi.org/10.1094/PHI-I-2004-0330-01. Updated 2010.

Ojanguren-Affilastro, A. O. (2005). Estudio monográfico de los escorpiones de la República Argentina. *Revista Ibérica de Aracnología, 11*, 75–241.

Olivero, P. A., González, A., Mattoni, C. I., & Peretti, A. V. (2015). Chemical caresses: Geographical variation of male sexual signals in a Neotropical scorpion. *Behavior, 152*(12–13), 1745–1763.
Olson, M. E. (2012). The developmental renaissance in adaptationism. *Trends in Ecology & Evolution, 27*(5), 278–287.
Oswald, M. E., & Grosjean, S. (2012). Confirmation bias. In R. F. Pohl (Ed.), *Cognitive illusions* (pp. 91–108). Routledge.
Paps, J. (2016). *These bizarre creatures defy what we think we know about plants and animals.* The Conversation. Retrieved from https://theconversation.com/these-bizarre-creatures-defy-what-we-think-we-know-about-plants-and-animals-69316. September 21.
Peretti, A. V. (1993). *Estudio de la biología reproductiva en escorpiones argentinos (Arachnida, Scorpiones): Un enfoque etológico.* (Doctoral thesis). Facultad de Ciencias Exactas, Físicas y Naturales, Universidad Nacional de Córdoba.
Pigliucci, M. (2014). Between holism and reductionism: A philosophical primer on emergence. *Biological Journal of the Linnean Society, 112*, 261–267.
Poggi, J., & Ohsawa, G. (1986). *The order of the universe*. George Ohsawa Macrobiotic Foundation.
Pozzo, M. I., & Soloviev, K. (2011). Culturas y lenguas: La impronta cultural en la interpretación lingüística. *Tiempo de Educar (Universidad Autónoma del Estado de México), 12*(24), 171–205.
Raghavender, U. S. (2021). *Ecological genomics*. Delve Publishing.
Ruíz, R., & Ayala, F. J. (1998). El método en las ciencias. In *Epistemología y darwinismo*. Fondo de Cultura Económica de España (Colección Ciencia y Tecnología).
Ruppert, E. E., & Barnes, R. D. (1996). *Invertebrate zoology* (6th ed.). Saunders College Publishing.
Sawyer, R. K. (2005). *Social emergence: Societies as complex systems*. Cambridge University Press.
Schaefer, H. M., & Ruxton, G. D. (2009). Deception in plants: Mimicry or perceptual exploitation? *Trends in Ecology & Evolution, 24*(12), 676–685.
Schierwater, B., & Desalle, R. (2021). Organizing Invertebrates. In B. Schierwater & R. DeSalle (Eds.), *Invertebrate zoology. A tree of life approach* (pp. 81–92). CRC Press.
Scott-Phillips, T. C. (2008). On the correct application of animal signaling theory to human communication. In A. D. M. Smith, K. Smith, & R. Ferreri Cancho (Eds.), *Proceedings of the 7th international conference on the evolution of language* (pp. 275–282). World Scientific.
Sherry, D. F. (2005). Do ideas about function help in the study of causation? *Animal Biology, 55*(4), 441–456.
Shimazaki, M., & Nakaya, K. (2004). Functional anatomy of the luring apparatus of the deep-sea ceratioid anglerfish *Cryptopsaras couesii* (Lophiiformes: Ceratiidae). *Ichthyological Research, 51*, 33–37.
Stephenson, A. G., & Bertin, R. I. (1983). Male competition, female choice, and sexual selection in plants. *Pollination Biology, 109*, 149.
Tavassoly, I., Goldfarb, J., & Iyengar, R. (2018). Systems biology primer: The basic methods and approaches. *Essays in Biochemistry, 62*(4), 487–500.
The American Heritage®, Dictionary of the English Language, 2003.
Tinbergen, N. (1963). On aims and methods of ethology. *Zeitschrift für Tierpsychologie, 20*, 410–433.
Trevors, J. T., & Saier, M. H., Jr. (2010). A tale of two worlds: The natural world and the artificial world. *Water, Air, and Soil Pollution, 205*, 37–38.
Van Straalen, N. M., & Roelofs, D. (2011). *An introduction to ecological genomics*. Oxford University Press.
Vieira, S., Biscoito, M., Encarnação, H., Delgado, J., & Pietsch, T. W. (2013). Sexual parasitism in the deep-sea ceratioid anglerfish *Centrophryne spinulosa* Regan and Trewavas (Lophiiformes: Centrophrynidae). *Copeia, 2013*(4), 666–669.
Wayne, J. (2020). Fundamental laws of biology. *SSRN*, 3722831.
Weaver, W. (1991). Science and complexity. In G. J. Klir (Ed.), *Facets of systems science* (pp. 449–456). Springer US.
Weller, J. M. (1955). Protista: Non-plants, non-animals? *Journal of Paleontology*, 707–710.

Wendt, H. (1982). *El descubrimiento de los animales. De la leyenda del unicornio hasta la etología*. Planeta.

Wilson, D. S., & Sober, E. (1989). Reviving the superorganism. *Journal of Theoretical Biology, 136*(3), 337–356.

Won, M. Y., Oyama, L. B., Courtney, S. J., Creevey, C. J., & Huws, S. A. (2020). Can rumen bacteria communicate to each other? *Microbiome, 8*(1), 1–8.

Yang, Z., & Rannala, B. (2012). Molecular phylogenetics: Principles and practice. *Nature Reviews Genetics, 13*(5), 303–314.

Chapter 2
A Small World of Exceptions?

Insisting on rules would not have improved things, they would have interrupted the process

—Paul K. Feyerabend, 1978

Contents

Abstract When rules are mentioned in biology, the opposite concept of exceptions immediately arises. In this chapter, we analyze the existing definitions of what an exception is in general and in life sciences. We give examples that invite us to reflect on the richness and complexity of exceptions since it is not just an absolute concept, but there can be gradients of types of exceptions depending on the type of rule to which they are linked. We mention what exceptions mean in everyday life and what value they have in the scientific field. We also comment on the concept of exception, which is extended to what is considered exceptional in terms of biological charac-

A. V. Peretti et al., *Rules and Exceptions in Biology: from Fundamental Concepts to Applications*, https://doi.org/10.1007/978-3-031-55382-0_2

teristics. In this context, single and multiple trait exceptions and the implications of population outliers are discussed.

2.1 Introduction

When we think of exceptions, even outside the scientific field, we think of rare events or situations that indirectly reinforce the pattern, that is, what is typical or normal. Something like this happens every time. Let us begin to illustrate this topic using a curious city anecdote. In the city of Córdoba, Argentina, we observe the sculpture carved in marble of a polar bear, which has, between its front paws, caught a fish, possibly a salmon. This sculpture, the "bear" as it is called there, has a long local history. It was commissioned by the Argentine painter Roberto Juan Viola and the Spanish sculptor Alberto Barral on the occasion of the inauguration of a bridge that crosses the Suquía River in the city at the end of the 1940s. The problem was that the bridge was named "Antártida Argentina," and it was planned to place an animal representative of the Antarctic region on one of the bridge's ends. There were applauses when the statue was unveiled, but one of the then Mayor's advisors came up to him and whispered in his ear that there were no bears in Antarctica! Well, that sculpture was removed from the place, the name of the bridge was not changed to "Ártico Argentino" (because it would also be inaccurate), and, unfortunately, they did not put a penguin sculpture instead. The bear statue began a journey of almost six decades through various city places and squares until finally falling into what is perhaps its final place, gardens outside the city museum of contemporary art. The statue's history has given rise to extensive notes and a recent book in which all the anecdotes summarized here are described in detail (Lavezzo, 2013).

This curious domestic history shows what an exception in the anecdotal history of a society is. First, bears inhabiting Antarctica would have been an exception. The presence of such a sculpture in a city of the Southern hemisphere, wandering for decades, was a total rarity, a strangeness for those who saw it without knowing the original story. How many rarities will there be in other cities? Minimal stories of local events marked in the popular unconscious, enrich the collection of domestic histories, which are no less striking for that. The truth is that the rare, the exceptional, always attracts or repels us, but it is not indifferent to us. As we will see later, this response is also observed in the scientific field, where the oddities, whatever their nature, attract our attention powerfully, or, in contrast, we directly ignore them and do not even notice them. It is like a "black and white" our position toward exceptions, a fact that is reflected in the value or not that we give to them in science, and in the case of biology, especially in crucial fields such as developmental biology, ecology, and the evolution of organisms and systems.

2.2 What Is an Exception?

The definition has many options, but they coincide with what is central. We could say that an exception is "*something excluded from, or not in accordance with the general rule, principle, or class*" (Collins English Dictionary, 2003) (Box 2.1). If we pay attention to this definition, it is self-evident that it has no independent or absolute entity since it is intimately linked to the definition of its opposite, i.e., the rule. Therefore, a definition of rule and a definition of exception become antonyms and, at the same time, "mutually necessary" opposites to give an entity and complete the understanding of what each of them alludes to. Interestingly, the definition of an exception, as in the case of rules, exists "before" any analysis of its existence in a given discipline. These basic definitions apply to the sciences, the arts, sports, and other areas of knowledge and human activities.

BOX 2.1 Glossary and synonyms of exception

Glossary

Anomaly: something that deviates from what is standard, normal, or expected.

Exception: a person or thing that is excluded from a general statement or does not follow a rule.

Extrem: not usual; exceptional. the highest or most extreme degree of something. furthest from the centre or a given point.

Rarity: a rare thing, especially one having particular value. unusually good or remarkable. [Rare: of an event, situation, or condition) not occurring very often.]

Teratology: the scientific study of congenital abnormalities and abnormal formations.

Synonyms

anormal unusual infrequent singular exceeding
remarkable irregular particular uncommon outlier
exceptional oddity minority special atypical
abnormal prodigious strange deformed monstrous

License: own creation.

2.2.1 *They Are in Our Daily Life*

The truth is that "rule" and "exception" are closely related concepts that scientists have in mind during their routine research. This "keeping in mind" can be formally embodied in expressions (laws, statements of patterns, and predictions) or more implicitly or intuitively. During our research, we use samples either in descriptive or experimental work, where we measure parameters of selected variables (e.g., traits such as size, length, width, height, and weight). Moreover, it is in these factors that we then see how they do or do not conform to a "norm" and whether we have outliers or "exceptional" cases. The problem is that if the latter is not conspicuous or "striking" to the observer, they may go unnoticed. There is a sensory bias that often allows us to appreciate what the rule is, perhaps because it is the majority and it may be the easiest thing to perceive. At the same time, if the oddities or exceptional cases are not aberrant, they are barely detected and often ignored. We will return to specific cases of extreme values in Chap. 5 when we analyze the nature and types of exceptions.

Interestingly, from areas such as positive psychology (e.g., coaching strategies and leadership), the value of feeling "exceptional," for example, to be out of the "norm" of the masses, is valuable and promoted. In addition to belonging to being part of a social group, here, the value of individuality is promoted, whether in personal life, at work, or in social life.

Therefore, we are not indifferent to exceptions if they are marked in the reference pattern and indicate a highly valued quality in society. This situation has been used by advertising, marketing, and fields nourished by striking oddities but synonymous with positive, healthy individuality above the common. This attracts us both to feel (and be?) different and to make a difference over the rest. In the history of humanity, not only in science but also in politics, art, and sport, the exceptional has been a synonym for "success", geniality, superlative, and promoter of resounding changes. Sports idols, i.e., the "best in the world at..." are exceptions to the rule. In other words, the rest of us are not the best in that measured parameter. Later, we will discuss the question of to what extent something or someone is exceptional as a "whole" or only because of a single "one variable." In Fig. 2.1, we see just a few media examples of athletes who are or were considered the best of the best, undisputed, i.e., "exceptional." For example, the fastest man in the world, the tennis player who has won everything, the Olympic super medalist, or the best in basketball are always highlighted. Each of these outstanding individuals represents an extreme of the best. They can be considered a stereotype of the ideal in that sport.

If we pay attention, everything goes hand in hand with the concept of "almost" perfection. Hence, idolatry because we see these "idols" as invincible. In other words, perfection is the opposite of mistakes and failures. For this reason, when these idols "fail" at what they excel at, society stops projecting its dreams onto them, often projects its failures, and may not forgive them for their mistakes. Society condemns how someone "exceptionally unique" can cease to be so from one moment to the next. The case of the great footballer Lionel Messi, for example,

Fig. 2.1 Exceptional athletes. Some of the best athletes in their disciplines. (**a**) Soccer player, Leonel Messi. (Credits: Tasnim News Agency). (**b**) Olympic sprinter, Usain Bolt (Credits: Erik van Leeuwen) (**c**) Olympic swimmer, Michael Phelps (in the middle) (Credits: White House photo by Eric Drape). (**d**) Olympic tennis player, Serena Williams. (Credits: Edwin Martinez from The Bronx).
Licenses via Wikimedia Commons and GNU free documentation. (**a**) Leonel Messi. License: CC BY 4.0. https://creativecommons.org/licenses/by/4.0, via Wikimedia Commons. https://es.wikipedia.org/wiki/Lionel_Messi#/media/Archivo:Argentina_vs_mexico_messi_vs_alvarado.jpg. (**b**) Usain Bolt. License: GNU free documentation license. https://es.wikipedia.org/wiki/Archivo:Usain_Bolt_100_m_heats_Moscow_2013.jpg. (**c**) Michael Phelps. License: Public Domain. https://es.m.wikipedia.org/wiki/Archivo:Michael_Phelps_Ryan_Lochte_Laszlo_Cseh_medals_2008_Olympics.jpg. (**d**) Serena Williams. License: CC BY 2.0 https://creativecommons.org/licenses/by/2.0, via Wikimedia Commons. https://commons.wikimedia.org/wiki/File:Serena_Williams_US_Open_2013.jpg

reflects this duality. In his golden era at FC Barcelona, Messi won many national and international trophies. In the Argentinian national team, however, he was unable to achieve the same level of performance, which is why he was fiercely attacked by fans and many critics in his country. It must have been a difficult time for Messi. The media forgot that this great player is not only an “idol,” but also a living being and not perfect.

Furthermore, the working environment at his club was different from that of the national team. So in this case, a behavior or conduct (i.e., in this case, the outstanding ability to play football) that was almost perfect in one context (at FC Barcelona) was not comparable in another (in the Argentinian national team). However, the criticism and attacks changed dramatically after he directed the Argentinian national team, which won three trophies, beating their former champions in the process: Brazil at the 2021 Copa América, Italy, at the 2022 Copa de Campeones Conmebol-UEFA Finalissima, and France at the 2023 FIFA World Cup. The environment at these tournaments was also like that of FC Barcelona and Messi could feel at home there. With his performance, he has once again shown that he is far more than just a normal footballer. We could list other cases of athletes whose peak performance depends heavily on the context; it is therefore not their sole responsibility to be or not to be "exceptional" in all situations.

In the case of women, it is even more remarkable. When talking about the female athlete and all the challenges she faces because she is a woman, all the social constructs about gender roles are included, and how these have generated stereotypes and expectations that "should" be followed. Although in recent years, there has been significant progress in gender parity in sports, the media continues to offer a stereotyped image of women, which also has an impact on public opinion. Even so, physical beauty is an essential characteristic of women in sports and not men (Mujika-Alberdi et al., 2021). Moreover, the same occurs in the case of motherhood (Ferrer et al., 2022). Therefore, we are talking about situations that could be considered double exceptions. For example, an outstanding woman in sports will not be an exception only because of her different sporting abilities but also in terms of a break in the "mandates" of her assigned role as a woman within society. However, we can still count on many exceptional women in this field, such as Martina Navrátilová, Serena Williams, Simone Biles, Annika Sörenstam, and Marta Vieira da Silva.

In the field of the arts, musicians such as Mozart, Beethoven, Wagner, Chopin, Clara Schumann, Fanny Mendelssohn, and Barbara Strozzi or painters such as Van Gogh, Dalí, Picasso, and Frida Kahlo are examples of the impact generated by being in front of "exceptionally good" beings in "that" which they do in an incredible way. The word "genius" implies, by definition, to be someone different from the rule, to be an exception. The Oxford Languages dictionary defines genius as *"exceptional intellectual or creative power or other natural ability"* or *"an exceptionally intelligent person or one with exceptional skill in a particular area of activity."* As we can see in both definitions, words such as *exceptional* and *exceptionally* undoubtedly refer to qualities outside the performance of individuals in the society from which this genius comes.

On the other hand, taking the Olympic Games as an example, the news highlights the fastest runner, such as the Great Usain Bolt, or the fastest swimmer, such as Michael Phelps, but not the slowest, the least skilled. That is, we do not know or are not made aware of the "other" extreme. In this case, the one who arrives last at the finish line can be considered a circumstantial extreme for that discipline and that edition of the Olympic Games but does not represent the true extreme. The bias is due to the implicit social condemnation that being last is equivalent to total failure, even if you have been to the Olympic Games. It may not be disclosed so as not to stigmatize that person. Are there people faster than Bolt running or Phelps swimming? We do not know, and the Olympic Games bring together a group that has passed numerous filters and has wanted to dedicate themselves professionally to these activities. In short, sports, arts, and sciences highlight the best, but we need to know or be interested in the "worst," the other side of the spectrum, according to the respective canons in each discipline. Someone might ask: why know and divulge who was the slowest? As scientists, we want to know or have it on par with the best. Considering also the "worst" would allow us to analyze, for example, the causes of the differences between the "excellent vs. poor" extremes at the level of genetic and environmental variables translated into physical or psychological qualities. Therefore, there is an impressive bias in humans to "show" the best and consciously or unconsciously hide the "worst" individuals; is not it interesting to highlight the weaknesses of our species?

Exceptions among humans can also have a connotation opposite to the extreme admiration or idolatry we have discussed. That is, the exception is linked to what produces "aversion" or abhorrence. There is a gradient of traits among humans, be they behavioral or physical, and canons established by societies of what is beautiful and what is ugly. Imagine a pendulum moving from the positive (admiration) to the other extreme of the negative (aversion). Then, we can see how fabled characters such as Frankenstein's monster, demons, malformations, and historical characters who have committed crimes can all fall into being socially considered negative exceptions, "rejectable" or certainly not "admirable" abnormalities.

Moreover, someone who is a "genius" in one variable or area of life may be unacceptable in others. In the past, oddities have been considered (consciously or unconsciously) "negative" in many societies. This bias has led, in many cases, to xenophobia, racism, discrimination, and crimes. In short, this has been experienced in many stages of human history, from Sparta (where according to Plutarch, sick or disabled babies were sacrificed—Penrose, 2015; but see Hurst, 2008) to Nazism, which combined the rejection of everything that did not conform to the concept of rule, which, according to the canons, was the Aryan stereotype.

We know that, fortunately, erroneous and fanatical look has been changing and our contemporary look as a society is far from that of yesteryear. Everything that represents not only minorities but even more true rarities is currently contemplated differently. In the novel "The Elephant Man" by Bernard Pomerance or the character Quasimodo (pun for "almost human") in "Notre Dame de Paris" by Victor Hugo, situations associated with the duality of the concept of physical beauty on the one hand and spiritual beauty on the other are reflected. These stereotypes draw attention to how society tends to focus mainly on the former. The "Elephant Man" was the nickname given to John Merrick. This English citizen became famous due to his terrible malformations (Proteus syndrome), as his moral, ethical, and intellectual values far surpassed the qualities of many of his contemporaries. Unfortunately, Merrick lived condemned to general rejection and was exhibited at numerous fairs as a "freak" until even after his death. Merrick wrote of himself and completed a poem by Watts: "*Is true that my form is something odd/But blaming me is blaming God; Could I create myself anew/would not fail pleasing you. If I could reach from pole to pole/Or grasp the ocean with a span, I would be measured by the soul/The mind's the standard of the man.*" (Adapted from Isaac Watts's "False Greatness.")

Exceptions shock us, and many atrocities have been committed against what we consider strange and different, often supported by pseudoscientific (or incomplete) arguments. This view is not only toward other human beings but also animals (e.g., toward animals taken as monsters, as associated with witchcraft, such as black cats, owls, spiders, and bats). In a neutral framework (i.e., neither negative nor positive), the exceptions often promote their gathering in written compendiums or public recognitions that include awards for what is extreme in some quality. One example is the famous Guinness Book, which includes thousands of disciplines, characteristics, facets, and the most unusual records. These refer to fields in which the protagonists, whether humans, animals, or plants, could fall into categories of teratologies or exceptions attractive to our anthropic bias. Everything is finally about surpassing sporting records, breaking records, expressing this or that ability or quality, and gathering everything that is "outstandingly exceptional."

2.2.2 Does the Scientific Method Ignore What Is Nothing or Even a Little "Repeatable?"

This fundamental question goes hand in hand with another, more specific question for this book: are exceptions significant for biology? We mean "important" both for biology itself and for the scientists who study it in its different disciplines. The answer is a resounding Yes! Exceptions are often the starting point for many changes in forms, behaviors in organisms, and systems in evolution. In other words, they are the initial kick-off or "raw material" for traits that will later be consolidated.

As we will see later, typically, we will not find only one type of exception but several. We can find levels or, instead, a gradient in the same way as when we referred to rules earlier. We can identify exceptional cases from interspecific or higher "interclade" degrees. Another more circumscribed level will be the intraspecific level. For example, individuals deviate from the characteristic pattern of their species. Even more, restricted in terms of level, there may be population patterns, and within them are the exceptions. It is like viewing the world using a telescope, a magnifying glass, or a microscope regarding the organismic levels of biological systems. Therefore, some exceptions may attract our attention more than others (conditioned by a particular sensory bias), while many exceptions may not even have been discovered. Biology, by its very nature of organismic diversity, from bacteria to plants and animals of all evolutionary levels, added to eco-geographical and temporal variability, offers us a wide range of these exceptions of different gradients or levels. If we add, as we will highlight later, the influence of a man on climate change, we will have the potential diversity for more variants to occur within a given pattern or norm, which can become very high.

The essence of exceptions is to be unusual, either in the frequency of occurrence with what is expected (something more quantitative within a continuum) or radically different forms of the pattern (something more qualitatively different). This characteristic implies that they should be taken into account in more detail by the scientific method. Based on statistics and probabilities of occurrence to define patterns, the scientific method does not necessarily contemplate highlighting what is rare as a fundamental piece of the descriptions, at least in daily practice judging by what is reflected in most scientific publications. The scientific method classically relies on experimentation and the "repeatability" of facts, phenomena, and data taken in control vs. experimental/s groups after this or that treatment until a sample size (n) is reached. However, it is obviously in the very nature of the exception not to be repeatable or to be repeatable in a very low or even not precisely the same way in case it depends on several variables. This characteristic finally results in the fact that exceptions "do not fit" in the logical foundations of the scientific method, at least the conventional one (i.e., positivism vs. a more holistic view which, on the contrary, does consider them more). We will address part of this problem when we discuss exceptions from the statistical analysis of biological data.

In a way, the scientist's gaze tends to recognize patterns, requiring a much more detailed look to detect exceptions. On many occasions, exceptions can be recognized, although their subsequent analysis is generally truncated, exposing their existence as marginal and without much biological relevance. Without devaluing the identification, study, and comparison of rules, the challenge is to become more committed as researchers to avoid overlooking or hiding (because they are not considered essential) biological exceptions. In this way, it would be possible to have integrative approaches for a better understanding of biological phenomena.

2.3 Waiting for Exceptions According to Rule Levels

In the previous chapter, we analyzed a certain degree of hierarchy regarding the level of "influence" between sciences such as mathematics, physics, chemistry, and biology. It is interesting to consider the "probability" of exceptions in each of these scientific areas (Fig. 1.2a). In principle, this probability will be inversely related to the level of influence of the science one refers to. Thus, they would be null within Physics, something more common in Chemistry, and even more so in Biology. These differences then translate into why we expect a higher occurrence of exceptions in specific sciences, particularly in the life sciences, and not so in others. This tendency is reflected in greater or lesser confidence in the enunciation of rules in these sciences.

Thus, in mathematics and physics, there would be no significant objections to the search for rules that can govern all physical phenomena. Chemistry also, although it may have some minor cases where specific chemical rules do not apply. Finally, biology challenges with the power or not to propose comprehensive laws (the most "universal") (e.g., those of Mendel or Hardy–Weinberg in genetics) due to the existence of exceptions, which by default are expected to occur in most cases. The exception to this would be the case of the level I rules and the fundamental laws of biology discussed in the previous chapter, which are worth reiterating. Within the framework of these, there would be no exceptions. However, as we have already shown, it has yet to be possible to postulate rules and the existence of recognizable patterns for many situations in various biological disciplines. In any case, the key is to contemplate that exceptions will occur, as we have already mentioned, according to the level of influence (I, II, and others) of the rule posited.

This greater feasibility of proposing rules, more generalizable and comprehensive laws in physics and much less in biology, could also be seen, on another level, for example, among the disciplines of biology in terms of the probability that exceptions can occur. Thus, those disciplines framed in molecular and cellular biology, having a more direct connection with physics–chemistry and at the same time being the bases of the disciplines of organismic biology, could be the ones that presentin comparison with the lattera lower number of exceptions. A level I rule of biology that we have already discussed that all organisms are defined by cells with membranes comes from molecular and cell biology. However, those rules that are more labile in terms of rule rigidity, such as those from the field of ecology and behavior, would be more subject to possible exceptions. Indeed, in these biological disciplines, interindividual and even intraindividual variabilities in time and space are a central aspect of their essence.

2.4 The "Exceptional" from the Point of View of Exceptions

When we refer to exceptions, rarities, and abnormalities, we limit ourselves to facts, groups, processes, and events that refer to something that deviates from the rule in a strict, visible sense. However, we could also include cases and situations related to this concept, although not strictly speaking. Thus, adverbs of frequency are used in examples such as "this organism always flies fast," "this process seldom, occasionally fast/often, frequently slow," or "this animal is rarely big among its group." Additionally, when mentioning something such as "this group is exceptional" or "that biological system is exceptional because of such a quality," we refer to the use of the term exceptional. However, we could replace it with related terms, such as rarely or abnormally, to describe the frequency of a particular characteristic or adjective (big or small, slow or fast, high or low).

Therefore, if we were to pay attention within the book to all these variants derived from the concept of exception, we could come across an endless universe of examples. This point does not invalidate the more limited analysis of striking, direct, or drastic cases of exceptions. In any case, this complexity makes the problem richer and shows an almost continuous gradient of which we may not yet be aware regarding the boundaries between rules and exceptions. Our flexibility in analyzing biological phenomena is necessary. Otherwise, reductionism will not allow a more holistic view to emerge.

2.4.1 An "Exceptional" Example? The Rapid Evolution of HIV

Concerning what has been said about terminology linked to exceptions, we can take a specific case close to humans at the evolutionary level. When illustrating cases of rapid evolution, perhaps the example of what happens with HIV is presented as an exceptional case. Reference is made to the fact that the evolution of this virus is "exceptionally" rapid (Hutchinson, 2001; Ridley, 2004). Indeed, the evolution of HIV takes place within the body of an individual and is, comparatively, exceptionally rapid compared to most other viruses.

This microcosm of mutations, natural selection, and finally adaptation (in this case, drug resistance) simulate what occurs at the macroevolutionary level in other organisms. Even a relatively small change at the mutational level quickly translates into an adaptive process of the virus to the 3TC drug. Elements such as having a high default mutation rate, a large population size in the host, and a high recombination rate strongly favor such exceptionally rapid evolution (Ridley, 2004). However, much remains to be clarified. For example, more needs to be known about the molecular mechanisms used in viral replication and the strategies adopted by the virus to evade host immune surveillance (Becerra et al., 2016). For example, a mutant "Tat" protein has recently been found to inhibit the infection of human cells by strains of various HIV-1 subtypes (Rustanti et al., 2017).

This rapid change due to a high level of mutations is standard in RNA viruses compared to DNA viruses (Holland et al., 1982; Steinhauer & Holland, 1987). Indeed, other RNA viruses, such as Ebola or influenza viruses, are also challenging to treat due to their high mutational rate (Hoenen et al., 2015; Harding & Heaton, 2018). These are rarities from the perspective of evolutionary speed that invite us to analyze why such a high mutational rate and associated aspects compare them with that of other viruses or pathogens. In this sense, these RNA viruses share the fact that they do not have a highly developed proofreading process (Elena & Sanjuán, 2005). This characteristic results in these "errors" due to mutations not being corrected, thus increasing the genetic diversity of the virus variants and their possible recombinations, which is essential for rapid evolution. Indeed, genetic variants coexist in the virus population, "cooperating" to cause a more significant effect, virulence, on the host than if only one existed (Fraser et al., 2014). Selection will occur in that population of coexisting variants as "a whole" rather than on an individual variant. These variants end up giving identity to a given population in terms of composition, functioning, and virulence. The term "quasispecies" has been applied to many of them due to their unique profile (Domingo et al., 1985; Najera et al., 1995; Eigen & Biebricher, 2018).

Therefore, HIV and many other RNA viruses are biological exceptions to other types of viruses because of their rapidity in evolving within an individual, resulting in increased drug resistance. In this sense, each one is an exception from a statistical approach. This situation does not imply that we are speaking here of an exception regarding the frequency of virus abundance but of a specific quality: speed of evolution. This particular case serves to warn us, once again, that just as when analyzing the evolution of traits by natural or sexual selection, it is always best when analyzing exceptions to specify which trait or group of traits we are referring to. In the example of HIV, we are dealing with a trait with a particular expression (rapid evolution) in a context (evolutionary time of only a few days) and space (within the host). We will return to HIV and examples of disease evolution later in the book when we analyze the dynamism between rules and exceptions.

2.5 Exceptions: Single-Trait or Multitrait?

When an individual is an exception in a group, is it as a "whole" or regarding a particular variable? Perhaps we inadvertently tend to speak of an exception based on a more relevant variable rather than a set of variables, either because of our sensory or conceptual bias or because the latter is critical from a biological point of view. However, it is worth asking whether we limit ourselves to a typified variable or variables when we have an individual who falls into the exception category. Alternatively, we can indirectly generalize it and involve more variables until we reach the whole individual without analyzing the latter. The implications of considering this are important. For example, an individual could be an exception within the population in terms of the variable "body size" (e.g., being tiny or giant), but this does not imply that other individual variables are associated or correlated, such as competitive ability or genetic quality. It may or may not, but we should check this and not generalize from this exception for one variable to the whole possible set of variables. In other words, we should not label the individual based on a single characteristic. This option limits expressing what a pattern, a rule, and what is not is an exception.

When, in Chap. 5, we look at the potential value of specific "variants" in a given trait, we will inevitably limit ourselves to a particular trait. Therefore, it is essential that, as far as possible, we take into account this limitation in the approach. We have already mentioned that when talking about selection processes in evolution, it is best to start by analyzing a particular trait before moving on to a "multitrait" complex. In this sense, it is important to avoid categorizing the decision to consider an individual solely based on the rarity or exceptionality of a particular characteristic as either "positive" or "negative" at a reductionist functional level. The same could be said if what is being analyzed is at a supraindividual level. Examples of this could be a population, a species, a larger taxonomic group, or even a more extensive process (e.g., not a trait in the strict sense but a process or mechanism). The concept of rarity is linked to the frequency of occurrence and abundance at the population or whole-species level. There are interesting studies on the biology of rarities (almost synonymizing this term with "infrequency").

On this subject, it is worth mentioning the book "The Biology of Rarity," edited by Kunin & Gaston (1997), which was a synthesis of what had been published up to that time on rare species, their dynamics in ecosystems, and reproduction systems.

It is mainly focused on the presence of rarities as the "opposite" of the abundance or density of a species. This point is also expressed as species-range size distribution in terms of geographical level (Gaston, 2014). These authors point out that these rare species, whether locally rare or geographically restricted, not only differ from the common ones by this "attribute" linked to their frequency in the number of individuals but also present other characteristics that differentiate them. These features exist, for example, at the reproductive level, such as low genetic incompatibility when crossing, a tendency to reproduce by chance, low reproductive effort, and low dispersal ability. They add that several mechanisms may be responsible for such differences, but in practice, they may be challenging to differentiate. However, the cataloging of rare-common differences that recur is of great importance, as it may allow us to partially compensate for the bias of the published literature toward reports with the most common taxa (Kunin & Gaston, 1993). The implications of considering rare species in conservation contexts are significant (Rey Benayas et al., 1999). We highlight the last sentence on the bias, perhaps logical, of having focused on the most common taxa. This bias has meant losing sight of several opportunities to learn about exceptions in terms of strategies, mechanisms, reproductive patterns, life histories, and morpho-physiological patterns, which could be critical to a much better understanding of "the rule," the dominant pattern, in these common species.

Are we, in science, to a greater or lesser degree, limited in experimental designs due to focusing our attention on a handful of variables to label what is and is not an exception? From the social point of view, we have already mentioned the high cost for human individuals of "falling" into the category of exception according to beauty standards or being labeled as "abnormal" in their cognitive or social capabilities. The opposite is also true, as we have already mentioned: cases in which individuals who are exceptionally good at a particular skill in the art, science, or sport are expected to be exceptionally "positive" in other cognitive, social, and human values and skills. Thus, we tend to be biased toward the exceptionally "negative" or "positive," attaching the rest of the qualities of that individual in the same line. This bias brings rejection, prejudice, and discrimination, in the first case, or excessive admiration, idolatry, and mythologizing in the second. Biology does not escape this bias, at least in terms of omitting to report exceptions, underestimating or overestimating them, as the case may be.

As we will see below, we can find a species that is taken as an exception either in its closest clade (e.g., genus or family) or more distant clade (order, class, phylum). This criterion may have been based on a particular trait or several traits. In many publications, this is reflected in the titles, where "a trait" is emphasized to mention that a species or individual is an exception. However, this unintentional reductionism may occur due to our bias toward the most striking aspects of an organism, focusing on exceptional traits such as extreme age, size, or weight, as we will explore later.

The point is that there may well be a very close association between more than one trait (e.g., between size and weight), and thus, the exception is no longer, strictly speaking at least, "uni-trait." Even so, what may still stand out as such is one trait (i.e., variable) over the associated other(s), even if the latter may also be labeled as an exception or at least fall in the minority of the variable distribution. Then, it is evident that in many cases, there appear before our eyes organisms that can undoubtedly be considered valid exceptions at the level of not one but several traits. These traits may not even be directly associated with each other. In general, the exceptional characteristics are presented in atomized form or as watertight behaviors at the level of "uni-traits."

2.6 "Invisible" Exceptions: Outliers in a Population

Thus far, we have mentioned two territories of exceptions, on the one hand, those linked to abnormalities of form or teratology, and on the other hand, exceptions of taxonomic nature in biological diversity. In the first case, they involve extreme rarities within a species, and in the second, between species (or the clade to which we refer). Both types will be discussed in more detail in the following chapters. Now, unlike these two options, some rarities could be more striking. We would say they are almost "invisible" to an observer, at least in a first approximation. We refer to extreme cases in the phenotypic expression of a given characteristic or trait. These cases fall in the extreme quantitative values, either maximum or minimum, well outside the "average" value within the population. Indeed, as shown in Fig. 2.2a, under standard conditions, the "norm" of the population (its normal distribution) shows us that such extreme values exist. However, as scientists, we generally pay little attention to these cases far from the mean. Thus, we express with means and

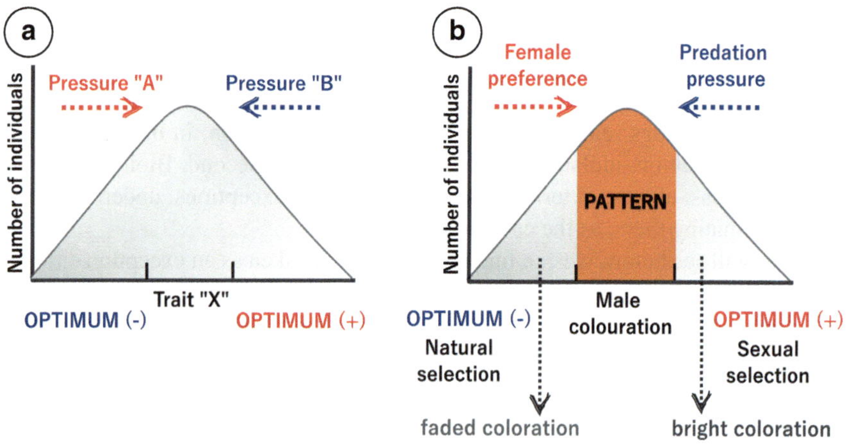

Fig. 2.2 Graphical representation of the distribution of individuals of a population with a certain trait "X". (**a**) This example shows how mean values relate to what is known as the pattern or rule and extreme negative or positive values to what is called the "exception." Certain selective pressures can generate changes in the distribution. (**b**) Example applied where the trait of interest is male coloration, and the selective pressures opposed are sexual and natural selection

their standard deviations what the "majority" of the individuals in the sample do, which is understandable if we want to characterize a given sample statistically. Statistics are designed to show this descriptively and comparatively since we can compare two or more samples according to these parameters. This is based on how concentrated or dispersed the majority is around a representative value.

This approach is necessary to know the "pattern" of what we are analyzing, studying, and observing at the intra- or intergroup level. It is not common, of course, that in a scientific article, the researcher stops in the results to show or analyze extreme values. For example, suppose the average height of individuals in a population is analyzed. In that case, the average value will be apparent in general terms or expressed according to the association with other variables, such as age, weight, and origin. However, no one usually stops saying anything about the tallest or shortest individual in height. According to this or that factor, this sample exhibits significantly higher or lower values in height. Nevertheless, this refers to the mean, not the individual point value.

Figure 2.2b introduces an additional component to consider: selection pressures. In a classical Gaussian bell, which hypothetically exhibits the typical normality in the distribution of a variable (e.g., trait "x"), a static scenario is reflected, a "snapshot" product of stabilizing selection on that variable. Therefore, we observe the outcome of the interplay between various selective forces that influence the expression of a trait to varying degrees. These forces can include abiotic factors such as temperature, latitude, and altitude, as well as biotic factors such as predation and feeding. The equilibrium resulting from this interplay is not static but rather fluctuating and modal. Life is characterized by the existence of these fluctuations, even though our analysis may often portray the "whole" as a snapshot in the punctual instant of our analysis, summarized by average values of a given variable. As we will see in the next chapter, those extremes can be the basis for unbalanced selective pressure changes. This situation can go from one stable state to a different one in a "directional" way (i.e., toward one extreme of the distribution) or to two different events in a "disruptive" way (i.e., toward both extremes of the original distribution). Therefore, it is in these extreme values that new possibilities for change, both at the ontogenetic and evolutionary levels, can potentially emerge as significant turning points.

However, even without relying on the potential value of evolutionary change to explain the exceptions, it is important to recognize that the outliers within a population provide valuable information about the prevailing "pattern" that is observed in the majority. We often fail to acknowledge that everything we analyze about the world around us is from an explicitly or implicitly rationalized comparative perspective. In other words, our assessments are far from absolute. When, for example, we describe an individual as being much taller or shorter than another, we are making a comparison. If we say this red color is much stronger than another but fainter than a third one, we are also engaging in comparison. Therefore, the comparison implicitly introduces the concept of extreme values.

On the other hand, going back to something previously said, if we have a mean value, it is interesting to ask what happened to those individuals who exhibit the expression of the trait in its extreme ranges (e.g., in statistics, those who fall in the maximum and minimum interquartile range). How many are there? What are their particularities? This point connects to the variance in the sample, i.e., how concentrated the sample is to the average value. By placing more emphasis on analyzing and explicitly describing extreme values, we can indirectly gain a deeper understanding of why other values that align closely with the mean exist. Neglecting the analysis of extreme values often leads to an incomplete explanation of the overall distribution and characteristics of the data. To that end, we cannot deal with absolutes at the variable level, and we will have to consider at least two or more variables. Thus, taking the example of height, we can add age, origin, or weight and thus, for example, connect that those who have the highest extreme values in terms of height are the ones who weigh more, and perhaps this is because they have eaten better. In contrast, those who measure less are those who may suffer from malnutrition and thus are the ones with the lowest weight. On the other hand, the average is those who weigh an average, corresponding to a stable, moderate diet—moreover, the examples in Fig. 2.2a, b are limited to a primary type of statistical data distribution, the normal distribution. However, we could have distributions that may be skewed in one direction, for example, toward the maximum or minimum.

Therefore, in our analysis, it would be advisable to keep in mind that by considering the extreme values, we reinforce the understanding of the statistical mean and, in addition, the influence of factors that could affect the trait analyzed in the individuals of the sample. Take, for instance, the case of the former athlete Usain Bolt. Scientists and doctors have extensively studied his exceptional speed, aiming to understand the factors that contribute to his extraordinary performance. In essence, they seek to uncover the underlying causes that allow them to achieve the maximum speed of human running speed. A combination of anthropometric characteristics, motor coordination, power generation capabilities, and effective running techniques would partly explain his performance. In addition, this indirectly sheds light on the fact that this could explain the lower speed of other competitors or people in general (e.g., poor development of motor coordination and power generation), added to anthropometric differences inherent to the athlete (Čoh et al., 2018). Thus, we can conclude that any extreme value in the distribution of a trait is "resignified" when contrasted. Otherwise, the extreme value remains an "ornament" and goes unnoticed or is considered only a mere curiosity.

2.7 Outliers: Those Underestimated Exceptions

Gladwell's book "Outliers: The Story of Success" (2008), which attempts to understand the secret of the success of great minds and "exceptional" athletes, was at the top of the New York Times bestseller list for 11 consecutive weeks. This work shows us that outliers and attempts to understand them fascinate the general public. However, the presence of "outliers" in scientific practice is far from being analogous to this example, as they tend to concern scientists. Outliers are defined as "*an outlying observation, or outlier is one characteristic that appears to deviate markedly from other members of the sample in which it occurs*" (Grubbs, 1969). Alternatively, an outlier is "*an observation that lies outside the overall pattern of a distribution*" (Moore & McCabe, 1999). Another typical formal statistical definition of an outlier is "*a point that falls more than 1.5 times beyond the interquartile range, either above the third or below the first quartiles*" (Renze, 2011). Whether considered for statistical analyses, outliers tend to go unnoticed, except as "nuisance" points to the researcher (Fig. 2.3a–c), in the vast cloud of data displayed on a plot such as a scatterplot or a histogram. In many cases, these values are labeled "suspicious," "ugly," and "nasty," or that can "ruin" or "destroy" our inferences (McClelland, 2000). Therefore, we see how an exceptional value can be considered from a biased negative valuation in our scientific work.

At the single-trait level, an outlier is a point that deviates from the "majority." Although we can quickly visualize several ways of detection in a histogram, the values are far from the frequent intervals and have low representativeness (Aguinis et al., 2013) (Fig. 2.3a). In a box plot, they can be seen as points more or less distant from the interquartile range boundary (Fig. 2.3c). Outliers can also appear when relating two or more variables or data sets (multitrait approach) (Fig. 2.3b). Points away from the fit line in scatter plots can be considered outliers. We can have an

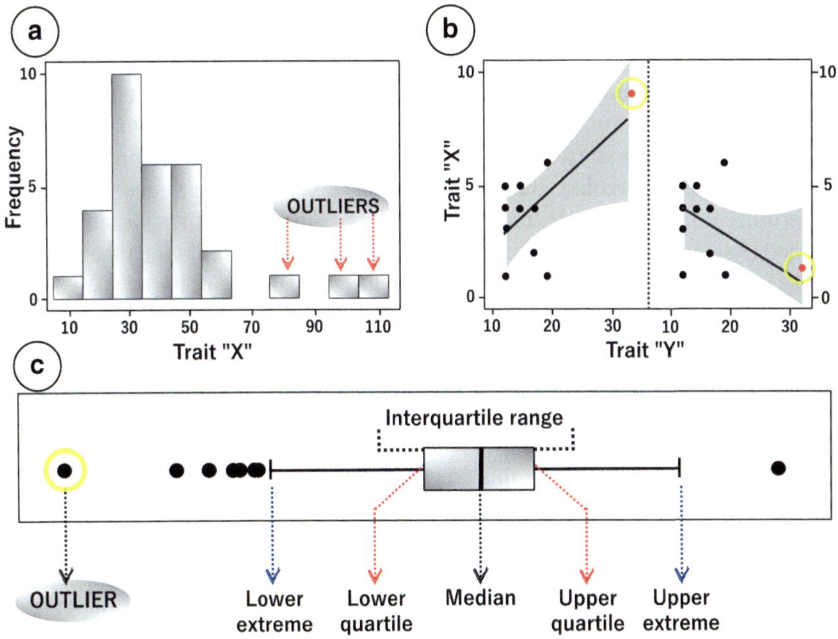

Fig. 2.3 Outliers in different statistical graphs. (**a**) Histogram showing the frequency of occurrence of trait "X" in the population, with extremes in positive values that can be considered statistical "exceptions" or "outliers." (**b**) Scatter plot showing relationships between variables "X" and "Y" with extreme values influencing the relationship patterns found. (**c**) Boxplot and its components, showing cases outside what is considered the pattern or norm (interquartile range)

outlier arising from the connection, or linkage, of two extreme values of variables but not necessarily in the same direction. For example, it may arise from the association between the minimum value of one variable and another variable's maximum value. Additionally, an outlier may occur between a medium value of one variable and a high value of another. In other words, the occurrence and characteristics of outliers and their linkage to the data pool of the respective variables that give rise to them cannot be predicted at all. However, they appear much more often than assumed or reported.

Outliers are usually more influential (adding more weight than others in the estimation) than data close to the mean. For example, if from a data set, we remove these values, our inferences or results may vary greatly; not so if we remove a point from the region of highest density. How much our results vary by removing these outliers gives us an idea of the "robustness" of our data set. Hypothesis testing is highly sensitive to the noncompliance of many model assumptions and the presence of these outliers. If data do not comply with the assumptions, the ability to detect true effects decreases (affects the p value, effect size, and estimated confidence intervals).

The presence of an outlier may also imply a type of problem. For example, a case that does not fit the model under study or perhaps a measurement error. In the case of the second option, outliers are good "alerts" of possible errors, so we can detect them and correct our data collection or discard them because of the impossibility of repetition. However, when faced with "real" outliers (not measurement errors), researchers can proceed in different ways or make different decisions. In many cases, the outlier (or several if more than one) is removed with some statistical techniques such as "least squares fitting to data" (Renze, 2011), or imputation techniques are applied to replace the outlier with one corresponding to the mean (Dang & Serfling, 2011). However, removing outliers or replacing them is not recommended in scientific practice. This reflects our inability to address the inherent variability of biological systems or our need to "fit" the data to our hypotheses. These manipulations (sometimes gross) may alter the inferences from the data since there is a bias, the sample size may be decreased, and the distribution or variances may be affected. Cortina (2002) warns, "*caution also must be used because, in most cases, deletion [of outliers] helps us to support our hypotheses (...) choosing a course of action post hoc that is certain to increase our chances of finding what we want to find is a dangerous practice.*" If this practice is carried out, it is always necessary to make it clear to the reader that outliers have been removed.

A more Solomonic approach is to perform statistical analyses with and without the outliers and report and discuss both results (Aguinis et al., 2013). In other cases, we can perform statistical manipulations in our statistical models to "deweight" those outlier observations using robust techniques (submit median, trimmed means, and winsorized mean) so that the estimates better represent the sample. We can try transforming the data to minimize the distorting effect of the extreme case (Zimmerman, 1995) (logarithmic or square root transformation) or use statistical models that have different assumptions (e.g., nonparametric statistics, bootstrapping, and generalized models). There are also specific robust methods for hypothesis testing samples with outliers, such as Yuen's robust test using trimmed means. One of the difficulties of outlier treatments is that there is no universally accepted theoretical framework for their treatment. In addition, the way to proceed varies greatly depending on the particular discipline (e.g., economics, psychology, pharmaceutics, and biology) (Rousseeuw & Leroy, 2003; Aguinis et al., 2013).

An evolutionary biologist may be interested in these plot points and will attempt to inspect them in light of other associated characteristics of the individual, including interpretations somewhat within the scope of the study. The extreme values of a particular trait within the natural distribution of the population, whether or not they are considered outliers, have potential value from an adaptive point of view. Charles Darwin saw this in his classic studies on the dynamics of variation and selection (Darwin, 1859, 1868, 1871), and William Bateson in his analysis of variability during development (Cock & Forsdyke, 2008; Peterson, 2008). Historically, evolutionary biology has shown that some rare expressions of a trait or set of traits can be influenced by directional selection in some contexts (Westneat & Fox, 2010). Thus, what used to be the tail end of the distribution may be the "middle" value of a particular trait now and vice versa. Indeed, this dynamism is at the heart of natural and sexual selection.

2.7.1 An Outlier Hiding a Spider's Secrets

The *Allocosa senex* is a wolf spider inhabiting coastal areas that possesses sex role reversal, as females compete for males and males exert mate choice (Aisenberg et al., 2007) (Fig. 2.4a). The male builds a particularly long burrow, compared to the one inhabited by the female, and remains in it until a receptive female finds it and enters for copulation. Before mating, the female evaluates the length of the burrow as an indirect indicator of the male "quality." Finally, the male withdraws from the burrow, and the female remains inside until the offspring are born. Interestingly, the granulometry of the sand on the coasts where this species lives influences the characteristics of the burrows. In this framework, De Simone et al. (2015) found that adult males and juveniles built longer burrows in coarser substrates, whereas in substrates with fine sand (to which they were not accustomed), burrows were significantly shorter (Fig. 2.4b).

There is an exciting point for our book. An outlier in the box plot corresponds to juvenile burrows in the fine-grained substrate (Fig. 2.4b). One individual built a burrow much longer than the rest, even larger than the burrow built in a coarse sand substrate. In other words, a "super burrow" is built by a juvenile male. If we were to analyze, as we said before, only this variable (i.e., "length of the burrow"), it would remain a simple curiosity. However, if we add other variables to understand this "atypical" value, for example, specific body characteristics, we can appreciate something very striking (Fig. 2.4c). This atypical individual in its burrow construction was the largest and heaviest, so it could be considered a "super individual" from

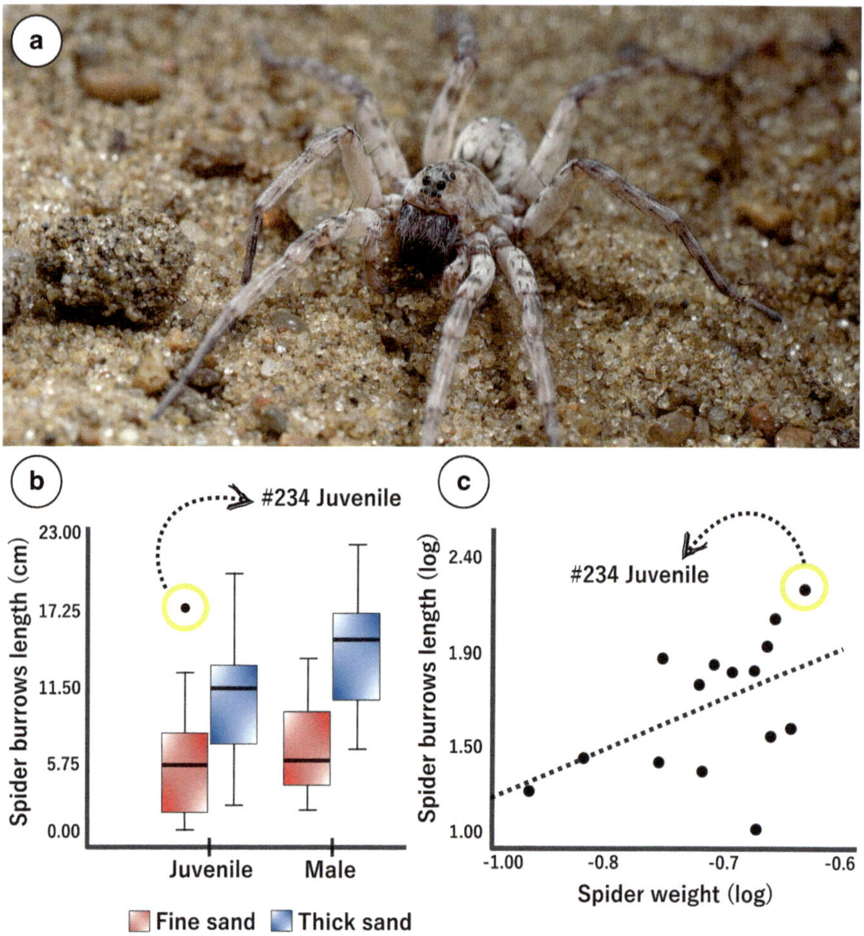

Fig. 2.4 Examples of how the analysis of outliers revealed new key information in a scientific study. (**a**) Burrow-digging wolf spider *Allocosa senex* (Image courtesy of Matías Izquierdo). (**b**) Comparison of the length of the burrows built in sand with different substrates, highlighting a very long burrow made by a juvenile in fine sand. (**c**) Relationship of the size of the burrow with the weight of the spiders, showing the heavy weight of the juvenile that made the long burrow

the point of view of traditional physical conditions. The "super burrow" was made precisely by the "super individual," considering size and body weight. As these individuals were collected in the field, we cannot discuss their origin or developmental conditions. This individual probably possessed sufficient reserves and strength to overcome the difficulties in building the burrow in an adverse substrate. Indeed, burrows in fine sand tend to collapse and require more silk covering by the building spider. This example shows how not limiting ourselves to a "simple" outlier can teach us more about the relationship between variables and possible implications for the biology of this species. We consider this example as "a grain of sand," metaphorically speaking, in the immense universe of outliers. Any researcher is confronted with these "annoying" points in the analysis. It would be a matter of paying much more attention to them and valuing them in our discussions.

2.8 Key Concepts and Ideas in This Chapter

- The definition of exception is strongly linked to the definition of rule. Both are antonyms and necessary to give entity to what each alludes to.
- In human history, exceptions have been seen as synonymous with success and what stands out positively from the average, not only in science but also in politics, art, sports, and other areas. However, they can also have the opposite effect when they are exceptions in a negative extreme such as deformities, teratologies, or monstrosities, which in this case produce aversion or rejection.
- In biology, exceptions can become a starting point that generates changes, whether in morphology, behaviors, different types of organisms, and systems of evolution.
- To detect exceptions in the scientific world, we must have a detailed look to avoid obviating, hiding, or considering them less important. This care gives us a more complete and comprehensive approach to all biological phenomena.
- Individuals who are exceptions within a population could be considered so by observing a single variable that makes them stand out from the rest or by a set of variables or traits, which may or may not be associated with each other.
- Extreme cases in the phenotypic expression of a given trait in a population may be less striking to the observer. However, these quantitative values that are far from the average value and that fall outside the norm are essential when we talk about the influence of factors affecting that trait.
- Outliers are essential data for detecting oddities in the natural world. Many times, they are eliminated or seen as a nuisance to the study. However, outliers can be an opportunity. After checking the nature of the data, a detailed analysis can help one to find interesting patterns in our study model.

References

Aguinis, H., Gottfredson, R. K., & Joo, H. (2013). Best-practice recommendations for defining, identifying, and handling outliers. *Organizational Research Methods, 16*(2), 270–301.

Aisenberg, A., Viera, C., & Costa, F. G. (2007). Daring females, devoted males and reversed sexual size dimorphism in the sand-dwelling spider *Allocosa brasiliensis* (Araneae, Lycosidae). *Behavioral Ecology and Sociobiology, 62*, 29–35.

Becerra, J. C., Bildstein, L. S., & Gach, J. S. (2016). R*ecent insights into the HIV/AIDS pandemic. Microbial Cell, 3*(9), 451.

Cock, A. G., & Forsdyke, D. R. (2008). *Treasure your exceptions: The science and life of William Bateson*. Springer.

Čoh, M., Hébert-Losier, K., Štuhec, S., Babić, V., & Supej, M. (2018). Kinematics of Usain Bolt's maximal sprint velocity. *Kinesiology, 50*(2), 172–180.
Collins English Dictionary. (2003). Glasgow, UK: HarperCollins.
Cortina, J. M. (2002). Big things have small beginnings: An assortment of "minor" methodological misunderstandings. *Journal of Management, 28*(3), 339–362.
Dang, X., & Serfling, R. (2011). A numerical study of multiple imputation methods using nonparametric multivariate outlier identifiers and depth-based performance criteria with clinical laboratory data. *Journal of Statistical Computation and Simulation, 81*(5), 547–560.
Darwin, C. (1859). *On the origin of species by means of natural selection, or the preservation of favoured races in the struggle for life*. John Murray.
Darwin, C. (1868). *The variation of animals and plants under domestication*. John Murray.
Darwin, C. (1871). *The descent of man, and selection in relation to sex*. John Murray.
De Simone, G. A., Aisenberg, A., & Peretti, A. V. (2015). Female and juvenile burrow digging in *Allocosa brasiliensis*, a South American sand-dwelling wolf spider. *Arachnology, 16*(8), 276–280.
Domingo, E., Martínez-Salas, E., Sobrino, F., de la Torre, J. C., Portela, A., Ortín, J., et al. (1985). The quasispecies (extremely heterogeneous) nature of viral RNA genome populations: Biological relevance—A review. *Gene, 40*(1), 1–8.
Eigen, M., & Biebricher, C. K. (2018). Sequence space and quasispecies distribution. In *RNA genetics* (pp. 211–245). CRC Press.
Elena, S. F., & Sanjuán, R. (2005). Adaptive value of high mutation rates of RNA viruses: Separating causes from consequences. *Journal of Virology, 79*(18), 11555–11558.
Ferrer, L., Stambulova, N., Borrueco, M., & Torregrossa, M. (2022). Maternidad en el tenis profesional: ¿es suficiente con cambiar la normativa? *Cuadernos de Psicología del Deporte, 22*(2), 47–61.
Fraser, C., Lythgoe, K., Leventhal, G. E., Shirreff, G., Hollingsworth, T. D., Alizon, S., & Bonhoeffer, S. (2014). Virulence and pathogenesis of HIV-1 infection: An evolutionary perspective. *Science, 343*(6177), 1243727.
Gaston, K. J. (2014). Species-range size distributions: Products of speciation, extinction and transformations. In F. A. Smith, J. L. Gittleman, & J. H. Brown (Eds.), *Foundations of macroecology: Classic papers with commentaries* (pp. 388–399). The University of Chicago Press.
Gladwell, M. (2008). *Outliers: The story of success*. Little.
Grubbs, F. E. (1969). Procedures for detecting outlying observations in samples. *Technometrics, 11*, 1–21.
Harding, A. T., & Heaton, N. S. (2018). Efforts to improve the seasonal influenza vaccine. *Vaccine, 6*(2), 19.
Hoenen, T., Safronetz, D., Groseth, A., Wollenberg, K. R., Koita, O. A., Diarra, B., et al. (2015). Mutation rate and genotype variation of Ebola virus from Mali case sequences. *Science, 348*(6230), 117–119.
Holland, J., Spindler, K., Horodyski, F., Grabau, E., Nichol, S., & VandePol, S. (1982). Rapid evolution of RNA genomes. *Science, 215*(4540), 1577–1585.
Hurst, R. (2008). Are disabled people human? In P. Healey & S. Rayner (Eds.), *Unnatural selection. The challenges of engineering tomorrow's people* (pp. 74–80). Routledge.
Hutchinson, J. F. (2001). The biology and evolution of HIV. *Annual Review of Anthropology*, 85–108.
Kunin, W. E., & Gaston, K. J. (1993). The biology of rarity: Patterns, causes and consequences. *Trends in Ecology & Evolution, 8*(8), 298–301.
Kunin, W. E., & Gaston, K. J. (1997). *The biology of rarity: Causes and consequences of rare*. Science+Business Media, B. V.
Lavezzo, F. (2013). *El oso antártico*. Cartografías. 124 pp.
McClelland, G. H. (2000). Nasty, ill-mannered observations can ruin your analysis. In H. T. Reis & C. M. Judd (Eds.), *Handbook of research methods in social and personality psychology* (pp. 393–411). Cambridge University Press.

Moore, D. S., & McCabe, G. P. (1999). *Introduction to the practice of statistics* (3rd ed.). Freeman, W. H.

Mujika-Alberdi, A., García-Arrizabalaga, I., & Gibaja-Martins, J. J. (2021). Mujeres deportistas: Poca visibilidad y mucho estereotipo. *Cuadernos de Psicología del Deporte, 21*(3), 269–283.

Nájera, I., Holguin, A., Quinones-Mateu, M. E., Muñoz-Fernández, M. A., Nájera, R., López-Galíndez, C., & Domingo, E. (1995). Pol gene quasispecies of human immunodeficiency virus: Mutations associated with drug resistance in virus from patients undergoing no drug therapy. *Journal of Virology, 69*(1), 23–31.

Penrose, W. D., Jr. (2015). The discourse of disability in ancient Greece. *Classical World*, 499–523.

Peterson, E. L. (2008). William Bateson: From Balanoglossus to materials for the study of variation: The transatlantic roots of discontinuity and the (un)naturalness of selection. *Journal of the History of Biology, 41*, 267–305.

Renze, J. (2011). *Outlier. From MathWorld—A Wolfram web resource, created by Eric W. Weisstein*. Retrieved from http://mathworld.wolfram.com/Outlier.html

Rey Benayas, J. M., Scheiner, S. M., García Sánchez-Colomer, M., & Levassor, C. (1999). Commonness and rarity: Theory and application of a new model to Mediterranean montane grasslands. *Conservation Ecology, 3*(1), 5.

Ridley, M. (2004). *Evolution* (3rd ed.). Blackwell Publishing.

Rousseeuw, P. J., & Leroy, A. M. (2003). *Robust regression and outlier detection*. Wiley.

Rustanti, L., Jin, H., Lor, M., Lin, M. H., Rawle, D. J., & Harrich, D. (2017). A mutant Tat protein inhibits infection of human cells by strains from diverse HIV-1 subtypes. *Virology Journal, 14*(1), 52.

Steinhauer, D. A., & Holland, J. J. (1987). Rapid evolution of RNA viruses. *Annual Reviews in Microbiology, 41*(1), 409–431.

Westneat, D. F., & Fox, C. W. (2010). *Evolutionary behavioral ecology*. Oxford University Press.

Zimmerman, D. W. (1995). Increasing the power of nonparametric tests by detecting and downweighting outliers. *The Journal of Experimental Education, 64*(1), 71–78.

Chapter 3
Associations and Dynamics Between Rules and Exceptions

Everything flows, nothing stands still

—Heraclitus

Contents

Abstract There is a complex relationship between rules and exceptions in different contexts. Rules and exceptions are not always distinct but rather dynamically related concepts. There are a variety of alternative terms, such as patterns, normalities, rarities, and abnormalities, that can be used interchangeably with rules and exceptions. This emphasizes the flexibility in the use of these terms. Singularity is often used as a synonym for exception, as something unique. This uniqueness can be both positive and negative and contributes to the value of individuality in biology. This feature demonstrates the possibility of mixed patterns and the importance of scale in analyzing

A. V. Peretti et al., *Rules and Exceptions in Biology: from Fundamental Concepts to Applications*, https://doi.org/10.1007/978-3-031-55382-0_3

them. Four scenarios of change and two types of dynamics between rules and exceptions can emerge. These scenarios cover situations where rules remain stable, exceptions persist, rules become exceptions, and vice versa. These dynamics can occur in a variety of contexts, including biological evolution. Changes in scientific knowledge and paradigms can be compared to the dynamics between rules and exceptions. Overall, this chapter explores the complex and interrelated nature of rules and exceptions in different domains, highlighting their dynamic and evolving relationships.

3.1 Introduction

We have referred to rules and exceptions according to their definitions and contexts to see how they can be linked. Specifically, we will analyze which rules and exceptions are not watertight compartments, but, on the contrary, there is a dynamic between them, both being important. However, before this, we will review alternative or similar expressions to the concepts of rules and exceptions. Some of them we have already mentioned, and the reader may wonder whether we can use them as synonyms.

3.2 Complementary Concepts When Focusing on Rules and Exceptions

3.2.1 On Terms and Their Synonyms Related to the Normal and the Abnormal

When we talk about rules, we can refer to them directly throughout the book as patterns or normalities. The same applies to exceptions and associated terms such as rarities or abnormalities. In this book, we focus mainly on the use of the terms rule and exception as antonyms, although we also use the meanings of patterns and rarities in some cases, and it is important to emphasize that we should be flexible in the use of these terms (Boxes 1.1 and 2.1). The concept of rule is more closely linked to that of law or may allude to establishing and identifying more flexible "mini laws." Patterns are also something that the observer identifies from the observation of nature. From a pattern, a rule may or may not be explicitly stated. We can state that when we speak of patterns, these are what exists as such, the morph "x," the functional type "x," that is, what "is" expressed from the genotype at the phenotypic level of this or that organism and that is shared, expressed, evidenced by the majority of the individuals of the population, species, clade. Thus, the pattern occurs at the level of the group in a given time and space. In a way, the rule can be synonymous, so we recommend flexibility on the part of the reader when we use rules or patterns indistinctly (or other options such as norm, standard, etc.).

As for exceptions, its entity will always be relative to something else (time, group, or space). Rarities correspond as an antonym to patterns (they allude to what is not common concerning the pattern), and hence this book could be entitled "patterns and rarities" instead of "rules and exceptions." In any case, the obsession with definitions should not distract us from what we are analyzing, the common versus the rare, whether the researchers of the discipline expose it as a rule versus exception or as pattern versus rarity or, what is hardly less evident, as frequent versus infrequent occurrence. This situation derives from the fact that nature is not discrete but rather continuous in many of its transitions. The discrete, deeply analyzed, finally becomes a continuum in our eyes. Let us always remember that definitions help, they are necessary, but they must not cage our minds.

3.2.2 Singularity: Each Living Being is Unique

Interestingly, one of the synonyms of exception is "singularity," which has the synonyms "unique, particular, special, remarkable, outstanding, rare, strange" and the antonym "common." Uniqueness, therefore, is the quality that distinguishes something from other things of its kind. For example, the concept of uniqueness has been used in mathematics for specific functions that exhibit unexpected behavior when certain values are assigned to the independent variable(s) (Arnold, 1981). In the case of "unique," it is said to be "singular, exclusive, one, pure, scarce, peculiar, characteristic," its antonym being "ordinary." Whether singular or unique, these terms connect with things that we will see in a broad, holistic view with the value of each organism in nature.

In humans, a "singular" person stands out (positively or negatively) above the rest. We have elaborated on this idea in Chap. 2. However, here the question arises, can we extrapolate this same thing to the rest of the living beings? We do not refer only to the explicitly striking ones but to some quality or several qualities that make them an exception in their group or according to a view of time or space. It is a paradox that what is the rule, for example, in a species, is the product of the sum of singularities. This "cryptic singularity" (i.e., not perceived or being underestimated by the observer) within the pattern is something that, from a holistic point of view, should call attention to the importance and value of the uniqueness of each living being in intra and interspecific interrelationships.

In this context, it is helpful to bring up the inspiring book *Life Trajectories: Biology, Freedom, and Determinism* by Rose (2001), where the author draws our attention to scientific reductionism when we only think of the determinism of genes to mark and define the identity of living organisms. Indeed, since the book's publication, this has fortunately been changing toward a view that pays more attention to the surrounding environment, epigenetics, the evolutionary biology of development, the social context, and, finally, to each individual in himself/herself. Rose (2001) shows and highlights the value of individuality by commenting that everyone has "his or her life trajectory." It is interesting to note that other authors, such as Fritjof

Capra, tend to focus more on the "search for the pattern," even analyzing similar aspects such as the phenomenon of emergence, patterns in nature, and body geometry. In other words, they aim to find a synthesis of the "whole" in life. Its vision is as a "network" or "intimately linked system" rather than emphasizing singularity (Capra, 1996; Capra & Luisi, 2014). It is not that Steven Rose denies these approaches but emphasizes more individuality, the freedom implicit in each individual to have, exhibit, and perhaps change his or her trajectory, as an undervalued aspect of modern biology. Perhaps because of the obsession to "order" and hierarchize the observed and the impressive biological diversity in patterns and rules, we are missing the richness of the uniqueness of each living being. For example, the value of the individual has been indirectly highlighted in the advancement of personality studies in animals, including "lower" groups such as arthropods, anemones, or even bacteria (Briffa & Weiss, 2010; Briffa & Greenaway, 2011; Carere & Maestripieri, 2013; Kralj-Fišer & Schuett, 2014; Kaiser & Müller, 2021).

Note that while Capra's vision is strongly connected with the search for rules, Rose's vision links more emphatically with the valuation and rescue of rarities or exceptions. The "holistic" context is very present in both visions, but they point to different scales of life organization. In the following, we will examine the dynamics between these "opposites." They are focused (unless otherwise stated) on rules and exceptions and their interchange in different perspectives and contexts.

3.2.3 *Are There Rules for Predicting the Occurrence or Type of Exception?*

It is interesting to consider whether it is possible to predict the occurrence of exceptions. That is, we can propose predictive rules for the occurrence of different types of exceptions according, for example, to a certain spatial context (e.g., surrounding environment, other organisms at the community level) or temporal context (e.g., ontogeny, evolutionary time). Therefore, if a particular factor exists, then the probability of occurrence of a particular rarity increases or decreases.

Theodosius Dobzhansky summarized in his writings on genetics some of the classic experiments carried out during the first half of the twentieth century on the effect of mutations to promote predictable changes (Dobzhansky, 1982; Ceccarelli, 2010). Indeed, today, we know that some specific pathologies and syndromes can appear because of certain carcinogenic factors related to age and nutrition. Therefore, if there are rare hereditary or environmental pathologies in biomedicine, the question arises as to whether we can predict their appearance. If we extrapolate this beyond biomedicine, we may find that certain teratological exceptions are not always due to chance but will be determined by spatial and temporal factors. Unfortunately, it is not easy for us to identify them to make accurate predictions. For example, the malformations detected in fossils of ammonites and other groups may have been the consequence of teratologies produced by extreme climatic changes or gene depletion. We will discuss some examples of this in Chap. 5 of this book.

3.3 Three Perspectives on the Dynamics Between Rules and Exceptions

Thus far, we have presented the main characters, rules and exceptions, their terminological variants, and fundamental theoretical options for their interpretation. We will now consider them in more detail regarding how they are sustained in time, space, and group. That is, we will not only limit ourselves to the variable "time," whether historical from the point of view of knowledge, or "evolutionary" within the framework of the evolution of a species. Indeed, these three contexts interact, and the subject is much more complex than simple linear temporal unidirectionality. We will address different scenarios and examples throughout the book.

3.3.1 The Possibility of Mixed Patterns

As we see in Fig. 3.1a, we can have situations where the three contexts (taxonomic group, spatial, and temporal) combine or interact to a greater or lesser extent. For example, a rule can change in time in general and then according to certain spaces. Additionally, all this can already be adjusted according to groups. This interaction can also refer to something evolutionary. For example, how a body structure changed over time within a clade (e.g., time-group combination) and, in turn, already in particular environments.

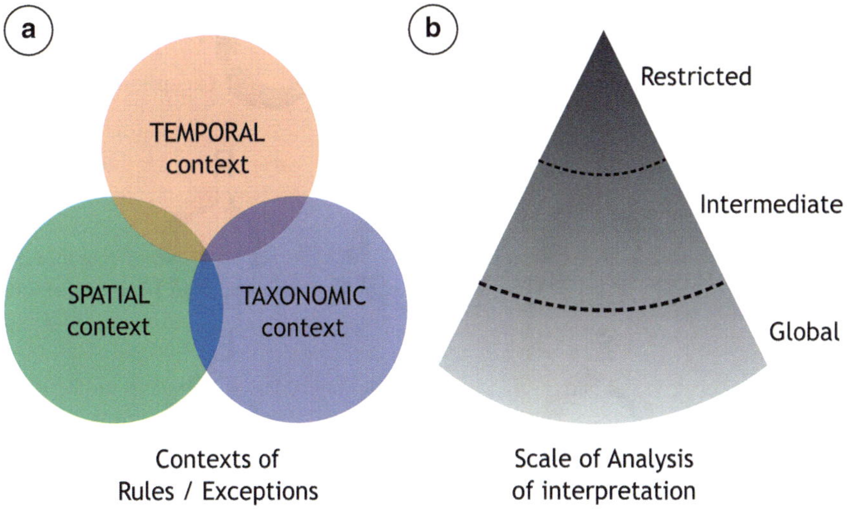

Fig. 3.1 Context and scale of analysis governing the dynamics between rules and exceptions. (**a**) Basic combinations in terms of interactions between time, space, and group contexts in the dynamics between rules and exceptions. (**b**) The scale of the analysis will be larger or smaller according to the level of the analyzed rule (e.g., type I, II, III, etc.)

On the other hand, the scale of the analysis may also be different (Fig. 3.1b). That is, if the study is focused on details of a phenomenon or is much more global. This is very important because, as we saw in Chap. 1, this will determine whether we are more or less likely to find exceptions to them. It is not the same to analyze major patterns, more inclusive of all living beings (e.g., the presence of a genetic code and how it works), as it is to analyze much more particular patterns (e.g., modifications in a unique phenotypic trait, such as a particular appendage, be it a fin, a leg, or a wing). More exceptions might be expected in the more comprehensive rules if they were "generalized" without discrimination and thus included "subrules" that were overlooked. However, exceptions would not be expected to occur if they were rules of type I. It is an apparent paradox that may lead to confusion, but these are two different situations that refer to the more globally applicable rules.

3.4 Main Directions of the Changes

We can find at least four situations of changes and two of change of dynamics between rules and exceptions in any of the contexts of time, space, and group (separate or mixed, Fig. 3.2). In the course of this book, we discuss and give examples of these situations in more detail: (A) *No change*: we can have (1) rules that remain the same (i.e., nothing or little change), regardless of the context (Chap. 4), and (2)

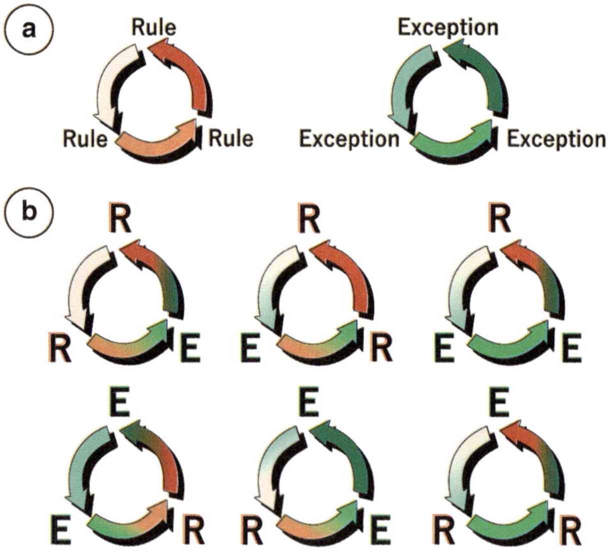

Fig. 3.2 Dynamics between rules and exceptions. (**a**) Situations in which both rules and exceptions remain stable, without significant modifications over time. (**b**) Situation in which rules and exceptions remain unstable. Transitions can be maintained over time or reoccur. *Abbreviations*: *E* exception, *R* rule

exceptions that persist relatively as such (i.e., nothing or little change), regardless of the context (Chap. 5) (Fig. 3.2a). (B) *Dynamic framework*: that is, specific to a changing world. Here, we find (1) exceptions that become rules (Chap. 6), as well as the opposite: (2) rules that become exceptions (Chap. 7) (Fig. 3.2b). These four scenarios serve as a guide for the examples we will discuss in the following chapters. Biology offers many examples of these four possibilities of dynamism between rules and exceptions, whether in time (historical or evolutionary), in place/space, or in group/clade (Fig. 3.2b).

3.4.1 "Conservative" Rules and Exceptions?

Indeed, whether dealing with a rule or an exception, the system can maintain very stable temporal, spatial, and group conditions. While minor variations are allowed, their significance is negligible. The constancy of these rules or exceptions is not a result of their lack of variation. The ability for rules or exceptions to remain unchanged depends partly on the scale of analysis, encompassing temporal, spatial, and group dimensions. Instability is more likely to be found when the scale is significantly large. Conversely, with a limited scale, it becomes less probable to detect changes, resulting in greater stability. This apparent paradox, which links the perception of change to the scale, can be resolved by recognizing that this "conservatism" is not universally upheld in the studied system. In other words, we observe a dynamic relationship between rules and exceptions in these three contexts. Depending on the specific case, we can observe changes in their importance and applicability.

3.4.2 Movements of Rules "With Continuous Distribution"

The rule (implicitly also the associated exceptions but with emphasis on the rule) undergoes changes when it "moves" toward ranges previously considered minorities or mere exceptions and may even "dilute" or disappear entirely as an initial rule. We can apply the look of the analysis of character analogies since the dynamics between rules and exceptions (either in one direction or the other) is a crucial aspect of explaining selective pressures (e.g., natural or sexual selection). We could start from a situation of (dynamic) equilibrium whose distribution is reflected in a way analogous to a stabilizing selection in evolution. That is, where the rule or norm is exhibited by an average number of individuals located at the center of the distribution, while progressively to the left and right, there will be fewer individuals exhibiting the rule at its maximum expression until reaching extreme values at the margins of the distribution. The latter will be, in a "combined" form, the sum of (1) an "extreme" expression of the trait, far from typical and, perhaps also (not always), (2) a very low frequency of occurrence (i.e., few individuals) compared to a high occurrence (many individuals) concerning that particular trait(s).

From this initial scenario, a shift may occur within the data distribution, favoring individuals who exhibit a specific variant of the overall spectrum of expression in the analyzed pattern. This shift resembles the effects observed under directional pressure. Therefore, as a result of this movement, the conformations of what will later become the new majorities and minorities will be modified. It is the same if we have a situation of disruption, where formerly extreme values become rules, or at least one becomes the rule. This situation may or may not imply the disappearance of the original rule.

We could also analyze a situation where the rule becomes ambiguous and unclear. An example could be the case of a scenario analogous to a dispersion of the expression of the feature to both sides by similar parts but at the expense of the midpoint. Therefore, the rule will be "fuzzy." This situation could occur if there are many exceptions to the rule that would paradoxically attain the entity of "mini rules," either individually or in groups, through potential associations between them. This scenario could lead to an extreme situation where it would already imply the "disappearance" of the rule as such within the framework of a new flat-type distribution. Here, there would no longer be any exceptions either. Minorities and majorities would no longer be distinguishable as such. Everything would be "equal."

The above leads us to attempt to plot the possible scenarios of "movements" of the rule. The options may imply, depending on the perspective of analysis, (1) movement due to biology's action, for example, due to evolutionary or ecological change, and (2) movement from the approach of the history of biology as a science, that is, due to a modification in the vision of man as a researcher of biological phenomena. The analysis of these movements will allow us to conclude that the evolution of rules follows the same patterns of biological evolution (i.e., evolutionary processes of traits and phylogenetic patterns of taxa). Furthermore, there are similar patterns of evolution of rules in forms of change in the distributions of the elements involved in all contexts of the universe, including abiotic, biotic, and even manifestations of man in interpreting these—for example, science, ideas, art, technology.

3.4.2.1 Analysis from Biological Evolution (i.e., From Scenario 1 of Rule Movement)

Not all traits are distributed evenly in a population. In some cases, there is a clear "zone" where the majority of individuals fall, while others fall outside of this range. Figure 3.3a shows a state according to a neutral or stabilized situation. Not necessarily toward both sides of the mean distribution, we will find real exceptions (e.g., it could well be something already biased toward one side). Within a particular "zone," there exists a specific range of trait expressions that is considered the typical pattern due to its prevalence among the majority of individuals (represented on the Y-axis as either an absolute number or proportion). For instance, in a species analyzed from a group context, we would observe a "continuous" rule that pertains to a character or event displaying continuous attributes, such as a specific color tone or size. Additionally, there would be "minorities" comprising individuals expressing a narrower range of the rule. Among these minorities, extreme values can be found.

The transition from the situation described in Fig. 3.3a to the scenario depicted in Fig. 3.3b involves a shift driven by directionality. In this new scenario, what was previously considered an exception (or a group of exceptions or their associations, to account for multiple possibilities) now becomes the new rule. This phenomenon can be observed in intraspecific analyses of traits. For instance, rare or extreme colorations, ornaments, or unusual body sizes may evolve into the new norm if they are progressively favored through natural or sexual selection over the previously predominant expressions conforming to the majority.

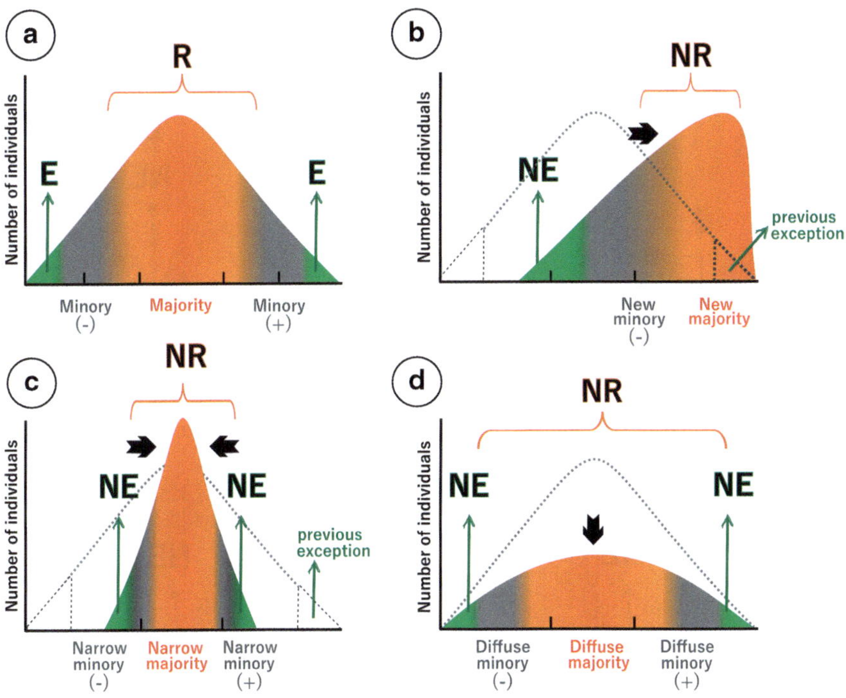

Fig. 3.3 Dynamics between rules and exceptions illustrated in individuals of a population with a certain trait "X." (**a**) A neutral or stabilized situation, where the majority of individuals fall in the "norm" or "rule," while others fall outside of this range (minorities and exceptions). (**b**) New scenario with a shift driven by directionality, which was previously considered an exception, now becomes the new rule. (**c**) The rule zones are maintained but with higher numerical representativeness, so we speak of "augmented" patterns, and a narrowed rule. (**d**) The areas of rules, minorities, and exceptions are maintained but with lower numerical representativeness, so we speak of "diluted" patterns. *Color references:* orange indicates the majority/rule zone; green indicates the exception zone; gray indicates the minority zone. *References:* Arrows indicate selective pressures that change distributions; dotted line, neutral pattern 3.3A

Another option for change in the rule would be the one shown in Fig. 3.4b, where the "old" rule is maintained, but another equally important one emerges from some previously extreme value. For example, if a range of coloration that was the "mean" persists in one sex but no longer in the other, it would shift to another coloration value. A minimum of the population previously expressed this value (e.g., now males are shifting to an intense body color due to advantages in a mate choice context). Alternatively, this disruptive situation could occur in two geographic spaces—for example, individuals of the same species in two populations isolated from each other. In one of these populations, the species changed to a more cryptic color for camouflage to avoid predators. In a phylogenetic case, imagine that an ancestral species would retain the classic pattern, the "old" rule, while a new species would present the new rule as a pattern. It is interesting to see how specific extreme values

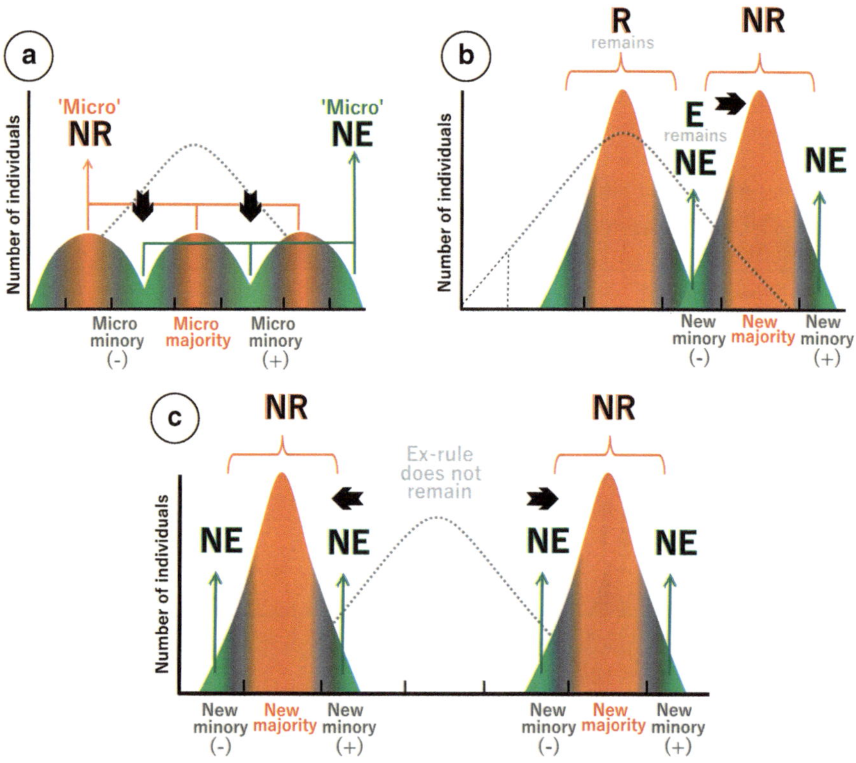

Fig. 3.4 Dynamics between rules and exceptions illustrated in individuals of a population with a certain trait "X". (**a**) Distribution of "mini rules" in a given distribution, with the appearance of exceptions and majorities in local regions. (**b**) The "old" rule is maintained, but another equally important rule emerges from some previously extreme value. (**c**) Two nonextreme values in a minority pool on either side of the mean become two rules, and the old rule does not persist. *Color references:* orange indicates the majority/rule zone; green indicates the exception zone; gray indicates the minority zone. *References:* Arrows indicate selective pressures that change distributions; dotted line, neutral pattern 3.3A

could overlap between the field of the two rules at the beginning and progressively reach a point where these margins would no longer be shared. We must remember that this can be analyzed in populations at the level of an intraspecific or interspecific analysis.

An intermediate situation occurs when two nonextreme values in a minority pool on either side of the mean become two rules (Fig. 3.4c). In this case, the old rule does not persist (it either no longer exists or is now an exception to one and the other new rule). A situation could even occur where the new rules on each side arise from the respective exceptions (i.e., not only from one of the minimum values, as in Fig. 3.4c). The rule tends to disappear in any of the options. This could occur, for example, because it could be more costly in the group, spatial, or temporal context (or a conjunction of all of them).

Another case is when there is polymorphism in the trait and some advantage in it. For example, populations with different geographic distributions may have distinct colorations because they are adapted to the specific environments where they live and to camouflage themselves from predators. The rule (i.e., a stable mean color to which the majority conformed in Fig. 3.3a) is diluted, and we could already have what is schematized in Fig. 3.3d.

There it could even be in the range from zero to "x" value of the trait expression as before or else spread toward new values (hence the "cuts" with two orange lines, to show that this "plateau" could well be more extensive than what is shown on the X-axis for reasons of graph space). In addition, we could have before this or at some time, space or group a situation of several smaller, smaller "mini rules" (Fig. 3.4a), with extreme cases with fewer individuals for each or in the "flatter" total of Fig. 3.3d, only two extremes. In either case, the gap will be smaller than that between the extreme cases and the mean in Fig. 3.3a.

Finally, the latter could directly disappear, as a low proportion of individuals exhibit any trait value. Here, it is impossible to identify the original rule. This situation would represent an extreme, almost evolutionary anarchism, at least temporarily. This case is illustrated with the symbol of infinity, not because it is so in reality but could extend further than is limited in the X-axis.

In short, every example of an evolutionary, ecological, and developmental nature could be expressed from a viewpoint such as the one mentioned change in the rule of biology itself. It is certainly not easy to frame options and examples so quickly, first, because, in many cases, we need to learn all the processes involved. These examples start from a normal distribution of traits and illustrate what would be defined as a rule. However, what would happen with other types of distribution? Could rules behave under these other types of statistical distributions, such as exponential ones? Moreover, as mentioned, we are assuming a "continuity," that is, continuous values, when it could be presence–absence or with gaps, intervals, without being continuous per se.

Of course, these graphs are limited and only schematize the evolution of a single trait. Changes could occur at a multidimensional level, with multiple variables intervening together. Moreover, graphically, "positive" or "negative" points are quantified as a function of values of expression of the pattern presented by individuals. It is limited to think that it is perhaps as simple as "above" or "below" the rule's core. In addition, the graphs aimed to consider the number of individuals as the determining parameter to establish a rule or an exception. We take the criterion of "how many" individuals express traits in a specific manner. Indeed, depending on that, we say whether or not that is a rule. In short, the baseline analysis presented here of general guidelines and representative examples of the dynamics between rules and exceptions may inspire mathematicians in biology to organize this potential better, considering its value and limitations.

The Sexual Conflict as an Example of the "Conflict of Interest" Between R and E

An interesting example of dynamics between rules and exceptions occurs in areas of evolutionary biology concerning the interests of the sexes. Specifically, we refer to sexual conflict and, in particular, to when intralocus type conflict has been analyzed. We take as an example the work of Cox & Caslbeek (2009), which is useful as a review with its graphs adapted from other studies on the subject (e.g., Rice & Chippindale, 2001; Day & Bonduriansky, 2004; Bedhomme & Chippindale, 2007).

First, we would have a theoretical "neutral" situation in which the conflict between males and females does not exist (or is at its minimum expression). This situation would coincide smoothly with the distribution in Fig. 3.3a of the relative fitness optima of each sex. Although this would be the theoretical distribution of an eventual phenotypic trait encoded by a gene, it could also happen that there are allelic variants with different optima in each sex. This would then imply that there is, in fact, an underlying "unresolved" conflict. In other words, as in dotted lines, we could imagine two bells, such as those in Fig. 3.4c, the presence of a conflict that underlies but does not yet manifest itself for different reasons.

On the other hand, the conflict could become "partially resolved" if, in one sex, a particular expression of the trait under analysis occurs and is exhibited to the left of the mean value of the previous distribution of the same. In contrast, in the other sex, the expression occurs to the right of that preceding mean. Finally, the conflict could be "totally resolved" if the "old" rule disappears entirely. Therefore, on both sides of the distribution, they have extreme values as patterns in relation to the previous situation and may not even overlap with the average zone of the previous distribution of the trait. In short, there are at least four possible situations in intersex conflict: no conflict, unresolved conflict, partially resolved conflict, and fully resolved conflict. All this is a function of different optima of the trait expression in each sex, a fact reflected in the distribution of expected and observed values.

From a broader perspective, when considering the theoretical level, this entire framework and context of sexual conflict prompt us to contemplate whether there might be "distinct optima" for implicit (or perhaps "evident" but unexpressed) rules and exceptions. This question could be posed both by action from biology and the history of science, as we explain below. These optima between a rule and its pool of exceptions would be quite opposed. However, they would not cease to be like "the two sides of the same coin." They mutually complement and rely on each other to establish their individual identities and strengthen their respective characteristics. Suffice it to recall how the respective definitions overlap and need to give entity to what is a rule and what is an exception. Seen from a dynamic approach, we find that they are different, with optima going in opposite directions (i.e., if you are a rule, you are not an exception and vice versa). Nevertheless, they can transmute to each other in any change that eventually occurs at the biological evolutionary level or of scientific knowledge in temporal, spatial, and group contexts.

On the other hand, we could make the situation more complex if we were to add to this "disruption in optimum" the fact that something can be a rule or an exception. This criterion will depend on the perspective or scale from which the situation is analyzed, so the conclusions should be limited to a given perspective and approach without ever generalizing. An example we will analyze further is the existence of traits that appear in phylogenetically derived groups within a clade and do not occur in the rest of the members. Such is the case that we will analyze later the presence of a true epithelium in a group derived from sponges. In them, the presence of epithelium is a rule. However, this characteristic is a remarkable exception in most of the phyla.

3.4.2.2 Analysis from the History of Biological Sciences (i.e., From Scenario 2)

The previously exposed view on the "movements" of rules in biology could also be helpful to us when analyzing changes in the historical rules of knowledge in biological science. We have commented on changes in the rules due to the contribution of new technologies that modify paradigms and advances in methodologies and science. In this context, Ruiz & Ayala (1998) highlight the analogy between the progress of scientific knowledge, characterized by its continual changes, and biological evolution. They emphasize that science, like species, is in a constant state of motion and transformation. There are also periods of quiescence without change and another new stage of upheaval and modification. Thus, science can be examined and understood in the same way as evolutionary epistemologies have proceeded. In that case, the objects of knowledge are not living organisms but scientific concepts and theories. First, the aforementioned epistemological perspectives of Lakatos, Feyerabend, and, mainly, Kuhn could be well embodied in such graphs.

For example, certain ideas may be the prevailing paradigm with counter ideas still scattered (Fig. 3.3a), followed by a moment of "chaos" where several options are possible, and the central paradigm or rule is destabilized (Fig. 3.3b, c). This situation may result in some new ideas becoming a new paradigm and replacing the old rule. Lakatos would follow the shift from Fig. 3.3a, b, where at least "all," 100% of the scope of the original rule does not fall away. However, the outermost belt of the idea, the initial paradigm, is modified, but the core (here, as the rule that continues) is still preserved. The new is not the opposite but complementary and belongs to an even more superficial level of the belt concerning the altered one. In contrast,

Feyerabend's anarchism could be embodied in Fig. 3.3d, where there are almost no rules, or they are no longer there. In this situation, every exception has a "double entity" of common and rare; this discrimination would no longer exist and would no longer make sense.

Second, more on the practical side, such as changes in scientific rules at the historical level on biological concepts, such as monogamy in birds, and molecular genetics, among other fields, serve this purpose. For example, an early rule postulates that "most birds are monogamous" (Fig. 3.3a), where the mean implies a low level of polyandry. We will see in more detail in the following chapters that this idea was modified by advances in molecular techniques that showed that polyandry values were higher than initially thought (e.g., this would be the change to Fig. 3.3b). From the "global" view for all birds, this change of interpretation has indeed occurred beyond a specific clade. Even if we were to analyze the issue of monogamy in the animal kingdom, the vague idea of monogamy as something virtually "ideal" prior to the theory of evolution and influenced by anthropocentrism has disappeared as such. Today, we know that is one of many options within reproductive systems and that these are highly context-dependent (which would lead to a figure of at least Fig. 3.3d if given equally). At least there can be at least two options in a given clade, such as monogamy or promiscuity, depending on particular taxa (this would be more toward, Fig. 3.4b or c, in a phylogenetic framework).

3.5 Key Concepts and Ideas in This Chapter

- When we speak of the concept of rule, it can be linked to that of law, although it does not imply that it is and may be more flexible. The same is true of the concept of exception when it is different or unique.
- Patterns that can be observed in nature can give rise to the enunciation or not of a rule, as can its antonym, rarity, related to the uncommon.
- One of the synonyms of rarity could be uniqueness, which distinguishes rare things from ordinary things. An interesting question is whether these singularities can be predicted, that is, whether predictive rules about the occurrence of exceptions can be proposed.
- The dynamics between rules and exceptions, in one direction or another, are key to their existence. The analysis of these movements will allow us to conclude that the evolution of rules follows the same patterns as biological evolution.
- There are also cases where the rules are more ambiguous or fuzzy. In such a case, the spread of the trait expression is to both sides by similar parts. This situation occurs when there are so many exceptions that they reach the same entity as the rules and could even make them disappear as such and generate a new flat-type distribution without rules or exceptions.
- Scientific paradigm shifts in biology can be analyzed in the same way as movements in rules in biological evolution. This view is interesting because it can

even transcend biology and allow us to see the dynamics of rules and exceptions from another perspective.

References

Arnold, V. I. (1981). *Singularity theory. Selected papers*. Cambridge University Press.

Bedhomme, S., & Chippindale, A. K. (2007). Irreconcilable differences: when sexual dimorphism fails to resolve sexual conflict. In D. J. Fairbairn, W. U. Blanckenhorn, & T. Székely (Eds.), *Sex, size and gender roles: Evolutionary studies of sexual size dimorphism* (pp. 185–194). Oxford University Press.

Briffa, M., & Greenaway, J. (2011). High in situ repeatability of behaviour indicates animal personality in the beadlet anemone *Actinia equina* (Cnidaria). *PLoS One, 6*(7), e21963.

Briffa, M., & Weiss, A. (2010). Animal personality. *Current Biology, 20*(21), R912–R914.

Capra, F. (1996). *The web of life*. Anchor Books.

Capra, F., & Luisi, P. L. (2014). *The systems view of life: A unifying vision*. Cambridge University Press.

Carere, C., & Maestripieri, D. (Eds.). (2013). *Animal personalities: Behavior, physiology, and evolution*. University of Chicago Press.

Ceccarelli, L. (2010). *Shaping science with rhetoric: The cases of Dobzhansky, Schrodinger, and Wilson*. University of Chicago Press.

Cox, R. M., & Calsbeek, R. (2009). Sexually antagonistic selection, sexual dimorphism, and the resolution of intralocus sexual conflict. *The American Naturalist, 173*(2), 176–187.

Day, T., & Bonduriansky, R. (2004). Intralocus sexual conflict can drive the evolution of genomic imprinting. *Genetics, 167*, 1537–1546.

Dobzhansky, T. (1982). *Genetics and the origin of species (No. 11)*. Columbia University Press.

Kaiser, M. I., & Müller, C. (2021). What is an animal personality? *Biology & Philosophy, 36*(1), 1.

Kralj-Fišer, S., & Schuett, W. (2014). Studying personality variation in invertebrates: Why bother? *Animal Behaviour, 91*, 41–52.

Rice, W. R., & Chippindale, A. K. (2001). Intersexual ontogenetic conflict. *Journal of Evolutionary Biology, 14*, 685–693.

Rose, S. (2001). *Trayectorias de vida: biología, libertad, determinismo*. Ediciones Granica SA.

Ruíz, R., & Ayala, F. J. (1998). *eEl método en las ciencias. Epistemología y darwinismo*. S. L. Fondo de Cultura Económica de España (Colección Ciencia y Tecnología).

Part II
Exploring Rules and Exceptions from Life Sciences

Chapter 4
When THAT Rule Almost Persists as THAT Rule

One never attains the total truth, nor is one ever totally far from it

—Aristotle

Contents

Abstract In biology, theories are constantly enriched and adapted by new examples and scientific advances. In this chapter, we focus on the debate about the rules in biology and their persistence over time as rules accepted by the scientific community. We exemplify this with Darwin's theory of natural selection, which has been nourished by new disciplines and technologies during the twentieth and twenty-first centuries and has undergone transformations, although it undoubtedly remains the dominant explanation of the evolution of biological systems. Biological rules are dynamic, although they tend to remain relatively stable over time, and their response to the appearance of exceptions can be variable, although the integration of new knowledge into the field is usually gradual until the accumulation of exceptions is so significant that the applicability of the rules is questioned (crisis *sensu* Kuhnian revolution of scientific knowledge). We illustrate this concept by analyzing

A. V. Peretti et al., *Rules and Exceptions in Biology: from Fundamental Concepts to Applications*, https://doi.org/10.1007/978-3-031-55382-0_4

the historical evolution of biological species concepts, and we draw attention to the extrapolation of rules originally proposed for certain organisms and applied indiscriminately to other groups without exhaustive testing (e.g., eco-geographic rules).

4.1 Introduction

Chapter 1 explored the concepts linked to rules and some basic examples. We have seen that rules can change completely, become an exception, or even disappear. Something much more subtle is that the rules are gradually modified and adjusted over time, space, or a group of living beings. With new information in these contexts, a rule is modified according to the novelties now included. This situation happened with many classical rules of biology in fields such as evolution, genetics, ecology, and biodiversity, among others. This "polishing" of this or that rule is so common that it often goes unnoticed, either because it was considered obvious or because we have not stopped examining the point. From a conservative perspective, the "rule–rule" situation will most likely evolve into dogmatism. However, it also provides relatively "stable" principles over time, allowing biological facts to be understood and interpreted. We have already seen that a rule can be general or restricted. The problem with any rule, especially in biology, is that it must inevitably be taken as a principle or norm with some degree of flexibility. Otherwise, the first exception encountered would justify rejecting the rule as inadequate when it should be restricted relative to its influence to include additional cases. Although biological rules generally explain the diversity of natural phenomena (e.g., ecology, genetics, paleontology), many "exceptions to the rule" have been found over time. However, the critical point is that new information does not necessarily mean that such rules are wrong. However, some of the original claims must be revised and discussed more carefully according to specific organisms and regions (Hone et al., 2008; Alho et al., 2011).

Therefore, when advancing knowledge on a given topic, we have seen how one reaches a point where the need arises to set basic principles to organize it, something inherent to the very essence of scientific thinking. One reason is to provide logical coherence to this vast set of new data, enabling future experimental or comparative work on its components within an integrated and unified context, rather than a confused, diffuse, or chaotic one. This plan can be challenging, especially in biology, where there are, by nature, patterns and their exceptions. owever, it is good to list some cases where these rules or principles have been proposed from evolutionary biology, developmental biology, ecology, and other fields. Many of these aspects will be addressed in the subsequent chapters, particularly those related to the dynamics between rules and exceptions. Therefore, whether in specific or broader fields, including biology, the integration of new knowledge necessitates the organization of clear guidelines and rules, as well as the re-evaluation of existing concepts and theories. This revitalization process involves incorporating novel findings to enhance their relevance. It is sometimes a process that takes time. Sometimes it is slow and, most of all, never-ending. In the following, we will examine some areas of biology to illustrate this.

4.2 Flexibilizing and Enriching Fixed Ideas: A Brief Historical Timeline in Genetics and Evolution

In the framework of the dynamics of rules, a body of ideas considered solid at a given time will only sometimes be overturned entirely or replaced by others in the future. In any case, what can happen is that new evidence, new information, or alternative options are incorporated into the broad umbrella of the initial rule. In this way, the rule or "super rule" (if it is too broad) will be more flexible and enriched with new data and perspectives. Indeed, this ultimately happens from the Kuhnian view of the advancement of scientific knowledge and mainly refers to the theory of organic evolution. This theory has been taken as a framework analogous to what occurs at the epistemological level concerning models on conceptual evolution (e.g., Popper, 1980, 1991; Ruíz & Ayala, 1998).

The occurrence of changes referring to knowledge about organic evolution is obviously to be expected since it is a scientific theory, which implies something vast that does not include a single principle or rule. Indeed, a classical definition of a scientific theory is a "systematic ideational structure of broad scope, conceived by the human imagination, which encompasses a family of empirical (experiential) laws regarding regularities existing in objects and events, both observed and posited." As we saw, the theory itself will include principles, laws, and rules (as we said, these terms can and are usually used as synonyms) that attempt to organize the observable regularities in objects and events.

Thus, it is expected that when new information about sublevels of rules and principles is incorporated into evolutionary biology, the theory is eventually renewed, and a new version emerges. As we observed in the first chapter, the "levels" of rules are generally varied. It is true, in this sense, that at the "macro" level of the theory of evolution, we have level I. Let us recall, in this case, what the third law of biology says: "All living organisms have arisen from an evolutionary process." While this principle remains constant, it prompts us to consider whether the accompanying body of data, hypotheses, ideas, pathways, and mechanisms remain unchanged. The history of science in evolutionary biology demonstrates, thankfully, that this theory is inherently flexible due to its subject of study.

Understanding all the details of the mechanisms and processes involved in how this universal level I law is expressed is not simple and has yet to be completely deciphered. We are dealing with more than a hundred-year-old theory which has undergone changes and has not remained unaltered or fixed. This theory has not been replaced by another but has been enriched, adapted, and modified over the years based on advances in related disciplines. The theory of natural selection, presented by Charles Darwin in 1859, marked a profound departure from traditional creationist beliefs. Additionally, within the scientific community, there was a prevailing emphasis on Lamarckism, which suggested the inheritance of acquired characteristics. Unlike Alfred Wallace, Darwin did not outright reject this concept (Caso, 2020). However, through the introduction of his new theory, he indirectly challenged and diminished the significance of the Lamarckian paradigm.

One of Jean-Baptiste Lamarck's notable contributions lies in that he was the first to propose an evolutionary theory (Lamarck, 1809) that, while not directly conflicting with religious beliefs, provided a groundbreaking conceptual framework in the scientific community of his time. Indeed, Lamarck's ideas were rooted in a solid scientific, observational, comparative basis, especially considering the early stages of biological knowledge when the fundamental principles were still being established. Lamarck proposed two challenging laws: 1. on modifications of organs during ontogeny of the individual by their use or disuse, and 2. on the nature of the heritability of such modification, as long as it lasted in time and was present in all progenitors. His second law, the inheritance of acquired characters, was later particularly questioned and rejected by most evolutionary biologists. However, in the forthcoming chapter on the transformation of rules into exceptions, we will delve into the resurgence of Lamarckism in recent decades (neo-Lamarckism). This resurgence suggests that the possibility of the second law regarding the inheritance of acquired traits cannot be entirely dismissed across all groups of living organisms.

The modifications of Darwin's original theory of natural selection that occurred during the twentieth and twenty-first centuries have come mainly from technological advances and knowledge associated with new disciplines that have helped to further "polish" the strictly Darwinian vision. In general, it is not a matter of rejecting the initial theory but of adding more aspects to it, some of which were very controversial at the time. Ultimately, they reinforce, to a greater or lesser extent, the central postulates originally enunciated by Darwin. The dimensions in which evolution operates are increased, taking into account more types of variations that occur in the life history of living beings (e.g., the genetic, epigenetic, behavioral, and symbolic ones analyzed by Jablonka & Lamb (2014) in their excellent book "Evolution in four dimensions"). Darwin himself relied on nineteenth-century rules governing growth and reproduction, inheritance, interindividual variability, and struggle for existence. It was from these rules that he developed his concepts on natural selection, through which divergence of characters involved in this struggle would occur, with the most successful form having more survival and success. In contrast, the least improved form would eventually become extinct. Thus, at that moment, he enriches and reorganizes the previously existing knowledge about living beings.

As we know, the incorporation of Gregor Mendel's studies on heredity was a key addition and modification to Darwin's original postulates. The publication of all his work (Mendel, 1866) in which he postulates the three laws of heredity (or two in his alternative vision if the first one is taken as implicit) was a fundamental pillar to give more sense over time to many of the conjectures and hypotheses that Darwin had drawn up. As we know, Mendel's great work was not seriously considered when he made it known in 1866. Indeed, it was around 1900 that it was rediscovered by several researchers, including William Bateson, who translated his work into English (Bateson, 1902), thus forming the new and revolutionary scientific discipline called "Genetics" (Gayon, 2016).

However, the relevance of Mendel's studies in evolution was questioned and debated for a long time. The subsequent contributions of W. Bateson, and particularly that of Thomas H. Morgan and his team, regarding the Mendelian basis of inheritance and the concept of chromosomes, solidified the gene theory (Morgan et al., 1915; Morgan, 1926) and established genetics as a crucial discipline for the study of evolution. The unit of variation was no longer the old concept of "gemmule" but the genes. However, studies on DNA structure would still be needed to better define the concept of "gene" at the molecular level, a term that, in turn, would undergo modifications in its meaning until today (Portin & Wilkins, 2017). From Morgan's work onwards, the focus at the hereditary level progressively concentrated on the structure of DNA as a unit of variation in evolution. With the contributions of the biologist and statistician Ronald Fisher (1930) and the geneticists Sewall G. Wright (1931) and John B. S. Haldane (1932), who would establish more firmly the knowledge on the genetic basis of evolution, in particular about the foundations of population genetics.

The discipline of evolutionary biology progressively arrived at a more integrative theory from Darwin's original; for this consolidation, the contributions of the geneticist Theodosius Dobzhansky (1937), the zoologist Ernst Mayr (1942), the paleontologist George G. Simpson (1944), and the botanist George L. Stebbins (1950), among other key researchers, were fundamental. This improvement originated the neo-Darwinism or synthetic theory of evolution (among several matching terminological options). Again, rather than contradicting Darwin's original ideas, what was done was to add new information from modern genetics, paleontology, and comparative morphology. In the early 1950s, with the knowledge of the chemical structure of DNA (Franklin & Gosling, 1953; Watson & Crick, 1953; Wilkins et al., 1953), the molecular basis of evolutionary theory was increasingly deepened. It involved a refinement in the studies of evolutionary relationships and the concepts of homologies based initially only on phenotypic characters. This integration led over the years to revisit and reformulate concepts linked to the phylogenetic relationships of many groups of organisms. This work involves comparative morphology and phylogenetics (Liò & Goldman, 1998; Whelan et al., 2001). Together, advances in genetics and biochemistry would progressively bring molecular biology to its apogee.

Advances in epigenetics have demonstrated that heritable changes do not always involve mutations in the DNA sequence. Changes could also be due to nucleotide methylation or histone protein acetylation. These events affect gene expression and RNA micromodifications (Gayon, 2016; Banta & Richards, 2018; Lind & Spagopoulou, 2018). Epigenetics recovered, in a way, the role of the environment in shaping phenotypes, relaxing the quasidogmatic idea of unidirectionality from DNA to phenotype existing up to that time (Jablonka, 2017; Cabej, 2018). Knowledge about long-standing concepts, such as developmental phenotypic plasticity, was revitalized and enriched, with Mary J. West-Eberhard at the forefront, giving it a more important role in evolution (West-Eberhard, 1989, 2003, 2005). The value traditionally given to random mutations was not discarded at all. However, it was relativized as the only way to generate variants in a framework of natural and sexual selection.

These new perspectives from modern molecular genetics, epigenetics, and phenotypic plasticity resulted in a more effective and better integration of developmental biology with evolution, subjects that always seemed curiously associated but not too influential on each other. It became clear that ontogeny could explain variants in traits that were not necessarily explained by the classical action of natural selection on mutations but rather by expressions during the development of the individual and of a common set of genes. Progressively, a new concept of evolutionary developmental biology, Evo-Devo (Hall, 2003; Carroll, 2008), emerged to summarize this type of association. The concept of ecological evolutionary developmental biology, Eco-Evo-Devo (Abouheif et al., 2014), was added to more strongly incorporate the role of ecology at the level of development and evolution. Eco-Evo-Devo pays attention to the influence of environmental signals in shaping phenotypes during the ontogeny of an organism and how this affects the evolutionary process of a species. Both views, particularly the first as the foundational one, represented a new paradigm in the vision of evolution, referring to the fact that it consists of changes in developmental processes. In the field of Evo-Devo, there is a focus on the recovery of macroevolutionary phenomena, which involves exploring the potential continuity between these phenomena and genetic-molecular and organismic changes. On the other hand, in Eco-Evo-Devo, the emphasis is primarily on recovering the characteristics of microevolution (Rendón & Folguera, 2014).

More recently, new innovative visions were raised with advances in neurosciences, knowledge of "networks" at the nervous system level, and extensive analogous concepts of "networks" in molecular biology and ecosystems. The study of gene (or genetic) regulatory networks (Thompson et al., 2015) is, in fact, an essential aspect of the Evo-Devo and Eco-Evo-Devo areas. This field is still in full development.

The addition of novel fields that are still in the process of consolidation, such as epigenetics, phenotypic plasticity, Evo-Devo, Eco-Evo-Devo, gene regulatory networks, and neuron-like networks, among others, is what has led many scientific leaders in these fields and evolution to call for a total and profound "update" of the theory of evolution (Müller, 2007; Pigliucci, 2007). It is proposed to formalize and not be limited to the traditional and unidirectional perspective of random mutations and their subsequent natural selection according to their adaptive functional value. This classical perspective is not denied, but other possibilities are added, complementary or alternative, depending on the context, level of analysis, and group. Thus, instead of weakening Darwinian ideas, the simple, straightforward, and practical scheme (still the heart of the theory) that was proposed more than 150 years ago is reinforced. This new "postsynthetic theory" stage is called extended evolutionary synthesis (Pigliucci & Müller, 2010; Laland et al., 2015). This new name tells us that there is no contradiction with the preexisting but a significant conceptual enrichment from the advances in these new areas (Pigliucci & Finkelman, 2014; Müller, 2017; Fábregas-Tejeda & Vergara-Silva, 2018).

Therefore, since evolutionary theory is a dynamic body of rules and principles, it is expected that many more instances of refinement of this key area of biological knowledge will emerge in the future. New "extensions" of the theory will continue to take place in line with advances in different fields of biology. The question also arises: could it be rejected at some point instead of being extended or enriched? This situation would lead to a crisis no longer only partial, as when it was seen as incomplete in the past, but to a level of a Kuhnian revolution of scientific knowledge. However, scientists have shown in the long run to be flexible and adaptable to perfect it and avoid falling into dogmatism, something that Darwin himself would surely expect. Again, in perspective, we see how evolutionary theory is becoming more complex. From a broad view at the level of time, it is a great rule that is becoming more flexible and enriched rather than fragmented or replaced in its foundations.

We can conclude that (1) a great initial rule (the evolutionary theory proposed by Darwin) varied but never ceased to be such; it was never replaced 100% but was made more flexible, enriched, and therefore improved; (2) the theory was perfected by the addition of other associated rules of another level, such as those of genetics; (3) in this process, there were phenomena that initially could have been considered somewhat circumscribed processes but that gradually became known to be quite generalized among organisms (e.g., phenotypic plasticity, epigenetics); and (4) indeed, some of these cases implied conceptual changes from considering an exception as a rule or vice versa, some of which will be analyzed in the following chapters.

4.3 What Is a Biological Species: Is There a Rule?

If there is a very controversial subject, diverse in approaches and criteria, then in an endless debate at the core of biology, it is referred to as the "species concept" (Kimbel & Martin, 1993; Wheeler & Meier, 2000). The "species problem" (as is known) of what a species is has dominated discussions in the biological sciences (Clarke & Okasha, 2013). Indeed, the recent detailed review by Zachos (2016) shows that this issue is still under continuous discussion. There is now a greater tendency toward a more holistic view, as recent advances in genetics, phenotypic plasticity, and phylogeny are taken into account, some of which we have exposed previously when discussing the extended theory of evolution. Indeed, we could almost be talking about a level I rule regarding the degree of inclusiveness involved in defining a species. From the strictly human, scientific interest, from our operational need as researchers we need to discern "without error" what is the "basic unit" of analysis, which will then be linked at the phylogenetic level in trees, evolutionary patterns, with other organisms. These relationships extend beyond interspecific connections to higher hierarchical levels, encompassing genera, families, and orders. It means how much we have wondered throughout the history of biology about how to define an "organism."

In addition to the challenge to define a species, we must consider both our external perspective as observers and the intrinsic nature of the species itself within the biological reality of organisms. One thing is our gaze as observers, and another is what the species represents by its existence in that biological reality. These two ideas are closely interconnected, yet they entail distinct implications. For example, when we talk about all organisms arising from an evolutionary process, this is a natural biological fact. However, this has yet to imply that in the history of biology, there is a definitive explanation of what evolution is and how it is carried out. Our view is more changeable than nature itself.

Thus, at least two types of perspectives arise. On the one hand, a transversal view, when analyzing and evaluating the differences between morphs that are potentially different. An illustrative example can be observed by examining the terminal branches of a phylogenetic tree, where we currently make transversal comparisons within a specific clade. This approach, even when dealing with fossils or the complexities of defining a species, has influenced definitions derived from phylogenetics. An alternative framework, focusing on a temporal perspective, adopts a more longitudinal approach by considering the transformation of a species over time. This viewpoint acknowledges that a species undergoes changes and ultimately ceases to be classified as the same species. This transformation has been extensively discussed (Wheeler & Meier, 2000). As summarized by Zachos (2016), the boundaries between both general perspectives, the first more focused on the pattern ("the picture"), while the second more on the process ("the film"), can be complicated to establish.

There are so many perspectives to define species that it is difficult to say, at least from the human methodological point of view, "what a species is." Indeed, we have multiple definitions (to explore this diversity of concepts, approaches, and implications, see Mallet, 2007; Zachos, 2016) (Box 4.1). One of the most widely used is the one defined at the population level, the so-called "Biological concept of species", enunciated by Mayr (1970) "*Species are groups of interbreeding natural populations reproductively isolated from other such groups.*" Indeed, multiple variants of this classic definition are commonly found on science education sites. For example, "*A species is often defined as a group of individuals that actually or potentially interbreed in nature*" (University of California, Berkeley, 2023), or "*A biological species is a group of organisms that can reproduce with one another in nature and produce fertile offspring*" (Nature Education, 2014). However, both sites clarify that these definitions have limitations since they do not contemplate, for example, organisms with asexual reproduction, something familiar in many invertebrates and plants.

Other species concepts are based on essential taxonomic criteria, particularly comparative morphological criteria, and more recently, evolutionary genetics has been added. Thus, there are typological definitions of species (including several, even ancient ones), evolutionary definitions of species (a more linear view of the status in evolutionary time as mentioned above), and phylogenetic and cladistic definitions (a more transversal view, a specific current branch) (Mallet, 2007) (Box 4.1). Alternatively, or as a complementary approach (depending on the researcher), there are two additional concepts to consider: the concept of recognition and the ecological concept of species. To decide in our book on these options is not the intention and would be superficial. What we find helpful in this controversial topic is how a rule can mutate over time. Additionally, how many definitions and criteria are presented in opposition to each other? The consideration of these options is not intended to imply a unidirectional choice between them. Instead, it can be viewed as a multilinear approach, with varying degrees of emphasis depending on the researcher's specific area of expertise. For instance, an ecologist, taxonomist, or geneticist may lean more toward one concept over the other. Suppose the focus is more on gene flow and interbreeding. In that case, the researcher will prefer a definition associated with an "inbreeding concept" (i.e., biological, recognition, isolation concepts) over others such as cohesion or monophyletic species concepts (Lee, 2003).

Therefore, in the face of so many options of definitions, it emerges that (1) we cannot say that there is "a single rule or law" for what is considered a species and (2) the understanding of species has undergone changes throughout history, influenced by advancements in knowledge regarding crucial aspects of organisms, including their evolution, genetics, ecology, and behavior. (3) Even today, there is still a diversity of definitions and concepts due to the perspective of each discipline and its biases (e.g., in evolution, ecology, taxonomy, behavior, genetics); (4) the definition of what is considered to be a species will always be relative and biased, depending on the approach of study. Thus, while some definitions will prevail over others (e.g., biological or evolutionary over many others), it will not imply the elimination of other concept options as "wrong."

Of particular interest in this issue are two definitions that, in some way, have tried to find a unifying concept of species. One of them is that coined by Templeton (1989, 1998). His definition is a summation of all or almost all the components implied in the other definitions: "*A taxon characterized by cohesion mechanisms, including reproductive isolation, recognition mechanisms, ecological niche, as well as by genealogical distinctness.*" Indeed, this definition is known as the cohesion concept (Box 4.1). On the other hand, the concise general lineage concept proposed by de Queiroz (1998, 2005): "Species are independent lineages." In this case, rather than adding several elements to the definition, the author highlights what will always exist, what will characterize any taxonomic entity that is to be called a "species," an element common to all of them. In essence, the only necessary property will be to have the general concept of species as separately evolving metapopulation lineages (de Queiroz, 2007). All the other properties considered by other concepts are not necessary but will depend on whether or not the species in question has adopted them throughout its existence. Indeed, the author emphasizes that lineages do not have to be "*phenetically distinguishable, diagnosable, monophyletic, intrinsically reproductively isolated, ecologically divergent, or anything else to be considered species. They only must evolve separately from other lineages*" (de Queiroz, 2007).

From the framework of our book, we see that the species problem is an excellent example (Box 4.1) of how a rule (the concrete answer to what is a species?) will depend on two-time contexts. On the one hand, we refer to historical time in the biological sciences as regard changes in concepts, as well as to evolutionary time as regard changes in the entity called species. On the other hand, the species concept will depend on the context of the group, i.e., the taxon being analyzed. For example, for an asexual taxon, a concept different from the classical biological definition would be easier to adopt. In turn, species concepts will depend on the spatial component. For example, the ecological definition of species focuses on the specific niche occupied by the organism within an ecosystem. In short, it is only possible to choose one rule if there are many partialized, interrelated, context-dependent rules.

Box 4.1 Alternative contemporary species concepts

Biological	Natural reproduction resulting in viable and fertile offspring 1. Intrinsic reproductive isolation 2. Shared specific mate recognition or fertilization system
Ecological	Same niche or adaptive zone
Evolutionary	Unique evolutionary role, tendencies, and historical fate
Cohesion	Phenotypic cohesion: genetic or demographic exchangeability
Phylogenetic	Heterogeneous 1. Hennigian: Ancestor becomes extinct when lineage splits 2. Monophyletic: Monophyly (an ancestor and all its descendants) 3. Genealogical: Exclusive coalescence of alleles 4. Diagnoseable: Diagnosability
Phenetic	Form a phenetic cluster
Genotypic cluster	Form a genotypic cluster: deficits of genetic intermediates

Modified from de Queiroz, 2007.

The truth is that everything is linked in biology. Although discussions about definitions concerning the species problem can be endless and fruitless (like most debates about "definitions"), its implications would not be minor. For example, there is an extreme definition of species based on the degree of extinction danger, which considers any isolated population a species (Waples, 1991). Therefore, we have that a species concept will impact fields that, in principle, may not be directly associated in theory, such as endangered species by human action ("Endangered Species Act"—George & Mayden, 2005). Here, the question arises: could the number of species in our biological lists and catalogs be reduced or increased depending on the concept adopted to define a species? Eventually, this would impact species management and conservation strategies, perhaps not only at the scientific level but also at the political level, for the preservation of natural areas where endangered species live (George & Mayden, 2005).

4.4 Biological Rules: Some Historical Changes and Limitations

If there is a field of biological sciences where we can appreciate in a didactic way the implications of the scope and limitations of some postulates, then it is the one concerning the so-called biological rules, which in many cases date back to the times of Aristotle. By biological rules, we refer to a set of type II level rules, which focus on ecological, eco-geographical, embryological, and genetic, among other areas. We make this clarification since, in Chap. 1, we mentioned the levels of rules and said that there was a level I that encompassed all biological organisms. From this level, we move on to more specific areas where other levels appear that are more restricted to certain organisms or ecological and/or geographical situations.

4.4.1 Analysis of Some Examples

Biology students come across many rules and laws during their careers. For instance, Table 4.1 lists examples of biological rules, their scope, and some references to their application. Many of them refer initially to a particular animal group. The problem, and not a minor one, is that some of these rules were later extrapolated to other groups of organisms, even quite distant phylogenetically from those originally studied. This led us to think that these rules had been poorly formulated from the beginning or that they arose from considering extremely restricted cases. In reality, the problem often lies in the extrapolation of the rule without taking into account the factors mentioned above or even "forcing" its extension to other organisms. Below, we highlight two illustrative cases of the problem, Bergmann's and Emery's rules.

4.4.1.1 Bergmann's Rule

A classic example of a biological rule of the eco-geographic type is Bergmann's rule. This rule states that within a widely distributed taxonomic clade, larger populations and species are found in colder environments, while smaller species are found in warmer regions (Bergmann, 1847). As depicted in Fig. 4.1, this particular rule has received extensive analysis and has been the subject of considerable controversy for over 150 years. There has been a significant amount of extrapolation from the original postulates, which were formulated for specific mammalian taxa, to phylogenetically distant groups such as insects (Shelomi, 2012). Furthermore, it is not a minor fact that the original paper was written in German; therefore, the English translations could have added some confusion to the topic.

Table 4.1 Rules in biology. Examples of biological rules (principles, laws, theories), year of postulation, biological field, and range of application

Year	Biological rule, law or principle	Biological field	Statement	Author	Application range	Citation
Dogmas/axioms						
1958	Central dogma of molecular biology	Physiology	DNA makes RNA, and RNA makes proteins. Once "information" has passed into protein it cannot get out again. The transfer of information from nucleic acid to nucleic acid, or from nucleic acid to protein may be possible, but transfer from protein to protein, or from protein to nucleic acid is impossible	F. Crick	Broad	Crick (1958)
1977	Woese's dogma	Evolution	The evolution of ribosomal RNA was a necessary precursor to the evolution of modern life forms	C. Woese	Broad	Woese & Fox (1977)
Rules						
1833	Gloger's rule	Ecology	Within a species, more heavily pigmented forms tend to be found in more humid environments (e.g., near the equator)	C. W. L. Gloger	Restricted (Endotherms, plants see Lev-Yadun, 2016)	Gloger (1833)
1847	Bergmann's rule	Ecology	Within a broadly distributed taxonomic clade, populations and species of larger size are found in colder environments, and species of smaller size are found in warmer regions	C. Bergmann	Restricted Endotherms, ectotherms see Mousseau, 1997; Vinarskiĭ (2014)	Bergmann (1847)
1877	Allen's rule	Ecology	Body shapes and proportions vary by climatic temperature by either minimizing exposed surface area to minimize heat loss in cold climates or maximizing exposed surface area to maximize heat loss in hot climates. Animals from colder climates have shorter appendages than animals from warmer climates	J. A. Allen	Intermediate Endotherms, ectotherms see Alho et al. (2011)	Allen (1877)
1884	Hertwig's rule, long axis rule	Cell Biology	A cell divides along its long axis. Predicts cell division orientation, which is important for tissue architecture, cell fate, and morphogenesis	O. Hertwig	Broad	Hertwig (1884)
1892	Jordan's rule	Ecology	There is an inverse relationship between water temperature and meristic characteristics (e.g., number of fin rays, vertebrae, or scale numbers) which are seen to increase with decreasing temperature	D.S. Jordan	Very restricted (Fishes)	Jordan (1892)
1894	Bateson's rule	Morphology	The long axes of the normal appendage and the two extra appendages lie in a single plane. One of the extra appendages is closer to the axis of the normal appendage, and the other is farther away. The closer extra appendage is structured and positioned as a mirror image of the normal appendage in a plane mirror placed between the normal appendage and the closer one. The farther appendage is the mirror image of the closer one in a plane mirror similarly positioned between the two extra appendages	W. Bateson	Broad	Bateson (1971)

Year	Biological rule, law or principle	Biological field	Statement	Author	Application range	Citation
1880	Deep-sea gigantism	Ecology	Deep-sea animals are larger than their shallow-water counterparts	H. Nottidge Moseley	Very restricted (Deep-sea animals)	Moseley (1880)
1887, 1907	Cope-Depéret rule	Evolution	Animal population lineages tend to increase in body size over evolutionary time	E. D. Cope, C. J. J, Depéret	Broad	Depéret (1907), Cope (1887)
1909	Emery's rule	Ecology	Insect social parasites are often closely related to their hosts, such as being in the same genus	C. Emery	Very restricted (Host-parasites)	Emery (1909)
1915	Harrison's rule	Ecology	Parasite body sizes covary with those of their hosts	L. Harrison	Very restricted (Host-parasites)	Harrison (1915)
1922	Haldane's rule	Genetics	If in a species hybrid only one sex is inviable or sterile, that sex is more likely to be the heterogametic sex	J. B. Sanderson Haldane	Very restricted (Hybrid species)	Haldane (1922)
1942	Eichler's rule	Evolution	Parasites tend to be highly specific to their hosts, and thus it seems reasonable to expect a positive covariation between the taxonomic richness of hosts and that of their parasites	W. Eichler	Very restricted (Host-parasites)	Eichler (1942)
1950	Rensch's rule	Ecology	Across species within a lineage, sexual size dimorphism increases with body size when the male is the larger sex, and decreases as body size increases when the female is the larger sex	B. Rensch	Broad (Animals, plants see Kavanagh et al., 2011)	Rensch (1950)
1959	Baker's rule	Population	Colonization by self-compatible organisms is more likely to be successful than colonization by self-incompatible organisms because of the ability for self-compatible organisms to produce offspring without pollination agents	H.G. Baker	Very restricted (Self-compatible organisms)	Baker (1959)
1963	The self-thinning rule, -3/2 power rule, Yoda's power law	Ecology	Beyond a certain density of sowing, the number of surviving plants is not related to the initial seed density; instead, a constant relationship (W) is evident between the density of survivors and their total biomass.	K. Yoda and colleagues	Broad (Plants)	Yoda et al. (1963)

(continued)

Table 4.1 (continued)

Year	Biological rule, law or principle	Biological field	Statement	Author	Application range	Citation
1964	Foster's rule, the island rule, or the island effect	Ecology	Members of a species tend to be smaller or bigger depending upon the resource availability in the environment. Insular dwarfism explains the shrinking of animals with the course of time due to limited food supply and more competition among the species. Insular gigantism is seen in animals that migrate to the islands from the mainlands. This is mainly because of reduced predation and lesser competition	J. B. Foster	Broad (Animals, plants see Biddick et al., 2019)	Foster (1964)
1964	Hamilton's rule	Genetics	Selection will favor altruistic behavior if the benefits to the recipient, multiplied by the relatedness between actor and recipient, outweigh the costs to the actor. This is a prediction from the Theory of kin selection	W. D. Hamilton	Restricted (Altruistic species)	Hamilton (1964)
1966	Hennig's progression rule	Evolution	As clades diversify, plesiomorphic traits dominate at centers of origin. The "most displaced" member of the clade will have the most apomorphic traits	W. Hennig	Broad	Hennig (1966)
1966	Deviation rule	Evolution	When two sister species split one deviates more strongly than the other from the common stem species	W. Hennig	Broad	Hennig (1950)
1967	Janzen's rule	Ecology	Tropical habitats with reduced thermal seasonality would select for species with narrow thermal tolerance, thereby limiting dispersal among sites of different elevations showing little overlap in temperature	D. H. Janzen	Restricted (Tropical insects)	Janzen (1967)
1967	Species–area relationship	Ecology	The number of species found in a given area increases with the size of the area sampled	R. MacArthur & E.O. Wilson	Broad	MacArthur & Wilson (1967)
1971	Thorson's rule	Ecology	At low latitudes, animals tend to produce large numbers of eggs developing to pelagic and widely dispersing larvae, whereas at high latitudes such organisms tend to produce fewer and larger lecithotrophic (yolk-feeding) eggs and larger offspring, often by viviparity or ovoviviparity, which are often brooded	S. A. Mileikovsky	Very restricted (Benthic marine invertebrates)	Mileikovsky (1971)
1975	Community assembly rules	Ecology	Competition, not random immigration, was the main force structuring the species composition of islands. Reduced niche overlap, species occurring together on islands should have less niche overlap than random assemblages because they have undergone specialization	J. Diamond	Very restricted (Island species)	Cody & Diamond (1975)
1989	Rapoport's rule	Ecology	Latitudinal ranges of plants and animals are generally smaller at lower latitudes than at higher latitudes	G.C. Stevens	Broad	Stevens (1989)
1996	Temperature–size rule	Ecology	Organisms exhibiting a plastic response are capable of allowing their body size to fluctuate with environmental temperature	D. Atkinson	Broad	Atkinson (1994)
1995	Phylogenetic niche conservatism	Ecology	Closely related species tend to occupy similar ecological niches	P.H. Harvey & M.D. Pagel	Broad	Harvey & Pagel (1991)

Year	Biological rule, law or principle	Biological field	Statement	Author	Application range	Citation
2019	Read's rule	Ecology	Trees that associate with arbuscular mycorrhizal fungi would dominate in warmer forests (such as in the tropics, where tree growth is limited by phosphorous) and ectomycorrhizal fungi in colder forests (where decomposition is slow and leaf litter is abundant)	B. S. Steidinger and colleagues	Very restricted (Mycorrhizal fungi)	Steidinger et al. (2019)
Principles						
1865	Krogh's principle	Epistemology	For such a large number of problems, there will be some animal of choice, or a few such animals, on which it can be most conveniently studied	C. Bernard	Broad	Bernard (1865)
1908	Hardy–Weinberg principle	Genetics	Allele and genotype frequencies in a population will remain constant from generation to generation in the absence of other evolutionary influences	G. Hardy & W. Weinberg	Broad	Hardy (1908)
1948	Bateman's principle	Evolution	Males typically experience more variance in reproductive success than females, due to their ability to mate with multiple partners	A. J. Bateman	Broad (Sexual species)	Bateman (1948)
1954	Lack's principle	Ecology	The clutch size has been adapted by natural selection to correspond with the largest number of young for which the parents can, on average, provide enough food	D. Lack	Restricted (Birds)	Lack (1954)
1975	Handicap principle	Evolution	Costly signals must be reliable signals, costing the signaler something that could not be afforded by an individual with less of a particular trait. It will be selectively advantageous for a female to choose as a mate a male with some conspicuous characteristic which is also a handicap as far as survival is concerned	A. Zahavi	Broad (Sexual species)	Zahavi (1975)
1934	Dale's principle, Dale's law	Physiology	A neuron performs the same chemical action at all of its synaptic connections to other cells, regardless of the identity of the target cell	H.H. Dale	Broad	Dale (1935)
Laws						
1816	Smith's law or principle of Faunal succession	Paleontology	Sedimentary strata contain different fossils (fauna and flora) that succeed each other vertically in a specific and predictable order that can be identified over a wide horizontal distance	W. Smith	Restricted (Fossils)	Smith (1816)
1824–1826	Meckel-Serres law of parallelism	Embriology	A higher animal in its embryological development recapitulates the adult structures of animals below it on the linear "scale of being." Conversely, lower animals represent permanent larval structures of the more advanced forms	J. Meckel & E. Serres	Broad	Serres (1827)

(continued)

Table 4.1 (continued)

Year	Biological rule, law or principle	Biological field	Statement	Author	Application range	Citation
1828	Von Baer's laws or laws of development	Embriology	Embryos start from a common form and develop into increasingly specialized forms so that the diversification of embryonic form mirrors the taxonomic and phylogenetic tree. Therefore, all animals in a phylum share a similar early embryo; animals in smaller taxa (classes, orders, families, genera, species) share later and later embryonic stages	K.E. von Baer	Broad	Von Baer (1828)
1835	Law of specific nerve energies	Physiology	The nature of perception is defined by the pathway over which the sensory information is carried. Hence, the origin of the sensation is not important. Therefore, the difference in perception of seeing, hearing, and touch is not caused by differences in the stimuli themselves but by the different nervous structures that these stimuli excite	J. P. Müller	Broad	Müller (1838)
1838	Surface law	Evolution	The heat production of different-sized species should be related to their surface area rather than their body masses if they were to maintain the same body temperature. Small animals should have greater mass-specific rates of metabolism than larger animals, just as relative surface area increases with decreased size of similar-shaped objects	P. Sarrus & J.F. Rameaux	Restricted (Endotherms)	Sarrus & Rameaux (1839)
1840	Liebig's law of the minimum	Ecology	Growth is dictated not by total resources available, but by the scarcest resource (limiting factor)	J. Von Liebig	Broad	Liebig (1840)
1865–1866	Laws of inheritance or Mendel's laws of inheritance	Genetics	A set of rules on the transmission by genetic inheritance of characteristics from parent organisms to their offspring. It includes the law of Independent Assortment: inheritance of one gene does not affect the inheritance of another gene unless the genes are linked on the same chromosome. Law of Dominance: one allele (the dominant allele) can mask the expression of another allele (the recessive allele). Law of Segregation: when two traits come together in one hybrid pair, the two characters do not mix with each other and are independent of each other. Each gamete receives one of the two alleles during meiosis of the chromosome	G. Mendel	Broad	Mendel (1866)
1904, 1934	Gause's law, competitive exclusion principle	Ecology	Two species competing for the same resource cannot coexist at constant population values. The competition leads either to the extinction of the weaker competitor or to an evolutionary or behavioral shift toward a different ecological niche	J. Grinnell, G. Gause	Broad	Grinnell (1904); Gause (1934)
1889	Galton's ancestral law	Genetics	The two parents contribute between them on average one-half of the total heritage of the offspring, the four grandparents one-quarter, and so on	F. Galton	Broad	Galton (1889)
1893	Dollo's law of irreversibility	Evolution	Once an organism has evolved in a certain way, it will not return exactly to its previous form	L. Dollo	Broad	Dollo (1893)

Year	Biological rule, law or principle	Biological field	Statement	Author	Application range	Citation
1905	Law of limiting factors	Ecology	Growth is limited by the factor that is in the shortest supply. When a process is conditioned as to its rapidity by several separate factors, the rate of the process is limited by the pace of the "slowest" factor	F. Blackman	Broad (Plants)	Blackman (1905)
1911	Shelford's law of tolerance	Ecology	An organism's success is based on a complex set of conditions and each organism has a certain minimum, maximum, and optimum environmental factor or combination of factors that determine success	V. E. Shelford	Broad	Shelford (1911)
1914	Williston's law	Evolution	Multipart structures in organisms tend to reduce the number of elements during evolution but, at the same time, provide a mechanism to increase diversity and functional specialization	W. Williston	Broad (Animals)	Williston (1914)
1922	Jordan's law	Ecology	Given any species in any region, the nearest related species is unlikely to be found in the same region nor in a remote region, but in a neighboring district separated from the first by a barrier of some sort	D. Jordan	Broad	Jordan (1922)
1928	Lewis' law	Cell Biology	The average apical area of a cell with n neighbors (divided by the average apical area of all cells) is proportional to its shape	F. T. Lewis	Very restricted (epithelial cells)	Lewis (1928)
1932	Kleiber's law, 3/4 rule	Evolution	The metabolic rate of an animal is proportional to its body mass raised to the 3/4 power	M. Kleiber	Restricted (Endotherms)	Kleiber (1932)
1936	Allometric scaling law	Evolution	A biological variable Y is related to the mass M of the organism by a power Law, Y=Y0M(exp=b), where b is the so-called allometric exponent	J. Huxley & G. Tessier	Broad	Huxley & Tessier (1936)
1942	Lindeman's law of trophic efficiency, ten percent law	Ecology	During the transfer of organic food energy from one trophic level to the next higher level, only about ten percent of the transferred energy is stored as flesh. The remaining is lost during transfer, broken down in respiration, or lost to incomplete digestion by higher trophic levels	R.L. Lindeman	Broad	Lindeman (1942)
1967	Ohno's law	Genetics	The gene content of the mammalian species has been conserved over species not only in the DNA content but also in the genes themselves. That is, nearly all mammalian species have conserved the X chromosome from their primordial X chromosome of a common ancestor	S. Ohno	Restricted (Mammals)	Ohno (1967)
1973	Van Valen's law, law of constant extinction	Evolution	The probability of extinction for species and larger evolutionary groups bears no relation to how long it may have already existed (Red Queen hypothesis). Species must constantly adapt, evolve, and proliferate to survive while pitted against ever-evolving opposing species	L. Van Valen	Broad	Van Valen Leigh (1973)
2000	Schmalhausen's law	Ecology	A population at the extreme limit of its tolerance in any one aspect is more vulnerable to small differences in any other aspect	R. Lewontin & R. Levins	Broad	Lewontin & Levins (2000)

(continued)

Table 4.1 (continued)

Year	Biological rule, law or principle	Biological field	Statement	Author	Application range	Citation
Theories						
1762	Germ theory of disease	Medicine	Microorganisms known as pathogens or "germs" can invade humans, other animals, and other living hosts. Their growth and reproduction within their hosts can cause disease	M. von Plenciz, etc.	Broad	von Plenciz (1705–1786)
1830s	Cell theory	Cell Biology	All living organisms are composed of one or more cells. The cell is the most basic unit of life	M. Schleiden & T. Schwann	Broad	Schleiden (1838)
1858–1859	Theory of evolution by natural selection	Evolution	Natural selection is the differential survival and reproduction of individuals due to differences in phenotype. It is a key mechanism of evolution, the change in the heritable traits characteristic of a population over generations	C. Darwin & A. R. Wallace	Broad	Darwin (1859)
1868	Ernst Haeckel's revised recapitulation theory, biogenetic law	Embriology	Development (size and shape changes during ontogeny) of an individual organism literally and faithfully recapitulates the evolutionary history of its species (phylogeny). The development of advanced species passes through stages represented by adult organisms of the more primitive species and, thus, is classified as a "strong recapitulation" theory	E. Haeckel	Broad	Haeckel (1868)
1932	Shifting balance theory	Evolution	Adaptive evolution may proceed most quickly when a population divides into subpopulations with restricted gene flow. In fitness landscapes (evolutionary landscapes) the populations may move across an adaptive valley to a higher adaptive peak	S. Wright	Broad	Wright (1932)
1944	Quantum evolution theory	Evolution	To explain the rapid emergence of higher taxonomic groups in the fossil record. Evolutionary rates differ from group to group and even among closely related lineages. These different rates of evolutionary change can be bradytelic (slow tempo), horotelic (medium tempo), and tachytelic (rapid tempo). The word "quantum" therefore refers to an "all-or-none reaction," where transitional forms are particularly unstable, and thereby perish rapidly and completely	G. G. Simpson	Broad	Simpson (1944)
1949	Hebbian theory, Hebb's rule, and cell assembly theory	Physiology	An increase in synaptic efficacy arises from a presynaptic cell's repeated and persistent stimulation of a postsynaptic cell. It is an attempt to explain synaptic plasticity, the adaptation of brain neurons during the learning process	D. Hebb	Broad	Hebb (1949)

Year	Biological rule, law or principle	Biological field	Statement	Author	Application range	Citation
1966, 1976	Selfish gene theory	Evolution	Gene-centered view of evolution (the more two individuals are genetically related, it makes for them to behave cooperatively with each other). A lineage is expected to evolve to maximize its inclusive fitness. As a result, populations will tend toward an evolutionarily stable strategy	R. Dawkins, G. C. Williams	Broad	Dawkins (1976)
1967	Island biogeography theory	Ecology	The number of species on an island is determined by a balance between immigration of new species to the island and extinction of existing species on the island	E.O. Wilson & R.H. MacArthur	Broad	MacArthur & Wilson (1967)
1967	r/K selection theory	Ecology	Selection of combinations of traits in an organism that trade-off between quantity and quality of offspring. The focus on either an increased quantity of offspring at the expense of individual parental investment of r-strategists, or on a reduced quantity of offspring with a corresponding increased parental investment of K-strategists, varies widely, seemingly to promote success in particular environments	E.O. Wilson & R.H. MacArthur	Broad	MacArthur & Wilson (1967)
1967	Serial endosymbiotic theory	Cell Biology, Evolution	States that eukaryotic cells evolved from a symbiotic relationship between different types of prokaryotic cells	L. Margulis	Broad	Sagan (1967)
1970s	Gaia theory, Gaia paradigm, and Gaia principle	Epistemology	Living organisms interact with their inorganic surroundings on the Earth to form a synergistic and self-regulating, complex system that helps to maintain and perpetuate the conditions for life on the planet	J. Lovelock & L. Margulis	Broad	Lovelock (1972)
1972	Punctuated equilibrium theory	Evolution	Once a species appears in the fossil record, the population will become stable, showing little evolutionary change for most of its geological history (stasis). When significant evolutionary change occurs, the theory proposes that it is generally restricted to rare and geologically rapid events of branching speciation called cladogenesis	N. Eldredge & S.J. Gould	Broad	Eldredge & Gould (1972)
1973	Evolutionary game theory	Evolution	Application of game theory to evolving populations in biology. It defines a framework of contests, strategies, and analytics into which Darwinian competition can be modeled	J. Maynard Smith & G. R. Price	Broad	Maynard-Smith & Price (1973)
1989	Contingency theory, wonderful life theory	Evolution	After hundreds of different phyla evolved during the Cambrian period, many of them subsequently became extinct, leaving the relatively few phyla that exist today	S. J. Gould	Broad	Gould (1990)

(continued)

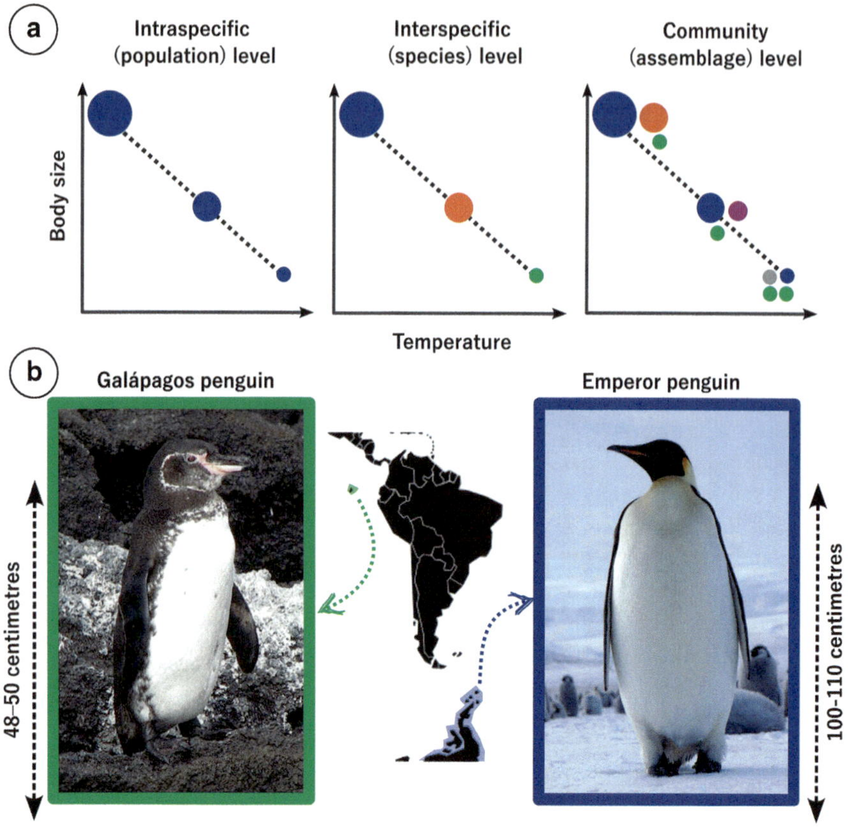

Fig. 4.1 Bergmann's rule. **(a)** Different levels of applicability of Bergmann's pattern in organisms (Modified from de Mungee et al., 2021). **(b)** Example of Bergmann's rule in penguin species (Credits: Mike Weston, Hannes Grobe/AWI).

Credits: **(a)** Galapagos penguin. License: CC BY 2.0 https://creativecommons.org/licenses/by/2.0, via Wikimedia Commons.https://commons.wikimedia.org/wiki/File:Gal%C3%A1pagos_Penguin_(Spheniscus_mendiculus)_-standing_on_rock.jpg. Emperor penguin. License: CC BY 3.0 https://creativecommons.org/licenses/by/3.0, via Wikimedia Commons. https://commons.wikimedia.org/wiki/File:Emperor-left_hg.jpg

In this situation, it is not surprising then that numerous exceptions began to appear, leading to the discussion about whether the rule truly worked, whether it was initially wrongly stated, or whether its reliable degree of scope was quite limited (Geist, 1987; Ashton et al., 2000; Olalla-Tárraga, 2011; Clauss et al., 2013). As some authors have pointed out, to accurately assess the link between predictions and empirical evidence for Bergmann's rule, it is key to revisit its original postulates (e.g., Pincheira-Donoso, 2010; Salewski & Watt, 2017). That is, to focus on its usefulness for predicting eco-geographics patterns specifically in certain homeotherm vertebrates. Adopting this conservative position, therefore, implies being cautious in extrapolating such a rule to, for example, ectothermic animals (Vinarskiĭ, 2014), something that would be discouraged (Salewski & Watt, 2017).

It has also been investigated whether Bergmann's rule could be applied to plants (Moles et al., 2009). This has been analyzed in Cactaceae, with articles that do not support the rule (Drezner, 2003) and others in which observations would indicate a concordant trend (Donati et al., 2016). However, this intention to examine the extent to which a rule originated for certain animals applies or not to another kingdom reflects extremely the "obsession" to make extrapolations that help to simplify the comprehension of the immense diversity of phenotypic patterns in biology. This is very understandable, although risky. A concrete attempt to "extend" Bergmann's rule was made by Hesse et al. (1937) through the "heart-weight rule." This holds that species inhabiting colder climates have a larger heart relative to body weight than close phylogenetic-related species inhabiting warmer climates. This rule can be an extension of the principles contemplated in Bergmann's rule for a specific organ of the body, the heart. As stated initially by Bergmann, this so-called Hessian rule also emphasizes the existence of a close phylogenetic relationship between the species being compared. This is an important point that, as we will exemplify later, if not taken into account, will inevitably bring much confusion to the assessment and scope of the rule.

The attempt to extend the scope of a rule or principle to other groups and/or situations is valid and inherent to the sciences. However, it can lead to a certain reductionism if we do not take the necessary precautions. It is well known that even today, there are still old debates about the degree of precision and relevance of this or that rule. Indeed, it is not uncommon to see titles heading articles such as "Rule A does not apply to situation X" or "Evidence rejecting rule B for group Y," etc. This is the first point to consider regarding what we will see in the next chapter, focusing on the exceptions. Here, an important question may arise: To what extent is the finding of an "exception" of the rule reliable, or is it simply an artifice of our bias in trying to forcefully apply a rule to something (e.g., taxon, context, etc.) not initially covered by the rule? To answer this question, it is essential to know whether we are dealing with a situation of "mixing" of statements. That is, to be clear about the original author's predictions of the rule and which statements were later added as extensions or extrapolations carried out by other researchers. Therefore, the groups that do not fit a specific rule, rather than being a real exception, may represent observations that were never contemplated initially. This difference is essential to not "waste" the word "exception" on something that does not apply.

4.4.1.2 Emery's Rule

Although not reaching the same level of study and controversy as Bergmann's rule, another example of a biological rule that highlights the debates regarding the group for which it was originally proposed is the Emery Rule. In 1909, the entomologist Carlo Emery of the University of Naples observed that social parasites (i.e., klepto-parasites) among insects tended to exert such parasitism on species or genera to which they were phylogenetically close (Emery, 1909). Their specialization to kleptoparasitism arises within a well-defined clade of Hymenoptera, among close groups, and not in taxonomically distant groups. It is important to note that the postulation of this rule arose from studies that Emery conducted specifically on distinct groups of ants that were closely related to each other. However, it was subsequently explored by meta-analysis whether this rule was valid for other ant groups where social parasitism was also present (Huang & Dornhaus, 2008). The original rule did not apply to all analyzed types of social parasitism in ants.

In a recent study by Lopez-Osorio et al. (2015), it was found that phylogenetic tests refute the applicability of this rule in explaining the evolution of social parasitism in certain hymenopterans, including yellowjackets and hornets. The point is that it was not Emery who generalized this rule. Emery explained social parasitism in a specific group of ants without even speaking of a rule. As is often the case, it was extrapolated and applied to other Hymenoptera and even within other completely distant organisms, such as fungi. Thus, if we adhere exclusively to Emery's original idea and his precise observations, this rule should never have been enunciated, or at least it should have been enunciated with appropriate caution and limitations.

4.4.1.3 More Rules, More Caution

Like these two preceding cases of biological rules, other rules have also been subject to criticisms related to the scope of their postulates. For example, in the field of eco-geographical rules, J. A. Allen described in 1877 that endothermic animals from colder places usually have shorter body appendages (limbs) compared to those endotherms living in warmer climates. However, Allen's rule has been criticized for its application in some groups of endotherms and/or in some areas of the world (Serrat et al., 2008; Alho et al., 2011). Something similar has occurred with Cope's rule. The rule is named after the paleontologist E. D. Cope, who contributed to evolutionary theory, including the "law of the unspecialized," which we will see in Chap. 6 (Raia & Fortelius, 2013). Cope's rule holds that population lineages increase in body size over evolutionary time (Rensch, 1948). Again, this principle is invalid for some taxonomic groups (Hone & Benton, 2005).

Many of these rules were actually described by the author after whom they were named (e.g., Bergmann's, Allen's). However, others were named in honor of the author who had provided the seminal observations, which were the foundation for the future rule (e.g., Cope's, Rapoport's). Sooner or later, papers appear that falsify the generalizations made of specific rules, circumscribing and better delimiting the rule's scope. Indeed, there is a growing interest in refining these rules, taking into account factors not considered in detail. Such is the case with Hamilton's rule (Dillard & Westneat, 2016). This rule states that genes should increase in frequency when the relatedness of a receptor to an actor, multiplied by the benefit to the receptor, exceeds the reproductive cost to the actor (this is a prediction of the theory of family selection formulated by W. D. Hamilton—Queller & Strassman, 2002). These conditions are fundamental to understanding this rule. It is interesting that, among other factors to take into account, Queller & Strassman (2002) analyze social and genetic monogamy, and even in the former, they see this link with cooperation, for example, in communities.

On the other hand, some works have also suggested the importance of considering more than one rule in an investigation, mainly when they could be linked to each other (Sargis et al., 2018). Such is the case of Island's rule and the aforementioned Bergmann's rule. The Island rule (or "Island effect" or "Foster's rule"—Foster, 1964) is an eco-geographical rule in evolutionary biology that states that members of a species decrease or increase depending on the resources available in the environment. In its statement, we see that there are common points with Bergmann's rule. Thus, it is necessary to evaluate multiple variables simultaneously in studies of eco-geographical norms. Sargis et al. (2018) studied these rules simultaneously in the common treeshrew (*Tupaia glis*), a group of small mammals. They observed a strong negative effect of latitude on body size, indicating the opposite of what is expected by Bergmann's rule. They also observed no difference in body size between the island and mainland populations. They found that among island populations, there was an effect of island area and maximum sea depth on body size. While there was a strong latitudinal effect on body size, the authors note that neither Bergmann's rule nor the Island rule would apply, at least clearly, for this group of small mammals.

This last example is very useful since it shows the importance of evaluating more than one rule in combination, especially if they could be linked, and then it leads us to the original statements of these rules. Indeed, the Island rule was extended in the framework of the so-called Island Biogeography Theory (Case, 1978). In addition, it was analyzed whether or not plants follow the statements of that rule (Biddick et al., 2019). As we said previously, the latter is challenging and scientifically valuable. However, it involves risks at the level of interpretation since it is indirectly "forcing" the applicability of a principle originally conceived for animals. In other words, it is not a universal type I rule but rather a type II rule, which implies that from its conception, it was circumscribed to specific taxa. In this context, it is necessary that in publications demonstrating "nullity," rejection, or nonapplicability of a rule, it should be stated whether or not the taxon and/or conditions are the same under which the rule was initially described. Making this point clear in scientific publications and discussions is essential to avoid confusion about the scope, clarity, and flexibility of biological rules. Otherwise, biologists themselves are "justifying" a certain discredit that is erroneously given to specific fields of biological sciences (mainly in diversity, ecology, and evolution) for the lack of laws, principles, and rules of reliable application.

The proposal of modifications to enhance the understanding, clarity, and applicability of these rules presents a valuable challenge for biologists. Again, it is key to keep in mind what the original author postulated and, vitally, what researchers ultimately "understand" by such a rule. In this context, Delhey's (2019) recent review of the Gloger rule provides an excellent example of the analysis that would be needed for most biological rules. First, Gloger (1833) pointed out that the birds' overall pigmentation and coloration patterns would be more intense at lower latitudes, which coincides with warmer regions. The effect of humidity was also highlighted since desert birds are paler even if they inhabit warmer areas. That is, the temperature–humidity combination would work together. Rensch (1929) expanded the scope of these ideas and framed them in what he defined as "Gloger's rule." He even expressed a "simple" version of the rule, not different from that already mentioned by Gloger himself, and a "complex" version of it. In this complex version, more details are given on the effect of temperature and humidity on certain types of pigments. Precisely, the deposition of phaeo-melanins should decrease with temperature, whereas only extreme cold will reduce the deposition of eu-melanin. In addition, the eu-melanin deposition should decrease with lower humidity, whereas phaeo-melanin deposition will increase.

Dehley's analysis of the literature on the subject (Delhey, 2019) concludes that Gloger's rule in its "complex" version has unfortunately not been analyzed in detail thus far. For the "simple version," the rule would not be satisfied for the temperature factor, being only consistent with the humidity factor. Therefore, he stresses the need to reformulate the rule, focusing only on humidity. However, he points out that more studies are needed and that to avoid biases in the work, it should always be clear which definition is being used and cite the appropriate references. He mainly suggests Rensch (1929) and not Gloger (1833) since the latter was not who actually coined the rule. In addition, he points out that because the original papers were written in German, later translations into English may have contributed to some confusion surrounding the rule. He concludes that a complete re-evaluation of Gloger's rule should identify and test mechanistic links between coloration and climate. For example, the likely mechanisms that would lead to greater melanin deposition in more humid-leafier-regions would be camouflage and protection against parasites and pathogens. This case exemplifies the changes and misinterpretations of a rule according to its original postulates and the importance of clarifying these changes and citing them correctly.

Finally, another relevant aspect is the theoretical basis underlying the formulation of biological rules. The formulation of rules explicitly—or even implicitly—has been based mainly on empirical methods, i.e., on the observation of facts and experimentation. Such is the case of the accumulation of information from comparative observational studies for eco-geographical rules, as well as experimental studies in others, such as those linked to genetics (e.g., Hamilton's rule). However, the theoretical bases may not have been strong in all cases. A pending is, therefore, to advance theoretical modeling to see the extent to which this or that rule is explained and supported. In particular, this is important when adding factors such as taxonomic group, region, age, and kinship index according to the type of rule being analyzed. In this context, we can highlight the study of Jørgensen (2002), who analyzes cases in ecology that demonstrate that it is possible to give theoretical explanations from the ecosystem: (1) the lower the mortality of mammals, the later the offspring are weaned; (2) lower mortality implies later first parturition with lower risk; and (3) exploitation competition can be explained by a slight advantage in the growth rate relative to mortality rate; (4) It is a good strategy for ectoparasites to have relatively high mortality in early stages. In addition, the optimal founder density is higher for relatively high mortality at an early stage. All these "rules," as stated by the author, would thus be supported by ecosystem theory, not only by empirical data.

On the other hand, as part of the "search for rules," there is still debate about how many species can coexist in a community. When the number of species reaches a certain threshold, the community may saturate. It is critical to be clear whether this occurs in all communities since, in the current times, with the influence and challenges of global climate change, biodiversity conservation studies and policies should be based on precise predictions. Recent studies indicate that many communities could be "unsaturated," i.e., not having a maximum of coexisting species. In this regard, Mateo et al. (2017) reconsider the problem to clarify the "rules" on this issue. They emphasize that perhaps a partial view prevails regarding the saturation of communities. They propose that biodiversity models could adequately address this problem by considering an existing "duality": it would be better to consider communities as unsaturated, but where species composition is influenced and conditioned by different drivers and scale-dependent biodiversity factors.

In conclusion, if we consider all the examples mentioned above of biological rules (whether consolidated, questioned, or in development), it will be good to continue advancing in comprehensive approaches that combine the empirical approach and theoretical modeling in their respective disciplines. Otherwise, as mentioned above, we biologists are proposing rules with imprecision or refuting them by a mere accumulation of observations without giving attention to the original statements of the rules. The challenge will be to contribute to the knowledge of the principles without neglecting biological diversity, which, ironically, may cause some confusion when the principles of the rules are extrapolated and generalized to nonrelated organisms and only from empirical data.

4.5 Key Concepts and Ideas in this Chapter

- In evolutionary biology, it is common for theories to be eventually renewed, enriched, and adapted by incorporating new information from advances in related disciplines, and thus, new versions of the original theory emerge.
- Darwin's theory of natural selection has undergone modifications during the twentieth and twenty-first centuries from new disciplines, which have helped to refine Darwin's strict vision and reinforce the central postulates initially enunciated.
- Molecular genetics, epigenetics, and phenotypic plasticity generated new perspectives to improve the integration of developmental biology with evolution.
- Progressively, the concepts of evolutionary developmental biology (Evo-Devo) and ecological evolutionary developmental biology (Eco-Evo-Devo) emerged, constituting a new paradigm in terms of the vision of evolution.
- A controversial topic refers to the concept of species because there is a diversity of definitions, approaches, and perspectives to define it. This means that there is no single rule in this case, and each discipline will relativize it according to the scope of the study.

- Many times, some rules have been extrapolated to other groups of organisms for which they were not initially conceived by their authors. This has led people to think that they have been poorly thought out; when in fact, the problem arises when this extrapolation forces their application.
- The rules should be based not only on data obtained from experiments but also on theoretical contributions from different disciplines.

References

Abouheif, E., Favé, M. J., Ibarrarán-Viniegra, A. S., Lesoway, M. P., Rafiqi, A. M., & Rajakumar, R. (2014). Eco-Evo-Devo: The time has come. In C. Landry & N. Aubin-Horth (Eds.), *Ecological genomics. Advances in experimental medicine and biology*. Springer. pp. 781.

Alho, J. S., Herczeg, G., Laugen, A. T., Räsänen, K., Laurila, A., & Merilä, J. (2011). Allen's rule revisited: Quantitative genetics of extremity length in the common frog along a latitudinal gradient. *Journal of Evolutionary Biology, 24*, 59–70.

Allen, J. A. (1877). The influence of physical conditions in the genesis of species. *Radical Review, 1*, 108–140.

Ashton, K. G., Tracy, M. C., & de Queiroz, A. (2000). Is Bergmann's rule valid for mammals? *The American Naturalist, 156*(4), 390–415.

Atkinson, D. (1994). Temperature and organism size: A biological law for ectotherms? *Advances in Ecological Research, 25*, 1–58.

Blackman, F. F. (1905). Optima and limiting factors. *Annals of Botany, 19*(74), 281–295.

Baker, H. G. (1959). Reproductive methods as factors in speciation in flowering plants. In *Cold Spring Harbor Symposia on Quantitative Biology* (pp. 177–191). Cold Spring Harbor Laboratory Press.

Banta, J. A., & Richards, C. L. (2018). Quantitative epigenetics and evolution. *Heredity, 121*, 210–224.

Bateman, A. J. (1948). Intrasexual selection in Drosophila. *Heredity, 2*(3), 349–368.

Bateson, W. (1902). *Mendel's principles of heredity: A defence*. University Press.

Bateson, G. (1971). A re-examination of "Bateson's Rule". *Journal of Genetics, 60*, 230–240.

Bergmann, C. (1847). Ueber die Verhältnisse der Wärmeökonomie der Thiere zu ihrer Grösse. *Göttinger Studien, 3*, 595–708.

Bernard, C. (1865). *Introduction à l'étude de la médecine expérimentale*. Baillière.

Biddick, M., Hendriks, A., & Burns, K. C. (2019). Plants obey (and disobey) the island rule. *Proceedings of the National Academy of Sciences, 116*(47), 201907424.

Cabej, N. (2018). *Epigenetic principles of evolution*. Academic Press.

Carroll, S. B. (2008). Evo-Devo and an expanding evolutionary synthesis: A genetic theory of morphological evolution. *Cell, 13*(1), 25–36.

Case, T. J. (1978). A general explanation for insular body size trends in terrestrial vertebrates. *Ecology, 59*(1), 1–18.

Caso, J. M. R. (2020). El "darwinismo puro" de Alfred Russel Wallace: Aportaciones a la teoría evolutiva moderna. *Asclepio, 72*(2), 324.

Clarke, E., & Okasha, S. (2013). Species and organisms: What are the problems? In P. Huneman & F. Bouchard (Eds.), *From groups to individuals. Evolution and emerging individuality*. pp. 55.

Clauss, M., Dittmann, M. T., Müller, D. W. H., Meloro, C., & Crodon, D. (2013). Bergmann's rule in mammals: A cross-species interspecific pattern. *Oikos, 122*(10), 1465–1472.

Cody, M. L., & Diamond, J. M. (1975). *Ecology and evolution of communities*. Belknap Press, Harvard University Press.

Cope, E. D. (1887). *The origin of the fittest: Essays on evolution*. D. Appleton.

Crick, F. H. C. (1958). On protein synthesis. In *Symposia of the Society for Experimental Biology; Number XII: The Biological Replication of Macromolecules* (pp. 138–163). Cambridge University Press.
Darwin, C. (1859). *On the origin of species by means of natural selection, or the preservation of favoured races in the struggle for life*. John Murray.
Dale, H. (1935). Pharmacology and nerve-endings. *Proceedings of the Royal Society of Medicine, 28*, 319–332.
Dawkins, R. (1976). *The selfish gene*. Oxford University Press.
de Queiroz, K. (1998). The general lineage concept of species, species criteria, and the process of speciation: A conceptual unification and terminological recommendations. In *Endless forms: Species and speciation* (pp. 57–75). Oxford University Press.
de Queiroz, K. (2005). Ernst Mayr and the modern concept of species. *Proceedings of the National Academy of Sciences, 102*, 6600–6607.
de Queiroz, K. (2007). Species concepts and species delimitation. *Systematic Biology, 56*(6), 879–886.
Delhey, K. (2019). A review of Gloger's rule, an ecogeographical rule of colour: Definitions, interpretations and evidence. *Biological Reviews, 94*, 1294–1316.
Depéret, C. (1907). *Les Transformations du monde animal*. Ernest Flammarion.
Dillard, J. R., & Westneat, D. F. (2016). Disentangling the correlated evolution of monogamy and cooperation. *Trends in Ecology & Evolution, 31*(7), 503–513.
Dobzhansky, T. (1937). *Genetics and the origin of species*. Columbia University Press.
Dollo, L. (1893). Les lois de l'évolution. Bulletin de la Société Belge de Géologie de Paléontologie & D'hydrologie, VII: 164–166.
Donati, D., Bianchi, C., Pezzi, G., Conte, L., Hofer, A., & Chiarucci, A. (2016). Biogeography and ecology of the genus *Turbinicarpus* (Cactaceae): Environmental controls of taxa richness and morphology. *Systematics and Biodiversity, 15*(4), 361–371.
Drezner, T. D. (2003). Revisiting Bergmann's rule for saguaros (*Carnegiea gigantea* (Engelm.) Britt. and Rose): Stem diameter patterns over space. *Journal of Biogeography, 30*(3), 353–359.
Eichler, W. D. (1942). Die Entfaltungsregel und andere Gesetzmäßigkeiten in den parasitogenetischen Beziehungen der Mallophagen und anderer ständiger Parasiten zu ihren Wirten. *Zoologischer Anzeiger, 137*, 77–83.
Eldredge, N., & Gould, S. J. (1972). Punctuated equilibria: an alternative to phyletic gradualism. *Models in Paleobiology, 82*, 115.
Emery, C. (1909). Über den Ursprung der dulotischen, parasitischen und myrmekophilen Ameisen. *Biologisches Centralblatt, 29*, 352–362.
Fábregas-Tejeda, A., & Vergara-Silva, F. (2018). The emerging structure of the extended evolutionary synthesis: Where does Evo-Devo fit in? *Theory in Biosciences, 137*(2), 169–184.
Fisher, R. A. (1930). *The genetical theory of natural selection*. Clarendon Press.
Foster, J. B. (1964). The evolution of mammals on islands. *Nature, 202*(4929), 234–235.
Franklin, R. E., & Gosling, R. G. (1953). Molecular configuration in sodium thymonucleate. *Nature, 171*, 740–741.
Galton, F. (1889). *Natural Inheritance*. Macmillan & Co.
Gause, G. F. (1934). *The Struggle for Existence*. Williams & Wilkins.
Gayon, J. (2016). From Mendel to epigenetics: History of genetics. *Comptes Rendus Biologies, 339*(7–8), 225–230.
Geist, V. (1987). Bergmann's rule is invalid. *Canadian Journal of Zoology, 65*(4), 1035–1038.
George, A. L., & Mayden, R. L. (2005). Species concepts and the endangered species act: How a valid biological definition of species enhances the legal protection of biodiversity. *Natural Resources Journal, 45*(2), 369–407.
Gloger, C. W. L. (1833). *Abänderungsweise der einzelnen, einer Veränderung durch das Klima unterworfenen Farben. Das Abändern der Vögel durch Einfluss des Klimas*. August Schulz.
Gould, S. J. (1990). *Wonderful life: The burgess shale and the nature of history*. W. W. Norton & Company.

Grinnell, J. (1904). The origin and distribution of the chest-nut-backed chickadee. *The Auk, 21*(3), 364–382.
Haeckel, E. (1868). *Natürliche Schöpfungsgeschichte*. Georg Reimer.
Haldane, J. B. (1922). Sex ratio and unisexual sterility in hybrid animals. *Journal of Genetics, 12*, 101–109.
Haldane, J. B. S. (1932). *The Causes of Evolution*. Longman, Green and Co. Princeton University Press.
Hall, B. (2003). Evo-Devo: Evolutionary developmental mechanisms. *International Journal of Developmental Biology, 47*, 491–495.
Hamilton, W. D. (1964). The genetical evolution of social behaviour II. *Journal of theoretical biology, 7*(1), 17–52.
Hardy, G. H. (1908). Mendelian proportions in a mixed population. *Science, 28*(706), 49–50.
Harrison, L. (1915). Mallophaga from Apteryx, and their significance; with a note on the genus Rallicola. *Parasitology, 8*(1), 88–100.
Harvey, P. H., & Pagel, M. D. (1991). The comparative method in evolutionary biology. Oxford: university press.
Hebb, D. O. (1949). *The Organization of Behavior*. Wiley & Sons.
Hennig, W. (1950). *Grundzüge einer Theorie der phylogenetischen Systematik*. Deutscher Zentralverlag.
Hennig, W. (1966). The Diptera fauna of New Zealand as a problem in systematics and zoogeography. *Pacific Insects Monograph, 9*, 81.
Hertwig, O. (1884). Das Problem der Befruchtung und der Isotropie des Eies, eine Theory der Vererbung. *Jenaische Zeitschrift fuer Naturwissenschaft, 18*, 21–23.
Hesse, R., Allee, W. C., & Schmidt, K. P. (1937). *Ecological animal geography*. Wiley.
Hone, D. W., & Benton, M. J. (2005). The evolution of large size: How does Cope's Rule work? *Trends in Ecology & Evolution, 20*, 4–6.
Hone, D. W. E., Dyke, G. J., Haden, M., & Benton, M. J. (2008). Body size evolution in Mesozoic birds. *Journal of Evolutionary Biology, 21*(2), 618–624.
Huang, M. H., & Dornhaus, A. (2008). A meta-analysis of ant social parasitism: Host characteristics of different parasitism types and a test of Emery's rule. *Ecological Entomology, 33*(5), 589–596.
Huxley, J. S., & Teissier, G. (1936). Terminology of relative growth. *Nature, 137*(3471), 780–781.
Jablonka, E. (2017). The evolutionary implications of epigenetic inheritance. *Interface Focus, 7*, 20160135.
Jablonka, E., & Lamb, M. J. (2014). *Evolution in four dimensions: Genetic, epigenetic, behavioral, and symbolic variation in the history of life*. The MIT Press.
Janzen, D. H. (1967). Why mountain passes are higher in the tropics. *The American Naturalist, 101*(919), 233–249.
Jordan, D. S. (1892). Relations of temperature to vertebrae among fishes. *Proceedings of the United States National Museum, 1891*, 107–120.
Jordan, D. S. (1922). Two volumes. In *The days of a man: Being memories of a naturalist, teacher and minor prophet of democracy*. World Book Company.
Jørgensen, S. E. (2002). Explanation of ecological rules and observation by application of ecosystem theory and ecological models. *Ecological Modelling, 158*, 241–248.
Kavanagh, P. H., Lehnebach, C. A., Shea, M. J., & Burns, K. C. (2011). Allometry of sexual size dimorphism in dioecious plants: Do plants obey Rensch's rule? *The American Naturalist, 178*(5), 596–601.
Kimbel, W. H., & Martin, L. B. (Eds.). (1993). *Species, species concepts and primate evolution*. Springer.
Kleiber, M. (1932). Body size and metabolism. *Hilgardia, 6*, 315–353.
Lack, D. (1954). *The natural regulation of animal numbers*. Clarendon.

Laland, K. N., Uller, T., Feldman, M. W., Sterelny, K., Müller, G. B., Moczek, A., Jablonka, E., & Odling-Smee, J. (2015). The extended evolutionary synthesis: Its structure, assumptions and predictions. *Proceedings of the Royal Society B, 282*(1813), 20151019.
Lamarck, J. B. (1809). *Philosophie zoologique*. New York: Hafner Publishing Company, 1963.
Lee, M. S. Y. (2003). Species concepts and species reality: Salvaging a Linnaean rank. *Journal of Evolutionary Biology, 16*, 179–188.
Lev-Yadun, S. (2016). *Defensive (anti-herbivory) coloration in land plants*. Springer International Publishing.
Lewis, F. T. (1928). The correlation between cell division and the shapes and sizes of prismatic cells in the epidermis of cucumis. *The Anatomical Record, 38*(3), 341–376.
Lewontin, R. C., & Levins, R. (2000). Schmalhausen's law. *Capitalism Nature Socialism, 11*, 103–108.
Liebig, J. V. (1840). *Die organische Chemie in ihrer Anwendung auf Agricultur und Physiologie (Organic chemistry in its applications to agriculture and physiology)*. Friedrich Vieweg und Sohn Publishing Company.
Lind, M. I., & Spagopoulou, F. (2018). Evolutionary consequences of epigenetic inheritance. *Heredity, 121*, 205–209.
Lindeman, R. L. (1942). The trophic-dynamic aspect of ecology. *Ecology, 23*(4), 399–417.
Liò, P., & Goldman, N. (1998). Models of molecular evolution and phylogeny. *Genome Research, 8*, 1233–1244.
Lopez-Osorio, F., Perrard, A., Pickett, K. M., Carpenter, J. M., & Agnarsson, I. (2015). Phylogenetic tests reject Emery's rule in the evolution of social parasitism in yellowjackets and hornets (Hymenoptera: Vespidae, Vespinae). *Royal Society Open Science, 2*, 150159.
Lovelock, J. E. (1972). Gaia as seen through the atmosphere. *Atmospheric Environment, 6(8):580*, 5–90076.
Macarthur, R. H., & Wilson, E. O. (1967). *The theory of island biogeography*. Princeton Univ. Press.
Mallet, J. (2007). Species, concepts of. In S. Levin et al. (Eds.), *Encyclopedia of biodiversity* (pp. 427–440). Elsevier.
Mateo, R. G., Mokany, K., & Guisan, A. (2017). Biodiversity models: What if unsaturation is the rule? *Trends in Ecology & Evolution, 32*(8), 556–566.
Maynard-Smith, J. M., & Price, G. R. (1973). The logic of animal conflict. *Nature, 246*(5427), 15–18.
Mayr, E. (1942). *Systematics and the origin of species*. Columbia University Press.
Mayr, E. (1970). *Populations, species, and evolution; an abridgment of animal species and evolution*. Belknap Press of Harvard University Press.
Mendel, G. (1866). *Versuche über Planzenhybriden, Verhandlungen des naturforschenden Vereines in Brünn*, Bd. für das Jahr 1865, Abhandlungen (pp. 3–47).
Mileikovsky, S. A. (1971). Types of larval development in marine bottom invertebrates, their distribution and ecological significance: a re-evaluation. *Marine Biology, 10*, 193–213.
Moles, A. T., Warton, D. I., Warman, L., Swenson, N. G., Laffan, S. W., Zanne, A. E., Pitman, A., Hemmings, F. A., & Leishman, M. R. (2009). Global patterns in plant height. *Journal of Ecology, 97*(5), 923–932.
Morgan, T. H. (1926). *The theory of the gene*. Yale University Press.
Morgan, T. H., Sturtevant, A. H., Muller, H. J., & Bridges, C. B. (1915). *The mechanism of Mendelian heredity*. H. Holt & Co..
Moseley, H. N. (1880). Deep-sea dredging and life in the deep sea. *Nature, 21*, 591–593.
Mousseau, T. A. (1997). Ectotherms follow the converse to Bergmann's rule. *Evolution*, 630–632.
Müller, J. (1838). *Handbook of human physiology* (Vol. 1). J. Hölscher.
Müller, G. B. (2007). Evo-devo: Extending the evolutionary synthesis. *Nature Reviews Genetics, 8*, 943–949.
Müller, G. B. (2017). Why an extended evolutionary synthesis is necessary. *Interface focus, 7*(5), 20170015.

Mungee, M., Pandit, R., & Athreya, R. (2021). Taxonomic scale dependency of Bergmann's patterns: a cross-scale comparison of hawkmoths and birds along a tropical elevational gradient. *Journal of Tropical Ecology, 37*(6), 302–312.

Nature Education. (2014). *Species*. Retrieved from https://www.nature.com/scitable/definition/species-312/

Ohno, S. (1967). *Sex Chromosomes and Sex-linked Genes*. Springer.

Olalla-Tárraga, M. A. (2011). "Nullius in Bergmann" or the pluralistic approach to ecogeographical rules: A reply to Watt et al. *Oikos, 120*, 1441–1444.

Pigliucci, M. (2007). Do we need an extended evolutionary synthesis? *Evolution, 61*(12), 2743–2749.

Pigliucci, M., & Finkelman, L. (2014). The extended (evolutionary) synthesis debate: Where science meets philosophy. *BioScience, 64*(6), 511–516.

Pigliucci, M., & Müller, G. B. (2010). *Evolution, the extended synthesis*. The MIT Press.

Pincheira-Donoso, D. (2010). The balance between predictions and evidence and the search for universal macroecological patterns: Taking Bergmann's rule back to its endothermic origin. *Theory in Biosciences, 129*(4), 247–253.

Popper, K. R. (1980). Selección natural y la emergencia de la mente. *Teorema: Revista Internacional de Filosofía, 10*(2/3), 191–213.

Popper, K. R. (1991). *Conjeturas y refutaciones: El desarrollo del conocimiento científico*. Paidós Ibérica.

Portin, P., & Wilkins, A. (2017). The evolving definition of the term "Gene". *Genetics, 205*(4), 1353–1364.

Queller, D. C., & Strassman, J. E. (2002). Quick guide: Kin selection. *Current Biology, 12*(24), R832.

Raia, P., & Fortelius, M. (2013). Cope's law of the unspecialized, Cope's rule, and weak directionality in evolution. *Evolutionary Ecology Research, 15*(7), 747–756.

Rendón, C. A., & Folguera, G. (2014). Evo-devo como disciplina integradora: La temporalidad de los procesos biológicos como estrategia de análisis. *Theoria, 81*, 395–415.

Rensch, B. (1929). *Das Prinzip geographischer Rassenkreise und das Problem der Artbildung*. Gebrueder Borntraeger.

Rensch, B. (1948). Histological changes correlated with evolutionary changes of body size. *Evolution, 2*(3), 218–230.

Rensch, B. (1950). Die Abhängigkeit der relativen Sexualdifferenz von der Körpergrösse. *Bonner Zoologische Beiträge, 1*, 58–69.

Ruíz, R., & Ayala, F. J. (1998). *El método en las ciencias. Epistemología y darwinismo*. Fondo de Cultura Económica de España (Colección Ciencia y Tecnología).

Sagan, L. (1967). On the origin of mitosing cells. *Journal of Theoretical Biology, 14*(3), 225-IN6.

Salewski, V., & Watt, C. (2017). Bergmann's rule: A biophysiological rule examined in birds. *Oikos, 126*, 161–172.

Sargis, E. J., Millien, V., Woodman, N., & Olson, L. E. (2018). Rule reversal: Ecogeographical patterns of body size variation in the common treeshrew (Mammalia, Scandentia). *Ecology and Evolution, 8*(3), 1634–1645.

Sarrus, F., & Rameaux, J. F. (1839). Application des sciences accessoires et principalement des mathématiques à la physiologie générale. Bulletin de l'Academie Royale de Medecine de Belgique, 3, 1094-1100.

Schleiden, M. J. (1838). Beiträge zur Phytogenesis. *Archiv für Anatomie, Physiologie und wissenschaftliche Medicin, 1*, 137–176.

Serrat, M. A., King, D., & Lovejoy, C. O. (2008). Temperature regulates limb length in homeotherms by directly modulating cartilage growth. *Proceedings of the National Academy of Sciences, 105*(49), 19348–19353.

Serres, E. R. A. (1827). Anatomie comparée du cerveau, dans les quatres classes des animaux vertébrés: appliquée à la physiologie et à la pathologie du système nerveux.

Shelford, V. E. (1911). Physiological animal geography. *Journal of Morphology, 22*(3), 551–618.

Shelomi, M. (2012). Where are we now? Bergmann's rule sensu lato in insects. *The American Naturalist, 180*(4), 511–519.
Simpson, G. G. (1944). *Tempo and mode in evolution*. Columbia University Press.
Smith, W. (1816). *Strata identified by organized fossils: containing prints on colored paper of the most characteristic specimens in each stratum*. W. Arding.
Stebbins, C. L. (1950). *Variation and evolution in plants*. Oxford University Press.
Steidinger, B. S., Crowther, T. W., Liang, J., Van Nuland, M. E., Werner, G. D., Reich, P. B., et al. (2019). Climatic controls of decomposition drive the global biogeography of forest-tree symbioses. *Nature, 569*(7756), 404–408.
Stevens, G. C. (1989). The latitudinal gradient in geographical range: how so many species coexist in the tropics. *The American Naturalist, 133*(2), 240–256.
Templeton, A. R. (1989). The meaning of species and speciation: A genetic perspective. In *Speciation and its consequences* (pp. 3–27). Sinauer Associates.
Templeton, A. R. (1998). Species and speciation: Geography, population structure, ecology, and gene trees. In *Endless forms: Species and speciation* (pp. 32–43). Oxford University Press.
Thompson, D., Regev, A., & Roy, S. (2015). Comparative analysis of gene regulatory networks: From network reconstruction to evolution. *Annual Review of Cell and Developmental Biology, 31*, 6.1–6.30.
University of California, Berkeley. (2023). *Defining a species. Understanding evolution*. Retrieved from https://evolution.berkeley.edu/evolution-101/speciation/defining-a-species/
Van Valen, L. (1973). A new evolutionary law. *Evolution Theory, 1*, 1–30.
Vinarskiĭ, M. V. (2014). On the application of Bergmann's rule to ectothermic organisms: The state of the art. *Biology Bulletin Reviews, 4*(3), 232–242.
Von Baer, K. E. (1828). *Über Entwickelungsgeschichte der Thiere*. Beobachtung und Reflexion.
Waples, R. S. (1991). Pacific Salmon, *Oncorhynchus* spp., and the definition of "Species" Under the Endangered Species Act. *Marine Fisheries Review, 53*(3), 11–22.
Watson, J. D., & Crick, F. H. (1953). A structure for deoxyribose nucleic acid. *Nature, 171*, 737–738.
West-Eberhard, M. J. (1989). Phenotypic plasticity and the origins of diversity. *Annual Review of Ecology and Systematics, 20*(1), 249–278.
West-Eberhard, M. J. (2003). *Developmental plasticity and evolution*. Oxford University Press.
West-Eberhard, M. J. (2005). Developmental plasticity and the origin of species differences. *Proceedings of the National Academy of Sciences, 102*(1), 6543–6549.
Wheeler, Q. D., & Meier, R. (2000). *Species concepts and phylogenetic theory: A debate*. Columbia University Press.
Whelan, S., Liò, P., & Goldman, N. (2001). Molecular phylogenetics: State-of-the-art methods for looking into the past. *Trends in Genetics, 17*(5), 262–272.
Wilkins, M. H. F., Stokes, A. R., & Wilson, H. R. (1953). Molecular structure of deoxypentose nucleic acids. *Nature, 171*, 738–740.
Williston, S. W. (1914). *Water reptiles of the past and present*. University of Chicago Press.
Woese, C. R., & Fox, G. E. (1977). Phylogenetic structure of the prokaryotic domain: The primary kingdoms. *Proceedings of the National Academy of Sciences, 74*(11), 5088–5090.
Wright, S. G. (1931). Evolution in Mendelian populations. *Genetics, 16*, 97–159.
Wright, S. G. (1932). The roles of mutation, inbreeding, crossbreeding, and selection in evolution. *Proceedings of the Sixth International Congress of Genetics*, 356–366.
Yoda, K., Kira, T., Ogawa, H., & Hozumi, K. (1963). Intraspecific competition among higher plants. XI. Self-thinning in overcrowded pure stands under cultivated and natural conditions. *Journal Biology Osaka City University, 14*, 107–129.
Zachos, F. E. (2016). *Species concepts in biology: Historical development, theoretical foundations and practical relevance*. Springer.
Zahavi, A. (1975). Mate selection: a selection for a handicap. *Journal of Theoretical Biology, 53*, 205–214.

Chapter 5
When THAT Exception Persists Almost as THAT Exception

Rarely do we realize that we are surrounded by the extraordinary

—Paulo Coelho

Contents

A. V. Peretti et al., *Rules and Exceptions in Biology: from Fundamental Concepts to Applications*, https://doi.org/10.1007/978-3-031-55382-0_5

Abstract We present the counterpart to Chap. 4 in Chap. 5, where we focus on the role of "permanent" exceptions and their importance in scientific knowledge and biology. These exceptions can be analyzed at any scale and are often not appreciated beyond anecdotal mentions. They can be aberrant or teratological groups, taxonomic groups unique for some features of their biology, "intermediate exceptions" with minorities of species or clades with certain characteristics compared to the majority, or organisms with exceptional distributions in space or time. In this chapter, we reviewed groups that have been difficult to classify because of their rarity, rarities that can be used for human benefit, exceptional groups that make us rethink phylogenetic relationships, rare and incredible biological phenomena, and evolutionary and ecological aspects of rare species. Our goal is to vindicate rarities and minorities, to highlight their importance for the understanding of evolution, and to begin to make these cases visible and treasured in the teaching of biology.

5.1 Introduction

As we saw in Chap. 2, when thinking about situations or organisms characterized as "permanent" exceptions (i.e., they have always been exceptions and are expected to remain as such), we probably first think immediately of outstanding individuals with extreme "positive" traits from an adaptive perspective. However, there are also "negative" extremes, and we find the norm or the majority between these positive and negative extremes. Indeed, there is a trend in biological sciences to use the so-called morphological "aberrations" to illustrate the extreme concept of exceptions. Occasionally, certain groups accumulate multiple "rarities" or exceptional features, and we said there are exceptional or "aberrant" among their group. The word "aberrant" may be more common than we think in the scientific literature or scientific communication. Our emotional perception may influence us to use the word "aberrant" subjectively. This word refers to something "*deviating from the normal or usual type, someone behaving in an abnormal or untypical way, or something deviating from truth, morality, etc.*" (Collins English Dictionary, 2003). Thus, although it is used more for human acts, the definition is connected to that of "exception," since it deviates from the regular pattern. If it is aberrant, it is because it presents a particular trait or "aberration." In biology, this word is the "*deviation from the normal type of a morphological or physiological character.*" Complementarily, in biology, we refer to something aberrant to an organism that presents one or more differences concerning those presented by the species to which it belongs.

We can also find aberrant organisms that show a mixture of several characters or a maximum reduction of the ancestral form. This includes groups that are usually rare in number, living fossils, or parasitic species (Wisnivesky, 2003). Viruses, for example, were considered "aberrant" forms of bacteria at the time M. W. Beijerinck began to study them (Jenkins, 2003), considering that there could be no smaller organisms. The term "aberrant" is also commonly used among paleontologists to highlight discoveries of strange fossils. For example, the case of the aquatic reptile *Sclerocormus* supported the idea of a rapid evolutionary divergence that occurred in many groups after the great Permian extinction (Jiang et al., 2016).

Charles Darwin (1859) refers to aberrant species when he mentions that "*as a small number of the old and intermediate forms have transmitted to the present day very little modified offspring, these constitute the so-called aberrant or osculating species.*" He adds that this perception of aberration is intensified as more related species become extinct. This makes them even more striking, unique, true exceptions among the larger clades that encompass them. He also says, "*...they are almost always represented by very few species, and these generally differ greatly from each other: which also implies extinctions*" and "*...leaving a small number of representatives which are still preserved under extraordinarily favorable conditions.*" He gives examples of aberrant species such as the platypus (*Ornithorhynchus anatinus*) and the South American lungfish (*Lepidosiren paradoxa*). He points out that even if species more closely related to them were still alive, they would still be aberrant. The reference to the small number of representatives is added to the rarity of the morphology, making a "cocktail" that attracted his attention.

These Darwinian concepts have not become obsolete in the lexicon of many authors in biology. One example is the phylum Gnathostomulida, mentioned as a small group of "aberrant" organisms in some invertebrate zoology textbooks (Marshall & Williams, 1985). These animals are tiny, vermiform-bodied, and live among particles on the seafloor, particularly in anaerobic environments, where they feed on fungi and microorganisms. Their evolutionary and systematic history is still not entirely clear, and they present a mix of features that a priori resemble groups of now obsolete categories, such as the "acoelomates," but at the same time present features of "pseudocoelomates" (Ruppert et al., 2003). Indeed, they were initially considered an "aberrant" group of turbellarians within the flatworms, and only in 1969 were they considered a separate phylum (Sterrer et al., 1985). Currently, they are included in the clade Gnathifera, grouping them together with rotifers, among others (Marlétaz et al., 2019; Giribet & Edgecombe, 2020).

Similar is the case with the so-called "sea spiders" of the class Pycnogonida. Pycnogonids are often said to possess "aberrant" characteristics compared to other representatives of the Subphylum Chelicerata (Ruppert & Barnes, 1994), a group they share with the horseshoe crabs (class Merostomata), and the arachnids (class Arachnida). These unique features include an articulated "thorax" instead of the typical prosoma and opisthosoma found in other chelicerates, well-developed ovigerous (egg-carrying) legs, particularly in males, and various additional traits that justify the use of the term "aberrant." Similar situations of use of this term are often repeated in any animal group. For example, in birds, certain anomalies in color, design, and development of flight feathers, rectrices, and remiges are considered "aberrant" (Sáez-Royuela et al., 1975). Additionally, some fish, such as abyssal fish, with large teeth and lures, even soles with both eyes on the same side of the body, can be labeled "aberrant."

Interestingly, the perception of aberration may be greater as we move up the zoological scale, and we are further impacted by this "departure from the common." Table 5.1 shows examples of various taxa in terms of so-called "aberrant" groups, highlighting the traits that make these groups extraordinary. We must acknowledge and consider emotional subjectivity and anthropocentric prejudices, which are generally unintentional in using the term "aberration." Here, we prefer the use of exception or rarity in many of these cases, but we do not deny the term "aberration" for some cases, as it is something we can still find in biological descriptions. What is clear is that exceptions or aberrations can occur in different taxonomic groups in different spatial and temporal contexts, ranging from morphological curiosities at phenotypic extremes to unique teratological cases.

Table 5.1 Examples of species considered aberrant in biology highlighting the aberrant traits

	Species	Trait that makes it aberrant	Reference
Vertebrate animals			
Axolotl	*Ambystoma mexicanum*	Able to regenerate limbs and internal organs	Monaghan et al. (2007); Satoh et al. (2008); Voss et al. (2009)
Platypus	*Ornithorhynchus anatinus*	Mammal that lays eggs, venomous, with a duck-billed snout and a beaver tail	Whittington et al. (2008)
Common vampire bat	*Desmodus rotundus*	Bats that feed on the blood of other animals	Greenhall (1972)
Naked mole-rat	*Heterocephalus glaber*	Hairless rodent	Tucker (1981); Daly & Buffenstein (1998)
Northern snakehead	*Channa argus*	Fish can breathe air and move on land to reach other bodies of water	Ishimatsu & Itazawa (1981, 1983)
Four-eyed fishes	*Anableps* sp.	Fish with eyes divided into two sections, allowing them to see above and below the water	Borwein & Hollenberg (1973); Sivak (1976); Zahl & Gomprecht (1977)
Invertebrate animals			
Vinegar fly	*Drosophila guttifera*	Flies in mating groups of males beat each other with their wings until some of them become unconscious	Niida & Koshikawa (2021)
Green ormer	*Haliotis tuberculata*	Mollusk that can regenerate the shell	Fleury et al. (2008)
Immortal jellyfish	*Turritopsis dohrnii*	Jellyfish can reverse their life cycle and return to their juvenile stage	Lisenkova et al. (2017); Martell et al. (2016)
Ground beetle	*Brachinus* sp.	Beetle that produces a chemical reaction to defend itself from predators	Lai (2010); McManus et al. (2018); Attygalle et al. (2020)
Plants			
Corpse flower	*Rafflesia arnoldii*	It produces the largest flower in the world but without leaves or roots, it obtains nutrients by parasitizing other plants	Nikolov & Davis (2017)
Venus flytrap	*Dionaea muscipula*	Carnivorous plants that feed on insects and other small animals	Poppinga & Joyeux (2011); Pavlovič et al. (2017)
Bacteria			
Extremophile Bacteria	*Deinococcus radiodurans*	Resists high radiation doses	Daly (2009); Krisko & Radman (2013)

5.2 Teratologies: Aberrations Revalorized

5.2.1 *Defining Teratologies, Subjectivities, and the Relationship Between Concepts*

Teratology is a scientific discipline that studies abnormal morpho-physiological developments in various taxa, employing terms such as "*individuals of a species that do not respond to the common pattern*" (Dicke, 1989). Other definitions of teratology are more emphatic, such as "*the scientific study of congenital abnormalities and abnormal formations*" (Oxford English Dictionary, 2017). Etymologically, it is derived from the French word *tératologie*, and this term in turn comes from the Greek τέρας, -ατος (téras, -atos), which is the genitive form of "téras," which refers to something extraordinary, monstrous, or prodigious, often denoting something that surpasses the boundaries of the natural. On the other hand, λογία "logos" signifies study, treatise, reason, or discourse. Consequently, cf. gr. τερατολογία "teratology" can be translated literally as the study or treatise of congenital malformations, deformities, or dysfunctions in living beings (Treviño-Rodriguez, 2023). Additionally, etymologically, the word teratological implies a dichotomy between something monstrous and prodigious, referring to a phenomenon that can cause fascination and/or terror. Another definition for teratology is "*a collection of tales about mythical or fantastic creatures and monsters*" (Collins English Dictionary, 2003). The term "monster" is undoubtedly striking and begs the question, what is a monstrosity? Definitions range from "*something very ugly and usually large*" (Cambridge Advanced Learner's Dictionary, 2023) to "*a person or animal that is markedly unusual or deformed*" (The American Heritage Roget's Thesaurus). As the reader will see, again, the emphasis is on something, someone, which departs from what is considered "regular," i.e., the pattern, the norm. The first definition is interesting because it includes a physical characteristic such as size and an attribution linked to our human perception of what is considered "ugly" as opposed to what is "beautiful" to our senses. This is linked to artistic representations and the valuation of beauty across different temporal and sociocultural contexts, which we will see in Chap. 8.

If we return to the meaning of teratology, we see that the intrinsic components that give it entity (e.g., monstrosity) or its antonyms (beauty *versus* ugliness) are based mainly on human perception. Something that pleases or does not please our senses, a sensory and emotional experience, similar to when we analyze the term "aberration." This suggestive bias does not detract from the value of this discipline, and it is important to bear in mind that it arose to highlight in some way biological abnormalities or aberrations that would otherwise be ignored, hidden, or denied. Nevertheless, it is crucial to acknowledge that science, or certain scientific fields, can be influenced by subjective human perspectives and biases. Ultimately, nothing we describe nature is without a human bias, given our sensory universe and its limitations.

One of the complementary meanings of teratology is the "relation of prodigies." This is interesting as the term "prodigy" (from the Latin *prodigium*) has an intrinsic duality, which ultimately confers a certain neutrality to the term, highlighting an oddity regardless of whether it is categorized as positive or negative. A prodigy could be considered "*an amazing or unusual thing, especially one out of the ordinary course of nature*" (Oxford English Dictionary, 2017). The complementary definitions of prodigy, such as "*an outstanding example of a particular quality*" or "*a young person with exceptional qualities or abilities*" (Oxford English Dictionary, 2017), are more subjective. In these cases, it is evident that the term prodigy is framed in positive attributions from a human perspective to refer to something or someone exceptional. Finally, a prodigy can be defined as a "miracle, an event of divine origin," which, in addition to being at the opposite pole to "monstrosity," appeals to a supernatural sense.

5.2.2 Utility of Teratologies in Evolution

"Monsters" or teratological forms have classically been viewed as nonfunctional entities, evolutionary dead ends not worthy of study. However, the recognition of "monsters" in studies such as Evo-Devo arises from the research for "logics" or "rules" to explain morphological development (Alberch, 1989; Reiss et al., 2008). Back in the 1700s, and after the contributions of some scientists based on studies with teratological humans (like the studies of two French anatomists, Jean Méry and Alexis Littré), the French Academy of Sciences proclaimed: "*Monsters are usually regarded as games of nature. But the philosophers are very much convinced that nature does not play, that she always follows the same rules and that all her works are, so to speak, equally serious. There can be extraordinary things but not irregularities. Furthermore, it happens very often that these very extraordinary things provide the path to discover the general rules in which they all are included*" (Histoire et Memoires de l'Academie Royale Des. Sciences, 1703 [1720]: 28.; Moscoso, 1998). This strategy of using teratologies to understand development goes back a long way (Paré, 1573; Aldrovandi, 1642), and since then, contributions to the field have been somewhat irregular, as the perspectives on which studies have focused. Modern versions include these "monsters" in fields such as evolutionary teratology or evolutionary developmental pathology (Evo-Devo-Path) (Moscoso, 1998; Blumberg, 2009; Nestawal, 2014; Guinard, 2015; Diogo et al., 2017; Diogo, 2020).

Why do some monster forms, despite being nonadaptive, appear repeatedly? Perhaps internal variations and rules of form–function relationships in development could answer this question. Teratologies can be generated in an organized or nonchaotic way, with certain forms being more likely than others (Alberch, 1980). This characteristic makes the description and classification of these morphological variants possible (Saint-Hilaire, 1836; Alberch, 1989). For example, two-headed monsters are relatively common in many vertebrate groups (Stockard, 1921; Saint-Hilaire, 1836), while three-headed monsters are extremely rare in nature. In turn, certain teratologies are more common in certain species than others (Stockard, 1921; Bulmer, 1970; Stephens & Shepard, 1983). In contrast, the recurrence of teratologies in different species suggests that the ontogenetic pathways that produce these morphologies may be maintained throughout the phylogeny (Alberch, 1989).

Indeed, Blumberg (2009), in his book "Freaks of nature," highlights multiple examples of the connections between teratology, developmental biology, and evolution. For example, from a biomedical perspective, it can be very useful to explain patterns of teratology by failures in physiological or developmental processes. This allows us, through the presence of specific malformations, to better understand certain physiological processes that fail (whether due to genetic and/or environmental factors) and to understand more clearly how these physiological processes function under normal conditions (e.g., in a healthy individual). Interestingly, both Alberch and Blumberg emphasize the role of "internal machinery" in determining form as a counterbalance to a pure adaptationist view with natural selection as an omnipotent external force. Stephen Jay Gould and Richard Lewontin had called for an appreciation of "*organisms as integrated wholes, fundamentally not decomposable into independent and separately optimized parts,*" also criticizing extreme adaptationism (Gould, 1977; Gould & Lewontin, 1979).

Different interpretations of morphological variability from diverse disciplines can be mixed, and the tools and explanations provided by each discipline are equally rich and valid. By conceptualizing potential morphological variations as a morphospace, we can identify unoccupied or sparsely represented regions as spaces associated with morphologies that may indicate lower fitness (Olson, 2012). This could be explained by the elimination of these morphologies by natural selection or by constraints inherent to development ("constraints," "scaling laws") (West & Brown, 2005). Olson (2012) suggests thinking about teratology in a framework of the possible and the impossible; they may be abnormal but possible morphologies and forms that are imaginable but unobserved. This means that to study whether specific data support biological "laws" or rules, the study of teratologies is fundamental. Uncommon variants give evolutionary biologists valuable information about what can and cannot develop ontogenetically (Reiss et al., 2008; Blumberg, 2009; Olson, 2012) and what certain morphological features would look like if they were evolutionarily selected. The empty spaces represent a wide range of morphological possibilities (Blumberg, 2009). The variations that occur naturally in species are the necessary substratum where selection acts. The "normal" to the "aberrant" are part of this continuum, and knowing this gradient is key to understanding evolutionary morphological trajectories.

Different approaches can study and explain empty spaces, but a developmental approach is necessary to understand the adaptive landscape of these morphological variants (Klingenberg, 2010; Olson, 2012). An embryological approach studies species' possible and not possible developmental pathways and their evolutionary implications (Alberch, 1989; Reiss et al., 2008; Blumberg, 2009). However, experimental manipulations can be performed to reach these empty spaces in morphospace (Sinervo & Litch, 1991; Conner et al., 2011). Finally, comparative studies allow the morphospace to be "filled" by characters from different species to analyze discontinuities and differences in their occupancy (Alberch, 1989; Polly, 2008;

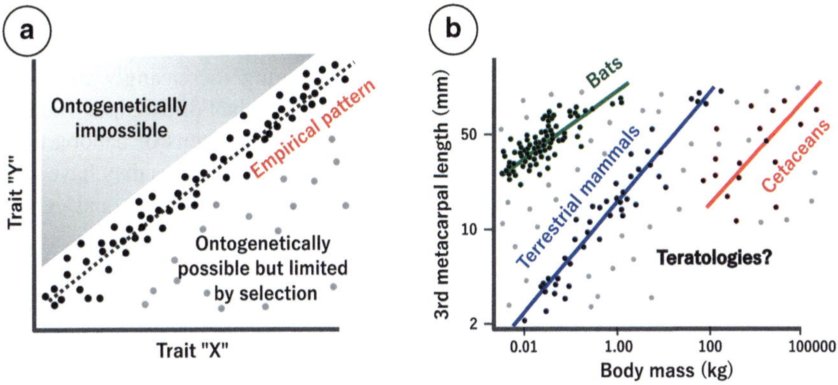

Fig. 5.1 Explanations for the empty space found in morphospaces. (**a**) Diagram illustrating an empirical pattern of a scaling relationship, on the upper diagonal, empty space would be ontogenetically impossible following a developmental constraint explanation. On the lower diagonal, the morphospace would be ontogenetically possible, though constrained by selection (gray dots), following an adaptationist explanation so the empirical pattern would represent the highest possible fitness. (**b**) Study of bone size scaling patterns in terrestrial mammals from a comparative (comparison with bats and cetaceans) and embryological–teratological approach by examining bone teratologies (Modified from Olson, 2012; Swartz & Middleton, 2007)

Minelli, 2009). An example is the study of limb bone proportions of terrestrial mammals, where the authors tested whether morphological relationships outside the allometric scaling line in the empty space of the morphospace were ontogenetically impossible or the result of selective processes (Swartz & Middleton, 2007) (Fig 5.1a). By comparison with nonterrestrial mammals, relationships outside the allometric scaling line were found to be ontogenetically accessible, and by an embryological-teratological study, bone disorders in bone size ratios and size–density relationships were found to occupy other empty spaces of the morphospace (Fig. 5.1b). Thus, by combining different approaches, it was possible to conclude that these relationships were not ontogenetic "constraints" but adaptive outcomes to life on land in mammals (Swartz & Middleton, 2007; Olson, 2012).

5.2.3 Factors That Can Generate Teratological Traits

As discussed earlier, teratologies can be categorized to identify a "teratological pattern," which involves investigating the underlying causes of specific teratological traits. This pursuit has been ongoing since the era of T. H. Morgan's classical genetics and subsequent studies exploring the physicochemical factors capable of inducing abnormalities, both genetic and phenotypic, in animal models such as *Drosophila* (Bianchini et al., 2018; Li & Bi, 2018; Mishra & Barik, 2018) and amphibians (Pownall & Saha, 2018). Over time, an increasing number of teratogenic elements have been identified. Teratogenic elements are defined as chemical compounds that can lead to developmental abnormalities in offspring if consumed during prenatal stages (Barik & Mishra, 2019). This field is becoming increasingly important because it can be extrapolated to possible effects on humans. For example, Barik & Mishra (2019) have recently reviewed the potential as teratogens of nanoparticles (thus in vogue promoting their use in cutting-edge biomedicine), as they have been found to cause defects at the behavioral, developmental, phenotypic, and genetic levels in *Drosophila*.

As we observed in animals, teratologies in plants and fungi can also be indicators of specific environmental variables such as lack of illumination (Michael et al., 1983) or the presence of harmful compounds that function as teratogens and produce individuals who deviate from the "normal" in diverse ways (Masters, 1869; Goebel, 1896; Bos, 1957). Studies on abnormal fungal fruiting bodies that develop in response to infection by other fungi or exposure to toxins have been very useful in analyzing the processes responsible for the appearance of the symmetrical form in fungi and for understanding their extreme sensitivity to environmental stress. An example is a phenomenon known as "rosecomb," which occurs in cultured fungi due to exposure to diesel vapor, kerosene, creosote, Lysol, and other petroleum-derived chemicals (Phillips, 1881; Lambert, 1930). Some of these exogenous chemical agents are associated with genetic instability, producing malformations and distortions in fruiting bodies (Umar & Van Griensven, 1999; Kearney & Kearney, 2009, 2011). The response of fungi is progressive depending on the severity and duration of exposure to teratogenic compounds (Flegg, 1983), and morphogenic changes begin to occur in the middle of fruiting structure development (Kearney & Kearney, 2009).

The study of teratogenic factors can also help examine the possible effect of some pollutants and climatic disturbances on the evolutionary history of organisms. For example, Munnecke et al. (2012) examined the occurrence of teratologies in a group of organisms, the acritarchs. These organisms are a collection of organic-walled microfossils characterized by their diverse shapes and sizes. They are considered an informal and polyphyletic group, meaning they consist of organisms from multiple evolutionary lineages with unknown biological affiliations (Servais et al., 2004). Munnecke and collaborators studied this group on a particular evolutionary time subject to major environmental perturbations: the late Ordovician–Silurian interval. This period was characterized by pronounced negative isotope excursions (of carbon $\delta 13C$ and oxygen $\delta 18O$) measured in marine and terrestrial substrates that indicate large-scale changes in the global carbon cycle and mass extinction events (Calner, 2008). The preceding stages of some of these excursions are marked by the high abundance of malformed acritarchs in various parts of the world (Delabroye et al., 2012). Although the mechanisms directly involved in the occurrence of teratologies in these organisms are still unknown, isotope excursions, as well as contamination by heavy metals, changes in temperature, salinity, and sea level may have been key factors involved in the teratogenesis of acritarchs (Munnecke et al., 2012).

A similar example occurs in the fossil palynological record that can provide clues to past climatic disturbances (Visscher et al., 2004; Foster & Afonin, 2005;

van de Schootbrugge & Wignall, 2016; Lindström et al., 2019; Gravendyck et al., 2020; Marshall et al., 2020). The process of pollen production is known as microsporogenesis and is one of the most vulnerable processes to environmental stress in the life cycle of seed plants (De Storme & Geelen, 2014). Deviations in microsporogenesis can influence the shape and size of pollen and spores (collectively termed palynomorphs), and due to the resistance of palynomorph walls to decomposition, these teratologies could appear in the fossil record. Therefore, morphological teratologies of palynomorphs are unique indicators of paleoenvironmental stress for terrestrial ecosystems (Benca et al., 2022). Aberrant spores and pollen grains were found in the fossil record from the Permian–Triassic and Triassic–Jurassic boundaries. Possible drivers of the malformations may have been geographic and temporal proximity to volcanic eruptions (Broadley et al., 2018; Lindström et al., 2019; Chu et al., 2021), ozone depletion, or elevated UV-B exposure (Visscher et al., 2004; Benca et al., 2022). The study of these fossil teratologies provides new insights into the different responses of the plant kingdom to ecological crises (Gravendyck et al., 2020).

5.2.4 Invertebrate Teratologies

There are reports of teratologies and deformities in vertebrates (Hendrickx & Binkerd, 1990; Kalter, 2003; Garcês et al., 2020), a notorious case being Janus, the longest-lived (25 years) two-headed turtle in the Geneva Natural History Museum. This turtle also has two hearts, two lungs, and a digestive tract, with the right head controlling the right legs and the left head the legs on the corresponding side. His keepers claim that Janus has two personalities and tastes and that they occasionally conflict with each other, for example, when walking (Farge & Mantovani, 2022). Beyond this lovely case, many reports of teratologies have occurred in invertebrates collected in the wild (Spanó et al., 2003; Asiain & Márquez, 2009; Eeva & Penttinen, 2009; Lesniewska et al., 2009; Ferrer et al., 2014; Chitimia-Dobler et al., 2017; Mitchell et al., 2020). In this group, mutants were crucial in discovering the role of homeotic genes in shaping the anteroposterior axis and conferring segments and appendages their specific identities (Rosenberg et al., 2009; Wanninger, 2015). In general, the systematics are good at detecting these morphological "aberrations" when analyzing characters for taxonomic descriptions. For example, in Coleoptera, multiple reports of abnormalities were reported, recurrently, some titled "*A remarkable teratological specimen of…*." Such is the case of a chrysomelid Coleoptera that, instead of a palpus, has an antenna (Clark & Belo Neto, 2010), a scarabaeid with an unusual malformation of the eyes (Gasca-Álvarez et al., 2018), and another of the same family with a deformation of the right elytra (Barria et al., 2020). These are typical examples of reports in which the teratological case is presented along with a brief analysis but is not intended to connect to causes, processes, or evolutionary implications.

In annelids, the bifurcation of the anteroposterior axis is common, i.e., worms with two or more heads and/or tails (Andrews, 1892; Dualan & Williams, 2011).

These bifurcations can be classified into different types according to anatomy (lateral, dorsoventral) or developmental origin (embryonic, postembryonic). Postembryonic bifurcations may occur during growth, regeneration, and reproduction. Lateral bifurcation can result in well-integrated phenotypes in adults, whereas dorsoventral bifurcation does not since it requires discontinuity of at least some internal organs (Andrews, 1894; Ponz-Segrelles et al., 2022). This phenomenon is especially notable in the Syllidae family (Polychaeta), specifically within the "Ribbon Clade." These organisms reproduce through stolon formation, which can exhibit abnormal patterns such as collateral stolon emergence (where new stolons cluster tightly together in a ventral to posterior arrangement) and successive gemmation (involving a dorsoventral bifurcation where each segment gives rise to only one stolon, following a posterior to anterior developmental gradient) (Johnson, 1902; Okada, 1937; Nogueira & Fukuda, 2008; Alvarez-Campos et al., 2013). Developmental anomalies in annelids emerge as a model that should be incorporated into the growing field of Evo-Devo and examined with modern techniques and perspectives (Ponz-Segrelles et al., 2022).

When delving into the underlying causes of these teratologies, it becomes evident that various factors, particularly during embryogenesis, can contribute to their development. Numerous studies with an experimental approach in arachnids have successfully identified the causative factors behind these morphological disorders (Buczek, 2000; Napiórkowska & Templin, 2017; Buczek et al., 2019). Among these factors, thermal shocks during embryogenesis have been found to play a significant role in generating different types of teratology, including oligomely (absence of one or more appendages), schistomely (bifurcation of appendages), heterosymely (fusion of adjacent appendages), symely (fusion of appendages from the same pair), and bicephaly (Jacunski et al. 2002; Jacunski & Templin, 1992, 2003; Napiórkowska & Templin, 2013, 2017, 2018) (Fig. 5.2a). The giant house spider *Eratigena atrica* is a great model used for experimental teratogenesis studies. In this species, deformities also include changes in the central nervous system, such as ganglion fusion (Jacunski et al., 2005; Napiórkowska & Templin 2017). Studying these teratological cases is extremely important, as they could teach us how certain environmental conditions affect arthropod development (Balazuc, 1948; Burke et al., 2018). This may also allow us to predict morphological changes and phenotypic consequences of the use of certain chemicals or temperature increases (e.g., linked to climate change) in the embryology of certain animal groups.

It is interesting, for example, that in spiders, there are body portions more prone to suffer teratologies, such as walking legs and ocular structures (Jiménez & Llinas, 2002; Napiórkowska et al., 2007; Gonzalez-Moliné, 2008; Mitchell et al., 2020). In contrast, in other structures, such as chelicerae, pedipalps, and genital structures, it is less common to find deformations (although see Muma, 1943; Kaston, 1962, 1963; Izquierdo, 2021; Nadolny et al., 2022) (Fig. 5.2b). Aberrant copulatory organs are usually more closely linked to cytogenetic (aberrant karyotypes) and genetic

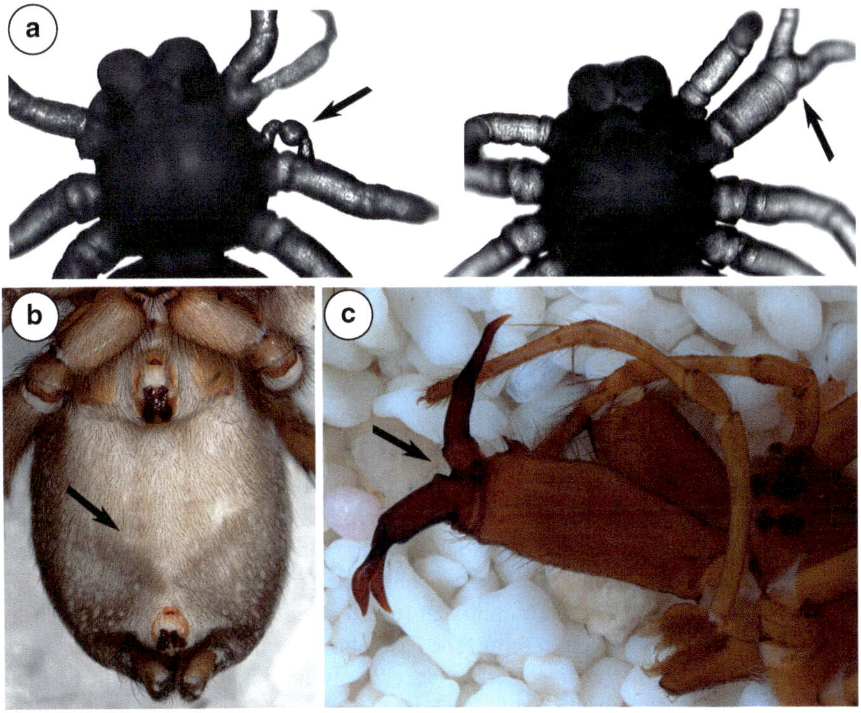

Fig. 5.2 Teratologies in spiders. (**a**) Ventral view of *Eratigena atrica* juveniles showing therato-genic changes after thermal shocks during embryogenesis (from Napiórkowska et al., 2017. (**b**) Ventral view of the opisthosoma of *Almafuerte peripampasica* (Image courtesy of Matías Izquierdo). (**c**) Teratological chelicerae of *Tetragnatha versicolor* (Image courtesy of Darko Cotoras). Black arroSws indicate teratological traits
(a) *Eratigena atrica* juveniles. License: Public Domain. (b) *Almafuerte peripampasica* teratology. License: Author permission. (c) *Tetragnatha versicolor* teratology. License: Author permission.

(aberrant patterns of gene expression in the subadult instar) disorders. This may result in "abnormal" gynandromorphic individuals or intersex cases (Roberts & Parker, 1973).

A noteworthy case was recently reported, where teratology was described in the long-jawed orb weaver *Tetragnatha versicolor* that possessed a duplicated left chelicera on its proximal–distal axis, with a terminal secondary schistomely on the fang (Cotoras et al., 2021) (Fig. 5.2c). This teratology could have occurred from an injury to the area where the duplication occurs, similar to what has been observed in the sea spider *Pycnogonum litorale* (Scholtz & Brenneis, 2016). The mechanism behind this could be the "boundary model" (Meinhardt, 1986), where an appendage develops from the boundaries between two cell populations (anterior and posterior), and a mechanical injury can cause the displacement of a third cell group, causing juxtaposition of morphogens that confer anterior, posterior, and ventral fates during regeneration. Knowing the possibility of the existence of this phenomenon behind the teratology of the proximal–distal axis of the chelicerae allows the study of the mechanistic basis of the induction of the proximal–distal axis of the chelicerae in chelicerates, which could be a key to corroborate the homology with other head appendages in Panarthropoda (primary antennae—mandibulates, chelicerae—euchelicerates, and chelifores—sea spiders) (Budd, 2002; Brenneis et al., 2008). The study of this teratology and its comparative approach (Olson, 2012) leads to elucidating shared patterns in the formation of chelicerae, opening the door to future functional and experimental studies of appendage development in different study models.

5.2.4.1 Reflections from the Discovery of Teratological Scorpions

Certainly, in each field, scientists come across rarities that fall into the teratological domain, which may not have evolutionary potential, but inform us a lot about what may be happening, for instance, at the physiological level. An example, derived from personal observations in the field of scorpion reproductive biology, highlights both the teratological occurrence and the subsequent resistance encountered within a segment of the scientific community in regard to acknowledging and reporting teratology. The fact referred to report the case of a male in a species of *Bothriurus bonariensis* scorpion that presented an excessively long courtship, with the male not being able to deposit the spermatophore. The courtship took almost 24 hours, when it usually occurs after no more than one hour (Peretti, 1992). At some point in this unusually long courtship, the couple was separated, and the male had a part of his spermatophore outside the genital opening. The male died within a few days, and the specimen was dissected. Surprisingly, it was found that this individual had only half of the reproductive system, so it can be considered an "aberrant" individual.

It was remarkable that despite lacking complete reproductive organs, this male scorpion engaged in courtship behavior and initiated the expulsion of a partial spermatophore through his genital opening, which did not fully exit. Reporting this curiosity as a scientific note in a local entomological journal (Peretti, 2000) was a difficult task. On the one hand, a strange case was exposed, and on the other hand, the implications of the lack of part of the reproductive apparatus were discussed. This involved more than three years of reviews and a journal referee labeling the report as of minor importance, as it was "merely teratological." That is what it was, something teratological, but apparently, this expression was taken as a synonym for something that was not worth being reported at a scientific level. This simple but intense experience made me reflect on the extent to which we researchers omit reporting rare situations regarding the morphology, physiology, or behavior of the animal group studied. It could simply be "self-censorship," considering that teratology is not worth reporting or that it is worth reporting, but we are too busy with more important things.

5.2.5 *Some Examples of Teratologies in Plants*

References to teratologies are more commonly found in relation to animals, including humans. The inclusion of teratology in other kingdoms, such as plants, algae, or fungi, is relatively rare and not as explicitly explored. This observation points to a certain bias among human scientists. However, it is important to recognize that abnormalities and aberrations also occur in these "nonanimal" groups. While the study of teratology in the plant and fungal kingdoms is not as advanced as in animals, there are multiple examples of such phenomena documented in these organisms (Phillips, 1881; Goebel, 1896; Bos, 1957; Meyerowitz et al., 1989).

5.2.5.1 Desired Malformations in Plants

"Malformations" in plants have attracted the attention of botanists for ages. Plants can exhibit mutations that result in various types of deformations. One notable example is fasciation, a phenomenon characterized by abnormalities in the growth of the apical meristem. These abnormalities lead to the formation of flattened, ribbon-like, crested, or sinuous shapes in the stems, roots, fruits, or flowers, deviating from their typical appearance (White, 1948; Iliev & Kitin, 2011) (Fig. 5.3b). This phenomenon is produced by mutations that may result from bacterial infection, insect or parasite attacks, or chemical or mechanical damage (Iliev & Kitin, 2011). Although these morphological alterations are not lethal, they can be inherited and are more common in families such as Amaranthaceae and Cactaceae (Piringer & Borthwick, 1961; Maliarenko & Mudrak, 2013).

Another teratology in plants appears as variegation and virescence, also due to genetic alterations that produce abnormal pigmentations in various portions of plants (Darwin, 1844; Kirk & Tilney-Bassett, 1978; Marcotrigiano, 1997) (Fig. 5.3a). A common cause of variegation is the masking of the green pigment by others, such as anthocyanins. This often extends to the entire leaf, turning it reddish or purple. Variegation is considered unstable and can be easily lost in plants that turn green again, but in some species, it is stable and does not change during the plant's life (Yu et al., 2007). Variegation mutants have played a prominent role in the history of genetics (Tilney-Bassett, 1975; Kirk & Tilney-Bassett, 1978). For example, it was observed that the mechanism of variegation transmission did not obey Mendel's laws (Baur, 1909; Correns & Correns, 1924), which opened the way for the discovery of non-Mendelian (maternal) inheritance (Granick, 1955; Rheinberger, 2013). We will also see later that some types of variegation can serve for mimicry in some plants.

Some cactus species that exhibit fasciation are highly valued by collectors of these species (Fig. 5.3c). This illustrates some of the ideas previously discussed regarding valuation and subjectivity when considering teratological individuals. Depending on the organisms in which the malformation occurs and the morphological outcome of the malformation, our subjectivity may appreciate these anomalies and intentionally seek them out. For example, some varieties are attractive and ornamental, and gardeners and plant cultivators tend to preserve and even replicate these

Fig. 5.3 Teratologies in plants. (**a**) Variegated leaves of *Calathea roseopicta* (Credits: Forest & Kim Starr) (**b**) Normal (on top) and fasciated (below) white-rayed mule's ear (*Wyethia helianthoides*) (Credits: Perduejn) (**c**) Fasciation of saguaro cactus (*Carnegiea gigantea*) showing a "crested" type (Credits: James G. Howes). (**d**) Witches' broom on birch caused by the fungus *Taphrina betulina* (Credits: Morten DD Hansen)

(**a**) *Calathea medallion.* License: CC BY 3.0 https://creativecommons.org/licenses/by/3.0, via Wikimedia Commons. https://commons.wikimedia.org/wiki/File:Starr_070906-8758_Calathea_roseopicta.jpg. (**b**) White-Rayed Mule's Ears (*Wyethia helianthoides*). License: CC BY 3.0, https://creativecommons.org/licenses/by/3.0 via Wikimedia Commons. https://commons.wikimedia.org/wiki/File:MulesEarFasciated_107393.jpg. (**c**) Crested saguaro cactus. License: Public Domain. https://upload.wikimedia.org/wikipedia/commons/4/43/Crested_Saguaro_cactus.jpg. (**d**) *Taphrina betulina.* License: No restrictions (CC0), via Wikimedia Commons. https://commons.wikimedia.org/wiki/File:Taphrina_betulina_31161103.jpg

cases (Datta, 2009). In this case, a certain "teratological degree" is accepted and sought after because that malformation is considered striking, beautiful, exotic, and extraordinary in a positive sense. Botanists even generate variegated plants through inoculation with certain viruses (Marcotrigiano, 1997). Such is the attraction for plants with these characteristics that there is in the United Kingdom the "Special Interest Group on Variegated Plants of the Hardy Plant Society" that, since 1980, shares and promotes the love for variegation.

5.2.5.2 Harmful Teratologies Could Bring Ecological Benefits

Another interesting example of teratology is the case of "witch broom," an unusual multiplication of branches on the upper part of the woody plant that resembles a broom or bird's nest (Hawksworth, 1961; Mathiasen, 1996) (Fig. 5.3d). This formation is caused by diseases produced by phytoplasmas, fungi, mites, aphids, and nematodes, causing what is known as "Witch Broom Disease." This disease can cause economic damage in plants such as cocoa (Evans, 2002, 2007) and cherry (Fujii et al., 1968), as well as in trees from which wood is extracted, such as pine, oak, and walnut (Gomez et al., 1996; Yun et al., 2001; Spicer, 2016). Forests with infestations of witch broom are more prone to forest fires (Harrington & Hawksworth, 1990). In the case of cocoa in Brazil, its production has been in decline for several years since a large outbreak of "witches' broom" in the late 1980s (Pereira et al., 1990; Evans, 2002; Griffith et al., 2003), which meant that thousands of hectares of cocoa farms were abandoned in Brazil and elsewhere in Latin America (Evans & Prior, 1987; Pereira et al., 1996; Griffith et al., 2003; Hebbar, 2007). Multiple lines of research are currently being developed to "combat" this teratology and mitigate its economic consequences (Purdy & Schmidt, 1996; Lopes et al., 2003; Medeiros et al., 2010; Tirado-Gallego et al., 2016). Here, we are at a negative extreme of teratology, which from the name already carries a bias associated with something "unpleasant" or "ugly" (since witches have had this reputation since ancient times).

Paradoxically, "witches' brooms" are ecologically important because they create and enhance the structural diversity of the forest canopy (Mathiasen, 1996; Tinnin & Forbes, 1999). Many wildlife species visit witches' brooms for nesting, foraging, latrines, hiding, or resting (Hawksworth & Wiens, 1996; Mathiasen, 1996; Shaw et al., 2004). For example, many threatened or endangered raptors use brooms in the western United States as nesting sites (Fletcher, 1990; Martin et al., 1992; Seamans & Gutiérrez, 1995; Forsman et al., 2015). Some mammals also nest on brooms, such as red squirrels at various sites in North America (Parks et al., 1999; Tinnin & Forbes, 1999; Hedwall & Mathiasen, 2006) and northern flying squirrels that form aggregations within brooms during cold winter periods, utilizing brooms twice as much as other nest sites (Mowrey & Zasada, 1984; Cotton & Parker, 2000). Maintaining conifers with "witches' brooms" in their habitat has been proposed to protect these animals (Mowrey & Zasada, 1984; Cotton & Parker, 2000; Garnett et al., 2006). This raises the complexity of categorizing teratology on certain occasions and reflects the subjectivity we may encounter when assessing these phenomena.

5.2.5.3 Teratologies That Inspired Theories of Plant Physiology

Especially interesting is how observations of certain plant teratologies inspired theories and generalizations that were fundamental in plant physiology. In this case, a unique or exceptional trait can lead to establishing a general rule. We have previously highlighted the case of variegation and the subsequent development of non-Mendelian laws, although we would like to highlight the case of "phyllody" in plants. The German poet and philosopher Johann Wolfgang von Goethe observed at the end of the eighteenth century some roses in which the floral organs were substituted by structures similar to leaves or stems, in his words he expressed: "*From our acquaintance with this abnormal metamorphosis, we are enabled to unveil the secrets that normal metamorphosis conceals from us, and to see distinctly what, from the regular course of development, we can only infer.*" These teratological roses led him to suggest the "leaf theory", where it was postulated that all parts of the flower are highly specialized leaves. When specific interferences occur in the early stages of growth (e.g., caused by phytoplasma or virus infections) (Mor & Zieslin, 1992; Sim et al., 2004; Hogenhout & Segura, 2009), these organs may develop, departing from the basic "construction plan" from the general pattern.

This type of anomaly is known as "phyllody" (Masters, 1869), i.e., the abnormal development of floral parts into leaf structures, and implies sterility of the plant (Mor & Zieslin, 1992). Goethe called this "metamorphosis" the main subject of his homonymous essay (von Goethe, 1790). Just as teratologies and exceptions were ignored or denied for a long time, the same has happened with certain theories or observations based on these anomalies, such as what happened with Goethe's proposal. However, this theory was fundamental for explaining homeotic genes, homology in comparative anatomy, the so-called ABC model of plants, and its implication in the field of Evo-Devo in plants (Coen & Meyerowitz, 1991; Weigel & Meyerowitz, 1994; Causier et al., 2010).

5.3 Treasure Your Exceptions

William Bateson was one of the first geneticists to emphasize the significant role that phenotypic rarities and exceptions play in both developmental and adaptive processes. One of his famous quotes was highlighted in Cock and Forsdyke's (2008) inspiring biography of the scientist: "*if I may throw out a word of counsel to beginners, it is: Treasure your exceptions! When there are none, the work gets so dull that no one cares to carry it further. Keep them always uncovered and in sight. Exceptions are like the rough brickwork of a growing building, which tells that there is more to come and shows where the next construction is to be*"—W. Bateson (1908). This highlights the importance of appreciating rare cases, minorities, and exceptions, including the value of novel traits within an evolutionary framework (e.g., West-Eberhard, 2003, 2005). We will comment on these works later in the chapter focused on changes from exceptions to rules.

In many groups of organisms, the "classic" definitions and descriptions in biology configure in part what we can understand as generalities, rules, or what the "majority" of species share, and this will be what identifies the group. However, as we go deeper into the details of these descriptions, rare cases appear occasionally, usually mentioned as friendly curiosities of natural history rather than as something of evolutionary interest. When there are "exceptions" that make up certain minorities, we are dealing with what we can call "intermediate exceptions," i.e., species that deviate from the general pattern in a not-so-drastic way and that include characters measured as continuous variables (e.g., number of groups, body size, intensity of coloration). In the descriptions, we can implicitly find these minorities, and it is important to keep in mind that these "intermediate" exceptions also have biological values.

Thousands of examples could be mentioned regarding these "intermediate" exceptions, implicit in the definitions, which tend to indirectly state the rule or pattern for a given trait under analysis. However, the deepening and valorization of these exceptions are not common. To learn about them, it is often necessary to go to specific scientific articles where we can go deeper into these exceptions and sometimes (not often) to discussions or explanations about their existence. Moving beyond the mere observation of patterns or "what" occurs, delving into the "why" enables a better understanding of the underlying causality behind the analyzed traits. This includes exploring "intermediate" exceptions that may arise at the group, temporal, spatial, or combined levels. Without intending to be deterministic, explanations of a functional-adaptive and/or evolutionary nature can be found for the existence of these "intermediate" exceptions.

A taxonomic curiosity or exception can be invaluable from a comparative point of view since, from a phylogenetic perspective, it provides valuable information on potential phylogenetic and evolutionary hypotheses. As we saw in Chap. 2, the basis of the beginning of an evolutionary process can be characterized by an exception, from the point of view of how common this pattern or trait is in the group under study, either from a global or a more specific perspective within a clade. Therefore, it is crucial to approach each definition cautiously to avoid falling into the reductionism of generalizations. We must understand that "intermediate" (i.e., not extreme but more subtle) exceptions are often prevalent in biological diversity, and therefore, we must recognize the importance of exploring these exceptions in biological research and education.

5.3.1 Temptation to Generalize When Encountering an Interesting Exception

The exceptional is in itself something neutral, neither useful nor useless, since everything will ultimately depend on the interpretation and contextualization of this morphological variant. It is as negative to quickly discard or omit them in the study of a species, as it is to fall into an incorrect generalization and interpretation of what the exception represents or its potential evolutionary consequences. Something very important in studying patterns is the detailed preliminary study with sufficient sample size to corroborate that specific observations are indeed norms or patterns or that they are exceptions. Failures or oversights in this initial determination can lead to misinterpretation of what a few data show with great repercussions for particular theoretical fields or paradigms.

A classic example, in the field of sexual selection, in particular the mechanisms of sperm competition, was that of the "killer" sperm also known as the "kamikaze" sperm hypothesis. This was postulated in a brief review by Baker & Bellis (1988), who argued that in mammals, males transfer part of their sperm agglutinated to slow down the sperm of rivals. Thus, in addition to its primary role in fertilization, spermatozoa would serve a secondary function of obstructing the success of competing males by destroying their immediate rival's spermatozoa. Later, it became evident that these speculative postulates were based on infrequent occurrences of sperm agglutination within the female reproductive tract. This so-called "reproductive tactic" was not a deliberate strategy in itself (Moore et al., 1999). Instead, sperm agglutination was recognized as a rare or anomalous event. The mistake here was attributing adaptive significance to an uncommon observation without thorough prior analysis and experimentation.

5.4 Unfolded Life Diversity: Some Rare Taxa and Amazing Strategies

As we are seeing, a whole book could be written on exceptions in different taxonomic groups alone. Here, the intention is to highlight some biological examples that would otherwise be omitted or ignored. Undoubtedly, in the present book, however, we give exceptions an important entity since it is an aspect of biology that is highly underestimated, often even underappreciated. If we are talking about exceptions, we can mention "group exceptions" or taxonomic exceptions in more precise terms. Here, we do not refer to exceptions within a species or a population but to the species, genus, or family as a "whole," a unit of taxonomic rarity. In several cases, we address evolutionarily derived groups within the clade under study, which exhibit many autapomorphies. We can also focus on phenotypic extremes within a species. That is, to consider cases that are part of the natural variability within the normal range of a particular phenotypic trait.

From a biodiversity standpoint, we can assess the degree to which supraspecific species or groups have been genuine rarities and taxonomic exceptions. In other words, they were or have been exceptions compared to other members of phylogenetically related taxa. Those taxa that are rare not only in having, for example, fewer species or individuals (this would be purely a numerical variable) but also in terms of body structure, physiology, ecology, etc., are true exceptions within the pattern or rule that, explicitly or not, characterizes that group of organisms. In other cases, we can highlight exceptions at the "spatial" level in terms of unique distributions or completely different from what is expected for that taxonomic group. In the following sections, we will try to include some examples in the context of rarities and exceptions that show that we can find amazing cases in all groups.

5.4.1 Exceptional Animal and Plant Groups

5.4.1.1 A Protozoan That Does Not Appear to Be One

A few years ago, a notable species within the protozoa was reported, *Spiculosiphon oceana* (Foraminifera) (Maldonado et al., 2013). This species combines several exceptions to the Order in terms of its size, appearance (and way of reaching it), and feeding strategy. *Spiculosiphon oceana* reaches a size of 4–5 centimeters in length (Maldonado et al., 2013), so it was initially considered a multicellular animal, more precisely, a sponge. Most foraminifera are only a few millimeters long and can form shells that they use for shelter (Ruppert et al., 2003). While these shells, in general, are formed by external mineral particles or calcium carbonate (Kaminski, 2004), the shell of *S. oceana* is composed of silica spicules of dead sponge cemented and bonded with a spongin-like organic substance (Maldonado et al., 2013). Thus, the peduncle is covered by regularly arranged spicules, while the globular region is irregularly arranged in an arborescent manner, allowing the extrusion of pseudopods through the cavities between the spicules. Spiculosiphon oceana collect and arrange spicules with high selectivity. This high selectivity also appears to be an exception within the genus (Christiansen, 1964). These organisms show chemoselectivity and can precisely discriminate siliceous materials from other common materials for the construction of shells. In addition, they can discriminate between silica materials of different nature and even between spicules of different shapes and sizes to locate them in different body portions (Maldonado et al., 2013).

Foraminifera are heterotrophic organisms that capture prey through their pseudopods, usually small organisms such as diatoms or bacteria. However, *S. oceana* has a peculiar feeding strategy related to its morphology and resemblance to carnivorous sponges. As we will explain below, some sponges have exceptional adaptations, such as a carnivorous feeding strategy (Demospongiae, Cladorhizidae) (Vacelet & Boury-Esnault, 1995; Vacelet & Custódio, 2007; Lee et al., 2012). For example, *S. oceana* exhibits remarkably similar size and morphology to sponges of the genus *Asbestopluma*. Both these sponges and *S. oceana* are carnivorous predators, feeding on other smaller calcareous foraminifera and planktonic organisms. This curious case could be considered a morphological and functional convergence between exceptions from different taxonomic groups. In these sessile organisms, prey is digested intracellularly since there is no digestive cavity per se. As previously discussed, these convergent exceptions would reflect adaptation to similar environmental conditions and pressures. Strikingly, some species of *Spiculosiphon* have been found in the same geographic location as cladorrhizid sponges (Vacelet et al., 1994; Vacelet & Boury-Esnault, 1995; Maldonado et al., 2013).

5.4.1.2 Tardigrades: The Most Resilient Organisms?

If we talk about exceptions, we must mention the tardigrades, a taxonomic group with unique characteristics in the animal world. These microscopic invertebrates (0.5–1 mm) were first described in 1773, and their discoverer was surprised by their bear-like morphology: "*Strange is this little animal, because of its exceptional and strange morphology and because it closely resembles a bear in miniature. That is the reason why I decided to call it little water bear.* "—J.A.E. Goeze (1773).

With more than a thousand species worldwide, these organisms are present in virtually every habitat on Earth (Guidetti & Bertolani, 2005): mountain peaks, underwater volcanoes, tropical rainforests, and even polar environments. However, tardigrades are exceptional because they are hyperresilient and extremophilic, i.e., they are resistant to extreme conditions. They can tolerate temperatures from -200 °C to 151 °C (Wright et al., 1992; Hengherr et al., 2008; Horikawa et al., 2012), high saline concentrations, the absence of oxygen, highly toxic chemicals (Baumann, 1922; Ramløv & Westh, 2001), and extremely high pressures (Seki & Toyoshima, 1998; Ono et al., 2008). Tardigrades tolerate extreme desiccation conditions through a reversible physiological process called "anhydrobiosis" (Alpert, 2006; Jönsson et al., 2008; Wełnicz et al., 2011). Through this process, they reduce their size and body water (going from 85% to 3% body water) and metabolism by entering a state of suspended animation (Crowe, 1975; Wright, 1989; Jönsson & Bertolani, 2001). The time they can spend in this state can be several years (Jönsson & Bertolani, 2001; Rebecchi et al., 2006; Guidetti & Jönsson, 2002), surviving up to three decades (Tsujimoto et al., 2016).

One of the most striking characteristics of tardigrades is their tolerance to high doses of ionizing radiation (May & Guimard, 1964; Jönsson et al., 2005; Horikawa et al., 2006). Both during anhydrobiosis and in the normal state, these organisms possess high radiation resistance (Jönsson et al., 2005; Horikawa et al., 2006), which could indicate that tolerance is not restricted to biomechanical protection of cells during anhydrobiosis. However, it is not all daisies and sunshine for irradiated tardigrades since, although they can survive, molt, and lay eggs, these eggs are not viable, so radiation produces sterility in these organisms (Jönsson et al., 2005; Horikawa et al., 2006). Recently, it was found that the protection of tardigrade cells from X-rays is due to a specific protein (Dsup = Damage suppression protein) that can be added to human cell cultures, providing them with partial protection from this type of radiation (Hashimoto et al., 2016; Chavez et al., 2019). Unraveling the mechanisms underlying anhydrobiosis and radiotolerance could contribute considerably to developing new technologies in radioprotection, cryopreservation, and preservation of biological material and in developing new technologies that protect humans from these types of stressors (Table 5.2).

Recently, along with the description of a new species of the genus *Paramacrobiotus*, their amazing ability to resist ultraviolet radiation was discovered. They possess a "fluorescence shield" (Suma et al., 2020). Although fluorescence has been described in many animals, very few studies have proven its functional significance (Lagorio et al., 2015; Jeng, 2019). In this case, it could be experimentally demonstrated that the presence of fluorescent pigments conferred protection from ultraviolet radiation

Table 5.2 Some examples of living organisms that represent exceptions or rarities within their clades in terms of specific functions that, from a human perspective, could serve to minimize, combat, or optimize industrial, agricultural, environmental, or medical production processes

Organism	Quality as an exception within a group	Function in nature	Potential application	References
Wax moth *Galleria mellonella*	Larvae can degrade polyethylene	The butterfly and its larva inhabit colony enclosures with beehives. The larva feeds on beeswax	Biorecycling/ degradation	Bombelli et al. (2017)
Bacterium *Ideonella sakaiensis*	Degradation of hydrocarbon-derived compounds	Use polyethylene terephthalate (PET) as the primary carbon source	Recycling and reuse of PET	Yoshida et al. (2016), Bornscheuer (2016)
Amazon mushroom *Pestalotiopsis microspora*	Polyurethane digestion under aerobic and anaerobic conditions	Use polyurethane as the only carbon source	Bioremediation	Russell et al. (2011)
Coelacanth *Latimeria chalumnae*	"Living fossil"	It retains primitive traits due to its slow evolution	Its notochord is sold on the illegal market as a fountain of youth. Many environmental associations point to well-known cosmetic companies that include coelacanth notochord in their formulas	Amemiya et al. (2013)
Rare microbes	Important role in biogeochemical cycles	Keystone species regulating the functioning of host-associated, terrestrial, and aquatic environments	These microbes have diverse enzymes, some key in industrial processes, such as thermophilic cellulases for biofuel applications	Rubin (2008), Jousset et al. (2017)
Tardigrades	Anhydrobiosis, resilience, UV resistance, resistance to ionization and radiation	"Superpredators" on flesh-eating bacteria and nematodes, regulating the food chain	Development of new technologies that protect humans from radioprotection, cryopreservation, and preservation of biological material	Jönsson (2007), Ono et al. (2008), Guidetti et al. (2012), Hashimoto et al. (2016), Weronika & Łukasz (2017), Chavez et al. (2019), Suma et al. (2020)

(continued)

Table 5.2 (continued)

Organism	Quality as an exception within a group	Function in nature	Potential application	References
Smith's mantis shrimp *Gonodactylus smithii*	Vivid colors, a high-range vision of the spectrum, powerful frontal appendages	Predators, regulators of the reef food chain	Potential models for studies of shooting mechanics, hydraulics, hunting apparatus, or the neuroscience of memory and learning	Patek et al. (2013), Chiao et al. (2000)
Dictyostelium discoideum	Cell movement, chemical signaling. Cell differentiation. Chemotaxis. Sociality	Decomposers in soils rich in organic materials	Useful in human cancer research, model for studying cell differentiation	Kay et al. (1978), Solomon et al. (2000), Hägele et al. (2000), Williams et al. (2006), Annesley & Fisher (2009), Jia et al. (2009), Clarke (2010), Sillo et al. (2011), Bozzaro & Eichinger (2011), Steinert (2011), Müller-Taubenberger et al. (2013), Hashimura et al. (2019), Pal et al. (2019), Schaf et al. (2019), Pears & Gross (2021)
Panellus stipticus	Bioluminescence	Wood decomposers	Bioindicator of environmental pollution	Stevani et al. (2013), Prasher et al. (2014), Ventura et al. (2021)

to these animals. Furthermore, Suma et al. (2020) were able to extract these pigments and transfer the protection to other UV-sensitive organisms, doubling their survival, so this ability of tardigrades could be universal and eventually have promising applications in industry or other applied fields (Table 5.2).

Anhydrobiosis not only allows tardigrades to survive in extreme terrestrial conditions but also allows these animals to survive unnatural conditions such as space vacuum (which imposes extreme dehydration) and solar/galactic cosmic radiation (Jönsson et al., 2008). Different species of tardigrades were sent to outer space twice (FOTON-M3 Mission—2007—Jönsson et al., 2008; Rebecchi et al., 2009,

2011a; Persson et al., 2011; Shuttle Mission STS-134 Endeavour—2011—Rebecchi et al., 2011b; Vukich et al., 2012), returning to Earth safe and sound and even being able to reproduce upon arrival. In space conditions, solar radiation imposes damage and changes in DNA structure by specific photoproducts. It is speculated that tardigrades may have special configurations of DNA or other cellular components that prevent cell damage and superefficient DNA repair systems (Jönsson, 2007; Rebecchi et al., 2011b; Vukich et al., 2012). Thus, tardigrades are promising models for Astrobiology, offering the unique opportunity to study the effects of space exposure on metazoans. Then, the strategies used by these incredible animals can be extrapolated to protect other organisms exposed to these environments, including humans (Jönsson, 2007; Ono et al., 2008; Guidetti et al., 2012; Weronika & Łukasz, 2017). Recently, using mathematical models, the exceptional resilience of tardigrades has been calculated, and it has been suggested that they would survive astrophysical catastrophes (such as ocean temperatures rising to boil), representing the most resilient life form in the animal world (Sloan et al., 2017).

5.4.1.3 Sponges Breaking Several Rules

Sponges are found within the phylum Porifera, which currently has more than 9000 species. This group of organisms has specific characteristics that make it unique: they are sessile, with an internal skeleton formed by calcareaous spicules or silica spicules, with or without spongin (collagen-like threads). Their body is perforated, forming an aquifer system through which water enters and exits through a larger perforation, the osculum. This water current is used for feeding, waste elimination, gamete release, and gas exchange. No less important is the fact that they lack true differentiated tissues. These characteristics are listed at the beginning of their description in any book on invertebrate zoology and phylogeny (e.g., Ruppert et al., 2003; Brusca et al., 2016; Giribet & Edgecombe, 2020).

In sponges, we can find multiple "intermediate" type exceptions. For example, it is common to think of a sponge as an immobile, medium-sized, brownish-colored animal. However, "*brownish color is the exception rather than the rule in Porifera*" (Ruppert & Barnes, 1994). Brownish sponges predominate in the class Calcarea but not in the others, where most species have bright colors, which are thought to serve as protection against solar radiation, as well as aposematic colors. Another example involves exceptions linked to the size of the organisms. Also within the class Calcarea, there is a very simple type of body structure called "asconoid," present, for example, in the genus *Leucosolenia*, where each individual of the colony (united at their bases by a stolon) has a size of 2.5 cm or even less. In this genus, it has been established that "*only rarely [each individual] is more than 10 cm tall*" (Ruppert et al., 2003). This would imply that such "rare occasions" do exist and lead us to ask: what are these exceptional cases? Do they occur in particular places or types of water currents? Indeed, the "rule" obviously is that the asconoid type is always small, but the exceptions of larger sponges should invite us to think about what lies behind them. In this case, the exception of the "implicit rule" for body size in

Leucosolenia is *L. somesii,* in which each member of the colony can exceed 10 centimeters, reaching 12 cm (Van Soest et al., 2007).

The absence of true tissues is so important that sponges have traditionally been grouped into a particular subkingdom known as Parazoa. This differs from the Eumetazoa subkingdom, where the rest of the animal groups are found, where we can already find true tissues (i.e., solid junctions between the cells of a given stratum). In parazoans, it was known that the cells of the outermost layer, the pinacoderm, laterally lacked cell–cell junctions between each other (e.g., the desmosomes of higher animals) and a basal lamina for seating (formed by a collagen layer in other animals). Therefore, although they form strata "analogous" to eumetazoans (e.g., pinacoderm equivalent to a "proto-epidermis"), they cannot be termed true tissues but as an ordered, but not cohesively united, aggregation of cells. This was the classical concept and rule for Porifera, and it is still the one that is referred to when defining them in their generalities. However, such a rule, almost of a level I for the group, presents surprising exceptions discovered and analyzed in depth not long ago.

Within the family Oscarellidae (class Homoscleromorpha), there is an exception that represents a very valuable evolutionary link between Porifera and the animals of the subkingdom Eumetazoa. We refer to what happens in the genera *Oscarella* and *Pseudocorticium* (Fig. 5.4b) (Van Soest et al., 2012). In them, we find something that departs from what is reported as a rule for the phylum Porifera, as many lack a skeleton and mainly possess a true epithelium. Ereskovsky et al. (2009) analyzed the ultrastructure of the pinacoderm layer of these sponges and found that at the base of this layer, there is a basal lamina formed by type IV collagen, as occurs in eumetazoans. Therefore, this pinacoderm can be considered an intimately associated cellular layer, i.e., a true tissue (Fig. 5.4c). Therefore, using the term epidermis instead of pinacoderm is correct. These discoveries were very important, to the point of suggesting that Homoscleromorpha would form a kind of "link" between the subkingdoms Parazoa (rest of Porifera) and Eumetazoa.

Now, what explanation could we have for the appearance of this "taxonomic exception" that *Oscarella* represents? The authors mentioned above have tried to provide one explanation that is quite logical: these sponges inhabit very shallow waters on the Mediterranean coasts, where they may be embedded in rocks or their interstices, that are continuously whipped by water waves. These conditions favor the absence of a skeleton. However, the lack of a skeleton also means having a firmer pinacoderm to resist the beating of the waves against the body of the sponge and prevent it from being destroyed. The presence of basal lamina would contribute to this, forming a proper epithelium. As we can see, two specific modifications (lack of skeleton and appearance of an epithelium) would be intimately linked to environmental characteristics. However, the latter alone would not explain why changes "should" be directed toward these morphological patterns; we are not appealing to determinism but rather to the selection of specific random mutations that, primarily, would have occurred in the ancestors of these sponges while they were trying to colonize the more coastal and agitated habitats of the Mediterranean. Thus, we see that the exceptions have their explanation, which may result from adaptive processes to the surrounding environment.

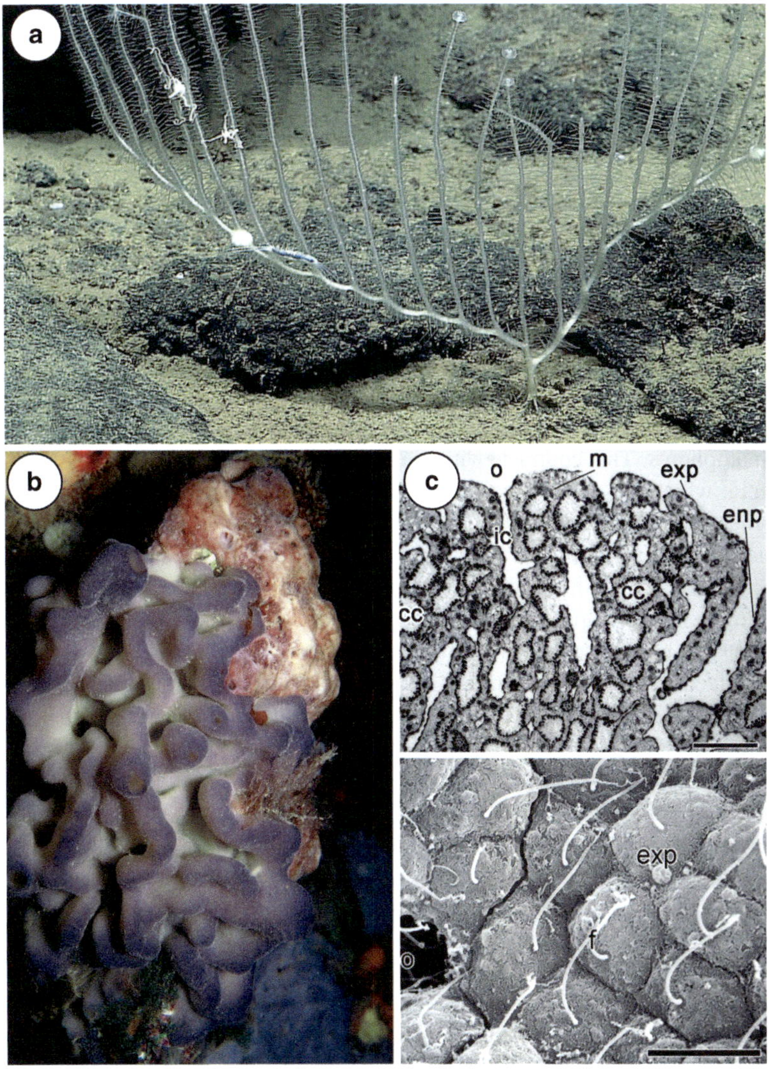

Fig. 5.4 Homoscleromorphae, a group full of exceptions among sea sponges. (**a**) Curious deep-sea carnivorous sponges (Credits: National Oceanic and Atmospheric Administration I Office of Exploration and Research, NOAA Photo Library). (**b**) Oscarellidae with its strange morphology (Credits: Géry PARENT.). (**c**) Histology of *Oscarella lobularis* showing the basal membrane (from Ereskovsky et al., 2015)

(**a**) *Chondrocladia lyra*. License: CC BY 2.0. https://creativecommons.org/licenses/by/2.0. https://www.flickr.com/photos/noaaphotolib/39695069932/. (**b**) *Oscarella lobularis*. License: Public domain. https://commons.wikimedia.org/wiki/File:Oscarella_lobularis_(Schmidt,_1862).jpg. (**c**) *Oscarella lobularis*. License: Public domain. https://doi.org/10.1371/journal.pone.0134566

Continuing with other examples of this wonderful group of invertebrates, we can contemplate another rarity. This time, it represents an extreme within a class, with so many autapomorphies that make it a well-derived group. We refer to a particular species of the so-called "carnivorous" sponges. Lee et al. (2012) described the fascinating *Chondrocladia lyra* (Cladorhizidae) (Fig. 5.4a), which has a particular body arrangement consisting of stems with spicules that become two rows of intersecting columns. Small animals such as crustaceans are entrapped within them and subsequently phagocytosed by the pinacoderm. This represents a distinct departure from the feeding mechanism commonly observed in other Porifera. In those species, nutrients are acquired through water currents via orifices and the capture of microparticles by choanocytes, specialized cells found in ducts or aquifer chambers. Without the meticulous examination of spicules and histology, it would be very challenging to categorize the sponges belonging to the Cladorhizidae family as part of the demosponges.

In these sponges, we also found globular apical zones in each column containing spermatophores. This contrasts with the male gamete-producing zones of the other Porifera, which are derived from choanocytes. Here one may ask "why" or "what for" these sponges exhibit these traits that confer an exceptional character to this species. *Chondrocladia lyra* lives at great depths (approximately 3,500 meters), with low food availability and less water flow, and in conditions of low illumination, resulting in less phytoplankton available to filter. Under these conditions, the "hunting" net-like body arrangement would prove more efficient than an in-stream system for feeding (Lee et al., 2012). Thus, this "body shape" would have developed due to the environmental characteristics of this particular environment that favor a shift in body morphological pattern from a microscopic water-filtering in-stream feeding system to one adapted for capture with a larger available food web (Lee et al., 2012). Additionally, the sperm production zone is moved to the apical zone of each capture column, optimizing the entire lateral surface of each column. That is, almost the entire body is now used exclusively as a capture net, while at the same time, the dispersal of gametes occurs from the top of the individual's body.

Again, we can, as in the case of sponges with epithelium, find plausible explanations for these exceptions on a comparative level with other members of their clade in terms of morphology, physiology, and ecology. These two examples from marine sponges illustrate how taxonomic exceptions highlight the remarkable extent to which evolution can manifest adaptive processes in surprising phenotypic patterns that are well suited to the environment.

5.4.1.4 Some "Worms" as Models of Exceptions

When we speak of exceptions, particularly at the taxonomic-relative level, animal models can be particularly useful. Examples are those known as "worms." This terminology encompasses several phyla, such as flatworms, nematodes, annelids, and xenacelomorphs, among other animals. Within them, some cases fall into the category of exceptions. In the following sections, we will take some of these cases as examples, detailing the features that make them taxonomic exceptions.

Using Foreign Weaponry: Flatworms and Nematocysts

Turbellaria, a class of flatworms of the phylum Platyhelminthes, is characterized as a basal group and is primarily free-living, generally predator and scavenger, although there are some commensal and ectoparasitic species (Ruppert et al., 2003). Within this diversity, there are many particular feeding strategies. Predators, for example, have hooks, secrete mucus, or use secretions from the epidermis to obtain food. The secretions of the epithelium and mucus help prey adhere and protect the worm from attack, as it can move easily, thanks to this lubrication. In this context, one tactic in the face of potential danger is the expulsion of "rhabdites," protein capsules, or acid mucopolysaccharides from glandular cells of the epidermis. In turn, the rhabdites protect against water stress and allow the cilia on the ventral side to move over a mucous base, facilitating locomotion.

However, although we find variants of these types of secreted products, an exceptional example is when foreign "weaponry" is used (i.e., that has been initially produced by another organism). We have as a maximum exponent of this rare species of the genus *Microstomum* that sometimes feeds on hydras, a typical freshwater cnidarian. The point is that they will use certain defense elements of these prey for self-protection. Among the species in which this phenomenon has been analyzed are *M. caudatum* and *M. lineare*, with long-standing studies on the subject (Kepner, 1911; Kepner & Nuttycombe, 1929).

To understand this, let us review that these cnidarians possess specialized structures called "nematocysts" in their tentacles as a protective mechanism (i.e., small coiled threads produced and stored inside cnidocytes). The nematocysts are discharged by the hydra to defend themselves and, at the same time, immobilize prey. They can be of different types, such as enveloping, sticky, and toxic (toxicists). This last variant is characterized by having a basal part that forms a vesicle charged with venom with a neurotoxic effect. It is precisely this type of nematocyst that *Microstomum* recycles. When ingesting the captured hydra, *Microstomum* does not digest the toxicists but makes them migrate from its digestive tract to the epidermis, where they will remain housed inside the cells of the simple epithelium (Kepner & Nuttycombe, 1929; Kepner, 1951; Krohne, 2018, 2020). In this way, *Microstomum* may use them for its defense or prey capture, similar to freshwater hydra. Interestingly, *Microstomum* tends to eat hydras until he fills that "nematocyst battery," after which it ceases to feed on them. This behavior suggests that its ingestion is primarily driven by the need to acquire defensive armaments. Once *Microstomum* has used them, it will return to feed on hydras and repeat this cycle of supplying itself with defensive nematocysts. Even after asexual reproduction, when the worm divides transversely, the nematocyst endowment will be passed on to the offspring, thus becoming an epigenetic type of inheritance in a more literal sense.

In the context of this book, there is no doubt that we are in the presence of true group exceptions. *Microstomum* represents an absolute rarity among the approximately 20,000 species of flatworms and 3500 turbellarian species. There are discussions about how this worm came to develop this strategy and its implications from an Evo-Devo point of view. What is certain is that by the 1950s, this example came to be widely used as evidence of acquired traits with a clear adaptive value. This was mainly seen as support for the twentieth-century "Neo-Lamarckian" school, with criticisms of Darwinism on unconventional forms of adaptations from a viewpoint prior to the development of epigenetics (Grassé, 1977).

On the other hand, it is interesting to know if this strategy is unique or if it also appears in some other groups. In other words, are more animals exhibiting this "armament-stealing" behavior? Gooheart & Bely (2016) note that while chemical sequestration is widespread among animals (e.g., subsequently using prey toxins to prey or for self-defense), the ability to seize entire structures is not common. While nematocyst sequestration remains rare, "Kleptocnidae" behavior (Krohne, 2018, 2020) also occurs exceptionally at the "intragroup" level in other organisms. In addition to Platyhelminthes, it appears in representatives of Ctenophora, Acoelomorpha, and Mollusca (Gooheart & Bely, 2016). Nematocyst sequestration is estimated to have evolved at least 9–17 times in these phyla. This ability is not restricted to *Microstomum* but is shared, analogously and with their respective particularities, among several species of different phyla. Thus, we see that an exception is not something that exists as an absolute but that even though it is a rarity, it can be characteristic of a pool of organisms, which within their respective clades are unique for this trait.

"Photosynthetic" Worms

Within the xenacelomorphs, we can find other exceptional cases. One of them is related to the "rule" that suggests that, apart from corals and other sedentary organisms, triblastic metazoans (i.e., which already possess mesoderm in addition to ectoderm and endoderm) have to ingest food to acquire nutrients. This is the opposite of what can occur with autotrophic organisms such as algae and plants. However, cases of symbiosis between animals and plants exist, although they are strange compared to the predominant pattern in the Animal Kingdom. In this framework, we find the worm *Symsagittifera roscoffensis*, which presents green algae symbionts (called *Zoochlorella* because it is inside an animal). This alga, *Tetraselmis convolutae*, photosynthesizes, and the worm feeds on the product of photosynthesis (Muscatine et al., 1974). This represents a clear example of photosymbiosis, in this case, between a metazoan and an alga. In effect, the alga is introduced into the early larval stages of the worm, which still feeds heterotrophically. When it reaches adulthood, it becomes totally dependent on the algae, to such an extent that its absence is lethal for the worm, so that the two organisms can be said to form a "photosymbiotic unit" (Bailly et al., 2014). It no longer presents a translucent color but is green due to the content of these green *Zoochlorella*. The worm is gregarious and even behaves by clustering toward light areas to facilitate photosynthesis. The adult of this worm is considered a true "solar-powered mature stage" due to this type of behavior (Bailly et al., 2014).

Several studies have been carried out on *S. roscoffensis*, as a model Acoela species for studies on this symbiotic relationship, at the level of respiration and photosynthetic physiology. Other more recent studies have focused on its phototaxis behavior concerning facilitating the photosynthesis of its symbiont (Nissen et al., 2015). In addition, it is an interesting organism for studying brain regeneration capacity (Sprecher et al., 2015). Definitely, this exceptional species has become a model for research on photosymbiotic relationships in marine animals due to this particular type of linkage with algae. In fact, the knowledge of *S. roscoffensis*. also allows extrapolation to other photosymbiotic animals, such as corals, which are vulnerable to climate change (e.g., water warming and acidification) (Bailly et al., 2014). Another species of the same class, *convoluta* partially prescinds from this symbiosis with an alga of the genus *Licmophora*, as it can obtain its food heterotrophically by consuming juvenile mussels, copepods, and other microorganisms (Byrnes & Witman, 2003), similar to other free-living worms. In a way, linked to these examples, we will later discuss the concept of symbiogenesis postulated by Lynn Margulis, among others, as an alternative—or at least as a complement yet to be synthesized as a whole—to the Synthetic or Neo-Darwinian Theory.

Atypical Parasites: Separate Sexes in a Hermaphroditic World

Among the invertebrate parasites, we found trematode flatworms of the digenean group, characterized by parasites that mainly have vertebrates as definitive hosts. The intermediate host, where the larval stages develop, is usually a gastropod mollusk. A typical characteristic of the digeneans is hermaphroditism. In fact, this particularity is reflected in a neophorous reproductive system, with great development and division of functions of the female part of the reproductive apparatus (an ovary and specialized vitelline glands) in such a way as to produce ectolecithal eggs or eggs that take up yolk from the vitelline glands (Adell & Riutort, 2021). In contrast, there is an Arcophore reproductive system where the gonad does not exhibit division of functions and produces endolecithal eggs (common in some Turbellaria). Thus, digeneans are very effective at producing thousands of eggs, which is crucial in the life cycle of a parasite due to the high mortality when moving from one host to another (Ruppert & Barnes, 1996). Among the digeneans, the family Schistosomatidae stands out for exhibiting particular characteristics. This family has approximately 100 species among the approximately 6000 species of the subclass Digenea (Loker & Brant, 2006). These organisms are parasites hosted in the portal-hepatic venous system of the definitive host (generally two) and have the "furcocercaria" as the vertebrate infecting larval stage, which is mobile and infests the definitive host. The best-known species of the group where sexual dimorphism is highly accentuated is *Schistosoma mansoni*, which causes schistosomiasis that affects Southeast Asia and part of the tropical and subtropical regions of the world.

What makes schizosomatids particularly unique is that they are dioecious. This synapomorphy, presenting separate sexes, is a rarity in the class Trematoda. Indeed, dioecious parasitic flatworms are rare but occur among the Digenea, Monogenea, and Eucestoda that inhabit the gills, intestines, and circulatory systems (Platt & Brooks, 1997). Interestingly, dioecy would not have occurred during specialization on homeothermic definitive hosts but would have been present when infecting ectothermic archosaurs (Brant & Loker, 2005). Today, there is a family of schizosomatids specialized in homeothermic hosts and two other families in ectotherms. From an evolutionary point of view within the class, what we call the "rule of hermaphroditism" was replaced in these groups by the "rule of dioecy" beyond the type of definitive host.

Why should these organisms abandon the successful hermaphroditic system? Platt & Brooks (1997) mention as one of the possibilities the adaptation to a blood medium, mainly venous, coupled with the need for females to oviposit in thin mesenteric veins associated with the intestinal wall. The combination of these factors could have favored the replacement of a depressed and extensive body by a filiform and elongated body. This change, however, represented a constraint to the further optimized development of a hermaphroditic reproductive apparatus. As a solution, this change was accompanied first by androdiocidity and progressively by marked sexual dimorphism. This resulted in flat, more muscular males that house the female within a "gynecophore" channel formed by the folding of the male at the time of fertilization. This relationship between the sexes allows the transport of females to oviposition sites. Females, on the other hand, specialized to be slimmer and longer to enter the subepidermal venous capillaries of the gut in which they oviposited (Platt & Brooks, 1997; Brant & Loker, 2005). It has been postulated that genetic changes, such as chromosomal inversion or translocation, mediate the appearance of dioecy (Platt & Brooks, 1997).

Interestingly, we can find different degrees of sexual dimorphism according to each host's characteristics of the arterial and mesenteric systems. As we named it, *S. mansoni* has a marked sexual dimorphism, while other species, such as *Dendritobilharzia pulverulenta*, have less sexual dimorphism, inhabiting the arterial system of ducks. In this species, the females remain in the large vessels of the heart and do not migrate to areas of thin subintestinal veins as the blood pressure is very high, ensuring the transport of eggs in the larger blood vessels. Consequently, these worms are poorly filiform, and while they remain dioecious, sexual dimorphism is slight (Loker & Brant, 2006).

An exciting concept can be illustrated with this example: "*a chain, associated concatenation, of exceptions to the rule that harbor other exceptions and so on*." In other words, schistosomatids are exceptions because they are dioecious among the Trematodes. In turn, this leads to great sexual dimorphism, which precisely characterizes "most" schistosomatids and becomes a rule pattern for the family. However, as the last example shows, there are exceptions to that family rule, according to the particularities of the punctual habitat within the circulatory system. This concept or "paradox" of a chain of exceptions and rules serves as a reminder that the classification of something as a rule or exception is relative and dependent on the context of comparison and the level of analysis. As we suggested before, rules and exceptions are not closed compartments but are intrinsic dynamic, depending on the comparative, evolutionary, spatial, and group-level perspective (or their combination) being considered. These mere examples serve to reinforce the need to always holistically and dynamically analyze the patterns and rarities at the evolutionary and phylogenetic levels.

5.4.1.5 Contributions from a Rare Crustacean

Smith's mantis shrimp, *Gonodactylus smithii*, is a Malacostracan crustacean inhabiting the Great Barrier Reef of Australia. This species represents an exception not only in its group but also among predatory organisms in general due to its "leverage and pressure" apparatus used for prey capture (Cox et al., 2014). This consists of a series of modified parts of its exoskeleton in the style of ossicles that articulate, resulting in movements of such speed and force that they produce something equivalent to the shot of a bullet on the shell of a clam, to the extent that it is said that this shot is like that of "a 22 revolver."

In connection with the aforementioned, and equally remarkable albeit extensively publicized, these organisms stand out due to their exceptional visual capabilities. They are considered one of the best and complete visions among animals (Marshall et al., 1991). They are endowed with two pedunculated eyes that can move independently and even in different directions from each other. Their vision is chromatic, so they are specialized for detecting colored signals from their conspecifics and even fish or other animals (Chiao et al., 2000). It has 16 photoreceptors, much more than the only 3 present in humans. In addition, it has infrared and ultraviolet vision and four linear and two circular polarization types (Kleinlogel & White, 2008). This allows it to better map objects to distinguish them properly. *Gonodactylus smithii*, as well as the peacock mantis shrimp, *Odontodactylus scyllarus*, represent the first cases of animals with dynamically polarized vision, where polarization information is maximized through rotational movement of the eyes (Daly et al., 2016). Their eyes are of the compound type, as in insects and some crustaceans, but are optically divided into three separate zones that produce overlapping fields of vision (Cronin et al., 2017). The ommatidia have different sizes in the central girdle relative to those that make up the rest of the eye, representing a more conventional compound eye. These central girdle ommatidia are usually larger and are believed to be used for color perception and polarization at different wavelengths as well as ultraviolet light. Finally, the presence of "pupils," up to three per eye, means that each eye zone can see the same point simultaneously from several viewing directions (Land et al., 1990). This feature could achieve stereoscopic vision from one eye (Exner, 1891).

More recently, the finding of another trait has added a reputation of true rarity to these species. These animals have neural memory and learning centers in their brain (Wolff et al., 2017), something not previously known to exist among crustaceans but in insects, identified as mushroom bodies. In fact, these appear in social insects such as bees, as they are complex centers associated with memory capacity and learning linked to having, for example, excellent spatial orientation at feeding time. In these shrimps, it could be linked to their ability to visit certain areas several times, remembering paths and places. The finding of these centers opens the debate as to whether these mushroom bodies were in the common ancestor of crustaceans and insects and were lost later in evolution in most of the former (with the exception of these shrimps) or whether it is an evolutionary convergence observed in *G. smithii*. Future studies at the transcriptomic level will then be able to see and compare the naturalness of specific proteins and thus answer this question.

Thus, as a whole, we have here a species that at least shows multiple striking characteristics that are rare in its group and even in comparison to other distant ones, making it a "multitrait" exception. The value of these exceptions as potential models for studies of shooting mechanics, hydraulics, hunting apparatus, or the neuroscience of memory and learning is invaluable (Table 5.2).

5.4.1.6 Cephalopods: The Most Intelligent Invertebrates?

Cephalopods have approximately 800 current species and are a remarkable class among invertebrates, as they possess many characteristics that undoubtedly position them as an exception to the group. First, these organisms are among the largest invertebrates, with the squid genera *Architeuthis* and *Mesonychoteuthis* being able to reach approximately 18 meters and weigh up to 1 ton (Clarke, 1966; Roeleveld & Lipinski, 1991; Ellis, 2006). These organisms inhabit the abyssal depths, and it is thought that this large size could be because they inhabit polar environments with high latitudes so that they would follow ecological rules, some of which we saw previously (e.g., temperature-size rule, Bergmann rule, oxygen availability coupled with low metabolic rates hypothesis) (Atkinson, 1994; Chapelle & Peck, 1999, 2004; Vermeij, 2016).

The eyes of these squids are the largest in the animal kingdom and have evolved to detect large predators rather than prey, although they are considered voracious predators themselves (Rosa et al., 2017), being key in marine food webs (Xavier et al., 2002, 2011; Cherel et al., 2004; Cherel & Hobson, 2007). It is believed that the rarity and extraordinary size of these creatures inspired legends from Scandinavian mythology, giving rise to the "kraken" or "leviathan" (probably from the genus *Architeuthis*), which may have originated from sightings of squid 33–45 m long (Philip, 2004). These creatures inspired Jules Verne in his work Twenty Thousand Leagues Under the Seas (Verne, 1870) and have appeared in popular films such as Pirates of the Caribbean: Dead Man's Chest (Verbinski, 2006) and several video games (Salvador & Tomotani, 2014). This highlights the human fascination with these extremes, ranging from mythological terror to literary inspiration (Ellis, 2006).

At the morphological level, cephalopods have other characteristics that distinguish them from other invertebrates: they have the foot next to the head diversified from 8 (*Octopus*) to 90 tentacles (*Nautilus*), they have appendages that are modified into reproductive structures to introduce spermatophores into the female paleal cavity, they have strong jaws that they use to tear prey, their circulatory system is closed with a network of capillaries, a systemic heart for the blood and two gill hearts that provide pressure to the circulation and the entrance of blood to the gills (Ruppert & Barnes, 1996).

However, the most exceptional feature of this group is certainly its complex nervous system that resembles that of vertebrates in relative size (Packard, 1972) and complexity (Young, 1991; Hochner et al., 2006; Shigeno et al., 2018). It consists of ganglia around the esophagus, forming a brain surrounded by a cartilaginous mass. This feature explains the existence of extremely fast behaviors, as they possess giant motor nerve fibers that control the contractions of the mantle muscles, allowing the outflow of pressurized water. In addition, they possess first- and second-order giant neurons connected to stellate ganglia and third-order giant neurons that innervate mantle muscle fibers. Due to the characteristics of their nervous system, cephalopods are considered the most "intelligent" invertebrates because they are able to open closed containers, find exits to mazes and learn specific behaviors, have long-term memory, and can rapidly change color (Darmaillacq et al., 2014; Mather & Dickel, 2017; Hanlon & Messenger, 2018). These exceptional traits have successfully allowed cephalopods to occupy multiple ecological niches by presenting complex inter and intraspecific interactions.

The intelligence of cephalopods and vertebrates with large brains appears to have evolved through different pathways at different pressures and in association with different life histories, despite sharing great sophistication (Amodio et al., 2019). Cephalopods' great flexibility, mate competition, and potential cooperative hunting with other species (Bayley & Rose, 2020) make it clear that their interactions with other individuals would not be as simple as previously thought (Boal, 1996; Hanlon & Messenger, 2018). In addition, octopuses may use coconut shells as a defensive tool and even bivalve shells and stones as shields against shark attacks (Jeffs & Brownlow, 2017), which could be related to planning abilities and the expression of behavioral innovation (Finn et al., 2009). Cephalopod research has great potential for understanding the evolution of intelligence in different animal groups (Amodio et al., 2019).

5.4.1.7 Exceptional Fungi

Challenges When Classifying Fungi

Fungi are extraordinary organisms and represent one of the cases where their multiple "outside the rules" characteristics have troubled naturalists and scientists throughout history, which has been reflected in the uncertainty in their taxonomic classification over time (Dube, 2015). This ambiguity is also reflected in the etymological root of "fungus" from the ancient Greek σφογγος (sponge), which refers to the macroscopic structures of molds and mushrooms. From this, the German terms "Schwamm" (sponge), "Schimmel" (mold), and the French "Champignon" have been derived. At first, fungi were considered closer to plants within the subkingdom Talobionta (the thallophytes, or thallus plants) (Scagel et al., 1977) and were believed to be descended from red algae (Rhodophyta) (Demoulin, 1985), perhaps because they possess plant-like cell walls composed of chitin rather than cellulose. In addition, they share other plant characteristics, such as vacuoles, reproduction by spores such as mosses or ferns, growth in the soil, and lack of locomotion.

Later, they were divided into a "separate" kingdom and grouped as "fungi" (Whittaker, 1959), although some were part of the kingdom Protozoa (amoeboid fungi and fungi with zoospores). In the Cavalier-Smith eight-kingdom scheme, they belonged partly to the kingdom Protozoa (the amoeboid fungi), partly to the kingdom Chromista (the Pseudofungi), and all others to the kingdom Fungi. Fungi produce L-lysine amino acids similar to euglenoids and bacteria (Zabriskie & Jackson, 2000). However, certain characteristics, such as the absence of chloroplasts (implying the impossibility of photosynthesis), heterotrophy (feeding by absorption), and the accumulation of glycogen as a cellular reserve substance (and not starch-like plants), make it evident that fungi are phylogenetically closer to animals (forming the clade Opisthokonta by possessing flagellated cells) (Cavalier-Smith, 1987; Stechmann & Cavalier-Smith, 2002). In other words, the rarity of these organisms as an "exquisite corpse" has made it difficult to include them in a single group or to group them with specific taxa.

Amoeboid Fungi

It is estimated that more than 5 million different species of fungi are on the planet, but less than 15% are currently known. When we think of fungi, the first thing that comes to mind is the charismatic "mushrooms" with prominent fruiting bodies, generally multicellular, with sexual reproduction. Of course, these are abundant and more showy and are called "true" mushrooms, belonging to the subkingdom Eumycota. However, there is a very peculiar group of fungi known as "nontrue," "amoeboid," "mucilaginous molds" or "pseudofungi," with approximately 1200 species belonging to the group of Myxomycota or Mycetozoa (Phylum Amoebozoa and Chromalveolata) (Hibbett et al., 2007). From an epistemological point of view, it is interesting how the "true" is more showy or conspicuous, while the exceptional is relegated to something "not true." However, we have already seen scientists' difficulty in determining what is and what is not a fungus. These organisms go through unicellular free-living phases as amoebae (phagotrophic phase without a cell wall), fusing into multicellular pseudoplasmodia within the same protoplasm (propagation phase). In favorable environments, they grow and live single-celled in isolation. Under environmental stress (e.g., lack of food, unfavorable edaphoclimate), these organisms change phase, aggregate by chemical signals, and form pseudoplasmodia that have defined anterior and posterior parts, respond to light and temperature gradients, and can migrate. Under appropriate conditions, mature pseudoplasmodia form a fruiting body (20% of cells die to form it) carrying one or more spore clusters (Raper, 1984; Kessin, 2001). Spores are released into the soil after programmed cell death of the fruiting body (Baldauf & Doolittle, 1997).

The simplicity of the life cycle of *Dictyostelium discoideum*, its cell movement, chemical signaling, and development apply to human cancer research, as well as other genetic, cellular, and biochemical processes in other organisms (Williams et al., 2006; Annesley & Fisher, 2009; Müller-Taubenberger et al., 2013; Pears & Gross, 2021) (Fig. 5.5a). This species is a model for studying cell differentiation, as different environmental factors or certain cell–cell interactions affect cell fate (Kay et al., 1978). It is a marvelous organism for studying chemotaxis, so it is used as a model in investigations of human inflammation, arthritis, asthma, lymphocyte trafficking, and axon guidance (Hashimura et al., 2019; Pal et al., 2019). The use of *D. discoideum* to study disease mechanisms and identify drug modes of action and toxicological effects, especially during cell development, is increasing (Schaf et al., 2019). It is also known that *Dictyostelium* can be infected with the bacteria *Legionella pneumophila* (Hägele et al., 2000; Solomon et al., 2000) and *Salmonella typhimurium* (Jia et al., 2009; Sillo et al., 2011), which has prompted its use as a model host in biomedical studies (Clarke, 2010; Bozzaro & Eichinger, 2011; Steinert, 2011) (Table 5.2).

It has been described that in colonies of *D. discoideum,* there is a "primitive breeding" behavior (Brock et al., 2011) of symbiotic bacteria within the fruiting bodies, an incredible example of evolutionary convergence concerning other agricultural organisms (e.g., insects). In general, this species is a solitary predator of bacteria by phagocytosis, even though these use strategies to avoid predation (Raper, 1937; Matz & Kjelleberg, 2005). However, some populations of *D. discoideum* have been found to stop feeding and incorporate the bacteria into their fruiting body. The "seeding" of bacteria at the site of spore dispersal is very advantageous in environments with few food resources (Brock et al., 2011). However, not all colonies perform this behavior because this practice carries associated costs: colonies that do not consume all the bacteria produce smaller spores or have less dispersal capacity. In addition, the benefit is much lower for spores containing bacteria that reach food-rich environments (Brock et al., 2011). Interestingly, they can also host bacteria, not for cultivation, but are defensive symbionts to inhibit the growth of nonfarmer strains of *D. discoideum* that might "prey" on other crops (Brock et al., 2013). This illustrates that microbes are capable of being prudent in the use of resources for future use, which may be important for microbial markets to evolve (Werner et al., 2014). Furthermore, successful breeding appears to be a complex evolutionary adaptation, as it requires additional strategies, such as acquiring other symbionts to defend and "privatize" crops (Brock et al., 2013).

However, within their multiple particularities, these species have proven to be unique with respect to their "sociality" (Raper, 1935). Cellular aggregation could be considered analogous to a social group or a chimeric multicellular organism (i.e., mixing of two or more genetically distinct clones) and is vulnerable to internal conflict since a "social dilemma" appears regarding cell fate. The expected "social contract" is that the frequency of each clone between the spores and the stem is the same. If this is not the case, we can say that the dominant clone cheated the minority clone by obtaining more representation in the spores, and cooperation can be put at risk when cheaters gain an advantage (Strassmann et al., 2000; Buttery et al., 2009).

Fig. 5.5 Examples of exceptional fungi. (**a**) Multicellular organism of *Dictyostelium discoideum* under the microscope (Credits: Bruno in Columbus). (**b**) The saprobe *Panellus stipticus* displaying bioluminescence (Credits: Ylem.). (**c**) Distribution of *Chorioactis geaster.* (**d**) *Chorioactis geaster* after spore dispersal. (**e**) Fruiting bodies of *Armillaria ostoyae* on a dead tree. (**f**) Rhizomorphic network of *Armillaria* spp (Credits: Henk Monster.)

(**a**) *Dictyostelium discoideum.* License: Public domain. https://commons.wikimedia.org/wiki/File:Dictyostelium_Aggregation.JPG. (**b**) *Panellus stipticus.* License: Public domain. https://commons.wikimedia.org/wiki/File:PanellusStipticusAug12_2009.jpg. (**c**) *Chorioactis geaster* distribution. (**d**) *Chorioactis geaster.* License: CC BY 3.0 https://creativecommons.org/licenses/by/3.0, via Wikimedia Commons. https://commons.wikimedia.org/wiki/File:Devil%27s_cigar_Chorioactis_geaster.jpg. (**e**) *Armillaria ostoyae.* License: Public domain. (**f**) Rizomorphen. License: CC BY 3.0 https://creativecommons.org/licenses/by/3.0, via Wikimedia Commons. https://commons.wikimedia.org/wiki/File:Rhizomorphen_or_Mycelial_chords_(rizomorfen_of_%22schoenveters),_mostly_from_the_Honey_mushroom_at_a_dead_poplar_at_Schellingwoude._-_panoramio.jpg

Because of these advantages, altruism is not expected to be evolutionarily stable (i.e., evolutionary pitfall), whereas the "cheating" strategy should become dominant (Medina et al., 2019). In this case, although cheating occurs, it is greatly reduced by various means, including high relatedness, kin recognition, pleiotropy, and other mechanisms (Gilbert et al., 2007; Ostrowski et al., 2008; Buttery et al., 2012). This system is ideal for studying the evolution of conflict versus cooperation or altruism and the molecular study of social evolution (Strassmann & Queller, 2011).

Glowing Fungi

The emission of light by living organisms is a phenomenon known as bioluminescence, produced by a chemical reaction of the enzyme luciferase with oxygen and the substrate luciferin, producing energy in the form of cold light. This phenomenon is distributed in several animal phyla, although it is rarely reported in sessile organisms. There are approximately 70 species of bioluminescent fungi (many of the genus *Mycena* and *Armillaria*), so it could be considered an exception of the fungi group, although it is a widespread phenomenon in almost 17 phyla (Desjardin et al., 2008) (Fig. 5.5b). All fungi that are bioluminescent share the same enzymatic pathways, suggesting that the biochemical pathway to bioluminescence appeared early in the evolution of mushroom-forming Agaricales (Oliveira et al., 2012). Although rare, this phenomenon in fungi has attracted the attention of humans since antiquity. More than two thousand years ago, Aristotle marveled at wood illuminated by what he called a "cold fire" (fox fire). The natives of Meghalaya in India used mushrooms as torches to not get lost in the forests. More than 100 years ago, naturalist George Gardner, watching children playing in the streets of Brazil with giant fireflies, noticed that they were mushrooms.

Fluorescing fungi emit green light with maximum intensity in the range of 520–530 nm (Chew et al., 2014) and can fluoresce in different parts, such as mycelia, fruiting bodies, or both, in rhizomorphs or spores (Herring, 1994; Weitz, 2004; Moore et al., 2021). One of the proposed functions for the existence of bioluminescence has to do with attracting pollinators. Glow-in-the-dark fungi would have evolved to attract nocturnal insects and thus ensure the dispersal of their spores (Desjardin et al., 2008). Sivinski (1981) showed that more arthropods were attracted to leaf litter with *Mycena mycelium* than to leaf litter without this fungus; this attraction was visual and did not involve chemical signals. Similarly, Oliveira et al. (2015) demonstrated this in *Neonothopanus gardneri* in the Brazilian coconut forest using plastic models of fungi with green LED-emitting light that attracted a wide variety of insects. Surprisingly, this bioluminescence followed a circadian rhythm, causing the mushrooms to "light up" at night (Oliveira et al., 2015).

Panellus stipticus is one of the brightest mushrooms in the world. Interestingly, this species is distributed worldwide, but only some strains, specifically those from North America, are bioluminescent. This species and other bioluminescent species have been tested as bioindicators of environmental contamination to heavy metals (such as Cd and Cu), as bioluminescence can be very sensitive to different chemical compounds in soil (Weitz et al., 2002; Mendes & Stevani, 2010; Stevani et al., 2013; Prasher et al., 2014; Ventura et al., 2020). Light emission appeared to decrease because these toxic compounds are related to the uncoupling of mitochondrial oxidative phosphorylation, leading to impaired ATP synthesis (Ventura et al., 2020). Although these lines of research are still under development, there is evidence that bioluminescence in these fungi can be considered an interesting variable as a bioindicator of environmental pollution (Soares et al., 2022) (Table 5.2).

An Unusual Distribution

The fungus of the species *Chorioactis geaster*, known as "devil's cigar," is considered one of the rarest on the planet (Rudy & Keller, 1996). It has a strange shape in itself, as the fruiting body resembles a dark brown cigar and grows on dead roots of elm, cedar, or oak trees. For dispersal of its spores, the fungus opens radially in a star shape (Otani, 1980), exposing the hymenium (Seaver, 1937) accompanied by the release of the spore cloud (hence its name cigar), which may be accompanied by a distinctive hissing sound audible up to a considerable distance (Rudy & Keller, 1996) (Fig. 5.5d).

However, it is not its morphology that makes this species so peculiar but its geographic distribution (Wolf & Wolf, 1947). We are faced here with a very particular spatial exception since the devil's cigar is only found in two places, separated by more than 10,000 km: in Texas (United States) (Seaver, 1937) and southern Japan (Imazeki, 1938) (Fig. 5.5c). These two localities on opposite sides of the globe are at similar latitudes with similar annual photoperiods. In 1939, mycologist Fred Jay Seaver wrote: "*It would be difficult indeed to account for [this distribution], and we merely accept the facts as they are*." Paradoxically, it is still unknown why this fungus has such an unusually disjunct distribution. Comparisons of the genomes of the Texas and Japanese populations showed divergence approximately 19 million years ago, revealing that the colonization of the new habitat was not related to *Homo sapiens* (Peterson et al., 2004). Meanwhile, due to deforestation and replacing oak with cedar trees, the devil's cigar is considered an endangered species in Japan.

The Oregon Monster

Few would imagine that when asked about the largest organism on the planet, the answer would be a fungus (Smith et al., 1992). However, with an extension of 9 km^2, weighing 650 tons (Stanosz et al., 1987) and glowing at night, the fungus known as "the Oregon monster" (*Armillaria gallica*) occupies a large portion of the forests of this state. This fungus has even become a tourist attraction in Michigan. This organism spreads its filaments or rhizomorphs (root-like multicellular structures of clonal dispersal) through the root structure of coniferous forests, bridging its host plants, as it is, in fact, a parasite (Baumgartner et al., 2011) (Fig. 5.5e, f). The expansion and stability of these environments have allowed the colonization and growth of this fungus between 2400 and 8650 years old. Rhizomorphs develop subways by aggregation and coordinated parallel growth of hyphae occupying the surrounding soil and colonizing new food resources (Heinzelmann et al., 2019). Rhizomorphs grow 1 meter per year and perform long-distance nutrient uptake and translocation functions (Morrison, 1975; Baumgartner et al., 2011). The form of dispersal and pathogenicity is unique to this species, involving ancestral genetic toolkits for wood decay, morphogenesis, and complex multicellularity (Sipos et al., 2017).

Some of the extraordinary qualities of this fungal species have led scientists to ask how these long-lived organisms can keep deleterious mutations to a minimum during somatic growth. Very low genetic variations (≅3 orders of magnitude lower than most filamentous fungi) and a very low mutation rate have been reported, making the genome of this fungus highly stable (Aanen, 2014; Anderson & Catona, 2014; Anderson et al., 2018). The number of cell divisions is thought to be reduced, and growth is similar to that of a plant shoot, with cell divisions occurring just below the apex. This would reduce the number of new mutations in the apical zone, preventing the accumulation of mutations in the population (Anderson & Catona, 2014). Most astonishingly, the "Oregon monster" characteristics could be useful in fighting cancer (Han et al., 2009; Chen et al., 2016). In addition to the extremely low mutation rate, *Armillaria* species could exhibit inherent resistance to DNA damage. Studying the genetic mechanisms underlying this stability in the genome could provide potential new ways to treat cancer (Table 5.2).

5.4.1.8 Carnivore Plants: Exceptions in the Group Show Us the Carnivore Rule

When we consider plants, we think of autotrophic and photosynthetic organisms, which transform inorganic matter from the environment into organic molecules and do not require the intervention of other organisms to obtain their energy (a rule I for the plant kingdom) (Curtis et al., 2008). This definition is a great plant rule, but as we saw above, some plants, such as parasitic plants, do not comply with this premise. Another exceptional example is carnivorous plants that acquire extra nutrients from other organisms (heterotrophy), such as insects and even small vertebrates (Adamec, 1997). The leaves of these plants have been modified to attract and trap their prey, after which they absorb nutrients from the prey by producing enzymes or digestive bacteria (Ellison, 2006; Barthlott et al., 2007) (Fig. 5.6a). These plants are thought to have evolved this ability because they inhabit certain nutrient-poor soils, especially nitrogen, such as acidic swampy soils and rocky bluffs (Feldman, 2009). Some, such as pitcher plants (Nepenthaceae) and many *Drosera*, expect passive prey, although they have intricate slippery edges and sticky-tipped hairs to secure insects (Darwin, 1875; Juniper et al., 1989). Others are more active, such as certain droseras that coil up, pushing prey into the sticky center of the trap, and others have an outer ring of fast-moving tentacles, the most sophisticated being the Venus flytrap (*Dionaea muscipula*), of which we will discuss later.

Murderous Plants Strange as Fiction

These plants have attracted the attention of botanists and ecologists for more than 200 years who have been trying to obtain answers about how these unique organisms evolve and how their anatomy works. Charles Darwin wrote the first known treatise on these plants in 1875 after spending more than 15 years studying them with fascination. He wrote to Charles Lyell in the late 1860s about a species of *Drosera*: "*...at this present moment I care more about Drosera than the origin of all the species in the world… Is it not curious that a plant should be far more sensitive to a touch than any nerve in the human body!'*. In his autobiography, he wrote: "*The fact that a plant should secrete, when properly excited, a fluid containing an acid and ferment, closely analogous to the digestive fluid of an animal, was certainly a remarkable discovery*." For Darwin, carnivorous plants were one of the most critical points of evolutionary theory (Fig. 5.6c).

However, just as these "plant monsters" inspired Darwin's fascination, they were long considered by the popular imagination to be "killer" plants. These man-eating plants appeared regularly in Anglo-American fictional accounts in the late nineteenth century, perpetuating pseudoscientific stories of explorers encountering horrors such as the "man-eating tree" of Madagascar or the "vampiric vine" of Namibia (Prior, 1939) (Fig. 5.6b). Sir Arthur Conan Doyle fanned this fiction by publishing "The American's Tale" (1880), a story in which an enormous Venus flytrap killed a man in the forests of Arizona. Additionally, H.G. Wells wrote about a "blood-sucking orchid" laying the groundwork for the man-eating plants appearing in future musical, film, and television adaptations in the twentieth century (Price, 2013). Interestingly, ideas and fables about killer plants began to emerge after Darwin popularized them with his treatise on these wonderful (and harmless) plants.

A Molecular Recipe Behind the Evolution of Carnivory

The carnivorous habit has evolved in at least 12 lineages independently (more than a dozen genera in 5 families and 810 species, so it is considered an example of morphological convergence) (Albert et al., 1992; Givnish, 2015; Fleischmann et al., 2018) (Fig. 5.6a). In each of these times, the evolutionary force that led to carnivory seems to have been the need to find alternative sources of vital nutrients such as nitrogen and phosphorus that they could not extract from the soil. The amazing thing is that this exceptional adaptation follows the same "recipe" or "rule" in all carnivorous plants! (Hedrich & Fukushima, 2021). The leaves of these plants have acquired new enzymatic functions: chitinases (digest chitin, the main component of the insect exoskeleton), proteases (breakdown proteins), and acid phosphatases (release phosphate groups from molecules and help mobilize phosphorus from prey). The production and control of these enzymes are genetically regulated and are acquired by a genetic phenomenon of co-option, where certain genes that these plants already had to adapt to these new functions, somewhat like a genetic "recycling" (Pavlovič & Saganová, 2015; Bemm et al., 2016; Pavlovič &

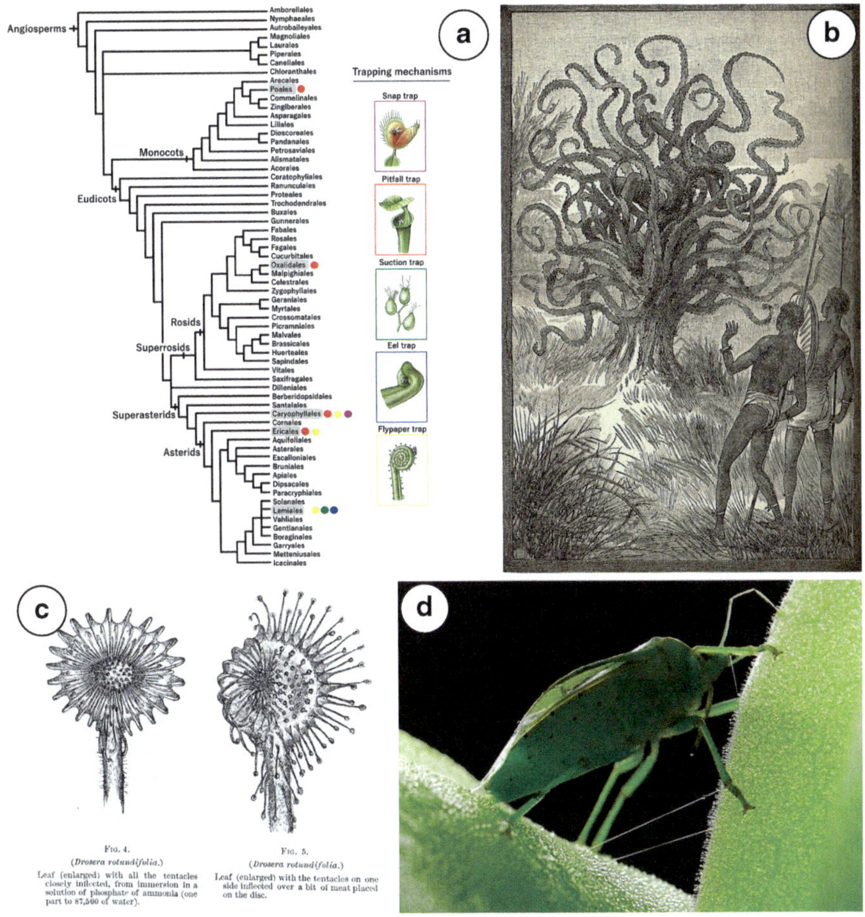

Fig. 5.6 Rare in the plant world: carnivorous plants. (**a**) Phylogenetic positions and trap types of carnivorous plants. Carnivorous plants evolved multiple times across the angiosperm lineage (modified from Hedrich & Fukushima, 2021). (**b**) Sacrificed to a man-eating plant (from Buel, 1887). (**c**) Selected image of the insectivorous plant *Drosera rotundifolia* described by Darwin (from Darwin, 1875). (**d**) A plant bug on a *Pinguicula gigantea* (Credits: Noah Elhardt).

(**a**) Phylogenetic positions and trap types of carnivorous plants. (**b**) Sacrificed to a man-eating plant. License: CC BY 4.0. https://creativecommons.org/licenses/by/4.0. https://commons.wikimedia.org/wiki/File:The_ya-te-veo.jpg (**c**) Selected image of the insectivorous plant *Drosera rotundifolia*. License: Public domain. https://commons.wikimedia.org/wiki/File:C._Darwin,_Insectivorous_Plants_Wellcome_L0031429.jpg. (**d**) Selected image of the insectivorous plant *Drosera rotundifolia*. License: Public domain. https://commons.wikimedia.org/wiki/File:Pinguicula_gigantea_ne.jpg

Mithöfer, 2019). These genes were initially involved in immune defense or stress response (jasmonate defense response), so they could be considered a case of exaptation (Bemm et al., 2016; Fukushima et al., 2017). The genes for these enzymes were reused when plants started to eat the organisms from which they were originally protected. Certain stimuli given by prey consumption increase the level of jasmonates, which transcriptionally activate the genes encoding digestive enzymes (Bemm et al., 2016; Yilamujiang et al., 2016; Pavlovič et al., 2017; Jakšová et al., 2020).

Digestion by prey enzymes is the first part of the process, as the plant must then transport these nutrients into the plant interior. In a typical plant, nutrients are transported through root transporter proteins. However, it has been recently discovered that the Venus flytrap possesses these transporter proteins in its modified leaves (Scherzer et al., 2013). Root genes were also co-opted to work at new sites, although these transporters are only activated upon uptake in the traps, whereas in the roots, they are always active. Co-optation appears to have originated from mutations related to fixed gene duplications because they were adaptive (Ibarra-Laclette et al., 2013; Palfalvi et al., 2020).

Vegetarian Carnivorous Plants

Gene co-option appears to be shared by many species; however, Kocáb et al. (2020) investigated the carnivorous butterwort *Pinguicula* sp., a small, herbaceous, rosette-like plant that uses sticky leaves with sticky mucilage to attack insects (Fig. 5.6d). The glands of this species would be like "sacs" that would undergo a type of autophagy and discharge enzymes onto prey, unlike the more precise mechanism of jasmonate-mediated digestive enzyme expression and secretion in carnivorous plants of the order Caryophyllales (Heslop-Harrison & Knox, 1971; Heslop-Harrison & Heslop-Harrison, 1980). Kocáb et al. (2020) found that enzyme production in *Pinguicula* does not appear to be regulated by the jasmonate pathway. Instead, they found alpha-amylase in its digestive fluid, which is unique among carnivorous plants.

Amylase is an enzyme that catalyzes the hydrolysis of starch, a polysaccharide produced by green plants as an energy store. The rosette shape of *Pinguicula* would serve to trap quite a bit of plant material (Darwin, 1875), and amylases can help digest this alternative carbon source (mixotrophic organism). This "vegetarianism" has been described for some *Utricularia* species (Peroutka et al., 2008; Koller-Peroutka et al., 2015), although until now, this compound has not been found in digestive fluids. This discovery could mean that other genes unrelated to defense have been coopted on the way to carnivory, although the mechanism triggering enzyme secretion remains a mystery.

5.4.2 *Extraordinary Phenomena*

5.4.2.1 Exceptions from the World of Deception

One fascinating phenomenon that often surprises us is the remarkable similarities observed among distantly related groups, be it in terms of morphology, behavior, or physiology. Evolutionary convergences have long captivated biologists, even before the times of Darwin himself. It was through his work that we started developing and organizing our understanding of shared ancestry, distinguishing between homologies and analogies, and exploring the occurrence of multiple convergences within a clade. In this context, some species deviate entirely from the typical pattern or rule of a taxon to "resemble" other organisms or environmental characteristics (camouflage), to go unnoticed, or to obtain some kind of advantage to ensure their survival. This remarkable capability is referred to as mimicry and encompasses various forms, such as aposematism, Batesian mimicry, and Müllerian mimicry. It is exemplified by iconic species such as the chameleon, stick insects, caterpillars that mimic branches or imitate the appearance of small snakes, butterflies that seamlessly blend with tree trunks, and seahorses whose bodies mimic the texture of algae. In Table 5.3, several examples are listed, with a summary of their main phenotypic characteristics, the percentage of mimetic species concerning their clade, and references. These examples have at least a couple of aspects in common: (1) they mimic something of the phenotype of a taxonomic group distantly related to them and (2) they are numerical rarities compared to the rest of the large clade that contains them. It is intriguing to integrate the investigation of mimicry cases, encompassing the description of mimicry itself, its associated behavioral ecology, and phylogeny, to gain a comprehensive understanding of its adaptive significance. Furthermore, with the progress of developmental biology, it becomes increasingly valuable to explore the ontogeny of mimicry in individuals displaying this trait.

Beetles That Mimic Ants

Army ants are ecologically dominant predatory insects in the tropics, with large nomadic colonies harboring a great diversity of parasitic myrmecophiles (i.e., organisms living in association with ants) (Rettenmeyer et al., 2011). An interesting case of mimicry is that of the army ant-parasitic rove beetles (Coleoptera). These organisms, known as ant-mimics, closely resemble their ant counterparts in terms of body shape (e.g., petiolate abdomen, geniculate antennae, and highly elongated legs) and behavior. In fact, they mimic the ants so well that they are readily accepted by the aggressive ants, joining their colonies and temporarily inhabiting their nests. These ant-mimics not only feed alongside the ant workers but also consume the workers' offspring (Akre & Rettenmeyer, 1966; Parker, 2016) (Fig. 5.7a, b). What is curious is that this phenomenon is not limited to a single species of staphylinid. Instead, multiple unrelated groups of ant-mimics within the subfamily Aleocharinae have independently developed this distinctive pattern at least 12 times (Maruyama

Table 5.3 Selected examples of mimetic species and models for mimicry, indicating the type of mimicry, the mimetic sex and trait, and the signal nature and receiver

Mimetic species (signaler)	Vulgar name	Model species	Vulgar name	Mimetic sex	Type of mimetism		Mimetic trait	Signal reciever	Signal nature
Automimicry, Browerian mimicry, Bakerian mimicry, or intraspecific mimicry: the same species of animal is imitated; (a) weakly defended members of a species with warning coloration are parasitic on more strongly defended members of their species, mimicking them to provide the negative reinforcement learning needed for warning signals to function; (b) a less vulnerable part of an animal's body resembles a more vulnerable part, providing an immediate selective advantage; (c) decieve prey so that it perceives that the attack is originating from where it is not (confounding effect); (d) intersexual mimicry occurs when individuals of one sex in a species mimic members of the opposite sex to facilitate sneak mating									
Chaetodon capistratus	Foureye butterflyfish	*Chaetodon capistratus*	Foureye butterflyfish	Both	Automimicry (b)	Deceptive mimicry	Body colors (ocelli)	Predators	Visual
Charina bottae, Hydrophis pachycercos	Rubber boa	*Charina bottae, Hydrophis pachycercos*	Rubber boa	Both	Automimicry (b, c)	Deceptive mimicry	Shape of tail-like head	Predators, preys	Visual
Danaus plexippus	Monarch butterfly	*Danaus plexippus*	Monarch butterfly	Both	Automimicry (a)	Deceptive mimicry	Warning coloration	Predators	Visual
Paracerceis sculpta	Marine isopod	*Paracerceis sculpta* (female and juveniles)	Marine isopod	Males (beta, gamma)	Automimicry (d)	Deceptive mimicry	Color and shape	Males	Visual
Subfamily Theclinae	Hairstreak butterflies	Subfamily Theclinae	Hairstreak butterflies	Both	Automimicry (b)	Deceptive mimicry	Shape of wing tail as antenna	Predators	Visual
Uta stansburiana	Common side-blotched lizards	*Uta stansburiana* (female)	Common side-blotched lizards	Males	Automimicry (d)	Deceptive mimicry	Color, behavior	Males	Visual, tactile

Aggressive mimicry: copying by a predator or parasitic species of the appearance of another harmless species, in order to disguise its nature and have easy access to its prey or host. The model suffers costs from being mimicked, due to the erosion of signal reliability and costs of being parasitized. Therefore, host phenotypes evolve toward increased discriminability from their parasites. *Pouyannian mimicry: flowers mimic a potential female mate visually, but the key stimuli are often chemical and tactile

Acherontia atropos	Death's-head hawkmoth	Honey bees colonies		Both	Aggressive mimicry	Deceptive mimicry	Epicuticular lipid pattern of the host	Prey	Chemical
Anomalospiza imberbis	Cuckoo-finch	*Prinia subflava*	Tawny-flanked prinia	Female	Aggressive mimicry	Deceptive mimicry	Egg color	Host	Visual
Carrion flowers, corpse flowers		Rotting flesh		Both	Aggressive mimicry (Pouyannian)	Deceptive mimicry	Color, shape, odor (chemicals)	Pollinator (necrophagous insects)	Visual, chemical, tactile
Chlorobalius leucoviridis	Spotted predatory katydid (males)	Tribe Cicadettini	Cicadas	Both	Aggressive mimicry (sexual)	Deceptive mimicry	Song, behavior	Prey (males)	Visual, Acoustic
Cuculus canorus	Common cuckoo	Several Passeriformes species		Female	Aggressive mimicry	Deceptive mimicry	Egg color	Host	Visual
Drakaea spp.	Hammer orchid	Female wasp		Both	Aggressive mimicry (Pouyannian)	Deceptive mimicry	visual and olfactory cues.	Pollinator (male wasps)	Visual, chemical
Epipactis helleborine	Broad-leaved helleborine orchid	Cabbage leaves infested with caterpillars		Both	Aggressive mimicry (Pouyannian)	Deceptive mimicry	Odor (chemicals: green-leaf volatiles)	Pollinator (social wasps)	Chemical
Hedychrum rutilans	Cuckoo wasps	*Philanthus triangulum*	European beewolf	Female	Aggressive mimicry	Deceptive mimicry	Composition of cuticular compound	Prey	Chemical

(continued)

Table 5.3 (continued)

Mimetic species (signaler)	Vulgar name	Model species	Vulgar name	Mimetic sex	Type of mimetism		Mimetic trait	Signal reciever	Signal nature
Hymenopus coronatus	Malaysian orchid mantis	Orchid flowers		Both	Aggressive mimicry	Deceptive mimicry	Color	Prey	Visual
Lanius excubitor	great grey shrike	Several small Passeriformes species		Both	Aggressive mimicry	Deceptive mimicry	Song	Prey	Acoustic
Macrochelys temminckii	alligator snapping turtle	Worms		Both	Aggressive mimicry	Deceptive mimicry	Shape, movement	Prey	Visual
Mastophora hutchinsoni	Bolas spider	Moth species		Both	Aggressive mimicry	Deceptive mimicry	Sex pheromone of prey	Prey (males)	Chemical
Misumena vatia	Crab spider	Flowers		Both	Aggressive mimicry	Deceptive mimicry	Color	Prey	Visual
Phengaris arion	Large blue	*Myrmica sabuleti*	Red ant	Both (larvae)	Aggressive mimicry	Deceptive mimicry	Chemical compounds and sound	Host, Prey	Chemical, Acoustic
Photuris spp.	Fireflies	*Photuris* spp.	Fireflies	Female	Aggressive mimicry	Deceptive mimicry	Visual light displays	Prey (males)	Visual
Plagiotremus rhinorhynchos	Bluestriped fangblenny	*Labroides dimidiatus*	Bluestreak cleaner wrasse (juvenile)	Both	Aggressive mimicry	Deceptive mimicry	Body colors, behavior	Prey	Visual
Stenolemus bituberus	Assassin bug	Prey of web-building spiders		Both	Aggressive mimicry	Deceptive mimicry	Vibrations on spider web	Prey (spider)	Vibrations
Stenoria analis	Ivy bee blister beetle	*Colletes hederae*	Ivy bee	Both (larvae)	Aggressive mimicry	Deceptive mimicry	Sex pheromone of host	Host (male)	Chemical

Synemosyna formica	Ant-mimicking jumping spider	Ants		Both	Aggressive mimicry	Deceptive mimicry	Shape, color and behavior	Prey	Visual
Volucella bombylans	Hoverfly bee	*Bombus lapidarius*	Red-tailed bumblebee	Both	Aggressive mimicry	Deceptive mimicry	Shape, color and behavior	Prey (larvae)	Visual
Batesian mimicry (defensive mimicry): an animal copies the appearance or behavior of a more dangerous or repugnant animal to discourage attacks by predators. Models are predicted to benefit from evolving away from Batesian mimics because mimicry erodes signal reliability, which could reduce the extent to which predators avoid the model									
Chionactis occipitalis	Northern shovel-nosed snake	*Micrurus fulvius*	Coral snake	Both	Batesian mimicry	Deceptive mimicry	Body colors	Predators	Visual
Cuculus spp.	Cuckoo species	*Accipiter nisus*	Eurasian sparrowhawks	Both	Batesian mimicry	Deceptive mimicry	Shape, color, behavior	Predators	Visual
Glaucidium californicum	Northern pygmy owl	Eyes of a predator		Both	Batesian mimicry	Deceptive mimicry	Body colors (ocelli)	Predator	Visual
Lampropeltis elapsoides	Scarlet king snake	*Micrurus fulvius*	Eastern coral snake	Both	Batesian mimicry	Deceptive mimicry	Body colors	Predators	Visual
Laniocera hypopyrra	Cinereous mourner	*Megalopygidae* caterpillars		Both (nestlings)	Batesian mimicry	Deceptive mimicry	Shape and color	Predators	Visual
Lycodon aulicus	Indian Wolf snake	*Bungarus caeruleus*	Common Krait Snake	Both	Batesian mimicry	Deceptive mimicry	Body colors	Predators	Visual
Onopordum sp., *Arctium tomentosum*		Spider Web		Both	Batesian mimicry	Deceptive mimicry	Trichomes in stems and leave	Herbivores	Visual, tactile
Mycalesis patnia, Caligo idomeneus, Haetera piera, Hipparchia semele	Butterflies with ocelli	Eyes of a predator		Both	Batesian mimicry	Deceptive mimicry	Body colors (ocelli)	Predator	Visual

(continued)

Table 5.3 (continued)

Mimetic species (signaler)	Vulgar name	Model species	Vulgar name	Mimetic sex	Type of mimetism		Mimetic trait	Signal reciever	Signal nature
Papilio polytes	Common mormon	*Pachliopta aristolochiae, Atrophaneura polyphontes*	Common rose butterfly, red-bodied swallowtails	Both	Batesian mimicry	Deceptive mimicry	Shape and color	Predators	Visual
Thaumoctopus mimicus	Mimic octopus	Various marine animals		Both	Batesian mimicry	Deceptive mimicry	Shape, color and behavior	Predators	Visual
Batesian food-source mimicry: in floral mimicry, rewardless flowers attract their pollinators through dishonest signaling of a resource. Rewardless flowers mimic the morphology, color, and/or scent of rewarding flowers									
Arisaema sikokianum	Japanese Jack-in-the-pulpit	Mushrooms		Both	Batesian food-source mimicry	Honest mimicry	Chemical compounds	Pollinator (mushroom-associated flies)	Chemical
Dracula spp.	Drac orchid	Mushrooms		Both	Batesian food-source mimicry	Deceptive mimicry	Chemical compounds	Pollinator (mushroom-associated flies)	Chemical
Epidendrum ibaguense	Epiphytic orchid	*Asclepias curassavica*	Tropical milkweed	Both	Batesian food-source mimicry	Deceptive mimicry	Flower shape and color	Pollinator (*Agraulis vanillae*)	Visual

Müllerian mimicry (defensive mimicry): this is a collaborative mimicry where species with the same defense mechanism (bad taste, poison, etc.) and the same predator develop a similar appearance to ensure greater survival. The predator will eat a single individual of either species and in the future will learn to avoid them both

Apheloria spp.	Millipedes	*Brachoria cedra*	Millipedes	Both	Müllerian mimicry	Honest mimicry	Body colors	Predator	Visual
Danaus plexippus	Monarch butterfly	*Limenitis archippus*	Viceroy butterfly	Both	Müllerian mimicry	Honest mimicry	Body colors	Predator	Visual
Euploea core	Common crow butterfly	Other toxic butterflies		Both	Müllerian mimicry	Honest mimicry	Body colors	Predator	Visual
Heliconius erato	Red passion flower butterfly	*Heliconius melpomene*	Postman butterfly	Both	Müllerian mimicry	Honest mimicry	Body colors	Predator	Visual
Morpho peleides	Blue morpho butterfly	Other blue butterflies		Both	Müllerian mimicry	Honest mimicry	Body colors	Predator	Visual
Pitohui dichrous	Hooded pitohui	*Pitohui uropygialis*	Southern variable pitohui	Both	Müllerian mimicry	Honest mimicry	Body colors	Predator	Visual
Ranitomeya imitator	Amazonian mimic poison frog	Dendrobatidae	Poison dart frogs	Both	Müllerian mimicry	Honest mimicry	Body colors	Predator	Visual

Fig. 5.7 The extraordinary world of mimicry. (**a**) Aleocharinae free-living beetle *Athetavaga* sp. (from Pentinsaari et al., 2019). (**b**) Army ant social parasites of Aleocharinae with myrmecoid morphology (*Pseudomimeciton* sp.). (**c**) Sea snake (*Laticauda* colubrina) mimicked by octopus. (Credits: Christian Gloor). (**d**) Octopus (*Thaumoctopus mimicus*) mimicking a sea snake. (Credits: Elias Levy.). (**e–f**) Mimetic fungus *Leratiomyces erythrocephalus* mimicking truffles of New Zealand (Credits: Bernard Spragg; F. Tony Wills).

(**a**) *Athetavaga* sp. License: CC BY 4.0 https://creativecommons.org/licenses/by/4.0, via Wikimedia Commons. (**b**) *Pseudomimeciton* sp. License: Public domain. https://commons.wikimedia.org/w/index.php?search=Pseudomimeciton+&title=Special:MediaSearch&go=Go&type=image. (**c**) *Laticauda colubrina*. License: CC BY 2.0. https://creativecommons.org/licenses/by/2.0, via Wikimedia Commons. https://commons.wikimedia.org/wiki/File:Laticauda_colubrina_(27567428624).jpg. (**d**) *Thaumoctopus mimicus*. License: CC BY 2.0. https://creativecommons.org/licenses/by/2.0, via Wikimedia https://commons.wikimedia.org/wiki/File:Mimic_Octopus_(14280646758).jpg. (**e**) *Leratiomyces erythrocephalus*. License: Public domain, via Wikimedia Commons. https://commons.wikimedia.org/wiki/File:Scarlet_Pouch_Leratiomyces_erythrocephalus_(52149653411).jpg. (**f**) *Leratiomyces erythrocephalus*. License: CC BY 2.5 https://creativecommons.org/licenses/by/2.5, via Wikimedia Commons. https://commons.wikimedia.org/wiki/File:Red_pouch_fungus_01.jpg

& Parker, 2017). Each independent clade is restricted to one zoogeographic region and is highly specific to a single ant genus. This could be an intermediate-type exception involving several species. However, considering the number of mimetic beetles throughout the order, it is clear that we are still dealing with a strategy that remains rare at the taxonomic level. However, mimicry generally involves a highly complex system of morpho-behavioral modifications. In the case mentioned, these changes occurred in diverse groups and at different points in time. Natural selection would have played a great role since it can hardly be argued that it resulted from a mere evolutionary convergence phenomenon.

Cephalopods as Masters of Camouflage

Octopuses, in particular, have an exceptional defensive system and are considered "masters of camouflage." Camouflaging is one of the most iconic examples of their behavioral flexibility. Their nervous system allows them to mimic their environment and to make extremely rapid changes in color and skin texture (Messenger, 2001) (Fig. 5.7c, d). They present pigment cells in the mantle called chromatophores that possess pigments that expand or condense using muscle contractions controlled by the nervous system (Messenger, 2001; Hanlon & Messenger, 2018). In addition, cephalopods can modify their strategies according to substrate and predator type (Langridge et al., 2007; Josef et al., 2012; Staudinger et al., 2013). For example, cuttlefishes may change coloration when facing visual predators but opt for flight strategies when dealing with predators that use chemical signals (Langridge et al., 2007).

Mushrooms That Look Like Berries

Fungal spores are primarily dispersed through anemochory, where the wind carries lightweight spores over significant distances. However, alternative methods of dispersal exist, including consumption by mammals, as seen in the case of truffles. Truffle-like fungi have rare and even unique forms of dispersal within the fungal world. The species *Leratiomyces erythrocephalus* grows in Australia and New Zealand, possessing a vivid red fruiting body with a "pouch" shape on a delicate stalk (Fig. 5.7e, f). These mushrooms mimic red berries, and in these geographic areas, they are believed to visually attract birds rather than attract mammals by smell (Medway, 2000). Because they closely resemble berries, these fungi are dispersed by birds over long distances and benefit from their nitrogenous resources in the digestive system.

Defensive Masquerade in Plants

As we have seen above, plant variegation could be considered teratology or chromatic aberration. Variegation involves a photosynthetic loss for the plant and has negative consequences. However, it has been found that in *Caladium steudneriifolium,* it could be a defensive strategy or a form of "masquerading" (i.e., organisms assimilate to inanimate or inedible objects in their environment) (Skelhorn, 2015). Leaf-mining moths frequently attack these plants, and some individuals have been found to develop variegated patterns similar to the damage produced by these insects to avoid further moth visits (Soltau et al., 2009). These authors painted nonvariegated leaves by copying the variegated patterns and found that within three months, these leaves were significantly less frequently damaged by moths than healthy leaves (Soltau et al., 2009). Moths were found to infest almost 10% of nonvariegated leaves, whereas they only mined approximately 2% of variegated leaves (Soltau et al., 2009). These herbivores may avoid leaves with visible markings, implying that damage-mimicry may reduce oviposition, consistent with the mimicry hypothesis (Smith, 1986; Lev-Yadun et al., 2002). Interestingly, in the populations where this variegation was found, there are also nonvariegated individuals, so it would seem that both morphs could be successful and maintained in the population (Smith, 1986; Soltau et al., 2009).

5.4.2.2 Strange Forms of Pollination

In angiosperms, pollination is the process by which pollen is transferred from the stamens to the stigma of the flowers, where it germinates and fertilizes the flower's ovules with the consequent production of seeds, fruits, and a new generation of plants. As we will see in detail in Chap. 6, this ecological process is of utmost importance in the evolution of flowering plants and typically involves a vector responsible for transport that can be biotic or abiotic (Kay & Sargent, 2009; Willmer, 2011). Pollen transfer can take place via wind (anemophilous) or water (hydrophilous), but in the vast majority of plants, it takes place by pollinating animals (zoophilous). In general, animals (e.g., butterflies, bees, hummingbirds, bats, mammals) seek to feed on the nectar and pollen of flowers, and in doing so, they become impregnated with the pollen that they carry to other flowers.

Darwin greatly contributed to this fascinating topic by studying orchids, their varied forms, and the relationship of these patterns with the interaction of pollinators (Darwin, 1862, 1877; Micheneau et al., 2009). Orchids present unique flowers that attract suitable pollinators for each species, being frequent examples of plant–insect coevolution. An iconic example discussed above was the moth predicted by Darwin when studying the Madagascar orchid *Angraecum sesquipedale*, which had a spur of more than 30 cm long with nectar only in its basal part. This unique morphology led Darwin to conclude that there must be a nocturnal moth with a long proboscis capable of releasing this nectar. It was not until 1903, more than 40 years later, that the moth with a 30 cm long proboscis was discovered: *Xanthopan morganii praedicta*, whose subspecific epithet alludes to Darwin's prediction of its existence. The case of the predicted moth is an example of the morphospaces imagined but not found (at that time). It is interesting how exceptional morphologies could be predicted from other coevolution examples.

"(…) I must say a few words on the Angraecum sesquipedale *(…). A whip-like green nectary of astonishing length hangs down beneath the labellum. (…) We shall, I think, see that the fertilization of the plant depends on this length and on nectar being contained only within the lower and attenuated extremity. It is, however, surprising that any insect should be able to reach the nectar: (…) in Madagascar there must be moths with probosces capable of extension to a length of between ten and eleven inches*"—Charles Darwin, 1862

Pollinators Not as Ridiculous as Previously Assumed

Pollination by snails and slugs is known as malacophily and is considered one of the rarest forms of pollination (Abrol, 2012) since these animals feed on soft vegetation and floral organs and are more likely to inhibit pollination than to promote it. Many researchers consider malacophily to be "notorious and obscure" (Faegri & van der Pijl, 1979) or even "ridiculous" (Atwood Jr., 1982). However, although infrequent, some cases of malacophily are known (Pammel & King, 1930; Sarma et al., 2007; Tandon & Ram, 2010; Abrol, 2012; Raju et al., 2014; Gardère, 2018). Plants generally grow in clusters, and flowers possess nonprotruding stigmas and anthers, which allows snails or slugs to crawl over the flowers, take up pollen, and deposit it on adjacent stigmas (Pammel & King, 1930). Strictly malacophilous plants have not been documented, although *Evolvulus nummularius* plants, for example, use snails (*Lamellaxis gracile*) as supplementary pollinators to bees (Sarma et al., 2007). Even on rainy days, snails are exclusive pollinators (Sarma et al., 2007). More recently, Raju et al. (2014) reported that *Euplecta subdecussata* pollinated *Ipomoea pes-caprae* during the flowering season, moving between flowers without damaging the style or stigma, being an essential pollinator and affecting self and cross-pollination. In this species, bees and wasps are pollinators, but snails are equally crucial as these insects (Raju et al., 2014).

Underwater Pollination

Unlike in terrestrial environments, animals do not participate in pollination under the sea (Vermeij & Grosberg, 2010). As a result, hydrophily is common to these plants and is considered a particular and near-exclusive adaptation of plants to life in the marine habitat. These plants usually have small, inconspicuous male flowers that release large numbers of pollen grains that drift into the water, where the large, feathery stigmas of the female flowers catch them. Hydrophily occurs in only 14 of the 14,000 genera of angiosperms, and 10 of these genera occur in marine environments (Ackerman, 2006). In the large group of flowering plants, hydrophilia could be considered a rarity, although it would be the rule for aquatic plants. However, a group of aquatic angiosperms presents a different form of pollination, being, in turn, a rarity among those of their group. As we saw in the case of schizomids, this could be a chain of exceptions and rules within a group.

Flowering plants in the marine environment are known as seagrasses and are generally hydrophilic (Cox, 1988; Ackerman, 2006). Flowers are ephemeral, female flowers have tentacle-like stigmas, and male flowers produce copious filiform or spherical pollen released in mucilage strands (Cox, 1988; Ackerman, 2006). However, the species known as turtle grass (*Thalassia testudinum*) can be considered an exception because, in addition to being hydrophilic, complementarily, its flowers are pollinated by marine invertebrates (zoobenthophily) (van Tussenbroek et al., 2012, 2016). These invertebrates, during the night, visit the flowers and transport and transfer the pollen-encrusted mucilage masses from male flowers to the stigmas of female flowers, and in turn, this pollen would be a food reward for these pollinators (van Tussenbroek et al., 2016). Like insect pollination, pollination by marine invertebrate fauna may have played a role in the transition from terrestrial to underwater abiotic pollination by water flow (van Tussenbroek et al., 2016).

These findings in seagrasses have prompted research on pollination in other marine organisms. Recently, the red algal species *Gracilaria gracilis* has been observed in symbiosis or mutualism with the marine isopod species *Idotea balthica* (Lavaut et al., 2022). The result of this symbiosis is isopod-mediated pollination of the algae, as biotic interactions dramatically increase the success of algal fertilization through the transport of spermatozoa in the animal's body. This is extremely interesting because the female gametes are not released, and the male gametes are not flagellated. This algal species would be ambophilous and could use both water and the isopod as pollinators (Lavaut et al., 2022). The isopod also engages in a double mutualism with the algae since it not only transports its pollen but also removes it by foraging epiphytic algae growing on the red algae (Ollerton & Ren, 2022). It appears that the symbiosis between the alga and the isopod was established approximately 300 million years ago. However, the alga is much older, and it is thought that perhaps pollination at that time was only water-mediated or perhaps by some other organism that we do not know about (Ollerton & Ren, 2022). This finding suggests that animal-mediated fertilization may have evolved independently in terrestrial and marine environments and raises the possibility of its appearance in the sea before plants moved onto land.

Intermediary Pollinators of "Armed" Plants

It is believed that plants, due to their immobility and lack of sensory perception, are not able to physically compete for reproductive opportunities (male–male competition) (Willson, 1990; Andersson, 1994; Arnold, 1994; Skogsmyr & Lankinen, 2002; Delph & Ashman, 2006). However, direct physical competition between individuals has been described in milkweeds (Oxypetalinae) during pollination (Cocucci et al., 2014) (Fig. 5.8a). In these species, large amounts of pollen aggregate in two pollinia connected by sterile accessory structures, forming a complex pollinarium (Harder & Johnson, 2008; Wiemer et al., 2012) (Fig. 5.8b). At the time pollinating insects visit the flowers, a process of "concatenation" occurs, where a pollinarium can join

Fig. 5.8 A strange case of pollination. (**a**) *Morrenia odorata* (Familia Apocynaceae). (**b–d**) Pollinaria and pollinators of *Morrenia odorata* (Credits: Cody H.)

Credits: (**a**) *Morrenia odorata*. License: CC BY 2.0 https://creativecommons.org/licenses/by/2.0, via Wikimedia Commons. https://commons.wikimedia.org/wiki/File:Morrenia_odorata_(Hook._%26_Arn.)_Lindl.jpg. (**b–d**) Pollinaria and pollinators of *Morrenia odorata* License: Author permission

others already present on the pollinator (Coombs et al., 2012) (Fig. 5.8c, d). Interestingly, pollinaria have evolved "horn-like" structures to physically compete with other pollinaria to gain a "privileged" place on the pollinator's body, as it would be advantageous to prevent new pollinaria from chaining together.

These horns would be botanical equivalents of the horns in many male vertebrates, and the reproductive performance of a pollinarium depends on its position in a chain, which in turn will depend on the characteristics of the horns of those pollinaria (Cocucci et al., 2014). In the phylogeny of Oxypetalinae, horns were significantly associated with loss of concatenation, whereas the loss of horns caused a return to concatenation (Cocucci et al., 2014). If horns are experimentally removed in *Oxypetalum solanoides*, pollinaria are reattached, whereby horns are evolutionarily acquired or lost to avoid or promote concatenations (Cocucci et al., 2014).

These studies suggest that self-propulsion and well-developed sensory perception are not necessary for sexual selection through intrasexual competition to take place (Cocucci et al., 2014). Only physical contact would be sufficient to influence competitors' reproductive success and promote the evolution of defensive and attack weaponry. Due to a scientific tradition, some research questions are more studied or considered more important in some taxa than others (Lankinen & Karlsson Green, 2015). With this in mind, we wonder if these "exceptional" cases are truly exceptional or if they represent a bias in studies of sexual selection or search for trait explanations in plants. What is clear is that fields such as botany could benefit greatly by incorporating and testing the impact of selective pressures generated by sexual selection and sexual conflict (Lankinen & Karlsson Green, 2015). Furthermore, the fact that there are intermediary "mediators" or "agents" such as pollinators that have coevolved with plants provides a new perspective to enrich studies of sexual selection in plants.

5.4.2.3 Reproductive Biology: Source of Valuable "Exceptions" at the Evolutionary Level

Reproductive biology is a field that exhibits numerous remarkable exceptions across morphology, behavior, and physiology. As we will see below, we discover that many of the so-called "exceptions" have become more like "cryptic" rules, perhaps not visible to the inexpert eye. This evolution stems partly from the profound influence of sexual selection on various reproductive traits and the continuous progress and evolving perspectives within the field (the example of how the real and widespread existence of monogamy in birds was relativized made it clear that these compartments of "rules" and "exceptions" in reproductive biology can be relatively weak). However, here, we will highlight some considerations regarding evolutionarily maintained minorities or extraordinary exceptions at the morphological and behavioral level, although we will also emphasize some phenomena that may be more common than we think.

Minorities That Remain: Alternative and Conditional Sexual Strategies

A context in which minorities can have an important evolutionary value is when we refer to tactics or strategies adopted by a part of the population. A strategy is the decision-making rules involved in certain reproductive behaviors. Individuals allocate resources to one or another (mutually exclusive) strategy to achieve the same functional end (Brockmann, 2001). These strategies can be alternative, conditional, or mixed (Gross, 1996; Taborsky et al., 2008). In the case of alternative reproductive tactics (ARTs), there is genetic polymorphism, with the same fitness for morphs given by frequency-dependent selection (i.e., a tactic is more successful when it is rarer but less successful when it is more common). Under strong selective pressures, for example, small males at a disadvantage in physical fights with other males or display against females adopt ARTs (Oliveira et al., 2008). This selective pressure sets the stage for the disruptive selection of morphologies (i.e., morphs) (Gadgil, 1972) or behaviors. There are numerous cases of these reproductive strategies. In these cases, we can consider that alternative tactics will not always be represented in equal frequency of occurrence in a given species or population of the same species. In some cases, the occurrence of "morphs" does not have a behavioral correlation, so it is necessary to be cautious with speculations about alternative strategies.

On the other hand, conditional strategies involve individuals' decisions based on the context or state of the individuals who optimize their fitness. The organisms are genetically monomorphic, and the fitnesses of the strategies are not the same (but the fitnesses of the alternatives at the switch point are equal). A well-known example is that of dung beetles of the genus *Onthophagus* that can exhibit a fighting or sneaking phenotype as adults depending on the body size that is determined by larval feeding (Emlen, 1994; Moczek & Emlen, 2000; Hunt & Simmons, 2001) (Fig. 5.9a). Adult body size distribution is discontinuous, reflecting different developmental trajectories (Eberhard & Gutiérrez, 1991). Switch points can be sensitive to ecological and demographic events (Eadie & Fryxell, 1992; Reynolds et al., 1993; Carroll & Corneli, 1995). For example, in addition to body size, population density affects the success of alternative tactics (fighter and nonfighter male phenotypes) (Radwan, 1993).

Individuals of the alternative male morph are usually called "sneakers," and in many cases, they do not invest in "armament" but rather in other adaptations to maximize fertilization (e.g., gonadal investment, sperm number, etc.) (Taborsky, 1998; Simmons et al., 1999). Because copulations of females with sneaker males and with males of the main morph (e.g., territorial, fighting, older) are common, it is frequent that ejaculates from both males compete for egg fertilization (Parker, 1990; Simmons, 2001). The example of male sneakers is very illustrative of ignoring or neglecting a less noticed phenomenon, such as sperm competition, either because of its range in conspicuousness to the bias of our sensory universe or its frequency of occurrence. This changed with the advancement of the conceptual framework of sexual conflict and postcopulatory sexual selection, under which tactics of copulation and sperm competition were more readily identified and interpreted in an evolutionary framework (Gross, 1982, 1985). Nor is it appropriate to take lower frequency tactics or strategies as the main pattern of a species, perhaps because they are so conspicuous. For example, it would be a mistake to overestimate a minority alternative or conditional strategy as the main pattern (Peretti & Córdoba-Aguilar, 2007).

In numerous instances, the origins of these strategies, whether they are alternative, conditional, minor, or even exceedingly rare, remain elusive. They may represent minority strategies today, but no one knows if they may represent the norm in the future. Many of them may represent differences at the genetic level or phenotypic plasticity of the trait we are analyzing, which could well form the tipping point for future evolutionary strategies of these types. There is much to be discovered in

Fig. 5.9 Uncommon and incredible examples in the world of animal reproduction. (**a**) Examples of alternative male morphs in *Onthophagus taurus* (top) and *O. nigriventris* (bottom). Large males are shown on the left and small males on the right (from Kijimoto et al., 2013). (**b**) Homosexuality in kudu males. (Credits: Brian Ralphs.) (**c**) A male Kobudai (*Semicossyphus reticulatus*) in Japan. In this species the females have the ability to turn into males, an example of a transgender fish. (Credits: Cinz) (**d**) Egg-carrying male of *Ibalonius* sp. with magnification of egg clutches (from Sharma et al., 2017)

Credits: (**a**) License: Used with permission granted by the publisher on 04 January 2023. (**b**) License: CC BY 2.0. https://creativecommons.org/licenses/by/2.0, via Wikimedia Commons. https://commons.wikimedia.org/wiki/File:BVR_8158_(33817940031).jpg. (**c**) License: CC BY 2.0. https://creativecommons.org/licenses/by/2.0, via Wikimedia Commons. https://commons.wikimedia.org/wiki/File:Semicossyphus_reticulatus.jpg. (**d**) License: used with permission granted by the publisher on January 4, 2023.

the world of reproductive strategies, but it will be challenging if we do not reinforce the focus on rare extreme cases or outliers as the basis for them. In other words, if we always wait for them to appear to a minimal but remarkable degree to be considered in our studies, we are missing the beginning of everything. The beginning of life may have been marked by something like this, rarities that later became regular, as we have already mentioned and will see in the following chapters in detail.

Animal Homosexuality

Homosexuality can be regarded as an exceptional occurrence when compared to the overall number of heterosexual individuals within a species with a dioecious reproductive system (Thornhill & Alcock, 1983; Maklakov & Bonduriansky, 2009). However, there are an increasing number of species where cases of homosexuality are reported, so we wondered whether this phenomenon could become a rule (Bailey & Zuk, 2009; Poiani, 2010) (Fig. 5.9b). In other words, it is not the pattern, the norm (although here, this word runs the risk of falling into the anthropological in terms of normal versus abnormal from the "moral" point of view, which we will not debate in this book). Strict homosexuality is limited to humans, domesticated sheep, and other unconfirmed species (Munévar, 2014). However, more than 1,500 species of animals present "atypical" sexual practices such as mating with individuals of the same sex between males (e.g., lions, giraffes, bison, dolphins, and elephants), females (e.g., seagulls, koalas, lizards, birds, beetles), bisexual behaviors (e.g., bonobos, bats), and long-term homosexual relationships (e.g., penguins, albatrosses) (Munévar, 2014).

Cases of homosexual behaviors have also been reported in animals taxonomically far from vertebrates, such as insects (e.g., fruit flies, weevils, seed and flour beetles, damselflies, field crickets) (McRobert & Tompkins, 1988; Castro et al., 1994; Harari et al., 2000; Van Gossum et al., 2005b; Bailey & French, 2012; Scharf & Martin, 2013; Olivero & Robillard, 2017) and arachnids (e.g., spiders, pseudoscorpions) (Scharf & Martin, 2013; R. Palen Pietri, pers. comm.) in more or less detail. Sometimes up to 50% of male mating attempts are directed at other males (Aiken, 1981; Serrano et al., 1991, 2000). At present, it is more often the case that "strange" reports of homosexuality can still be found in these groups, but these are increasingly embedded in an evolutionary context (Scharf & Martin, 2013). For example, YouTube videos have reported that a male of the cricket *Xenogryllus marmoratus* has mounted another male (Olivero & Robillard, 2017). In this case, the male appears to be pretending to be a female in order to consume a nutritious glandular secretion from the other male when mounting him (generally in this species the female mounts the male). In many insect species, the explanation for these behaviors has to do with recognition failures, lack of experience, or benefits of increased overall mating frequency to males (Thornhill & Alcock, 1983; Serrano et al., 1991, 2000; Harari et al., 2000; Switzer et al., 2004; Scharf & Martin, 2013).

The phenomenon of homosexuality has been cited or studied tentatively since the eighteenth century by George Edwards, who sought the causes of the "abnormal" behaviors he saw in birds and attributed them to captivity (Sommer & Vasey, 2006). Darwin himself made some observations of behaviors between individuals of the same sex, although he was evasive in his explanations of them, describing them as superficial and transitory (Brooks, 2021). For example, when faced with birds courting same-sex individuals after the loss of a mate, he said: "*How is it that so many birds are ready immediately to replace a lost mate? Magpies, jays, carrion-crows, partridges (...) offer at first sight the most perplexing case. But birds of the same sex, although of course not truly paired, sometimes live in pairs or in small parties (...) With partridges two females have been known to live with one male, and two males with one female. In all such cases, it is probable that the union would be easily broken*" (Darwin, 1871). In the years that followed, numerous observations of homosexual behavior or intersexuality were reported as curiosities by various naturalists. These observations in that historical context led to strong academic disputes, such as the one that followed the publication of "*Perversion sexuelle chez des Coléoptères mâles*" by the French entomologist Henri Gadeau de Kerville in 1896. After several observations, Gadeau asserted the existence of a "homosexual instinct," which led to heated debate and detailed studies of the behavior of these insects (Brooks, 2009, 2021).

Classical interpretations attribute these behaviors to failures in recognizing the other sex, although gradually, the study of animal homosexuality has gone from being a simple "curiosity" to becoming more relevant because of its socioecological implications. In this context, with the publication of the controversial but challenging book "Evolution's Rainbow" by Joan Roughgarden in 2004, this topic was discussed concerning sexual diversity in nature, including men, from a more "social" perspective within biology. Roughgarden illustrates multiple and fascinating examples within the animal kingdom and from cultures beyond occidental ones, showing a "range" of options in terms of sexuality, which the author analyzes as a natural variation. In this framework, it is necessary to look not only subjectively but also objectively at homosexuality more broadly and more integrated with the social aspects of the species. Evaluating homosexuality with a precise approach, free of prejudices, is difficult, and the attempts made thus far are inconsistent in their results.

In recent decades, evidence has begun to accumulate that homosexuality is a behavior that arises primarily from biological or genetic factors (although there is a role for environmental factors in the expression of a homosexual phenotype) (Pillard & Bailey, 1998; Swaab, 2008; Bao & Swaab, 2010; Ganna et al., 2019). If facultative homosexual organisms reproduce at a lower rate than heterosexual organisms, how can the maintenance of homosexuality be explained? This question is what makes the issue of homosexuality an "evolutionary puzzle" or known as "Darwin's paradox" since it supposedly contradicts the theory of evolution by implying sexual behavior without reproductive purposes, or at least it is adaptive value not immediately apparent (Gavrilets & Rice, 2006; Bailey & Zuk, 2009; Poiani, 2010). There are multiple hypotheses to explain the evolution and maintenance of homosexuality, and many are not mutually exclusive (Sommer & Vasey, 2006; Bailey & Zuk, 2009; Dukas, 2010; Poiani, 2010). Among these are kin selection (i.e., the reproductive success of relatives who might pass on the genes for homosexuality is encouraged), genetic overdominance (i.e., a heterozygous version of a gene provides an organism with some reproductive advantage), maternal effects (i.e., the fetus is influenced by the environment of the mother's womb, leading to changes that predispose to homosexuality), and antagonistic sexual selection (i.e., genes are spread through a population giving a reproductive advantage to one sex and disadvantaging another, e.g., the mother may increase her fertility at the expense of her child's ability to procreate). There is also evidence in birds and mammals that homosexuality can be a form of “practice” for juveniles and a way to maintain alliances within social groups (Sommer & Vasey, 2006; Bailey & Zuk, 2009). Of course, the causes and consequences of homosexuality can vary greatly among species and contexts, although genetic studies in model organisms have helped to understand common patterns in the evolution of this phenomenon (Bailey & Zuk, 2009). As we previously stated, particularly in humans, we must be very cautious with the search for explanations for homosexuality, and here, it is argued that sexual orientation is the product of a relationship between human biology and the sociocultural environment and that many times the cultural meaning of sexual orientation surpasses biological explanations (Roughgarden, 2004; Alonso, 2014, Ganna et al., 2019; but see Hamer et al., 2021; Ganna et al., 2019).

On Female Andromorph Dragonflies

Color polymorphism (i.e., the presence of two or more genetically determined color morphs within a population) is widespread in diverse taxa, such as fish, mammals, birds, and insects (Roulin, 2004). When polymorphisms are observed in a single sex (intrasexual), their evolution and maintenance are explained by sexual conflict (Schluter, 2001; Roulin, 2004; Svensson et al., 2009) or by differences in immunocompetence (i.e., different morphs adapt to environments with different exposures to parasites, leading to differential resistance to parasites) (Roulin, 2004; McKinnon & Pierotti, 2010; Sánchez-Guillén et al., 2013). As mentioned above, these polymorphisms are maintained by frequency-dependent negative selection (Bots et al., 2015). Among insects, odonates in particular are characterized by color polymorphism, especially in females. (Fincke et al., 2005). Females usually show a “gynochrome” morph (i.e., female coloration) and an “androchrome” morph (i.e., male coloration). Females mimic males not only in coloration but also in some behaviors (e.g., territoriality, aggressiveness) (Sherratt 2001; Robertson, 1985; Hinnekint, 1987; Cordero, 1992; Cordero & Pérez, 1998; Andrés et al., 2002; Sirot et al., 2003) and even in other morphological traits (Abbott & Gosden, 2009). In this way, these females remain relatively unnoticed, at least initially, avoiding sexual “harassment” by males and minimizing the cost of unwanted mating and the risk of physical injury. (Svensson et al., 2005; Takahashi & Watanabe, 2010; Takahashi et al., 2010; Iserbyt et al., 2013). However, these females may suffer increased predation and parasitism (Robertson, 1985; Svensson et al., 2005), as well as decreased fecundity due to morphological changes (e.g., narrowing of the abdomen) (Gosden & Svensson, 2009; Sánchez-Guillén et al., 2017). As an evolutionary response to this confusion of the multiple morphs, males may develop a reversible preference for mating and harassing the most common female type in the population (Van Gossum et al., 2001).

The damselfly genus *Ischnura* has species with polymorphic females and is an evolutionary and ecological model system for studying the selective forces that maintain this polychromatism (Wellenreuther et al., 2014). This genus has females with one androchrome morph and one or more gynochrome morphs controlled by three alleles with complete and hierarchical dominance (Sánchez-Guillén et al., 2005). In this genus, Sánchez-Guillén et al. (2017) wondered whether these morphs might signal different alternative reproductive tactics. These authors found that behavior differed significantly between morphs but not within morphs, while in males, the tactics were affected by the abundance of female morphs in the population to avoid lost mating opportunities (Van Gossum et al., 2005a; Sánchez-Guillén et al., 2013, 2017).

Fluid Sex in Fish

Among the possible reproductive strategies, we find animals with "fluid sex," i.e., that change sex under certain conditions (at the embryonic stage or as adults) (Avise & Mank, 2009). Among fishes, 2% of species (approximately 500 species) exhibit this phenomenon to some degree (Devlin & Nagahama, 2002; De Mitcheson & Liu, 2008; Avise & Mank, 2009). For example, the kobudai fish, *Semicossyphus reticulatus*, transforms from female to male (i.e., protogynous sequential hermaphrodite) by increasing in size and bulbous forehead, developing testes and an aggressive character (Ochi et al., 2017) (Fig. 5.9c). Others are more extreme, such as the angelfish *Centropyge potteri*, in which all fish begin life as females and all males were once females (Bauer Jr & Bauer, 1981). Although less common, some fishes may change from males to females (i.e., sequential protandric hermaphrodites), such as the famous clownfish *Amphiprion clarkii* (Moyer & Nakazono, 1978). In these species, a pair of reproductive females and males live together with subordinate nonreproductive fishes inside sea anemones. When the dominant female dies, the larger male becomes a female. These sex changes of both a protogynous and a protandrous system seem to occur because of the "size advantage hypothesis," as it would be advantageous for some fishes to be of different sexes once they have reached a specific size (Ghiselin, 1969; Warner, 1975). Other species, such as gobies (*Gobiodon* and *Paragobiodon*), can adopt both sexes depending on the circumstances (i.e., bidirectional hermaphrodites) (Nakashima et al., 1996; Munday et al., 1998; Cole & Hoese, 2001). These species live in crevices within coral reefs, are poorly mobile and have little opportunity to find mates. It is therefore an advantage that two fish that meet by chance can form a pair regardless of their sex (Nakashima et al., 1996; Munday et al., 1998).

These sex-switching strategies in fish testify to a remarkable plasticity in this group of vertebrates, which is probably due to the great diversity of species adapted to a variety of environments. These strategies would be highly adaptive, especially in changing environments, as they would in many cases allow fish to double their reproductive output (as both males and females) (Leonard, 2006). "Fluid sex" has occurred several times in different fish families over the course of evolution, and although a number of genetic and hormonal triggers come into play in all of them, the one responsible for sex change would be the enzyme aromatase (Piferrer et al., 1994). It is important to remember that temperature can determine sex in fish (Conover & Kynard, 1981). This enzyme in particular is temperature sensitive and can be activated or deactivated depending on the temperature (e.g., male development is more frequent in warm waters) (Devlin & Nagahama, 2002; Uchida et al., 2004; Ospina-Alvarez & Piferrer, 2008). In addition, certain chemical pollutants, water acidity, and parasites, such as *Wolbachia*, affect aromatase in developing fish embryos (Jobling et al., 1998; Kidd et al., 2007; Brown et al., 2015), which could have serious consequences in a climate change scenario (Devlin & Nagahama, 2002; Ospina-Alvarez & Piferrer, 2008).

Speculating Behavior from Morphological Observations

When observing a "rarity," it is important to be cautious in assessing its significance and evolutionary or phylogenetic implications. This is particularly true when attempting to predict specific behavioral patterns based on morphological observations, especially if those predicted behaviors are infrequent compared to the more common behaviors within those groups. For example, if we never saw a butterfly fly, could we claim that wings are for flight? Intuitively, we appeal to the comparison with other animals that possess this trait, and therefore, from these "models," we deduce that since wings serve for flight, they most likely have the same function in butterflies. The problem occurs when there are no models against which to compare or when what exists is phylogenetically very distant, which makes extrapolation between groups riskier. These extrapolations have occurred numerous times in the history of ethology, and associations between traits and behaviors have been made without direct observation. This has led to premature "paradigms" with implications for phylogenetic interpretations in certain taxonomic groups. Remember the categorization of Porifera, cnidarians, or fungi as plants in the past because many of them are sessile.

An example that illustrates this topic comes from a very interesting group of arachnids of the family Podoctidae, order Opiliones known as harvestmen. A characteristic feature of this family is the males' peculiar paternal care of the eggs (Martens, 1993; Kury & Machado, 2003). In the vast majority of harvestmen, males take care of the eggs deposited by females on substrates such as plant leaves, under logs, or stones (Machado & Raimundo, 2001; Machado & Macías-Ordóñez, 2007). However, the difference and particularity presented by this family are that males carry eggs attached (individually or grouped) to their hind legs. This was reported long ago (Martens, 1993) and echoed by other authors as a unique case among harvestmen (Tallamy, 2001; Kury & Machado, 2003; Requena et al., 2014). In an evolutionary context, this showed an interesting convergence with other arthropods, such as insects, such as water bugs, where males carry the eggs laid by the inseminated female on their dorsum (Munguía-Steyer et al., 2019; Ohba, 2019), or other chelicerates, such as pycnogonids (Lee & Kim, 2020).

We had a case of a rarity from the perspective of the order Opiliones and arachnids in general and even in an arthropod context. In that "relative" perspective, this may be considered a rarity or exception (i.e., the rule would be that when there is egg care, this occurs on an "extrabodily" substrate). However, in these harvestmen, there are no observations or studies describing female egg-laying behavior in this male body zone or behavior where the male picks up eggs from the substrate. There is no "direct" evidence that the eggs are from harvestmen. Unexpectedly and as part of a molecular phylogenetic study of this family of harvestmen, a DNA analysis of these eggs indicated that they did not belong to harvestmen but to spiders (Sharma et al., 2017) (Fig. 5.9d). The publication of this study sparked a substantial debate between the authors and proponents of the paternal care hypothesis, such as Machado & Wolff (2018), regarding the presence of eggs in this particular male body zone. Machado and Wolf base their arguments on knowledge of the morphology, behavior, and specific characteristics of both harvestmen and spiders. Against this, Sharma et al. (2018) performed new and more detailed DNA analyses and reconfirmed that the eggs belonged to spiders and not harvestmen. This debate seems to be still ongoing and highlights the relative importance of different sources and methodologies to sustain specific hypotheses and postulates. We must be cautious in inferring a particular behavior from a trait that would "indicate" or "suggest" it.

5.4.3 *Evolutionary and Ecological Exceptions*

5.4.3.1 Evolutionary Simplification

O'Malley et al. (2016) reviewed the situation of increasing complexity in living organisms, evaluating the role of simplification (e.g., morphological, physiological, etc.) in macroevolution. These authors emphasize that simplification should be considered together with complexification in the generation of macroevolutionary patterns. The evolutionary importance of simplification is evident in several eukaryotic taxa, including protists adapted to parasitism, in addition to algae, plants (e.g., by loss of the flagellum), and metazoan animals (O'Malley et al., 2016). We could draw an association between some examples of simplification and exceptional cases in biology since this simplification process can generate true rarities.

How To Be a Good Parasite?

In groups with parasitism, a simplification or organic reduction is common, leading to the loss of genes, characteristics, and/or functions essential for survival outside the host. An example of simplification could be the loss of photosynthetic capacity that led to the emergence of particular groups of eukaryotes. Particularly notable are some parasites derived from green algae, which do not photosynthesize and parasitize insects (Blouin & Lane, 2012; Pombert et al., 2014). One of the best-known cases is the apicomplexan protist Plasmodium, which is vectored by mosquitoes and hosted by humans, causing Malaria disease. This apicomplexan has plastids (apicoplast) derived from red algae (McFadden et al., 1996; Kohler et al., 1997). In particular, the evolution of the parasite lifestyle is associated with a genomic reduction with a loss of functions no longer necessary in their novel environment. Surprisingly, some parasites, such as *Helicosporidium* derived from green algae, undergo some nuclear genome reduction, but most ancestral metabolic functions are retained, including some photosynthesis-related functions, such as those of the Calvin cycle. While some gene families lost complexity, others, such as those associated with chitinases, expanded in response to parasitism. Helicosporidium represents an example of a transition from an autotrophic to a heterotrophic parasitic organism where certain functions independent of the new lifestyle are preserved (Pombert et al., 2014).

Another example linked to parasitism occurs in insects within the superfamily of dipteran Hippoboscoidea, which contains tsetse flies (Glossinidae), louse flies (Hippoboscidae), and two families of bat flies (Streblidae and Nycteribiidae). Except for tsetse flies that are free-living and contact their host during feeding, all hypobioscids are highly specialized obligate ectoparasitic flies that spend all or most of their adult lives on their hosts. Several species, particularly within the bat parasites of the family Nycteribiidae, are notable for their simplified morphologies, which are extreme specializations of their life form. These flies are similar to spiders, with completely reduced wings, dorsoventrally flattened thorax, and dorsally inserted legs. Because of these simplifications and "extreme" modifications, homology determination has been extremely difficult, and phylogenetic relationships based on morphology have been almost impossible to elucidate (Petersen et al., 2007).

Taken together, all these examples highlight how the path of simplification to become a "good parasite" would be a common evolutionary process, as a "rule," in groups that are phylogenetically very distant from each other. However, we again encounter the "paradox" of a chain of exceptions and rules. With the exception of having reduced morphologies, we find the rule, which would be a "common" way in parasites. However, returning to the last example, if morphological simplification is beneficial for parasitism (Andersen, 1997), why are not all species of Hippoboscoidea apterous? This is because extreme simplification may entail certain costs, which in this case would have to do with the dispersal of the organisms. As we said, Nycteribiidae species parasitize bats that, in general, have gregarious habits, and this is why more than 50% of the species of this family could have faced this extreme simplification since dispersal could be achieved in the same way due to the proximity between individuals (Petersen et al., 2007).

On De-evolution and When the Simple Deceives Us

Another interesting example occurs in geckos, where an innovation such as the adhesive system suffers a secondary loss or simplification. This may sound counterintuitive since many structural innovations have enabled many animal groups' diversification and ecological specialization, so why lose it? (Hale, 1996; Garber & Rehg, 1999; Zippel et al., 1999). Sometimes, however, the simplification or loss of these innovations can be even more advantageous (Wiens & Slingluff, 2001; Bejder & Hall, 2002; Skinner et al., 2008). In geckos, the adhesive system evolved independently at least 11 times, allowing these organisms to occupy inaccessible regions due to increased climbing effectiveness (Autumn & Peattie, 2002). However, the adhesive system appears to have been simplified or lost at least nine times (Gamble et al., 2012). This could be due to certain constraints imposed by these adhesive structures to run, making these organisms much slower (Russell & Higham, 2009). Higham et al. (2014) found that the rate of morphological evolution was higher for gecko species that had lost or simplified their adhesive systems, so these species appear to have overcome these constraints. The African gecko lineages of Phyllodactylidae suffered morphological simplifications in their adhesive systems associated with a shift from a rupicolous lifestyle to burrowing in loose sand (Lamb & Bauer, 2006; Gamble et al., 2012).

Interestingly, sometimes these "simplifications" resulted in derived morphologies different from the ancestral states without the character, such as *Chondrodactylus angulifer*, which lost the adhesive system but retained structures such as tendons, muscles, and associated skeleton (Lamb & Bauer, 2006; Johnson & Russell, 2009). The "at first sight" morphological simplifications turned out to be reductions of functionally integrated modules rather than dissolutions of them (Von Dassow & Munro, 1999). This interesting pattern of secondary loss leading to derived morphologies that are not necessarily "simpler" can be framed in what is known as Dollo's law (Law of Phylogenetic Irreversibility or the Law of Irreversible Evolution) (Dollo, 1893; Gould, 1970). This law states that "*an organism never returns exactly to a former state, even if it finds itself placed in conditions of existence identical to those in which it has previously lived... it always keeps some trace of the intermediate stages through which it has passed*" (Gould, 1970). A well-known example is the loss in many animals, including humans, of gulonolactone oxidase, the final enzyme in the vitamin C biosynthetic pathway, necessitating its exogenous supplementation (Chaudhuri & Chatterjee, 1969; Nishikimi et al., 1994; Cui et al., 2011). During vertebrate evolution, many species lost gulonolactone oxidase enzyme-encoding genes, never reverting to ancestral gene expression states (Yang, 2013). Phylogenetic reconstructions generally follow parsimony methods, and relationships can be reconstructed by "Dollo parsimony," i.e., making it impossible to revert to ancestral states after losses (Quesne, 1974; Farris, 1977; but see Fitch, 1971).

However, as we have stressed repeatedly in this book, there are no absolutes, and all rules have their exceptions. In the case of this law, there are examples where structures appear to have effectively "reverted" to "simpler" character states. An iconic case occurs in certain gastropods that have lost their spiral shell to be replaced by limpet-like morphology (Collin & Cipriani, 2003). Genera of the family Calyptraeidae (*Trochita, Sigapatella*, and *Zegalerus*) appear to have recovered a coiled shell from limpet morphologies (Pagel, 2004). Although it could be argued that these gastropods have evolved a different type of coiling from their coiled ancestor, Collin & Cipriani (2003) hypothesize that the genes responsible for coiling were silenced and that the genetic architecture was maintained by pleiotropy, as has been proposed in other examples (West-Eberhard, 2003; Whiting et al., 2003;

Wiens et al., 2007; Lynch & Wagner, 2010; Lynch, 2022). Many of these species have larval stages with spiral shells (Collin, 2003), so the expression of genes involved in spiralization may have been maintained and "reactivated" in adult stages as a form of heterochrony. This juvenilization would indicate that the "spiral shell" state that could be considered "more complex" is simpler since it involves the retention of juvenile stages (heterochrony) (Collin & Cipriani, 2003).

5.4.3.2 Exceptions at First Sight

"Nonanimal" Eyes

Initially, photoreceptor structures were considered exclusively animal characteristics, although we will see some exceptions. Eye structures in the form of spots are considered the simplest form of photoreception. These are known as "eye spots" and are formed by specific photoreceptor proteins and a signal transduction mechanism that results in a photovoltaic-type response (Hegemann, 1997). Eye spots are present in certain flagellated microorganisms and algae. We will return to the idea of senses in plants in Chap. 6. One of the best-known examples of vision outside animals is *Euglena*. This flagellate belonging to the phylum Euglenozoa is famous for having both heterotrophic (via osmotrophy) and autotrophic (photosynthesis) feeding facultatively, depending on the light conditions of the environment. The more light there is, the more autotrophic feeding tends to occur.

Euglena has a red-colored stigma or "eyespot" in its cytoplasm and is close to the flagellum, which consists of a cell organelle made up of a concentration of carotenoid pigment granules (flavoproteins). The function of the *Euglena* eyespot is to filter sunlight falling on a light-detecting structure at the base of the flagellum (the paraflagellar body), allowing only certain wavelengths of light to reach it. In other words, the eye spot is not a photoreceptor but a filter or selector of light waves. As the cell rotates with respect to the light source, the eye spot partially blocks the source, allowing *Euglena* to find the light and move toward it. In this case, there is positive phototaxis, but in other flagellates, it can be negative. Eye spots are also found in some algae. For example, in *Chlamydomonas*, the eye spot is part of the chloroplast, and the pigment is formed by rhodopsins (Suzuki et al., 2003). Kreimer (2009) refers to the presence of eyespots in algae as constituting a "primordial visual system."

Cave Dwellers: Blind Fish

The sensory world and how organisms process the information surrounding them is a wonderful field of biology. Many species have adapted to specific environmental conditions by showing drastic modifications of their phenotype regarding their sensory organs and structures. Just as the appearance of a new sensory structure is important in this context, the loss of sensory structures and/or their modification to the minimum expression are also relevant. This occurs in many subterranean species of troglofauna (cavities and caves) and stygofauna (aquifers, underwater caves, and fissures). Many species inhabiting these regions share characteristics such as loss of vision and pigmentation and enhancement of other senses such as hearing, touch, and smell (Chapman, 1982). Since the ideas of Lankester (1893), who postulated that cave dwellers were descendants of eyeless mutants that had become trapped and unable to leave the caves, much progress has been made in the study of troglobitic species and the functional and evolutionary significance of their morphological adaptations.

Examples could be blind cavefish, blind snakes, and moles. These animals are taxonomically rare, standing out from the other species in their clade. If we focus on fish, it is interesting to note that there are 111 species of blind fish out of 35000 total species worldwide. A well-studied example is the Mexican tetra (*Astyanax mexicanus*), which has an epigean form (surface fish) and at least 30 populations of eyeless hypogean forms (cavefish) (Jeffery, 2001, 2020). Therefore, they are considered excellent models of parallel and convergent evolution. Current alternative hypotheses suggest that adaptation to cave environments may have occurred either once or multiple times during the evolutionary history of this species (Dowling et al., 2002). The loss of eyes in cave populations has provided an excellent model for studying the "plasticity-first" hypothesis. This suggests that previously hidden variations in plastic responses may be exposed following the colonization of a new environment in which different induced phenotypes are beneficial, thereby exposing this variation to selection and facilitating adaptive evolution. Indeed, the cave environment is evolutionarily novel, and the populations that inhabit it are derived from surface populations, in which vision was highly evolved. There is evidence that embryos begin to develop eyes, but then this process stops, and a layer of skin develops over the eye. On the other hand, another theory states that there is a disruption of genes necessary for eye development that would then be maintained by neutral selection.

A very interesting experiment was conducted by Yamamoto & Jeffery (2000), who transplanted crystallins from embryos of surface individuals into cave embryos and vice versa. They found that the surface tetras failed to develop eyes, while the cave forms could fully develop eyes. This would indicate that the lens of cave forms produces signals indicative of eye degeneration (Yamamoto & Jeffery, 2000). However, why do cavefish lenses undergo this "cell death"? Yamamoto et al. (2004) seem to have found the answer by analyzing the expression of genes involved in organ- and tissue-specific cell differentiation. These fish produce large amounts of Sonic hedgehog (Shh) genes in the area that will become the head, leaving few (and not enough) cells for eye formation (Yamamoto et al., 2004; Pottin et al., 2011). As previously mentioned, the loss of specific morphological features or sensory abilities in tetras may result in the acquisition of other traits. Among the losses, in addition to vision and eyes, these fish had modifications and losses of melanin pigmentation, heart regenerative capacity, changes in circadian rhythms, and social behaviors (Jeffery, 2020). New traits included increased olfactory lobes, hypothalamus, jaws, taste buds, teeth, neuromasts, and body fat. The trade-off between the loss of eyes and the increase in other senses is controlled by Shh signaling, which is overexpressed in the cave tetra form (Yamamoto et al., 2009; Pottin et al., 2011; Hinaux et al., 2016).

The Whites of Your Eyes Are More Unique Than You Think

"*Don't fire till you see the whites of their eyes!*" was one of the most iconic phrases supposedly exclaimed in June 1775 by Colonel William Prescott at the Battle of Bunker Hill. On that occasion, the British army was approaching, and the American militia had them in their sights, waiting until they were close enough to attack. However, beyond the historical fact, we are concerned here with an interesting issue, something that makes us unique among animals and that is precisely the whites of our eyes. This portion of our eyes is known as the "sclera," an opaque protective layer that surrounds the eyeball, in addition to the transparent cornea at the front of the eye. Humans are characterized by a white sclera, while most animals have dark or iris-colored sclera (except for goats and crows) (Kobayashi & Kohshima, 1997, 2001). In addition, humans have the highest proportion of exposed sclerae in the eye contour, which is horizontally elongated (Kobayashi & Kohshima, 1997).

However, why is this marked difference that makes us unique? It is thought that the nonwhite sclera in most mammals may have evolved so that a flash of white would not attract the attention of predators. The risk of predation may have decreased with the evolution of increased body size and the use of tools and fire. Thus, in humans, the white sclera may be an adaptation to expand the visual field by allowing greater eye movement, especially in the horizontal direction, and to improve the detection of the gaze direction of another individual (Kobayashi & Kohshima, 1997). This is known as the "cooperative eye hypothesis" (framed in the "gaze-signaling hypothesis"), which makes it easier to see in which direction we are looking (Kobayashi & Kohshima, 2001; Tomasello et al., 2007; Kano et al., 2022). This would be a fundamental part of nonverbal communication in humans, as the gaze is used in many social activities, such as cooperation, teaching, and language learning (Tomasello et al., 2007; Kano et al., 2022). This ability in our ancestors might have been advantageous during hunts, where particular eye movements and gazes might have been key to alerting other hunters to potential prey or dangers.

Eyes with white sclerae, on the other hand, are rare among other great apes (Kobayashi & Kohshima, 2001). In general, great apes have dark sclerae that make it difficult to locate the edge of their irises and, thus, to estimate their line of sight (Kobayashi & Kohshima, 2001). This may be because their societies are less cooperative, and it is more important for them to hide the direction of their gaze from others, for example, so that they do not reveal the location of food. The "gaze camouflage" hypothesis predicts that dark pigments serve to camouflage gaze direction in competitive social environments. One study compared the reaction of nonhuman apes and human infants to an experimenter's head and eye movements, finding that while human infants followed the experimenter's gaze, apes only followed head movements (Tomasello et al., 2007). In canids, for example, it has been found that species that hunt in groups are more likely to have eyes with white sclera than solitary ones.

On the other hand, the "self-domestication" hypothesis posits white sclerae as a by-product of this process. During domestication, selection for tolerance and against aggression is linked to a reduction in the number and speed of migration of neural crest cells in early embryogenesis (Wilkins et al., 2014). The melanocytes that color the sclera are derived from neural crest cells, so reducing melanocytes in white sclerae could be a by-product of selection for increased docility during domestication (Wilkins et al., 2014). Another theory explains the white coloring of the sclera as a fundamental part of transmitting emotional signals in combination with the fine musculature around the eyes (Jessen & Grossmann, 2014). It could also indicate the health status of individuals through color variations (e.g., people associate yellowish sclera as less healthy) (Provine et al., 2013a, 2013b). In one study, humans with more reddish or yellowish sclerae were found to be rated as less healthy, attractive, and less young (Provine et al., 2013a, 2013b). In addition, human adults and infants show preferences for animals with white sclera (Segal et al., 2016). These hypotheses are not mutually exclusive, and highly visible eyes may serve all or several of these functions and have evolved under various selection pressures (Tomasello et al., 2007; Mearing et al., 2022).

5.4.3.3 Value of "Rare" Species in Ecosystems

Due to significant levels of disturbance, many ecosystems are currently experiencing what is widely recognized as the "sixth mass extinction" (Barnosky et al., 2011). In this context, "rare" species are more vulnerable than widespread or abundant species (Purvis et al., 2000; Davies et al., 2004; Sekercioglu et al., 2008). However, what is the basis for considering a species as "rare"? It is important to clarify that rarity is always context dependent, as rare species in one system may not be rare in other geographically or temporally different systems (Lawler et al., 2003). There are many ways to define rarity, but the most common are considering species with very small population sizes, restricted geographic ranges, or narrow habitat or feeding tolerances (Rabinowitz, 1981; Harnik et al., 2012) (Fig. 5.10a). Thus, rare species have received attention in conservation biology, although little is known about the functional consequences of their decline or extinction (Lyons et al., 2005).

Rarity can result from different processes, from stochastic processes to inherent trade-offs in life-history strategies (Ai et al., 2013). In addition, rarity can be a consequence of biotic and abiotic interactions. Highly specialized species with very restricted environmental niches may be abundant in a few environments but rare in many others. Alternatively, for example, noncompetitive species or species that depend on the abundance of others (frequency-dependent selection) may remain rare (Meyer & Kassen, 2007). In the second case, it is known that, for example, the greater the frequency dependence of another species, the more persistent the rare species will be (Yenni et al., 2012). Finally, rare species may arise and be maintained by abiotic changes, as they are known to be more sensitive to environmental fluctuations and more prone to extinction (Gaston, 2008) (Fig. 5.10a).

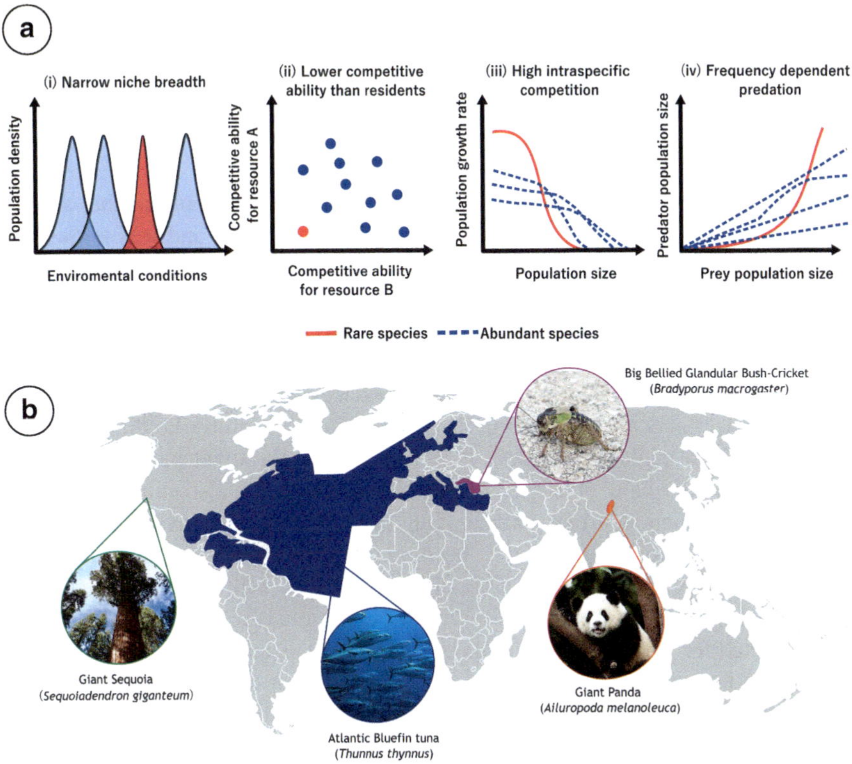

Fig. 5.10 Species rare in their distribution and abundance and processes that generate these patterns. (**a**) Potential mechanisms that can drive local rarity: (i, ii) mechanisms related to species characteristics and (ii, iv) mechanisms related to local biotic and abiotic conditions (modified from Jousset et al., 2017). (**b**) Geographically restricted species and species with small population sizes: geographically restricted giant sequoia; Atlantic bluefin tuna is widespread but has a small population size; giant panda and glandular bush-cricket are rare in number and are geographically restricted (modified from Dee et al., 2019) (*Bradyporus macrogaster.* Image courtesy of Otto Bylén Claess on)
(**a**) License: own creation. (**b**) *Bradyporus Macrogaster*: https://colombia.inaturalist.org/observations/119406275. License: Author permission. Bluefin tuna: License: Public domain. https://pixabay.com/es/photos/pez-submarino-mar-oceano-agua-5999306/; Giant sequoia: License: Public domain. https://pixabay.com/es/photos/secoya-secuoya-gigante-2406954/; Giant panda: License: Public domain. https://www.pexels.com/es-es/foto/animal-pelo-fauna-mamifero-7619818/

The study of rare species can often lead to difficulties regarding the limited ecological knowledge about them (Lyons et al., 2005). The commonness-dominance paradigm argues that the focus on common species is justified because they possess greater biomass and use more energy in communities (mass–ratio hypothesis) (Grime, 1998). Meanwhile, rare species that use marginal resources or habitats play a minimal role in ecosystems (resource availability hypothesis, Venier & Fahrig, 1996) (Gaston, 2011). However, multiple studies have found that rare species can have a significant value and contribute to various ecosystem services (Mouillot et al., 2013; Jain et al., 2014; Leitão et al., 2016; Dee et al., 2019).

Ecosystem Services of Rare Species

The focus of conservation has broadened to protect the contributions that animals and plants make to humans, known as "ecosystem services," and it is known that many rare species can provide these services (Dee et al., 2019). Rare species can make significant contributions through their mere presence, regardless of their abundance. This can be particularly evident in various experience contexts, such as safaris, natural sightings, or visits to zoos (Booth et al., 2011; Daniel et al., 2012). Additionally, rare species provide cultural services, fostering cultural identity, a sense of place, and fulfilling iconic or spiritual roles, thereby holding relational value (e.g., bald eagles *Haliaeetus leucocephalus* or the giant panda *Ailuropoda melanoleuca*) (Garibaldi & Turner, 2004). Notably, they can also enhance biodiversity in settings such as urban gardens (Ong et al., 2022). Paradoxically, other rare species possess more "value" as they become rarer. In this case, humans value them for being "unique" (Courchamp et al., 2006; Hall et al., 2008) as species harvested for luxury goods (e.g., sturgeons for caviar, bluefin tuna) (Gault et al., 2008), animals targeted for trophy hunting (Palazy et al., 2012), used for traditional medicine (e.g., rhino horns), ornamental species (e.g., orchids), or exotic pets (e.g., Banggai cardinal fish) (Courchamp et al., 2006; Dee et al., 2019) (Fig. 5.10b).

In contrast, other rare species will provide greater services if they are more abundant. This is the case for the Giant Sequoia *Sequoiadendron giganteum*, which is geographically restricted but locally abundant and contributes greatly to its local community services through carbon storage (Fauset et al., 2015). This case shows that previously abundant and keystone species in their communities that suffered declines or disappeared led to the loss of ecosystem structure and services. Due to the introduction of a pathogen from Asia (*Cryphonectria parasitica*), the American chestnut *Castanea dentata*, which was a foundation species in eastern North American forests and which previously sequestered carbon and affected stream hydrology, suffered a decline due to chestnut blight (Paillet, 2002; Ellison et al., 2005). Fundamental ecosystem services (decomposition, nutrient cycling, productivity) changed drastically following chestnut replacement by other species, altering terrestrial and aquatic processes. For example, in many cases, chestnut was replaced by oak trees (*Quercus* spp.) with leaves with lower nutritional quality, which, when falling into the water, produced a decline in leaf processing and consumption, decreasing growth rates and adult body mass in macroinvertebrate shredder communities (Ellison et al., 2005). We will return to the role of alien introductions and species declines in Chaps. 6 and 7.

On the other hand, there are species that, despite being rare, contribute disproportionately through indirect interspecific interactions, providing unique functional roles or facilitating them (Lyons et al., 2005; Mouillot et al., 2013;

Delgado-Baquerizo et al., 2016; Jousset et al., 2017; Ziegler et al., 2018). For example, some rare or threatened nitrogen-fixing species and their microbial mutualists enhance soil fertility, which significantly benefits biomass production in nitrogen-limited grasslands associated with higher biofuel yields (Leach & Givnish, 1996; Tilman et al., 2006). Mouillot et al. (2013) studied the role of rare species and their functional traits in ecosystems such as coral reef fishes, alpine plants, and tropical trees. They found that rare species supported a diverse combination of functional traits. In addition, rare species can stabilize food webs (O'Gorman et al., 2011), reduce competition between service providers (Holt et al., 1994; Tilman et al., 1997), or be keystone species or predatory species with disproportionate roles in structuring communities (Lyons et al., 2005; Mouillot et al., 2013). This is the case for rare sea otters (*Enhydra lutris*) in Aleutian kelp forests that limit the presence and abundance of sea urchins, which indirectly increases kelp biomass and carbon available for carbon storage (Estes et al., 1998; Wilmers et al., 2012).

Rare Species Loss Affects Functional Assemblages in Tropics

Leitão et al. (2016) conducted a study that explored the impact of the loss of rare species in the tropics on the structure of functional assemblages (i.e., diversity and distribution of functional traits). The tropics are characterized by numerous rare species (Hercos et al., 2013; Hubbell, 2013), with a high rate of species extinction (Brook et al., 2006). These authors found that several rare species support unique roles in assemblages with common species, so their loss would be critical for the functional structure of these communities, leading to declines in ecological processes at both local and regional scales (Leitão et al., 2016). The effect they found on the communities was disproportionately high, leading to a drastic decrease in functional richness, specialization, and originality.

Surprisingly, these results were found in three taxonomic groups that are very different in terms of their ecology or evolutionary history. For example, the cassowary *Casuarius casuarius*, classified as a vulnerable species by the International Union for Conservation of Nature (IUCN), is considered rare, being the only large-bodied disperser of plants over long distances in Australian tropical rainforests. Its loss would impact plant population dynamics and in turn, strongly affect the functional structure of bird assemblages in the Australian Wet Tropics. Another example was the periphyton-grazing fishes, with superrestricted geographic ranges in small Amazonian Forest streams. The importance of these fishes is key, as they are the only species involved in the incorporation of carbon along the river continuum through the exploitation of periphyton (complex mixture of algae, heterotrophic microbes, cyanobacteria, and detritus attached to submerged surfaces in aquatic ecosystems).

The last example these authors addressed was that of the rainforest trees *Brosimum acutifolium* and *Protium giganteum* from French Guiana. The first species has leaves with a milky latex, associated with the ability to resist herbivores and fungal infections. *Protium giganteum* has a particular bark that protects it from the wildfires that may occur in the region. These species provide resistance and resilience to various disturbances and environmental fluctuations in the region. These results allow us to conclude that beyond taxonomic or esthetic losses, the extinction of rare species can result in the loss of irreplaceable functions, a decrease in ecological niches in the communities, and a disruption of interactions between species, drastically impacting the integrity of many ecological processes (Leitão et al., 2016).

The Rare Biosphere

In general, microbial communities are composed of a large number of rare species (with low abundances), which is globally known as "the rare biosphere" (Sogin et al., 2006; Skopina et al., 2016). Between 1.5 and 28% of all microbes are "conditionally rare taxa" (Shade et al., 2014), meaning that under favorable conditions, these species can disproportionately affect ecosystem functions such as biogeochemical cycles (Falkowski et al., 2008; Lynch & Neufeld, 2015; Jousset et al., 2017). Rare species are important in nutrient cycling, greenhouse gas emissions, and pollutant degradation. They also prevent the invasion of new species by supporting the structure of specific communities and stabilizing community functions in fluctuating environments. Other rare microbes are essential in associations with plants and animals, preventing infection by pathogens and stimulating host immunity (Jousset et al., 2017). These microbes have diverse enzymes, some of which are key in industrial processes, such as thermophilic cellulases for biofuel applications (Rubin, 2008). The rare biosphere is not exempt from the consequences of global climate change and land-use alterations, so it is susceptible to these impacts and changes in its ecosystem functions (Rodrigues et al., 2013). This is why the "rare biosphere" is considered at the frontier of microbial ecology research (Jousset et al., 2017).

Microbial systems are characterized by high functional redundancy (Rousk et al., 2009); that is, many species share similar functions, so rare species are not as important in functional diversity. However, species that are not relevant in certain conditions may be relevant in changing conditions, offering a pool of genetic resources and providing necessary functions or facilitating particular interspecific interactions (Shade et al., 2014; Fetzer et al., 2015). An interesting example is that of microbial communities with a role in the degradation of organic compounds such as pollutants. Removing rare species in activated sludge and freshwater dramatically reduces the ability to degrade pollutants and toxins (Hernandez-Raquet et al., 2013; Delgado-Baquerizo et al., 2016). On the other hand, the removal of rare species in the soil can lead to an increase in new unwanted pathogenic species (Mallon et al., 2015), which demonstrates the crucial role of rare species in biotechnological applications such as crop protection (Xiong et al., 2021).

An example illustrating the importance of the rare biosphere is found in associations between corals and photosynthetic dinoflagellates (*Symbiodinium* spp.). Although reef ecosystems cover a small percentage of the world's oceans, nearly one-third of the world's marine biodiversity depends on their functional integrity

(Reaka-Kudla, 1997). Association with photosynthetic dinoflagellates is crucial in coral reefs but in turn, possesses a high vulnerability to local stressors and the effects of climate change (Hoegh-Guldberg et al., 2007) since ocean warming and other impacts destroy coral–symbiodinium symbiosis (coral bleaching). The dominant *Symbiodinium* type is the most studied and stress-susceptible in the associations, but new molecular techniques are allowing the discovery of many rare *Symbiodinium* species in low abundances that would be stress-resistant and would allow a replacement of the dominant types (Berkelmans & van Oppen, 2006; Boulotte et al., 2016). These rare *Symbiodinium* taxa contribute to host–symbiont community stability and promote the environmental resilience of the coral holobiont (Ziegler et al., 2018). Interestingly, these "microbiome" switches or flexibility of change could facilitate relatively rapid responses to environmental changes in other organisms (Ziegler et al., 2019; Voolstra & Ziegler, 2020). We will return to this symbiotic association later in the book.

5.5 Key Concepts and Ideas in This Chapter

- Aberrant species represent both teratological cases and exceptional organisms. These species that captivated Darwin and still amaze us today may be unique in their taxonomic groups, rare in their geographic range, hyperderived groups, or "living fossils."
- Teratology examines abnormal morpho-physiological developments in various taxa and can be useful for understanding possible morphological pathways by linking internal developmental rules, as well as adaptive and functional explanations. This has led to a vindication of the reporting of teratological cases, as well as the study of their causes and their inclusion in fields such as Evo-Devo.
- Teratologies have inspired great theories such as homeotic genes or the ABC model in plants. Their study can alert us to current or past environmental pollution, certain teratologies in plants are intentionally produced, and other teratologies are combated because of the ecological damage they cause.
- Exceptions and rarities are invaluable in the study of biology and should be more emphasized in the definitions of taxonomic groups. Sometimes there are particular rarities, but in other cases, there are "intermediate exceptions" with minorities of species or clades with certain characteristics compared to the majority.
- We can find exceptions in many animal and plant groups; we just have to know how to observe, relativize, and value these cases. In some cases, we face a "paradox" of a chain of exceptions and rules, demonstrating how interesting and complex the analysis of these cases can be.
- We review groups that, due to their rarities, have been difficult to classify, rarities that can be used to human benefit, exceptional groups that make us reconsider phylogenetic relationships, phenomena with incredible exceptions, animals and plants that are not what they seem or do not do what is expected, and evolutionary and ecological aspects of rare species.

References

Aanen, D. K. (2014). How a long-lived fungus keeps mutations in check. *Science, 346*(6212), 922–923.

Abbott, J. K., & Gosden, T. P. (2009). Correlated morphological and colour differences among females of the damselfly *Ischnura elegans*. *Ecological Entomology, 34*(3), 378–386.

Abrol, D. P. (2012). *Pollination biology: Biodiversity conservation and agricultural production*. Springer.

Ackerman, J. D. (2006). Sexual reproduction of seagrasses: Pollination in the marine context. In A. W. D. Larkum, R. J. Orth, & C. M. Duarte (Eds.), *Seagrasses: Biology, ecology and conservation* (pp. 89–109). Springer.

Adamec, L. (1997). Mineral nutrition of carnivorous plants: A review. *The Botanical Review, 63*, 273–299.

Adell, T., & Riutort, M. (2021). Phylum Platyhelminthes. In B. Schierwater & R. DeSalle (Eds.), *Invertebrate zoology: A tree of life approach* (pp. 219–230). CRC Press.

Ai, D., Chu, C., Ellwood, M. F., Hou, R., & Wang, G. (2013). Migration and niche partitioning simultaneously increase species richness and rarity. *Ecological Modelling, 258*, 33–39.

Aiken, R. B. (1981). The relationship between body weight and homosexual mounting in *Palmacorixa nana* Walley (Heteroptera: Corixidae). *Florida Entomologist*, 267–271.

Akre, R. D., & Rettenmeyer, C. W. (1966). Behavior of Staphylinidae associated with army ants (Formicidae: Ecitonini). *Journal of the Kansas Entomological Society, 39*(4), 745–782.

Alberch, P. (1980). Ontogenesis and morphological diversification. *American Zoologist, 20*(4), 653–667.

Alberch, P. (1989). The logic of monsters: Evidence for internal constraint in development and evolution. *Geobios, 22*, 21–57.

Albert, V. A., Williams, S. E., & Chase, M. W. (1992). Carnivorous plants: Phylogeny and structural evolution. *Science, 257*(5076), 1491–1495.

Aldrovandi, U. (1642). Monstrorum Historia cum Paralipomenis historiae omnium animalium. Bartholomaeus Ambrosinus. labore, et studio volumen composuit. In *Marcus Antonius Bernia in lucem edidit. Proprijs sumptibus. cum indice copiosissimo*. typis Nicolai Tebaldini.

Alonso, D. L. (2014). *Biología de la homosexualidad*. Síntesis.

Alpert, P. (2006). Constraints of tolerance: Why are desiccation-tolerant organisms so small or rare? *Journal of Experimental Biology, 209*(9), 1575–1584.

Alvarez-Campos, P., San Martin, G., & Aguado, M. T. (2013). A new species and new record of the commensal genus *Alcyonosyllis* Glasby & Watson, 2001 and a new species of *Parahaplosyllis* Hartmann-Schröder, 1990, (Annelida: Syllidae: Syllinae) from Philippines Islands. *Zootaxa, 3734*(2), 156–168.

Amemiya, C. T., Alföldi, J., Lee, A. P., Fan, S., Philippe, H., MacCallum, I., et al. (2013). The African coelacanth genome provides insights into tetrapod evolution. *Nature, 496*(7445), 311–316.

Amodio, P., Boeckle, M., Schnell, A. K., Ostojic, L., Fiorito, G., & Clayton, N. S. (2019). Grow smart and die young: Why did cephalopods evolve intelligence? *Trends in Ecology & Evolution, 34*(1), 45–56.

Andersen, N. M. (1997). Phylogenetic tests of evolutionary scenarios: The evolution of flightlessness and wing polymorphism in insects. In P. Grandcolas (Ed.), *The Origin of Biodiversity of Insects: Phylogenetic Tests of Evolutionary Scenarios* (pp. 91–108). Memoires du Museum d'Histoire Naturelle.

Anderson, J. B., & Catona, S. (2014). Genomewide mutation dynamic within a long-lived individual of *Armillaria gallica*. *Mycologia, 106*(4), 642–648.

Anderson, J. B., Bruhn, J. N., Kasimer, D., Wang, H., Rodrigue, N., & Smith, M. L. (2018). Clonal evolution and genome stability in a 2500-year-old fungal individual. *Proceedings of the Royal Society B, 285*(1893), 20182233.

Andersson, M. (1994). *Sexual selection*. Princeton University Press.

Andrés, J. A., Sánchez-Guillén, R. A., & Rivera, A. C. (2002). Evolution of female colour polymorphism in damselflies: Testing the hypotheses. *Animal Behaviour, 63*(4), 677–685.

Andrews, E. A. (1892). Bifurcated annelids. *The American Naturalist, 26*(309), 725–733.

Andrews, E. A. (1894). Memoirs: Some Abnormal Annelids. *Journal of Cell Science, 2*(144), 435–460.

Annesley, S. J., & Fisher, P. R. (2009). *Dictyostelium discoideum*—A model for many reasons. *Molecular and Cellular Biochemistry, 329*, 73–91.

Arnold, S. J. (1994). Is there a unifying concept of sexual selection that applies to both plants and animals? *The American Naturalist, 144*, S1–S12.

Asiain, J., & Márquez, J. (2009). New teratological examples in Neotropical Staphylinidae (Insecta: Coleoptera), with a compilation of previous teratological records. *Revista Mexicana de Biodiversidad, 80*, 129–139.

Atkinson, D. (1994). Temperature and organism size: A biological law for ectotherms? *Advances in Ecological Research, 25*, 1–58.

Attygalle, A. B., Xu, S., Moore, W., McManus, R., Gill, A., & Will, K. (2020). Biosynthetic origin of benzoquinones in the explosive discharge of the bombardier beetle Brachinus elongatulus. *The Science of Nature, 107*, 1–11.

Atwood, J. T., Jr. (1982). How is *Paphiopedilum* pollinated? *American Orchid Society Bulletin, 51*, 1057–1058.

Autumn, K., & Peattie, A. M. (2002). Mechanisms of adhesion in geckos. *Integrative and Comparative Biology, 42*(6), 1081–1090.

Avise, J. C., & Mank, J. E. (2009). Evolutionary perspectives on hermaphroditism in fishes. *Sexual Development, 3*(2–3), 152–163.

Bailey, N. W., & French, N. (2012). Same-sex sexual behaviour and mistaken identity in male field crickets. *Teleogryllus oceanicus. Animal Behaviour, 84*(4), 1031–1038.

Bailey, N. W., & Zuk, M. (2009). Same-sex sexual behavior and evolution. *Trends in Ecology & Evolution, 24*(8), 439–446.

Bailly, X., Laguerre, L., Correc, G., Dupont, S., Kurth, T., Pfannkuchen, A., et al. (2014). The chimerical and multifaceted marine acoel *Symsagittifera roscoffensis*: from photosymbiosis to brain regeneration. *Frontiers in Microbiology, 5*, 498.

Baker, R., & Bellis, M. (1988). "Kamikaze" sperm in mammals? *Animal Behaviour, 36*(3), 936–939.

Balazuc, J. (1948). La tératologie des coléoptères et expériences de transplantation sur *Tenebrio molitor* L. *Mémoires du Muséum National d'Histoire Naturelle, 25*, 1–293.

Baldauf, S. L., & Doolittle, W. F. (1997). Origin and evolution of the slime molds (Mycetozoa). *Proceedings of the National Academy of Sciences, 94*(22), 12007–12012.

Bao, A. M., & Swaab, D. F. (2010). Sex differences in the brain, behavior, and neuropsychiatric disorders. *The Neuroscientist, 16*(5), 550–565.

Barik, B. K., & Mishra, M. (2019). Nanoparticles as a potential teratogen: a lesson learnt from fruit fly. *Nanotoxicology, 13*(2), 258–284.

Barnosky, A. D., Matzke, N., Tomiya, S., Wogan, G. O. U., Swartz, B., Quental, T. B., et al. (2011). Has the Earth's sixth mass extinction already arrived? *Nature, 471*(7336), 51–57.

Barria, M. D., Quirós, D. I., & Emmen, D. (2020). Reporte de un caso teratológico en *Macraspis chrysis* (Linnaeus) (Coleoptera: Scarabaeidae: Rutelinae: Rutelini) de Panamá. *Revista Chilena de Entomología, 46*(1), 47–50.

Barthlott, W., Porembski, S., Seine, R., & Theisen, I. (2007). *The curious world of carnivorous plants*. Timber Press.

Bateson, W. (1908). *The Methods and scope of genetics, an inaugural lesson delivered 23 October 1908*. At the University Press.

Bauer, J. A., Jr., & Bauer, S. E. (1981). Reproductive biology of pigmy angelfishes of the genus *Centropyge* (Pomacanthidae). *Bulletin of Marine Science, 31*(3), 495–513.

Baumann, H. (1922). Die anabiose der tardigraden. *Zoologischer Jahrbucher, 45*, 501–556.

Baumgartner, K., Coetzee, M. P., & Hoffmeister, D. (2011). Secrets of the subterranean pathosystem of *Armillaria*. *Molecular Plant Pathology, 12*(6), 515–534.

Baur, E. (1909). Das Wesen und die Erblichkeitsverhältnisse der "Varietätes albomarginatae hort." von Pelargonium zonale. *Zeitschrift für induktive Abstammungs- und Vererbungslehre, 2*(1), 81–106.

Bayley, D. T. I., & Rose, A. (2020). Multi-species co-operative hunting behaviour in a remote Indian Ocean reef system. *Marine and Freshwater Behaviour and Physiology, 53*(1), 35–42.

Bejder, L., & Hall, B. K. (2002). Limbs in whales and limblessness in other vertebrates: Mechanisms of evolutionary and developmental transformation and loss. *Evolution & Development, 4*(6), 445–458.

Bemm, F., Becker, D., Larisch, C., Kreuzer, I., Escalante-Perez, M., Schulze, W. X., et al. (2016). Venus flytrap carnivorous lifestyle builds on herbivore defense strategies. *Genome Research, 26*(6), 812–825.

Benca, J. P., Duijnstee, I. A., & Looy, C. V. (2022). Fossilized pollen malformations as indicators of past environmental stress and meiotic disruption: insights from modern conifers. *Paleobiology, 48*(4), 677–710.

Berkelmans, R., & Van Oppen, M. J. (2006). The role of zooxanthellae in the thermal tolerance of corals: A 'nugget of hope' for coral reefs in an era of climate change. *Proceedings of the Royal Society B: Biological Sciences, 273*(1599), 2305–2312.

Bianchini, M. C., Portela, J. L. R., Puntel, R. L., & Ávila, D. S. (2018). Cellular responses in *Drosophila melanogaster* following teratogen exposure. In L. Félix (Ed.), *Teratogenicity testing: Methods in molecular biology* (pp. 243–276). Humana Press.

Blouin, N. A., & Lane, C. E. (2012). Red algal parasites: models for a life history evolution that leaves photosynthesis behind again and again. *Bioessays, 34*(3), 226–235.

Blumberg, M. S. (2009). *Freaks of nature: And what they tell us about development and evolution*. Oxford University Press.

Boal, J. G. (1996). Absence of social recognition in laboratory-reared cuttlefish, *Sepia officinalis* L. (Mollusca: Cephalopoda). *Animal Behaviour, 52*(3), 529–537.

Bombelli, P., Howe, C. J., & Bertocchini, F. (2017). Polyethylene bio-degradation by caterpillars of the wax moth Galleria mellonella. *Current Biology, 27*(8), R292–R293.

Booth, J. E., Gaston, K. J., Evans, K. L., & Armsworth, P. R. (2011). The value of species rarity in biodiversity recreation: A birdwatching example. *Biological Conservation, 144*(11), 2728–2732.

Bornscheuer, U. T. (2016). Feeding on plastic. *Science, 351*(6278), 1154–1155.

Borwein, B., & Hollenberg, M. J. (1973). The photoreceptors of the "four-eyed" fish, Anableps anableps L. *Journal of Morphology, 140*(4), 405–441.

Bos, L. (1957). Plant teratology and plant pathology: Met een samenvatting: Plantenteratologie en Planteziektenkunde. *Tijdschrift Over Plantenziekten, 63*(5), 222–231.

Bots, J., Iserbyt, A., Van Gossum, H., Hammers, M., & Sherratt, T. N. (2015). Frequency-dependent selection on female morphs driven by premating interactions with males. *The American Naturalist, 186*(1), 141–150.

Boulotte, N. M., Dalton, S. J., Carroll, A. G., Harrison, P. L., Putnam, H. M., Peplow, L. M., & Van Oppen, M. J. (2016). Exploring the *Symbiodinium* rare biosphere provides evidence for symbiont switching in reef-building corals. *The ISME Journal, 10*(11), 2693–2701.

Bozzaro, S., & Eichinger, L. (2011). The professional phagocyte *Dictyostelium discoideum* as a model host for bacterial pathogens. *Current Drug Targets, 12*(7), 942–954.

Brant, S. V., & Loker, E. S. (2005). Can specialized pathogens colonize distantly related hosts? Schistosome evolution as a case study. *PLoS Pathogens, 1*(3), e38.

Brenneis, G., Ungerer, P., & Scholtz, G. (2008). The chelifores of sea spiders (Arthropoda, Pycnogonida) are the appendages of the deutocerebral segment. *Evolution & Development, 10*(6), 717–724.

Broadley, M. W., Barry, P. H., Ballentine, C. J., Taylor, L. A., & Burgess, R. (2018). End-Permian extinction amplified by plume-induced release of recycled lithospheric volatiles. *Nature Geoscience, 11*(9), 682–687.

Brock, D. A., Douglas, T. E., Queller, D. C., & Strassmann, J. E. (2011). Primitive agriculture in a social amoeba. *Nature, 469*(7330), 393–396.

Brock, D. A., Read, S., Bozhchenko, A., Queller, D. C., & Strassmann, J. E. (2013). Social amoeba farmers carry defensive symbionts to protect and privatize their crops. *Nature Communications, 4*(1), 2385.

Brockmann, H. J. (2001). The evolution of alternative strategies and tactics. *Advances in the Study of Behavior, 30*, 1–51.

Brook, B. W., Bradshaw, C. J., Koh, L. P., & Sodhi, N. S. (2006). Momentum drives the crash: Mass extinction in the tropics. *Biotropica, 38*(3), 302–305.

Brooks, R. (2009). All too human: responses to same-sex copulation in the common cockchafer (*Melolontha melolontha* (L.)), 1834–1900. *Archives of Natural History, 36*(1), 146–159.

Brooks, R. (2021). Darwin's closet: The queer sides of the descent of man (1871). *Zoological Journal of the Linnean Society, 191*(2), 323–346.

Brown, A. R., Owen, S. F., Peters, J., Zhang, Y., Soffker, M., Paull, G. C., et al. (2015). Climate change and pollution speed declines in zebrafish populations. *Proceedings of the National Academy of Sciences, 112*(11), E1237–E1246.

Brusca, R. C., Moore, W., & Shuster, S. M. (2016). *Invertebrates* (3rd ed.). Oxford University Press.

Buczck, A. (2000). Experimental teratogeny in the tick *Hyalomma marginatum marginatum* (Acari: Ixodida: Ixodidae): effect of high humidity on embryonic development. *Journal of Medical Entomology, 37*(6), 807–814.

Buczek, A., Bartosik, K., Buczek, A. M., Buczek, W., & Kulina, D. (2019). Abnormal development of *Hyalomma marginatum* ticks (Acari: Ixodidae) induced by plant cytotoxic substances. *Toxins, 11*(8), 445.

Budd, G. E. (2002). A palaeontological solution to the arthropod head problem. *Nature, 417*(6886), 271–275.

Buel, J. W. (1887). *Sea and land: An illustrated history of the wonderful and curious things of nature existing before and since the Deluge* (pp. 475–477). Historical Publishing Company.

Bulmer, M. G. (1970). *The biology of twinning in man*. Clarendon.

Burke, A. F., Sole, C. L., & Scholtz, C. H. (2018). A remarkable teratological case for *Eucymatodera parva* Schenkling, 1908 (Cleridae: Tillinae) from Namibia. *African Entomology, 26*(1), 250–253.

Buttery, N. J., Rozen, D. E., Wolf, J. B., & Thompson, C. R. (2009). Quantification of social behavior in *Dictyostelium discoideum* reveals complex fixed and facultative strategies. *Current Biology, 19*(16), 1373–1377.

Buttery, N. J., Jack, C. N., Adu-Oppong, B., Snyder, K. T., Thompson, C. R., Queller, D. C., & Strassmann, J. E. (2012). Structured growth and genetic drift raise relatedness in the social amoeba *Dictyostelium discoideum*. *Biology Letters, 8*(5), 794–797.

Byrnes, J., & Witman, J. D. (2003). Impact assessment of an invasive flatworm, *Convoluta convoluta*, in the Southern Gulf of Maine. *Journal of Experimental Marine Biology and Ecology, 293*(2), 173–191.

Calner, M. (2008). Silurian global events—At the tipping point of climate change. In A. M. T. Elewa (Ed.), *Mass Extinctions* (pp. 21–58). Springer-Verlag, Cambridge University Press.

Cambridge Advanced Learner's Dictionary (2023). Cambridge, UK: Cambridge University Press.

Carroll, S. P., & Corneli, P. S. (1995). Divergence in male mating tactics between two populations of the soapberry bug: II. Genetic change and the evolution of a plastic reaction norm in a variable social environment. *Behavioral Ecology, 6*(1), 46–56.

Castro, L., Toro, M. A., & López-Fanjul, C. (1994). The genetic properties of homosexual copulation behaviour in *Tribolium castaneum*: Artificial selection. *Genetics Selection Evolution, 26*(4), 361–367.

Causier, B., Schwarz-Sommer, Z., & Davies, B. (2010). Floral organ identity: 20 years of ABCs. *Seminars in Cell & Developmental Biology, 21*(1), 73–79.

Cavalier-Smith, T. (1987). The origin of fungi and pseudofungi. In A. D. M. Rayner, C. M. Brasier, & D. M. Moore (Eds.), *Evolutionary biology of the fungi, british mycological society symposium* (Vol. 13, pp. 339–353). Cambridge University Press.

Chapelle, G., & Peck, L. S. (1999). Polar gigantism dictated by oxygen availability. *Nature, 399*(6732), 114–115.

Chapelle, G., & Peck, L. S. (2004). Amphipod crustacean size spectra: New insights in the relationship between size and oxygen. *Oikos, 106*(1), 167–175.
Chapman, P. (1982). The origin of troglobites. *Proceedings of the University of Bristol Speleological Society, 16*(2), 133–141.
Chaudhuri, C. R., & Chatterjee, I. B. (1969). L-ascorbic acid synthesis in birds: Phylogenetic trend. *Science, 164*(3878), 435–436.
Chavez, C., Cruz-Becerra, G., Fei, J., Kassavetis, G. A., & Kadonaga, J. T. (2019). The tardigrade damage suppressor protein binds to nucleosomes and protects DNA from hydroxyl radicals. *eLife, 8*, e47682.
Chen, Y. J., Chen, C. C., & Huang, H. L. (2016). Induction of apoptosis by *Armillaria mellea* constituent armillarikin in human hepatocellular carcinoma. *OncoTargets and Therapy, 9*, 4773–4783.
Cherel, Y., & Hobson, K. A. (2007). Geographical variation in carbon stable isotope signatures of marine predators: A tool to investigate their foraging areas in the Southern Ocean. *Marine Ecology Progress Series, 329*, 281–287.
Cherel, Y., Duhamel, G., & Gasco, N. (2004). Cephalopod fauna of subantarctic islands: New information from predators. *Marine Ecology Progress Series, 266*, 143–156.
Chew, A. L. C., Desjardin, D. E., Tan, Y. S., Musa, M. Y., & Sabaratnam, V. (2014). Bioluminescent fungi from Peninsular Malaysia—A taxonomic and phylogenetic overview. *Fungal Diversity, 70*(1), 149–187.
Chiao, C. C., Cronin, T. W., & Marshall, N. J. (2000). Eye design and color signaling in a Stomatopod Crustacean *Gonodactylus smithii*. *Brain, Behavior and Evolution, 56*, 107–122.
Chitimia-Dobler, L., Bestehorn, M., Bröker, M., Borde, J., Molcanyi, T., Andersen, N. S., et al. (2017). Morphological anomalies in *Ixodes ricinus* and *Ixodes inopinatus* collected from tick-borne encephalitis natural foci in Central Europe. *Experimental and Applied Acarology, 72*(4), 379–397.
Christiansen, B. O. (1964). *Spiculosiphon radiata*, a new Foraminifera from northern Norway. *Astarte, 25*, 1–8.
Chu, D., Corso, J. D., Shu, W., Song, H., Wignall, P. B., Grasby, S. E., et al. (2021). Metal-induced stress in survivor plants following the end-Permian collapse of land ecosystems. *Geology, 49*(6), 657–661.
Clark, S., & Belo Neto, L. A. (2010). A remarkable teratological specimen of *Pseudoluperus longulus* (LeConte) (Coleoptera: Chrysomelidae) from Utah, USA. *The Coleopterists Bulletin, 64*(4), 383–385.
Clarke, M. R. (1966). A review of the systematics and ecology of oceanic squids. *Advances in Marine Biology, 4*, 91–300.
Clarke, M. (2010). Recent insights into host–pathogen interactions from *Dictyostelium*. *Cellular Microbiology, 12*(3), 283–291.
Cock, A. G., & Forsdyke, D. R. (2008). *Treasure your exceptions: The science and life of William Bateson*. Springer.
Cocucci, A. A., Marino, S., Baranzelli, M., Wiemer, A. P., & Sérsic, A. (2014). The buck in the milkweed: Evidence of male-male interference among pollinaria on pollinators. *New Phytologist, 203*(1), 280–286.
Coen, E. S., & Meyerowitz, E. M. (1991). The war of the whorls: Genetic interactions controlling flower development. *Nature, 353*(6339), 31–37.
Cole, K. S., & Hoese, D. F. (2001). Gonad morphology, colony demography and evidence for hermaphroditism in *Gobiodon okinawae* (Teleostei, Gobiidae). *Environmental Biology of Fishes, 61*, 161–173.
Collin, R. (2003). Worldwide patterns in mode of development in calyptraeid gastropods. *Marine Ecology Progress Series, 247*, 103–122.
Collin, R., & Cipriani, R. (2003). Dollo's law and the re–evolution of shell coiling. *Proceedings of the Royal Society of London Series B: Biological Sciences, 270*(1533), 2551–2555.
Collins English Dictionary (2003). Glasgow, UK: HarperCollins.

Conner, J. K., Karoly, K., Stewart, C., Koelling, V. A., Sahli, H. F., & Shaw, F. H. (2011). Rapid independent trait evolution despite a strong pleiotropic genetic correlation. *The American Naturalist, 178*(4), 429–441.
Conover, D. O., & Kynard, B. E. (1981). Environmental sex determination: interaction of temperature and genotype in a fish. *Science, 213*(4507), 577–579.
Coombs, G., Dold, A. P., Brassine, E. I., & Peter, C. I. (2012). Large pollen loads of a South African asclepiad do not interfere with the foraging behaviour or efficiency of pollinating honey bees. *Naturwissenschaften, 99*, 545–552.
Cordero, A. (1992). Morphological variability, female polymorphism and heritability of body length in *Ischnura graellsii* (Rambur) (Zygoptera: Coenagrionidae). *Odonatologica, 21*(4), 409–419.
Cordero, A. C., & Pérez, F. E. (1998). Mating frequency, population density and female polychromatism in the damselfly *Ischnura graellsii*: An analysis of four natural populations. *Etología, 6*, 61–66.
Correns, C. E., & Correns, C. E. (1924). *Vererbungsversuche mit buntblättrigen Sippen. II: Vier neue Typen bunter Periklinalchimären* (pp. 989–1023). Springer.
Cotoras, D. D., Castanheira, P. D. S., & Sharma, P. P. (2021). Implications of a cheliceral axial duplication in *Tetragnatha versicolor* (Araneae: Tetragnathidae) for arachnid deuterocerebral appendage development. *Development Genes and Evolution, 231*(5–6), 131–139.
Cotton, C. L., & Parker, K. L. (2000). Winter activity patterns of northern flying squirrels in subboreal forests. *Canadian Journal of Zoology, 78*(11), 1896–1901.
Courchamp, F., Angulo, E., Rivalan, P., Hall, R. J., Signoret, L., Bull, L., & Meinard, Y. (2006). Rarity value and species extinction: The anthropogenic Allee effect. *PLoS Biology, 4*(12), e415.
Cox, P. A. (1988). Hydrophilous pollination. *Annual Review of Ecology and Systematics, 19*(1), 261–279.
Cox, S. M., Schmidt, D., Modarres-Sadeghi, Y., & Patek, S. N. (2014). A physical model of the extreme mantis shrimp strike: Kinematics and cavitation of Ninjabot. *Bioinspiration & Biomimetics, 9*(1), 016014.
Cronin, T. W., Marshall, N. J., & Caldwell, R. L. (2017). *Stomatopod vision*. Oxford University Press.
Crowe, J. H. (1975). The physiology of cryptobiosis in tardigrades. *Memorie dell'Istituto Italiano di Idrobiologia, 32*, 37–59.
Cui, J., Pan, Y. H., Zhang, Y., Jones, G., & Zhang, S. (2011). Progressive pseudogenization: Vitamin C synthesis and its loss in bats. *Molecular Biology and Evolution, 28*(2), 1025–1031.
Curtis, H., Schnek, A., Barnes, S., & Massarini, A. (2008). *Biología de Curtis* (7th ed.). Editorial Médica Panamericana.
Daly, M. J. (2009). A new perspective on radiation resistance based on Deinococcus radiodurans. *Nature Reviews Microbiology, 7*(3), 237–245.
Daly, T. J., & Buffenstein, R. (1998). Skin morphology and its role in thermoregulation in mole-rats, Heterocephalus glaber and Cryptomys hottentotus. *The Journal of Anatomy, 193*(4), 495–502.
Daly, I. M., How, M. J., Partridge, J. C., Temple, S. E., Marshall, N. J., & Roberts, N. W. (2016). Dynamic polarization vision in mantis shrimps. *Nature Communications, 7*, 12140.
Daniel, T. C., Muhar, A., Arnberger, A., Aznar, O., Boyd, J. W., Chan, K. M., et al. (2012). Contributions of cultural services to the ecosystem services agenda. *Proceedings of the National Academy of Sciences, 109*(23), 8812–8819.
Darmaillacq, A. S., Dickel, L., & Mather, J. (2014). *Cephalopod cognition*. Cambridge University Press.
Darwin, C. R. (1844). Variegated leaves. *The Gardeners' Chronicle and Agricultural Gazette, 37*, 621.
Darwin, C. (1859). *On the origin of species by means of natural selection, or the preservation of favoured races in the struggle for life*. John Murray.
Darwin, C. (1862). *On the various contrivances by which British and foreign orchids are fertilised by insects, and on the good effect of intercrossing*. John Murray.

Darwin, C. (1871). *The descent of man, and selection in relation to sex*. John Murray.
Darwin, C. (1875). *Insectivorous plants*. John Murray.
Darwin, C. (1877). *The various contrivances by which orchids are fertilised by insects*. John Murray.
Datta, S. K. (2009). Plant chimeras and their role in development of new ornamental varieties. *Journal of Ornamental Horticulture, 12*(2), 75–94.
Davies, K. F., Margules, C. R., & Lawrence, J. F. (2004). A synergistic effect puts rare, specialized species at greater risk of extinction. *Ecology, 85*(1), 265–271.
De Mitcheson, Y. S., & Liu, M. (2008). Functional hermaphroditism in teleosts. *Fish and Fisheries, 9*(1), 1–43.
De Storme, N., & Geelen, D. (2014). The impact of environmental stress on male reproductive development in plants: Biological processes and molecular mechanisms. *Plant, Cell & Environment, 37*(1), 1–18.
Dee, L. E., Cowles, J., Isbell, F., Pau, S., Gaines, S. D., & Reich, P. B. (2019). When do ecosystem services depend on rare species? *Trends in Ecology & Evolution, 34*(8), 746–758.
Delabroye, A., Munnecke, A., Servais, T., Vandenbroucke, T. R. A., & Vecoli, M. (2012). Abnormal forms of acritarchs (phytoplankton) in the upper Hirnantian (Upper Ordovician) of Anticosti Island, Canada. *Review of Palaeobotany and Palynology, 173*, 46–56.
Delgado-Baquerizo, M., Giaramida, L., Reich, P. B., Khachane, A. N., Hamonts, K., Edwards, C., et al. (2016). Lack of functional redundancy in the relationship between microbial diversity and ecosystem functioning. *Journal of Ecology, 104*(4), 936–946.
Delph, L. F., & Ashman, T. L. (2006). Trait selection in flowering plants: how does sexual selection contribute? *Integrative and Comparative Biology, 46*(4), 465–472.
Demoulin, V. (1985). The red algal-higher fungi phylogenetic link: the last ten years. *BioSystems, 18*(3–4), 347–356.
Desjardin, D. E., Oliveira, A. G., & Stevani, C. V. (2008). Fungi bioluminescence revisited. *Photochemical & Photobiological Sciences, 7*, 170–182.
Devlin, R. H., & Nagahama, Y. (2002). Sex determination and sex differentiation in fish: An overview of genetic, physiological, and environmental influences. *Aquaculture, 208*(3–4), 191–364.
Dicke, J. M. (1989). Teratology: principles and practice. *Medical Clinics of North America, 73*(3), 567–582.
Diogo, R. (2020). Introduction to evolutionary developmental pathology, or evo-devo-path: On neodarwinism, natural mutants, hopeful monsters, syndromes, genomics, variations, humans, apes, chameleons, and dinosaurs. *Current Molecular Biology Reports, 6*(2), 11–15.
Diogo, R., Guinard, G., & Diaz, R. E., Jr. (2017). Dinosaurs, chameleons, humans, and evo-devo path: Linking Étienne Geoffroy's teratology, Waddington's homeorhesis, Alberch's logic of "monsters," and Goldschmidt hopeful "monsters". *Journal of Experimental Zoology Part B: Molecular and Developmental Evolution, 328*(3), 207–229.
Dollo, L. (1893). Les lois de l'évolution. *Bulletin de la Société Belge de Géologie, de Paléontologie et D'hydrologie, VII*, 164–166.
Dowling, T. E., Martasian, D. P., & Jeffery, W. R. (2002). Evidence for multiple genetic forms with similar eyeless phenotypes in the blind cavefish, *Astyanax mexicanus*. *Molecular Biology and Evolution, 19*(4), 446–455.
Dualan, I. V., & Williams, J. D. (2011). Palp growth, regeneration, and longevity of the obligate hermit crab symbiont *Dipolydora commensalis* (Annelida: Spionidae). *Invertebrate Biology, 130*(3), 264–276.
Dube, H. C. (2015). *An introduction to fungi* (4th ed.). Scientific Publishers.
Dukas, R. (2010). Causes and consequences of male–male courtship in fruit flies. *Animal Behaviour, 80*(5), 913–919.
Eadie, J. M., & Fryxell, J. M. (1992). Density dependence, frequency dependence, and alternative nesting strategies in goldeneyes. *The American Naturalist, 140*(4), 621–641.
Eberhard, W. G., & Gutiérrez, E. E. (1991). Male dimorphisms in beetles and earwigs and the question of developmental constraints. *Evolution, 45*(1), 18–28.
Eeva, T., & Penttinen, R. (2009). Leg deformities of oribatid mites as an indicator of environmental pollution. *Science of the Total Environment, 407*, 4771–4776.

Ellis, R. (2006). *Monsters of the Sea*. Universidad de California.

Ellison, A. M. (2006). Nutrient limitation and stoichiometry of carnivorous plants. *Plant Biology, 8*(6), 740–747.

Ellison, A. M., Bank, M. S., Clinton, B. D., Colburn, E. A., Elliott, K., Ford, C. R., et al. (2005). Loss of foundation species: Consequences for the structure and dynamics of forested ecosystems. *Frontiers in Ecology and the Environment, 3*(9), 479–486.

Emlen, D. J. (1994). Environmental control of horn length dimorphism in the beetle *Onthophagus acuminatus* (Coleoptera: Scarabaeidae). *Proceedings of the Royal Society of London Series B: Biological Sciences, 256*(1346), 131–136.

Ereskovsky, A. V., Borchiellini, C., Gazave, E., Ivanisevic, J., Lapebie, P., Perez, T., et al. (2009). The Homoscleromorph sponge *Oscarella lobularis*, a promising sponge model in evolutionary and developmental biology: Model sponge *Oscarella lobularis*. *BioEssays, 31*(1), 89–97.

Ereskovsky, A. V., Borisenko, I. E., Lapébie, P., Gazave, E., Tokina, D. B., & Borchiellini, C. (2015). *Oscarella lobularis* (Homoscleromorpha, Porifera) regeneration: Epithelial morphogenesis and metaplasia. *PLoS One, 10*(8), e0134566.

Estes, J. A., Tinker, M. T., Williams, T. M., & Doak, D. F. (1998). Killer whale predation on sea otters linking oceanic and nearshore ecosystems. *Science, 282*(5388), 473–476.

Evans, H. C. (2002). Invasive neotropical pathogens of tree crops. In R. Watling, J. C. Frankland, A. M. Ainsworth, S. Isaac, & C. H. Robinson (Eds.), *Tropical mycology* (pp. 83–112). CABI Publishing.

Evans, H. C. (2007). Cacao diseases—The trilogy revisited. *Phytopathology, 97*, 1640–1643.

Evans, H. C., & Prior, C. (1987). Cocoa pod diseases: causal agents and control. *Outlook on Agriculture, 16*, 35–41.

Exner, S. (1891). *Die Physiologie der Facettirten Augen von Krebsen und Insecten*. Deuticke.

Faegri, K., & Van der Pijl, L. (1979). A short history of the study of pollination ecology. *The Principles of Pollination Ecology*, 1–77.

Falkowski, P. G., Fenchel, T., & Delong, E. F. (2008). The microbial engines that drive Earth's biogeochemical cycles. *Science, 320*(5879), 1034–1039.

Farge, E., & Mantovani, C. (2022, September). Double celebration: two-headed tortoise Janus turns 25. *Reuters*. Retrieved from https://www.reuters.com/lifestyle/oddly-enough/double-celebration- two-headed-tortoise-janus-turns-25-2022-09-02/

Farris, J. S. (1977). Phylogenetic analysis under Dollo's Law. *Systematic Biology, 26*(1), 77–88.

Fauset, S., Johnson, M. O., Gloor, M., Baker, T. R., Monteagudo, M. A., Brienen, R. J., et al. (2015). Hyperdominance in Amazonian forest carbon cycling. *Nature Communications, 6*(1), 1–9.

Feldman, L. J. (2009). Insectivorous Plants. In V. H. Resh & R. T. Carde (Eds.), *Encyclopedia of insects* (pp. 511–514). Academic Press.

Ferrer, J., Fernández-López, J., González, A. F., Menéndez, D. F., & Torralba-Burrial, A. (2014). Unusual teratology for a tenebrionid: Antennal schistomelia in *Probaticus granulatus* (Allard, 1876) (Coleoptera: Tenebrionidae). *The Coleopterists Bulletin, 68*(1), 139–142.

Fetzer, I., Johst, K., Schäwe, R., Banitz, T., Harms, H., & Chatzinotas, A. (2015). The extent of functional redundancy changes as species' roles shift in different environments. *Proceedings of the National Academy of Sciences, 112*(48), 14888–14893.

Fincke, O. M., Jödicke, R., Paulson, D. R., & Schultz, T. D. (2005). The evolution and frequency of female color morphs in Holarctic Odonata: why are male-like females typically the minority? *International Journal of Odonatology, 8*(2), 183–212.

Finn, J. K., Tregenza, T., & Norman, M. D. (2009). Defensive tool use in a coconut-carrying octopus. *Current Biology, 19*(23), R1069–R1070.

Fitch, W. M. (1971). Toward defining the course of evolution: Minimum change for a specific tree topology. *Systematic Biology, 20*(4), 406–416.

Flegg, P. B. (1983). Response of the sporophores of the cultivated mushroom (*Agaricus bisporus*) to volatile substances. *Scientia Horticulturae, 21*, 301–310.

Fleischmann, A., Schlauer, J., Smith, S. A., Givnish, T. J., Ellison, A., & Adamec, L. (2018). Evolution of carnivory in angiosperms. In A. Ellison & L. Adamec (Eds.), *Carnivorous plants: Physiology, ecology, and evolution* (pp. 22–42). Oxford University Press.

Fletcher, K. W. (1990). *Habitats used, abundance and distribution of the Mexican spotted owl Strix occidentalis lucida on national forest system lands*. USDA Forest Service, Southwestern Region.

Fleury, C., Marin, F., Marie, B., Luquet, G., Thomas, J., Josse, C., & Lebel, J. M. (2008). Shell repair process in the green ormer Haliotis tuberculata: A histological and microstructural study. *Tissue and Cell, 40*(3), 207–218.

Forsman, E. D., Sovern, S. G., Taylor, M., & Biswell, B. L. (2015). Home range and habitat selection by northern spotted owls on the eastern slope of the Cascade Mountains, Washington. *Journal of Raptor Research, 49*(2), 109–128.

Foster, C. B., & Afonin, S. A. (2005). Abnormal pollen grains: An outcome of deteriorating atmospheric conditions around the Permian–Triassic boundary. *Journal of the Geological Society, 162*(4), 653–659.

Fujii, S., Aoki, H., Kômoto, M., Munakata, K., & Tamura, T. (1968). Studies on the hypertrophic disease of Cherry (Genus *Prunus*), So-called "Witch's Broom" Caused by Taphrina wiesneri: Part I. On the Metabolites of *T. wiesneri* Part II. On the phenolic acids and their related compounds, especially on melilotic acid in infected Japanese Flowering Cherry Leaves. *Agricultural and Biological Chemistry, 32*(7), 810–815.

Fukushima, K., Fang, X., Alvarez-Ponce, D., Cai, H., Carretero-Paulet, L., Chen, C., et al. (2017). Genome of the pitcher plant *Cephalotus* reveals genetic changes associated with carnivory. *Nature Ecology & Evolution, 1*(3), 1–9.

Gadgil, M. (1972). Male dimorphism as a consequence of sexual selection. *The American Naturalist, 106*(951), 574–580.

Gamble, T., Greenbaum, E., Jackman, T. R., Russell, A. P., & Bauer, A. M. (2012). Repeated origin and loss of adhesive toepads in geckos. *PLoS One, 7*(6), e39429.

Ganna, A., Verweij, K. J., Nivard, M. G., Maier, R., Wedow, R., Busch, A. S., et al. (2019). Large-scale GWAS reveals insights into the genetic architecture of same-sex sexual behavior. *Science, 365*(6456), eaat7693.

Garber, P. A., & Rehg, J. A. (1999). The ecological role of the prehensile tail in white-faced capuchins (*Cebus capucinus*). *American Journal of Physical Anthropology, 110*(3), 325–339.

Garcês, A., Pires, I., & Rodrigues, P. (2020). Teratological effects of pesticides in vertebrates: A review. *Journal of Environmental Science and Health, Part B, 55*(1), 75–89.

Gardère, M. L. (2018). Snails in the flowers of *Campanula jacobaea* C. Sm. ex Webb (Campanulaceae) from the island of São Nicolau, Cabo Verde archipelago. *Adansonia, 40*(2), 183–186.

Garibaldi, A., & Turner, N. (2004). Cultural keystone species: Implications for ecological conservation and restoration. *Ecology and Society, 9*(3).

Garnett, G. N., Chambers, C. L., & Mathiasen, R. L. (2006). Use of witches' brooms by Abert squirrels in ponderosa pine forests. *Wildlife Society Bulletin, 34*(2), 467–472.

Gasca-Álvarez, H. J., Skelley, P. E., & Deloya, C. (2018). A remarkable teratological specimen of *Trichiotinus rufobrunneus* (Casey) (Coleoptera: Scarabaeidae: Cetoniinae: Trichiini). *Insecta Mundi, 0680*, 1–5.

Gaston, K. J. (2008). Biodiversity and extinction: the importance of being common. *Progress in Physical Geography, 32*(1), 73–79.

Gaston, K. J. (2011). Common ecology. *Bioscience, 61*(5), 354–362.

Gault, A., Meinard, Y., & Courchamp, F. (2008). Consumers' taste for rarity drives sturgeons to extinction. *Conservation Letters, 1*(5), 199–207.

Gavrilets, S., & Rice, W. R. (2006). Genetic models of homosexuality: Generating testable predictions. *Proceedings of the Royal Society B: Biological Sciences, 273*(1605), 3031–3038.

Ghiselin, M. T. (1969). The evolution of hermaphroditism among animals. *The Quarterly Review of Biology, 44*(2), 189–208.

Gilbert, O. M., Foster, K. R., Mehdiabadi, N. J., Strassmann, J. E., & Queller, D. C. (2007). High relatedness maintains multicellular cooperation in a social amoeba by controlling cheater mutants. *Proceedings of the National Academy of Sciences, 104*(21), 8913–8917.

Giribet, G., & Edgecombe, G. D. (2020). *The invertebrate tree of life*. Princeton University Press.

Givnish, T. J. (2015). New evidence on the origin of carnivorous plants. *Proceedings of the National Academy of Sciences, 112*(1), 10–11.

Goebel, K. (1896). Teratology in modern botany. *Science Progress (1894–1898), 6*(1), 84–100.

Goeze, J. A. E. (1773). *Herrn Karl Bonnets Abhandlungen aus der Insektologie aus d. Franz. übers.u. mit einigen Zusätzen hrsg. v. Joh.* August Ephraim Goeze. Gebauer.

Gomez, G. G., Conci, L. R., Ducasse, D. A., & Nome, S. F. (1996). Purification of the phytoplasma associated with China-tree (*Melia azedarach* L.) Decline and the Production of a polyclonal antiserum for its detection. *Journal of Phytopathology, 144*(9–10), 473–477.

Gonzalez-Moliné, A. L. (2008). Anomalía ocular en *Macrothele calpeiana* (Walckenaer, 1805) (Araneae, Hexathelidae). *Boletin de la Sociedad Entomológica Aragonesa, 42*, 397–398.

Gooheart, J. A., & Bely, A. E. (2016). Sequestration of nematocysts by divergent cnidarian predators: mechanism, function, and evolution. *Invertebrate Biology, 136*(1), 75–91.

Gosden, T. P., & Svensson, E. I. (2009). Density-dependent male mating harassment, female resistance, and male mimicry. *The American Naturalist, 173*(6), 709–721.

Gould, S. J. (1970). Dollo on Dollo's law: Irreversibility and the status of evolutionary laws. *Journal of the History of Biology, 3*(2), 189–212.

Gould, S. J. (1977). *Ontogeny and phylogeny*. Harvard University Press.

Gould, S. J., & Lewontin, R. C. (1979). The spandrels of San Marco and the Panglossian paradigm: A critique of the adaptationist programme. *Proceedings of the Royal Society of London. Series B. Biological Sciences, 205*(1161), 581–598.

Granick, S. (1955). Die plastiden und chondriosomen. In W. Ruhland (Ed.), *Encyclopedia of plant physiology* (pp. 507–564). Springer Verlag.

Grassé, P. P. (1977). *Evolution of living organisms: evidence for a new theory of transformation*. Academic Press.

Gravendyck, J., Schobben, M., Bachelier, J. B., & Kürschner, W. M. (2020). Macroecological patterns of the terrestrial vegetation history during the end-Triassic biotic crisis in the central European Basin: A palynological study of the Bonenburg section (NW-Germany) and its supra-regional implications. *Global and Planetary Change, 194*, 103286.

Griffith, G. W., Nicholson, J., Nenninger, A., Birch, R. N., & Hedger, J. N. (2003). Witches' brooms and frosty pods: Two major pathogens of cacao. *New Zealand Journal of Botany, 41*(3), 423–435.

Grime, J. P. (1998). Benefits of plant diversity to ecosystems: Immediate, filter and founder effects. *Journal of Ecology, 86*(6), 902–910.

Gross, M. R. (1982). Sneakers, satellites and parentals: Polymorphic mating strategies in North American sunfishes. *Zeitschrift für Tierpsychologie, 60*(1), 1–26.

Gross, M. R. (1985). Disruptive selection for alternative life histories in salmon. *Nature, 313*(5997), 47–48.

Gross, M. R. (1996). Alternative reproductive strategies and tactics: Diversity within sexes. *Trends in Ecology & Evolution, 11*(2), 92–98.

Guidetti, R., & Bertolani, R. (2005). Tardigrade taxonomy: An updated check list of the taxa and a list of characters for their identification. *Zootaxa, 845*(1), 1–46.

Guidetti, R., & Jönsson, K. I. (2002). Long-term anhydrobiotic survival in semi-terrestrial micrometazoans. *Journal of Zoology, 257*(2), 181–187.

Guidetti, R., Rizzo, A. M., Altiero, T., & Rebecchi, L. (2012). What can we learn from the toughest animals of the Earth? Water bears (tardigrades) as multicellular model organisms in order to perform scientific preparations for lunar exploration. *Planetary and Space Science, 74*(1), 97–102.

Guinard, G. (2015). Introduction to evolutionary teratology, with an application to the forelimbs of tyrannosauridae and carnotaurinae (Dinosauria: Theropoda). *Evolutionary Biology, 42*(1), 20–41.

Hägele, S., Köhler, R., Merkert, H., Schleicher, M., Hacker, J., & Steinert, M. (2000). *Dictyostelium discoideum*: A new host model system for intracellular pathogens of the genus *Legionella*. *Cellular Microbiology, 2*(2), 165–171.

Hale, M. E. (1996). Functional morphology of ventral tail bending and prehensile abilities of the seahorse, *Hippocampus kuda*. *Journal of Morphology, 227*(1), 51–65.
Hall, R. J., Milner-Gulland, E. J., & Courchamp, F. (2008). Endangering the endangered: The effects of perceived rarity on species exploitation. *Conservation Letters, 1*(2), 75–81.
Hamer, D., Mustanski, B., Sell, R., Sanders, S. A., & Garcia, J. R. (2021). Comment on "Large-scale GWAS reveals insights into the genetic architecture of same-sex sexual behavior". *Science, 371*(6536), eaba2941.
Han, S. S. R., Cho, C. K., Lee, Y. W., & Yoo, H. S. (2009). Antimetastatic and immunomodulating effect of water extracts from various mushrooms. *Journal of Acupuncture and Meridian Studies, 2*(3), 218–227.
Hanlon, R. T., & Messenger, J. B. (2018). *Cephalopod behaviour*. Cambridge University Press.
Harari, A. R., Brockmann, H. J., & Landolt, P. J. (2000). Intrasexual mounting in the beetle *Diaprepes abbreviatus* (L.). *Proceedings of the Royal Society of London. Series B: Biological Sciences, 267*(1457), 2071–2079.
Harder, L. D., & Johnson, S. D. (2008). Function and evolution of aggregated pollen in angiosperms. *International Journal of Plant Sciences, 169*(1), 59–78.
Harnik, P. G., Simpson, C., & Payne, J. L. (2012). Long-term differences in extinction risk among the seven forms of rarity. *Proceedings of the Royal Society B: Biological Sciences, 279*(1749), 4969–4976.
Harrington, M. G., & Hawksworth, F. G. (1990). Interactions of fire and dwarf mistletoe on mortality of southwestern ponderosa pine. In J. S. Krammes (Ed.), *Effects of fire management of southwestern natural resources: Symposium proceedings* (pp. 234–240). USDA Forest Service, Rocky Mountain Forest and Range Experiment Station General Technical Report RM-191.
Hashimoto, T., Horikawa, D. D., Saito, Y., Kuwahara, H., Kozuka-Hata, H., Shin-i, T., et al. (2016). Extremotolerant tardigrade genome and improved radiotolerance of human cultured cells by tardigrade-unique protein. *Nature Communications, 7*(1), 12808.
Hashimura, H., Morimoto, Y. V., Yasui, M., & Ueda, M. (2019). Collective cell migration of *Dictyostelium* without cAMP oscillations at multicellular stages. *Communications Biology, 2*(1), 34.
Hawksworth, F. G. (1961). *Dwarf mistletoe of ponderosa pine in the Southwest (No. 1246)*. US Department of Agriculture.
Hawksworth, F. G., & Wiens, D. (1996). *Dwarf mistletoes: Biology, pathology, and systematics*. In *United States Forest Service Agricultural Handbook 709*.
Hebbar, P. K. (2007). Cacao diseases: a global perspective from an industry point of view. *Phytopathology, 97*(12), 1658–1663.
Hedrich, R., & Fukushima, K. (2021). On the origin of carnivory: Molecular physiology and evolution of plants on an animal diet. *Annual Review of Plant Biology, 72*, 133–153.
Hedwall, S. J., & Mathiasen, R. L. (2006). Wildlife use of Douglas-fir dwarf mistletoe witches' brooms in the Southwest. *Western North American Naturalist, 66*(4), 450–455.
Hegemann, P. (1997). Vision in microalgae. *Planta, 203*(3), 265–274.
Heinzelmann, R., Dutech, C., Tsykun, T., Labbé, F., Soularue, J. P., & Prospero, S. (2019). Latest advances and future perspectives in *Armillaria* research. *Canadian Journal of Plant Pathology, 41*(1), 1–23.
Hendrickx, A. G., & Binkerd, P. E. (1990). Nonhuman primates and teratological research. *Journal of Medical Primatology, 19*(2), 81–108.
Hengherr, S., Brümmer, F., & Schill, R. O. (2008). Anhydrobiosis in tardigrades and its effects on longevity traits. *Journal of Zoology, 275*(3), 216–220.
Hercos, A. P., Sobansky, M., Queiroz, H. L. D., & Magurran, A. E. (2013). Local and regional rarity in a diverse tropical fish assemblage. *Proceedings of the Royal Society B: Biological Sciences, 280*(1751), 20122076.
Hernandez-Raquet, G., Durand, E., Braun, F., Cravo-Laureau, C., & Godon, J. J. (2013). Impact of microbial diversity depletion on xenobiotic degradation by sewage-activated sludge. *Environmental Microbiology Reports, 5*(4), 588–594.

Herring, P. J. (1994). Luminous fungi. *Mycologist, 8*(4), 181–183.

Heslop-Harrison, Y., & Heslop-Harrison, J. (1980). Chloride ion movement and enzyme secretion from the digestive glands of *Pinguicula*. *Annals of Botany, 45*(6), 729–731.

Heslop-Harrison, Y., & Knox, R. B. (1971). A cytochemical study of the leaf-gland enzymes of insectivorous plants of the genus *Pinguicula*. *Planta, 96*, 183–211.

Hibbett, D. S., Binder, M., Bischoff, J. F., Blackwell, M., Cannon, P. F., Eriksson, O. E., et al. (2007). A higher-level phylogenetic classification of the Fungi. *Mycological Research, 111*(5), 509–547.

Higham, T. E., Russell, A. P., Birn-Jeffery, A., Collins, C. E., & Hulsey, C. D. (2014). Use it or lose it: Locomotor evolution associated with the loss of adhesion in geckos. *Integrative and comparative biology, 54*, E91.

Hinaux, H., Devos, L., Blin, M., Elipot, Y., Bibliowicz, J., Alié, A., & Rétaux, S. (2016). Sensory evolution in blind cavefish is driven by early embryonic events during gastrulation and neurulation. *Development, 143*(23), 4521–4532.

Hinnekint, B. O. N. (1987). Population dynamics of *Ischnura elegans* (Vander Linden)(Insecta: Odonata) with special reference to morphological colour changes, female polymorphism, multiannual cycles and their influence on behaviour. *Hydrobiologia, 146*, 3–31.

Hochner, B., Shomrat, T., & Fiorito, G. (2006). The octopus: A model for a comparative analysis of the evolution of learning and memory mechanisms. *The Biological Bulletin, 210*(3), 308–317.

Hoegh-Guldberg, O., Mumby, P. J., Hooten, A. J., Steneck, R. S., Greenfield, P., Gomez, E., et al. (2007). Coral reefs under rapid climate change and ocean acidification. *Science, 318*(5857), 1737–1742.

Hogenhout, S. A., & Segura, M. (2009). Phytoplasma genomics, from sequencing to comparative and functional genomics: What have we learnt. *Phytoplasmas: Genomes, Plant Hosts and Vectors*, 19–36.

Holt, R. D., Grover, J., & Tilman, D. (1994). Simple rules for interspecific dominance in systems with exploitative and apparent competition. *The American Naturalist, 144*(5), 741–771.

Horikawa, D. D., Sakashita, T., Katagiri, C., Watanabe, M., Kikawada, T., Nakahara, Y., et al. (2006). Radiation tolerance in the tardigrade *Milnesium tardigradum*. *International Journal of Radiation Biology, 82*(12), 843–848.

Horikawa, D. D., Yamaguchi, A., Sakashita, T., Tanaka, D., Hamada, N., Yukuhiro, F., et al. (2012). Tolerance of anhydrobiotic eggs of the Tardigrade *Ramazzottius varieornatus* to extreme environments. *Astrobiology, 12*(4), 283–289.

Hubbell, S. P. (2013). Tropical rain forest conservation and the twin challenges of diversity and rarity. *Ecology and Evolution, 3*(10), 3263–3274.

Hunt, J., & Simmons, L. W. (2001). Status-dependent selection in the dimorphic beetle *Onthophagus taurus*. *Proceedings of the Royal Society of London Series B: Biological Sciences, 268*(1484), 2409–2414.

Ibarra-Laclette, E., Lyons, E., Hernández-Guzmán, G., Pérez-Torres, C. A., Carretero-Paulet, L., Chang, T. H., et al. (2013). Architecture and evolution of a minute plant genome. *Nature, 498*(7452), 94–98.

Iliev, I., & Kitin, P. (2011). Origin, morphology, and anatomy of fasciation in plants cultured in vivo and in vitro. *Plant Growth Regulation, 63*(2), 115–129.

Imazeki, R. (1938). A rare fungus, *Urnula geaster* Peck grows in Kyusyu, Japan. *Journal of Japanese Botany, 14*, 680–684.

Iserbyt, A., Bots, J., Van Gossum, H., & Sherratt, T. N. (2013). Negative frequency-dependent selection or alternative reproductive tactics: Maintenance of female polymorphism in natural populations. *BMC Evolutionary Biology, 13*(1), 1–11.

Ishimatsu, A., & Itazawa, Y. (1981). Ventilation of the air-breathing organ in the snakehead Channa argus. *Japanese Journal of Ichthyology, 28*(3), 276–282.

Ishimatsu, A., & Itazawa, Y. (1983). Difference in blood oxygen levels in the outflow vessels of the heart of an air-breathing fish, Channa argus: Do separate blood streams exist in a teleostean heart? *Journal of Comparative Physiology, 149*, 435–440.

Izquierdo, M. A. (2021). A teratologic spider with duplicated reproductive organs. *The Journal of Arachnology, 48*(3), 343–345.
Jacunski, L., & Templin, J. (1992). Anatomical examination of the anterior prosoma in bicephalous monsters of *Tegenaria atrica* CL Koch. *Bulletin of the Polish Academy of Sciences. Biological Sciences, 40*(2).
Jacuński, L., & Templin, J. (2003). Morphology of prosoma in bicephalous monsters of *Tegenaria atrica* CL Koch. *Journal of Thermal Biology, 28*(5), 393–396.
Jacunski, L., Templin, J., & Napiórkowska, T. (2002). Doubling of prosoma in larva of *Tegenaria atrica* CL Koch [Agelenidae]. *Bulletin of the Polish Academy of Sciences. Biological Sciences, 50*(1).
Jacunski, L., Templin, J., & Napiórkowska, T. (2005). Changes in the neuromerism of the subesophageal part of the nervous system in oligomelic individuals of *Tegenaria atrica* (Arachnida). *Biologia, 60*(5), 589–592.
Jain, M., Flynn, D. F., Prager, C. M., Hart, G. M., DeVan, C. M., Ahrestani, F. S., et al. (2014). The importance of rare species: A trait-based assessment of rare species contributions to functional diversity and possible ecosystem function in tall-grass prairies. *Ecology and Evolution, 4*(1), 104–112.
Jakšová, J., Libiaková, M., Bokor, B., Petřík, I., Novák, O., & Pavlovič, A. (2020). Taste for protein: Chemical signal from prey stimulates enzyme secretion through jasmonate signaling in the carnivorous plant Venus flytrap. *Plant Physiology and Biochemistry, 146*, 90–97.
Jeffery, W. R. (2001). Cavefish as a model system in evolutionary developmental biology. *Developmental Biology, 231*(1), 1–12.
Jeffery, W. R. (2020). *Astyanax* surface and cave fish morphs. *EvoDevo, 11*(1), 1–10.
Jeffs, K., & Brownlow, M. (2017). *Green Seas, Blue Planet II.*
Jeng, M. L. (2019). Biofluorescence in terrestrial animals, with emphasis on fireflies: A review and field observation. *Bioluminescence—Analytical Applications and Basic Biology*, 1–16.
Jenkins, J. B. (2003). *Genética.* Editorial Reverté.
Jessen, S., & Grossmann, T. (2014). Unconscious discrimination of social cues from eye whites in infants. *Proceedings of the National Academy of Sciences, 111*(45), 16208–16213.
Jia, K., Thomas, C., Akbar, M., Sun, Q., Adams-Huet, B., Gilpin, C., & Levine, B. (2009). Autophagy genes protect against *Salmonella typhimurium* infection and mediate insulin signaling-regulated pathogen resistance. *Proceedings of the National Academy of Sciences, 106*(34), 14564–14569.
Jiang, D. Y., Motani, R., Huang, J. D., Tintori, A., Hu, Y. C., Rieppel, O., et al. (2016). A large aberrant stem ichthyosauriform indicating early rise and demise of ichthyosauromorphs in the wake of the end-Permian extinction. *Scientific Reports, 6*(1), 1–9.
Jiménez, M. L., & Llinas, J. (2002). Revision of ocular anomalies in epigean spiders (Arachnida: Araneae) with notes on four new records. *Anales del Instituto de Biología. Serie Zoología, 73*(2), 241–250.
Jobling, S., Nolan, M., Tyler, C. R., Brighty, G., & Sumpter, J. P. (1998). Widespread sexual disruption in wild fish. *Environmental Science & Technology, 32*(17), 2498–2506.
Johnson, H. P. (1902). Collateral budding in annelids of the genus *Trypanosyllis*. *The American Naturalist, 36*(424), 295–315.
Johnson, M. K., & Russell, A. P. (2009). Configuration of the setal fields of *Rhoptropus* (Gekkota: Gekkonidae): Functional, evolutionary, ecological and phylogenetic implications of observed pattern. *Journal of Anatomy, 214*(6), 937–955.
Jönsson, K. I. (2007). Tardigrades as a potential model organism in space research. *Astrobiology, 7*(5), 757–766.
Jönsson, K. I., & Bertolani, R. (2001). Facts and fiction about long-term survival in tardigrades. *Journal of Zoology, 255*(1), 121–123.
Jönsson, I. K., Harms-Ringdahl, M., & Torudd, J. (2005). Radiation tolerance in the eutardigrade *Richtersius coronifer*. *International Journal of Radiation Biology, 81*(9), 649–656.
Jönsson, K. I., Rabbow, E., Schill, R. O., Harms-Ringdahl, M., & Rettberg, P. (2008). Tardigrades survive exposure to space in low Earth orbit. *Current Biology, 18*(17), R729–R731.

Josef, N., Amodio, P., Fiorito, G., & Shashar, N. (2012). Camouflaging in a complex environment—octopuses use specific features of their surroundings for background matching. *PLoS One, 7*(5), e37579.

Jousset, A., Bienhold, C., Chatzinotas, A., Gallien, L., Gobet, A., Kurm, V., et al. (2017). Where less may be more: how the rare biosphere pulls ecosystems strings. *The ISME Journal, 11*(4), 853–862.

Juniper, B. E., Robins, R. J., & Joel, D. M. (1989). *The carnivorous plants*. Academic Press.

Kalter, H. (2003). Teratology in the 20th century: Environmental causes of congenital malformations in humans and how their environmental causes were established. *Neurotoxicology and Teratology, 25*, 131–282.

Kaminski, M. A. (2004). *The year 2000 classification of the agglutinated foraminifera*. Grzybowski Foundation.

Kano, F., Kawaguchi, Y., & Hanling, Y. (2022). Experimental evidence that uniformly white sclera enhances the visibility of eye-gaze direction in humans and chimpanzees. *Elife, 11*, e74086.

Kaston, B. J. (1962). Ocular anomalies in spiders. *Bulletin of the Brooklyn Entomological Society, 57*, 17–21.

Kaston, B. J. (1963). Deformities of external genitalia in spiders. *Journal of the New York Entomological Society, 71*, 30–39.

Kay, K. M., & Sargent, R. D. (2009). The role of animal pollination in plant speciation: Integrating ecology, geography, and genetics. *Annual Review of Ecology, Evolution, and Systematics, 40*, 637–656.

Kay, R. R., Tilly, R., & Garrod, D. (1978). Requirement for cell differentiation in *Dictyostelium discoideum*. *Nature, 271*(5640), 58–60.

Kearney, R., & Kearney, E. (2009). *Waxcaps fungi (Hygrocybe)*. Abnormalities in Lane Cove Bushland Park.

Kearney, R., & Kearney, E. (2011). The listing of an Australian Hygrocybeae community and its new species under State and Commonwealth legislations. *Fungal Conservation, 1*, 13–28.

Kepner, W. A. (1911). Concerning the nematocysts of *Microstoma*. *Science, 34*, 213–214.

Kepner, W. A. (1951). Food reactions of amoebas and the manipulation of nematocysts of *Hydra* by *Microstomum*. *Science, 114*(2969), 556–557.

Kepner, W. A., & Nuttycombe, J. W. (1929). Further studies upon the nematocysts of *Microstomum caudatum*. *Biological Bulletin, 57*(2), 69–78.

Kessin, R. H. (2001). *Dictyostelium: evolution, cell biology, and the development of multicellularity*. Cambridge University Press.

Kidd, K. A., Blanchfield, P. J., Mills, K. H., Palace, V. P., Evans, R. E., Lazorchak, J. M., & Flick, R. W. (2007). Collapse of a fish population after exposure to a synthetic estrogen. *Proceedings of the National Academy of Sciences, 104*(21), 8897–8901.

Kijimoto, T., Pespeni, M., Beckers, O., Moczek, A. P., et al. (2013). Beetle horns and horned beetles: Emerging models in developmental evolution and ecology. *Wiley Interdisciplinary Reviews: Developmental Biology, 2*(3), 405–418.

Kirk, J. T. O., & Tilney-Bassett, R. A. E. (1978). *The plastids* (2nd ed.). Elsevier.

Kleinlogel, S., & White, A. G. (2008). The secret world of shrimps: Polarisation vision at its best. *PLoS One, 3*(5), e2190.

Klingenberg, C. P. (2010). Evolution and development of shape: Integrating quantitative approaches. *Nature Reviews Genetics, 11*(9), 623–635.

Kobayashi, H., & Kohshima, S. (1997). Unique morphology of the human eye. *Nature, 387*(6635), 767–768.

Kobayashi, H., & Kohshima, S. (2001). Unique morphology of the human eye and its adaptive meaning: comparative studies on external morphology of the primate eye. *Journal of Human Evolution, 40*(5), 419–435.

Kocáb, O., Jakšová, J., Novák, O., Petřík, I., Lenobel, R., Chamrád, I., & Pavlovič, A. (2020). Jasmonate-independent regulation of digestive enzyme activity in the carnivorous butterwort *Pinguicula× Tina*. *Journal of Experimental Botany, 71*(12), 3749–3758.

Kohler, S., Delwiche, C. F., Denny, P. W., et al. (1997). A plastid of probable green algal origin in Apicomplexan parasites. *Science, 275*, 1485–1148.
Koller-Peroutka, M., Lendl, T., Watzka, M., & Adlassnig, W. (2015). Capture of algae promotes growth and propagation in aquatic *Utricularia*. *Annals of Botany, 115*(2), 227–236.
Kreimer, G. (2009). The green algal eyespot apparatus: A primordial visual system and more? *Current Genetics, 55*, 19–43.
Krisko, A., & Radman, M. (2013). Biology of extreme radiation resistance: the way of Deinococcus radiodurans. *Cold Spring Harbor perspectives in biology, 5*(7), a012765.
Krohne, G. (2018). Organelle survival in a foreign organism: *Hydra* nematocysts in the flatworm *Microstomum lineare*. *European Journal of Cell Biology, 97*(4), 289–299.
Krohne, G. (2020). Hydra nematocysts in the flatworm *Microstomum lineare*: In search for alterations preceding their disappearance from the new host. *Cell and Tissue Research, 379*(1), 63–71.
Kury, A. B., & Machado, G. (2003). Transporte de ovos pelos machos: uma sinapomorfia comportamental para os opiliões da família Podoctidae (Opiliones: Laniatores).IV Encontro de Aracnólogos do Cone Sul, , p 127).
Lagorio, M. G., Cordon, G. B., & Iriel, A. (2015). Reviewing the relevance of fluorescence in biological systems. *Photochemical & Photobiological Sciences, 14*(9), 1538–1559.
Lai, C. (2010). *Potential applications of the natural design of internal explosion chambers in the bombardier beetle (Carabidae, Brachinus)*. (Doctoral dissertation,. Massachusetts Institute of Technology).
Lamb, T., & Bauer, A. M. (2006). Footprints in the sand: independent reduction of subdigital lamellae in the Namib–Kalahari burrowing geckos. *Proceedings of the Royal Society B: Biological Sciences, 273*(1588), 855–864.
Lambert, E. B. (1930). Two new diseases of cultivated Mushrooms. *Phytopathology, 20*(11).
Land, M. F., Marshall, N. J., Brownless, D., & Cronin, T. W. (1990). The eye-movements of the mantis shrimp *Odontodactylus scyllarus* (Crustacea: Stomatopoda). *Journal of Comparative Physiology., 167*, 155–166.
Langridge, K. V., Broom, M., & Osorio, D. (2007). Selective signalling by cuttlefish to predators. *Current Biology, 17*(24), R1044–R1045.
Lankester, E. R. (1893). Blind animals in caves. *Nature, 47*(1221), 486–486.
Lankinen, Å., & Karlsson Green, K. (2015). Using theories of sexual selection and sexual conflict to improve our understanding of plant ecology and evolution. *AoB Plants, 7*.
Lavaut, E., Guillemin, M. L., Colin, S., Faure, A., Coudret, J., Destombe, C., & Valero, M. (2022). Pollinators of the sea: A discovery of animal-mediated fertilization in seaweed. *Science, 377*(6605), 528–530.
Lawler, J. J., White, D., Sifneos, J. C., & Master, L. L. (2003). Rare species and the use of indicator groups for conservation planning. *Conservation Biology, 17*(3), 875–882.
Leach, M. K., & Givnish, T. J. (1996). Ecological determinants of species loss in remnant prairies. *Science, 273*(5281), 1555–1558.
Lee, D., & Kim, W. (2020). New species of *Pycnogonum* (Pycnogonida: Pycnogonidae) from Green Island, Taiwan, with an additional note on the holotype of *P. spatium*. *Zootaxa, 4750*(1), zootaxa-4750.
Lee, W. L., Reiswig, H. M., Austin, W. C., & Lundsten, L. (2012). An extraordinary new carnivorous sponge, *Chondrocladia lyra*, in the new subgenus *Symmetrocladia* (Demospongiae, Cladorhizidae), from off of northern California, USA. *Invertebrate Biology, 131*(4), 259–284.
Leitão, R. P., Zuanon, J., Villéger, S., Williams, S. E., Baraloto, C., Fortunel, C., et al. (2016). Rare species contribute disproportionately to the functional structure of species assemblages. *Proceedings of the Royal Society B: Biological Sciences, 283*(1828), 20160084.
Leonard, J. L. (2006). Sexual selection: Lessons from hermaphrodite mating systems. *Integrative and Comparative Biology, 46*(4), 349–367.
Lesniewska, M., Bonato, L., Minelli, A., & Fusco, G. (2009). Trunk anomalies in the centipede *Stigmatogaster subterranea* provide insight into late-embryonic segmentation. *Arthropod Structure & Development, 38*, 417–426.

Lev-Yadun, S., Inbar, M., Izhaki, I., & Dafni, A. (2002). Colour patterns in vegetative parts of plants deserve more research attention. *Trends in Plant Science, 7*(2), 59–60.

Li, D., & Bi, X. (2018). Effect of teratogens on development of *Drosophila melanogaster*. In L. Félix (Ed.), *Teratogenicity testing: Methods and protocols* (pp. 233–241). Humana Press.

Lindström, S., Sanei, H., Van De Schootbrugge, B., Pedersen, G. K., Lesher, C. E., Tegner, C., et al. (2019). Volcanic mercury and mutagenesis in land plants during the end-Triassic mass extinction. *Science. Advances, 5*(10), eaaw4018.

Lisenkova, A. A., Grigorenko, A. P., Tyazhelova, T. V., Andreeva, T. V., Gusev, F. E., Manakhov, A. D., & Rogaev, E. I. (2017). Complete mitochondrial genome and evolutionary analysis of Turritopsis dohrnii, the "immortal" jellyfish with a reversible life-cycle. *Molecular Phylogenetics and Evolution, 107*, 232–238.

Loker, E. S., & Brant, S. V. (2006). Diversification, dioecy and dimorphism in schistosomes. *Trends in Parasitology, 22*(11), 521–528.

Lopes, U. V., Monteiro, W. R., Pires, J. L., Rocha, J. B., & Pinto, L. R. M. (2003). On farm selection for witches' broom resistance in Bahia, Brazil—A historical retrospective. In *Proceedings of the 14th International Cocoa Research Conference*.

Lynch, V. J. (2022). Is there a loophole in Dollo's law? A DevoEvo perspective on irreversibility (of felid dentition). *Journal of Experimental Zoology Part B: Molecular and Developmental Evolution*, 1–9.

Lynch, M. D., & Neufeld, J. D. (2015). Ecology and exploration of the rare biosphere. *Nature Reviews Microbiology, 13*(4), 217–229.

Lynch, V. J., & Wagner, G. P. (2010). Did egg-laying boas break dollo's law? Phylogenetic evidence for reversal to oviparity in sand boas (eryx: boidae). *Evolution, 64*, 207–216.

Lyons, K. G., Brigham, C. A., Traut, B. H., & Schwartz, M. W. (2005). Rare species and ecosystem functioning. *Conservation Biology, 19*(4), 1019–1024.

Machado, G., & Macías-Ordóñez, R. (2007). Social behavior. In R. Pinto-da-Rocha, G. Machado, & G. Giribet (Eds.), *Harvestmen: The biology of opiliones* (pp. 400–413). Harvard Univ. Press.

Machado, G., & Raimundo, R. L. G. (2001). Parental investment and the evolution of subsocial behaviour in harvestmen (Arachnida Opiliones). *Ethology Ecology & Evolution, 13*(2), 133–150.

Machado, G., & Wolff, J. O. (2018). The assassination of a hypothesis by non-critical interpretation of molecular data: A comment on. *Molecular Phylogenetics and Evolution, 129*, 346–348.

Maklakov, A. A., & Bonduriansky, R. (2009). Sex differences in survival costs of homosexual and heterosexual interactions: Evidence from a fly and a beetle. *Animal Behaviour, 77*(6), 1375–1379.

Maldonado, M., Lopez-Acosta, M., Sitja, C., Aguilar, R., Garcia, S., & Vacelet, J. (2013). A giant foraminifer that converges to the feeding strategy of carnivorous sponges: *Spiculosiphon oceana* sp. nov. (Foraminifera, Astrorhizida). *Zootaxa, 3669*, 571–584.

Maliarenko, V. M., & Mudrak, T. P. (2013). Cactus viruses in fasciated plants. *Biologija, 59*(2).

Mallon, C. A., Poly, F., Le Roux, X., Marring, I., van Elsas, J. D., & Salles, J. F. (2015). Resource pulses can alleviate the biodiversity–invasion relationship in soil microbial communities. *Ecology, 96*(4), 915–926.

Marcotrigiano, M. (1997). Chimeras and variegation: Patterns of deceit. *HortScience, 32*(5), 773–784.

Marlétaz, F., Peijnenburg, K. T. C. A., Goto, T., Satoh, N., & Rokhsar, D. S. (2019). A new spiralian phylogeny places the enigmatic arrow worms among gnathiferans. *Current Biology, 29*(2), 312–318.

Marshall, A. J., & Williams, W. D. (1985). *Zoología. Invertebrados* (Vol. 1). Reverté.

Marshall, N. J., Land, M. F., King, C. A., & Cronin, T. W. (1991). The compound eyes of mantis shrimps (Crustacea, Hoplocarida, Stomatopoda) I. Compound eye structure: the detection of polarized light. *Philosophical Transactions of the Royal Society of London. Series B: Biological Sciences, 334*(1269), 33–56.

Marshall, J. E., Lakin, J., Troth, I., & Wallace-Johnson, S. M. (2020). UV-B radiation was the Devonian–Carboniferous boundary terrestrial extinction kill mechanism. *Science Advances, 6*(22), eaba0768.

Martell, L., Piraino, S., Gravili, C., & Boero, F. (2016). Life cycle, morphology and medusa ontogenesis of Turritopsis dohrnii (Cnidaria: Hydrozoa). *Italian Journal of Zoology, 83*(3), 390–399.

Martens, J. (1993). Further cases of paternal care in Opiliones (Arachnida). *Tropical Zoology, 6*(1), 97–107.

Martin, S. K., Beatty, J. S., & Hawksworth, F. G. (1992). Douglas-fir dwarf mistletoe brooms and Spotted Owl nesting habitat: eastern Cascade Range, Washington. In *Proceedings of the 40th Annual Western International Forest Disease Work Conference* (pp. 77–79). USDA Forest Service, Pacific Southwest Research Station.

Maruyama, M., & Parker, J. (2017). Deep-time convergence in rove beetle symbionts of army ants. *Current Biology, 27*(6), 920–926.

Masters, M. T. (1869). *Vegetable teratology: An account of the principle deviations from the usual construction of plants*. Ray Society.

Mather, J. A., & Dickel, L. (2017). Cephalopod complex cognition. *Current Opinion in Behavioral Sciences, 16*, 131–137.

Mathiasen, R. L. (1996). Dwarf mistletoes in forest canopies. *Northwest Science, 70*, 61–71.

Matz, C., & Kjelleberg, S. (2005). Off the hook—How bacteria survive protozoan grazing. *Trends in Microbiology, 13*, 302–307.

May, R. M., & Guimard, J. (1964). The radiation resistance of some micro-animals. *International Journal of Radiation Biology and Related Studies in Physics, Chemistry, and Medicine, 8*(6), 479–485.

McFadden, G. I., Reith, M. E., Munholland, J., & Lang-Unnasch, N. (1996). Plastid in human parasites. *Nature, 381*(6582), 482–482.

McKinnon, J. S., & Pierotti, M. E. (2010). Colour polymorphism and correlated characters: genetic mechanisms and evolution. *Molecular Ecology, 19*(23), 5101–5125.

McManus, R., Ravenscraft, A., & Moore, W. (2018). Bacterial associates of a gregarious riparian beetle with explosive defensive chemistry. *Frontiers in Microbiology, 9*, 2361.

McRobert, S. P., & Tompkins, L. (1988). Two consequences of homosexual courtship performed by *Drosophila melanogaster* and *Drosophila affinis* males. *Evolution, 42*(5), 1093–1097.

Mearing, A. S., Burkart, J. M., Dunn, J., Street, S. E., & Koops, K. (2022). The evolutionary drivers of primate scleral coloration. *Scientific Reports, 12*(1), 1–10.

Medeiros, F. H. V., Pomella, A. W. V., De Souza, J. T., Niella, G. R., Valle, R., Bateman, R. P., et al. (2010). A novel, integrated method for management of witches' broom disease in Cacao in Bahia, Brazil. *Crop Protection, 29*(7), 704–711.

Medina, J. M., Shreenidhi, P. M., Larsen, T. J., Queller, D. C., & Strassmann, J. E. (2019). Cooperation and conflict in the social amoeba *Dictyostelium discoideum*. *International Journal of Developmental Biology, 63*, 371–382.

Medway, D. G. (2000). Mycophagy by north island robin. *Australasian Mycologist, 19*, 102.

Meinhardt, H. (1986). The threefold subdivision of segments and the initiation of legs and wings in insects. *Trends in Genetics, 2*, 36–41.

Mendes, L. F., & Stevani, C. V. (2010). Evaluation of metal toxicity by a modified method based on the fungus *Gerronema viridilucens* bioluminescence in agar medium. *Environmental Toxicology and Chemistry, 29*(2), 320–326.

Messenger, J. B. (2001). Cephalopod chromatophores: neurobiology and natural history. *Biological Reviews, 76*(4), 473–528.

Meyer, J. R., & Kassen, R. (2007). The effects of competition and predation on diversification in a model adaptive radiation. *Nature, 446*(7134), 432–435.

Meyerowitz, E. M., Smyth, D. R., & Bowman, J. L. (1989). Abnormal flowers and pattern formation in floral. *Development, 106*(2), 209–217.

Michael, E., Henning, B., & Kreisel, H. (1983). *Handbuch jür Piltfreunde* (Vol. 5). G. Fischer Verlag.

Micheneau, C., Johnson, S. D., & Fay, M. F. (2009). Orchid pollination: From Darwin to the present day. *Botanical Journal of the Linnean Society, 161*(1), 1–19.

Minelli, A. (2009). *Perspectives in animal phylogeny and evolution*. Oxford University Press.

Mishra, M., & Barik, B. K. (2018). Behavioral teratogenesis in *Drosophila melanogaster*. In L. Félix (Ed.), *Teratogenicity testing* (pp. 277–298). Humana Press.

Mitchell, S. M., Sole, C. S., & Lyle, R. L. (2020). Teratological cases of the ocular patterns in the South African endemic trapdoor spider genus *Stasimopus* Simon (1892) (Araneae, Mygalomorphae, Stasimopidae). *African Zoology, 55*(4), 363–367.

Moczek, A. P., & Emlen, D. J. (2000). Male horn dimorphism in the scarab beetle, *Onthophagus taurus*: do alternative reproductive tactics favour alternative phenotypes? *Animal Behaviour, 59*(2), 459–466.

Monaghan, J. R., Walker, J. A., Page, R. B., Putta, S., Beachy, C. K., & Voss, S. R. (2007). Early gene expression during natural spinal cord regeneration in the salamander Ambystoma mexicanum. *Journal of Neurochemistry, 101*(1), 27–40.

Moore, H. D. M., Martin, M., & Birkhead, T. R. (1999). No evidence for killer sperm or other selective interactions between human spermatozoa in ejaculates of different males in vitro. *Proceedings of the Royal Society of London Series B: Biological Sciences, 266*(1436), 2343–2350.

Moore, D., Robson, G. D., & Trinci, A. P. (2021). *21st century guidebook to fungi*. Cambridge University Press.

Mor, Y., & Zieslin, N. (1992). Phyllody malformation in flowers of *Rosa* × hybrida cv. Motrea: effects of rootstocks, flower position, growth regulators and season. *Journal of Experimental Botany, 43*(1), 89–93.

Morrison, D. J. (1975). Ion uptake by rhizomorphs of *Armillaria mellea*. *Canadian Journal of Botany, 53*(1), 48–51.

Moscoso, J. (1998). Monsters as evidence: The uses of the abnormal body during the early eighteenth century. *Journal of the History of Biology, 31*(3), 355–382.

Mouillot, D., Bellwood, D. R., Baraloto, C., Chave, J., Galzin, R., Harmelin-Vivien, M., et al. (2013). Rare species support vulnerable functions in high-diversity ecosystems. *PLoS Biology, 11*(5), e1001569.

Mowrey, R. A., & Zasada, J. C. (1984). Den tree use and movements of northern flying squirrels in interior Alaska and implications for forest management. In W. R. Meehan, T. R. Merrell, & T. A. Hanley (Eds.), *Fish and wildlife relationships in old growth forests: Proceedings of a symposium, Juneau, Alaska* (pp. 351–356). American Institute of Fishery Research Biologists.

Moyer, J. T., & Nakazono, A. (1978). Protandrous hermaphroditism in six species of the anemonefish genus *Amphiprion* in Japan. *Japanese Journal of Ichthyology, 25*(2), 101–106.

Müller-Taubenberger, A., Kortholt, A., & Eichinger, L. (2013). Simple system–substantial share: The use of *Dictyostelium* in cell biology and molecular medicine. *European Journal of Cell Biology, 92*(2), 45–53.

Muma, M. H. (1943). Spider deformities. *Bulletin of Natural History Society, 13*, 80–81.

Munday, P. L., Caley, M. J., & Jones, G. P. (1998). Bi-directional sex change in a coral-dwelling goby. *Behavioral Ecology and Sociobiology, 43*, 371–377.

Munévar, G. (2014). Darwinism and homosexuality. In F. de Sousa & G. Munévar (Eds.), *Sex, reproduction and Darwinism* (pp. 95–108). Pickering & Chatto.

Munguía-Steyer, R., González-García, E., Castaños, C. E., & Córdoba-Aguilar, A. (2019). Costly parenting: physiological condition over time and season in males of the giant waterbug *Abedus dilatatus*. *Physiological Entomology, 44*(3–4), 236–244.

Munnecke, A., Delabroye, A., Servais, T., Vandenbroucke, T. R., & Vecoli, M. (2012). Systematic occurrences of malformed (teratological) acritarchs in the run-up of Early Palaeozoic δ13C isotope excursions. *Palaeogeography, Palaeoclimatology, Palaeoecology, 367*, 137–146.

Muscatine, L., Boyle, J. E., & Smith, D. C. (1974). Symbiosis of the acoel flatworm *Convoluta roscoffensis* with the alga *Platymonas convolutae*. *Proceedings of the Royal Society of London B: Biological Sciences, 187*(1087), 221–234.

Nadolny, A. A., Marusik, Y. M., Kronestedt, T., Kovblyuk, M. M., & Zamani, A. (2022). New cases of teratological deformities in wolf spiders (Araneae: Lycosidae). *Arachnology, 19*(2), 585–590.

Nakashima, Y., Kuwamura, T., & Yogo, Y. (1996). Both-ways sex change in monogamous coral gobies, *Gobiodon* spp. *Environmental Biology of Fishes, 46*, 281–288.

Napiórkowska, T., & Templin, J. (2013). Symely, a seldom occurring developmental anomaly in the spider *Tegenaria atrica*. *Invertebrate Reproduction & Development, 57*(2), 95–100.

Napiórkowska, T., & Templin, J. (2017). Teratological changes on the prosoma of *Eratigena atrica* spiders caused by alternating temperatures. *Invertebrate Survival Journal, 14*(1), 480–487.

Napiórkowska, T., & Templin, J. (2018). Heterosymely and accompanying anomalies in the spider *Eratigena atrica* (CL Koch, 1843) (Araneae: Agelenidae). *Annales Zoologici, 68*(4), 909–914.

Napiórkowska, T., Jacuński, L., & Templin, J. (2007). Epimorphosis and repair processes of schistomelic pedipalps and walking appendages in *Tegenaria atrica* (Araneae, Agelenidae). *Biologia, 62*, 756–762.

Nestawal, S. (2014). Semantics of monstrosity in the nineteenth century. *Medicina nei secoli: Journal of History of Medicine and Medical Humanities, 26*(1), 167–196.

Niida, T., & Koshikawa, S. (2021). No evidence for contribution of sexually monomorphic wing pigmentation pattern to mate choice in Drosophila guttifera. *Ethology, 127*(7), 527–536.

Nikolov, L. A., & Davis, C. C. (2017). The big, the bad, and the beautiful: Biology of the world's largest flowers. *Journal of Systematics and Evolution, 55*(6), 516–524.

Nishikimi, M., Fukuyama, R., Minoshima, S., Shimizu, N., & Yagi, K. (1994). Cloning and chromosomal mapping of the human nonfunctional gene for L-gulono-gamma-lactone oxidase, the enzyme for L-ascorbic acid biosynthesis missing in man. *Journal of Biological Chemistry, 269*(18), 13685–13688.

Nissen, M., Shcherbakov, D., Heyer, A., Brümmer, F., & Schill, R. O. (2015). Behaviour of the plathelminth *Symsagittifera roscoffensis* under different light conditions and the consequences for the symbiotic algae *Tetraselmis convolutae*. *The Journal of Experimental Biology, 218*(11), 1693–1698.

Nogueira, J. M. D. M., & Fukuda, M. V. (2008). A new species of *Trypanosyllis* (Polychaeta: Syllidae) from Brazil, with a redescription of Brazilian material of *Trypanosyllis zebra*. *Journal of the Marine Biological Association of the United Kingdom, 88*(5), 913–924.

O'Gorman, E. J., Yearsley, J. M., Crowe, T. P., Emmerson, M. C., Jacob, U., & Petchey, O. L. (2011). Loss of functionally unique species may gradually undermine ecosystems. *Proceedings of the Royal Society B: Biological Sciences, 278*(1713), 1886–1893.

O'Malley, M. A., Wideman, J. G., & Ruiz-Trillo, I. (2016). Losing complexity: The role of simplification in macroevolution. *Trends in Ecology & Evolution, 31*(8), 608–621.

Ochi, Y., Fukui, Y., Sakai, Y., & Hashimoto, H. (2017). Age, growth and reproduction of the Asian sheephead wrasse *Semicossyphus reticulatus* in the western Seto Inland Sea. *Japan. Ichthyological Research, 64*(4), 415–422.

Ohba, S. Y. (2019). Ecology of giant water bugs (Hemiptera: Heteroptera: Belostomatidae). *Entomological Science, 22*(1), 6–20.

Okada, Y. K. (1937). La stolonisation et les caractères sexuels du Stolon chez les Syllidiens Polychètes: (Etudes sur les Syllidiens III). *Japanese Journal of Sanitary Zoology, 7*, 441–490.

Oliveira, R. F., Taborsky, M., & Brockmann, H. J. (2008). *Alternative reproductive tactics: An integrative approach*. Cambridge University Press.

Oliveira, A. G., Desjardin, D. E., Perry, B. A., & Stevani, C. V. (2012). Evidence that a single bioluminescent system is shared by all known bioluminescent fungal lineages. *Photochemical & Photobiological Sciences, 11*, 848–852.

Oliveira, A. G., Stevani, C. V., Waldenmaier, H. E., Viviani, V., Emerson, J. M., Loros, J. J., & Dunlap, J. C. (2015). Circadian control sheds light on fungal bioluminescence. *Current Biology, 25*(7), 964–968.

Olivero, P., & Robillard, T. (2017). Same-sex sexual behavior in *Xenogryllus marmoratus* (Haan, 1844) (Grylloidea: Gryllidae: Eneopterinae): Observation in the wild from YouTube. *Journal of Orthoptera Research*, 1–5.

Ollerton, J., & Ren, Z. X. (2022). Did pollination exist before plants? *Science, 377*(6605), 471–472.

Olson, M. E. (2012). The developmental renaissance in adaptationism. *Trends in Ecology & Evolution, 27*(5), 278–287.

Ong, T. W., Lin, B. B., Lucatero, A., Cohen, H., Bichier, P., Egerer, M. H., et al. (2022). Rarity begets rarity: Social and environmental drivers of rare organisms in cities. *Ecological Applications, e2708.*

Ono, F., Saigusa, M., Uozumi, T., Matsushima, Y., Ikeda, H., Saini, N. L., & Yamashita, M. (2008). Effect of high hydrostatic pressure on the life of the tiny animal tardigrade. *Journal of Physics and Chemistry of Solids, 69*(9), 2297–2300.

Ospina-Alvarez, N., & Piferrer, F. (2008). Temperature-dependent sex determination in fish revisited: Prevalence, a single sex ratio response pattern, and possible effects of climate change. *PLoS One, 3*(7), e2837.

Ostrowski, E. A., Katoh, M., Shaulsky, G., Queller, D. C., & Strassmann, J. E. (2008). Kin discrimination increases with genetic distance in a social amoeba. *PLoS Biology, 6*(11), e287.

Otani, Y. (1980). Sarcoscyphineae of Japan. *Transactions of the Mycological Society of Japan, 21*, 149–179.

Oxford English Dictionary (2017). Oxford, UK: Oxford University Press.

Packard, A. (1972). Cephalopods and fish: the limits of convergence. *Biological Reviews, 47*(2), 241–307.

Pagel, M. (2004). Limpets break Dollo's law. *Trends in Ecology & Evolution, 19*(6), 278–280.

Paillet, F. L. (2002). Chestnut: history and ecology of a transformed species. *Journal of Biogeography, 29*(10-11), 1517–1530.

Pal, D. S., Li, X., Banerjee, T., Miao, Y., & Devreotes, P. N. (2019). The excitable signal transduction networks: movers and shapers of eukaryotic cell migration. *The International Journal of Developmental Biology, 63*(8–10), 407.

Palazy, L., Bonenfant, C., Gaillard, J. M., & Courchamp, F. (2012). Rarity, trophy hunting and ungulates. *Animal Conservation, 15*(1), 4–11.

Palfalvi, G., Hackl, T., Terhoeven, N., Shibata, T. F., Nishiyama, T., Ankenbrand, M., et al. (2020). Genomes of the venus flytrap and close relatives unveil the roots of plant carnivory. *Current Biology, 30*(12), 2312–2320.

Pammel, L. H., & King, C. M. (1930). Honey plants of Iowa. *Iowa Geological Survey Bulletin, 7*.

Paré, A. (1573). *Des monstres tant terrestres que marins avec leurs portraits*. Paris.

Parker, G. A. (1990). Sperm competition games: Raffles and roles. *Proceedings of the Royal Society of LondonSeries B: Biological Sciences, 242*(1304), 120–126.

Parker, J. (2016). Myrmecophily in beetles (Coleoptera): Evolutionary patterns and biological mechanisms. *Myrmecological News, 22*, 65–108.

Parks, C. G., Bull, E. L., Tinnin, R. O., Shepherd, J. F., & Blumton, A. K. (1999). Wildlife use of dwarf mistletoe brooms in Douglas-fir in northeast Oregon. *Western Journal of Applied Forestry, 14*(2), 100–105.

Patek, S. N., Rosario, M. V., & Taylor, J. R. A. (2013). Comparative spring mechanics in mantis shrimp. *Journal of Experimental Biology, 216*(7), 1317–1329.

Pavlovič, A., & Mithöfer, A. (2019). Jasmonate signalling in carnivorous plants: Copycat of plant defence mechanisms. *Journal of Experimental Botany, 70*(13), 3379–3389.

Pavlovič, A., & Saganová, M. (2015). A novel insight into the cost–benefit model for the evolution of botanical carnivory. *Annals of Botany, 115*(7), 1075–1092.

Pavlovič, A., Jakšová, J., & Novák, O. (2017). Triggering a false alarm: Wounding mimics prey capture in the carnivorous Venus flytrap (*Dionaea muscipula*). *New Phytologist, 216*(3), 927–938.

Pears, C. J., & Gross, J. D. (2021). Microbe profile: *Dictyostelium discoideum*: model system for development, chemotaxis and biomedical research. *Microbiology, 167*(3), 001040.

Pentinsaari, M., Anderson, R., Borowiec, L., Bouchard, P., Brunke, A., Douglas, H., et al. (2019). DNA barcodes reveal 63 overlooked species of Canadian beetles (Insecta, Coleoptera). *ZooKeys, 894*, 53.

Pereira, J. L., Ram, A., de Figueiredo, J. M., & de Almeida, L. C. C. (1990). First occurrence of witches' broom disease in the principal cocoa-growing region of Brazil. *Tropical Agriculture, 67*(2), 188–189.

Pereira, J. L., De Almeida, L. C. C., & Santos, S. M. (1996). Witches' broom disease of cocoa in Bahia: Attempts at eradication and containment. *Crop Protection, 15*(8), 743–752.

Peretti, A. V. (1992). El espermatoforo de *Bothriurus bonariensis* (CL Koch)(Scorpiones, Bothriuridae): Morfología y funcionamiento. *Boletín de la Sociedad de Biología de Concepción, 63*, 157–167.

Peretti, A. V. (2000). Existencia de cortejo en el campo de machos de *Bothriurus bonariensis* (Scorpiones: Bothriuridae) que carecen de un órgano paraxial. *Revista de la Sociedad Entomológica Argentina, 59*(1–4).

Peretti, A. V., & Córdoba-Aguilar, A. (2007). On the value of fine-scaled behavioural observations for studies of sexual coercion. *Ethology, Ecology & Evolution, 19*, 77–86.

Peroutka, M., Adlassnig, W., Volgger, M., Lendl, T., Url, W. G., & Lichtscheidl, I. K. (2008). *Utricularia*: A vegetarian carnivorous plant? Algae as prey of bladderwort in oligotrophic bogs. *Plant Ecology, 199*, 153–162.

Persson, D., Halberg, K. A., Jönsson, K. I., & Ricci, C. (2011). Survival of tardigrades after exposure to space in low Earth orbit. *Journal of Zoological Systematics and Evolutionary Research, 49*, 1–7.

Petersen, F. T., Meier, R., Kutty, S. N., & Wiegmann, B. M. (2007). The phylogeny and evolution of host choice in the Hippoboscoidea (Diptera) as reconstructed using four molecular markers. *Molecular Phylogenetics and Evolution, 45*(1), 111–122.

Peterson, K. R., Bell, C. D., Kurogi, S., & Pfister, D. H. (2004). Phylogeny and biogeography of *Chorioactis geaster* (Pezizales, Ascomycota) inferred from nuclear ribosomal DNA sequences. *Harvard Papers in Botany, 9*(2), 141–152.

Philip, N. (2004). *Mythology of the world. Kingfisher*. Publications.

Phillips, W. (1881). Monstrosities in fungi. *Transactions of the Woolhope Naturalists' Field Club*, 103–106.

Piferrer, F., Zanuy, S., Carrillo, M., Solar, I. I., Devlin, R. H., & Donaldson, E. M. (1994). Brief treatment with an aromatase inhibitor during sex differentiation causes chromosomally female salmon to develop as normal, functional males. *Journal of Experimental Zoology, 270*(3), 255–262.

Pillard, R. C., & Bailey, J. M. (1998). Human sexual orientation has a heritable component. *Human Biology*, 347–365.

Piringer, A. A., & Borthwick, H. A. (1961). Effects of photoperiod and kind of supplemental light on growth, flowering, and stem fasciation of *Celosia*. *American Journal of Botany, 48*(7), 588–592.

Platt, T. R., & Brooks, D. R. (1997). Evolution of the schistosomes (Digenea: Schistosomatoidea): the origin of dioecy and colonization of the venous system. *The Journal of Parasitology, 83*(6), 1035–1044.

Poiani, A. (2010). *Animal homosexuality: A biosocial perspective*. Cambridge University Press.

Polly, P. D. (2008). Developmental dynamics and G-matrices: Can morphometric spaces be used to model phenotypic evolution? *Evolutionary Biology, 35*, 83–96.

Pombert, J. F., Blouin, N. A., Lane, C., Boucias, D., & Keeling, P. J. (2014). A lack of parasitic reduction in the obligate parasitic green alga *Helicosporidium*. *PLoS Genetics, 10*(5), e1004355.

Ponz-Segrelles, G., Ribeiro, R. P., & Aguado, M. T. (2022). Monsters reveal patterns: bifurcated annelids and their implications for the study of development and evolution. *Biological Reviews, 97*(3), 896–922.

Poppinga, S., & Joyeux, M. (2011). Different mechanics of snap-trapping in the two closely related carnivorous plants Dionaea muscipula and Aldrovanda vesiculosa. *Physical Review, 84*(4), 041928.

Pottin, K., Hinaux, H., & Rétaux, S. (2011). Restoring eye size in *Astyanax mexicanus* blind cavefish embryos through modulation of the Shh and Fgf8 forebrain organising centres. *Development, 138*(12), 2467–2476.

Pownall, M. E., & Saha, M. S. (2018). Histological observation of teratogenic phenotypes induced in frog embryo assays. *Methods in Molecular Biology, 1797*, 309–323.
Prasher, I. B., Chandel, V. C., & Chauhan, R. (2014). Effect of Carbon and Nitrogen Sources on Growth and Luminescence of *Panellus stipticus* (Bull.) P. Karst. *Journal of Advanced Botany and Zoology, 2*(1), 32–36.
Price, C. B. (2013). Vegetable monsters: man-eating trees in Fin-de-Siècle fiction. *Victorian Literature and Culture, 41*(2), 311–327.
Prior, S. (1939). *Carnivorous plants and "the man eating tree"*. Field Museum of Natural History.
Provine, R. R., Cabrera, M. O., & Nave-Blodgett, J. (2013a). Red, yellow, and super-white sclera. *Human Nature, 24*(2), 126–136.
Provine, R. R., Nave-Blodgett, J., & Cabrera, M. O. (2013b). The emotional eye: Red sclera as a uniquely human cue of emotion. *Ethology, 119*(11), 993–998.
Purdy, L. H., & Schmidt, R. A. (1996). Status of Cacao witches' broom: biology, epidemiology, and management. *Annual Review of Phytopathology, 34*(1), 573–594.
Purvis, A., Agapow, P. M., Gittleman, J. L., & Mace, G. M. (2000). Nonrandom extinction and the loss of evolutionary history. *Science, 288*(5464), 328–330.
Quesne, W. J. (1974). The uniquely evolved character concept and its cladistic application. *Systematic Biology, 23*(4), 513–517.
Rabinowitz, D. (1981). Seven forms of rarity. In H. Synge (Ed.), *The biological aspects of rare plant conservation* (pp. 205–217). John Wiley and Sons.
Radwan, J. (1993). The adaptive significance of male polymorphism in the acarid mite *Caloglyphus berlesei*. *Behavioral Ecology and Sociobiology, 33*, 201–208.
Raju, A. S., Raju, P. S., & Ramana, K. V. (2014). Melittophily and malacophily in *Ipomoea pes-caprae* (Convolvulaceae). *The Journal of Asian Biodiversity, 6*(2).
Ramløv, H., & Westh, P. (2001). Cryptobiosis in the eutardigrade *Adorybiotus* (*Richtersius*) coronifer: tolerance to alcohols, temperature and de novo protein synthesis. *Zoologischer Anzeiger: A Journal of Comparative Zoology, 240*(3–4), 517–523.
Raper, K. B. (1935). *Dictyostelium discoideum*, a new species of slime mold from decaying forest leaves. *Journal of Agricultural Research, 50*, 135–147.
Raper, K. B. (1937). Growth and development of *Dictyostelium discoideum* with different bacterial associates. *Journal of Agricultural Research, 55*, 289–316.
Raper, K. B. (1984). *The Dictyostelids*. Princeton University Press.
Reaka-Kudla, M. L. (1997). Global biodiversity of coral reefs: a comparison with rainforests. In M. L. Reaka-Kudla, D. E. Wilson, & E. O. Wilson (Eds.), *Biodiversity II: Understanding and protecting our biological resources* (pp. 83–108). Joseph Henry Press.
Rebecchi, L., Guidetti, R., Borsari, S., Altiero, T., & Bertolani, R. (2006). Dynamics of long-term anhydrobiotic survival of lichen-dwelling tardigrades. *Hydrobiologia, 558*, 23–30.
Rebecchi, L., Altiero, T., Guidetti, R., Cesari, M., Bertolani, R., Negroni, M., Rizzo, A. M., & Giovannini, C. (2009). Tardigrade resistance to space effects: first results of experiments on the LIFE-TARSE mission on FOTON-M3 (September 2007). *Astrobiology, 9*, 581–591.
Rebecchi, L., Altiero, T., Guidetti, R., Cesari, M., Bertolani, R., Negroni, M., Rizzo, A. M., & Giovannini, C. (2011a). Resistance of the Tardigrade *Richtersius coronifer* to Space Flight (LIFE–TARSE mission on FOTON-M3). *Astrobiology, 11*, 551–556.
Rebecchi, L., Altiero, T., Guidetti, R., Cesari, M., Bertolani, R., Rizzo, A. M., & Giovannini, C. (2011b). Tardigrade resistance to space effects: first results of experiments on the LIFE-TARSE mission on FOTON-M3 (September 2007). *Astrobiology, 11*, 551–556.
Reiss, J. O., Burke, A. C., Archer, C., De Renzi, M., Dopazo, H., Etxeberría, A., et al. (2008). Pere alberch: Originator of evodevo. *Biological Theory, 3*, 351–356.
Requena, G. S., Munguía-Steyer, R., & Machado, G. (2014). Paternal care and sexual selection in arthropods. In R. H. Macedo & G. Machado (Eds.), *Sexual selection: perspectives and models from the Neotropics* (pp. 201–233). Academic Press.
Rettenmeyer, C. W., Rettenmeyer, M. E., Joseph, J., & Berghoff, S. M. (2011). The largest animal association centered on one species: the army ant *Eciton burchellii* and its more than 300 associates. *Insectes Sociaux, 58*, 281–292.

Reynolds, J. D., Gross, M. D., & Coombs, M. J. (1993). Environmental conditions and male morphology determine alternative mating behaviour in Trinidadian guppies. *Animal Behaviour, 45*(1), 145–152.

Rheinberger, H. J. (2013). Re-discovering mendel: the case of Carl Correns. *Science & Education, 24*(1–2), 51–60.

Roberts, M. J., & Parker, J. R. (1973). Gynandry and intersexuality in spiders. *Bulletin of the British Arachnological Society, 2*(9), 177.

Robertson, H. M. (1985). Female dimorphism and mating behaviour in a damselfly, *Ischnura ramburi*: Females mimicking males. *Animal Behaviour, 33*(3), 805–809.

Rodrigues, J. L., Pellizari, V. H., Mueller, R., Baek, K., Jesus, E. D. C., Paula, F. S., et al. (2013). Conversion of the Amazon rainforest to agriculture results in biotic homogenization of soil bacterial communities. *Proceedings of the National Academy of Sciences, 110*(3), 988–993.

Roeleveld, M. A. C., & Lipinski, M. R. (1991). The giant squid *Architeuthis* in southern African waters. *Journal of Zoology, 224*(3), 431–477.

Rosa, R., Lopes, V. M., Guerreiro, M., Bolstad, K., & Xavier, J. C. (2017). Biology and ecology of the world's largest invertebrate, the colossal squid (*Mesonychoteuthis hamiltoni*): A short review. *Polar Biology, 40*(9), 1871–1883.

Rosenberg, M. I., Lynch, J. A., & Desplan, C. (2009). Heads and tails: Evolution of antero-posterior patterning in insects. *Biochimica et Biophysica Acta (BBA)-Gene Regulatory Mechanisms, 1789*(4), 333–342.

Roughgarden, J. (2004). *Evolution's Rainbow*. University of California Press.

Roulin, A. (2004). The evolution, maintenance and adaptive function of genetic colour polymorphism in birds. *Biological Reviews, 79*(4), 815–848.

Rousk, J., Brookes, P. C., & Bååth, E. (2009). Contrasting soil pH effects on fungal and bacterial growth suggest functional redundancy in carbon mineralization. *Applied and Environmental Microbiology, 75*(6), 1589–1596.

Rubin, E. M. (2008). Genomics of cellulosic biofuels. *Nature, 454*(7206), 841–845.

Rudy, K. C., & Keller, H. W. (1996). The rare and fascinating devil's cigar. *Chorioactis geaster. Mycologist, 10*(1), 33–35.

Ruppert, E. E., & Barnes, R. D. (1994). *Invertebrate Zoology*. Harcourt Brace & Company.

Ruppert, E. E., & Barnes, R. D. (1996). *Invertebrate zoology* (6th ed.). Saunders College Publishing, Ft. Worth.

Ruppert, E. E., Fox, R. S., & Barnes, R. D. (2003). *Invertebrate Zoology: A functional evolutionary approach* (6th ed.). Cengage Learning.

Russell, A. P., & Higham, T. E. (2009). A new angle on clinging in geckos: Incline, not substrate, triggers the deployment of the adhesive system. *Proceedings of the Royal Society B: Biological Sciences, 276*(1673), 3705–3709.

Russell, J. R., Huang, J., Anand, P., Kucera, K., Sandoval, A. G., Dantzler, K. W., et al. (2011). Biodegradation of polyester polyurethane by endophytic fungi. *Applied and Environmental Microbiology, 77*(17), 6076–6084.

Sáez-Royuela, R., Bernis, F., Fernández Cruz, M., & Maluquer, S. (1975). *Ornitología y conservación de la naturaleza hoy. Estudio de la avifauna ibérica en homenaje a José A. Valverde Gómez.* Centro de Publicaciones. Ministerio de Educación y Ciencia, .

Saint-Hilaire, G. I. (1836). *Histoire générale et particulière des anomalies... ou Traité de Tératologie*. Librarie de l'Académie Royale de Médecine.

Salvador, R. B., & Tomotani, B. M. (2014). The Kraken: When myth encounters science. *História, Ciências, Saúde-Manguinhos, 21*, 971–994.

Sánchez-Guillén, R., Van Gossum, H., & Cordero Rivera, A. (2005). Hybridization and the inheritance of female colour polymorphism in two ischnurid damselflies (Odonata: Coenagrionidae). *Biological Journal of the Linnean Society, 85*(4), 471–481.

Sánchez-Guillén, R. A., Martínez-Zamilpa, S. M. J., Jiménez-Cortés, J. G., Forbes, M. R. L., & Córdoba-Aguilar, A. (2013). Maintenance of polymorphic females: Do parasites play a role? *Oecologia, 171*(1), 105–113.

Sánchez-Guillén, R. A., Wellenreuther, M., Chávez-Ríos, J. R., Beatty, C. D., Rivas-Torres, A., Velasquez-Velez, M., & Cordero-Rivera, A. (2017). Alternative reproductive strategies and the maintenance of female color polymorphism in damselflies. *Ecology and Evolution, 7*(15), 5592–5602.

Sarma, K., Tandon, R., Shivanna, K. R., & Ram, H. M. (2007). Snail-pollination in *Volvulopsis nummularium*. *Current Science*, 826–831.

Satoh, A., Graham, G. M. C., Bryant, S. V., & Gardiner, D. M. (2008). Neurotrophic regulation of epidermal dedifferentiation during wound healing and limb regeneration in the axolotl (Ambystoma mexicanum). *Developmental Biology, 319*(2), 321–335.

Scagel, R. F., Bandoni, R. J., Rouse, G. E., Schofield, W. B., Stein, J. R., & Taylor, T. M. C. (1977). *El Reino Vegetal. Los grupos de plantas y sus relaciones evolutivas*. Omega: Barcelona.

Schaf, J., Damstra-Oddy, J., & Williams, R. S. (2019). *Dictyostelium discoideum* as a pharmacological model system to study the mechanisms of medicinal drugs and natural products. *International Journal of Developmental Biology, 63*, 541–550.

Scharf, I., & Martin, O. Y. (2013). Same-sex sexual behavior in insects and arachnids: Prevalence, causes, and consequences. *Behavioral Ecology and Sociobiology, 67*, 1719–1730.

Scherzer, S., Krol, E., Kreuzer, I., Kruse, J., Karl, F., von Rüden, M., et al. (2013). The *Dionaea muscipula* ammonium channel DmAMT1 provides NH4+ uptake associated with Venus flytrap's prey digestion. *Current Biology, 23*(17), 1649–1657.

Schluter, D. (2001). Ecology and the origin of species. *Trends in Ecology & Evolution, 16*(7), 372–380.

Scholtz, G., & Brenneis, G. (2016). The pattern of a specimen of *Pycnogonum litorale* (Arthropoda, Pycnogonida) with a supernumerary leg can be explained with the "boundary model" of appendage formation. *The Science of Nature, 103*, 1–9.

Seamans, M. E., & Gutiérrez, R. J. (1995). Breeding habitat of the Mexican spotted owl in the Tularosa Mountains, New Mexico. *The Condor, 97*(4), 944–952.

Seaver, F. J. (1937). Photographs and descriptions of cup-fungi—XXV. Urnula Geaster. *Mycologia, 29*(1), 60–65.

Segal, N. L., Goetz, A. T., & Maldonado, A. C. (2016). Preferences for visible white sclera in adults, children and autism spectrum disorder children: Implications of the cooperative eye hypothesis. *Evolution and Human Behavior, 37*(1), 35–39.

Sekercioglu, C. H., Schneider, S. H., Fay, J. P., & Loarie, S. R. (2008). Climate change, elevational range shifts, and bird extinctions. *Conservation Biology, 22*(1), 140–150.

Seki, K., & Toyoshima, M. (1998). Preserving tardigrades under pressure. *Nature, 395*(6705), 853–854.

Serrano, J. M., Castro, L., Toro, M. A., & López-Fanjul, C. (1991). The genetic properties of homosexual copulation behavior in *Tribolium castaneum*: Diallel analysis. *Behavior Genetics, 21*, 547–558.

Serrano, J. M., Castro, L., Toro, M. A., & López-Fanjul, C. (2000). Inter-and intraspecific sexual discrimination in the flour beetles *Tribolium castaneum* and *Tribolium confusum*. *Heredity, 85*(2), 142–146.

Servais, T., Li, J., Stricanne, L., Vecoli, M., & Wicander, R. (2004). Acritarchs. In B. Webby, F. Paris, & I. Percival (Eds.), *The great ordovician biodiversification event* (pp. 348–360). Columbia University Press.

Shade, A., Jones, S. E., Caporaso, J. G., Handelsman, J., Knight, R., Fierer, N., & Gilbert, J. A. (2014). Conditionally rare taxa disproportionately contribute to temporal changes in microbial diversity. *MBio, 5*(4), 10–1128.

Sharma, P. P., Santiago, M. A., Kriebel, R., Lipps, S. M., Buenavente, P. A. C., Diesmos, A. C., Janda, M., Boyer, S. L., Clouse, R. M., & Wheeler, W. C. (2017). A multilocus phylogeny of Podoctidae (Arachnida, Opiliones, Laniatores) and parametric shape analysis reveal the disutility of subfamilial nomenclature in armored harvestman systematics. *Molecular Phylogenetics and Evolution, 106*, 164–173.

Sharma, P. P., Oberski, J. T., Santiago, M. A., Kriebel, R., Lipps, S. M., Buenavente, P. A., et al. (2018). There is no evidence that Podoctidae carry eggs of their own species: Reply to Machado and Wolff (2017). *Molecular Phylogenetics and Evolution, 129*, 349–353.

Shaw, D. C., Watson, D. M., & Mathiasen, R. L. (2004). Comparison of dwarf mistletoes (*Arceuthobium* spp., Viscaceae) in the western United States with mistletoes (*Amyema* spp., Loranthaceae) in Australia—Ecological analogs and reciprocal models for ecosystem management. *Australian Journal of Botany, 52*(4), 481–498.

Sherratt, T. N. (2001). The evolution of female-limited polymorphisms in damselflies: A signal detection model. *Ecology Letters, 4*(1), 22–29.

Shigeno, S., Andrews, P. L., Ponte, G., & Fiorito, G. (2018). Cephalopod brains: An overview of current knowledge to facilitate comparison with vertebrates. *Frontiers in Physiology, 9*, 952.

Sillo, A., Matthias, J., Konertz, R., Bozzaro, S., & Eichinger, L. (2011). *Salmonella typhimurium* is pathogenic for *Dictyostelium* cells and subverts the starvation response. *Cellular Microbiology, 13*(11), 1793–1811.

Sim, S., Rowhani, A., & Golino, D. (2004). Phyllody in roses. *American Rose, 39*(18), 32–34.

Simmons, L. G. (2001). *Sperm competition and its evolutionary consequences in insects*. Princeton University Press.

Simmons, L. W., Tomkins, J. L., & Hunt, J. (1999). Sperm competition games played by dimorphic male beetles. *Proceedings of the Royal Society of London Series B: Biological Sciences, 266*(1415), 145–150.

Sinervo, B., & Licht, P. (1991). Proximate constraints on the evolution of egg size, number, and total clutch mass in lizards. *Science, 252*(5010), 1300–1302.

Sipos, G., Prasanna, A. N., Walter, M. C., O'Connor, E., Bálint, B., Krizsán, K., et al. (2017). Genome expansion and lineage-specific genetic innovations in the forest pathogenic fungi *Armillaria*. *Nature Ecology & Evolution, 1*(12), 1931–1941.

Sirot, L. K., Brockmann, H. J., Marnis, C., & Muschett, G. (2003). Maintenance of a female-limited polymorphism in *Ischnura ramburi* (Zygoptera: Coenagrionidae). *Animal Behaviour, 66*(4), 763–775.

Sivak, J. G. (1976). Optics of the eye of the "four-eyed fish" (Anableps anableps). *Vision Research, 16*(5), 531–IN6.

Sivinski, J. (1981). Arthropods attracted to luminous fungi. *Psyche: A Journal of Entomology, 88*(3–4), 383–390.

Skelhorn, J. (2015). Masquerade. *Current Biology, 25*(15), R643–R644.

Skinner, A., Lee, M. S., & Hutchinson, M. N. (2008). Rapid and repeated limb loss in a clade of scincid lizards. *BMC Evolutionary Biology, 8*(1), 1–9.

Skogsmyr, I. O., & Lankinen, Å. (2002). Sexual selection: An evolutionary force in plants? *Biological Reviews, 77*(4), 537–562.

Skopina, M. Y., Vasileva, A. A., Pershina, E. V., & Pinevich, A. V. (2016). Diversity at low abundance: The phenomenon of the rare bacterial biosphere. *Microbiology, 85*(3), 272–282.

Sloan, D., Alves Batista, R., & Loeb, A. (2017). The resilience of life to astrophysical events. *Scientific Reports, 7*(1), 5419.

Smith, A. P. (1986). Ecology of a leaf color polymorphism in a tropical forest species: Habitat segregation and herbivory. *Oecologia, 69*, 283–287.

Smith, M. L., Bruhn, J. N., & Anderson, J. B. (1992). The fungus *Armillaria bulbosa* is among the largest and oldest living organisms. *Nature, 356*(6368), 428–431.

Soares, D. M., Procópio, D. P., Zamuner, C. K., Nóbrega, B. B., Bettim, M. R., De Rezende, G., et al. (2022). Fungal bioassays for environmental monitoring. *Frontiers in Bioengineering and Biotechnology, 10*.

Sogin, M. L., Morrison, H. G., Huber, J. A., Welch, D. M., Huse, S. M., Neal, P. R., et al. (2006). Microbial diversity in the deep sea and the underexplored "rare biosphere". *Proceedings of the National Academy of Sciences, 103*(32), 12115–12120.

Solomon, J. M., Rupper, A., Cardelli, J. A., & Isberg, R. R. (2000). Intracellular growth of *Legionella pneumophila* in *Dictyostelium discoideum*, a system for genetic analysis of host-pathogen interactions. *Infection and Immunity, 68*(5), 2939–2947.

Soltau, U., Dötterl, S., & Liede-Schumann, S. (2009). Leaf variegation in *Caladium steudneriifolium* (Araceae): A case of mimicry? *Evolutionary Ecology, 23*(4), 503–512.

Sommer, V., & Vasey, P. L. (2006). *Homosexual behaviour in animals: An evolutionary perspective*. Cambridge University Press.

Spanó, N., Ragonese, S., & Bianchini, M. L. (2003). An anomalous specimen of *Scyllarides latus* (Decapoda, Scyllaridae). *Crustaceana, 76*(7), 885–889.

Spicer, D. (2016). Witch's brooms in conifers. *Plantsman, 15*(3), 146–153.

Sprecher, S. G., Bernardo-Garcia, F. G., van Giesen, L., Hartenstein, V., Reichert, R., Neves, R., et al. (2015). Functional brain regeneration in the acoel worm *Symsagittifera roscoffensis*. *Biol Open, 4*(12), 1688–1695.

Stanosz, G. R., Patton, R. F., & Spear, R. N. (1987). Structure of *Armillaria rhizomorphs* from Wisconsin aspen stands. *Canadian Journal of Botany, 65*(10), 2124–2127.

Staudinger, M. D., Buresch, K. C., Mäthger, L. M., Fry, C., McAnulty, S., Ulmer, K. M., & Hanlon, R. T. (2013). Defensive responses of cuttlefish to different teleost predators. *The Biological Bulletin, 225*(3), 161–174.

Stechmann, A., & Cavalier-Smith, T. (2002). Rooting the eukaryote tree by using a derived gene fusion. *Science, 297*(5578), 89–91.

Steinert, M. (2011). Pathogen-host interactions in *Dictyostelium, Legionella, Mycobacterium* and other pathogens. *Seminars in Cell & Developmental Biology, 22*(1), 70–76.

Stephens, T. D., & Shepard, T. H. (1983). A review of limb defects in a large fetus collection. *American Journal of Human Genetics, 35*(3), 508.

Sterrer, W., Mainitz, M., & Rieger, R. M. (1985). Gnathostomulida: Enigmatic as ever. *The origins and relationships of lower invertebrates* (pp. 181–199).

Stevani, C. V., Oliveira, A. G., Mendes, L. F., Ventura, F. F., Waldenmaier, H. E., Carvalho, R. P., & Pereira, T. A. (2013). Current status of research on fungal bioluminescence: biochemistry and prospects for ecotoxicological application. *Photochemistry and Photobiology, 89*(6), 1318–1326.

Stockard, C. R. (1921). Developmental rate and structural expression: an experimental study of twins, 'double monsters' and single deformities, and the interaction among embryonic organs during their origin and development. *American Journal of Anatomy, 28*(2), 115–277.

Strassmann, J. E., & Queller, D. C. (2011). Evolution of cooperation and control of cheating in a social microbe. *Proceedings of the National Academy of Sciences, 108*(2), 10855–10862.

Strassmann, J. E., Zhu, Y., & Queller, D. C. (2000). Altruism and social cheating in the social amoeba *Dictyostelium discoideum*. *Nature, 408*(6815), 965–967.

Suma, H. R., Prakash, S., & Eswarappa, S. M. (2020). Naturally occurring fluorescence protects the eutardigrade *Paramacrobiotus* sp. from ultraviolet radiation. *Biology Letters, 16*(10), 20200391.

Suzuki, T., Yamasaki, K., Fujita, S., Oda, K., Iseki, M., Yoshida, K., et al. (2003). Archaeal-type rhodopsins in *Chlamydomonas*: model structure and intracellular localization. *Biochemical and Biophysical Research Communications, 301*(3), 711–717.

Svensson, E. I., Abbott, J., & Härdling, R. (2005). Female polymorphism, frequency dependence, and rapid evolutionary dynamics in natural populations. *The American Naturalist, 165*(5), 567–576.

Svensson, E. I., Abbott, J. K., Gosden, T. P., & Coreau, A. (2009). Female polymorphisms, sexual conflict and limits to speciation processes in animals. *Evolutionary Ecology, 23*, 93–108.

Swaab, D. F. (2008). Sexual orientation and its basis in brain structure and function. *Proceedings of the National Academy of Sciences, 105*(30), 10273–10274.

Swartz, S. M., & Middleton, K. M. (2007). Biomechanics of the bat limb skeleton: scaling, material properties and mechanics. *Cells Tissues Organs, 187*(1), 59–84.

Switzer, P. V., Forsythe, P. S., Escajeda, K., & Kruse, K. C. (2004). Effects of environmental and social conditions on homosexual pairing in the Japanese beetle (*Popillia japonica* Newman). *Journal of Insect Behavior, 17*, 1–16.

Taborsky, M. (1998). Sperm competition in fish: bourgeois' males and parasitic spawning. *Trends in Ecology & Evolution, 13*(6), 222–227.

Taborsky, M., Oliveira, R. F., & Brockmann, H. J. (2008). The evolution of alternative reproductive tactics: concepts and questions. In R. F. Oliveira, M. Taborsky, & H. J. Brockmann (Eds.), *Alternative reproductive tactics: An integrative approach* (pp. 1–21). Cambridge University Press.

Takahashi, Y., & Watanabe, M. (2010). Female reproductive success is affected by selective male harassment in the damselfly *Ischnura senegalensis*. *Animal Behaviour, 79*(1), 211–216.

Takahashi, Y., Yoshimura, J., Morita, S., & Watanabe, M. (2010). Negative frequency-dependent selection in female color polymorphism of a damselfly. *Evolution, 64*(12), 3620–3628.

Tallamy, D. W. (2001). Evolution of exclusive paternal care in arthropods. *Annual Review of Entomology, 46*(1), 139–165.

Tandon, R., & Ram, H. M. (2010). The saga of pollination biology. In V. P. Sharma (Ed.), *Nature at work: The ongoing Saga of evolution* (pp. 219–250). National Academy of Sciences.

Thornhill, R., & Alcock, J. (1983). *The evolution of insect mating systems*. Harvard University Press.

Tilman, D., Lehman, C. L., & Thomson, K. T. (1997). Plant diversity and ecosystem productivity: theoretical considerations. *Proceedings of the National Academy of Sciences, 94*(5), 1857–1861.

Tilman, D., Hill, J., & Lehman, C. (2006). Carbon-negative biofuels from low-input high-diversity grassland biomass. *Science, 314*(5805), 1598–1600.

Tilney-Bassett, R. A. E. (1975). Genetics of variegated plants. In C. W. Birky, P. S. Perlman, & T. J. Byers (Eds.), *Genetics and biogenesis of mitochondria and chloroplasts* (pp. 268–308). Ohio State University Press.

Tinnin, R. O., & Forbes, R. B. (1999). Red squirrel nests in witches' brooms in Douglas-fir trees. *Northwestern Naturalist*, 17–21.

Tirado-Gallego, P. A., Lopera-Álvarez, A., & Ríos-Osorio, L. A. (2016). Estrategias de control de *Moniliophthora roreri* y *Moniliophthora perniciosa* en *Theobroma* cacao L.: revisión sistemática. *Ciencia y Tecnología Agropecuaria, 17*(3), 417–430.

Tomasello, M., Hare, B., Lehmann, H., & Call, J. (2007). Reliance on head versus eyes in the gaze following of great apes and human infants: The cooperative eye hypothesis. *Journal of Human Evolution, 52*(3), 314–320.

Treviño-Rodriguez, J. G. (2023, July). Etimología de Teratología. *Etimologías*. https://etimologias.dechile.net/?teratologi.a.

Tsujimoto, M., Komori, O., & Imura, S. (2016). Effect of lifespan and age on reproductive performance of the tardigrade *Acutuncus antarcticus*: minimal reproductive senescence. *Hydrobiologia, 772*, 93–102.

Tucker, R. (1981). The digging behavior and skin differentiations in Heterocephalus glaber. *Journal of Morphology, 168*(1), 51–71.

Uchida, D., Yamashita, M., Kitano, T., & Iguchi, T. (2004). An aromatase inhibitor or high water temperature induce oocyte apoptosis and depletion of P450 aromatase activity in the gonads of genetic female zebrafish during sex-reversal. *Comparative Biochemistry and Physiology Part A: Molecular & Integrative Physiology, 137*(1), 11–20.

Umar, M. H., & Van Griensven, L. J. (1999). Studies on the morphogenesis of *Agaricus bisporus*: the dilemma of normal versus abnormal fruit body development. *Mycological Research, 103*(10), 1235–1244.

Vacelet, J., & Boury-Esnault, N. (1995). Carnivorous sponges. *Nature, 373*(6512), 333–335.

Vacelet, J., & Custódio, M. R. (2007). Diversity and evolution of deep-sea carnivorous sponges. In M. R. Custódio, G. Lôbo-Hajdu, E. Hajdu, & G. Muricy (Eds.), *Porifera research: Biodiversity, innovation and sustainability* (pp. 107–115). Série Livros.

Vacelet, J., Boury-Esnault, N., & Harmelin, J. G. (1994). Hexactinellid cave, a unique deep-sea habitat in the scuba zone. *Deep Sea Research Part I: Oceanographic Research Papers, 41*(7), 965–973.

van de Schootbrugge, B., & Wignall, P. B. (2016). A tale of two extinctions: Converging end-Permian and end-Triassic scenarios. *Geological Magazine, 153*(2), 332–354.

Van Gossum, H., Stoks, R., & De Bruyn, L. (2001). Frequency-dependent male mate harassment and intra-specific variation in its avoidance by females of the damselfly *Ischnura elegans*. *Behavioral Ecology and Sociobiology, 51*, 69–75.

Van Gossum, H., De Bruyn, L., & Stoks, R. (2005a). Reversible switches between male–male and male–female mating behaviour by male damselflies. *Biology Letters, 1*(3), 268–270.
Van Gossum, H., Stoks, R., & De Bruyn, L. (2005b). Lifetime fitness components in female colour morphs of a damselfly: Density-or frequency-dependent selection? *Biological Journal of the Linnean Society, 86*(4), 515–523.
Van Soest, R. W., De Kluijver, M. J., Van Bragt, P. H., Faasse, M., Nijland, R., Beglinger, E. J., et al. (2007). Sponge invaders in Dutch coastal waters. *Journal of the Marine Biological Association of the United Kingdom, 87*(6), 1733–1748.
Van Soest, R. W., Boury-Esnault, N., Vacelet, J., Dohrmann, M., Erpenbeck, D., De Voogd, N. J., Santodomingo, N., Vanhoorne, B., Kelly, M., & Hooper, J. N. (2012). Global diversity of sponges (Porifera). *PLoS One, 7*(4), e35105.
Van Tussenbroek, B. I., Monroy-Velazquez, L. V., & Solis-Weiss, V. (2012). Meso-fauna foraging on seagrass pollen may serve in marine zoophilous pollination. *Marine Ecology Progress Series, 469*, 1–6.
Van Tussenbroek, B. I., Villamil, N., Márquez-Guzmán, J., Wong, R., Monroy-Velázquez, L. V., & Solis-Weiss, V. (2016). Experimental evidence of pollination in marine flowers by invertebrate fauna. *Nature Communications, 7*(1), 12980.
Venier, L. A., & Fahrig, L. (1996). Habitat availability causes the species abundance-distribution relationship. *Oikos*, 564–570.
Ventura, F. F., Mendes, L. F., Oliveira, A. G., Bazito, R. C., Bechara, E. J., Freire, R. S., & Stevani, C. V. (2020). Evaluation of phenolic compound toxicity using a bioluminescent assay with the fungus *Gerronema viridilucens*. *Environmental Toxicology and Chemistry, 39*(8), 1558–1565.
Ventura, F. F., Soares, D. M., Bayle, K., Oliveira, A. G., Bechara, E. J., Freire, R. S., & Stevani, C. V. (2021). Toxicity of metal cations and phenolic compounds to the bioluminescent fungus Neonothopanus gardneri. *Environmental Advances, 4*, 100044.
Verbinski, G. (2006). *Pirates of the Caribbean: Dead man's chest*. Walt Disney Pictures.
Vermeij, G. J. (2016). Gigantism and its implications for the history of life. *PLoS One, 11*(1), e0146092.
Vermeij, G. J., & Grosberg, R. K. (2010). The great divergence: When did diversity on land exceed that in the sea? *Integrative and Comparative Biology, 50*, 675–682.
Verne, J. (1870). *Twenty thousand leagues under the seas*. Oxford University Press.
Visscher, H., Looy, C. V., Collinson, M. E., Brinkhuis, H., Van Konijnenburg-Van Cittert, J. H., Kürschner, W. M., & Sephton, M. A. (2004). Environmental mutagenesis during the end-Permian ecological crisis. *Proceedings of the National Academy of Sciences,* 101(35), 12952–12956.
Von Dassow, G., & Munro, E. (1999). Modularity in animal development and evolution: elements of a conceptual framework for EvoDevo. *Journal of Experimental Zoology, 285*(4), 307–325.
von Goethe, J. W. (1790). *Herzoglich Sachsen-Weimarischen Geheimraths Versuch die Metamorphose der Pflanzen zu erklären*. Ettinger.
Voolstra, C. R., & Ziegler, M. (2020). Adapting with microbial help: Microbiome flexibility facilitates rapid responses to environmental change. *BioEssays, 42*(7), 2000004.
Voss, S. R., Epperlein, H. H., & Tanaka, E. M. (2009). Ambystoma mexicanum, the axolotl: a versatile amphibian model for regeneration, development, and evolution studies. *Cold Spring Harbor Protocols, 2009*(8), pdb-emo128.
Vukich, M., Rettberg, P., & Parpart, K. M. (2012). Tardigrades in space research. *Proceedings of the Ninth International Symposium on Tardigrada. Zool. Anz., 251*, 345–348.
Wanninger, A. (2015). Morphology is dead–long live morphology! Integrating MorphoEvoDevo into molecular EvoDevo and phylogenomics. *Frontiers in Ecology and Evolution, 3*, 54.
Warner, R. R. (1975). The adaptive significance of sequential hermaphroditism in animals. *The American Naturalist, 109*(965), 61–82.
Weigel, D., & Meyerowitz, E. M. (1994). The ABCs of floral homeotic genes. *Cell, 78*(2), 203–209.
Weitz, W. H. J. (2004). Naturally bioluminescent fungi. *Mycologist, 18*(1), 4–5.
Weitz, H. J., Campbell, C. D., & Killham, K. (2002). Development of a novel, bioluminescence-based, fungal bioassay for toxicity testing. *Environmental Microbiology, 4*(7), 422–429.

Wellenreuther, M., Svensson, E. I., & Hansson, B. (2014). Sexual selection and genetic colour polymorphisms in animals. *Molecular Ecology, 23*(22), 5398–5414.
Wełnicz, W., Grohme, M. A., Kaczmarek, Ł., Schill, R. O., & Frohme, M. (2011). Anhydrobiosis in tardigrades—The last decade. *Journal of Insect Physiology, 57*(5), 577–583.
Werner, G. D., Strassmann, J. E., Ivens, A. B., Engelmoer, D. J., Verbruggen, E., Queller, D. C., et al. (2014). Evolution of microbial markets. *Proceedings of the National Academy of Sciences, 111*(4), 1237–1244.
Weronika, E., & Łukasz, K. (2017). Tardigrades in space research-past and future. *Origins of Life and Evolution of Biospheres, 47*, 545–553.
West, G. B., & Brown, J. H. (2005). The origin of allometric scaling laws in biology from genomes to ecosystems: Towards a quantitative unifying theory of biological structure and organization. *Journal of Experimental Biology, 208*(9), 1575–1592.
West-Eberhard, M. J. (2003). *Developmental plasticity and evolution*. Oxford University Press.
West-Eberhard, M. J. (2005). Developmental plasticity and the origin of species differences. *Proceedings of the National Academy of Sciences, 102*(1), 6543–6549.
White, O. E. (1948). Fasciation. *The Botanical Review, 14*(6), 319–358.
Whiting, M. F., Bradler, S., & Maxwell, T. (2003). Loss and recovery of wings in stick insects. *Nature, 421*, 264–267.
Whittaker, R. H. (1959). On the broad classification of organisms. *The Quarterly Review of Biology, 34*(3), 210–226.
Whittington, C. M., Papenfuss, A. T., Bansal, P., Torres, A. M., Wong, E. S., Deakin, J. E., & Belov, K. (2008). Defensins and the convergent evolution of platypus and reptile venom genes. *Genome Research, 18*(6), 986–994.
Wiemer, A. P., Sérsic, A. N., Marino, S., Simões, A. O., & Cocucci, A. A. (2012). Functional morphology and wasp pollination of two South *American asclepiads* (Asclepiadoideae–Apocynaceae). *Annals of Botany, 109*(1), 77–93.
Wiens, J. J., & Slingluff, J. L. (2001). How lizards turn into snakes: A phylogenetic analysis of body-form evolution in anguid lizards. *Evolution, 55*(11), 2303–2318.
Wiens, J. J., Kuczynski, C., Duellman, W. E., & Reeder, T. W. (2007). Loss and re-evolution of complex life cycles in marsupial frogs: Does ancestral trait reconstruction mislead? *Evolution, 61*, 1886–1899.
Wilkins, A. S., Wrangham, R. W., & Fitch, W. T. (2014). The "domestication syndrome" in mammals: A unified explanation based on neural crest cell behavior and genetics. *Genetics, 197*(3), 795–808.
Williams, R. S., Boeckeler, K., Gräf, R., Müller-Taubenberger, A., Li, Z., Isberg, R. R., et al. (2006). Towards a molecular understanding of human diseases using *Dictyostelium discoideum. Trends in Molecular Medicine, 12*(9), 415–424.
Willmer, P. (2011). *Pollination and floral ecology*. Princeton University Press.
Willson, M. F. (1990). Sexual selection in plants and animals. *Trends in Ecology & Evolution, 5*(7), 210–214.
Wilmers, C. C., Estes, J. A., Edwards, M., Laidre, K. L., & Konar, B. (2012). Do trophic cascades affect the storage and flux of atmospheric carbon? An analysis of sea otters and kelp forests. *Frontiers in Ecology and the Environment, 10*(8), 409–415.
Wisnivesky, C. (2003). *Ecología y epidemiología de las infecciones parasitarias*. Editorial Tecnológica de Costa Rica.
Wolf, F. A., & Wolf, F. T. (1947). *The fungi* (Vol. II). John Wiley & Sons, Inc..
Wolff, G. H., Halkinrud Thoen, H., Marshall, J., Sayre, M. E., & Strausfeld, N. J. (2017). An insect-like mushroom body in a crustacean brain. *eLife, 6*, e29889.
Wright, J. C. (1989). Desiccation tolerance and water-retentive mechanisms in tardigrades. *Journal of Experimental Biology, 142*(1), 267–292.
Wright, J. C., Westh, P., & Ramløv, H. (1992). Cryptobiosis in tardigrada. *Biological Reviews, 67*(1), 1–29.

Xavier, J., Rodhouse, P., Purves, M., Daw, T., Arata, J., & Pilling, G. (2002). Distribution of cephalopods recorded in the diet of the Patagonian toothfish (*Dissostichus eleginoides*) around South Georgia. *Polar Biology, 25*, 323–330.

Xavier, J. C., Phillips, R. A., & Cherel, Y. (2011). Cephalopods in marine predator diet assessments: why identifying upper and lower beaks is important. *ICES Journal of Marine Science, 68*(9), 1857–1864.

Xiong, C., He, J. Z., Singh, B. K., Zhu, Y. G., Wang, J. T., Li, P. P., et al. (2021). Rare taxa maintain the stability of crop microbiomes and ecosystem functions. *Environmental Microbiology, 23*(4), 1907–1924.

Yamamoto, Y., & Jeffery, W. R. (2000). Central role for the lens in cave fish eye degeneration. *Science, 289*(5479), 631–633.

Yamamoto, Y., Stock, D. W., & Jeffery, W. R. (2004). Hedgehog signalling controls eye degeneration in blind cavefish. *Nature, 431*(7010), 844–847.

Yamamoto, Y., Byerly, M. S., Jackman, W. R., & Jeffery, W. R. (2009). Pleiotropic functions of embryonic sonic hedgehog expression link jaw and taste bud amplification with eye loss during cavefish evolution. *Developmental Biology, 330*(1), 200–211.

Yang, H. (2013). Conserved or lost: molecular evolution of the key gene GULO in vertebrate vitamin C biosynthesis. *Biochemical Genetics, 51*, 413–425.

Yenni, G., Adler, P. B., & Ernest, S. M. (2012). Strong self-limitation promotes the persistence of rare species. *Ecology, 93*(3), 456–461.

Yilamujiang, A., Reichelt, M., & Mithöfer, A. (2016). Slow food: Insect prey and chitin induce phytohormone accumulation and gene expression in carnivorous *Nepenthes* plants. *Annals of Botany, 118*(2), 369–375.

Yoshida, S., Hiraga, K., Takehana, T., Taniguchi, I., Yamaji, H., Maeda, Y., et al. (2016). A bacterium that degrades and assimilates poly(ethylene terephthalate). *Science, 351*(6278), 1196–1199.

Young, J. Z. (1991). Computation in the learning system of cephalopods. *The Biological Bulletin, 180*(2), 200–208.

Yu, F. E. I., Fu, A., Aluru, M., Park, S., Xu, Y., Liu, H., et al. (2007). Variegation mutants and mechanisms of chloroplast biogenesis. *Plant, Cell & Environment, 30*(3), 350–365.

Yun, L., Wang, Y., Tian, Y., & Haoyuan, S. (2001). Advances in tissue culture and eliminating the Pathogeny of witch broom disease (MLO) of Chinese Jujube. *Journal of Fruit Science, 2*.

Zabriskie, T. M., & Jackson, M. D. (2000). Lysine biosynthesis and metabolism in fungi. *Natural Product Reports, 17*(1), 85–97.

Zahl, P. A., & Gomprecht, R. J. (1977). Visual versatility and feeding of the four-eyed fishes, Anableps. *Copeia, 4*, 791–793.

Ziegler, M., Eguíluz, V. M., Duarte, C. M., & Voolstra, C. R. (2018). Rare symbionts may contribute to the resilience of coral–algal assemblages. *The ISME Journal, 12*(1), 161–172.

Ziegler, M., Grupstra, C. G., Barreto, M. M., Eaton, M., BaOmar, J., Zubier, K., et al. (2019). Coral bacterial community structure responds to environmental change in a host-specific manner. *Nature Communications, 10*(1), 1–11.

Zippel, K. C., Glor, R. E., & Bertram, J. E. (1999). On caudal prehensility and phylogenetic constraint in lizards: the influence of ancestral anatomy on function in *Corucia* and *Furcifer*. *Journal of Morphology, 239*(2), 143–155.

Chapter 6
Focusing on Dynamics: When an Exception Becomes a Rule

The historian of science may be tempted to exclaim that when paradigms change, the world itself changes with them

—Thomas Khun

Contents

A. V. Peretti et al., *Rules and Exceptions in Biology: from Fundamental Concepts to Applications*, https://doi.org/10.1007/978-3-031-55382-0_6

Abstract In this chapter, we deal with the change from the exception to the rule in biological systems, both by the action of nature and by the changes that occur due to human action. We talk about the origin of life on planet Earth, the first organisms that colonized primitive environments and changed the atmosphere, giving rise to new forms of life, the appearance of eukaryotic, multicellular organisms, and the different forms of reproduction. We focus on events and changes that were initially considered teratological and that are familiar to our current vision. We also mention adaptations, plasticity, and different phenotypes that became advantages and allowed organisms to continue living in different environments. On the other hand, we point to global processes that affect humans and that in many cases are caused by humans. We discuss examples of diseases that turn into pandemics, the processes of environmental pollution, and accelerated climate change. Finally, we will discuss the changes in scientific ideas, which are closely linked to the social context at each moment in human history, the changes in the different fields of study and within society itself.

6.1 Introduction

At the beginning of the section that covers this chapter, we saw that stability and dynamism are states that can be ephemeral in many situations at the scale of time, space, or group. Accordingly, in the two previous chapters, we have seen cases in which stability, the "conserved," prevails over change at any scale or level, depending on the point of comparison. In this chapter, we focus on the situation that we could describe as a change from "exception to rule," when the uncommon becomes the pattern. It is one of the most exciting and striking situations for scientists and nature. The least interesting would be that nothing changes or is simply a matter of changes from rules to other rules (e.g., modulating their scope), but without implying that this change is for something previously considered a rarity.

There is perhaps a solid epistemological view whenever the word change is used in the dynamics of rules, often inherently implying the concept of exceptions. As discussed in the chapter on rules, Kuhn's epistemology emphasizes the significance of change and revolution, which are typically preceded by a crisis. However, it is important to note that this perspective does not solely or exclusively attribute the emergence of exceptions to the evolution of rules. Indeed, the idea of scientific revolutions implies the fall of a paradigm from the growth of a new body of arguments and data that refute that preexisting paradigm. In the beginning, the new idea was like a bottle in the sea, an undervalued exception for its acceptance in the community. Kuhn's epistemology inherently embodies the essence of change. In addition to the advancements made in refining his fundamental concepts, contemporary contributions by scholars such as Lakatos have helped to provide a clearer understanding of what aspects are more prone to change. Furthermore, Feyerabend's notion of nonacceptance of rules in a strict sense adds to the understanding of change. Nevertheless, one undeniable fact remains: change does indeed exist.

The phrase by Kuhn (1976) that heads this chapter alludes to that strange sensation of perceiving that the world has also changed because the scientific paradigm with which we understood it (or at least tried to understand it) has changed. This new perception is understandable since the change in the scientific gaze influences the way we perceive the surrounding world. As he argues, to the extent that scientists' only recourse to that world is what they see and do, we can say that they will respond to a different world after a revolution. It is a somewhat circular argument that reflects what often happens to us. In other words, although the world has not changed, by changing our gaze, it is presented to us as if it has changed. Technological advances are certainly promoters of these changes and new perspectives. It is feedback between science and technology. For example, the development of electron microscopy has allowed the exploration of cell ultrastructure and the structure of materials. It promoted exploring previously uninvestigated aspects with new discoveries, which again pushed for improving that technology. Additionally, as Kuhn argues, during revolutions, scientists see new and different things when they look at new instruments in places where they have looked before. This metaphor masterfully illustrates this concept of looking again at the known with new eyes: *"It is as if the professional community has suddenly been transported to another planet where familiar objects are seen in a different light and join the unknown. Of course, nothing of the sort happens: there is no geographical transplantation; outside the laboratory, everyday affairs usually continue as before"*—Kuhn, 1976.

However, it is crucial to avoid the temptation of solely engaging in an epistemological exercise focused on examining the dynamics between rules and exceptions within the context of the history of biological knowledge. This narrow approach only considers changes resulting from shifts in scientific paradigms, such as alterations in schools of thought and technological advancements. Although this is extremely interesting and important, the truth is that it responds to a specific dynamic framework: our history of knowledge. We have to balance our approach and assessment of exceptions from the other significant dynamics framework: the one based on changes in biology both in the short or medium term (e.g., at the physiological, behavioral, and ecological level) and in the long term (e.g., at the genetic, evolutionary level). The dynamics of the history of knowledge could be considered analogous to the Darwinian approach to biological evolution (Ruíz & Ayala, 1998). Thus, considering an evolutionary approach, we have already seen that, from an adaptive functional point of view, oddities can become a rule to be generalized. There may be a mixture of biological changes and changes in the scientific community's interpretation and view of the "knowledge" of these changes. Hence, the confusion or being able to easily fall, or be trapped, in the strictly epistemological.

Therefore, this chapter will address some "witness" cases of dynamism from exception to rule in biological systems. These will begin with the changes from exceptions to rules occurring at the biological level by the action of nature itself or the man on it. Subsequently, we will examine these changes within the framework of scientific knowledge of biology, driven by changes in schools of thought and technological advances.

6.2 E-R Changes by Nature's Action

6.2.1 At Crucial Moments of Biological Evolution

6.2.1.1 Origin of Life

Currently, the existence of life on planet Earth is the rule. Therefore, we find life forms adapted even to extreme environments of temperature, altitude, pressure, and humidity (Rampelotto, 2013). Therefore, the rarity on our planet is not to find traces of life somewhere. Life is massive and present everywhere, from large vertebrates, invertebrate organisms, and plants to simple microorganisms such as thermophilic bacteria. However, we know that life was not always the rule on Earth. Life was initially an absolute exception because it was a singular event on our planet. Moreover, the abundance of the initial organisms would logically have been low until a new stage was reached, in which the density of descendants progressively increased.

If someone from outside the Earth had observed our planet at that remote time, perhaps so great would be the magnitude of exception or rarity in the existence of life that they might not even have detected it. The question is whether something similar might happen to us when looking for life elsewhere in the solar system or the universe. Thus, that exception (i.e., initial life on Earth) could have remained as such. However, life began to develop, biological systems diversified, and the number of representatives increased and evolved (Solomon et al., 2018; Curtis et al., 2008). Given these circumstances, it is not surprising that while this book explores the concepts of rules and exceptions within the realm of life, including animals, plants, and other organisms, it is noteworthy that life itself, in its early origins, represented a remarkable exception marked by genuine innovation.

We will not discuss the hypotheses explaining the origin of life. It has been addressed in classic books, such as the pioneering work of Aleksandr Oparin (1924) and John B. S. Haldane (1929) and the subsequent writings of Francis Crick (1981), Robert Shapiro (1988), Leslie Orgel (1998a, 1998b), and Freeman Dyson (1999), among others. Laboratory experiments such as those of Stanley Miller (1953) have become classic examples in the field, despite the inherent limitations that their results may possess. The results obtained by Miller can also be obtained under conditions of atmospheric neutrality closer to an environment dominated by volcanic gases, as prevailed on the early Earth (Harold, 2001; Fitz et al., 2007). In addition, discussions on the possibility of the extraterrestrial origin of life through panspermia (Tepfer, 2008; Wickramasinghe, 2012; Cockell, 2016) or the bombardment of prebiotic foreign organic compounds (Lazcano-Araujo & Oró, 1981; Chyba et al.,

1990; Whittet, 1997; Callahan et al., 2011) also add exciting ideas to the hypotheses of the origin of life.

What is certain is that for our examples for the present book, the origin of life or "abiogenesis" combines the three contexts analyzed in this book. The context of a group, or individual, is represented by the appearance of the first living being (FUCA aka First Universal Common Ancestor—Prosdocimi et al., 2018). Meanwhile, the context of space refers to a particular place where life began, for example, a specific site in the primordial sea, rather than happening massively at various points. However, no one can say for sure! (Raup & Valentine, 1983, Shapiro, 2007; Velasco, 2018). Finally, the time context alludes to the millions of years it would have taken for the first living thing to appear since the formation of the Earth. Its appearance took place approximately between 3000 and 4000 million years (Schopf, 2006; Schopf et al., 2007; Raven et al., 2002; Dodd et al., 2017; Freeman et al., 2017; Betts et al., 2018) from precursors that were not categorized as living beings (e.g., complex and clustered organic molecules but not organized within a protective layer). The cell membrane was crucial for giving entity to that unit, the primitive cell, which progressively began to evolve and diversify.

Chaos and entropy ruled in that primitive world. Unlike today, on primitive Earth, there is no atmospheric oxygen and no ozone layer to filter out dangerous UV rays (Towe, 1981; Cockell, 2016). Thus, the mere existence and maintenance of certain complex organic compounds was already an exception (Solomon et al., 2018). Not only were these compounds challenging to generate, but they were difficult to maintain and replicate once they were generated by chance. Thus, once these molecules were formed and stabilized, it was possible to make further progress and improve the work, catalyzing processes that were previously unnecessary. Take, for example, the formation of a molecule such as RNA or a related primitive precursor. The RNA world theory (Gilbert, 1986) assumes that in the beginning, RNA was responsible for carrying genetic information and catalyzing protein formation processes (Cech, 1990; Altman, 1990, 2000). This model today can be seen as an exception, contrasting with the current model in living systems where the RNA/protein/DNA complex is in charge of performing such a task and with substantial improvements (Bernhardt, 2012). The rule at that primitive time of biotic origin was the circulation of molecules dissolved in the medium (Kauffman, 2011). The formation of a compound such as RNA requires the polymerization of a long chain of nucleotides. This chain is very unstable in an aquatic environment (Shapiro, 2007), requiring extra help from other coexisting compounds in the environment. The appearance of simple, selectively permeable membranes, combined with ions and the presence of amino acids, likely played a crucial role in preserving, stabilizing, and facilitating the evolution of this organic compound. Otherwise, it would have been exposed to bombardment and eliminated in the primordial broth (Carny & Gazit, 2005; Higgs & Lehman, 2015; Pressman et al., 2015). These future membranes were perhaps in the medium as soap droplets (micelles) in an aqueous solution (Luisi, 1996; Benner et al., 2010). The use of these structures to shelter molecules, especially with genetic information, may have been rare. However, it has become the rule maintained today in all living organisms and other biological entities, such as viruses. A pivotal transition occurred, leading to a change in the level of complexity commonly referred to as life. At the heart of this transformation lies the fundamental ability to reproduce, serving as a key defining characteristic. We see that the sum of oddities of a particular moment generated the landscape we now see as usual.

Among the oldest organisms (3.4–3.8 BYA), both autotrophs and heterotrophs were represented (Solomon et al., 2018). For a long time, it was thought that the first living cell was a heterotroph, consuming molecules and metabolizing them through fermentation since the absence of oxygen was the rule at that time (Dismukes et al., 2001). Gradually, however, more recent discoveries have turned that hypothesis toward chemoautotrophs or photoautotrophs (Ward & Kirschvink, 2015; Solomon et al., 2018). Notably, various groups of chemosynthetic bacteria have been discovered, which were likely well adapted to thrive under harsh environmental conditions, including extreme temperatures, prevailing on Earth during that period (Raven et al., 2002; Martin & Russell, 2003; Curtis et al., 2008; Martin et al., 2018; Freeman et al., 2017). These primitive organisms laid the foundation for subsequent Earth changes. Primitive autotrophs, such as photosynthetic cyanobacteria, gradually oxygenated the atmosphere and changed the environment forever (Des-Marais, 2000). The presence of a sufficient quantity of oxygen in the atmosphere allows the appearance and evolution of heterotrophic organisms. This represents a new change where something once regarded as an exception has now become the rule. In the next section of this chapter, we will review that critical moment, which is linked to a chain of energetic expansions in life.

In the context related to the origins of living beings, we cannot fail to mention the concept of LUCA (an acronym for Last Universal Common Ancestor), as it is known to be the immediate ancestor of all living beings (Woese, 1987; Forterre et al., 2005; Glansdorff et al., 2008; Weiss et al., 2018). LUCA is not considered the first life form on the planet (as we said, that is FUCA—Prosdocimi et al., 2019). LUCA represents the only kind of organism of that time that still has living descendants and gave rise to the three domains of life (Prosdocimi et al., 2019). In a framework similar to the one posed for the first living organism, LUCA's emergence, propagation, and diversification are framed in a transition from exception to rule; of course, a rule is represented by an expansion in descendants that reach up to the present epoch. Initially, it was also an exception that this organism, and perhaps earlier forms of it, were subject to natural selection (Darwin, 1859). Some studies suggest that prebiotic organic compounds may have been subject to selective forces (Ponnamperuma, 1978; Anderson, 1983; Lazcano & Miller, 1996; Subramanian et al., 2020). The ability of early organisms to undergo changes over time was a fundamental breakthrough that, as we discussed, currently stands as one of the fundamental principles defining all living entities: the capacity for evolution (Chodasewicz, 2014; Schulze-Makuch & Irwin, 2018).

A question that elicits a sense of fear or desolation within humanity is whether life on Earth is an exception in relation to the rest of the vast universe. Exobiology, research, and space missions seek to decipher this big question (Cockell, 2020; Schulze-Makuch & Irwin, 2018). Enrico Fermi once wondered "*where is everyone*?," alluding to the absence of contact with other extraterrestrial organisms or civilizations (Gray, 2015). Thus far, the existence of intelligent life forms and the technological development to make contact seems scarce or absent in the universe. However, the hypothesis that simple or inferior life forms are present on different exoplanets located within the habitability zone of their stars or even the possibility that life exists on planets or moons within the solar system would not seem at all far-fetched (Cockell, 2016). For now, the rule remains the general absence of life outside the Earth. We will return to this enigmatic field toward the end of Chaps. 7 and 8.

6.2.1.2 "Energy Expansions" of Evolution

When hypotheses about the origin of life are discussed, there is one point in common among them, the existence at that time of macro and microclimatic conditions very different from those that today form the pattern on earth. The combination of water, which is an essential component, along with specific critical chemical elements, serves as the building blocks for the formation of more intricate molecules. These molecules act as the foundation for cellular structures, even in their simplest form, thus laying the groundwork for the emergence of minimal yet living organisms. However, until recently, evolutionary biologists have not placed much emphasis on the effect of the type of energy prevailing at "x time" in favoring or not favoring the emergence and evolution of certain life forms. A look at the feedback between abiotic and biotic conditions, referring mainly to the association in "energetic" terms at each moment, has been missing.

The approach finally arrived very recently, in the hand of an inspiring study by Olivia P. Judson (Judson, 2017), an author well known in the field of science communication. She has proposed to pay attention to the "energy expansions" associated with the evolution of living beings, considering the type of source and factors that propitiate their appearance. She argues that the life system on earth can be divided into five "energy" epochs, each of which is conducive to the evolution of life forms that can exploit new energy sources. These sources are geochemical energy, sunlight, oxygen, flesh, and fire. Judson argues that both geochemical energy and sunlight were present from the beginning but that the others are already a consequence of evolutionary events (i.e., produced by the concrete activity of living organisms, such as the effect of cyanobacteria on the production of oxygen). If we consider that none of these energy sources have disappeared over time, this has directly resulted in an expansion of available energy resources for living organisms. Consequently, this has fostered an associated increase in the diversity and complexity of ecosystems. In addition, these expansions have implied transformations of crucial aspects of the terrestrial environment, which in turn have mediated future evolutionary changes. Incorporating this energetic perspective potentially empowers us to anticipate the trajectories of life systems on planetary scales, regardless of their location. Indeed, in the current era of exploring exoplanets similar to Earth throughout the universe, this approach holds significant value, as it enables us to make predictions about the expected conditions on a particular planet based on its known energetic state. We do not intend here to make an exhaustive presentation of their arguments, but we express the value of this approach for our book.

On the other hand, something that derives from this new vision is that many of the sources first alluded to did not exist, or if they did, they were infrequent. We can trace geochemical and solar energy from the beginning (although initially billions of years old, these were still in the process of becoming a rule). However, for example, oxygen in its free state is initially the product of the action of organisms such as cyanobacteria. Free oxygen initially represented an exceptional occurrence but gradually emerged as the prevailing pattern from an energetic perspective. The previous energy sources were incorporated to varying degrees alongside the rise of free oxygen. The same happened with the age of meat. Predators appeared when there were enough autotrophic organisms, and the ingestion of "meat" became generalized. When predators appeared, it was necessary for a minimum of autotroph photosynthesis systems or detritivorous organisms. The idea is going from a nonexistent fact or strategy to exception, from exception to rarity, from rarity to something in between, and then to something more generalized up to a specific limit (e.g., availability of "prey" at each time and site, *sensu* the predator–prey model). The fire epoch, also an exception at the beginning, appears only when atmospheric oxygen (produced by life) is sufficient and shapes regions by its drastic effect on plant and animal communities. Additionally, when humans appear, they learn to use it. Again, in the beginning, it was an exception, but little by little, it became a habit. It has already become the rule for cooking food and modifying the environment (e.g., burning pastures and forests to favor the growth of certain species humans wish to exploit for their use). Thus, we have at least three epochs, that of oxygen, that of meat, and that of fire, that at the beginning were exceptions but gradually became rules.

Judson also argues that all five forms of energy have been recovered. In the globality of the terrestrial life system, an additive effect of energy sources predominated. However, today, species have not been evenly distributed in the exploitation of energy sources, but rather those linked to oxygen and meat predominate (for example, consider the million species of insects vs. the few species that live on geochemical energy alone), or so it is believed. Some estimates suggest that the biosphere operating below the surface still rivals in terms of mass that of the surface (Whitman et al., 1998). That is, the magnitude of the contribution of each energy source to sustaining life activities in the taxa is different. In terms of direct contribution, from a relative point of view, specific energy sources, such as geochemical sources, became the exception. Geochemical energy went from an almost "absolute" rule billions of years ago to a more circumscribed rule, at least as a direct energy source in the strict sense. Finally, this new perspective that emphasizes the role of energy sources and feedback in evolution represents, in a way, a modification of the current paradigm regarding our understanding of evolution, as has happened with so many theories and hypotheses. This view does not challenge the classical neo-Darwinian framework but rather adds additional elements that enrich our previous ideas.

6.2.1.3 Origin of Eukaryotes (and Protist Radiation)

For almost 2000 million years, the only living organisms on earth were prokaryotes (Alegado & King, 2014). It is postulated that eukaryotic cells appeared approximately 1500 million years ago (Knoll et al., 2006). This "appearance" is, in itself, an enigma. Initially, the "first" eukaryote was not of the standard type or pattern. We could even say it was a "quasi" teratology in a prokaryotes' world. The appearance of an organism, still a cell, containing its genetic material enclosed within a double membrane, along with a prokaryote within its cytoplasm (which would eventually become the mitochondrion), might initially appear as an anomaly, deviating from the prevailing understanding of what constitutes an organism (cell) based on common knowledge. Its functionality made this appearance "monstrous" from the point of view of a "prokaryotic observer."

There have been many models to explain the origin of eukaryotes, and to date, this subject is still much debated. Among them, autogenous theories postulate that a prokaryotic cell is the ancestor of eukaryotes, and some organelles arise from invaginations of the plasma membrane (Reid, 1985). However, following these theories, it is still impossible to explain the origin of organelles, such as mitochondria and chloroplasts, with critical functions for modern eukaryotes. Thus, the endosymbiosis theory between two prokaryotic organisms has gained relevance (Margulis, 1981). While it may not provide a comprehensive explanation for the origin of all cellular structures, it remains a highly remarkable theory. Some authors have even developed theories explaining the origin and evolution of eukaryotes in two phases. These phases involve autogenous and symbiogenesis processes to explain the complexity of the eukaryotic cell, its organelles, and even the nuclear membrane's structure and cytoskeleton's structure (Poole & Penny, 2001; Cavalier-Smith, 2002a).

The role of endosymbiosis (from phagocytosis of a cell by its host) was and is questioned by many authors (Martin et al., 2001). However, there is more consensus to support the notion that the mitochondrion and chloroplast were indeed acquired through such a process, whereby they were originally separate organisms ingested by the host (López-García et al., 2017). Whether the host was a prokaryote, an intermediate prokaryote-half-eukaryote with a defined nucleus but still without mitochondria, or an archaeon is a matter of debate. Some authors suggest that the process of endosymbiosis that gave rise to mitochondria occurred after an earlier fusion between bacteria and archaea (Zillig et al., 1989). However, other works propose that more than one endosymbiosis originated the mitochondria and the nucleus from a bacterium and archaea phagocytosed by a host (Lake & Rivera, 1994; Horiike et al., 2001). However, the origin of the nucleus by endosymbiosis has been firmly rejected (Martin, 1999; Poole & Penny, 2001). Finally, it has also been postulated that the origin of mitochondria was one of the last steps in the evolution toward eukaryogenesis (Karlberg et al., 2000; Cavalier-Smith, 2002a; Andersson et al., 2003). The significant problem is that there is no agreement on what constitutes a valid hypothesis, but it is most widely accepted to postulate that the initial phagocytosed element was a bacterium (McFadden & van Dooren, 2004; Zimorski et al., 2014; López-García et al., 2017).

Recent advancements in genetics, our understanding of marine microfauna, research on viruses and the emergence of new domain classifications have proved to be crucial in the scientific interpretation of this ongoing discussion, which continues to be of great significance today. For example, the conservation of choanocytes in choanoflagellates and primitive animals such as sponges suggests that the animal progenitors were bacterivorous (Alegado & King, 2014). On the other hand, the diversification of animals coincides with a rapid reduction in the abundance of stromatolites, and from this, it could be thought that the bacteria involved in those formations were preyed upon by primitive animals (Garrett, 1970; Awramik, 1971; Walter & Heys, 1985). These phagocytosed bacteria could have impacted genome evolution by lateral gene transfer, providing new genetic material in animals and their origins (Alegado & King, 2014). There is a continuous stream of intriguing news claiming the discovery of the long-awaited "key" piece, whether through fossil findings, theoretical advancements, or breakthroughs in genetics and physiology. These findings aim to unravel the mystery surrounding the origin of eukaryotes (e.g., Keeling & Doolittle, 1995; Rivera & Lake, 2004; Gribaldo et al., 2010; Spang et al., 2015). In any case, the initial appearance of this "proto-eukaryote" was not a massive event. Even if it occurred more than once, it remained something scarce. One could well say "teratological," as we have already emphasized. The idea that the process of phagocytosis of another organism could eventually lead to a close relationship, such as the one observed between a whole cell and the mitochondrion, may initially seem far-fetched. However, it is undeniable that this process underwent numerous trials and errors before the emergence of a primitive "eukaryote" where it proved successful. In these early stages, the relationship likely began as a commensal one, without any adverse effects on the host organism.

Since the Paleozoic era, approximately 550 million years ago, eukaryotes have been extraordinarily successful. In addition to the multicellular plants, animals, and fungi that define the biosphere we know today, many simple groups of protists still contribute fundamentally to ecosystem function (Butterfield, 2015). However, this was not always the case. Eukaryote evolution represents one of the most elemental transitions in the history of life on Earth, yet there is considerable debate and a lack of consensus regarding the exact timing of this event. The increase in cellular complexity from prokaryotic cells would inevitably have generated explosive radiation of protists into every available ecosystem niche (Cavalier-Smith, 2002a). The cytoskeleton, molecular motors, and cellular endomembrane system exhibited the most far-reaching changes. Without these innovations, the existence of complex protozoa, as well as the diverse array of animals, plants, and fungi that we observe today, would not have been possible. Their functional diversification was accentuated by their new ability to ingest and digest other cells as food or keep them in place as "slaves" that provide energy and valuable metabolites (Cavalier-Smith, 2002b). The abundance, diversity, and evolutionary revolution of protists and their genetic material undoubtedly characterize them as transitional organisms (Lynch & Conery, 2003). The manner in which selection and genetic drift influence mutations in protists can be compared to the processes observed in large populations of evolutionarily significant prokaryotes, as well as in small populations of multicellular eukaryotes. This is due to this intermediate state in which protists can be considered precursors to other eukaryotes, as described by O'Malley et al. (2013).

In summary, the origin of eukaryotes represents an actual example of the passage from the abnormal to something that later resulted in the appearance of protists and then to the rest of the animal, plant, and fungal phyla. The controversy about this origin does not prevent us from tracing the reasoning that we could pose as a rule: "*everything that became massive, expanded, generalized, began as a rarity.*" Is this something that sounds logical, obvious, or even naive, although it is something that we do not value and do not think about in our view of biological evolution, at least not enough.

6.2.1.4 Symbiotic Evolution: From the Restricted to the Rule?

As we have seen, the origin of eukaryotes is closely connected to the concept of symbiogenesis, which is defined as an evolutionary mechanism that generates variation and produces new species through symbiotic associations between two individuals (Mereschkowsky, 1910; Margulis, 2010). This process has been recognized as a significant means of generating "biological novelties." While it has considerable support, it also faces criticism, particularly from proponents of neo-Darwinism (Speidel, 2000).

There are currently many symbiotic associations, and we have reviewed some of them in Chap. 5. An exceptional case is represented by lichens, which are the result of symbiosis between a fungus and an alga or a cyanobacterium. One-fifth of all known fungal species form symbiotic associations with green algae, cyanobacteria, or both to generate lichens as new organisms. It allows fungi to fulfill their carbohydrate requirements (Hawksworth, 1991; Hawksworth et al., 1995). This symbiotic association confers exceptional resistance to various environmental conditions, enabling them to occupy ecological niches characterized by conditions that individual organisms involved in the symbiosis would be unable to survive in (Schne et al., 2008). Lichens are believed to have arisen independently, which explains their high diversity and the occurrence of lichenized and nonlichenized fungal species within the same phylogenetic group (Lutzoni et al., 2001). Furthermore, independent losses of symbiotic associations may have occurred during the evolution of fungi, and thus, there are nonlichen-forming species derived from lichen-forming ancestors. However, the typical pattern observed in all these cases, as in many others, is similar, and symbiosis facilitates evolutionary innovation, although it may be lost later.

In addition, fungi can establish symbiotic associations with plant roots, forming mycorrhizae. Fungi play a crucial role in converting soil minerals, such as phosphorus, into a form that is readily available and usable by plant roots. They facilitate the transfer of nutrients from the soil to the plant (Selosse et al., 2006). Certain groups of cnidarians, such as corals, can also form symbiotic associations with algae. These algae provide them with energy through photosynthesis, enabling corals to meet their energy needs for processes such as calcification (Inoue et al., 2018). In the following chapters, we will see that variations in temperature, light, and nutrients can affect this symbiosis, modifying the abundance of microalgae and thus having secondary effects on coral physiology (Dubinsky & Jokiel, 1994; Anthony et al., 2008). The study of coral reefs can provide insights into fundamental processes in symbiotic relationships between species and demonstrate how they can adapt to environmental changes (Apprill, 2017). In addition, bacteria and fungi contribute to the cycling of organic matter, forming these characteristic multimembered marine microbiomes. Interestingly, many of these cases form an initial exception to a rule in their respective groups. By evolution, we move from an initial (fortuitous?) association between an organism and its endosymbiont to a persistent linkage over time. Thus, group and time contexts are present. For example, the colonization of land by plants approximately 400 million years ago was made possible by symbiosis with fungi (Lewis, 1987; Selosse & Le Tacon, 1998). These fungi formed associations with roots in approximately 90% of higher plant species and even with ferns and some mosses (Selosse & Le Tacon, 1998). The first plants that colonized terrestrial life probably lacked roots and therefore needed an association with fungi for water and mineral uptake (Aanen & Eggleton, 2017). However, this symbiosis was gradually lost or replaced in some lineages by other symbionts, such as nitrogen-fixing bacteria or ectomycorrhizal fungi, or by new adaptations such as advanced root morphologies.

Everything concerning endosymbiosis is also connected with changes in scientific interpretations. As mentioned earlier, the endosymbiotic hypothesis was promoted by Lynn Margulis. The postulates of this framework make it appear as an evolutionary theory in itself, holistic, cohesive to all living beings (Margulis & Sagan, 2002). Margulis's advocacy for this theory was so influential that it was often regarded as an alternative to the classical neo-Darwinian synthetic theory of evolution (i.e., classical neo-Darwinism). More recently, building upon the concept of symbiogenesis, Gilbert et al. (2012) proposed a "symbiotic" perspective on life. According to these researchers, the concept of a "biological individual" was useful in studying various aspects, such as anatomy, physiology, and genetics, which contributed to the development of specific biological subdisciplines, each with its own understanding of individuality. Nevertheless, with the advancements in genetics and physiology that have unveiled the close association between animals, plants, and their respective symbiotic microorganisms, there is a growing need to consider the "holobiont" as the natural biological unit. The term "holobiont" refers to the multicellular eukaryote and its persistent symbiont colonies, acknowledging the integrated nature of these organisms and their symbiotic partners. In other words, the barrier of the "individual" has long been crossed not only in scientific interpretation but also in organic evolution. Consequently, the concept of the "biological individual" is considered outdated. This shift in perspective leads to the understanding that symbiotic associations are no longer viewed as exceptions but rather as a widespread phenomenon. It prompts the exploration of a new notion of biological unity, wherein a group of multiple organisms comes together to achieve functional integration (Gilbert, 2014). Symbiont diversity plays a crucial role in the metabolism and development of individuals, as observed in symbiont microbes within the animal kingdom. These symbionts contribute to a second mode of gene inheritance, introducing variability to the host organism (Gilbert, 2011).

In the same line of analysis and curiously linked to the terminology of the present book, we highlight the statement of Munzi et al. (2019). Indeed, symbiosis is now recognized as a widespread phenomenon that was previously regarded as an "exception," limited to certain organisms such as corals, lichens, or mycorrhizae, as mentioned earlier. However, with the advancement of knowledge, it has become evident that symbiotic relationships are not exclusive to a few organisms but rather a "rule" applicable across all living systems and entities. The microbiome has emerged as an integral part of all life forms, underscoring the ubiquity and importance of symbiotic associations. Therefore, they emphasize that symbiosis could be seen as a fundamental strategy for life and that it represents biodiversity in action, where communication, information exchange, and regulation between the parts guarantee stability and adaptability.

Toward a Symbiogenic Phylogenetics (SYMPHY)?

In this framework of the implications of more integrative views claimed by symbiogenesis theory, with greater or lesser emphasis, it is interesting to note the recent study by Tripp et al. (2017) on their claim that current phylogenetics should take a more integrative view in terms of considering symbiomes in the tree of life. The title of their paper, "*Reshaping Darwin's Tree: Impact of the Symbiome*," refers to that claim about the hitherto overlooked or at least underestimated the impact of symbiomes. That is, he aims to leave behind a classical systematic vision, more focused on taking each species as a "unit" and instead be aware that in many cases (if not all?), it is a complex of species forming a unit, as is the case for symbiomes.

The term symbiome refers to taxa intimately associated in physical space (colocalized) and coevolve together under selection, i.e., as a consortium. It is different from the term holobionts. Although they can be taken as synonyms (as many of the already cited and discussed previously in this chapter), it is better to use holobionts when referring to physically associated taxa that do not coevolve or are necessarily codependent (obligate). Examples of symbiomes are intimate interactions such as those we observed above between plants with fungi, plants-cyanobacteria, algae-fungi, such as in the case of lichens, and even the intestinal insect-microbiome. For example, phylogenetic studies in aphids showed that endosymbiont bacteria of the genus *Buchnera* are one of the fundamental causes of aphid diversification. This indicates that the acquisition of a free-living ancestor of the bacterium was followed by the coevolution of both host and symbiont and the specialization of the host to different feeding niches (Moran et al., 1993). Thus, each aphid species harbors a strain of bacteria suitable for providing specific nutrients that are deficient in its diet (Gil & Latorre, 2019). With the availability of known host and symbiont genomes, it is now evident that it is the aphid that controls nutrient production by the bacteria in finely tuned metabolic complementation (Eisen, 2010; Wilson & Duncan, 2015). Another example is observed in most terrestrial plants, including cereals, which, as mentioned, can establish symbiotic associations with fungi and generate arbuscular mycorrhizae. These mycorrhizae help plants acquire nutrients and protect them from biotic and abiotic stresses (Harrison et al., 2002; Govindarajulu et al., 2005). Fossil records suggest that this symbiosis coincides with the emergence of the first land plants (Redecker et al., 2000). In recent years, genetic studies have identified genetic components necessary for establishing this symbiosis in model plants (Harrison, 2005; Bonfante & Genre, 2010).

In this context, it is imperative to consider the symbiotic phylogenetic vision of these authors, considering that much of the diversity not yet described on Earth is microbial, whether in mutualistic or parasitic associations. Indeed, for prokaryotes, the classical model of species as a single exclusive unit would not be appropriate or sufficient. The implications of considering a symbiotic phylogenetics (SYMPHY) vision would allow better decisions and management of species conservation tasks, invasive species, alternative fuels, technology, and biomedicine applications, among others (Tripp et al., 2017).

6.2.1.5 Origin of Sex

One aspect of biology that was an exception millions of years ago but evolved to become a rule in almost all organisms is that of sex. Countless books have been written on this subject, from the time of Darwin (1871) to the essay by Margulis & Sagan (1986), or now classic books on evolution by sexual selection (e.g., Andersson, 1994). In all cases, addressing how the first organisms reproduced and when sex appeared, its advantages and disadvantages have been the subject of multiple views, which have not always been in joint agreement with each other. We know that one of the fundamental properties of life is reproduction, which has evolved from simple to more complex forms. Initially, the rule was asexual reproduction, a process that involves the replication of genetic information so that the offspring are the same as the original individual. This mode of reproduction is still used by many unicellular species and some multicellular organisms (Cook, 1979). The evolution of sexual reproduction is an adaptive and specific characteristic of eukaryotic organisms. Here, the offspring's genetic material came from two different individuals, involving the combination of genetic traits where the gametes of both parents combined to form the offspring (Grützner et al., 2004; Avise, 2011).

The origin of sexual reproduction has long been a mystery. However, the form of reproduction used by organisms may have depended on the environmental circumstances surrounding them. For example, sexual reproduction could have been favored under stressful conditions (Baker & Parker, 1973). Indeed, considering that a very changing environment prevailed millions of years ago, high mutation rates could have promoted an asexual organism to become sexual. Sexual reproduction has numerous advantages that make it the rule. Among them, sex evolved as a mechanism to produce genetic variation through allelic recombination, which allowed organisms to adapt to changing environments (Baker & Parker, 1973; Bernstein et al., 1984, 1985, 1987). Other advantages of sex are that sex facilitates the propagation of advantageous traits (Crow, 1994), eliminates deleterious mutations from the genome (Kondrashov, 1988; Griffiths et al., 1999), and facilitates germline DNA repair (Crow, 1994). One of the most discussed hypotheses about sex maintenance (a fundamental fact for it to be constituted as a rule) is that it helps organisms resist parasites, a statement known as the Red Queen hypothesis (Barrière & Félix, 2005; Otto, 2008; Morran et al., 2011). Although sexual reproduction has advantages, it also entails certain associated costs. For example, when reproducing asexually, organisms do not need to invest time and energy in searching for and choosing a mate. At the level of population expansion, the disadvantage of sexual reproduction is that it grows less than an asexually reproducing population in each generation. In addition, an individual who reproduces sexually passes 50% of its genes to its offspring, whereas it would have passed 100% of its genes to the progeny if reproduction had been asexual. Therefore, an organism that reproduces sexually transmits fewer copies of its genes than an organism that reproduces asexually (Otto, 2008).

The origin of sexual reproduction dates back to the first prokaryotes, approximately 2 billion years ago, when bacteria began to exchange genes through conjugation, transformation, and translation. Although these processes differ from sexual reproduction, they share basic similarities (Otto, 2008). In eukaryotes, it is believed that actual sex emerged approximately 1.2 billion years ago in the Proterozoic Eon in algal species, so sex through different successful processes managed to persist over time (Butterfield, 2000). If environmental changes that produced DNA damage were maintained over time, then selection would have acted continuously in the transition from prokaryotes to eukaryotes, and adaptive mechanisms would have caused asexual reproduction to give way to sexual reproduction (Lodé, 2012; Bernstein & Bernstein, 2013).

Both types of reproduction coexist in numerous invertebrates and plants, known as heterogamy. For example, when population density is low in rotifer species, females reproduce asexually, and at higher densities, they accumulate a chemical signal that induces the transition to sexual reproduction (Snell et al., 2006). Within insects, ants, bees, and wasps exhibit sexual and asexual reproductive strategies. Aphids are one group that has this dual type of reproduction. By asexual reproduction, they quickly create winged offspring that colonize new plants and then use sexual reproduction in the fall to lay eggs for the next season (Stern, 2008). The water flea *Daphnia* reproduces by parthenogenesis to quickly populate ponds and then uses sexual reproduction as competition and predation increase (Burns, 2013; Walsh, 2013). Meanwhile, at the level of higher organisms, including vertebrates, most arthropods, and a significant portion of other invertebrates, sexual reproduction has become the main method of reproduction, predominantly the rule. Curiously, this does not rule out the presence of "reversed" exceptions, i.e., the presence of those primitive reproductive options in some of them (Table 6.1). There are cases where the costs of sexual reproduction are very high, for example, because of the difficulty of finding a mate; here, asexual reproduction, particularly parthenogenesis, is a common strategy in many species. Parthenogenetic organisms can be divided into facultative (where females can reproduce sexually and asexually) and

Table 6.1 Hypotheses about the advantages of sexual reproduction and examples of sexual and asexual reproduction

Hypotheses about the advantages of sexual reproduction			
Hypothesis	Description	Examples	References
Evolutionary advantage	Sexual reproduction increases genetic variability and facilitates adaptation to environmental changes	Sexual: Humans, birds, mammals, reptiles, amphibians Asexual: Bacteria, some plants, Acari	Bell (1982); Kondrashov (1982); Rice (2002); Otto & Lenormand (2002); Lenormand (2003); Otto (2009)
Quick evolution of resistance	Genetic recombination of sexual reproduction allows populations to evolve rapidly resistance to diseases and parasites	Sexual: Humans, insects, amphibians, birds Asexual: Some virus, fungus, Acari	Bell (1982); Hedrick (2005); Hughes et al. (2008); Frankham (2010); Frankham (2015)
Sexual selection	The competition to mate and the selective choice by females leads to an improvement in the genetic quality of offspring	Sexual: Birds, mammals, some insects Asexual: None	Andersson (1994); Tregenza & Wedell (2000); Brooks & Kemp (2001); Møller & Jennions (2001); Kokko & Jennions (2008)
Types of sexual and asexual reproduction in different living organisms			
Types of reproduction	*Description*	*Examples*	*Reference*
Sexual reproduction: Combination of (usually haploid, or having a single set of unpaired chromosomes) reproductive cells from two individuals to form a third (usually diploid, or having a pair of each type of chromosome) unique offspring. Sexual reproduction produces offspring with novel genetic combinations			
Syngamy	Takes place in multicellular organisms. It involves the fusion of two haploid gametes to create a genetically distinct diploid offspring	Humans and other animals. Plants. Algae as *Chlamydomonas*. Fungi such as *aspergillus*	Kondrashov (1993); Otto & Lenormand (2002); Campbell & Reece (2005); Bernstein & Bernstein (2010); Raven et al. (2002); Tortora et al. (2016)
Conjugation	Takes place in single-celled organisms. It involves the exchange of genetic material between two individuals	Bacteria. Unicellular organisms such as *paramecium*. Fungi such as *saccharomyces*. Algae such as *spirogyra*	

(continued)

Table 6.1 (continued)

Hypotheses about the advantages of sexual reproduction			
Hypothesis	Description	Examples	References
Asexual reproduction: Produces offspring that are genetically identical to the parent because the offspring are all clones of the original parent. This type of reproduction occurs in prokaryotic microorganisms (bacteria) and in some eukaryotic single-celled and multi-celled organisms			
Fission	The splitting of a parent cell into two or more daughter cells	Binary fission Bacteria *E. Coli*, multiple fission protozoa *plasmodium*, Schizogamy in apicomplexa *toxoplasma*	Steele (2002); Ruppert et al. (2004); Sánchez Alvarado (2006); Popham (2013); Pechenik (2015); Winfield & Hoare (2015); Booth & Deimel (2019); Kim et al. (2019)
Budding	The growth of a new individual from a small projection on the surface of the parent	Cnidarians *hydra*, unicellular fungus yeast, sponges *Ephydatia*	
Fragmentation	The breaking of the parent into pieces, each of which can grow into a new individual	Flatworms *Dugesia*, echinoderms starfish, algae *Ulva*	
Parthenogenesis	The development of an embryo from an unfertilized egg	Arachnids scorpions. Insects aphids, reptiles lizards, Rotifera	

obligate (where females reproduce only asexually). Among vertebrate species with facultative parthenogenesis, we can name scorpionfish species (Seiter & Stockmann, 2017), zebra sharks (Dudgeon et al., 2017), blacktip sharks (Feldheim et al., 2010), and Komodo dragon (Watts et al., 2006). In species with obligate parthenogenesis, we can name the lizard genus *Cnemidophorus* (Crews & Fitzgerald, 1980).

Sex Chromosomes and Their Entity as an Exception

It is important to note that the emergence of sex chromosomes, carrying genetic information associated with traits defining one sex (male or female), represents a unique novelty within the framework of what an exception is. Indeed, sex chromosomes can be regarded as an "exception to the rule" since they differ from all other chromosomes in the organism, not containing the same genes; they deviate from the pattern followed by homologous chromosomes (Audesirk et al., 2008). Before the appearance of sex chromosomes, all chromosomes shared the same genes in their pairs, maintained by mitotic cell division and subsequent regeneration. Nothing changed. In contrast, meiosis, as a new form of chromosome division, is closely linked to the appearance of sex chromosomes (Grützner et al., 2004). If, as the evidence indicates, sexual reproduction arose very early in eukaryotic evolution, possibly the essential features of meiosis were already present in prokaryotic ancestors (Bernstein & Bernstein, 2010, 2013). In genetic sex determination systems, the sex of an organism is determined by the genome it inherits. Different sex determination systems depend on the sex chromosomes, e.g., XY in mammals (Piprek, 2009; Fusco & Minelli, 2019), ZW in birds (Smith et al., 2003), and X0 in insects (Charlesworth, 2003; Kaiser & Bachtrog, 2010), among others.

Sex has evolved along with the reproductive system, starting with similar gametes (isogamy) and progressing to systems that have different types of gametes (anisogamy) (Gee, 1999; Freeman & Herron, 2007). The most morphologically simple organisms in any lineage tend toward isogamy, and the most complex organisms tend toward anisogamy (Margulis & Sagan, 1986). Therefore, the transition from isogamy to anisogamy represented a significant and crucial step in the evolution of sex, moving from exception to rule. This transition also played a vital role in the progressive definition of distinct phenotypic traits between the two sexes (Stearns, 1987). Classical schemes illustrate in a hypothetical Gaussian bell how disruptive selection favored the most extreme varieties from isogamy. On the one hand, a heavy, less mobile cell containing nutrients (e.g., yolk); on the other hand, a cell with fewer nutrients is smaller and more mobile. As a consequence, eggs and sperm emerged and became established as definitive gametes (Gee, 1999; Freeman & Herron, 2007).

In summary, we can say that when sex first evolved, isogamy was the rule. Sexual partners were nearly identical in appearance. However, many organisms and their reproductive mechanisms have evolved, giving rise to anisogamy, which is now the most widespread mechanism (Margulis & Sagan, 1986).

A Pluralistic Approach to Hypotheses

To the classical discussions of considering only one hypothesis or another to explain the reason for sex, we now bet on a more "pluralistic" approach. That is, several hypotheses should not be interpreted as mutually exclusive. The recent review by Neiman et al. (2017) alludes to this point as to why sex is maintained in evolution. The authors discuss the mechanisms underlying the maintenance of sexual reproduction from the perspective that different mechanisms could combine to prevent the replacement of sexual populations by asexuals. Although leading evolutionary biologists, such as R. A. Fisher, H. J. Muller, and A. Weismann, have already shown the advantages of sexual reproduction, the question is, what maintains sex as a widely used strategy as a reproductive mechanism? Numerous hypotheses are based on it, for example, Muller's ratchet (Muller, 1964), Lottery (Williams, 1975), Red Queen (Jaenike, 1987), and deterministic mutation accumulation (Kondrashov, 1984), among others. At this point, numerous articles published in the 1990s support the idea that several hypotheses complement each other to explain sex maintenance (Peck, 1993; Howard & Lively, 1998; Neiman et al., 2017). In this sense, in

1999, West and collaborators developed what is known as the "pluralistic approach to sex," which posits the combination and interaction between different hypotheses. Thus, the idea that a single mechanism is not sufficient to be able to explain the maintenance of sex is widely accepted in the scientific community. West et al. (1999) propose that different hypotheses coincide (some of them named above) and raise the possibility that different mechanisms may vary in their influence among different taxa (Neiman et al., 2017).

6.2.1.6 Origin of Multicellularity

Another remarkable example in biological evolution is the emergence of multicellularity, marking a transition from rarity to a rule. Perhaps one of the most significant examples of "generalized pattern." In the previous paragraphs, we have commented on the basics, that is, the origin of eukaryotes, which led to their radiation at the unicellular level through protists. However, evolution would take another even more significant step by forming closely associated groupings of cells and thus producing more complex beings. It would lay the foundations for the appearance of representatives of the various kingdoms that make up life on Earth. Interestingly, despite the advantages that multicellularity provides, it has not hindered the coexistence of unicellular organisms alongside multicellular life forms. This phenomenon continues to persist even after millions of years of evolution.

The origin of eukaryotic multicellular organisms represents a significant evolutionary milestone (Butterfield, 2000; Schlichting, 2003; Niklas & Newman, 2013, 2016). Interestingly, this milestone occurred not as a unique event but independently in numerous groups. In its simplest sense, it has been observed at least 25 times, even when considering simple cell aggregation (Grosberg & Strathmann, 2007; Niklas & Newman, 2013; Niklas et al., 2019; Claessen et al., 2014; Hammerschmidt et al., 2014). In a narrower sense, the definition of complex multicellularity suggests that the activities of cells must be coordinated effectively (Kaiser, 2001; Mian & Rose, 2011). Thus, this complex multicellularity would have appeared in separate events (Whittaker, 1969): two times in eukaryotes (once in Amoebae and Animalia [Metazoa]), three times in fungi, and six times in photosynthetic eukaryotes (Knoll, 2011; Niklas & Newman, 2020). What is clear from current phylogenetic analyses is that metazoans all derive from a common Precambrian unicellular ancestor (Sebé-Pedrós et al., 2017; Giribet & Edgecombe, 2020).

Let us consider a hypothetical primordial environment dominated by unicellular organisms (prokaryotes and eukaryotes). In this environment, a new organism arises either due to a failure to separate cells during division or through simple cell aggregation (Rokas, 2008). Regardless of the specific mechanisms, this new organism would have been a rarity during its initial stages. At that time, given our current knowledge of the evolution and success of multicellularity, this phenomenon might have been regarded as an abnormal form or a "teratology of unicellularity," particularly if we consider the hypothesis of cell separation failure.

Let us review what the "rules" were thus far. First, there was no persistent grouping of individuals. While unicellular organisms can form groups over time, their communication is limited, eventually relying on chemical signaling in their environment (Knoll, 2011). Second, cells lacked specialization or differentiation, meaning that all cells performed all functions. This lack of specialization may have hindered their exploration of specific niches (Umen, 2014). Third, physiological limitations and constraints were present during this period. For instance, the size of single-celled organisms was determined, in part, by diffusion distances within the cell (Niklas & Newman, 2013). Finally, size also made them prey for other single-celled predatory organisms and limited the prey size to be ingested (Niklas & Newman, 2016).

With this panorama, we see that a novel and unique aspect at that time is the intimate cooperation between cells, which goes beyond a simple momentary and random interaction. Then, and here is the rare or at least surprising, it jumps to a symbiotic relationship between different types of founder cells (Margulis & Chapman, 1998; Niklas & Newman, 2013). The cells stay together and begin to specialize, each doing a particular job and getting better and better as they evolve. It is another process that can be taken as a rarity since the rule was that all the work was concentrated in one individual, the same cell. After the appearance of multicellularity, the work was distributed. Each cell group was specialized in performing better actions than the individual cell because, as the famous saying goes: "*Jack of all trades, master of none.*" Among several examples, the specialization and differentiation of vegetative or somatic and reproductive cells are notorious (Michod et al., 2006; Knoll, 2011; Umen, 2014). We previously observed that sex appeared in metazoans. Among unicellular organisms, reproduction by cell division is the rule. Some authors even argue that a modification in the cell division phenomenon could have given way to multicellularity (Bonner, 2003; Niklas & Newman, 2016).

Returning to the division between cell lines, multicellularity brought something novel to reproduction: restricting it to specific cells. In addition, reproduction was not through mitosis but through the process of meiosis (Grosberg & Strathmann, 2007) (was it an aberrant process at some point?). As mentioned earlier, meiosis introduces genetic recombination, leading to a significant increase in diversity among organisms. This process helped prevent the existence of clonal individuals who could be easily wiped out by unfavorable changes (Curtis et al., 2008). Additionally, as we saw, sexual reproduction in unicellular organisms was mainly by isogamy but changed with the advent of multicellularity that incorporated oogamy (Wiese et al., 1979; Umen, 2014; Ruiz-Trillo & Nedelcu, 2015). Oogamy is a feature that, as we saw, brings many advantages over isogamy. For example, the nutritional investment in the progeny and the increased chances of collision between gametes due to the presence of numerous small gametes associated with the males.

Phylogeny and genomics propose a "generalized" trajectory for the evolution of multicellularity, which initially involves the cooption of existing genes for adhesion (for further details, refer to the comprehensive review by Knoll, 2011). However, upon closer examination, this explanation fails to account for the diverse transitions observed among different groups. For example, plants and fungi followed different trajectories for their respective multicellularity patterns from the initial elements they had in common (Niklas & Newman, 2013). Indeed, the genetics of organisms underwent significant changes. To illustrate, imagine the genome as a blueprint outlining the execution of routine cellular functions. However, as multicellular organisms evolved with division of labor and cell differentiation, these blueprints had to be modified since the functions differed from their unicellular counterparts or were possibly absent in them (Rokas, 2008). This adaptation was necessary to accommodate the new complexities arising from multicellularity. Thus, the cells of a multicellular organism need communication and coordination between different cell groups, cohesion between cells to form the individual, and differential development of cell lines that compose it (Rokas, 2008). All aspects are little exploited, totally absent, or exceptional in unicellular individuals.

Most strikingly, studies in unicellular counterparts show the presence of genes needed to generate proteins involved in cell–cell junctions and cell–cell signaling (King et al., 2008; Sebé-Pedrós et al., 2017; Paps & Holland, 2018; Richter et al., 2018). For example, choanocytes are cells present in sponges (Porifera). These cells closely resemble free unicellular forms called choanoflagellates. For a long time, the rule was to associate them directly, but recent studies have suggested that these are analogous forms that seek to solve the water flow problem at the microscopic level in a similar way. Indeed, this solution appears in several groups outside Porifera (Giribet & Edgecombe, 2020). However, choanoflagellates have demonstrated the ability to coalesce and generate small colonies and are considered the closest single-celled relatives of animals (King, 2005). The ability to form colonies suggests that these organisms potentially have the necessary genetic elements. Studies in a choanoflagellate species, *Monosiga brevicollis*, demonstrate that the genome possesses genes ready to encode the protein domains necessary for multicellularity (King et al., 2008; Fairclough et al., 2013; Suga et al., 2013; Sebé-Pedrós & de Mendoza, 2015).

How can we interpret this from the orbit of rules and exceptions? Imagine that one would have expected the appearance of these genes to be something fortuitous that triggered the appearance of the characters linked to multicellular organization. However, the presence of that information may have been the rule in unicellular organisms, but what may have been the exception was the manifestation of those characters that remained deactivated. While there was innovation in the genetic repertoire at the origin of metazoans, the unicellular ancestor had the basis for cell adhesion, signaling, and transcriptional regulation (Sebé-Pedrós et al., 2017). In contrast, some groups lose entirely some of the genes linked to multicellularity. A case present in current cnidarians is that of myxozoans, an exception among modern cnidarians. Myxozoans are highly modified cnidarians of parasitic life (Lom & Dyková, 2006) and are composed of a few cells. They were long considered protists and presented some of the smallest genomes in the animal kingdom (Chang et al., 2015). These authors claim that genes that are considered the stars of animal multicellularity were eliminated within this massive genome reduction. This elimination drastically modified the structural plan of the organism, which is very convenient for parasitic life. It is conceivable that this should have been the beginning among unicellular organisms, the absence of the genes for multicellularity and that a modification in the genome would generate these genes, whose expression would result in the characteristics that multicellularity, as a phenomenon, needs. However, we saw that the rule was probably the presence of these genes from the beginning, and then, they were lost in more derived groups, as in the case of the group of parasites mentioned above.

It is exciting that currently, we can use analysis models and somehow see the evolutionary path different organisms could have taken to go from unicellular to multicellular life. In this sense, sponges are organisms that would present the first level of organization within eumetazoans. As we saw, the presence of choanocytes makes Porifera a good candidate for the closest ancestor of metazoans, a place disputed with sea acorns (Ctenophora) (Giribet & Edgecombe, 2020). This cellular organization is more complex than other unicellular groupings but less complex than the tissue and organismic types of organizations present, for example, in cnidarians and Platyhelminthes. Sponges, unlike other animals, can dissociate their cells and regroup them, forming a new adult organism (Lavrov & Kosevich, 2014). Although the rule in this group is the absence of actual tissues, it does not prevent some kind of integration, communication, and differentiation in the cells that make up the individual. As we saw, the exception to this group, in terms of poriferous body constitution, is represented in the class Homoscleromorpha. Recall that what stands out most in this group is the presence of an actual basement membrane with type IV collagen underlying the pinacoderm and choanoderm and more complex cellular connections than in the rest of the sponges (Chap. 5). These features connect Porifera directly to actual animals (Gazave et al., 2012).

On the other hand, the green algae Volvocales, also called Chlamydomonadales, represent a fascinating group from an evolutionary point of view (Umen, 2014). They represent a rarity for the study of multicellularity because they are one of the few groups whose representatives can reconstruct the theoretical evolutionary steps from a unicellular to a multicellular organism. This evolutionary path is shown very clearly by Umen (2014) and by Niklas & Newman (2013). It starts with the genus *Chlamydomonas*, where no cell grouping can be seen, and unicellularity is the rule. Then, we have *Gonium*, where we already see cell polarity, cell number modulation, and cell junctions. *Pandorina* and *Eudorina*, on the other hand, are two genera that show slightly more complex and extensive colonies. Even greater complexity can be seen in *Pleodorina,* as it possesses specialization between somatic cells and germ cells and the presence of anisogamy but not yet oogamy. Finally, the whole level of multicellular organization is found in *Volvox*, where oogamy already appears, along with some other more specific characteristics that determine the conformation of a multicellular individual (Umen, 2014). The study of this algae group also showed an "innate" tendency toward multicellularity. What do we mean by this? In this scenario, the initial explanation does not always explicitly involve natural selection. Imagine, for example, that there was already a combination of multiple cells, primordial cell mixing, which may have occurred through cooperation, symbiogenesis, or a combination of various mechanisms. The role of natural selection may come into play at later stages of evolution, once these multicellular entities have formed and begun to undergo further changes and adaptations. It is interesting the comment of Furusawa & Kaneko (2002) in the conclusion of their work when they state that rather than emphasizing the "emergence" of multicellularity as "a difficult problem" in evolution, they consider that it arose as a natural consequence of a primordial colony of cells that could grow continuously. In other words, they perceive this event (i.e., the origin of multicellular organisms) as an expected and inevitable consequence of dynamical systems (Niklas & Newman, 2013; Lyons & Kolter, 2015). With a precise computer simulation analysis, Furusawa & Kaneko (2002) showed that cell differentiation could be triggered spontaneously mainly due to the intrinsic composition of the cell pool. While the internal composition of cells is, in principle, very similar, subtle differences ultimately cause differences to exist and generate cell types.

Thus, transitions between different cell types follow restrictive rules on transition dynamics. A clear example is transitioning from a totipotent cell to another cell type of lower rank. The totipotent cell can generate another cell of the same type and one of lower rank but once differentiated. A cell can only continue to generate its same type without being able to return to the possibility of generating any cell type (i.e., it loses totipotency). In this case, it is the growth of a cell colony over time, where sooner or later, finer cooperation, with a division of tasks, will occur. At least, this could be seen in the experiment of Boraas et al. (1998) in *Chlorella vulgaris*, where after a hundred generations of a unicellular colony, multicellularity emerged in the presence of a predator.

It showed the adaptive value of cooperating in a multicellular manner, in this case through colonies of eight cells acting as a functional unit (Stearns & Hoekstra, 2005).

6.2.1.7 Multicellularity and Cancer

In biology and medicine, there is often a necessity to analyze situations where multicellularity deviates from a controlled program (Trigos et al., 2018). As we have seen, the rule in multicellularity is that the cells that compose the organism reproduce within a plan and perform the assigned tasks for the organism's good. Sometimes these cells get out of control, stop reproducing and working for the organism's good, and start doing it selfishly for themselves (Aktipis et al., 2015). It is the case of cancer cell proliferation. Cancer is rare for multicellular organisms, an exception for a group of cells that do not differentiate as they should in the tissue where they are located. The oddity is that these cells selfishly seize control of resources and do not respond to the communication patterns established within the organism (Nedelcu, 2020). Therefore, they grow out of control, exhausting resources, invading tissues, and ultimately killing the organism. Nondifferentiation as an isolated error may occur in the tissue, but when it becomes widespread (becoming the pattern or rule within an organ or location within the individual), it is classified as a disease.

Evolutionarily, cancer is closely linked to the phenomenon of multicellularity (Davies & Lineweaver, 2011). Aktipis et al. (2015) argue that the generation of a multicellular organism requires cooperation between cells and mechanisms to suppress somatic deception, as cellular-level fitness must be restrained to promote the overall fitness of the organism. Many genes identified in cancer cells are primitive and conserved (Weinberg, 1983; Hartl et al., 2010; Srivastava et al., 2010). It is interesting to note where the exception lies in how these cells work. Studies suggest that the properties of multicellularity are lost or deregulated as genes associated with unicellularity are turned on specifically for cancer proliferation (Trigos et al., 2018). The fact that they are evolutionarily highly conserved suggests that they serve essential functions in cellular physiology and that their role in carcinogenesis represents a total exception or an aberrant deviation from the original functions (Weinberg, 1983). Thus, the genes needed for multicellular cooperation are the same genes that function incorrectly in cancer cells (Weinberg, 2007).

Currently, cancer is an exception from a natural point of view. However, some works ponder that very primitive metazoans are tumor-like neoplasms (Davies & Lineweaver, 2011) since they are colonies where cooperation, adhesion, and exchange of chemical information are relatively poor. Then, evolutionarily, these protozoa advanced by improving signaling and coordination. Thus, in the beginning, colonies that behaved analogously to "tumors" may have been the rule.

6.2.1.8 Changes Linked to the "Initial" Invasion of the Earth

In the history of life, there was a time when no organisms inhabited Earth. It is listed in any textbook of paleontology and evolution. Indeed, it is believed that most life forms initially evolved in marine habitats. Until the Ediacaran period (ca. 600 million years ago), many single-celled organisms inhabited the seas, such as eukaryotes, prokaryotes, and archaea (Cavalier-Smith, 2006; DeLong & Pace, 2001). From that time on, a wide variety of multicellular organisms began to develop, including algae, plants, fungi, and animals that evolved in the sea and, to a lesser extent, in freshwater habitats (Heckman et al., 2001).

The invasion, conquest, or colonization of land was one of the most fundamental events in evolution. This new environment without any competitors allowed the diversification of many organisms, leading to the evolution of new forms, behaviors, and life strategies. At that particular moment, immediately before this colonization began, the prevailing rule was "marine life" or, in other words, the sea as the "only" environment for life. As the invasion of land began, it was evident that, from a group perspective (in terms of diversity) and the number of individuals per group (density), living on land was only an exception rather than the norm. At this moment, life would be unnoticed outside water environments.

It is intriguing to consider that the first individuals to colonize the land may have been members of a relatively "marginal" population within their species. For example, if we were to imagine arthropods that went to land, it would be hard to imagine individuals well adapted to a deeper life in the sea or freshwater. Instead, it seems more plausible to think of individuals accustomed to living in intertidal zones or temporary pools, where water availability is subject to evaporation or runoff. That is, zones where amphibious life could be the intermediate "nexus" to move from water to land. The data also suggest that marine organisms may have undergone changes and tolerated brackish waters, such as estuaries. The salt content of the water is an essential initial aspect to control, so adapting to an estuary may have been a first step in colonizing terrestrial environments (Sánchez, 2007). Those individuals of the species accustomed to intermediate, amphibious areas would be the most suitable for colonizing areas already dominated by air rather than water. Once again, we highlight the significance of intrapopulation variability, particularly the individuals exhibiting "extreme" traits (e.g., internalized gills or cutaneous respiration in animals, plant cuticles, and well-defined stomata in plants, which are beneficial for terrestrial life). These specific traits render them better adapted for the successful colonization of the new environment.

Furthermore, in this context, the initial colonization of land is also linked to a concept we will discuss later in this book, known as the founder's principle, which explores at the genetic level how exceptions can become the rule. Specifically, we refer to how rare genes within the gene pool of the original population (e.g., marine organisms) may become prevalent in the population that has already established itself in the new terrestrial environment (Templeton, 1980; Provine, 2004). Indeed, if such genetic shifts can occur during the colonization of uninhabited islands, it is reasonable to assume that even more significant genetic changes must have taken place during the critical moments of colonization in such an extensive environment entirely devoid of living organisms. These rare genes (in frequency), which later become dominant in descendants, likely had significant implications for subsequent evolutionary patterns in plant and animal lineages. However, the implications of what was implicitly present, such as the founder principle, during the initial colonization of a new environment are often overlooked or not frequently mentioned. It could have been further enhanced if those original pioneer populations suffered a bottleneck effect by temporarily reducing those new gene pools due to environmental events or contingency phenomena, such as volcanic eruptions, solid and sudden climatic changes, etc. In this world of gradual colonization of new environments and associated processes, such as the founder principle, we should pause to contemplate the profound impact exceptions can have on shaping future patterns and rules in evolution. It is just a sample of the multidimensionality linked to exceptions (e.g., exceptions of different types, at different levels—time, space, group—of patterns, processes, etc.). This book may not fully illustrate all these variants. Nevertheless, the initial spark will ignite, paving the way for interdisciplinary collaboration among researchers and readers.

The first living organisms to colonize the land are believed to have been photosynthetic prokaryotes that inhabited humid environments near ocean shores, probably 2.6 billion years ago (Horodyski & Knauth, 1994; Prave, 2002). Thus, despite the lack of water, terrestrial environments offer an ideal location, with plenty of sunlight and carbon dioxide for photosynthesis, which makes the colonization of these spaces possible (Pisani et al., 2004; Sanford et al., 2012; Wu et al., 2014).

Subsequently, the ancestors of present-day plants were able to colonize drier environments. It is postulated that a group of freshwater plants gradually advanced to muddy shores, and actual land plants evolved from there (Heckman et al., 2001). The flora during the Cambrian was fully marine (Gradstein & Kerp, 2012). Evidence for the appearance of the first, still rare, land plants occurs in the Ordovician period (ca. 450 million years ago) in the form of fossil spores, and they began to diversify later in the late Silurian, approximately 430 million years ago (Wellman et al., 2003). Conditions improved for plant colonization. The climate was colder and wetter, coastal zones were wide and stable, and soil formation occurred. Spore production is considered key in the evolution of land plants (Gradstein & Kerp, 2012; Lei, 2017). It is interesting to see lignin as something that could be considered teratological in a marine aquatic environment, but not terrestrial.

Fungi seem to have colonized the soil at the same time as plants. This association between an autotroph and a fungus probably favored the colonization and permanence of eukaryotes in the terrestrial environment (Selosse & Le Tacon, 1998; Heckman et al., 2001). One of the first associations that favored the colonization of land by eukaryotes may have been the formation of lichens derived initially from a fungus and a cyanobacterium. This association would have resulted in another type of symbiosis between autotrophs and fungi, arbuscular mycorrhizae. Today, lichens together with arbuscular mycorrhizae are the most common symbiotic associations between autotrophs and fungi (see Sect. 6.4 on plant communication in this chapter). In these associations, fungi assist in the uptake of nutrients, metabolites, and water while benefiting from nutrients provided by the autotroph (Plackett & Coates, 2016; Audesirk et al., 2008). The early colonization of the terrestrial environment by plants and fungi likely contributed to the rise in atmospheric oxygen concentration, promoting the Cambrian explosion of animals (Heckman et al., 2001; Plackett & Coates, 2016). This point directly relates to the previously mentioned "energetic expansions" in the planet's history.

The colonization of the earth by plants was preceded in the early Cambrian by invertebrate animals. According to the fossil record, the first was *Apankura*, an Euthycarcinoid arthropod found in Argentina (Vaccari et al., 2004). This organism is thought to be the precursor of terrestrial mandibulates, and traces of this animal found in Canadian deposits also confirm its presence on land (Sánchez Alvarado, 2006). Arthropods were well adapted to life on land because they had an almost impermeable shell, presumably used for protection (Waters, 2003; Mayhew, 2007; Pisani et al., 2004). In addition to fungi, the evolution of plants is always closely related to arthropods, mainly insects. During the early phase of land colonization, arthropods were wingless, a condition known as "apterous." This characteristic did not significantly impact their interactions with plants since, at that time, plants were generally short in height (Sánchez Alvarado, 2006).

An important evolutionary event associated with the first steps of vertebrates onto land was the transformation of fins into limbs, thus moving from swimming to walking and initiating the enormous radiation of terrestrial vertebrates. If we were to view this event from an approach similar to that of Blumberg (2009), these were changes that, if they occurred in a stable marine environment, could well be viewed as simple teratologies. In contrast, this is not the case if that change, as we know, was linked to the colonization of a new environment where the adaptive value is evident.

On the other hand, the evolution of limbs was preceded by a significant change in sensory abilities (Nilsson, 2017). During the finned fish to early amphibious tetrapod transition, there was an increase in eye size and a shift in position toward the top of the head. The authors explain this change in the visual system as an adaptation to have better vision in the air (Nilsson et al., 2012; MacIver et al., 2017; Nilsson, 2017). Again, the reflection on how a series of changes can be, to a greater or lesser extent, considered teratological versus adaptive will depend on the context in which it occurs. However, in a time and space context, sometimes the boundaries could be more precise, as they will depend on the medium- and long-term fitness of the organism that presents that change.

At this point, it is interesting to slow down and discuss a particularity in the context of rules and exceptions. This refers to the fact that land invasion by plants is considered to have been a "single event" in the history of life. At the same time, animal groups colonized the land at different times during their evolution. Land plants form a monophyletic group that evolved as an adaptive response to migrating from a subaerial freshwater habitat to a terrestrial environment. Phylogenetic analysis of extant plants suggests that green algae are related to land plants, i.e., land plants likely evolved from a multicellular freshwater green alga (Wellman et al., 2003; Pisani et al., 2004; Qiu et al., 2006; Rubinstein et al., 2010; Hanschen et al., 2016). This monophyly strongly supports a single event of soil colonization by plants (Plackett & Coates, 2016). Nonetheless, the intricate morphologies that enabled animals to survive on land originally evolved in aquatic environments and independently spread to land on multiple occasions. For instance, in the case of arthropods, it is believed that at least four colonization events occurred, leading to the independent terrestrial colonization of "myriapods" (centipedes, millipedes, and related groups), insects, arachnids, and crustaceans (Pisani et al., 2004).

Thus, this unique event in the evolutionary history of plants represents, even more, a great rarity, an exception, of that time. One so important that it later implied crucial changes in ecosystems. For example, it facilitated the settlement of other groups, such as herbivorous animals, as they were already present as the raw food material (Benton, 2010). In other words, the invasion of animals was presented as recurrent, multiple patterns, while in plants, everything was born from a single situation.

In these early stages of colonization, those individuals with rapid growth and short life spans could have been the ones that were initially successful in colonization (Connell & Slatyer, 1977). At this time, these traits, perhaps not common in the ancestral core population but common in the marginal population, could have been later replaced by other species with a long life cycle and slower growth. Here, the rare thing at that early stage of invasion would have been the latter (a stage that could have lasted thousands of years or more—Connell & Slatyer, 1977). Again, we can always see what was familiar and rare at "x time," "x space," and "x taxonomic group." Indeed, whether we are discussing rules and exceptions—not solely based on our perspective of what "should be," but on the actual occurrences in nature—or patterns and rarities, the initial colonization of the Earth encompassed a multitude of situations where this duality was evident, as we have explored in this section.

Changes Associated with Ecological Succession Mechanisms Among Land Colonizers

Initial colonization by a given group, e.g., plants, will then impact successive colonizers of other groups, various animals, and so on. In this sense, three models have been proposed to try to explain the mechanisms that determine the sequence of species: facilitation, tolerance, and inhibition (Connell & Slatyer, 1977; Finegan, 1984; Huston & Smith, 1987; Pickett et al., 1987), something central when we think of a passage from exception to rule in the framework of colonization processes of new environments. In the "facilitation model," early successional species have specific characteristics that allow them to colonize first, such as the ability to produce a large number of propagules (Marks, 1974) and improve the conditions of the environment (food, shelter, soil, and microclimate, for example) to make them more suitable for later colonization by other species so that they can invade and grow to maturity (Connell & Slatyer, 1977; Read et al., 2016; Koffel et al., 2018). For example, among the roots of plants on aquatic margins, a microhabitat was created that was conducive to many arthropods that were thus able to colonize the land as the first animals on it (Ridley, 2004). Second, a "tolerance model" has been proposed, which states that the modifications produced in the environment by the first species that colonized do not increase or reduce the probability that new species can colonize the site, nor do they affect the growth of the new species. Species that develop later are simply those that arrive later. The sequence of species is determined solely by their life history characteristics (Connell & Slatyer, 1977; Finegan, 1984; Huston & Smith, 1987; Pickett et al., 1987). Finally, the third model, the "inhibition model," proposes that once the first colonizing species secures space and other resources, they inhibit the invasion of later species or suppress the growth of those already present. New species can invade only when the first species are damaged or have died. In contrast to the first two models, here, the species replacing the previous one need not be a different species adapted to conditions modified in a particular way by the previous species (Connell & Slatyer, 1977; Finegan, 1984; Huston & Smith, 1987; Pickett et al., 1987). Newly settled species may well have defined the disappearance or caused the disappearance of previously settled species, i.e., pioneers, to become rarities. However, it is currently suggested that these three models are not mutually exclusive and that one becomes more important than the other at different stages of succession. What is certain is that what was initially rare on earth acted as an indirect driving force for other groups that were also rare in their beginnings. That is, they were exceptions that established scenarios as rules for those that would come later. These rarities, whether still in their rarity state or gradually turning into patterns, influenced what others could or could not do to a certain extent, regardless of whether they belonged to the same clade (e.g., plants to plants) or not (e.g., fungi such as mycorrhizae–plants; plants–animals; animals–animals).

6.2.2 Perspectives and Cases of when "Oddities" Become the Pattern

6.2.2.1 Nonrepresentative Individuals Are Settlers

In population genetics, random events play a crucial role in the passing of genes from one generation to the next. These events, often referred to as genetic drift, are considered significant in evolution. For instance, in biological evolution, the existence and strength of random sampling events involving a small number of individuals as founders of a new population are well documented. This process presents changes in the frequency of genes, including some of them associated with diseases. Genetic drift encompasses processes such as the founder effect and bottleneck. These processes occur when a new population is established from a small group of randomly selected individuals (Templeton, 1980). The founder effect occurs when a small number of individuals break away from the original population and colonize new territory. These individuals may possess rare genes in the original population. Thus, these rare genes in the population will now be more common among the descendants, affecting the original genetic variation (Templeton, 1980; Clegg et al., 2002; Peter & Slatkin, 2015). The full founder effect requires not only a founder event but also that the founder individuals are not genetically representative of the original population.

On the other hand, a bottleneck effect occurs when an environmental catastrophe, such as an earthquake or a tsunami, eliminates most of the population at random, and only a few individuals survive regardless of the genes they have (Templeton, 1980; Cornuet & Luikart, 1996; Peter & Slatkin, 2015). Human behaviors can also act and produce bottleneck effects. For example, a species that suffered from bottleneck effects is the northern elephant seal (*Mirounga angustirostris*), which was hunted to near extinction. When the population recovered, less genetic variation was observed among individuals (Hoezel, 1999). It is interesting to note that both processes produce variations in allele frequencies, resulting in a decrease in the population's genetic diversity (Hoezel, 1999). Some genes may be eliminated from the population, while others that were rare now become more common in the new population. The link between founder and bottleneck events is the randomness with which individuals are selected.

Gene drift processes differ from natural selection in that gene drift eliminates individuals randomly. For example, a pest can only kill individuals lacking a particular gene (Lande, 1976). The genes with the best survival qualities are passed onto the next generation in natural selection. Thus, we could discuss whether the choice of the founding individuals of a new population was random or actually by natural selection, e.g., they are the strongest, better able to expand and travel to the new site, etc. The truth is that we have illustrative examples (mainly from the world of flies) that show how few individuals can initiate a new population. This population's allele frequencies will differ from the ancestral allele frequencies due to the founder effect and may or may not contribute to speciation processes. Such is the case for *Drosophila hawaiana* (Carson & Johnson, 1975; Templeton, 1979a, 1980), *D. mercatorum* (Templeton & Rankin, 1978; Templeton, 1979b), *D. silvestris,* and *D. heteroneura* (Carson & Kaneshiro, 1976). In other cases, however, the effects alone are not as decisive as those observed in other *Drosophila* species that have undergone founding events without speciation. This is the case, for example, of *D. melanogaster* itself (Powell, 1978; Templeton, 1979b, 1980). Additionally, within vertebrates, in some island-colonizing birds, the genetic variability of populations did not show a significant decrease (Charlesworth & Charlesworth, 1987; Moya et al., 1995; Clegg et al., 2002). Thus, only a minority of the processes linked to genetic drift lead to speciation. It could depend on specific characteristics of the original population, such as its original structure, genome, mating system, and mutations (Templeton, 1980). For example, concerning the structure of the original population, different ancestral population models have been described that can influence the selection of founder individuals, including the panmictic population model, Wrights Island model, and "Yanomama" model (Nei et al., 1975; Templeton, 1980).

Interestingly, more clarity is needed as to whether the emergence of speciation derived from gene drift events is expected. However, even so, it should not be underestimated since it does exist. In this regard, the words of Templeton (2008) stand out: "*Although rare, founder speciation can have a disproportionate importance in adaptive innovation and radiation, and examples are given to show that 'rare' does not mean 'unimportant' in evolution.*" It is worth noting that the terms "rare" and "exceptional," as applied to speciation by founder events, do not diminish its significance within an evolutionary framework.

On the other hand, the individuals who originate the new population often come from a marginal (i.e., noncore) subpopulation of the species. This "sample" in itself may not be representative of the genes of the ancestral population. Thus, individuals colonizing previously uninhabited spaces are descendants of those living near the "expansion front" of the original population (Hallatschek et al., 2007; Klopfstein et al., 2006; Peter & Slatkin, 2015). In some populations, a phenomenon known as the "spatial ordering of individuals" may occur. This happens when individuals with traits that enhance dispersal tend to accumulate at distribution edges (Phillips et al., 2010). This arrangement promotes the frequency of dispersal-promoting genotypes to be higher in expanding ranges of populations (Phillips et al., 2010; Szűcs et al., 2017a). Indeed, in this case, we encounter a situation that can be considered an exception both quantitatively and qualitatively. These populations may carry rare alleles that will become the most common after the foundational event concludes. Thus, we see that while the genes would be mainly those of the ancestral population, the allele frequency would not. We now have a significantly high frequency of alleles that were originally rare in the average population but prevailed in the "extreme" sample that eventually formed the foundation of the new population. The characteristics of that sample, which will be the ones that migrate and finally colonize the new territory, will be crucial. Implicitly, if the sample of a few individuals came from the middle, more representative population, we would not see such a marked and evident "effect" of the founder. This effect is seen in cases where the members of the founder sample present a disproportionate number of these "rare" alleles.

It is worth mentioning that homozygosity as such could not be expected from a founder effect but from restriction in allele frequency. Changes in allele frequency can be very dramatic and can be modeled (Slatkin & Excoffier, 2012). What is certain is how critical the founder effect may have been for the expansion of populations (Peter & Slatkin, 2015). Range expansion refers to the expansion of a population from a small region to a much larger region. These expansions are common in many species and occur on different time scales depending on the species (Peter & Slatkin, 2015). For example, two theoretical population genetic models refer to the range of population expansion. The first describes the change in allele frequency over space due to differences in dispersal and fitness in a continuously distributed population.

In contrast, the serial founder model assumes discrete locations that can be colonized by the population and not so extreme differentiation (Ramachandran et al., 2005; Slatkin & Excoffier, 2012; Peter & Slatkin, 2015). This links directly to the concept of variability we discussed in the previous sections regarding the extremes of expression of a trait and, thus, the individuals carrying that expression. Furthermore, it is possible for founders to include either multiple individuals or a single fertilized female [as per Mayr's definition from Mayr (1963)]. In some cases, these individuals may belong to an extreme quartile of the trait within the original population (Nei et al., 1975; Cornuet & Luikart, 1996). It could, as we said, be even stronger if it is already in a marginal distribution and, therefore, already exhibits a previous state of "low genetic representativeness." This founder could also possess a novel character associated with an adaptive function *per se* (see next section). It could even be an individual with a quasi/or fully heritable teratological trait and/or a carrier of diseases, something seen, for example, in some human colonizations (see Fix, 1997—on cline formation of gene frequencies produced by the effect of the founder).

There are multiple examples of founder effects in biology. As well compiled by Ridley (2004), a "natural laboratory" has been the colonization processes of humans, where medical anomalies can be seen in cases of foundational processes by scarce colonists, where the progenitors were carriers of those alleles. For example, Huntington's disease (a genetic disease with an abnormal dominant allele that disrupts the function of its nerve cells) is more common in Venezuela and South Africa, with Dutch ancestry, than in most other populations. This is because the gene carrying this disease was widespread among the small group of original Dutch settlers (Zeegers et al., 2003). Another example is the disease known as porphyria variegata (where an autosomal dominant gene produces a defective form of the enzyme protoporphyrinogen oxidase). Individuals carrying this gene suffer a severe reaction to anesthetics. This disease is widespread among the white population of South Africa with Dutch ancestry (Dean, 1982). To these examples, we can add that of the Pennsylvania Amish with German and Swiss ancestry, which was established in the United States in the eighteenth century (Curtis et al., 2008). This group has a high incidence of dwarfism and extremities with supernumerary fingers (polydactyly) called Elis-van Creveld Syndrome (McKusick, 2000). By chance, the members who migrated to the United States carried that rare recessive allele that, in homozygosis, manifests the syndrome. The Amish are a very close group with a high rate of inbreeding, which makes homozygosity prevalent. The incidence in the group is so high that their numbers resemble those reported for the rest of the world (Curtis et al., 2008). We will see other examples later when discussing disease incidences and rare diseases in Chap. 8.

From the above, it is then clear how the rare could become common because of the founder effect. However, implicitly (although it is not usually stated in the texts), we have that the common can become rare or even disappear, as we will explain in the next chapter. Following the example of allelic frequency, common alleles would become rare now in the population once settled in the new place. Interestingly, the spatial context is crucial here since it refers, even implicitly, to founders coming from elsewhere (see above for range expansion). Founding individuals encounter a different environment from where they originated. Therefore, it is crucial to understand the processes that will determine the successful colonization of new environments (Szűcs et al., 2017b). The number of individuals founding the new population appear to be a determinant of colonization success in various taxa, as it affects demographic and genetic processes (Colautti et al., 2006; Blackburn et al., 2015). Thus, whether the founding individuals are genetically and phenotypically diverse increases the establishment success of the new population (Forsman, 2014; Szűcs et al., 2017b).

However, in the framework of random sampling, we could alternatively refer to cases where the context of time, rather than space, would be the most gravitating factor in the changes in allele frequencies (making the rare familiar and vice versa). This is because the change can occur *in situ* in the same space but is progressively more and more marked over time. We already mentioned this when a population passes through a bottleneck. Indeed, when an initial population in a new territory or even a perfectly established population suffers a significant reduction in the number of individuals at random, those that survive will then expand their alleles in the descendants (Nei et al., 1975; Cornuet & Luikart, 1996; Hoezel, 1999). For our book, the common denominator is the dynamics between R and E. For example, we would pass an R (an x allele) to be an E after the bottleneck and perhaps an E (z allele) to R after the bottleneck. Then, by starting its expansion, we already have a change from what was once a common genotypic-phenotypic manifestation, the pattern, the norm, the rule, to an exception or perhaps disappearance.

Conversely, something that was rare will now become common, the rule. Everything will depend on whether the survivors associated with the genetic drift processes (founder effect and bottleneck) had alleles or allele frequency markedly different from the predecline population mean. Thus, we saw how the founder and bottleneck effects provide us with examples of something going from "rare" (in fact, this is used for "rare alleles" Ridley, 2004—p 176) to common. Finally, we now see the connection with the ideas of evolution by contingency discussed in Chap. 1, when specific evolutionary processes are initiated by contingency events (understanding contingency effects as what may or may not happen) rather than processes linked to natural selection (Fig. 6.1a, b).

What has been exposed in this section comes from something very classic in evolutionary biology, where it is appreciated that rarities exist in variability and that they can and have a great value for understanding what is later seen in derived populations. In the next section, we will analyze more about how trait rarities can be the critical element in establishing, in populations and species, novelties within the framework of phenotypic plasticity.

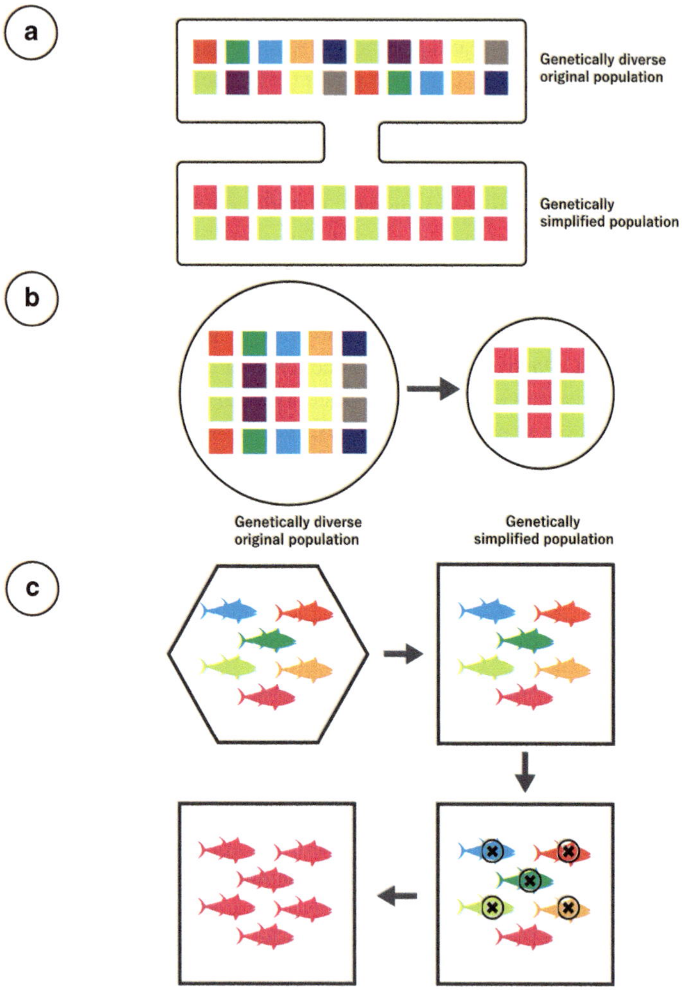

Fig. 6.1 Processes that can cause rare alleles to become common in the population. **(a)** Bottleneck effect. **(b)** Founder effect. **(c)** Environmentally induced phenotypic changes. A genetically diverse population undergoes a disturbance in the environment, where different genotypes differ in their response to the new environment. Selection acts and favors the genotype that produces the phenotype that best adapts to the new environment (Modified from Levis et al., 2018)

6.2.2.2 Minorities Channeled Via Phenotypic Plasticity

This extreme case of the appearance of a new trait, not linked to anything homologous to an ancestor, is a rarity in its initial stage. Both this situation and the previous one (of a phenotypic trait within the distribution range of the trait and present in a few individuals) connect to underlying genetic level variations. These variations may be minor or hidden. This links directly to cryptic genetic variation (CGV), a concept applied to the appearance of morphs within a species adapted to particular environments. CGV is a "*genetic variation that normally has little or no effect on phenotypic variation except under atypical conditions*" (Levis & Pfennig, 2016). The conditions that induce these cryptic genetic variations are usually rare or absent in population history. It limits the opportunities for selection to act on that variation and allows it to accumulate. The CGV then provides a pool of permanent genetic variation, facilitating adaptation when a rare condition becomes common (Paaby & Rockman, 2014).

When a trait emerges solely due to environmental induction, the genetic variation linked to the appearance of this novel trait becomes evident when there are changes in environmental conditions. In that context, the variation is manifested by an increase in the heritability of the trait, which will then be evolutionarily selectable (Queitsch et al., 2002; Ledón-Rettig et al., 2014). However, the way variability is experienced is not uniform across all organisms. For instance, animals can adapt to varying resource availability through their mobility, minimizing heterogeneity. On the other hand, plants, being sessile, begin their life cycle under specific conditions that may change over time, and they lack the ability to counteract these changes. Thus, plant survival and reproduction depend on tolerance to environmental extremes, which may be made possible by plastic responses (Bradshaw, 1965). For example, crucial adaptive variation in plants involves differences in leaf morphological, anatomical, and physiological characteristics along environmental gradients (light, water availability, etc.). These variations can be observed between species, populations of the same species, and even between leaves produced by a single plant. They can be maintained over time according to environmental characteristics, allowing tolerance to change and maintaining that new trait as a commonality (Wells & Pigliucci, 2000).

Indeed, this hidden variation, which may have minimal or no impact on phenotypic variation under typical conditions, can play a crucial role in "atypical" circumstances, where it becomes evident and acts as a cornerstone for later refinement of the novel trait through genetic accommodation. This refined expression of the trait can then become adaptive in a specific context. This is what is proposed by the "plasticity-first hypothesis", a mechanism of adaptive evolution in which environmental perturbation leads, through phenotypic plasticity, to developmental reorganization (through, e.g., of altered gene expression) and uncovers "cryptic genetic variation" to, and ultimately the production of, a novel developmental variant that immediately undergoes "phenotypic accommodation" and is subsequently refined through "genetic accommodation" (Levis & Pfennig, 2016) (Fig. 6.1c).

Here, we see two curious elements in the framework of the present book. How something that is "hidden" is not hidden under certain conditions, which could be the tip of something novel, which in turn is initially a rarity. In addition, the concept appeals to "atypical" conditions in which this could occur so that this GCS is discovered. As we see, concepts of rare and atypical are again ingredients at play in the beginnings of an adaptive process, from minor variations of atypical environmental change (in a broad, external sense). The novel will have the potential to continue polishing and adjusting to that new environment (i.e., have more sensitivity, adjustment, fit, coupling, and bound to that environment). Darwin's own words come to mind (1859) when referring to the degree to which the organization tends to progress, which sounds in a certain way indirectly linked to this subject: "*...the final result is that every being tends to perfect itself more and more concerning the conditions*" (pp. 153). This "perfecting" to which he referred well is, in this framework, given by that adjustment to the new environment (e.g., via genetic accommodation—Levis & Pfennig, 2016).

It is striking that plasticity will be the requirement to be less sensitive (i.e., less adjusted, limited to a single environment) but may then lead to new traits that may eventually become more sensitive, adjusted to that change that occurred. Moreover, this new morphotype, also adapted to its primitive environment, will coexist or not coexist with the ancestor if it is present. An example is the spadefoot toads, common in southern Canada, USA, and Mexico, of the genus *Spea*, which presents a new carnivorous morphotype (*Spea multiplicata*), something unusual in this group. This trait is presumed to have arisen via plasticity-first evolution. Today, two morphs coexist, an ancestral, omnivorous morph and a new, carnivorous morph. The dynamics between them are interesting. The tadpoles of this species consume mostly plant material. However, they can also be facultative carnivores, depending on the circumstances of the environment in which they grow, and can even eat smaller tadpoles of their species. It ensures the growth and metamorphosis of at least part of the offspring (Levis et al., 2015). Today, we cannot say that one is the rule and the other the exception, but perhaps the minority is, in any case, the carnivore.

It is intriguing to observe that at the initial stages of the evolutionary process leading to the emergence of this new morph, there was an inherent situation of exceptionality. This is true unless one postulates that it was a massive generalized explosion of the same cryptic genetic variation, manifested at the same time in the majority of the population subjected to an atypical environment (i.e., the one that led to the selection of some morphs). Following the line of Levis et al. (2015), this seems unlikely, as in many or most cases in evolution. The exception, initial rarity, in terms of frequency of distribution and abundance of that morph (i.e., individuals possessing that new trait or expression of an already existing trait such as body size), is in itself—from the statistical point of view—an exception. Thus, we quickly connect the above with the chapter on exceptions.

Importantly, the examples mentioned regarding the role of phenotypic plasticity in the emergence of new morphs refer to an intraspecific phenomenon. In other words, this does not necessarily imply the formation of a new species or a process of speciation; rather, it focuses on variations and adaptations within the same species. Nevertheless, it is evident that the appearance of new traits via phenotypic plasticity will connect with speciation processes in the event of subsequent changes in more extreme environments. Especially at the level of sexual preferences and sexual recognition, in short, pre- or postzygotic barriers of isolation between the new morph and the morph that gave rise to it (precedent, ancestor, as it is better to say). It has been argued that adaptive phenotypic plasticity can promote diversification and speciation through its effect on selection and gene flow (West-Eberhard, 2003; Wund, 2012). Populations with plastic phenotypes that arrive in a new environment may be more likely to persist because they are more responsive to environmental demands (West-Eberhard, 2005; Pfennig et al., 2010). When this plasticity is adaptive, the plastic population will be closer to optimal fitness, which may result in directional selection and consequently reduce gene flow and lead to speciation (Ghalambor et al., 2007; Crispo, 2008).

In addition, other examples of plasticity can also be observed in nematodes, where in some species, there is an evolution of food structures, promoting novel traits (Susoy et al., 2015). In some crustaceans, such as *Daphnia melanica*, there is a regulation of the melanin gene that allows for reduced pigmentation and adaptation to environments with novel predators (Scoville & Pfrender, 2010). In reptiles, for example, snake populations that frequently encounter large prey can accumulate mutations that result in larger head sizes (developmental adaptive plasticity) (Aubret et al., 2004). In birds, several species inhabiting noisy environments develop the ability to show changes in spectral settings and higher frequency songs (Slabbekoorn, 2013). (see Levis & Pfennig, 2016 for details). On the other hand, phenotypic plasticity can also be observed in plants, allowing some populations' novel traits to become pattern traits. In some acacias, resistance traits are induced and expressed in response to herbivore attacks (see Sect. 6.4.2.3). However, in other Central American species, which ants obligately inhabit, extrafloral nectar is secreted. This nectar is an indirect resistance, attracting ants that defend the plants from herbivores. The phylogeny of *Acacia* and other closely related genera indicates that induced nectar production is the "original" state, while constitutive extrafloral nectar flow is derived within the *Acacia* genus. A constitutive resistance trait has evolved from an inducible trait in response to particular functional demands (Heil et al., 2004). It is interesting to mention here one of the open questions left by these authors to explore in the future: are there particular taxonomic groups and traits more likely to undergo evolution by plasticity-first than others, and if so, why? This question is common to other problems or situations in evolutionary biology and connects to others posed earlier in this book, such as whether transitions from E to R are more common in certain groups than others and, if so, why.

Furthermore, numerous instances of plasticity, whether behavioral or morphological at the trait level, in general can enhance adaptation to new environments. Interestingly, these plastic responses can be further reinforced through hybridization processes, which aid in rapid expansion and adaptation. It may involve species exotic to that particular site (see below on introduced species) or range expansion in an area where it was already present. The combination of several factors, such as personality, communication, metabolism, hybridization, etc., means that the rapid evolution of multiple traits can occur during the range expansion processes of the organism in question (see Canestrelli et al., 2016). Range expansion requires the colonization of a new environment, which was previously inhabited by the focal species. During this expansion phase, essential traits favoring dispersal are those remaining at the dispersal front, being maintained in new generations and, in turn, accelerating the establishment of populations in colonized areas (Travis & Dytham, 2002; Phillips et al., 2010; Shine et al., 2011).

Recent eco-evolutionary experiments on expanding populations support the prediction that rapid evolution can accelerate the rate of population expansion (Williams et al., 2019). Experiments conducted, for example, on red flour beetles (*Tribolium castaneum*), demonstrated that the rapid evolution of existing traits increases the speed of population expansion in individuals at the expanding edge (Weiss-Lehman et al., 2017). Experimental populations moved faster where evolution was operating.

Several animal personality traits have a degree of plasticity that can vary between individuals in response to factors related to both ontogeny and adulthood (Dingemanse et al., 2010). It adds a level of greater complexity but also more significant variation available during range expansion. On the other hand, phenotypic changes appearing during the expansion phase could also strengthen species recognition and increase divergence in selected traits, e.g., sexually, promoting selective mating (Cardoso et al., 2014). It would contribute, for example, to the construction of isolation barriers before mating.

In sexual selection, traits linked to choice must be preferred by the chosen sex within a population. Additionally, these traits may change if the preference is altered or changed. For example, traits that enhance dispersal will increase the frequency at the expansion front (Shine et al., 2011), potentially increasing mating rates. Females may become more flexible in their choice due to a shortage of potential mates.

Thus, the overall process is more complex (i.e., not so univariate in terms of factors). Indeed, something as rare as hybridization could be used to fuel the emergence of more genetic diversity. It is crucial and widespread, for example, in plants (e.g., Rieseberg et al., 2007) when adapting to a new environment, and even more so if it is entirely exotic (e.g., see below cases of introductions in Australia). When related species live in sympatry and hybridization occurs, genes from one species are added to the genetic material of the other, or new hybrid gene combinations can be established (Barton, 2001). However, regardless of the mechanism, if these new combinations improve growth and the ability to compete in novel environments, invasion success may increase (Ellstrand & Schierenbeck, 2000; Blumler, 2003). However, increased invasiveness could be due to increased genetic variability and the fixation of new gene combinations and phenotypes (Arnold, 1997).

To this scenario, we should add behavioral plasticity as something "key" so that new morphological variants can finally become established. Body modifications are not dissociated (West-Eberhard, 2003) but are linked to specific behaviors—for example, using a new food resource, avoiding a new predator, or maintaining body temperature (Zuk et al., 2014). In the case of flexibility in antipredator behavior, this allows animals to respond quickly to changes in the predation environment (Lima & Dill, 1990). Response behavior to new predation threats may lead to the establishment of novel traits. It may allow populations to persist long enough for evolutionary adaptations to occur or expose them to new selection pressures. Plasticity in this type of behavior has been observed, for example, in populations of *Sceloporus undulatus* lizards, which perform body contractions and flee from attack by invasive ants of the species *Solenopsis invicta*. These behaviors have not been observed or are rare in these populations (Langkilde, 2009), but when present, they cause them to increase survival, both by reducing encounters with these ants (Freidenfelds et al., 2012) and by eliminating encounters if attacked by them (Graham et al., 2012; Langkilde & Freidenfelds, 2010). In addition, lizards inhabiting sites invaded by these ants have also evolved to have longer hind legs, which allow for increased efficiency in behavior against these predators (Langkilde, 2009).

In essence, any trait that undergoes modifications at the somatic level should also, by implication, be associated with ethological plasticity. This means that for a newly expressed genetic variant to be effectively utilized and beneficial in an animal's life, it needs to be appropriately coupled with plastic behavior. The alignment between the genetically expressed trait and the corresponding plastic behavior is crucial for successful adaptation to the new environment. If there is an incompatible match or a lack of alignment between the two, it may lead to difficulties in long-term persistence or a challenging mismatch when attempting to adapt to the new conditions. However, there may be certain limitations to innovation in terms of behavior. Although some changes can be advantageous, too rapid a change can have negative consequences (Berends et al., 2010). One example is the inconvenience of making changes in the face of transient ecological circumstances. Innovation may involve high costs. Therefore, a new solution may be equally beneficial as detrimental. It could justify limitations in accepting novel behavior (Brosnan & Hopper, 2014). Thus, innovative individuals do this primarily when their pattern behavior (their rule) is no longer beneficial, and therefore, change would bring them a more significant advantage. Brosnan & Hopper (2014) suggest that individuals only explore alternative strategies when there is an obligation to go through unavoidable circumstances. The interaction between the individual and the environment will determine whether those new strategies are maintained.

In this way, if sufficiently substantial, changes in the environment could lead to the emergence or expression of new traits, perhaps initially represented by a minority in populations. This implied a rarity, a rare trait. Indeed, these seemingly "rare" traits can, in certain instances, serve as a means to counteract environmental changes that might currently be unfavorable. Over time, these traits can acquire value and become selectable or heritable, providing a mechanism for organisms to adapt and thrive in their changing environment. It is how so-called minorities sometimes end up becoming, over time and as a result of evolutionary processes, the rule or pattern that allows populations to continue to exist, albeit in a different way.

6.2.2.3 Beginning of Coevolution Between Plants and Pollinators

As we started commenting in the previous chapter, the role of animal pollination is a point today undisputed in the evolution of angiosperms, even considered a key factor for the diversification of the group and specialization of many of its species

(Kay & Sargent, 2009; Sadava et al., 2011; Willmer, 2011). The ecology and evolution of plant–pollinator interactions can be very complex, depending on the degree of generalization versus specialization of the animals involved (Herrera, 1996; Mitchell et al., 2009; Willmer, 2011). How this level of coevolution was reached is somewhat controversial. The origin of such an interaction may be no different from that of any animal–animal or plant–plant interaction. However, the importance of animal-pollinator function, to the point of being the engine of plant evolution, is of great magnitude. It can be argued that, like all origins, this interaction began as a rarity.

A Little History

Initially, pollination by abiotic agents was the rule. In the Carboniferous, ferns and related plants had dust-like spores dispersed by wind (anemophilous dispersal). The heights of many of the tree ferns must have allowed them to use the wind very efficiently. The first seed-bearing plants were the conifers and related groups (gymnosperms) that dominated alongside the dinosaurs at the end of the Paleozoic period. Pollination was commonly wind-mediated without flowers but with seeds, and today, that is still the norm for most. However, as early as the Paleozoic and projected in some species that survived to the present, we can see insect-mediated pollination, particularly featuring some flies, beetles, and moths. The fossil record shows that in the beginning, pollination was very much linked to predation by pollinating agents and that the process was poorly articulated at the time. Today, the exceptions to anemophilous pollination within gymnosperms are Cycadales and Gnetales. *Gnetum*, *Ephedra*, and *Welwitschia* stand out in a particular group of gymnosperms. Cycadales and Gnetales are pollinated by dipterans, noctuid moths, or beetles (Willmer, 2011; Fogden & Fogden, 2018). Flowering plants evolved relatively recently in the Cretaceous. With them came numerous adaptations that coevolved with pollinators, resulting in an explosion of diversity in both groups (Sadava et al., 2011). Many novel nutritious foods have appeared, such as pollen, nectar, fruits, and seeds (Fogden & Fogden, 2018). The first to interact and take advantage of this new world of nutritional resources were insects, followed by vertebrates, particularly mammals and birds. The truth is that this condition, still not so common at the time, was expanding. Flowering plants directed the shift from abiotic-mediated pollination to biotic-mediated pollination. As an exception to this rule, some groups became anemophilous again (Charlesworth, 1993; Schoonhoven et al., 2005).

An exciting aspect arises when evaluating the planetary scale. The rule seems that anemophilous species of flowering plants increase with latitude and elevation. It is crucial because it is directly related to the precision of pollination between abiotic and biotic factors. Pollination mediated by biotic factors is much more effective and accurate. Imagine that the wind is less intense in dense and humid forest areas, so that this system could be more efficient in such places. The target is generally well localized, and the interacting structures are shaped by years of evolution

(Schoonhoven et al., 2005). There will be a progressive tendency toward the improvement of the plant to "take better advantage of its animal vector" to carry its pollen to another flower, from the resident, or donor, to the recipient, located in the same or different individual, depending on the type of reproduction (Graham et al., 2006).

In this sense, the current rule in flowering plants is hermaphroditism (both sexes in the same flower) (Renner & Ricklefs, 1995; Vamosi & Vamosi, 2004; Schoonhoven et al., 2005; Jarne & Auld, 2006). While there are exceptions, with many groups having separate sexes (in distinct monoecious flowers or distinct dioecious individuals, Graham et al., 2006), more than 80% of plants have flowers with both male and female reproductive organs (Willmer, 2011). Common sense would say this is a good strategy because they can self-fertilize (Jarne & Auld, 2006) and thus ensure seed production (Schoonhoven et al., 2005). While it is vital to ensure fertilization, it is more critical to ensure variation in the progeny in case of coping or a sudden environmental change (Darwin, 1859; Futuyma, 2013). Thus, many flowering plant species avoid self-fertilization with physiological, temporal, and spatial mechanisms (Fogden & Fogden, 2018). For example, like many other plants, orchids avoid self-fertilization (Darwin, 1869; Proctor et al., 1996). However, there are examples of orchids pollinated by insects that may become facultatively self-pollinating in areas where insects do not carry out pollination or are directly absent (Proctor et al., 1996). Thus, the orchid turns to self-fertilization because seed production with little genetic variability is much better for plant fitness than no seed.

The Flower and Its Rewards

The flower as a unit can be taken as a natural billboard. It provides much information to pollinators with its structure, shape, colors, etc. (Graham et al., 2006). Fossil evidence shows that primitive flowers present many reproductive structures and spiral arrangements. This rule, which probably existed from the beginning of the appearance of flowering plants (Angiosperms), has changed and is currently only represented in basal groups such as the Magnoliaceae. The current rule is regular radial symmetry and structures shaped according to the interaction with pollinators (Willmer, 2011). There is a tendency toward three-dimensional flowers that favor recognition by the pollinator, and the reproductive structures are better positioned to interact with the pollinator (Schoonhoven et al., 2005). If we discuss evolutionary primitives and take the example of Magnoliaceae, the culmination of evolutionary change would be orchids. We will see how evolution can mold curious but effective floral forms.

A significant milestone for the interaction success between pollinators and plants is the appearance of rewards for the pollinator. It all started with pollen rewards for the transfer of pollen itself. While one might think this would result in an enormous production of pollen, compared to the production in wind-pollinated species, it is considerably less. However, the evolutionary response to optimize the reward process was to switch from pollen consumption to nectar consumption. This change resulted in a benefit for the plant since pollen is much more costly from an energetic perspective than nectar. This is a nice strategy since nectar is sugary fluid and recall that a plant to run short of carbohydrates would be an exception.

In rare cases, some plants produce repellent pollen or nectar (Adler, 2000; Schoonhoven et al., 2005; Patiny, 2011; Stevenson, 2020). It could be posited as an exception since rewards are a significant part of the pollinator-plant relationship. In this context, it is difficult to explain why generating rewards that would be detrimental to the pollinator is an essential intermediary for plant reproduction. Several authors agree that the modularity of the plant–animal interaction (pollinator or herbivore) could partly answer this question (Adler, 2000; Stevenson, 2020). Many hypotheses claim benefits for the plant, such as favoring specialist pollinators, scaring away nectar thieves or low-quality pollinators, and modifying pollinator behavior, among others (Adler, 2000; Rivest & Forrest, 2020). It is also believed that toxic pollen may result from past selective pressures that do not operate today. However, the adaptive validity of this phenomenon has yet to be demonstrated experimentally (Adler, 2000). Unfortunately, to date, the origin and function of toxic nectar have yet to be fully understood.

Other rewards are comparatively more exceptional than pollen or nectar, such as oils, resins, perfumes, and egg deposition sites (Patiny, 2011). These exceptional rewards limit the diversity of pollinators ready to do the job. Within these examples of exceptional rewards is the case of butterflies (Ithomiini) that use alkaloids from the flowers of certain species of Boraginaceae and Asteraceae to produce pheromones. Males use these pheromones to attract females and generate receptivity-inhibitory compounds that are transferred after copulation (Fogden & Fogden, 2018).

One rule applicable to pollinators is redundancy (Willmer, 2011). This suggests that a pollinator can pollinate more than one type of plant, and a plant can receive more than one type of pollinator. Thus, if one of the two were to become extinct, the remnant would not *a priori* be in danger. On the side of the exception to this phenomenon are the cases of extreme codependency, where we can cite the iconic case of wasps and the fig tree (*Ficus* spp.), and moths, and the cassava plant (*Manihot esculenta*). The punctual case of wasps and fig trees is a classic example of 1:1 specificity and represents a firm rule in this interaction. However, this may be a somewhat reductionist approach to the subject (Cook & Rasplus, 2003), and there are papers suggesting exceptional cases of this rule (Rasplus, 1996; Michaloud et al., 1996).

Mutualism and Deception

While the plant–pollinator interaction is considered mutualistic, there is an underlying conflict: exceptionally, some plants cheat their pollinators (Dafni, 1983; Willson & Ågren, 1989; Ellis & Johnson, 2010; Ackerman et al., 2011). An alternative to this norm of reward production is the production of deception structures or substances, a situation where mutualistic equality is twisted to the side of the plant (Proctor et al., 1996). Overall, deception in the plant–pollinator interaction is rare, but it is not as rare as one might expect. These resources bet on pollinators being mistaken, but as a resource, they are less effective since these plants tend to receive relatively few visits (Baker, 1976; Little, 1983). There are different levels of deception. Simple levels, where plants falsify the amount of pollen available (Proctor et al., 1996; Shrestha et al., 2020), and more complex levels of deception, where mates and egg deposition sites are simulated. It occurs, for example, with saprophytic organisms and plants that simulate decaying matter or where the flower is modified in the form of a prison where insects are trapped, loaded with pollen, and then released (Urru et al., 2011). Finally, there are extreme levels of deception where plants mimic so accurately that they can simulate, smell, and even look like females of certain insect groups and even generate false copulations with young or unsuspecting males (Gaskett et al., 2008).

Concrete examples of these "extreme" adaptations in the shape of the floral area reached by the pollinator frequently appear among some species of orchids. Charles Darwin (1862) devoted an entire book to studying orchids. He described with his well-known fascination and thoroughness his observations and interpretations of pollination processes in British species. In a second edition (Darwin, 1877), he incorporated more tropical material based on observations that were not his own (Micheneau et al., 2009). Darwin's aim with these trials was to add evidence of the importance of cross-fertilization in natural selection and to use orchids as a model since they show complex morphologies that generally avoid self-fertilization and favor pollination (Fay & Chase, 2009). Darwin's optics produced a change of vision in the study of plants, especially flowers, in the Victorian society of the time. Darwin managed to "reinvent" plants, especially orchids, before public opinion by giving them a more vibrant tone with strategies and means to achieve them (Endersby, 2016). From the point of view of this book, Darwin added evidence to demonstrate that the rule was not that plants were static inanimate agents but that with their strategies, they could achieve incredible things with a will often assigned to animals (Endersby, 2016).

There are several morphologies of orchids. Some are simple with a more common "flower" appearance, such as *Eriaxis rigida* or *Stenorrhynchos speciosum* (see photos in Fay & Chase, 2009), and others with much more complex aspects, such as *Caladenia longicauda*. However, the climax of complexity and precision is found in species such as *Ophrys insectifera* or *O. apifera,* where the entire floral structure simulates an insect. These orchids use a tactic that is not the norm among flowering plants but is very common among orchids: deception (Proctor et al., 1996; Jersáková et al., 2006). These orchids of the genus *Ophrys* attract pollinators by simulating sexual substances produced by females (Gaskett et al., 2008). In effect, the males of certain insects, such as wasps, are the sexes deceived by the orchid. They are attracted and driven to copulate with the flower that mimics odor and even shapes their female, ending up loaded with pollen after the unfruitful mating. However, we have different levels of deception among orchids (see Phillips et al., 2014 for a detailed analysis of criteria for sexual deception in orchids). For example, some orchids mimic only the odor of the sex pheromones of receptive female insects to attract males (e.g., spider orchids, genus *Caladenia*). Additionally, as seen before, some species mimic a shape similar to the body of the female insect (e.g., warty hammer orchids, one is *Drakaea livida*, the aforementioned *O. insectifera*, and *O. apifera*) (Alcock, 2000; Gaskett et al., 2008; Fay & Chase, 2009; Bohman et al., 2014). Other species, along with scent, trapping mechanisms are added to ensure that the male insect carries pollen attached to his body (e.g., red-banded greenhood orchid *Pterostylis sanguinea*) (Peakall, 1990; Lehnebach et al., 2005; Gaskett et al., 2008; Phillips et al., 2014). Each example acts on males of a species or at least on particular genera of wasps (the first two examples) or dipterans (in the last one).

It is thought that the rule is that with this strategy, the plant biases mutualism to increase its fitness, and the deceived pollinator male gains nothing but loses nothing either (Schiestl, 2005). However, many exceptions are found where this deception leads to fitness costs for both males and females of the pollinator (Gaskett et al., 2008). These costs are linked to the loss of copulations with genuine females and especially to the loss of costly ejaculate due to pseudocopulation (Coleman, 1929; Blanco & Barboza, 2005), as occurs, for example, in *Cryptostylis erecta* and *Cryptostylis leptochila* and the male pollinator orchid wasp (*Lissopimpla excelsa*), where there is a high rate of ejaculation (Gaskett et al., 2008; Brunton-Martin et al., 2020). Thus, in orchids pollinated by pseudocopula, species pollinated by essence-collecting insects, and other groups that show an intense relationship with their pollinators, the rule is the attraction of the pollinator through olfactory chemical signals (essences) (Proctor et al., 1996). As we saw, this chemical attraction can be accompanied by morphological and functional changes.

Many species of orchids are pollinated by insects, which are rewarded with nectar, but the exceptions to this typical pattern in plants are, as we have seen, very numerous among orchids (Proctor et al., 1996). Thus, in this framework, we can ask ourselves how these deception strategies arose. We know that they involved adjustments over time that were not minor, reinforced by the success of the novelty in its beginnings on the part of the plant. While more experienced males may tend to avoid them and not want to "copulate" with these deceptive orchids (Peakall, 1990; Alcock, 2000; Wong & Schiestl, 2002; Schiestl, 2005), what is certain is that it has worked and continues to do so.

Thus, an important question arises: how can a plant reach such a degree of complexity or weave such a strategy? How can the plant be able to generate a prison, mimic an oviposition site, or even mimic the shape and signals of the female of its pollinator to attract it to pseudocopulation and benefit from the transfer of pollen? The functional question is obvious, the "what for." However, it is difficult to understand, even with advances in pollinator–plant interactions, especially insect–plant interactions, how to reach such perfection in the design of deception. The tip of the iceberg encloses hundreds of options that characterize the linkage between plants and their pollinators. From the perspective of our book, the example mentioned above leads us to see how something that was indeed a rarity at the beginning, perhaps something subtle, gradual, from a natural selection approach, became in certain groups of plants the common, the norm.

Moreover, if we have not learned this modification's value or adaptive function to date, we could consider it a clear teratology. Having insect-like flowers would have no meaning *per se* for the plant. Thus, if we look closely, we see how the functionally adaptive loses strength, and we can easily mislabel something.

For our analysis, it is interesting to note that from the intraspecific point of view, these cases began as rarities and ceased to be so. They emerged as novelties, rarities in the occurrence frequency, or the trait's conspicuousness. They may have involved an even subtle level of preadaptation in color, shape, or a type of "bouquet" (odor), as noted by Vereecken et al. (2012) in their comprehensive comparative study in orchids of the genus *Ophrys*. In that work, the authors even overturn the "Cope's rule of the unspecialized," that very specialized groups cannot "reverse" in the evolution toward more generalist forms; in this case, transitions in both directions between generalists (shelter mimicry) and specialists (sexual deception).

The "duality according to the perspective of the analysis" is that, at the level of the supraspecific group, these groups of plants are species, genera rare for the clade to which they belong (i.e., the deception of these types is not a "rule" or pattern widespread among plants). Moreover, many insects are also somewhat different from their relatives. Thus, from the intraspecific view, these exceptional cases of evolution are fantastic examples of the passage from rarity to a pattern (i.e., depending on the clarification, it may be rather supra-familial). They are still rarities since they are restricted to a delimited group. Whether we call a phenomenon a rule or an exception depends very much on the level of analysis we are carrying out.

6.2.3 A More Bidirectional E-R Dynamic in the Short Term: Facultative Organisms

When we have mentioned examples of the passage from exception to rule, whether we refer to a pattern (e.g., form, behavior, etc.), process (its causes and mechanisms), or both, it does not imply that this is necessarily unidirectional. It is expected that in many circumstances, this will have an apparent bidirectional character, i.e., that the passage from exception to rule may occur in reverse. It is not only evolutionary time, as would be a regression or change in a trait that will not always be equal to the ancestral one, but where the "spatial" context (i.e., environment, surrounding environment, geography, territory) is the trigger for that change from E to R or from R to E, even during the life of the organism.

Where then, can we easily see this relatively fluid dynamic in one direction or the other? For this purpose, we can use so-called "facultative organisms." This concept refers to organisms (animals, plants, bacteria, etc.) that exhibit different behaviors or traits in response to varying environmental conditions. In one spatial context (x1), they may display specific characteristics, while in another spatial frame (x2), they might exhibit entirely different traits. This modality of change means that what is a rule, even implicit, in one setting is not so in another; therefore, this feature is different in greater or lesser magnitude. This trait may be morphological, physiological, behavioral, or a combination. It is worth remembering that "optional" is defined as something that can be done or not done; that can develop or function but is not obligatory (The American Heritage® Dictionary of the English Language, 2003). Concerning living things, that term is pointedly applied to an organism that manifests its genotype under defined environmental or biological conditions, for example, a greater or lesser degree of parasitism or the ability to reproduce sexually or vegetatively (Burke & Bonduriansky, 2017). Many bacteria illustrate the above (Table 6.2), as in some cases, they can survive both in the presence of oxygen and in anaerobic systems or environments.

When we discussed the origin of life and how the evolution of the first organisms that existed on earth occurred, we said that the rule in that first context was to be anaerobic, and when some molecular oxygen arose in the atmosphere, it came directly from the photolysis of water or other means. However, it was still

Table 6.2 Some examples of facultative organisms where the rule operates according to the environmental context in which the species develops are highlighted

Organism	Facultative characteristic	Context 1	Context 2	References
Escherichia coli	Respiratory metabolism	Anaerobic	Aerobic	Stanier & Villanueva (1996)
Salmonella	Respiratory metabolism	Anaerobic	Aerobic	Stanier & Villanueva (1996)
Vibrio cholerae	Respiratory metabolism	Anaerobic	Aerobic	Stanier & Villanueva (1996); Fernandez & Alonso (2009)
Klebsiella sp.	Metabolism	Nitrogen fixation	Aerobic	Bagley (1985)
Hamiltonella defensa	Endosymbiosis with aphids	Presence: Resistance to natural enemies, increased thermal tolerance	Ausence	Castaneda et al. (2010)
Bombus sp. Bumblebee	Mutualism (pollination)	Legitimate pollination	Nectar robbing	Inouye & Ogilvie (2001); Lichtenberg et al. (2018)
Hypericum perforatum St. John's wort	Reproduction	Apomixis	Sexual reproduction	Pank et al. (2003)
Salmo salar Atlantic salmon	Life in fresh or saltwater	Lives in the ocean for most of its life	Reproduction in freshwater	McCormick et al. (1998)
Anemonia sulcata Sea anemone	Mutualism	Solitaries	Symbiosis with decapods	Calado et al. (2007)
Aphids	Reproduction	Sexual reproduction in times of low resources	Parthenogenesis in times of high resources	Simon et al. (2010)

too little to have a relevant role in microbial evolution. Indeed, in early evolution, one could disregard oxygen and consider that it was the age of anaerobes (Martin & Sousa, 2016). Then, as oxygen increased notably in the atmosphere, organisms began to adapt to metabolize it. However, some maintained the ability to perform different types of metabolism depending on the context of the presence or absence of oxygen (Curtis et al., 2008). These bacteria are the so-called facultative anaerobes (or also facultative aerobes). They develop an aerobic respiratory metabolism when they use oxygen in the environment or a fermentative metabolism in the case of its absence (Stanier & Villanueva, 1996). Thus, facultatively anaerobic bacteria can obtain energy through both types of metabolism, depending on environmental conditions. One of the main groups of this type of bacteria is Proteobacteria, which include some important pathogens, such as *Escherichia coli*, *Salmonella*, *Vibrio*, and *Helicobacter*. There are also free-living bacteria with facultative metabolism, such as those represented by nitrogen-fixing bacteria. This mechanism allows them to adapt to the environment, and the passage from exception to rule would occur in a bidirectional manner to meet the metabolic needs of the organism (Unden et al., 1994; Unden et al., 1995; Murashko & Lin-Chao, 2017).

From an ecological perspective, concerning bacteria, there are cases of organisms that can live with different patterns according to the environment in which they are found. Facultative endosymbionts constitute one case. These are common among arthropods (Duron et al., 2008; Henry et al., 2015), and they can provide, among other things, a defense against natural enemies, such as parasitoids or fungi (Burke et al., 2009). The functions of these symbionts can range from an obligate nutritional role to an optional role in protecting their hosts against environmental stress. For example, almost all aphids are infected with the obligate nutritional symbiont *Buchnera aphidicola*, which they generally need for survival (Shigenobu et al., 2000). However, many aphids also possess additional symbionts that are facultative from the host's perspective (Moran et al., 2005). This means that although they are not necessary for the aphid to survive, they give it an advantage in defending itself. These bacteria can be transmitted from mother to offspring. There may even be several types of bacteria in a host associated with the aphid's host plant, geographical area, and climatic characteristics, which influence the evolutionary biology of aphids (Barrenchea-Mellado et al., 2015).

Other species exhibit a form of facultative mutualism, i.e., they benefit from a symbiotic relationship with another species but can live without it. Myrmecophilous plants, such as some species of *Passiflora*, whose living tissues are occupied by ants, are an example of this type of mutualism (Apple & Feener, 2001). Facultative myrmecophiles have a more significant number of ant species associated with them than obligates, as they do not require ants for survival but would be expected to associate opportunistically with numerous species (Glasier et al., 2018). The opportunistic nature of these associations implies that there must be chance encounters of symbionts interacting with a host (Rodrigues et al., 2010). The ability of hosts to interact with more ant species would increase the likelihood of symbiosis and thus benefit facultative myrmecophiles (Glasier et al., 2018).

Similarly, facultative mutualism is observed in some insects and birds pollinating certain plants, depending on the environment (Ollerton, 2006). Sometimes, as a flexible rule, facultative mutualists may behave cooperatively or in a host-exploitative manner. This is the case, for example, in individuals of the yucca moth, which sometimes skip behaviors used to pollinate flowers (Tyre & Addicott, 1993). These behaviors exist in numerous pollinators, which may use flower-pollinating tactics to consume nectar or make holes in their base instead of "legitimately" visiting flowers (Inouye, 1980). Nectar "stealing" is widespread, exhibited by insects, birds, and mammals, and is seen in most flowers with a tubular corolla or nectar spur (Irwin et al., 2010). Whether an individual visits a flower legitimately or steals depends on the costs needed to access nectar with each tactic and the amount of nectar it can extract. The most efficient flower management tactic under a set of ecological conditions would be the prevailing tactic or, in this framework, the prevailing rule (Lichtenberg et al., 2018).

Another example of a facultative organism is epiphytic plants, which live perched on other plants but do not have metabolic contact with them (Benzing, 1990). However, they establish interactions and intervene in the nutrient cycle of the host, in addition to being necessary for the functioning of the ecosystem (Ceja et al., 2008). Some epiphytes are also facultative and can indistinctly inhabit both the canopy and the ground (Sartí, 2012). This means that they can use different strategies according to their needs in the environment they are in. These include hemiparasitic plants. These plants generally live attached to the xylem of a host plant and can be facultative or obligate, depending on their degree of dependence. Obligate parasites depend on the host throughout their life, while facultative parasites may live part of their life independently (Westwood et al., 2010).

Facultative hemiparasites can complete their life cycle without host plants because they are photosynthetic, but when they encounter the root of a host, they form connections called haustorial. Thus, they extract water and minerals directly (Press et al., 1990). Some examples are found within the families of Scrophulariaceae, Santalaceae, Oleaceae, etc. Even in some species, seedlings can be independent of the host for up to one year, as in *Atkinsonia* or *Rhinanthus* (Sosa & Tressens, 2002).

Therefore, in all these cases and others, we cannot speak of exceptions to the rule but rather that the rule is very flexible in an optional context. As we have seen, sometimes the obligate was initially the rule, as in the example of anaerobic organisms. In the same way, evolution can be observed in parasitic organisms that come from free-living organisms. The potential transition in this evolution can be seen in facultative parasites, which, according to environmental conditions, can behave in one way or another (Luong & Mathot, 2019). Facultative parasites can complete the life cycle without a host but exploit them under certain conditions. For example, in facultative parasitic nematodes, parasitic juveniles form under stressful conditions (high density, low food availability, high temperature), but under favorable conditions, they can complete their life cycle without a host (Viney, 1996; Stasiuk et al., 2012). In mites, for example, the reproductive state of the host, its availability, and the parasite's state are necessary conditions that dictate whether organisms will behave parasitically or not (Luong et al., 2017). These context conditions, in all cases, make it the rule to be free-living, as was the pattern in the beginning, or to parasitize a host when this way of life represents an advantage to them.

At the reproductive level, there are also cases of some organisms that follow specific guidelines according to the best strategy. That is to say. They cannot be pigeonholed as "always" following this principle or pattern in reproduction mode. In botany, for example, there is facultative apomixis, through which individuals can reproduce asexually, but without leaving aside sexual reproduction, in a balanced system. Facultative apomixis is widespread in the Rosaceae family (Richards, 1997). For animals, facultative parthenogenesis is common, and we have already discussed that it appears, among other groups, in many insects, such as bees, ants, or aphids. In the latter, reproduction by parthenogenesis occurs when the environment presents favorable conditions and abundant food. Otherwise, they produce males and females who reproduce sexually (Lanteri et al., 2010).

The "facultative" component may be discovered late in many organisms and may even have been considered a mere rarity initially. This could occur due to the bias of studying only a restricted space or group of species in a clade. Cases of these late discoveries are other groups to those already mentioned where cases of parthenogenesis are also still found, which were previously totally unknown or taken as mere exceptions. For example, in the reproduction of the spiny leaf stick insect *Extatosoma tiaratum*, it was observed that females could reproduce asexually when sexual reproduction represented high costs (Burke et al., 2015). Another example is the case of eastern copperhead snakes (*Agkistrodon contortrix*) and their venomous cottonmouth cousins (*A. piscivorus*), which until relatively recently were thought to reproduce asexually only under captive conditions. However, it was discovered that some litters of wild snakes had also been born by parthenogenesis and that females could reproduce in this way when males were absent, when the population was low, or when migrating to new habitats (Booth et al., 2012). A couple of years ago, the case of asexual reproduction was observed for the first time in an endangered sawfish species (*Pristis pectinata*), documenting the birth of seven asexually conceived offspring. This behavior, as in those snakes, has been observed only in captive individuals (Fields et al., 2015). Therefore, all these "late" discoveries indicate that there could be many more cases of groups of animals and plants that follow more than one strategy. It would make them facultative organisms in that part of their biology. In conclusion, this bias mentioned above may mean that some of these cases remain outside our knowledge, but far from being exceptions, they may be evidence of the existence of patterns that differentially show themselves according to needs throughout the life cycle.

6.3 E-R Changes Due to Human Activity on Nature

6.3.1 Exotic Groups that Become Local Pests

6.3.1.1 Introduced Animals

Most biological introductions, which may later become invasions, appear as a direct result of human activity (Sakai et al., 2001). Plants and animals have long been mobilized to different places around the globe, and invasive species have caused a great deal of change worldwide, often harming native communities (Simberloff, 2001). In the framework of this book, we could say that native species, in any place, would represent the rule, the common, and adapted group to their environment. Moreover, when a new and exotic species is introduced, it would represent an exception. The problem arises when exotic species present a great competitive ability and capacity to spread. With these characteristics, it can become an invasive species, with the capacity to exclude and displace native species, even leading to extinction (Gurevitch & Padilla, 2004; Bøhn et al., 2008). Here, we would have a change in the pattern, where now the alien species could become the ruler.

Biological invasions have become a significant threat to biodiversity, with main mechanisms such as competition and niche substitution, which can make invaders dominant in their new ecosystem (Cheng et al., 2009). The introduction of species into new ecosystems was generally intentional and human-facilitated. Many times, species in captivity manage to escape, and they can become established in the new environment, competing with native species. One of the main motives for intentionally introducing species is economic gain. There are numerous examples of species introduced for agricultural or livestock purposes. Indeed, the most numerous and abundant introduced species are agricultural plants and animals. Cattle, sheep, goats, horses, and chickens are examples of introduced animals in many regions. The domestic bee (*Apis mellifera*) is another example of a highly adaptable species that can naturalize in a wide variety of environments. On the other hand, other species have been introduced for recreational purposes, such as hunting and fishing, such as boards, salmon, and trout.

One crazy example is the introduction of hippopotamuses to the shores of the Magdalena River in Colombia (Buriticá, 2014). This story relates to drug trafficking, as the infamous Colombian drug trafficker Pablo Escobar was obsessed with exotic animals and set up a private zoo in one of his mansions. He brought in a couple of young hippopotamuses, among many other rare wild animals (he loved birds). After his death, his entire empire collapsed, and many animals were released into the wild. It is how a megaherbivore and important ecosystem engineer was introduced into a place far from its original environment. Hippos have a relatively high reproduction rate and no major predator or disease in the area. These characteristics will lead to a significant increase in individuals by 2050, with a projected population of 1,500 hippos. It is currently the largest herd outside Africa. Thus, hippos are not only invasive species that compete for resources, alter ecosystem processes, and shape the landscape but also, due to their high aggressiveness, endanger human settlements that share their new area of distribution (Monsalve, 2014). The Colombian government is currently implementing strategies to limit the growth and expansion of the hippo population to mitigate future social and ecological implications.

In South America, particularly in southern Chile and Argentina, the wild boar (*Sus scrofa*) is considered one of the most noxious invasive ungulates (Jaksic et al., 2002; Novillo & Ojeda, 2008; Ballari et al., 2015). Wild boars are native to Eurasia and North Africa and have invaded almost the entire world (Wehr, 2021). It is a successful invader because it has a high reproductive rate and remarkable plasticity of adaptation to a wide range of habitats, climates, and food (Coblentz & Baber, 1987; Podgórski et al., 2013; Ballari & Barrios-García, 2014). As an invasive species, it damages animal and plant communities and modifies ecosystem processes and habitats. For example, it favors the dispersal of invasive plant species, which generates a double disturbance in the invaded local ecosystem (Ballari et al., 2015). The wild boar was introduced into Argentina from Europe early in the last century for sport hunting. As often happens with this practice, some individuals escaped from confinement and spread throughout the territory (Navas, 1987). It is currently exploited as an economic and cultural resource in several provinces, but this makes it a sustainable activity far from controlling and eradicating this invasive species (Ballari et al., 2015).

Australia is the perfect case of an isolated geographical area in the world that has suffered the introduction of exotic animals in large numbers that displaced many unique local representatives. There are many examples of cases ranging from insects to mammals and plants, some of them for more than 100 years. Perhaps the worst of the invasive species in Australia and many parts of the world is the European rabbit (*Oryctolagus cuniculus*), which, both for its ability to proliferate and its adverse effects, is included in the 100 lists of the world's most harmful invasive alien species (Lowe et al., 2000). It was introduced as a game but adapted very well to the new conditions and without the danger of natural predators. The rabbit accelerated the process of erosion and desertification in Australia by consuming natural vegetation. In addition, this lack of food affected many native animals and livestock. In addition, introducing a natural enemy, the red fox, was a disaster, as it preferred to attack birds or marsupials. A solution for this invasive species was the introduction of some viruses, such as myxomatosis and later hemorrhagic disease, which managed to stabilize populations (Coman, 2010).

An important aspect to highlight in this scenario is the rapid expansion of some of the introduced species. Hybridization processes have favored it with locals in some cases, as well as a great capacity to adapt to new environments. We have as examples the "invasive cane toads" or cane toads (*Rhinella marina*) (satirized in the comic strip The Simpsons) and the invasion of fruit flies in eastern Australia (Canestrelli et al., 2016). The cane toad was introduced to Queensland from Hawaii in 1935 and spread rapidly across Australia. After that, the cane toad has become a plague and a threat to native species because of its ability to reproduce and its venom (Tyler, 1994). In addition, during expansion, toad populations underwent significant morphological changes that promoted the dispersal of the invasion front, such as longer legs, as well as behavioral changes, such as straighter dispersal trajectories. On the other hand, invasion front toads could reduce parasite load and positively regulate enzymes involved in energy metabolism and response to oxidative stress (Brown et al., 2015; Hudson et al., 2015). All this allowed this exotic species to become highly invasive and to spread rapidly, competing with native species.

The other example mentioned above is the fruit fly (*D. melanogaster*), introduced in northern Australia approximately 100 years ago. Researchers observed that this species underwent several morphological changes during its dispersal. These changes included alterations in wing size and shape, as well as modifications in life history traits, stress tolerance, and metabolic enzymes (Frentiu & Chenoweth, 2010; Ray et al., 2016). All these changes allowed rapid adaptation to new environmental conditions and enhanced dispersal and reproduction.

In the framework of this book, we could state that these species, due to their particular characteristics, whether accidentally or voluntarily introduced to new environments, have become the rule, competing with and even displacing native species in Australia, as is the case in many other places. Many of the competing native species still persist but have changed the rate of appearance. They are now becoming the exceptions or the rare species that these new invaders displaced.

6.3.1.2 Introduced Plants

As with animals, invasive plants can compete with native vegetation and displace or drive native species to extinction. In addition, one effect of invasive plants that become established in a new ecosystem is that they can significantly affect soils and their fertility, water availability, erosion, etc. (Giorgis et al., 2006). All these modifications caused by invasive plants can affect the ecosystem's biodiversity and have negative economic consequences (Pimentel et al., 2005). A typical example is *Mimosa pigra*, native to neotropical regions but which can become a significant invader. This species was introduced into Australia more than 100 years ago and has spread remarkably throughout northern Australia. Although it allows some endangered species to find refuge in its canopy, it is generally harmful to Australian wildlife, as their diet is still based on native plants, which are displaced by the rapid growth of the *Mimosa* (Miller & Lonsdale, 1987; Lonsdale et al., 1995). The same is true of the fig cactus belonging to the *Opuntia* genus (*Opuntia ficus-indica*) that is native to Mexico but has spread to the United States and Argentina. From the United States, it reached Australia to be used in gardening and then naturalized and spread, adapting to the climate. It was also introduced in Europe and from there to North Africa. If the conditions are right, this plant quickly spreads because it can reproduce asexually by fragmentation and grow extremely fast, making it easy to become a pest (Cañas, 2017).

The success of introduced species has many hypotheses that postulate the need for unique ecological characteristics to explain their dispersal. However, recent evidence suggests that native and introduced species have similar successional trajectories (Meiners, 2007), as do other characteristics, such as defenses (Lind & Parker, 2010), enemy release (Schultheis et al., 2015), and grouping mechanisms (Lemoine et al., 2016). This suggests that introduced and native species thrive for very similar reasons. Interestingly, the latter authors state: "*we show that native and introduced species appear to follow the same 'rules' for becoming abundant.*" In other words, introduced species are not so different from their native counterparts, at least regarding ecological behavior. An invasive species is more successful than a native species in competing with it and displacing it to become the new ruler of the ecosystem.

This may be due to particular characteristics, such as rapid growth or dispersal inherent to the plant, and not to the geographical origin or the ecosystem's characteristics.

Other Witness Cases in Different Regions of the World

In addition to the cases mentioned above, invasive species exist in all regions of the globe, and in the same way, they can be distributed and become pests. The important thing about these ecological explosions is that they can sometimes have severe effects, even affecting human life. An example mentioned in the valuable and interesting book "*The Ecology of Invasions by Animals and Plants*" (Elton, 2020) is that of the *Anopheles gambiae* mosquito, the carrier of malaria. In 1929, some mosquitoes accidentally arrived in northeastern Brazil and were probably transported from Dakar, Senegal. Although not much attention was initially paid to them, the insects spread silently along the coastal region. It caused one of the worst malaria epidemics in Brazil due to its rapid reproduction in open ponds. Fortunately, control of the species was possible, and *Anopheles gambiae* was exterminated from the South American continent.

The same is true for many other species in different parts of the world, introduced as a source of food, the fur industry, the timber industry, and other economic or recreational purposes such as hunting and fishing. The increase in the speed of transportation in recent years has maintained and intensified this bombardment of foreign species in all countries, brought accidentally or intentionally from places that used to be isolated. Although not all plants and animals carried worldwide succeed in becoming established, they can somehow change the balance of native species. This process is gradually changing the distribution of species around the globe. In this list, we can cite the case of the Burmese python (*Python bivittatus*), a snake species native to India, Pakistan, and Indochina (Uetz et al., 2019), introduced in Florida, United States to be marketed as a pet. Subsequently, many of these snakes were released into the Everglades, generating a problem by impacting the ecosystem through heavy predation on mammals, including some endangered species, resulting in a considerable decline in previously common species (Reichert et al., 2017; Sovie et al., 2016). Another species threatening the local species in the United States is the tiger salamander *Ambystoma tigrinum*, which was introduced for use in fisheries as bait and then dispersed, competing with the locally endemic Californian species *A. californiense* (Naylor et al., 2001; Riley et al., 2003). Exotic species were also introduced to the rest of the world for use in various industries, such as red foxes from Europe to the United States and Australia for the fur industry and recreational hunting. The European red fox is probably responsible for the decline of some small canids and ground-nesting birds in North America and numerous small and medium-sized rodents and marsupials in Australia (Global Invasive Species Database, 2021 (http://www.issg.org/database).

Witness Cases from the South of the World

Most of the invasive species in South America come from Europe. This is commonly observed, for example, in places such as Argentina related to the arrival of European immigrants in the late nineteenth and early twentieth centuries (Speziale et al., 2012). In general, exotic species in Argentina were introduced for ornamental reasons (Rejmánek & Richardson, 2013) but are also related to the timber industry, forestry functions, and erosion control (Giorgis & Tecco, 2014). Each region has some characteristic exotic species. The occurrence of exotic species greatly depends on the proximity to urban centers and human presence (Viano, 2014). There are many cases cited for Argentina, and although most of them were introduced for commercial purposes, their dispersal and reproduction characteristics, among others, made them established in the country.

An Edible Fruit Whose Thorny Plant Displaces Not Only Plants: The Case of the Blackberry

The blackberry (*Rubus ulmifolius*) is an exotic species in the province of Córdoba, Argentina belonging to the family Rosaceae. Due to its rapid growth and the fact that it reproduces successfully through its seeds and stems (Giorgis et al., 2006), it has been a severe problem in the province for several years. Although it is very commonly used for the construction of fences or the consumption of its fruits, it forms thorny thickets, causing the landscape and habitat structure to change drastically (Comparatore & Mazzolari, 2009). Despite the invasiveness of this species, its ability to rapidly produce fruit and the possibility of asexual reproduction have made it valuable for consumption and exploitation. Its fruit is used for gastronomy and is part of the group of red fruits. However, it also has interesting properties due to its high content of tannins, sugars, and other organic compounds. Therefore, in addition to being used for food preparations, it also has medicinal purposes (Ortiz, 2007). Despite this, wild varieties of the species have spread rapidly in some regions of the country. Therefore, measures have had to be taken to eradicate it through the interdisciplinary work of biologists, agronomists, and foresters to reverse the deterioration of the natural environment caused by this plant (Comparatore & Mazzolari, 2009) (Fig. 6.2a, b).

A Colorful Tree for the Landscape However, Harmful to Native Species: The Crataegus (Pyracantha angustifolia)

It is a shrub native to Africa, with long thorns and white flowers, whose fruits enable the species' rapid dispersal and invasion. These are fleshy fruits eaten by birds, which then disperse their seeds. Crataegus was introduced to different regions of Argentina for its ornamental value. Because of its rapid growth and reproduction, it is one of the main invasive species in many ecosystems today (Giorgis et al., 2006). Its dense foliage generates much shade, an attraction for ungulates such as goats, horses, or sheep, which also disperse its seeds. Crataegus also helps the growth of

Fig. 6.2 Invasive blackberry in mountain environments of Argentina**.** The blackberry (*Rubus ulmifolius*) invades various mountainous regions, including high altitudes such as Los Gigantes, Córdoba, Argentina. It competes with high-altitude grasslands at 1800 m and progressively fills the canyons and temporary watercourses inhabited by native species, displacing them. Credit: Gabriel Laufer; Javier martin (**a**) *Rubus ulmifolius.* License: Public domain. https://www.inaturalist.org/observations/61610193. (**b**) Rubus ulmifolius fruit. License: Public domain, via Wikimedia Commons. https://commons.wikimedia.org/wiki/File:Rubus_ulmifolius_Fruits_23August2009_SierraMadrona.jpg

another invasive exotic species, the broad-leaf privet (*Ligustrum lucidum*), which grows more densely under its foliage and replaces native trees (Giorgis et al., 2006; Viano, 2008). Other species that also represent major invasions in Córdoba are sweet briar and dog rose (*Rosa rubiginosa* and *R. canina*), cotoneaster (*Cotoneaster franchetii*), pine (*Pinus* sp.), thistle (*Carduus* sp.), etc. All of these factors are already established, which makes their eradication much more costly.

Some Examples of Exotic Animals

Among the invasive animals, we already cited the case of the wild board. However, one of the most important invasive species that can be cited for Argentina is the European hare (*Lepus europaeus*), which can inhabit different environments, colonizing quickly due to its high number of offspring in each gestation. Trout and salmon have also been introduced, especially for recreational sport fishing activities, and have affected several native species in Patagonia by competing for the same resources (Giorgis et al., 2006). The abundance and quantity of exotic species are constantly increasing. Those introduced as "rare" species can slowly take the place of native species if the necessary measures are not taken, knowing each particular species in each region where they have become established.

In conclusion, this pool of examples and witness cases mentioned from various parts of the planet, as well as hundreds of others that exist of the exact nature (which would require a book for that purpose), serve to illustrate how the concept of exception was initially present for the introduced species. Moreover, from a frequency of occurrence perspective, it gradually or drastically became the most abundant, the new rule for the region in question. In contrast, native species may become rare species, representing new exceptions, which, unfortunately, in many cases, represent the "previous step" to their extinction.

6.3.2 A Look at the Disease Cycle and the Role of Man

6.3.2.1 General Epidemiological Context

As a rule, humans have been dealing with diseases since the beginning. Diseases have a relatively basic development cycle within the host body. However, they can also have a cycle more related to the general population with sudden and abrupt changes in onset and incidence. At the end of this chapter, it will be apparent that both diseases and more global processes such as pandemics start as an exception that, with time, and helped mainly by the constant changes in human habits and relationships, can become the rule in a specific time and space.

We must first turn to epidemiology to rescue some concepts that will help better understand these cycles and patterns shown by diseases (Bonita et al., 2006; Rothman, 2012; Gordis, 2013). For an infectious disease to occur, three components should interact: the infectious agent or pathogen, the host, and the environment that makes these two components coincide. On the other hand, noninfectious diseases are multifactorial, so this simple model can be made more complex (Rothman, 1976). Infectious diseases have a cycle called natural history. It shows the development of the disease at the individual level, over time and without treatment. As a rule, the cycle begins with a susceptible host exposed to an infectious agent or conditions that initiate the process in the case of noninfectious diseases such as cancer. After exposure, there is an asymptomatic incubation or latency period of the disease, which can vary widely depending on the agent. At this stage, some pathological changes may be detected in the patient. When symptoms appear, the clinical stage of the disease is reached. Finally, the disease ends, and the host may recover, but the disease may also leave sequelae or disability, or in the worst case, death. In addition, disease cycles can be analyzed from a macrohealth point of view and not so much from an individual point of view as described above. As a rule, diseases have levels of presence in the community. The base level is the endemic level, which may be zero or not, and the incidence of the disease may be rare or very common but within a range considered normal. Thus, there are diseases whose base level is rare, and a single case sets off health alarms (e.g., rabies, poliomyelitis). In other cases where diseases are common, people wait for an abrupt change in the norm before becoming alarmed (e.g., influenza).

On this basis, we have different denominations according to the behavior and rarity of the disease. Sporadic diseases are rare and irregular. Endemic diseases are common in the population in a geographical area. When the incidence level of a disease suddenly increases above the expected baseline level, it is called an epidemic. When the epidemic spreads to other geographical areas, such as countries or continents, and affects many people generating community circulation, we are dealing with a pandemic. The emergence of an epidemic may be related to changes in exceptions to rules. For example, an agent may increase its virulence, which is the proportion of severe or fatal cases. Additionally, the agent may be rare or absent in one place and may be introduced. It can change the mode of transmission by making it more infective. It can also change the susceptibility of the host to be more likely to respond to the agent. However, these cycles also end naturally due to a compromise between host resistance and parasite virulence, artificially due to technological developments or a better understanding of the causes of infection or the development of sanitary measures.

It should be noted that the progressive eradication of the pathogen (or its decrease and/or disappearance due to seasonality) generates the reverse step, i.e., that a disease that is a rule becomes an exception. Thus, developing a vaccine for an agent that has become more virulent may cause the disease to become rarer, chronic, or disappear. Although terms such as epidemic were intended for infectious agents, they are now applied to noninfectious diseases such as obesity and diabetes (Centers for Disease Control and Prevention, 2004; Mitchell et al., 2011). McNeill (1998) strongly emphasizes that the human being can be seen as a disease and, to some degree, an epidemic for other organisms, which may engage in less virulent forms of behavior but constantly generate a disturbance to the system. This view was brought to the cinema by the 1999 film The Matrix. Thus, the human being was an exception when it was in harmony with the environment and became, as a rule, the predominant organism on the planet, becoming a "macroparasite" (McNeill, 1998). Diseases are also seen as a means of population control since, for example, in Africa, many parasites and microorganisms do not trigger the formation of immune reactions. As density increases, infection increases, and the system escalates into an epidemic that returns to equilibrium once people die or are unable to generate resources due to levels of disability (McNeill, 1998).

6.3.2.2 Historical Review and Some Emblematic Cases

Historically, disease cycles followed humanity as civilization progressed. Human actions such as trade routes, social groupings, dietary modifications, improved health practices, and armed conflicts influence the appearance and disappearance of diseases (Bollet & Jay, 2004). In the beginning, the rule was that humans were scattered in small groups of hunter-gatherers separated from each other. Epidemics and significant pandemics (human plagues) were rare, if not nonexistent, since the parasite or infectious agent could not easily find a host to perpetuate the ailment (Kenny, 2021). Paleontological analyses suggest that humans who reached adulthood had an average level of health (Bollet & Jay, 2004). Additionally, parasites are at risk of eventually killing their host before being transmitted (McNeill, 1998), which is evolutionarily not suitable for the pathogen and is generally an exception in an efficient parasite (McNeill, 1998; Méthot, 2012). However, this is not to say that humans did not get sick. Indeed, many diseases affected primates, from which we evolved but also common diseases that we know, such as herpes, hepatitis, or yellow fever, and bacterial diseases linked to hunting and consumption of wild animals, such as anthrax and brucellosis (Aberth, 2011).

Moreover, in these times, human diseases were the rule in tropical areas because that is where the ancestors of human beings came from. We can give as an example sleeping sickness produced by a unicellular organism (*Trypanosoma brucei*). This parasite is found as a rule in antelopes and is transmitted by tsetse flies. It is harmless to the antelope and the fly but not to the human host. It generates extreme weakness and can be lethal. It is a case where the parasite is, as a rule, in equilibrium with the vector and the wild animal host but, as an exception, generates severe disease in humans, perhaps because it has not yet had time to reach a mutualistic equilibrium with them (McNeill, 1998). However, as humans spread throughout the world, they left these tropical latitudes to reach more temperate or colder zones and brought many diseases. They also escaped from others that afflicted their predecessors (McNeill, 1998; Kenny, 2021). The ability to create shelters, use clothing, and cook food contributed to avoiding many diseases, resulting in improved health, increased vigor, and population growth (Birdsell, 1957, 1958; McNeill, 1998). Cultural evolution generated a significant change in the environment and nature, and the relationship with diseases also changed considerably (McNeill, 1998; Pagani-Balletti, 2020).

During the first great human revolution, the agricultural revolution, humans went from hunter-gatherers to more stable groups where agriculture was practiced. This change drastically favored the spread of diseases that, until then, were considered rare in humans or absent. It favored the emergence of "crowd diseases" (Bollet & Jay, 2004). It was also a period of domestication of wild animals, which favored the appearance of common diseases in animals that were rare in humans (zoonoses). It is possible to identify common diseases that are associated with illnesses observed in domestic animals. For example, mumps is related to rinderpest, canine distemper, smallpox to cowpox, and of course, influenza is shared by humans, pigs, and poultry (McNeill, 1998). Examples of diseases common in wild animals and, therefore, rare at first in humans are bubonic plague (burrowing rodents), yellow fever (monkeys), and rabies (bats). Furthermore, as humans continue to invade wild places or come into contact with wild creatures, the occurrence of such diseases becomes more common. For instance, the hypothesis of exotic species displayed in Asian markets has been linked to the appearance of diseases such as COVID-19.

Thus, settlement in cities, population increase, interaction with other individuals (people and animals), increased waste and pollution, and trade routes were the triggers that favored the appearance of significant plagues and epidemics (Aberth,

2011; Pagani-Balletti, 2020; Kenny, 2021). Ancient texts, from Mesopotamia and Far Eastern texts to Egyptian papyri and the Bible, speak of many diseases. However, the descriptions are not detailed enough, and plagues are often attributed to superior forces (Aberth, 2011). Trade and conquest routes, together with armed conflicts, led to the appearance of great pandemics that wiped out a large part of the world's population (McNeill, 1998; Pagani-Balletti, 2020). Thus, places where diseases developed a normal cycle of epidemics or local circulation were connected and allowed diseases to affect a wider geographical range, similar to what has happened recently with the COVID-19 pandemic. Examples include the Justinian plague (camel route), the bubonic plague that was brought from the East, or the plague of Amwâs and probably the plague of Athens, which are related to armed conflicts since fighting troops were the first to be infected (Dols, 1974; Conrad, 1998).

Some Witness Cases

The Black Plague

Brought from southern Russia, the disease is spread by rats. These rats are infested with fleas, which in turn are infected with a bacterium (*Yersinia pestis*). Once the rats die, the fleas parasitize other animals and humans, thus spreading the disease (McNeill, 1998; Pagani-Balletti, 2020). It can also be transmitted from human to human when it causes pneumonia, but this way of spreading the disease is rarer (Bollet & Jay, 2004). Geographically, travel and trade routes spread the disease throughout much of the known world at that time (McNeill, 1998; Bollet & Jay, 2004; Aberth, 2011). The term bubonic comes from the buboes or swollen lymph nodes it produces, and the term black derives from the gangrenous patches that severely ill people show in some regions of their body (Bollet & Jay, 2004).

Many of the epidemics of antiquity are attributed to the bubonic plague. However, many accounts are vague, and historians often do not agree with epidemiologists in identifying diseases, such as the plague of Athens (typhus) (Pagani-Balletti, 2020) and the Antonine plague in Rome (smallpox, typhus). The Justinian plague is very likely to have been bubonic plague. The Black Death was the most famous plague epidemic in history (fourteenth century). Here is a good model explaining how human activities (such as travel and trade) spread a disease. It is thought to originate from the hunting of marmots, which is believed to be the natural reservoir for infected fleas. The fur trade extended from Crimea to the Mediterranean. Many sick or dead animals were easily captured, and their skin was contaminated with hungry fleas. In addition, the Crimean port was used by Genoese and Venetian merchants who allied with rival Mongol tribes. One of these tribes attacked the Genoese and allies and forced them to retreat to Kaffa. Here, history documents one of the first examples of biological warfare, as the Mongols catapulted bodies of Black Death victims over the city walls. Although the bodies themselves could not contaminate people directly, it is believed that the rats in the city, along with the infected fleas that infested the bodies, played a significant role in spreading the disease. This led to the downfall of the city besieged by the Mongols. Thus, the survivors returned to their origin cities, and on the way, they spread the plague throughout the Mediterranean, helped in turn by the rats they carried as stowaways on their ships. Then, the maritime routes between infected cities caused the plague to spread throughout most of the known world. Often, the ships arrived at ports with many of the crew dead. An example of a practice introduced at that time that became the rule, especially in ports, is the establishment of quarantine. Beyond religious beliefs, it was said that in 40 days, a disease had time to appear and manifest itself. Quarantine centers were thus created. One of the best options was quarantines inside the ship since, on land, people complied with the quarantine, but rats and fleas did not.

Therefore, we can see how something restricted, endemic to Crimea (exception), which would have killed comparatively a handful of people, was transformed into a killing machine, unstoppable (The rule in medieval Europe). This was due to the convergence of various specific conditions, such as hunting and selling wild animal skins, the proliferation of rats in cities and ships, and trade routes that favored the spread and proliferation of the disease. It is estimated that the Black Death wiped out approximately 1/3 of the population of Europe and the Middle East. Because almost 7 out of 10 infected individuals died, many agricultural areas were deserted. The lack of labor generated famines and malnutrition that opened the door to other diseases, such as smallpox and typhus. Then, several epidemics of the Black Death occurred in Europe. It became endemic in England from the fourteenth to seventeenth century. At this time, there was a change in the social paradigm, and hygienic norms were established in the population, leading in the next century to the construction of sewers and drains to avoid massive contagions (Pagani-Balletti, 2020).

It is also thought that the pathogen in question, *Yersinia pestis*, has become less virulent, the hosts have become more resistant and there is a possibility that infection with another related strain, *Y. pseudotuberculosis*, could generate cross-resistance with *Y. pestis*. This is in line with the previously mentioned idea that pathogens evolve to be less virulent and can coexist with the host without killing it (Smith, 1934). The dynamics between the host and parasite are complex. Although a rat extermination program would favor a decrease in the frequency of the disease, in the short term, the incidence would increase. Hungry fleas desperately seek another host, and humans would be one of the safest options. Historically, during many of the bubonic plague epidemics, when large numbers of dead rats were found, it was a harbinger of an increase, an outbreak in human cases of the Black Death. However, many suggest that the extraordinary appearance of rats in European cities was due to the great slaughter of cats related to witchcraft and the devil (Pigna, 2020). There are currently some outbreaks of bubonic plague, suggesting that it can be maintained in natural reservoirs until it infects people and is reintroduced to new areas (Keeling & Gilligan, 2000; Stenseth et al., 2008; Gascuel et al., 2013).

As we saw, trade routes were vehicles for large-scale pests capable of wiping out entire people. We have a clear example of the colonization of the American territory and how many groups were decimated by rare or absent diseases in those populations. On the other hand, for European conquerors, these diseases were common, and their bodies already produced the immunity that Native Americans lacked

(Pagani-Balletti, 2020). It is another case of biological warfare that some of the most renowned conquerors used. Examples of this are the great Aztec and Inca civilizations that perished mainly with the introduction of smallpox (McNeill, 1998). Inevitably, the industrial revolution also led to the spread of diseases due to poor hygienic conditions and overcrowding at that time. In addition, high industrial pollution levels appeared and continued to increase exponentially. Here, the dynamics of diseases became chronic.

HIV and the Aids Pandemic

One of the most compelling cases in recent times of something that began as an absolute rarity, an exception from the human perspective for man and his health, concerns the emergence of HIV and its consequent production of AIDS. Indeed, the origin of HIV and its two types have been the subject of much debate. Both have been considered to be the result of multiple cross-species transfers of primate immunodeficiency viruses (SIVs) that naturally affect African primates (Holmes, 2001; Sharp & Hahn, 2010, 2011; Peeters et al., 2013). HIV can be categorized as a rarity because of the number of exceptional characteristics it possesses. It belongs to the retrovirus family. These viruses are exceptional because they can reverse the current standard order of cell biology we saw in the origin of life chapter (transcription of DNA into RNA) (Weiss, 1996). Retroviruses generally do not break the cell membrane to release their copies (Aberth, 2011). In addition, retroviral DNA can remain dormant within the cell for years and suddenly become activated. Within retroviruses, HIV is particular in that it can specifically target crucial cells of the immune system that present the CD4 receptor to which the virus interacts (Aberth, 2011). Compared to other viruses that wage battles with the immune system to see who prevails, HIV takes it hostage by using its cells to replicate. It also belongs to the subclass of retroviruses that cause a slow infection (Lentivirus) (Haase, 1986). Patients, in general, can go a long time being asymptomatic, which gives the virus an advantage from the point of view of propagation. Finally, patients are prone to other diseases, as AIDS is not the one that kills but leaves the way open for the body with no immune system to perish under the action of other diseases, which would otherwise be contained (Aberth, 2011). Something that was initially restricted to monkeys, then passed to apes and from them to man, generating a pandemic process. Considering the more than 26 species of nonhuman primates involved, hypotheses on the nonspecific transmission of viruses, their mutations, and consequent transformations between species, it has been discussed whether the appearance of AIDS then constituted a zoonosis (see debates between Hahn et al., 2000 & Marx et al., 2004).

The truth is that today, in terms of epidemiology, incidence, and prevalence of HIV, it can be stated that this is a disease that can no longer be pigeonholed into a social group as it was done in the early 1980s. Today, the incidence of the virus is related to culture and the level of awareness of people belonging to different groups. For example, today, in first-world countries, the disease affects mainly white heterosexuals and women. However, the virus continues to have an incidence within the homosexual community. The myth that this was a disease only associated with homosexuals is now a thing of the past. On the other hand, in Africa, there is a higher incidence in people of color and heterosexuals, with a high incidence in women. Beliefs and rituals have a lot to do with this. On a global scale, AIDS cases have decreased, and patients now live with the disease chronically, without it being a death sentence. These individuals have a life expectancy that is close to the average (Becerra et al., 2016). Unfortunately, due to the complexity of the virus, developing an effective vaccine is difficult, but the effects of antiviral therapy allow the disease to be controlled.

The AIDS pandemic serves as a clear example of a transition from something rare and confined to something widespread. Initially, specific groups such as homosexuals, sex workers, hemophiliacs, or Haitian immigrants were disproportionately affected and were at the forefront of detected cases (Aberth, 2011). However, as time progressed, the perception and understanding of the disease evolved, and the social taboo surrounding the subject began to diminish. Like other pandemics, human beings initially discriminated against those with the disease. It was only with the realization that anyone could be exposed and at risk that the collective perspective gradually changed. This shift in perspective has led to a broader and more empathetic approach to addressing the AIDS pandemic.

Influenza and Covid-19

Influenza viruses are classified according to their host and the subtype of surface proteins that allow them to interact with the host cell they infect. These proteins are highly variable due to the high rate of mutations and/or recombinations. Thus, the nomenclature of the virus has the letters A, B, and C according to the host and then H (Hemagglutinin) N (Neuraminidase) with a subscript indicating what type of protein it is (e.g., AH1N1 is influenza that infects humans and animals and has the subtypes 1 of membrane proteins). The natural reservoirs are animals such as birds, pigs, and bats. Some strains share one of the receptors between both species, which makes the infection less effective in humans, but recombination may make the infection more specific in humans (Webster, 2002; Rodríguez Lozano, 2009; Ozawa & Kawaoka, 2013).

Different types of influenza have caused many epidemics and pandemics. Perhaps the worst influenza pandemic known to date is the misnamed "Spanish flu" (Spain was the neutral nation in World War I, which was the first to report its occurrence) (Flecknoe et al., 2018). The virus is believed to have appeared explosively in Kansas, USA, at a military camp. The US military carried it to Europe in World War I (Patterson & Pyle, 1991; Oxford et al., 2005; Foss, 2020). It is estimated to have killed 27% of the world's population. The second wave is said to have been more deadly and coincided with the end of WWI. It was an exceptional moment in human history where the devastation of war, the mutation and increased virulence of the virus, and the relaxation of social measures combined, a cocktail that proved deadly. There were other influenza epidemics, some of which were classified as potentially pandemic pathogens (PPPs) according to the WHO. Several different viruses cause seasonal influenza. It is a chronic seasonal disease and is controlled by the vaccination of people at risk.

At the end of 2019, a respiratory syndrome generating pneumonia appeared in the Chinese city of Wuhan. It was detected as a coronavirus that was named SARS CoV-2, and the disease was named COVID-19 (Lango, 2020; Singhal, 2020; Al-Qahtani, 2020). It was thought that the outbreak could be controlled locally, and no immediate action was taken, allowing the progressive spread of the epidemic to become a global pandemic. Air routes and ease of travel for much of the world's population facilitated the rapid spread of the disease. Additionally, ignorance about the pathogen and poor political, social, and sanitary decisions made it difficult to control the disease in many countries.

As a rule, quarantine and lockdown systems were implemented globally. Many countries did not adequately comply with the primary control and prevention measures and served as a substrate for virus mutation and the emergence of new strains. By the mid-2020s, vaccines began to be applied, and they helped to control the spread of the disease, thus establishing a new "normality" thus far. This description is real-time, and there is still much to learn because it is fully developed and cannot be compared with the other pandemics we have analyzed (Pigna, 2020).

We see it very easy to relate the disease cycle to human activities (McNeill, 1998; Aberth, 2011). Many diseases common in the past are becoming rare today thanks to health policies, changes in people's habits, and global vaccination campaigns (Cockburn, 1963; McNeill, 1998).

Change in the Incidence of "Civilization Diseases"

Some diseases or conditions are entirely related to the development of modern civilization, especially Western civilization, mainly with its current diet and lifestyle (Carrera-Bastos et al., 2011; Kopp, 2019). Thus, these conditions were rare or nonexistent in ancient times but changed the status to frequent or common diseases (Bollet & Jay, 2004) and are called "Civilization diseases" (Carrera-Bastos et al., 2011; Kopp, 2019). Here, there is a clear shift between what was the rule and what was the exception. Until the late twentieth century, most deaths were due to infectious diseases. That was the rule. Today, many deaths are from noninfectious and degenerative diseases. It is even more common and notorious in Western civilized societies and infrequent in places where more primitive systems, such as hunting and gathering, and less westernized societies, are followed.

We will take as an example obesity, which is currently considered a pandemic with many health implications (Roth et al., 2004; James, 2008; Kopp, 2019). What is striking is the change of conception about this condition, which went from merely an esthetic and monocausal norm to being cataloged as a chronic, dangerous disease with multiple causes. What used to be seen as a rarity has now become commonplace. Its causes have much to do with the change from exception to rule in human habits. For example, the poor quality of sleep and/or lack of sleep causes circadian rhythms and metabolism to be altered. As a result, the individual spends more time awake and eats more outside regular eating schedules. Throughout history, we went from eating a few times a week or a few times a day to eating at all hours. Another cause is related to our stable environment, where physical activity is not the same as our ancestors did. Again, a change from exception to the implicit rule, where now the intake in portion size is much higher but commonly nutritionally poor (Kaidar-Person et al., 2008; Khanna et al., 2012; Lieskovská & Megyesiová, 2015; Bitar, 2018). In addition, another linked change is that houses are temperature conditioned, which results, for example, in the body not expending energy trying to conserve heat (Kotradyová et al., 2013). The type of food in modern societies has undergone significant changes at a much faster pace than human beings have been able to adapt to. As a general trend, we have moved away from a Paleolithic type of nutrition, where protein and fat were predominant over carbohydrates, and refined sugars and dairy products were absent (Kopp, 2019). Let us remember that our metabolism and genetics are still adapted to that nutrition. The deviation from this ancestral diet to modern dietary habits can contribute to nutritional issues and health problems in contemporary times (Rendel, 1970).

Fat has a particular behavior in obese people. It represents a tissue that usually fulfills the function of energy storage. However, adipocytes (cells that form fat tissue) also have a secretory function and can release hormones and fatty acids that impact metabolism and interact with other tissues (Roth et al., 2004; Berggren et al., 2005). When this tissue grows too much, it invades other tissues, causing fatty organs, coronary problems, diabetes, and inflammation. It is an exciting paradigm shift that changes how obesity is currently perceived and treated. Obesity is directly associated with two other common civilization diseases, coronary heart disease, and diabetes (Rana et al., 2007). Obesity is prone to the occurrence of these two conditions that were not as common in ancient times. In addition, the change in lifestyle that we cited for obesity predisposes individuals to the appearance of these and other associated diseases (Betlejewski, 2007). Obesity is a complex and challenging disease that requires a comprehensive approach to treatment. It is not just about relearning eating habits and incorporating exercise but also addressing psychological and emotional factors. A holistic approach that encompasses mental well-being and emotional support is crucial for successful management.

Regarding modern campaigns that promote obesity acceptance, there is a fine line between body acceptance and normalizing a serious health condition. While promoting body acceptance and self-love is essential, it is equally important to address the health risks associated with obesity. Sending messages that trivialize or normalize obesity can be dangerous, as it may undermine efforts to combat this health issue and compromise the overall well-being of individuals affected by it. Balancing body acceptance with the recognition of the importance of maintaining a healthy weight is essential to promote both physical and mental health.

Additionally, stress deserves a separate paragraph. Stress is a beneficial mechanism for survival. It keeps us alert and triggers the flight or fight reflex (Chrousos, 2009) that has served us so well from an evolutionary point of view. However, a mechanism that did not evolve as a disease today is an intermediary that generates many deaths worldwide or has undesirable consequences. The passage from exception to rule is apparent here.

Alzheimer's: Rare in Young People, Common in the Elderly?

Alzheimer's syndrome is a neurodegenerative disease that became prominent at the beginning of the twentieth century (Petsko, 2006). The disease was probably not new but very rare and had a very low incidence before this date (Mahandra, 1984). It is a rare disease in other species and predominantly in humans (Gunn-Moore et al., 2018). Historical analysis agrees that it had a low prevalence prior to the twentieth century (Bondi et al., 2017). Many think it was due to low life expectancy. It is a common perception that the rule in ancient times was to live to age 40. However, it is recognized that people have reached average ages similar to today without problems (Griffin, 2008). The rule in those times was that the adult survival rate was low (Robson & Kaplan, 2003; Burger et al., 2012; Kua et al., 2014). Therefore, it is not rare for humans to live more than 60 years. However, it was perhaps rare because they did not survive birth or infancy (Montagu, 1994). In addition, they had to endure wars and killings, severe famines, or, as we said before, plagues or diseases that some are today easily curable as superficial bacterial infection that today is treated with antibiotics but that in the nineteenth century was the most frequent cause of death (Zaffiri et al., 2012). Likewise, it is not ruled out, as with other syndromes, that it was not accurately diagnosed in ancient times, and its low incidence is partly due to this phenomenon (see Chap. 8).

This age-related view is associated with the low incidence of Alzheimer's disease in young people and its prevalence in older adults (Petsko, 2006). However, some studies associate the disease's spread with substantial physical activity declines. Imagine our hunter-gatherer ancestors, the rule was to spend a lot of time wandering and hunting, with all the physical exertion that entails. They performed an average of four times more physical activity than people do today. Currently, despite the growing fitness trend, the overall level of physical exercise in modern society has decreased significantly compared to that in the past.

People who perform moderate to high physical activity lower the probability of developing the disease by 30%. Therefore, whether it is because of the increase in human longevity or the decrease in physical activity, it is clear that Alzheimer's is a disease that has gone from being rare to very common and with a high incidence among adults.

Anecdotal Case in History

A rare and curious historical case was the dancing epidemic that ravaged Strasbourg, France, in the sixteenth century (Pigna, 2020). Infected people danced uncontrollably and died of exhaustion. It could have been an outbreak of collective psychosis, or it could have been associated with a rare genetic disease with a high prevalence rate due to the high consanguinity of the region. There is also talk of massive intoxication with hallucinogenic substances with precursors of LSD (Matossian, 1989; Waller, 2008, 2009; Pigna, 2020).

In conclusion, the passage from the exception to the rule has been given mainly by the change in the human population's lifestyle. Thus, the change in relationships, economy, science, technology, and food has impacted human life from the health point of view, opening doors to new diseases and conditions that, from being restricted or absent, have become part of people's daily lives. Some, such as contagious diseases, may be relatively easy to attack. However, others, such as diseases related to contemporary Western civilization, are much more challenging to combat and will again require a significant change from rule to an exception in human activities and customs. Just as we saw that we are responsible for their emergence, we are also responsible for their combat and eradication. In Chap. 8, we will continue to discuss rules and exceptions in medicine, giving a general overview of the subject's current state and the horizons we can expect.

6.3.3 Influence of Environmental Pollution and Climate Change

Humans have dramatically modified the environment without ceasing to be part of it. However, many human activities generate drastic modifications of the environment that tend to break the dynamic equilibrium between the two (see the analogy between humans and viruses in the previous section). A term has been coined to describe a new geological era known as the Anthropocene that began in 1950 with the appearance of general-purpose plastic and became a key indicator of human activity (Zalasiewicz et al., 2016; Geyer, 2020). While many may talk about many animals changing the environment drastically, such as beavers building dams, few animals have reached the level of change that humans have generated. One of the environmental concerns caused by human interference is environmental pollution.

Environmental pollution is the phenomenon by which harmful substances of various kinds are introduced into the environment, damaging the entire system, including the living organisms that inhabit it (Merriam-Webster, 2010). This pollution can be natural or artificial (Farmer, 2002). Thus, planet Earth has seen many similar instances in which harmful substances were produced that harmed or even changed the environment during its long life. We have already mentioned the origin of life and oxygen production, which may have been considered harmful at the time but was later used beneficially by organisms adapted for that purpose. Additionally, significant catastrophes that produced massive extinctions can be considered events that pollute the planet, harming many living organisms. It occurred in the Permian (Benton & Twitchett, 2003) and in the Cretaceous–Tertiary boundary, an event that ended with dinosaurs (Officer & Drake, 1983; Shaw, 2018). Certainly, in these cases, the primary causes were volcanoes, abrupt climate changes (Crowley & North, 1988), and/or space asteroids (Alvarez et al., 1980). However, it is rare for a living organism to modify its environment to jeopardize its own existence. While certain examples of species modifying the environment exist in nature, these species generate modifications while maintaining a sustainable balance between their population and the surrounding environment (Wright et al., 2002).

Like any other animal, human activities have historically produced manageable amounts of waste and residues without introducing or generating elements that were not naturally present in the environment. As humanity moved away from wilderness and clustered in cities, environmental change increased, as did water and land pollution (See Markham, 2019). The industrial revolution is one of the critical points in humanity that could be taken as a point of no return. While it sparked numerous advances that ultimately benefited society, the environment has endured and continues to endure significant suffering as a consequence. Air pollution levels escalated exponentially (e.g., Klimont et al., 2013). The levels peaked in the 1970s and dropped slightly until our time. This modification of the air brought with it problems with the ozone layer related to greenhouse gases. These gases concentrate in the atmosphere and generate a layer that does not allow heat dissipation, producing the famous greenhouse anomaly linked to climate disruption (Kweku et al., 2018). This phenomenon contributes to a larger-scale issue known as global warming. Despite attempts by various conspiracy theories and unfounded claims to divert attention and cast doubt on the reality of global warming, its validity is evident and undeniable to anyone who examines the evidence.

As with pollution, climate change is not a rare phenomenon on the planet (Crowley & North, 1988; Huber et al., 2000). There have been many periods of abrupt climate change (Crowley & North, 1988). For example, in The Middle Ages (mid-fourteenth to mid-nineteenth centuries), humanity witnessed "The Little Ice Age" in which the European climate was impacted, resulting in glacial expansions and altered climatic conditions (Mann, 2002; Matthews & Briffa, 2005). However, this was not a global-scale phenomenon but was confined to the northern hemisphere (Bradley & Jonest, 1993; Mann et al., 1999; Mann, 2002). However, none of these climate changes had a human component or even resulted from any other organism on the planet and are considered regular climate cycles.

Thus, we see that although environmental pollution and climate change may have a purely natural component, they are processes that have been greatly accelerated in recent decades by human action. Many patterns and processes they trigger were rare before and are very common now. We will cite a few examples to illustrate this idea.

6.3.3.1 Water and Air Pollution

Contamination of water, air, and soil has been extensively studied. For example, water can be contaminated with organic matter, mainly phosphorus and nitrogen, derived from nutrients from high loads of wastewater and fertilizers from agricultural activities. It results in cultural (human) eutrophication, which generates the pollution of marine coasts, estuaries, and rivers (Seitzinger et al., 2002 Glibert et al., 2005; Burkholder et al., 2007). It is one of the most significant pollution problems worldwide (Howarth et al., 2002; Glibert et al., 2005). Although eutrophication can occur naturally (e.g., upwelling) (Richardson & Jørgensen, 1996), it occurs in a more controlled manner compared to human-generated eutrophication, mainly because of the magnitude of the impact. Eutrophication, in turn, leads to uncontrolled growth of certain species of toxic harmful algal bloom species (Smayda, 1990; Anderson et al., 2002; Glibert et al., 2005), oxygen deficits, and death of various organisms (Bricker et al., 1999; Burkholder, 2001; Burkholder et al., 2007). There is a dynamic balance between light, nutrient availability, and temperature in nature. These are the factors that may limit plant biomass production. When nutrients condition biomass increase, their exogenous incorporation via eutrophication can affect the biomass in the ecosystem (Richardson & Jørgensen, 1996). Thus, naturally, this eutrophication phenomenon is infrequent, so the effects on algal and plankton populations are rare. At the same time, their influence on other organisms is rare when there is an uncontrolled increase in biomass (blooms), generating hypoxia and even changing the diversity and abundance of species. This behavior may be natural in the organism, but the magnitude of population growth and the exacerbated action on other individuals is often the exception in nature. Studies report that blooms are more significant, toxic, and prolonged than they used to be (e.g., Anderson et al., 2002) and that the presence of soot and smog in large cities is not normal. An interesting case of microevolution and the influence of the industrial revolution on air pollution and its impact on organisms is reflected in a species of butterfly associated with trees in the industrial regions of England. Peppered moths (*Biston betularia*) are nocturnal Lepidoptera that can have two morphs (*insularia* [light] and *carbonaria* [dark]) (Cook et al., 1986).

The light morph (*insularia*) mimics very well with the tree trunk, which was also light and contained lichens. Increased pollution and soot deposition resulted in the darkening of the tree trunk and elimination of lichens, making light butterflies highly visible and favoring the camouflage of dark butterflies (Majerus, 2009; Curtis et al., 2008). Although this phenomenon occurred in real time while Darwin was developing the theory of evolution, he did not explore it (Kettlewell, 1959). Here, we see how a phenotype that was considered rare changed its status to standard as a result of environmental change due directly to artificial air pollution. Furthermore, we know that, in general, with the exception of rare natural cases such as volcanoes or giant meteorite impacts, the air should be pure and easy to breathe. Unfortunately, this is currently the exception in certain megacities worldwide, whose pollution and toxicity make it difficult to breathe such air (Meteosim, 2021).

6.3.3.2 Spill

Natural disasters related to oil were also uncommon, especially when it spilled into the sea. As with other pollutants, oil spills exist naturally. They are cited in the bible and are common in places where oil is produced naturally (Nelson-Smith, 1971). The pollution we are referring to, which was exceptional until the middle of the nineteenth century, is marine or coastal pollution linked to the transport of crude oil or the use of gasoline by ships, pollution directly linked to human activity (Blumer, 1971). From the biological point of view, it is a disaster of enormous magnitude, affecting the ecosystem permanently for an extended period. Organisms are not accustomed to or adapted to this type of contamination, which causes many to perish, a situation that would be uncommon in systems without such contamination.

6.3.3.3 Soil

When we refer to the soil, the most abrupt change that man has made is the change in land management and the production of large arable areas (Power & Follett, 1987). Humans have modified the landscape and produced extensive monocultures that deviate from the natural patterns found in nature. The norm is that different plant species are combined (Hooper et al., 2012). Monoculture brings about significant pest challenges from insects (Altieri et al., 1984), fungi (Fisher et al., 2012), and rodents (Howard, 2018), necessitating the use of pesticides for their control and fertilizers to enhance crop production (Epstein, 1997). All of these examples are exceptions and exaggerations of what happens naturally.

6.3.3.4 Nuclear Energy

Concerning nuclear energy, both the war tests and the accidents related to nuclear power plants generated evident changes from exception to rule. Take, for example, the Chernobyl nuclear disaster (Haynes & Bojcun, 1988). It is evident that this event led to a disproportionate rise in diseases and malformations associated with elevated levels of artificial radiation (Lendon & Martin, 2007; Völkle, 2015), a trend that persists even today (Nesterenko & Yablokov, 2009). Additionally, animals had to adapt to that radioactive environment. Imagine the changes that must occur in nature and the adaptations that organisms must develop (Mousseau, 2021). Although many studies are controversial mainly due to the hermeticism of the former Soviet regime, some works clarify that species fall in abundance and biodiversity in these contaminated areas (Mousseau & Møller, 2012). Thus, the change from exception to rule is seen in the current presence of many radioactive substances or derivatives that are not commonly found in the environment (Kholosha et al., 2008). In addition, there are other cases where spider communities, for example, grow proportionally to the level of contamination, probably due to the disappearance of other predators of their food or the malformation and incapacity of prey (Møller et al., 2012).

6.3.3.5 Micro and Nanoplastics

Plastics are generally synthetic polymers. Polymers are not a rarity in nature. They appear in many natural materials that we have studied in biology. Basic examples are DNA, spider silk, plant cellulose, hair, and muscle fibers (Geyer, 2020). World War II triggered the production and use of plastics worldwide. Although plastics have been around for approximately 100 years, their production and use were rare prior to WWII (Geyer, 2020). Different types of plastics with different conformations are directly related to their recycling ability. In a gradient from the easiest to recycle to the most difficult, we would find thermoplastics such as PVC or PET capable of being melted and, therefore, potentially recyclable. Thermosets that are more complex in their molecular structure, such as plastic resins or melamine, are more difficult to recycle. Finally, the combination of thermosets with fibers makes fiberglass or carbon fiber-reinforced polymers even more difficult to recycle but resistant. The biodegradation of these materials is a complex mechanism involving microbes that breakdown the material into its essential components. It is a process that takes different rates that can range from weeks to decades.

Plastic became an indicator of human activity on a geological scale. Before 1950, it was absent from sediments, but it was present from that time onward (Zalasiewicz et al., 2016). We see how the very presence of the material went from being a rarity in nature to being a conspicuous constituent in nature. We will now look at the magnitude and impact of the occurrence of this material in the environment.

Plastic produces pollution on different scales. This material generates waste during production and disposal of the article after it has been used. Subsequently, recycling produces gas emissions into the atmosphere, as does disposal by burning. Both are emblematic cases of plastic impact mitigation that generate even more impact. The only use of waste that does not generate impact, at least for a time, is the reuse of the product (e.g., reusing plastic containers, bottles, or cups). However, at some point, the plastic product inevitably ends up being recycled, thermally destroyed, or, in the worst case, discarded. Due to improper disposal or negligence, plastic enters marine habitats at a dizzying rate (Steer & Thompson, 2020). Various analyses show that plastics of various sizes (5 mm and over to less than 0.1 mm) are now the rule in nature, appearing in freshwater, marine, and terrestrial environments as a potentially hazardous pollutant (Besseling et al., 2017; Li et al., 2019). The most prevalent forms of waste are micro and nanoplastics. An intriguing phenomenon is the formation of a 1.6 million km^2 synthetic "island" entirely made of waste known as "The Great Pacific Garbage Patch" (Kaiser, 2010; Gabrys, 2017; Lebreton et al., 2018). This immense patch, predominantly composed of plastics, floats through the Pacific Ocean. The rule was thus far that islands were generated by geological processes such as volcanic islands or biological processes such as coral islands. However, with their waste, humans have become agents that can generate something similar. Plastics can also interact with marine biota and bioaccumulate in living organisms in the ocean (Toussaint et al., 2019; Wang et al., 2021) or become entangled in them, causing disability or even death (Steer & Thompson, 2020).

As we have seen throughout this section, the examples are many and varied and work at different scales. What must be clear from this section is that man has become a potent agent of environmental change, even competing with critical geological processes that have shaped the planet in ancient times. By shifting the environmental balance, he produces many changes from exceptions to rules. Although the earth is suffering at the moment, in the long run, it is used to these cycles of contingency or stress (e.g., catastrophes and extinctions), and we have seen that it takes time, but it knows how to recover well. The one whose days may be numbered if it continues on the current path, is the human species. Humans may eventually join other large groups, such as dinosaurs, in the annals of species that succeeded and were able to transcend but ended up dying out abruptly. The rule is that no matter how long it takes, the Earth always recovers and is the species that goes extinct. If we do not change course, we will not be the exception. In the next chapter, we will discuss how processes linked to human-induced climate change impact species diversity, in particular, changes from rules to exceptions, the reverse order to the one just discussed.

6.3.4 Manipulating Organisms to a Destination

6.3.4.1 Artificial Selection: Turning Exceptions into Rules

By understanding the principles of natural selection, we can apply them to enhance the breeds and varieties of plants and animals. It is already known since Darwin himself (in fact, he devoted chapters on this) that artificial selection was initiated by man long before he had any knowledge of the existence of natural selection. The knowledge of natural selection has further refined these artificial processes, allowing for a more detailed consideration of rarities and exceptions in their entirety. This optimization enables more efficient and ethical applications in various fields, such as biomedicine, agronomy, and animal production. In this context, it is interesting to note that artificial selection is restricted to what is deliberately done by man, for example, for the domestication of animals or plants for agricultural or ornamental purposes. Clearly, its definition points exclusively to it: *"a process in the breeding of animals and in the cultivation of plants by which the breeder chooses to perpetuate only those forms having certain desirable inheritable characteristics"* (Collins English Dictionary, 2003). In other words, artificial selection is a method by which humans intervene in the reproduction of animals and plants to favor the development of selected traits, whether for productivity, resistance, docility, esthetics, or other purposes. In addition to these techniques based on sexual reproduction, genetic engineering techniques are also used, such as the creation of transgenic and other genetically modified organisms (Hill & Caballero, 1992; Conner, 2003).

Indeed, the definitions of artificial selection often overlook unguided human actions, which can have a similar impact on nature. These "involuntary" actions can also significantly influence and shape natural processes. In other words, we are referring to artificial selection that is not conscious, not directed to an objective, but that invariably acts at a selective level on certain organisms. Thus, depending on the type of planning, two types of artificial selection can be considered: conscious, when there is an objective, a determined selection plan, and unconscious or involuntary, when there is no previously determined objective.

In addition, conscious artificial selection can be differentiated between stabilizing type selection when those selected are individuals with intermediate characteristics, where genetic variance is expected to be reduced (Bulmer, 1971), or disruptive type selection when those selected are individuals with extreme characteristics and genetic variance is expected to increase (Hill & Caballero, 1992). Numerous examples can be included in involuntary artificial selection, and they occur when animals face an anthropic environment that we impose. They change to better adjust to it. Some classic illustrative cases are as follows: (1) due to the effect of the urban environment, for example, swallows in United States highways. Now short-winged individuals are more abundant because they are the ones that better maneuver to avoid being run over by trucks or cars, and this is reflected in less and less mortality (Brown & Brown, 2013); (2) due to hunting and fishing. For example, Chesapeake turtles are evolving to a larger size, as the smaller ones can enter the traps fishermen use for crabs (Santos et al., 2018); (3) because of environmental pollution, as some aquatic organisms survive or overcome challenges imposed by us. For example, fish and invertebrates adapted to PVC spilled throughout rivers and lakes (Lusher et al., 2020; Hammer et al., 2012); (4) due to the effect of the introduction of exotic species (either for productive purposes or to control some local pest) that finally force certain native organisms to adapt. A clear example appears in an already mentioned invasive species, the cane toad introduced in Australia (Sect. 6.3.1). Their influence reduced the head size of the native snakes, as the smaller snakes ate smaller, less toxic toads and were the surviving individuals (Phillips et al., 2003; Phillips & Shine, 2006); (5) due to the effect of organisms that are released into habitats other than their natural habitat and must adapt to survive. For example, on Pod Mrcaru Island in the Adriatic Sea, a population of lizards of the species *Podarcis sicula* was released. The lizards came from an island abundant in insects, so their diet was practically insectivorous. However, on the newly colonized island, the abundance of insects was much lower, and plants were more abundant. This species of lizard changed the basis of their diet and adapted to eating leaves. Among other adaptive characteristics, they developed a larger head to give a better bite, and their legs became shorter as there was no need to run to hunt (Herrel et al., 2008).

All these cases are ultimately events of natural selection in the broad sense. However, in a narrower sense, they represent examples of artificial selection since they involve modified scenarios and conditions imposed directly (with an objective) or indirectly (unconsciously) by man. These clarifications are important, as they expand the concept of artificial selection beyond its strict definition as deliberately oriented by humans. Whether we are aware of it or not, in all these cases, we start with exceptions that potentially lie at the extremes of the trait distribution, which is often continuous. For example, short wings in the US highway swallows, extralarge body size in the turtles mentioned above, and the small head in the snakes of northern Australia (Fig. 6.3a, c). In short, something analogous to natural selection on extreme variants is done, mainly by directional selection.

Indeed, the birth of agriculture serves as a compelling example of conscious artificial selection. During this transformative period, humans made a significant decision to choose wheat spikes in which the seeds (grains) would not naturally fall and disperse to the ground, as was the "rule" in the species. By deliberately selecting these monodispersing spikes, humans altered the natural mechanism of seed dispersal, which allowed for better control over the crop and facilitated cultivation. This act of conscious artificial selection marked a pivotal moment in human history

Fig. 6.3 Species under involuntary artificial selection. (**a**) *Petrochelidon pyrrhonota*, American cliff swallow (Credits: Channel City Camera Club from Santa Barbara). (**b**) *Malaclemys terrapin*, diamondback terrapin. (**c**) *Acanthophis praelongus*, northern death adder. (Credits: Valentin Moser) (**a**) *Petrochelidon pyrrhonota*. License: CC BY 2.0 <https://creativecommons.org/licenses/by/2.0>, via Wikimedia Commons. https://commons.wikimedia.org/wiki/File:Cliff_Swallow_(51120316295).jpg. (**b**) *Malaclemys terrapin*. License: Public domain, via Wikimedia Commons. No machine-readable author provided. Ltshears assumed (based on copyright claims). https://commons.wikimedia.org/wiki/File:Diamondback_Terrapin.jpg. (**c**) *Acanthophis praelongus*. License: CC BY 4.0 <https://creativecommons.org/licenses/by/4.0>. https://www.inaturalist.org/photos/29424801

and laid the foundation for modern agriculture. These seeds were sown. If the trait was inheritable and not merely ontogenetically acquired, a proportion of the offspring replicated that trait, which was in itself an exception. That is, something even teratological, since that characteristic of "not opening" of the spike was an attempt against the reproduction of the species. By using this to domesticate the plant, it is considered man-initiated agriculture. Different wheat species are used for different purposes. For example, the flour we use today also results from an artificial selection process. *Triticum monococcum* and *Aegilops speltoides* were crossed and resulted in *T. dicoccum,* and when crossed with *A. squarrosa,* they resulted in *T. aestivum*, known as common wheat, from which many products for human consumption are made (Gregory, 2009). Another notable example, because of its importance in human consumption, is maize. Almost all maize harvested in the world today is "hybrid maize," i.e., a product of artificial selection. We will discuss this in the following chapters. Improvement after domestication has led to changes in morphology, yield, plant habit, and biochemical composition (Gregory, 2009; Wright et al., 2005; Gage et al., 2017). Among other examples that increased their productivity thanks to artificial selection, we can name them rice (Izawa, 2007) and rye (Cooper, 1960; Robertson, 1961). The creation of different species of cabbage plants (*Brassica oleracea*) to obtain a greater variety of vegetables has given rise to different varieties grouped in cultivar groups, such as brussels sprouts, cauliflower, and broccoli (Gregory, 2009). The example of bananas (*Musa paradisiaca*) is an exceptional case of conscious artificial selection and how an exception becomes a rule. The domestic banana, whose fruit is consumed worldwide, is characterized by being fleshy and small-seeded and has the right length and thickness to fit in hand. This fruit descends from hybridization and selection between species of the Musaceae family, native to Indonesia. The fruit of this wild plant is small and oval, not so fleshy, and its seeds are enormous compared to those of its artificial descendants (Li & Ge, 2017). On the other hand, varieties of stunning beauty in ornamental plants have also developed through artificial selection for desired characteristics that, in most cases, were simple exceptions in ancestral populations (Zhang & Liu, 1998).

For example, of conscious artificial selection involving animals, we can mention a few among a large number. For example, the domestic dog (*Canis lupus familiaris*) originated from populations of the gray wolf (*C. lupus*); all current varieties of domestic dogs are the product of artificial selection. Current dog varieties are task-oriented, such as guarding, companionship, and esthetic purposes, or species that humans consider "cute," e.g., breeds such as poodles and bulldogs (Frank, 1980; Hedrick, 2015).

Today's dairy cows are the result of many years of selective breeding, where those individuals who produced more milk, the exceptions that differed from the average, were selected, and these were the ones that, in turn, were made to breed (Flori et al., 2009; Hedrick, 2015). In addition, excellent varieties have been created, such as the "Patagonian Dynasty," transgenic cows capable of producing insulin in their milk (Linares & Maciel, 2017). Other examples, such as farm animals for higher body weight and greater body fat thickness, have also been under artificial selection for higher productivity and profitability (Hetzer & Harvey, 1967; Verghese & Nordskog, 1968). Racehorses are bred under artificial selection to create a higher probability of producing faster and stronger horses in subsequent generations, which generates a business with significant economic benefits (Hanot et al., 2018). In the laboratory, mice, for example, have been selected to be able to run at a much faster speed than mice from control lines. These mice have body adaptations that give them higher levels of endurance (Swallow et al., 1998). Additionally, species such as guppies (*Poecilia reticulata*), freshwater fish native to Central America, are under artificial selection processes where those with the most attractive colors are selected to reproduce in search of the most "beautiful" ones (Endler, 1980; Kotrschal et al., 2014).

Indeed, these examples are repeated with some similarity in the process in many other examples of agriculture, forestry, fish farming, livestock, ornamentation, and biomedicine, among others. Every innovation arises from the deliberate selection of an unusual variety, either in its entirety or specifically with a unique trait "x." This trait may be rare and unremarkable in the organism's natural environment, but due to some desirable quality, it becomes the ideal choice for a particular human interest. That variety can indeed be a single specimen or two before the beginning of the process of artificial selection. Therefore, this context resembles the teratological from a realistic perspective of the species. Just think of the extravagant dog breeds from an adaptive point of view; there are teratologies everywhere.

An interesting point to mention is that many times it has been demonstrated how to go back from something that by artificial selection became the rule to the ancestral pattern. Let us think of cases where more than 10,000 years ago man "chose" an exception of interest in a certain organism and transformed it into a rule, perhaps having completely transformed what the initial rule was for that species. It should be noted that variants often appear in artificially selected organisms that a breeder may see as teratologies but which in reality are nothing more than ancestral traits that refuse to disappear. There are plans that seek to vindicate ancestral and/or lost traits. In this scenario prevails the quest for exceptions in the genes, phenotypes of some specimens, or dispersed breeds. It is aimed to "put it all together" to go toward that lost ancestral rule. In other words, to recover the appearance, the morphology of the original organism. The results of this selection can then be placed in nature to test the aptitude of these "new" (original?) variants (Conner, 2003). In this regard, we use two examples. The first is that of the Aurochs (*Bos primigenius*). This species of animal, similar to bulls and bison, existed thousands of years ago in Europe. Humans began to domesticate them and select various genetic traits to make them increasingly docile and tame, creating several different species of cattle. A group of researchers set out to "resurrect" the original aurochs. By crossing different species of cattle, they have achieved the original genetic characteristics of the aurochs and have managed to obtain 80% of the original genome (Kyselý, 2008). The second example conveys the common pigeon (*Columba livia*), where wild phenotype reconstruction could now occur (Sol, 2008).

It will be interesting to see in these projects if the current rule (traits that have already been artificially selected) remains as mere exceptions in these future "retro"—artificial secondary organisms, returning to the old version of the organism via reversion. This connects with an additional idea, implicit but worth emphasizing, which is that the varieties we have created by artificial selection are abnormalities for the species itself. This is clearly seen, as we exposed above, in some examples, in breeds of dogs, fish, cattle, and plants, with exacerbated traits of size, length, height, body color, appendages, or others. If these traits appeared in the original population, it would not be simply an extreme value of natural variability but true teratologies. Indeed, this would even imply a break in the continuum of trait distribution, something even more discrete and strikingly different. In other words, humans "create teratologies" for our benefit. This is a not insignificant point, which involves us directly in the dynamics between rules and exceptions with living beings, as no other organism on Earth has ever done before. Many varieties of plants and breeds of animals are "beautiful" from our point of view or even appear so to us because of their bizarre characters (this is what we are attracted to show). However, from the natural perspective of the species, they may be "monsters." Indeed, in some species, artificial selection has gone too far, favoring traits with which the species would not have been able to survive in nature for different reasons, for example, difficulty in feeding, moving, or reproducing, among others. In other words, this exception would hardly have become the rule. There are many illustrative cases in this respect of birds (e.g., Mitchell, 2014). For example, the common pigeon (*Columba livia*), which has almost no beak, bulging eyes, and a tiny skull, is one of the pigeon breeds we consider "beautiful" (Parrott-Holden, 1987); the Gibber italicus (*Serinus canaria*) domestic canary, which has been selectively bred for a few hundred years and where breeders can control everything from skeletal formation, the number of feathers, to the length and style of the bird's song (Bartels, 2003); chickens bred to develop larger breasts, the giant Junglefowl (*Gallus gallus*) have great difficulty moving and hearts that have not been adapted to support the additional mass (Paxton et al., 2010).

Among the fish, bubble eye goldfish (*Carassius auratus*) is a species that has been created through artificial selection, especially for the delight of the human observer, resulting in an incredible variety of shapes and sizes. This fish grows large fluid-filled sacs under the eyes and has malformed, upturned eyes, a curved spine, and no dorsal fin (Komiyama et al., 2009). The Rhodesian Ridgeback dog breed was bred for hunting and is large, fast, solid, and has excellent eyesight. However, it may possess a hole toward the end of the spine that leaves nerve tissue exposed to bacteria (Hedrick, 2015). Something similar occurs with the German Shepherd (Deutscher Schäferhund), a beautiful, intelligent, and protective dog with cinnamon and black fur widely used by security forces and the military. Among other diseases, it is common for large specimens to suffer from hip dysplasia, which causes deformities and problems in the posterior area and much pain (Hedhammar et al., 1979; Rossi, 2018).

On the other hand, it should be noted that most plant and animal species domesticated through artificial selection suffer a similar effect to the bottleneck (see Sect. 6.2.2.1) as their genetic diversity decreases concerning their wild ancestor (Wright et al., 2005). This results in increased vulnerability to disease and an increased incidence of morphological problems that would otherwise be minimized or eliminated by natural selection (Robertson, 1961; Conner, 2003). In addition, it can shorten the life span of selected organisms, create new mutations that it is not known how they will be routed later, and cause sterility (see the iconic example of sterility in bananas, Pearce, 2003, 2008).

All we have mentioned and exemplified in the present section provide clear evidence of the artificial selection of dynamics between exception and rule. If we as humans become more aware of this, it would allow us to be more responsible for the impact of creating breeds, "quasi" teratological variants, which could bring to a directional limit the genetics and phenotype of a given species (e.g., to the detriment of its ancestral genotype and morphotype). Most importantly, we must be more responsible for clearly foreseeing the direct and indirect effects of such manipulation and this "new organism" on other species and ecosystems. Time will tell how this will affect nature worldwide in a much more generalized way and what our role will be in this challenging scenario.

6.3.4.2 "Artificial" Cloning, the Transition from Exceptions to Rules?

On the subject of artificial cloning, it is necessary first to be clear about what cloning is in a natural context. In this sense, cloning obtains a biological entity (gene, chromosome, cell, or organism) genetically identical to another by asexual reproduction mechanisms where only one parent is involved (Chuaire et al., 2004). Cloning appeared at the early stages of the evolutionary process, when microorganisms began to produce copies of themselves by mitosis. Thus, a somatic cell replicates, and the genetic material of the daughter cells is identical to that of the mother cell. We could consider mitosis a type of "natural" cloning (Chuaire et al., 2004). In nature, asexual reproduction or clonal reproduction is widespread in "lower" organisms, such as bacteria, protists, and other unicellular organisms (Logares et al., 2009), as well as in multicellular animals, such as sponges, coelenterates, flatworms, annelids, and echinoderms (Ereskovsky & Tokina, 2007; Kostyuchenko et al., 2016; Rubilar et al., 2005). Asexual reproduction only occurs in organisms whose cells retain the property of embryonic totipotency (the ability to multiply and differentiate into different cell types). Indeed, cloning is an intrinsic element of perpetuation in the offspring of more organisms than we initially imagined. Thus, although the primary form of reproduction of eukaryotic organisms is sexual reproduction, many groups also use asexual reproduction as a mechanism for producing offspring. Most fungi can reproduce sexually and asexually (Elliott & Maheshwari, 1994; Taylor et al., 2015). Yeasts, for example, reproduce asexually by budding, and other fungi can reproduce by spore production. Examples of asexual reproduction are also observed in plants, which occurs when a plant part divides and develops separately into a new plant (Cook, 1979; Bengtsson & Ceplitis, 2000). Among the most common forms is the production of propagules and apomixis, among others. Examples are northern highbush blueberry (*Vaccinium corymbosum*), common hazel (*Corylus avellana*), and quaking aspen (*Populus tremuloides*).

Let us look at cloning from a macroevolutionary perspective, i.e., considering the whole tree of life. We could say that in terms of the number of species and branches derived in the evolution of living beings (e.g., toward newer animals and plants), it went from being a rule as such (i.e., reproduction only or mainly by cloning) to an exception (i.e., reproduction only by sexual means). However, if we focus on the microevolutionary level, we see that there is a dual dynamic with sexual reproduction in many groups. For example, some insect species can switch their reproductive mode between sexual and asexual reproduction, depending on seasonal changes in temperature and day length (Razmjou et al., 2010; Peng et al., 2017; Kokko, 2020). These two reproductive patterns in one species have been observed in aphids (Le

Trionnaire et al., 2008; Ollivier et al., 2012; Shuo et al., 2020), termites (Matsuura et al., 2009; Yashiro & Matsuura, 2014), ants (Pearcy et al., 2004; Fournier et al., 2005), bees (Beekman et al., 2011), beetles (Perotti et al., 2016), and locusts (Hamilton, 1953). Another example where this occurs is flatworms (Ramm, 2017). Among vertebrates, we can name whiptail lizards (*Cnemidophorus* sp.), a species of Teidae where asexual reproduction is favored in the peripheral zones of the distribution (Wright & Lowe, 1968).

In contrast to natural cloning, artificial cloning techniques require a manipulation process to obtain copies of the mother cell. Within artificial cloning, we can mention reproductive cloning, which aims to duplicate complete living beings (the embryo is transferred to a uterus for implantation in the endometrium, and thus a new organism develops) (Hochedlinger & Jaenisch, 2003; Rideout et al., 2002). On the other hand, there is therapeutic cloning, which aims to become an alternative for the treatment of some diseases and the replacement of injured tissues or organs (here, the embryo is transferred to a culture medium and will give rise to embryonic stem cells with the potential to differentiate into different cell types) (Keller & Snodgrass, 1999; Cibelli et al., 2001). Several methods have been developed for artificial cloning, including somatic cell nuclear transfer (SCNT) cloning (Pardo, 1997; Keefer, 2015). In this technique, two types of cells are used: somatic and female sex cells. In the procedure, genetic material is extracted from both cell types, and the nucleus of the somatic cell to be cloned (donor) is injected into the female sex cell (recipient) (Gurdon & Byrne, 2003; Li et al., 2003; Hochedlinger & Jaenisch, 2003).

Artificial cloning has been applied without much discussion in plants, which, possibly, being "phenotypically alien" to a human, did not produce ethical and moral controversy. Artificial cloning in plants is highly relevant (Giacometti, 1990). Initially, focused on cultivating flowers, most studies on artificial cloning in plants are currently focused on species with desirable agronomic characteristics, for example, to increase their production. With the application of the latest plant cloning techniques, it is possible to regenerate an entire plant from a single plant cell, tissue, or organ (González Jiménez, 1998; Torres et al., 2000).

The panorama of artificial cloning changes drastically when talking about animals, where numerous preliminary tests, and failed experiments, to others close to its realization, such as the pioneering work of J. Gurdon in the early 1960s when he successfully cloned the frog *Xenopus laevis* (Elsdale et al., 1960; Gurdon, 1996; Gurdon, 2006). We then have the iconic example of "Dolly" the sheep, which represented the first cloned mammal; more precisely, in July 1996, it was a sheep of the Finn Dorsert breed, with many previous failed attempts (a rarity in a pool of 278 attempts in total 1/277) (Campbell et al., 1996). Other examples of artificial cloning in vertebrates exist in monkeys of the genus *Macaca* (Meng et al., 1997; Liu et al., 2018), domestic cats (Shin et al., 2002; Gómez et al., 2008), and different species of goats, sheep, and cattle (to improve their productivity) (Smith et al., 2000; Kühhölzer et al., 2001; De Sousa et al., 2002). The artificial cloning of equines also has a tremendous economic impact since these animals are used for sports such as horseback riding and polo.

There are many advantages to be gained from artificial cloning, such as the cloning of endangered animal species, animals with important livestock characteristics (e.g., more productive cows), or genetic manipulation to produce transgenic animals that produce substances used in the pharmacological and sanitary fields. However, despite the advantages of artificial cloning, there are also ethical objections. The main ones refer to the fact that it could endanger the biodiversity of the animal world and the ecological balance itself since it could lead to a decrease in the genetic variability of species, with irreversible consequences for the ecosystem (Tsunoda & Kato, 2002; Wells, 2005). In addition, many cloned embryos die after implantation (Rideout III et al., 2001; Hochedlinger & Jaenisch, 2002), and if they survive, they show more significant than average body size (Hill et al., 2000; Humpherys et al., 2001; Rideout III et al., 2001), obesity (Tamashiro et al., 2003), and defects in the kidneys, heart, liver, lungs, and brain (Cibelli et al., 2002). As already mentioned, the cloning of the famous Dolly sheep (Campbell et al., 1996) resulted in an apparently healthy individual. However, one of the most significant concerns raised was that Dolly seemed to age at a faster rate compared to a typical sheep. Additionally, the cloning process needed a total of 277 unsuccessful embryo attempts before achieving a successful single birth.

Furthermore, artificial cloning has triggered numerous discussions as the potential for its future application in humans became evident. Ian Wilmut, the leader of the team that created Dolly, suggested that this scientific advance could serve as a foundation for combating diseases, marking a potential starting point in the pursuit of disease eradication. As previously mentioned, therapeutic cloning offers the potential for creating tissues or organs that can be used for transplantation, presenting an efficient alternative for treating specific pathologies. One significant advantage of this technique is the avoidance of immunological rejection (Pardo, 1997; Singec et al., 2007; Brown, 2009). We will return to this subject in Chap. 8 of this book. The researcher also highlighted the usefulness of dealing with infertility problems (e.g., through reproductive cloning, which is the most questioned), something that was a rarity, an impossibility, at the time and is now a feasible option. Undoubtedly, the prospects for nuclear transfer involving the artificial cloning process are manifold, and the potential is yet to be exploited (Colman & Kind, 2000; Dinnyes & Szmolenszky, 2005).

In this context, the possibility of cloning human beings has given rise to an intense ethical and political debate on handling reproductive cloning. It is argued that cloned individuals will not benefit from possessing a unique and unrepeatable identity (Bellver, 2002). Legislation varies widely among countries and tends to distinguish embryo cloning for therapeutic purposes from reproductive cloning, which aims to create complete human beings. The latter is legally prohibited in many countries. The Catholic Church has also spoken out against human artificial cloning techniques, arguing that a few individuals would appropriate the right to create other individuals by manipulating their biological identity (Nabavizadeh et al., 2016). In recent years, a misconception about artificial cloning has been gaining popularity, fueled by the media and cinema. This misconception suggests that a clone is an exact replica of another individual. However, clones defined in such a way do not exist, as an individual's phenotype is influenced not only by their genotype but also by epigenetic and developmental mechanisms involving other factors. Consequently, much of the societal rejection of artificial cloning lacks a solid fundamental basis (Kahn, 1997; Shapiro, 1997; Gracía, 1999).

From all of the above and within the framework of our book, it is clear that artificial cloning was a total "rarity" in its beginnings, something even unthinkable but which has been progressively advancing toward the transition from exception to rule. Are we in an "intermediate zone" from rarity to the pattern? In reality, it will continue to be considered a rarity or a pattern depending on what it is compared against. Generally, seeing artificial cloning as a rarity is already quite outdated. The applications of genetics in various fields, as already mentioned, invite us once again to reflect on the fact that everything rare can become a pattern or at least an "intermediate" but essential option and that it has both beneficial and harmful consequences that still need to be evaluated in depth. In this transition, it is expected that the techniques will improve over time and that the success rate will be increasingly higher, allowing artificial cloning processes to become more frequent and safer. In conclusion, artificial cloning is a highly illustrative case of the change from exception to rule by human action, and where advances in science and technology clearly show us that reality will surpass (or has already surpassed?) what seemed pure fiction until a couple of years ago.

6.3.4.3 Synthetic Biology: On the Way to Becoming Something Common?

Connected to artificial selection and cloning, we can also incorporate the field of synthetic biology into the framework of artificial exception-to-rule dynamism. Synthetic biology is defined as the application of science, technology, and engineering to the chemical synthesis of biological components and the design and use of those synthetic components as interchangeable parts to produce new organisms designed to perform specific functions (O'Malley et al., 2008; El Karoui et al., 2019). As a rule, synthetic biology uses artificially constructed parts such as amino acids, ribosomes, synthetic DNA sequences, artificial cells, or biological systems that are entirely alternative to those found in nature (Arkin, 2008; Cameron et al., 2014). Thus, by adding or deleting specific genes, synthetic biology produces genetically modified organisms (Keasling, 2007; O'Malley et al., 2008; Cameron et al., 2014). The basis of synthetic biology are biobricks, synthetic DNA fragments that code for a biological characteristic and can be combined to form complex circuits. Biobricks form the basic standard and modular unit that performs a simple function and allows a cell's behavior to be modified (Endy et al., 2010; de Lorenzo & Schmidt, 2017). In this context, the focus is on creating, modifying, or evolving biological systems with the purpose of controlling and utilizing them for direct biotechnological applications to meet human needs (Endy, 2005; Ruiz Mirazo & Moreno Bergareche, 2012).

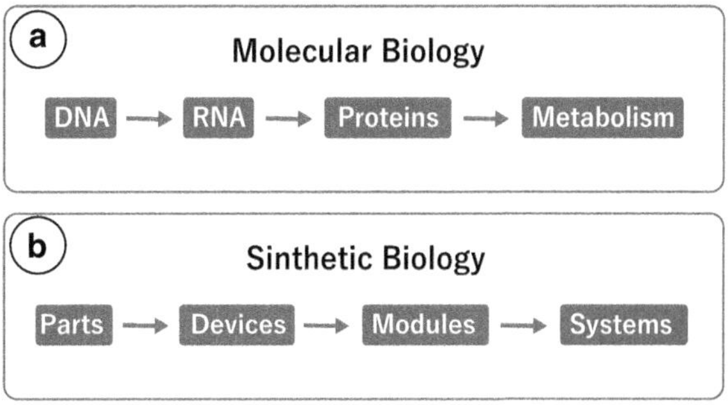

Fig. 6.4 Different approaches from molecular biology (**a**) and synthetic biology (**b**) (Modified from Lorenzo, 2014)

If we pause for a moment, we see that comparing synthetic and traditional biology may be interesting. As we said in Chap. 1 (Sect. 1.2.1), traditional biology (here, we will call it traditional to differentiate it without confusion from synthetic biology) has three fundamental laws, laws that undoubtedly must be fulfilled and that do not accept exceptions. These are the laws of thermodynamics (laws of physics and chemistry), systems defined by membranes, and living beings subject to a constant process of biological evolution. In this sense, although synthetic biology is a discipline that does not focus on the study of living beings but, as we have already said, has as its objective the study of systems that do not exist in nature, we can find similarities and differences between them. Thus, synthetic biology transcends manipulation and brings to the table some of the laws of traditional biology since processes of reproduction and evolution are also involved in biological systems. However, this evolution may take different paths from those that nature would take. Thus, a distinguishing feature of synthetic biology from traditional biology is its ability to alter the basic organizational structure and processes inherent to life (Ruiz Mirazo & Moreno Bergareche, 2012).

From a synthetic biology perspective, two main aspects come into play. The first aspect involves design, which is accomplished through layers of increasing complexity, encompassing the construction of parts, devices, modules, and systems (Endy, 2005; de Lorenzo & Schmidt, 2017). The second aspect is the flow of information where one of the principal dogmas of molecular biology, in which information flows from DNA to RNA, proteins, and then different specific functions, is now replaced by the compositional and quantitative principle of synthetic biology in which biological parts give rise to devices and these to modules and systems (Fig. 6.4a, b) (de Lorenzo, 2014; de Lorenzo & Schmidt, 2017; de Lorenzo, 2020).

In synthetic biology, a fundamental premise is to standardize biological components in a manner that separates them from their natural context (de Lorenzo, 2020). The behavior of these components becomes a crucial requirement for engineering new devices, regardless of the environment in which they operate. This is due to the intrinsic ability of biological functions to evolve together as a whole and thus behave in a highly context-dependent manner (de Lorenzo & Danchin, 2008; de Lorenzo,

2020). Furthermore, by way of rules, synthetic biological systems must meet particular characteristics to be considered as such, for example: (a) be computationally predictable, (b) measurable, (c) controllable, and (d) transformable (add and regulate previous functions) (Lee et al., 2013; Li et al., 2015; Muñoz-Miranda et al., 2019). Additionally, three stages are proposed for the construction of a synthetic biological system: (1) decoding and analysis of the genomes of biological systems, (2) synthesis of the parts that make up the genome and their assembly, and (3) the use of genome editing tools such as CRISPR–Cas9, which has been key to synthetic biology (Lee et al., 2013; Doudna & Charpentier, 2014). Other techniques used by synthetic biology include genetic code expansion, minimal genome, sequence assembly, and synthetic messenger RNA, among others (Lammoglia et al., 2016; Adli, 2018; Charlesworth et al., 2019).

The origin of this discipline dates back to the 1970s. Throughout its history, synthetic biology has gone through three stages. Therefore, its character of total rarity can be appreciated in its beginnings and then gradually assimilated as an increasingly predominant field. The first stage was the era of genetic engineering or biotechnology, which modified the genome of microorganisms (for example, the genome of *Escherichia coli* was altered to produce insulin). The second was the elaboration and development of synthetic genomics associated with producing drugs, biofuels, and genetically modified foods. Moreover, the third phase, which we are currently in, aims to achieve the complete synthesis of genomes (Alonso, 2013; Garro, 2016). In 2010, Venter and collaborators developed a new bacterium to self-replicate. This was the first time a cell's genetic material was replaced entirely by synthetic genes. This synthetic genome dictated the instructions for the bacterial cell's activity, growth, and division. With this new field of research, two different implementation strategies are proposed. From a more extreme perspective, synthetic biology aims to control and direct life, redesigning it without further contemplation (Endy, 2005; Endy et al., 2010). On the other hand, a more subtle implementation is proposed, taking advantage in some way of the properties that we recognize as inherent to living beings (Chen et al., 2004; Mansy et al., 2008).

We now focus on the applications that synthetic biology has or may have in the not-so-distant future. In that case, human society demands many products and resources to satisfy needs. These are constantly increasing due to the growing human population. It is then that synthetic biology could have a prominent role (Serrano, 2007; Muñoz-Miranda et al., 2019).

Some Fields where Synthetic Biology Could Become Very Common

Let us start by listing the fields and areas where synthetic biology can have significant participation and stop being a mere curiosity, a rarity. Medicine will be one of the fields that will benefit the most from advances in synthetic biology, where different projects are being developed in numerous areas, and even others are already being implemented. For example, (1) creating artificial cells called CARs (CAR T cells). Such cells can be detected through biosynthetic sensors, activate or inhibit themselves, and attack cancer cells without attacking healthy cells and thus decrease, at least in part, the adverse effects that current techniques have on patients (June et al., 2018; Chandran & Klebanoff, 2019; Ventura González et al., 2020). (2) Tissue repair is based on the design of molecular systems that can recognize tissue damage. (3) Cellular reprogramming, where stem cells are altered to have new characteristics. This therapy could significantly impact the immune system (Satoshi et al., 2018; El Karoui et al., 2019). (4) 3D bioprinting of organs and tissues. This method will address the significant lack of organ donors and better understand certain diseases. Tests and prototypes of different organs, such as heart, kidney, liver, corneas (to prevent blindness), and ovaries (against infertility), among others, have already begun (Murphy & Atala, 2014; Zhang & Zhang, 2015; Mandrycky et al., 2016). Amazingly, researchers in Germany have succeeded in creating transparent organs, making them possible to study. The SHANEL technique, described in the journal Cell, makes organs transparent and makes it possible to identify the different cell types to map them within a 3D representation of the organ (Zhao et al., 2020). Technology is significant in studying diseases since it allows a better understanding of their anatomy. (5) In addition, optimizing the production of antibodies or vaccines, to be able to reduce the cost and increase the speed of production, and give greater ease of application and distribution. (6) Genetic engineering of pigs to be resistant to viruses and have immune profiles similar to humans to allow the development of xenotransplantation (transplantation of cells, tissues or organs from one species to another, it is sought that the species are close to avoid rejection) and thus cover the significant lack of organs for transplantation (Burkard et al., 2018). (7) Synthetic constructs can also be designed to limit the spread of infections by attacking the vectors of certain viruses as in the case of mosquitoes (Ruder et al., 2011).

Another field that could benefit significantly from synthetic biology is drug production, where it will be possible to develop specific drugs. In this way, it will no longer be a rarity, an exception, for drugs to be tailored to each patient's specific, finely personalized needs (see also Chap. 8). For example, synthetic antimalarial drugs and malaria vaccines are under development (Vennerstrom et al., 2004; Keasling, 2007). The food industry is another field where great benefits could be obtained by implementing synthetic biology. This is a field in which an increasing number of innovations and methods are being sought through which the processes of obtaining different foods can be optimized, both in quantity and quality (Tyagi et al., 2014; Jullesson et al., 2015). On the other hand, synthetic biology offers the opportunity to create responsive materials that could be particularly useful for improving building materials (Le Feuvre & Scrutton, 2018).

Finally, bioremediation uses bacteria and fungi to remove toxic or noxious compounds from the environment. For example, molecular techniques targeting specific genes involved in key metabolic processes have been employed to study methanotrophs, which are microorganisms crucial in reducing methane emissions (a greenhouse-effect gas) from soil and sediment (Strong et al., 2015). These approaches have successfully revealed novel methanotrophs capable of actively degrading methane under environmental conditions. Additionally, methanotrophs possess the ability to cometabolize trichloroethylene (TCE), a common groundwater contaminant (Guo et al., 2001). Methane injection is often used to stimulate the TCE-degrading activity of indigenous methanotrophs in TCE bioremediation.

In other studies, in marine petroleum degradation, molecular ecological approaches have been utilized to study bacterial populations in petroleum-contaminated marine environments (Shi et al., 2019). Research has demonstrated that specific bacterial groups, such as those associated with α-Proteobacteria and γ-Proteobacteria, are frequently present in oil-contaminated marine environments and likely play a significant role in the bioremediation of spilled oil (Chang et al., 2000). Moreover, studies have highlighted the accelerated growth of bacterial populations belonging to the genus *Alcanivorax* when seawater is supplemented with petroleum and inorganic fertilizers. These findings suggest that certain bacterial populations hold potential for effective bioremediation of oil spills in marine environments. Indeed, numerous relevant examples (Watanabe, 2001; He et al., 2016) highlight the significance of these microorganisms, which are expected to play a crucial role in the biological bioremediation of our planet.

Synthetic Biodiversity Conservation

Despite decades of conservation effort, current indicators show that it has not been possible to reduce the rate of biodiversity loss (Butchart et al., 2010; Tittensor et al., 2014; Corlett, 2017; Piaggio et al., 2017), even with the increase in terrestrial and marine protected areas (Butchart et al., 2010; Butchart et al., 2015). Furthermore, several factors contribute to this decline: the negative impact of invasive species spreading deleterious genes within their populations (Webber et al., 2015; Faber et al., 2021), the prevalence of infectious diseases, the effects of climate change, and pollution (Redford et al., 2021; Hoegh-Guldberg et al., 2008), all collectively contributing to this troubling decrease. As a result, it is becoming increasingly apparent—an idea that was once inconceivable—that the field of conservation requires close collaboration between conservation biologists and synthetic biologists. This joint effort is essential to apply new techniques aimed at preventing biodiversity decline (Piaggio et al., 2017; Redford et al., 2013; Johnson et al., 2016). On another note, certain aspects related to reintroducing nearly extinct species (Seddon et al., 2014; Keefer, 2015; Saragusty et al., 2016) and utilizing gene editing to enhance resistance against diseases and the effects of climate change, such as higher temperatures or droughts, hold potential for safeguarding endangered plant and animal species. The implementation of synthetic biology techniques (Hoegh-Guldberg et al., 2008; Corlett, 2017) could enable these advancements to become more prevalent in the near future, possibly becoming standard practices.

Ethical Aspects

Because of the relevance of the multiple elements mentioned above that make up synthetic biology, it is important to highlight other disciplines that involve ethical, legal, philosophical, and economic aspects (Sandler, 2020). Synthetic biology is the subject of great debate worldwide due to the impact of its different applications. Its principles are too challenging for our conception of what is or is not the norm. Synthetic biology is an example of dual-use technology that promises numerous beneficial applications but can also generate significant detriments (El Karoui et al., 2019). Therefore, it is perceived both positively and negatively in society, which is naturally expected when faced with something unknown, and these new techniques being developed are no exception. The concepts of nature, life, and biodiversity can be significantly affected by genetic modification of individuals and the exploitation of genetic resources. Without ensuring fair sharing of benefits, this could exacerbate inequality in resource access, among other potential consequences. However, it is undeniable that genetic modification holds great potential, particularly in the field of medicine, where it may encounter less resistance (Sha et al., 2017; Muñoz-Miranda et al., 2019).

In this sense, there is an exponential annual increase in projects and patents linked to synthetic biology. When observed by area of application, the primary patents are for the medical, energy, and industrial sectors (Van Doren et al., 2013; Carlson, 2016). The countries that are leading patent applicants are the United States and Japan. Synthetic biology is a tool with high expectations and socioeconomic potential. Its field of action will likely grow exponentially in the next decade based on the new developments, patents, and industries being created and developed (Muñoz-Miranda et al., 2019).

In conclusion, synthetic biology has evolved into an essentially interdisciplinary science, which already has its principles and guidelines, under which it studies the interactions of multiple components, and which promises to be of great help in various fields where the main objective is to increase or improve human needs (Liu et al., 2013). Although for the moment, humanity still sees synthetic biology as a rarity or curiosity compared to traditional biology, it is unquestionable that it is advancing at an incredible speed and that it is progressively being used more frequently in various fields, where the application of its innovative techniques can provide more significant benefits. It is not venturesome to foresee then that, on a par with traditional biology (or perhaps above it in some cases?), synthetic biology will become something regular, even indispensable, in our daily lives. All the applications we have mentioned were unthinkable or barely dreamed in the past. They were, in any case, rarities for traditional biology but also for synthetic biology itself in its beginnings. However, we have observed that synthetic biology holds immense potential for change, and the term "rare or exceptional" no longer carries the same connotation as it does in traditional biology. In other words, because of its dynamism, almost anything seems possible for synthetic biology (if it cannot/is not now, it can/will be sooner or later). Definitely, weirdness as a synonym for "watertight" does not seem to be in their lexicon.

6.4 E-R Changes Due to Modifications in Scientific Interpretations

The history of science is an essential field where changes from exception to rule have continuously occurred and continue to do so. That is, how our perceptions of the surrounding world have been changing according to changes in the knowledge of biodiversity and technological advances applied to research. Our vision of the biological sciences is dynamic, which is the basis for permanently modifying what we consider a rarity or a pattern. Depending on the area of biology, what we now consider a rule, the typical pattern differs from what was considered a hundred years ago or even just a couple of years ago. Technological advances and the models of organisms used, new statistical and analytical models or newly acquired data can generate a bias in the observations and results, and above all, generalizations must be cautious.

6.4.1 Key Factors Associated with Change

These changes in the history of biology respond to different factors that generally work in combination. We could highlight the following:

6.4.1.1 Scientific and Sociocultural School Biases

Scientific trends have always been closely linked to the prevailing sociocultural context. Perhaps the most recognized case is how the social context and the nineteenth-century market and competition theories from Malthus served as inspiration for the concepts of competition and struggle for survival in Darwin's Theory of Natural Selection (Malthus, 1969; Herbert, 1971; Bowler, 1976). Among other aspects, in this context, sociocultural conditioning factors likely hindered the consideration of learning capacity and the presence of intelligence and emotions in "lower animals." Moreover, it has been much more challenging to consider plants beyond their decorative or complementary roles in an ecosystem. Instead of seeing them merely as simple primary producers, a more comprehensive and enriched perspective is needed. Additionally, the female role in sexual selection processes and mechanisms was accepted only a couple of decades ago (Smith, 1991; Andersson & Iwasa, 1996; Jennions & Kokko, 2010; Hosken & House, 2011; Ah-king, 2023). In short, a mixture of anthropocentrism and sociocultural bias has prevailed. It has prevented us from admitting that living beings exhibit characteristics previously limited to the human species and, even more, from relegating the female role as an evolutionary engine.

Moreover, science has tended to be reductionist in some arguments. For example, there has been a predominant tendency toward a unidirectional vision of phenomena as a linear cause–effect when many processes can feed back and multiple variables and factors can participate. We have an example of how epigenetics modified the unidirectional view of the mechanism of genome functioning (see below). Another case is the discovery of "flexibility" at the hormonal level, which is connected with flexibility at the level of animal behavior and, thus, the ability to be adaptable to new environments. In other words, according to Taff and Vitousek's (2016) review, flexibility at the endocrine level in a world that is dynamic. Thus, such a unidirectional view of nature becomes obsolete.

6.4.1.2 Knowledge of Diversity and Evolutionary Framework

Within the context of changes due to modifications in scientific visions or perceptions, the knowledge of biodiversity and its understanding within an evolutionary framework has undoubtedly played a preponderant role. The discovery over the centuries of organisms, the understanding of how they function, and the existence of an evolutionary theory to unite them were crucial for changes in what was considered as rules at the time. The iconic case of considering some invertebrates, such as Porifera and cnidarians, as plants in the early days of biological sciences, particularly during the Aristotelian era, can be attributed to a lack of scientific knowledge at that time (Ruppert & Barnes, 1996; Bavestrello & Sarà, 2009). The prevailing "rule" defining "being an animal" was movement, leading to the misconception that organisms with somewhat greenish colors and fixed to the substrate could only be plants. Furthermore, it is worth noting that before the theory of natural and sexual selection emerged, understanding apparent aspects such as the sexual dimorphism of traits, as well as many morphophysiological adaptations of organisms to their environment, was not possible. Prior to the advent of Darwin's ideas, in many cases when the sexes were highly dimorphic due to differential pressures of sexual selection, scientists proceeded, as we saw previously, to classify them as two distinct species (Kelley, 1988; Delph, 1999; Mori et al., 2017). During those periods in the history of biology, a significant mixture of scientific (or "proto-scientific") interpretations with obscurantism and religious interpretations played a crucial role, hindering scientific progress (Barbour, 1990; Evans & Evans, 2008; Rowbury, 2009).

6.4.1.3 Related to Current Equipment and Methodologies

Few data, insufficient technology, and a strong influence of a paradigm may lead to the omission of elements that could indicate that observed patterns do not represent rare phenomena. These factors also apply to the inverse situation (shift "from exception to rule"). Indeed, before the invention of key instruments that significantly advanced scientific knowledge and underwent further improvements, such as the telescope (from the initial eyepiece to the current sophisticated versions) and the microscope (starting from optical and then advancing to electronic ones), what was once considered the norm or standard quickly gave way to what could be perceived as an exception in its early stages. From the change in the view of the universe, our solar system in particular, to the understanding of microscopic life and its components, everything implied, explicitly or implicitly, constant changes in what was or was not "the rare vs. the common." In this context, there are also improvements in histochemical techniques in biochemistry, including DNA and RNA in particular. We also have an improvement in behavioral filming techniques and techniques for ecological and climatic studies. Undoubtedly, behind every technology, no matter how small it may seem, there was a part of the history of biological sciences linked to its progress. Mario Bunge's phrase summarizes this scenario very well: "*Whereas science elicits changes in order to know, technology knows in order to elicit changes*" (Bunge, 1979). As far as our book is concerned, we do not intend to bombard you with an exhaustive analysis of all the cases where this occurred; we leave that for a treatise on the history of science from its origins. Within this context, it is essential to emphasize recent examples of modifications. This serves as a reminder that current scientific interpretations will inevitably undergo changes in many cases. Being aware of these shifts helps us realize that our present understanding may evolve over time.

6.4.2 Some Case Histories of E - R Change by Scientific Vision

6.4.2.1 Learning and Animal Personality

Among the various topics where our interpretations have changed drastically in recent years, we can mention the field of study of animal learning and personality, both in general and, more specifically, concerning groups that were underestimated in the past, such as invertebrates. Let us first analyze what refers to learning, reviewing some concepts we will use later in the examples.

Animal Learning

Today, we know that the ability to learn is not a process exclusive to humans, and at this point, a paradigm shift occurred, which we will refer to throughout this section. Learning itself is an adaptive process. It is acquired through subtle and incremental changes (Moreno, 1994). Learning allows animals to adjust their behaviors according to the changes they detect in nature, a clear example being flight response in the face of the sudden appearance of a predator. In this sense, the learning of individuals must be plastic. By this, we mean that it must allow individuals to modify and adapt their behaviors and skills to changing conditions, i.e., the original learning must be adapted to the new conditions to which the individuals are exposed. The types of learning of a species can range from very elementary to very complex processes (Aguado-Aguilar, 2001). This variation in the complexity of learning processes will depend to a great extent on the development of the nervous system of organisms, and here, it was essential to have a better knowledge of the existing biodiversity since it allowed us to understand the variation in the development and complexity of the nervous system on an evolutionary scale.

Learning has been classified into nonassociative and associative. Nonassociative learning is considered the simplest and refers to changes in the response of individuals to repeated exposure to the same stimulus (Perry et al., 2013). Nonassociative learning comprises habituation and sensitization. Habituation involves a decrease in response to repeated stimuli that are not associated with a reward, while sensitization involves an increase in response to repeated stimuli (Moreno, 1994; Perry et al., 2013).

On the other hand, in associative learning, the individual learns that the response to a stimulus is associated with a given event (Perry et al., 2013). Here, it is critical to highlight the influential contribution of Ivan Pavlov through what was called classical conditioning (Pavlov, 1927; Perry et al., 2013). Pavlov was a great revolutionary in formulating the theory of the conditioned reflex, or classical conditioning, a significant step in understanding animal learning. In his experiments, he discovered that dogs' salivary and gastric secretions could be stimulated not only by food but also by images and sounds regularly presented to them before feeding as a result of associative learning (Castro & Wasserman, 2010). Another researcher who brought about a change in our conception of the learning capacity of animals was Edward Thorndike. He recorded the time it took for an animal to escape from a box and used dogs and cats as experimental models. Thorndike found that the time it took the animals to get out of the box decreased as a function of the repeated times they were subjected to this experience (Castro & Wasserman, 2010; Dorey, 2020).

Based on these experiments, the Law of Effect was developed, where if a positive outcome follows a behavior, then it is more likely to occur; however, if a negative response follows a behavior, then it is less likely to occur again (Castro & Wasserman, 2010; Dorey, 2020). To be sure, discoveries of this type in the sociocultural context of the time were revolutionary since it was rare to demonstrate with clear evidence that animals could learn and respond to certain stimuli for their benefit. A central point to highlight is that when in the past, reference was made to learning processes, it was common to think only of humans and animals such as rats (Santín et al., 2000; Vicens et al., 2003; Honey et al., 2014), monkeys (Waelti et al., 2001), and dogs (Dorey, 2020), where learning processes are fundamental for behaviors associated, for example, with foraging, reproductive behaviors, offspring care or escape (Kessels et al., 2001; Vicens et al., 2003). However, we did not even consider the idea that animals, wrongly called "inferior," could learn. Here, a paradigm shift occurred, accepting that animals with a less developed nervous system could also learn.

The implementation of new technologies was significant in this paradigm shift. Until not so long ago, descriptions of the behavior of individuals, associated with learning as with many other processes, were qualitative descriptions, in many cases with great detail, by the way. However, implementing new technologies and new forms of analysis allowed the study of behaviors to be a quantitative process, i.e., in a much more "strict" way. Regarding new technologies for behavior evaluation, the use of cameras and high-speed video cameras has proven valuable. They enable us to observe behaviors in slow motion, facilitating the detection of actions that occur at high speeds, which are not easily detectable by the naked eye (Dell et al., 2014; Egnor & Branson, 2016); development of computerized programs for the analysis of behaviors, which allow us to detail not only which behaviors individuals show in a given situation but also their frequency, duration, chronology with respect to other behaviors, making diagrams of relationship between behaviors, among others (Kyzar et al., 2012; Jones et al., 2016); technologies for bioacoustic studies, such as the use of microphones to analyze the behaviors of certain species remotely (Liedvogel et al., 2013); microclimatic sensors to estimate temperature and humidity; the use of radiofrequency techniques to monitor certain species such as pollinators (Barlow & O'Neill, 2020); and simple or very sophisticated statistical analyses depending on the need for the data obtained. These are just a few examples of the new technologies that have been implemented in recent years and have greatly benefited countless investigations, making it possible to change our interpretations and thus consider that numerous processes, such as learning and personality in invertebrates, are no longer rarities but on the contrary processes present in numerous species.

We will now present examples to illustrate that species of "lower" animals are no longer only exceptions where some form of learning has been demonstrated to exist. Indeed, both associative and nonassociative learning have been demonstrated in many invertebrates (Perry et al., 2013). Although they have simpler nervous systems than vertebrates, they possess a great variety in complexity, ranging from simple nervous networks to more complex cephalized systems (Giurfa, 2013; Perry et al., 2013). For several years, specific invertebrate models, with insects historically being prominent, have become crucial for studying learning. These models offer easier experimental conditions, making them valuable in research (Giurfa & Sandoz, 2012). The great use of insect species in studies on learning and memory is based on linking associative learning processes to specific neural circuits (Giurfa, 2015).

In the insect world, K. Takeda pioneered the study of learning ability, using the common honeybee, *Apis mellifera*, as a model. He developed a specific guideline called the olfactory conditioning protocol for the proboscis extension response (PER) (Takeda, 1961; Bitterman et al., 1983). With this protocol, he proposed evaluating learning and memory in studying olfactory discrimination at different levels of analysis ranging from molecular to behavioral (Takeda, 1961; Menzel, 1999). Thus, the foundations were laid for what would later become one of the most widely used tools for studying learning and memory in invertebrates (Bitterman et al., 1983; Giurfa & Sandoz, 2012). The original protocol proposed that the bees were first given an odor stimulus in the conditioning tests. After a few seconds, the antennae were allowed to touch a sugar solution, which caused the proboscis to extend, and the bee could lick the solution. This protocol was intended to demonstrate that bees learned to associate the olfactory stimulus with feeding (Takeda, 1961; Bitterman et al., 1983; Menzel, 1999).

Most behavioral biologists have historically underestimated the learning ability of arthropods, except for a few cases (e.g., Thorpe, 1943; Monteith, 1963). However, several studies have shown that crustaceans, spiders, and insects can modify their behavior based on previous experiences (e.g., Reaka, 1980; Dukas, 1998, 2008). The influence of experience is not limited to simple tasks but also involves complex behaviors, such as patterns of elaborate predatory and sexual behavior (e.g., Cronin et al., 2006; Dukas et al., 2006; Jakob et al., 2011). Species of ants, butterflies, and spiders have shown learning based on associations between colors and food presence (Perry et al., 2013). *Drosophila melanogaster* can be visually conditioned to modify its flight behavior in a simulator to avoid heat shock (Brembs & Heisenberg, 2000; Giurfa & Sandoz, 2012; Perry et al., 2013).

The common honeybee was also used as a model for the study of classical conditioning (Hammer, 1997; Menzel, 2001; Giurfa, 2007), whereas already mentioned, proboscis extension behavior is a function of olfactory conditioning (Bitterman et al., 1983; Giurfa & Sandoz, 2012). There are also analyses, with a focus on associative learning, to determine how bees learn different floral odors (Wright & Schiestl, 2009; Reinhard et al., 2010; Giurfa & Sandoz, 2012) and learning behaviors where bees show an association between thermal stimuli and feeding (Menzel et al., 2001; Hammer et al., 2009).

In addition, the classical conditioning process of the proboscis extension response was also applied to blowfly species, flies, and bumblebees (McGuire, 1984; Fresquet, 1999; Laloi et al., 1999). Another example is crayfish; when threatened, they produce flexions of the abdomen to move away from danger, and it can be said that they are under a habituation-type learning process since the probability of escape decreases with repeated stimulation (Krasne & Wine, 1987; Krasne & Teshiba, 1994).

Learning about invertebrates is not limited to the arthropod world. Perry et al. (2013) show a wide variety of different types of learning in nonarthropod invertebrates and protists. For example, among the primitive organisms where the existence of learning has been experimentally proven, we can name the amoeba *Physarum polycephalum*. They found here that after successive exposures, this species of amoeba stops avoiding certain substances it has repeatedly encountered; this is linked to the habituation type of learning (Pershin et al., 2009). This amoeba also overcame movement tests associated with changes in temperature and humidity (Pershin et al., 2009). On the other hand, rotifers, microscopic invertebrates containing only approximately 200 neurons, showed that their contractile capacity decreases in the face of repeated tactile stimuli over time (Applewhite, 1968). Ascidian larvae that possess only approximately 100 neurons (Meinertzhagen & Okamura, 2001) also show forms of nonassociative learning. Larvae modify their swimming behavior according to repeated light exposures. The behaviors showed habituation or sensitization depending on the light intensity to which they were subjected (Perry et al., 2013). Studies in flatworms and cnidarians, groups containing elementary nervous systems, have also demonstrated the existence of a type of classical conditioning. For example, flatworms contract the body upon exposure to light (Thompson & McConnell, 1955), and cnidarians can respond to olfactory and touch stimuli (Perry et al., 2013). In nematodes, it was shown within a now-famous model, *Caenorhabditis elegans*, that it can learn and remember stimuli it encounters around it (Rankin, 2004). It has also been shown that this nematode can learn based on mechanosensory (Rose & Rankin, 2001) and thermosensory stimuli (Mori, 1999).

One invertebrate group to highlight in learning is the cephalopods, which include octopuses, cuttlefish, and squid. These specialized mollusks form a fascinating group in which a significant development of cognitive capacity has been observed, including processes such as learning and memory (Schnell et al., 2021). Cephalopods show different forms of learning, mainly associative (Darmaillacq et al., 2004, 2014; Cole & Adamo, 2005). For example, the octopus *Octopus cyanea* can learn to associate a neutral stimulus, such as a sound, with a reward, such as obtaining food (Papini & Bitterman, 1991). The octopus *Amphioctopus marginatus* surrounds itself for protection, with coconut shells as a shield, a behavior that individuals are thought to associate with a reduced risk of predation (Finn et al., 2009). In addition, cephalopods possess spatial learning, such as exploratory learning (Karson et al., 2003; Jozet-Alves et al., 2008). On the other hand, in cognitive processes linked to memory, studies show that cephalopods have short- and long-term memory (Jozet-Alves et al., 2013; Zepeda et al., 2017). Cuttlefish, for example, can remember what they ate and where they consumed it. These processes allow them to undoubtedly optimize foraging behavior (Jozet-Alves et al., 2013; Schnell et al., 2021). In recent years, it has been suggested that various evolutionary pressures, such as predation, foraging, and mating competition, have significantly influenced the development of cephalopod cognitive processes that facilitate behavioral flexibility (Schnell & Clayton, 2019; Amodio et al., 2019). Undoubtedly, the most salient and recognized example is the behavioral flexibility of cephalopods in predation avoidance behaviors, where they can change their body coloration in mere seconds. This behavior has been associated with associative learning based on the individual's previous experience (Hough et al., 2016; Schnell et al., 2021).

Animal Personality

Linked to learning processes, we can refer to animal personality (Sih et al., 2004a, 2004b; Schuett et al., 2010), which is currently not considered a pattern, a rule, but no longer appears as just exceptions in the literature. The concept of personality refers to the fact that individuals in a population may show differences in behavior that are maintained over time and in different situations (Réale et al., 2007). Although psychologists have paid particular attention to the study of personality in both humans and animals for many years (Wilson, 1994; Gosling, 2001), it is only recently that animal personality has been included as a topic of study by ecologists and evolutionary biologists (Sih et al., 2004a, 2004b; Dall et al., 2004). We now have a large body of evidence that certain personality traits can affect essential processes such as dispersal, colonization, and reproduction, among others (Réale et al., 2007; Dingemanse et al., 2010; Stamps et al., 2012; Keiser et al., 2018). Personality in animals involves different traits that have been classified into shyness-boldness (reaction of individuals to nonnew risk situations), exploration (reaction of an individual to a new situation), activity (general level of activity of an individual), sociability (reaction to the presence or absence of conspecifics), and aggressiveness (agonistic reaction to conspecifics) (Réale et al., 2007).

When we estimate the repetitiveness of a trait, we consider that the environment is kept constant. This is to avoid any new variable influencing the analysis of the repetitiveness of the behavior we are evaluating. However, even if we achieve this, learning processes inherent to each individual, such as habituation or sensitization of the ones we talked about before, could affect the performance of the individuals (Réale et al., 2007). This idea that animals can have a determined personality changes the traditional conception where it was considered that animals could be plastic and consistently exhibit the optimum of their behaviors according to the different situations to which they were exposed (Sih et al., 2004a). Thus, although an individual will not show the same behaviors in successive events, intraindividual variations are expected to be smaller than interindividual variations (Sih et al., 2004a, 2004b). One hypothesis posits that individuals may show differences in certain behavioral traits because, for example, they may differ in their physiological and morphological characteristics (Stamps, 2007). Furthermore, it has been proposed that personality traits are genetically inherited (van Oers & Sinn, 2011). However, environmental factors, such as temperature and food availability, may also influence individuals' personalities. Undoubtedly, this field awaits further study (Groothuis & Trillmich, 2011; Kralj-Fišer & Schuett, 2014).

We widely accept that humans have different personality traits. For example, some are shyer than others, some are more active than others, and some are even more aggressive than others (Sih et al., 2004a, 2004b). As far as animals are concerned, studies of differences in personality traits have also been conducted in a wide variety of vertebrates, including mammals (Koolhaas et al., 1999; Gosling, 2001), birds (Drent et al., 2003), reptiles (Stapley, 2006), amphibians (Sih et al., 2003), and fish (Sih et al., 2004a, 2004b). For example, in the bird *Parus major*, differences are observed in behavioral traits associated with exploration, where some individuals are much more active than others, and it has been shown, thanks to artificial selection (see Sect. 6.3.4.1), that these traits are heritable (Drent et al., 2003; van Oers et al., 2004). In fish species such as *Lepomis gibbosus* (Sih et al., 2004a, 2004b) and *Poecilia reticulata* (Godin & Dugatkin, 1996), individuals show differences in personality traits associated with boldness, while others exhibit traits of shyness.

However, as we have already said, vertebrates are just a sample of animals that can show traits of a given personality. Even so, at present, we find it difficult to accept that animals such as invertebrates are also capable of showing differences in traits associated with personality, possibly because we interpret that their nervous system, which is much less developed, has a limited capacity. In this sense, the lack of studies on invertebrate species could be related to the fact that many consider them to be "mini-robots" that respond similarly to certain stimuli. Therefore, it is not expected to be a difference in the responses of different individuals (Kralj-Fišer & Schuett, 2014). These authors reviewed that there is personality at the level of exploratory activity, aggression, and boldness in numerous species of Arthropoda, Mollusca, and Cnidaria.

Within arthropods, differences in personality traits have been observed, for example, in insect species such as *Aquarius remigis*, where some males are consistently more active and aggressive than others. On average, more active males tend to encounter more females and mate more frequently (Sih et al., 2002; Sih & Watters, 2005). In fly species such as *Drosophila*, there are differences in levels of aggression between individuals (Zwarts et al., 2012; Alekseyenko et al., 2013). In *Apis mellifera*, individuals show differences in levels of boldness (Wray et al., 2011), and the same occurs in *Gryllus integer* (Niemela et al., 2012).

Within arachnids, spiders have also been found to exhibit personality traits. For example, females of *Agelenopsis aperta* showed different levels of aggressiveness among themselves (Riechert & Hedrick, 1993). In females of *Holocnemus pluchei*, there are interindividual differences in stridulation behavior in sexual interactions (Calbacho-Rosa et al., 2019). Some females stridulate more (possibly associated with more demanding females), while others stridulate less (perhaps associated with less demanding females). Males of *Argiope radon* show consistent differences in behaviors in sexual interactions that reflect differences in male quality (Wignall et al., 2014).

On the other hand, within mollusks, consistent differences in behaviors associated with boldness and activity traits have been shown in the species *Euprymna tasmanica* (Sinn & Moltschaniwskyj, 2005; Sinn et al., 2006), and in the species *Octopus rubescens,* individuals show different levels of boldness (Mather & Anderson, 1993). For Cnidaria, we can name the species *Actinia equina*, where different levels of boldness among individuals have also been observed (Rudin & Biffa, 2012).

From all mentioned in this section, we can conclude that learning and personality are not only processes exclusive to humans or higher animals but also to the vast majority of animals, if not all of them. Perceiving lower animals as organisms with the capacity to learn and show different personalities has remarkably changed our perception of what occurs in nature. Here, we again emphasize the importance of the sociocultural changes that have been developing throughout history, allowing us to become involved in this study. As we saw, another aspect that brought about changes in our interpretations was the critical advances in technology, which made it possible to carry out a large number of analyses and interpretations of processes that we had previously considered oddities simply because we could not observe them. Of course, as has been said throughout the book, the fact that an exception becomes a rule does not imply that it should be generalized to all organisms. However, as exemplified by the case of the learning ability and differentiated personality traits demonstrated among invertebrates, it does mean that researchers should take this new information into account in new experimental designs.

6.4.2.2 Neuronal Plasticity: Regeneration and New Connections

The conventional notion regarding the animal brain, including the human brain, is that its defining cells, neurons, lack the ability to regenerate, unlike the cells found in other organs. Advancements in technology and novel approaches have challenged long-standing beliefs, including the idea that neuronal regeneration is impossible. Presently, we have evidence to support the existence of neuronal regeneration, and while it may not be universally the rule for all organisms, it is no longer considered a mere exception.

First, let us clarify what we mean by the term regeneration. Regeneration, in biological terms, refers to restoring the structure and function of a damaged part of the organism (Bely & Nyberg, 2010; Poss, 2010). This regeneration can occur at different levels, cellular, tissue, organ, or, in the most extreme cases, of some part of the body (Bely & Nyberg, 2010). Three distinct mechanisms have been identified to facilitate the process of regeneration: (1) epimorphosis (differentiation of a blastema and then production of new tissue), (2) morphalaxis (transformation of existing tissues or body parts into new structures), and (3) compensatory growth (tissue that is not damaged enlarges to compensate for the loss of tissue) (Gilbert, 2003; Martínez et al., 2006).

Some examples of organisms showing regeneration processes in different tissues include invertebrates such as planarians, which can regenerate half of their missing body when cut in half (Reddien & Sánchez Alvarado, 2004; Gilbert, 2003). Hydra, another invertebrate, has shown the ability to regenerate into complete individuals when cut into several parts (Gilbert, 2003; Bosch, 2007). Additionally, the worm *Hofstenia miamia* can regenerate the missing part of its body after a longitudinal cut (Gehrke et al., 2019).

Among vertebrates, several examples of regeneration exist across different types of organisms. For instance, the zebrafish (*Danio rerio*) is capable of regenerating fins, as well as certain organs such as the heart and pancreas (Poss et al., 2003). In the case of amphibians, salamanders are notable for their ability to regenerate their limbs (Gilbert, 2003). Among reptiles, lizards have the capacity to autotomize their tails and subsequently regenerate them, and in some turtles, regeneration of at least some part of the shell has been observed (Alibardi, 2010). Finally, in mammals, regeneration processes also occur. For example, hair, nails, skin, mucous membranes, liver, and bone can undergo regenerative processes (Belkind-Gerson & Suárez-Rodríguez, 2004).

Regarding neuronal regeneration, the prevailing dogma until the late 1990s was that there was no formation of new neurons in adult organisms. Cells of the nervous system are widely believed to be one of the few, if not the only, cell types incapable of regeneration (Brain & Walton, 1969; Zupanc, 2008; Gao et al., 2017). However, this concept has been progressively changing. In a broad sense, we now know that neuronal regeneration in adult organisms exists and requires several coordinated processes, such as proliferation and migration of neuronal precursors to sites where some injury has occurred and the integration of the new cells into already existing circuits (Kernie & Parent, 2010).

Among invertebrates, neuronal regeneration has been demonstrated in several groups, for example, in the flatworm *Notoplana acticola* (Faisst et al., 1980) and among annelids in the earthworms *Lumbricus terrestris* and *Allolobophora caliginosa* (Hulsebosch & Bittner, 1980; Ortega & Olivares-Bañuelos, 2020). Among arthropods, cricket species have shown regeneration of sensory neurons (Murphey et al., 1983), and regeneration of motor fibers has been observed in cockroaches

(Stern et al., 1997). In Orthoptera, such as *Schistocerca gregaria*, neuronal regeneration has also been demonstrated (Pätschke et al., 2004; Lüdke & Lakes-Harlan, 2008). Echinoderms, including species such as the serpent star *Ophioderma longicauda* (Dupont & Thorndyke, 2006; Biressi et al., 2010) and the sea cucumber *Holothuria glaberrima* (Ortega & Olivares-Bañuelos, 2020), serve as excellent models for studying neuronal regeneration. In these organisms, there is compelling evidence of neuronal regeneration following injury, which is associated with the involvement of various neurotrophic factors.

In vertebrates such as fish, reptiles, and birds, neurogenesis has been shown to occur in various parts of the brain (Doetsch & Scharff, 2001; Zupanc, 2001; Kishimoto et al., 2012). In teleost fish, there is a high capacity for neuronal regeneration after injury, as occurs in *Apteronotus leptorhynchus* (Zupanc, 2008; Kishimoto et al., 2012). In amphibians, particularly in frog species, such as *Rana pipiens*, it has been seen that after a cut of the olfactory nerve, there is a process of mitosis of basal cells of the olfactory mucosa that then differentiate into neurons (Graziadei & DeHan, 1973; Zupanc, 2006). Adult reptiles, including certain turtles such as *Trachemys scripta* and lizards such as *Podarcis hispanica*, exhibit a remarkable characteristic: their telencephalon generates new cells almost every day. Subsequently, a significant portion of these newly generated cells differentiate into neuronal cells, serving as replacements for those that have been damaged (Font et al., 2001; Zupanc, 2006). Birds present an extraordinary situation, particularly in songbirds, where environmental changes have a notable impact on neuronal regeneration in adults. In species that reproduce seasonally, the lengthening of days significantly influences the growth of song control nuclei. This behavioral change, observed in males, is associated with the attraction of females (Tramontin & Brenowitz, 2000; Font et al., 2001; Zupanc, 2006).

In mammals, neurogenesis occurs in two specific areas of the brain: the anterior part of the subventricular zone (Eriksson et al., 1998; Curtis et al., 2007; Zupanc, 2008; Kishimoto et al., 2012) and the subgranular zone of the dentate gyrus in the hippocampus (Kempermann, 2002; Zupanc, 2008; Kishimoto et al., 2012). In some species of mice, and even in humans, some neurotrophic substances, such as retinoic acid, have been shown to enhance neuronal regeneration rates in both the central nervous system and the peripheral nervous system in vitro (Quinn & De Boni, 1991). However, the mechanisms by which these neurotrophic substances have such an effect are still unclear. One of the proposed hypotheses is that adult neurogenesis could be linked to learning (see Sect. 6.4.2.1) and memory processes (Font et al., 2001). In this sense, this process in vertebrates also involves structural modifications in the brain (Alvarez-Buylla, 1992; Kolb & Whishaw, 1998). It was also shown that although neurogenesis occurs throughout life, it decreases with the organism's age (Kuhn et al., 1996). Thus far, we have named a few examples to show that neuronal regeneration is widely distributed among various animal species. These critical advances in our knowledge are undoubtedly due mainly to the development of novel technologies that are continually advancing in their development.

For example, toward the end of the 1980s, a radiolabeling technique involving the use of dyes, such as thymidine (3H), which is incorporated into the DNA of dividing cells and can be detected by autoradiography, began to be used in studies of neuronal regeneration in adults of fish, reptile, and bird species. Thus, monitoring this dye made it possible to corroborate neuronal regeneration processes (Font et al., 2001; Zupanc & Sîrbulescu, 2011; Hulsebosch & Bittner, 1980; Gage & Temple, 2013). It was determined in mammals that a growth factor, IGF-1, and its receptor appear to be essential in the growth of new axons in mature cells (Dupraz et al., 2013). The PI3K enzyme activates different signaling pathways within the neuron that result in axon growth. Other essential advances that allowed changing our interpretations about the existence of neuronal regeneration were the advances in immunohistology techniques together with the use of confocal microscopy, which also allowed the study of neurogenesis in adult organisms (Gage & Temple, 2013). In addition, the development of methodologies that allowed the isolation, maintenance, and differentiation of precursor cells of nervous system cells under in vitro culture conditions (Kilpatrick & Bartlett, 1993; Reynolds & Weiss, 1992; Gage & Temple, 2013) have also contributed significantly to the increased knowledge on the subject and consequently to the change of our conceptions in terms of considering neuronal regeneration as an exception.

Furthermore, the capacity for neuronal regeneration itself merits further discussion, as it is closely connected to the continuous process of forming and refining neuronal connections throughout an organism's life. This vital issue for modern medicine is linked to an old idea that neuronal connections in the brain take place only in the early stages of development. It is now understood that this is not true. There is a phenomenon of "neuronal plasticity" that persists throughout life, contradicting the previously mentioned notion of a fixed, nonplastic system with no regenerative potential. Moreover, if such a nonplastic system were to exist, it would likely be limited to certain exceptions. Precisely, it is worth mentioning the theory of neuronal connections (Kandel et al., 2000), which proposes that even when the organism is an adult, there is a need for permanent contact between a neuron and its target, which can be another neuron, a muscle fiber, or a gland (García, 2003). Although this contact is more critical during development than in the adult organism, the cells of the nervous system need to interact with their targets to carry out their functions (De la Cruz et al., 1996; García, 2003). Although the adult mammalian nervous system does not have the capacity for the significant reorganization of its neural pathway connections, this has been observed after neuronal injury and motor learning (Cotman & Nieto-Sampedro, 1984; Malenka, 1995; Tsukahara, 1981).

Again, the data corroborate that neuronal regeneration and neural connections are not exceptions. In addition, it is now known that neurons can survive some types of axon injury and that the axon can reinnervate another target tissue (De la Cruz et al., 2000; García, 2003). They can also recover their functional capacity, that is, the capacity to produce synaptic potentials (Font et al., 2001; García, 2003), something that was not considered possible in the past. In addition, we also know that axonal regeneration is different between the central and peripheral nervous systems. This is due to differences in the environment in which axons regenerate. For example, it is believed that myelin in the central nervous system is an inhibitor of axonal regeneration (Liu et al., 2011; Zou et al., 2013; Pernet & Schwab, 2014). Among some of the mechanisms, we can highlight calcium waves, where for example, in invertebrates such as the nematode *Caenorhabditis elegans* and the mollusk *Aplysia* sp., it has been found that axon regeneration depends on calcium influx (Ghosh-Roy et al., 2010; Doron-Mandel et al., 2015). Among mammals, in rat species, for example, epigenetic modifications have been studied as another mechanism that may influence neuronal growth, as they may affect changes in genes that are associated with axon growth (VandenBosch & Reh, 2020), and local protein synthesis mechanisms, based on the ability of axons to synthesize proteins as mRNA, have been recorded along axons (Font et al., 2001; Zheng et al., 2001).

If we shift our attention to human beings as a study model, the remarkable advancements in neuroscience, a discipline currently at its peak, have significantly influenced our interpretations of what we previously considered as exceptions or rules. As we have already said, studies in neuroscience have discovered that the brain changes every time we learn something (Kandel et al., 2000; García, 2003). This occurs due to a process known as neuronal plasticity, which gives the nervous system the capacity to change its structure and function, ultimately producing changes in our way of thinking and acting (Carlin & Siekevitz, 1983; García, 2003; Gruart & Delgado-García, 2004). However, it has also been proposed that a more appropriate term is neural elasticity since this involves not only changing but also being able to return to the initial state (García, 2003). Currently, a considerable body of research in humans is devoted to determining the frequency of neuronal regeneration in adults and identifying the underlying mechanisms that facilitate this process. This investigation holds immense significance due to its potential application in therapies for patients who have suffered brain injuries or neurodegenerative diseases. Thus, this regeneration could be responsible for the improvement in many of them (Belkind-Gerson & Suárez-Rodríguez, 2004; Gao et al., 2017).

Neurotechnology stands out as one of the technological advancements that have revolutionized our understanding of the mechanisms operating within the nervous system. Neurotechnology is a developing discipline based on neuroscientific knowledge, including for example, the simulation of neuronal models, the interconnection of the brain with electronic systems, and sophisticated apparatus for analyzing brain activity. Neurotechnology, according to its mode of implementation, can be classified as invasive when it requires surgery for, for example, the placement of transmitters or receptors that capture signals from the brain or nervous system or noninvasive when it does not require surgery for its placement. Furthermore, new appliances available today have facilitated substantial progress in our comprehension of the nervous system, encompassing the phenomenon of neuronal regeneration. Let us consider that until the nineteenth century, studying the brain was limited to post-mortem examinations, as the available technology, such as X-rays, only allowed the visualization of the skull bones. However, a significant breakthrough occurred with the implementation of advanced technologies such as computed tomography (CT Scan) and magnetic resonance imaging (MRI), which revolutionized neuroscience by enabling the acquisition of detailed neuroimages. CT Scan made it possible to obtain slice images to analyze the object of interest; indeed, Allan McLeod Cormack and Godfrey Newbold Hounsfield received the Nobel Prize in Physiology for its development in 1979. For the development of magnetic resonance imaging, in 1978, an image of the human brain was obtained for the first time. Magnetic resonance imaging is currently one of the most widely used techniques for neuroimaging since it is a noninvasive technique thanks to which soft tissues can be seen. The ability to differentiate between gray matter and white matter within images has opened up new possibilities for monitoring vascular diseases of the brain, including neurodegenerative processes that lead to dementia and Alzheimer's disease. This advancement in neuroimaging has significantly enhanced our understanding and diagnosis of these conditions.

In addition, Parkinson's disease, a neurodegenerative disorder, causes the loss of different types of cells in the central and peripheral nervous systems, including dopaminergic neurons in the midbrain, which produces most of the symptoms in patients. In this sense, the analysis of the possible replacement of dopamine neurons has been closely studied for its possible implementation as a therapeutic alternative (Guardo-Gómez et al., 2020; Tabar & Studer, 2014). Thus, developing and implementing new and innovative techniques for regenerating nervous system tissues promises to obtain methods for the recovery of motor and sensory function of damaged cells (An et al., 2006; Guardo-Gómez et al., 2020). This is an example of the great benefits provided by the advancement of science and the possibility of having access to more excellent knowledge, thanks to which our interpretations about a certain subject change over time. However, their implementation requires a great deal of development and fine-tuning to be used "massively" among patients with a given pathology.

Moreover, while it has long been acknowledged that physical activity enhances overall health and well-being, emerging evidence indicates that it activates specific mechanisms that safeguard neurons. These mechanisms have been termed the "physiological neuroprotection system" (Herrero & Ferradaz, 2011). When the body starts moving, the brain coordinates a series of mechanisms, such as muscle movement, increased blood flow, and alterations in breathing and heart rate. Different nerve centers regulate everything, so physical activity causes several areas of the brain to be activated (Herrero & Ferradaz, 2011). For all these reasons, there is now concrete evidence that a sedentary lifestyle is a risk factor for developing neurodegenerative diseases. All these discoveries have led to significant changes at the social level. For example, many people have already included physical activity as part of their daily routines to contribute to a better general state of our organism, something that was not common before. We will return to this fact in Chap. 8.

In conclusion, our changes in the interpretations regarding neuronal regeneration and associated topics have been possible thanks to the tremendous technological advances that have been developed over time, some of which we have named above. Many neurological analyses or studies of various types are now considered routine, including those of high complexity, which was unthinkable only a couple of years ago. Neuronal regeneration, along with the generation of new neuronal connections, is no longer perceived as a mere curiosity or an exception in the outdated conception of "neurons do not regenerate" (an implicit rule of the past). As research expands to encompass a diverse range of model species, more studies consistently reveal that neuronal cells can regenerate in adults. This discovery is progressively shifting the perception of neuronal regeneration from an exception to a principle, and it is anticipated that further investigations will solidify this notion as a widely accepted rule.

6.4.2.3 Communication in Plants

In nature and science, plants are often overlooked and marginalized (Hoekstra, 2000; Jose et al., 2019). Many perceive plants as tedious compared to animals, as they may seem monotonous, relatively similar, and nonmobile, merely serving as a source of food for other organisms. This is a vague and reductionist view of what plants are. Indeed, it appears peculiar to categorize these organisms as almost "inanimate" and attribute no greater purpose to them beyond existence, reproduction, and being preyed upon. It is not easy to believe they do not have a goal and means to achieve it. Every living organism has one. Therefore, it is not strange to think that plants can perceive the environment, respond to changes, and transmit information to other individuals. The inability to perform these tasks would surely make the group disappear (Gagliano, 2015).

In biology, and in simple terms, communication is often described as a system in which at least two individuals exchange signals that lead to conditional and reversible changes (Karban, 2015). In this context, the term has been applied mainly to multicellular animals. Progressively more primitive groups, such as unicellular protists, were incorporated, and now we also speak of communication in bacteria (kingdom Bacteria) (Taga & Bassler, 2003; Federle & Bassler, 2003; Greenberg, 2003; Keller & Surette, 2006; Blackwell & Fuqua, 2011). The understanding of animal communication has evolved in recent years, especially with advances in the fields of neuroethology (e.g., Schöneich, 2020; Mariette et al., 2021), behavioral ecology (Higham & Hebets, 2013), and behavioral genetics (Ferveur, 2005; Robinson & Huang, 1998), moving on from past debates about instinct versus learning (Thorpe, 1956). However, a significant gap has remained almost constant regarding other kingdoms, such as fungi and plants.

Undoubtedly, the diversity of plants pervades every ecosystem. The importance of plant organisms in the diversification of life, together with cyanobacteria, by providing oxygen through photosynthesis in the ancient environments of Earth has never been denied (see Sect. 6.2.1.1). In addition, the interdependence of plant–animal interactions has been recognized and accepted since ancient times. This is most evident in Darwin's studies on pollinators and orchids (Micheneau et al., 2009). Additionally, various ecological aspects, such as seed dispersal, evolutionary studies of herbivory, and the evolution and ecology of carnivorous plants, have been subjects of investigation (Abrahamson, 1989). These studies collectively highlight the profound connections and contributions of plants in shaping the natural world and understanding the intricate relationships between organisms in diverse ecosystems.

However, amidst all these conceptual changes, there remains an aspect that has not received the attention it deserves, or perhaps has been underestimated: the existence of plant communication. In other words, the real presence of information transmission between plant individuals, where at least one sender and one receiver will share information in the form of signals, and these signals generate a response in the receiver (Karban, 2015), as in animals, there are diverse forms (Gagliano et al., 2012; Awan et al., 2019). In this context, current researchers broke paradigms, implicit biases in our view in the past in science, that plants could not communicate. Advances have been astonishing in the last hundred years. Although much remains to be explored, today, we can say that plant communication exists and that, rather than an exception of a particular species or case, it is the rule in plants (Baluška & Mancuso, 2007, 2009; Karban, 2021).

Certainly, communication is not the sole outcome. Nevertheless, researchers propose that plants may possess more advanced capabilities beyond perceiving and responding to their environment. There is a suggestion that plants exhibit electrical transmission (Bose, 1925; Scott, 1962; Mackie, 1970; Dziubinska, 2003) akin to the nervous system in animals, along with memory, intelligence (Trewavas, 2003, 2014; Baluška et al., 2018), and responses to circadian rhythms, similar to our need for "sleep" (Mancuso & Shabala, 2006; Mondal & Parui, 2013), among other traits previously associated only with animals. Indeed, all of this falls under the umbrella of the development of a relatively new field, the field of plant neurobiology (Brenner et al., 2006). The findings strongly suggest that animals and plants might not be as fundamentally different as traditionally believed. It appears that they could share evolutionary convergence, enabling them to interact with the environment and evolve in response (Trewavas, 2005). Here are some examples, many of them surprising, of communication in plants.

Herbivory, Communication and Reaction in Plants

There is a clear example to illustrate the communication between plants resulting from a stimulus and the subsequent response of individuals. In the 1980s, there was a significant drought in South Africa. Kudus (*Tragelaphus strepsiceros*) are large antelopes that inhabit this region and feed on various types of grass and tender shoots throughout the season (Dörgeloh, 2001). During droughts, kudus exhibited a preference for feeding on leaves from various Acacia species. There were many reports of healthy animals found dead, and deaths were directly associated with density (Van der Waal, 2005). In the rumen of dead animals, the concentration of tannins was too high (Van Hoven, 1984). Tannins are chemical compounds (secondary metabolites) produced by plants for defense against herbivory (Hagerman & Butler, 1981; Hassanpour et al., 2011) and can prevent digestion in the rumen of ungulates (Van Hoven, 1984). Ungulates can neutralize the tannin content at average concentrations, but these chemicals can be lethal at significant and prolonged concentrations. Acacias could quadruple the average concentration of tannins, resulting in kudu's death. The trigger for this response appears to be herbivory pressure in fenced areas of Game Ranches. The plants reacted by increasing tannin levels in their leaves. They "did not want" to be eaten or lose their precious leaves and responded chemically to overgrazing. Thus, the plants evaluated the environment and took action on the signal.

It was later found that plants not under high herbivory pressure (e.g., outside of the ranch boundaries) had the same elevated tannin concentrations. This was no coincidence. The researchers sought to elucidate whether the plants could give other individuals an "alarm signal" about the herbivory pressure to which they were being subjected. The hypothesis was that "there was something in the air." Therefore, the air around the Acacia branches was analyzed, and it was found that when the kudu ate the leaves, the plant emitted a light gas called ethylene, which is used for the ripening process of stored fruits (see below). This wind-borne gas reached other leaves, signaling the production of enzymes catalyzing the production of tannins. Thus, the plant communicates chemically with another plant and warns it of potential dangers. In turn, the plant that receives the warning responds in advance by preparing the production of herbivore repellents. This clearly illustrates the ability of plants to evaluate the environment and respond to changes and pressures, as well as to warn or alarm other individuals by transmitting information. These communication concepts were not uncommon in nature; however, scientists only occasionally identified them clearly. This serves as a clear example of exceptions that eventually become rules, reflecting the evolving perception of science regarding certain phenomena.

Let us now go deeper and explore the potential for underground communication among plants. It is crucial to understand how our perception of how plants interact with each other below the ground has evolved over time. In the past, it was thought that organisms were independent in a forest and did not communicate through their roots. Today, it is known, for example, that forests are a continuous network linked and communicated underground. In this sense, as we have already seen, plants can communicate via the air with other individuals and with different parts of the same organism, but we will also see that communication can flow through the soil. Roots are known to be selective in places with plenty of nutrients and water, but they also form interspecific relationships with fungi and bacteria. This is known as the rhizosphere (Berendsen et al., 2012). This rhizosphere is a mutualistic relationship where the plant is provided with nutrients from the soil that would be difficult to obtain, and in return, the symbionts obtain sugars from the plant. Obviously, and this is no exception, some organisms take advantage of and parasitize this association (Bouwmeester et al., 2007). Regardless, the association with fungi is very beneficial and, in some cases, is believed to be vital for plant survival (Van der Heijden et al., 2006). There are two leading symbiotic associations with fungi, arbuscular mycorrhizae (AM), and ectomycorrhizae (EM) (Lodge, 2000; Mohammadi et al., 2011). This relationship is and has been very important for plants to such an extent that many believe that plant radiation or the ability to colonize dry environments would not have been possible without a mutualistic relationship with fungi and other soil microorganisms. Mycorrhizae play a vital role in fostering connections within the forest, where the signals exchanged are beneficial rather than competitive, contrary to the longstanding belief (Castro-Delgado et al., 2020). Fungi, through mycelia, interconnect the roots of different organisms that would otherwise never touch (Mohammadi et al., 2011). Scientists have named this phenomenon "Wood Wide Web" in allusion to acronyms related to the internet (Simard et al., 1997; Work & Flowers, 2011; Castro-Delgado et al., 2020; Hooper, 2021). This is wise since this interplant network functions as if it were the computer internet, interconnecting nodes (organisms), and exchanging information and substances such as metabolites, water, compounds, and signals (Castro-Delgado et al., 2020). However, although the rule is that the exchange is beneficial for different individuals, there are some exceptions. As with the Internet, the network can steal information, hack, or attack another organism. For example, several representatives of a group of orchids (genus *Corallorhiza*) steal resources through the mycorrhizal network. Others, such as eastern American black walnut (*Juglans nigra*), disseminate toxic chemicals through the network, such as spreading malicious code on the Internet.

This network that interconnects plants is a valuable form of communication that goes unnoticed by most of us and can be taken as an exception when studied. However, it is a rule in the plant world. This network interconnects the organisms in the forest and allows intraspecific but also interspecific synergy.

Plant Senses

At this point, it is worth mentioning that the definitions with which we deal in science are very important when cataloging a phenomenon. For example, if we want to find senses in plants exactly like those we see in animals, it will be a difficult task. Similarly, if we seek to define mechanisms or structures in animals and plants in the same way, we will also be often contradicted, and we may draw false conclusions. If the definition of the phenomenon we are studying is clear, the results we obtain will also be clear to interpret. The senses in plants have long been neglected due to the discussed perception that they were almost inanimate sessile organisms that had no purpose beyond serving as food for other living things (Gagliano, 2013). However, that conception changed substantially from the last decades of the nineteenth century onward thanks to the research carried out by numerous researchers, especially many tenacious observations and questions made by Darwin himself. It is complex to think about senses in plants because, in the collective unconscious, the idea of senses is deeply rooted in what the senses are in vertebrates. However, we should keep an eye on the other side of the spectrum of living beings.

The senses have always been related to animals. For example, someone has eagle vision, lynx hearing, dog smell, and so on. However, only a few times do we hear similar phrases with plants as protagonists. It may have to do with our idea that the senses are associated with a more or less complex nervous system. Thus, any living organism lacking it or in poorly developed stages is likely to lack senses. Nevertheless, examples of minor invertebrates such as jellyfish or planarians with distinctive sense organs, the visual acuity of an octopus comparable to that of a vertebrate (Mathger et al., 2009; Hanke & Osorio, 2018), or the proprioception of a scorpion capable of assembling a map of the terrain with high precision (Wolf, 2017) are ignored. Something similar happens with plants.

To be precise in this respect, when we talk about senses, we have to think about the capacity of the organism to capture the signal and respond to it. To the surprise of many, plants can develop the most well-known senses, such as vision, smell, touch, hearing, proprioception, and even memory (Chamovitz, 2020). Indeed, most of the senses in plants do not align precisely with our current understanding. However, as mentioned earlier, adopting a broader definition allows for a better comprehension of how senses function in plants. We will even realize that they are similar to how they work on animals.

Sight

Perhaps plants do not form images, nor can they discern the color of a predatory larva or the type of fur of a savannah ungulate. However, what if we define sight as the ability to detect electromagnetic waves of different lengths and respond to them? We know that light is essential for them and that they can not only detect it but also grow toward it. In 1880, Darwin, with his son Francis, studied motion in plants (Kutschera & Briggs, 2009). They conducted an elegant experiment where they hid a plant in the dark and left a candlelight burning. They observed that the plant grew toward the light. They documented positive phototropism in plants (Holland et al., 2009). This discovery seems simple, but it is gigantic. Not only did the plant detect the light stimulus, but it sent signals to the stem to grow unevenly so that it could direct itself to where the stimulus was coming from. Darwin, curious as he was, sought to discern which was the detector of the stimulus in the plant (Kutschera & Briggs, 2009; Chamovitz, 2020). Thus, he covered different areas of the stem with opaque and transparent materials and exposed them to the same light source. Thus, they realized that the shoot tip participated in light harvesting. This is just one example of what plants can "see," but it is not just a matter of detecting the faint light of a candle.

Plants can also see different frequencies of electromagnetic waves, detect the time of day they are in, detect how much light they are receiving, and indirectly detect if something is blocking their light source (Chamovitz, 2020). Many of these abilities are given by different photoreceptors. Humans have five photoreceptors, one for lights and shadows, three for colors (RGB), and a fifth that regulates the internal clock (Kolb, 2012). Plants possess many photoreceptors. For example, studies in *Arabidopsis thaliana* showed that it has more than ten photoreceptors that tell the plant when to germinate, when to give flowers, and the intensity of light it receives, among others (Smith, 1982; Izawa, 2008; Babla et al., 2019; Chamovitz, 2020). The rule is that the proteins that function as photoreceptors differ between plants and animals but fulfill the same function (e.g., two photoreceptors that detect the color red). The exception to this rule is that both animals and plants share cryptochromes, the photoreceptors involved in circadian rhythm (Chaves et al., 2011). As Chamovitz (2020) quotes, "*in this sense, animals and plants 'see' in the same way.*"

Touch

Plants feel touch but do not feel pain (Chamovitz, 2020). Plants possess receptors that are analogous to those found in humans and other organisms. These receptors provide plants with information about their environment, enabling them to adjust their metabolism and respond accordingly. Although the idea of touch, proprioception, or perhaps the feeling of pain is closely related to the presence of a nervous system, we have examples that show how plants use a similar mechanism to "feel" without the need for a nervous system. Here, the common ground and what would seem to be the rule in sensing is electricity in the form of an electrochemical reaction, something shared by plants and animals (Baluška & Mancuso, 2009; Chamovitz, 2020). In humans and animals, ions play a crucial role in the transmission of signals to the brain. They enter and exit the cell membranes, propagating the signals to the brain. Something similar happens in plants without a nervous system to mediate the process.

To illustrate, there are two iconic examples of rapid plant movement. These plants have structures that move in a very short period of time. This type of movement is different from the standard movements in plants called tropisms (Forterre, 2013). Tropisms are slow movements; remember that the times of plants are not our desperate times (Trewavas, 2005). Plants are much slower than animals and can afford to be because they are also among the longest-lived organisms (Chamovitz, 2020). Technology has helped analyze normal plant movement thanks to time-lapse photography systems (e.g., Attenborough, 1995) that allow us to generate videos with perceptible speeds. Within this classification of rapid plant movement, there are three main groups: plants that catch and consume prey (carnivorous plants), plants that move leaves under a stimulus, and plants that use sudden movement to disperse seeds. In this section, we will address the first two examples. The number of plant species that show these characteristics are few, and they would be an exception within the large group of plants.

Among the trapping plants, we have seen the fantastic venus flytrap (*Dionaea muscipula*). Anyone with one of these plants will spend long periods (and many insects) to see them in action. Therefore, the insect prowls between the two trap-like lobes at the tips of the leaves. The trap is triggered upon stimulus, and the lobes quickly close and trap the organism (Hodick & Sievers, 1988; Volkov et al., 2008). Subsequently, digestive juices are responsible for dissolving and absorbing the prey (Scala et al., 1969). During his studies in this species, Darwin noticed that dark hairs inside the lobes function as triggers to activate the trap. Scientists have noted that at least two of these hairs must be stimulated within no more than 20 seconds for the trap to activate (Volkov et al., 2008; Böhm et al., 2016). This is very interesting as the plant has to retain somehow the activation information of one of the hairs for that period, something like a "short term memory" (see below). The hairs on these plants are not only sensitive but also selective, as they are not activated by any prey size. Burdon-Sanderson & Page (1877) and Burdon-Sanderson (1882) found that activation of the hairs generated an electrical signal similar to muscle contraction in animals.

The second iconic example of how touch affects plants is that of *Mimosa pudica*. This plant features a compound leaf, and its parts retract upon touch (Ahmad et al., 2012). Additionally, it exhibits responses to various stressors, including damage, vibrations, wind, and drought (Volkov et al., 2010). At the electrophysiological level, it functions similarly to the venus flytrap (Allen, 1969). Touch produces an electrical signal that results in leaf closure (Bose & Das, 1925). *Mimosa pudica* has specialized motor cells at the base of the leaves that cause the leaves to move. The basic principle is the movement of ions in and out of the cell that produces water movement and changes in the turgor that causes the leaves to move.

One effect that might be considered rare is that touching can harm plants (Jaffe & Forbes, 1993; Chamovitz, 2020). It has been found that touching plants can stunt their growth and kill certain parts. This knowledge came from an experiment investigating growth in plants. The researchers took leaves and measured them with a ruler every day. They noticed with amazement that these leaves never grew to a standard size, and many died. In contrast, the surrounding leaves did not show any problem. They realized that this was due to the manipulation of the leaves when measuring.

Finally, many climbing plants have mobile tendrils that are very sensitive to touch, and the slightest pressure they feel makes them grow around the object. The tendrils generally make a circular movement trying to contact some structure that serves as support. Once they touch something, they automatically begin to curl. Here, we see an association between movement and touch to find an anchoring place for the branches (Scorza & Dornelas, 2011). As we have seen thus far, plants feel a lot. They are capable of equating with us and other vertebrates in that sense. Regardless of the outcome, the plant rule seems to be electrical signals modulated by ions and some proteins but without a coordinating nervous system.

Proprioception.

The answer is a resounding yes. Plants can "know" where it is down, up, and sideways. They know how to orient themselves and where each of their parts is (Moulia et al., 2019, 2021). The plant "knows" where to put its roots and where the stems go. Experiments have shown that both zones orient themselves using gravity (Levernier et al., 2021). The roots have cells with statoliths and spherical structures that are heavy and respond to gravity. The aerial zones have a zone called the endodermis associated with the phloem, which also contains these statoliths. The position of the statoliths tells the plant where it is below. In addition, as we have seen, plants have an affinity for light or chemical cues that make them move toward the stimulus. Similar to gravity, these stimuli are detected, and the plant responds by bending to the position where it needs to go according to the stimulus (Vandenbrink & Kiss, 2019). It uses hormones called auxins that help in various processes. However, they are responsible for accumulating in certain areas, informing the surrounding cells that they must increase in size, which will cause the stem or root to bend (Levernier et al., 2021).

Smell

Again, if we define olfaction as the ability to perceive odors or scents, it is easy to fit with plants. We already had a preview of how plants use volatile compounds to communicate, where we talked about pollinators and plants. As we previously saw, ethylene is a gas that serves to ripen fruits and communication between individuals, among other functions (Grierson, 2013). Plants detect ethylene particles in the air and trigger a cascade of signals that produce, e.g., fruit softening (Barry & Giovannoni, 2007). Fruits can communicate via ethylene, i.e., they can "smell" each other (Chamovitz, 2020). This is demonstrated when you want to soften a fruit and put it next to another already ripe fruit. Ethylene production helps uniform fruit ripening and indirectly synchronizes the timing of seed dispersal. This gas is also associated with plant senescence processes in leaves and fruits (Iqbal et al., 2017).

Hearing

Here, we find an exception to the pattern cited in this section. There are, thus far, no rigorous and conclusive scientific studies on hearing (Chamovitz, 2020). The rule or paradigm at this time is that plants do not hear, and many agree that, unlike other senses, plants would not need to hear from an evolutionary perspective. However, it may also be possible that plants do hear, but we still need to perform the proper experiment to identify it. As happened historically with the other senses. Perhaps the rule that plants are deaf will change to the exception in the future. Time and scientific advances will tell.

Memory and Vegetable Intelligence?

Memory is closely linked to the senses. In our case, images, smells, or sounds can trigger happy or ugly memories. In the case of plants, we have seen cases such as the venus flytrap that can retain previous events and bring that information to process at the moment (differentiated trichome stimulation). This delayed response to a previous action requires "memory." Plants can present a basic level of memory called procedural memory (*sensu* Tulving, 1985). This simple memory results from sensing the environment and reacting to the stimulus. Nevertheless, a memory, however simple, must have three essential components (Klein, 2015). It must have the ability to form the memory (encode the information), retain the memory (store the information), and retrieve the memory (obtain the information again). Let us test this system with the example of the venus flytrap. Something unknown touches a hair (encodes information) and then retains the information for a few seconds until another hair is stimulated (stores the information). Once a second hair is touched, it retakes the previously stored information and closes the trap (retrieves the stored information). One hypothesis that explains how this happens is related to action potentials and calcium concentrations (Hodick & Sievers, 1988). The charge and ion concentration of the first stimulus are retained for a few seconds, and the second stimulus triggers another potential that, together with the previous one, exceeds a threshold and closes the trap. If this second stimulus does not arrive, the charge and ion concentration are lost.

Memory can be long-term, as in the case of some plants that need certain conditions to flower. Researchers believe these plants remember the conditions, especially the seasons where they should flower (Sung & Amasino, 2005; Amasino, 2010; Xu & Chong, 2018). Stress-mediated memory is also transgenerational and linked to epigenetics (Molinier et al., 2006; Iwasaki & Paszkowski, 2014; Kinoshita & Seki, 2014). Thus, plants that suffer some environmental stress change the composition of their DNA and transfer this information so that the progeny makes the same modification in their DNA without being exposed to the stressor.

Regarding intelligence in plants, the first thing to note is that there needs to be a consensus on the definition of intelligence (Manetas, 2012). It can also be believed that intelligence is an exception presented in humans and some groups, such as dolphins, elephants, and cephalopods (Manetas, 2012). For plants, the term is more elusive. When we compile a list of properties associated with intelligence, such as information sensing, processing, and integration; decision-making and behavior control; learning; memory; choice; self-recognition; and predictive modeling to optimize resource acquisition efficiently, we can observe that many of these traits are shared by plants (Trewavas, 2005).

As we have seen in the previous paragraphs, plants can sense the environment and respond accordingly. They can solve problems to be more efficient and have overall higher fitness. Beyond the issue of intelligence, let us keep in mind the current rule that plants are aware of their environment (Chamovitz, 2020). Let us recap, they are aware of electromagnetic waves (light), where it reaches, if it is reaching them, how much they should reach and what to do to get the most out of it. They are aware of the chemicals around them and in the air. They are aware of the touches they receive, gravity, their position relative to other plants, structures, the sun, and their environment. Plants are more aware of their environment than most of us! All the rest is widely speculative at the moment. Better development of the scientific paradigm coupled with new ways of assessing intelligence may give us a better view of the phenomenon in the future. The new branch of study called plant neurobiology is very suggestive and invites serious scientific discussion (Chamovitz, 2020).

From what we have seen in this whole section devoted to communication in plants, we should be generous when discussing rules and exceptions. Once more, we should remember that the results and discussions at a given time are restricted by the scientific paradigms and social thoughts of that particular moment, in addition to the technical and mental capabilities.

6.4.2.4 Epigenetics: When Environment Matters

In science, particularly in biology, there has been a well-known belief, often referred to as a "dogma," regarding the "gene–environment association." According to this traditional view, information flows solely from genes to the external environment through the production of proteins via RNA (a concept known as the genetic "unidirectionality" dogma) (Crick, 1958). As a result, it is commonly believed that if an organism is born with a specific genotype, it cannot be altered or modified. However, recent developments have shown the necessity to correct this perception. While this idea-dogma might have been useful in the early days of modern genetics, it no longer stands unchallenged today, especially in light of the growing evidence suggesting a significant influence of the environment on genes. Consequently, this notion can no longer be sustained without facing criticism similar to that encountered by ideas proposing a strong environmental impact on genetic expression. In his theory of evolution (previously discussed in Sect. 4.2 of this book), Lamarck postulates two laws. The first law addresses the fortification or weakening, and even disappearance, of organs based on their respective use or disuse. The second law is more controversial and proposes that nature preserves these changes in organs by passing them on to the next generation through the inheritance of acquired characters (Lamarck, 1809). This theory was discarded in the 1960s, and Charles Darwin's theory of evolutionary changes produced by natural selection was favored. However, it is now being realized that some evolutionary changes are close to the Lamarckism of that time (Ward, 2018). As mentioned earlier, the Lamarckian theory has regained attention and interest due to the progress in genetics and our enhanced understanding of DNA. This renewed fascination has been termed "neo-Lamarckism." The reason behind this resurgence is that several scientific findings can no longer be fully explained by Darwin's evolutionary theory alone; instead, they align more closely with Lamarck's theory. This necessarily means that evolutionary theory is being modified and updated, with the addition of new aspects that reinforce it. Within this framework, epigenetics stands out, defined as the mechanisms that regulate gene expression, which occurs without modifying the DNA sequence (Zulet et al., 2017). Epigenetics causes evolutionary changes; it is a process still poorly understood by some but of much importance for evolutionary theory (Ward, 2018). The most critical part of epigenetics, or at least the most controversial, is behavior inheritance. This theory was refuted many times for decades. However, several studies in mice have shown that acquired behavior can be inherited from one generation to the next (Francis et al., 2003; Darnaudery & Maccari, 2008; Champagne, 2014). This type of study caused evolutionists to reconsider adding these "neo-Lamarckian" findings to the current theory of evolution (Ward, 2018).

Epigenetics provides insight into how certain acquired proteins or life changes can be passed onto future generations through inheritance. In other words, it highlights the significance of recognizing that the conventional "rule" stating that information is solely transmitted from genes to the external environment is not entirely valid. The opposite exists, perhaps not only for the mere regulation of a gene or genes but also for modifying it, and this modification is inherited (epigenetically, this change). Today, for example, it is recognized that the nervous system possesses enormous plasticity. This flexibility can be translated into morphological and functional changes resulting from epigenetic mechanisms in response to some environmental signals. The processes observed in learning, memory, cognitive aging, addiction, depression, etc., can be regulated by epigenetic mechanisms, which were unthinkable some time ago (Mahgoub & Monteggia, 2013; Colciago et al., 2015). It has recently been shown that chronic stress, for example, predicts accelerated epigenetic aging (Juster et al., 2010; Zannas et al., 2015) and that meditation can significantly decrease these stress levels and age-related neuroplastic changes; hence, it has become a focus of scientific interest to promote healthy aging (Chaix et al., 2017).

The first discovery where unusual genome results were observed, in which the responses were not predictable (and thus initially thought to be rare), was made in the 1940s by Barbara McClintock (McClintock, 1942). In her dissertation upon receiving the Nobel Prize in 1983, McClintock stated that, building upon those early results, it was necessary to subject the cell to the same challenge repeatedly to observe the nature of the changes. She stated that the responses of genomes to unforeseen challenges are not precisely programmed. However, these challenges are detected, and the genome responds in a discernible but initially unthinkable way. These responses, now known to occur in many organisms, are essential for appreciating how a genome can reorganize itself when faced with a difficulty for which it is unprepared. Our current knowledge of genome responses suggests that these reorganizations originated from some "shock" that forced the genome to restructure itself to overcome a threat to its survival. Cells can "feel" what is happening. From this, in a certain way, they make decisions and act on them so that the DNA molecule is correct after accidents that occur inside the cell or outside, such as infections by viruses, poisons, and altered environments (McClintock, 1984). An example studied in the framework of epigenetics and its action is that of DNA methylation processes and others of this type (see table in Rey et al., 2016), which increasingly encompasses all living beings.

DNA methylation (the addition of a methyl group to a DNA nucleotide) modifies the accessibility of DNA for protein synthesis, which most often has consequences for gene expression (Bird, 2002). Some of these methylations can be reestablished during meiosis, but they can also be maintained between generations and lead to heritable phenotypic changes (Daxinger & Whitelaw, 2012). Regardless, methylation is one of many chromatin marking systems. This is obvious if we consider invertebrates, which have almost no methylation in their DNA. However, cellular phenotypes are transmitted to daughter cells. In the fruit fly *Drosophila*, for example, there are practical reasons to suggest that they may do so because protein complexes that bind to DNA may act as heritable chromatin marks; however, this is still unclear, and how that mechanism of inheritance works is not yet understood (Jablonka & Lamb, 2005). Another process within this framework is histone modifications in DNA, which can also affect the accessibility of its sequence to transcription enzymes and ultimately affect genes and their expression (Strahl & Allis, 2000). Some of these modifications are heritable and associated with changes in the organism's life history (Baulcombe & Dean, 2014).

Advances in cell and molecular biology have allowed and continue to allow discoveries in the processes of DNA regulation and demonstrate how specific mechanisms in cells, which could be seen as rare or which were unknown or understudied, can now be considered as a pattern of cell behavior in terms of gene expression mechanisms and therefore of (often heritable) changes in phenotype. Recent progress in transcriptomics, for example, revealed a staggering diversity of RNAs that do not encode proteins but are involved in the control of transcription and post-transcription gene expression, being integral parts of epigenetic regulation—even being inherited from one generation to the next through parental gamete tissues as a source of nongenetic inheritance (Luteijn & Ketting, 2013; Holoch & Moazed, 2015).

As the field of epigenetic studies expands, so does the interest in exploring new technologies. To date, numerous methods have been developed to provide a better and more accurate understanding of the functions of epigenetic regulation. Thus, the advancement in new methodologies allows the use of multiple strategies, such as high-quality antibodies, functional chromatin assays, imaging tools, and high-throughput sequencing technologies (Zheng & Bernstein, 2015). A breakthrough occurred with bisulfite sequencing, which is now a standard method for DNA methylation detection. This method led to numerous subsequent technologies, such as methylation-specific PCR (MSP) or combined bisulfite restriction (COBRA). New protocols have also been developed, combining basic techniques to provide a complete understanding of the overall dynamics of epigenetic processes in modulating the phenotypic expression of the genome (Tollefsbol, 2011). Along with these advances in methods to unravel the epigenome, challenges have also arisen in managing massive amounts of information. Therefore, new developments in bioinformatics and computational sciences are critical in managing datasets and interpreting epigenomic functions, as well as multiomics data derived from combinatorial integration analyses, to further advance the rapid development of epigenetics (Li, 2021).

Recently, there has been increasing interest in the study of whether environmental factors (such as chemical pollutants, dietary components, and temperature changes) modulate the establishment and maintenance of epigenetic modifications and, therefore, whether they could influence gene expression and phenotype (Feil & Fraga, 2012). To date, the central focus in "Environmental Epigenetics" is centered on DNA methylation. A classic example in plants is *Linaria vulgaris*, in which the flower can present inheritable bilateral or radial symmetry. The change from one type of flower to the other occurs occasionally and involves alterations in DNA methylation in a gene encoding a transcription factor without apparent genetic mutations (Cubas et al., 1999). In addition, there are studies on wild plant populations in which differential phenotypes have been identified that are associated with specific DNA methylation profiles, which could appear because of selection (Herrera & Bazaga, 2010; Lira-Medeiros et al., 2010).

In animal models, it has been shown that nutrition and exposure to certain environmental factors during development can lead to specific changes in particular loci of the epigenome (Rosenfeld, 2010). Concerning human studies, in addition to what has already been mentioned above regarding the nervous system, it is known that nutrition and exposure to toxic components at the gestational stage can have long-term phenotypic effects involving changes in DNA methylation (Heijmans et al., 2008; Waterland et al., 2010).

Most epigenetic variation is restored between generations. However, in some cases, environmentally induced epigenetic variation can result in heritable phenotypic plasticity that invokes Lamarckian-like inheritance. However, it can also result from random epimutations that may have higher mutation and reversion rates than DNA sequence mutations (Robertson & Wolf, 2012). Epigenetics today represents an additional source of variability, acting as an evolutionary force alongside natural selection and genetic drift (Schrey et al., 2012). Future research is still needed to understand the interactions of genotype, epigenotype, and environment and how they relate to selection. However, epigenetics provides a clear link between a gene's genotypic and phenotypic nature (Bolondi et al., 2017). As Jablonka & Lamb (2005) discuss in their book Evolution in four dimensions (genetics, Epigenetics, Behavioral, and Symbolic Variation in the History of Life), heritable epigenetic differences may also play an essential role in the origin of species. The authors argue that although speciation is a hotly debated topic, it is generally agreed that new species appear when populations become isolated from each other by a geographic or ecological barrier, which may cause populations to change. These changes prevent individuals from interbreeding successfully if they meet again. Although we generally assume that the changes are genetic, they can sometimes be thought of as epigenetic because while populations are isolated, they experience different conditions, use different resources, or experience a different climate. In this case, new epigenetic marks could be induced in both somatic and germline cells, which may affect the functioning of the organisms in the new environment and their ability to interbreed between populations. Then, if two previously isolated populations have acquired different epigenetic variations, the incompatibility between the marks transmitted by the gametes may prevent a hybrid embryo from developing typically. Thus, the initial reproductive barrier formation may result from epigenetics rather than genetic changes (Jablonka & Lamb, 2005).

In this framework, the genetic and epigenetic aspects of the gene can be combined structurally and functionally to include its different levels uniquely (Bolondi et al., 2017). The latter reinforces our idea that what was previously thought of as a unidirectional pathway of gene transcription can now be viewed differently and that in the pathway of selection, not only the gene as a transcribed DNA segment itself acts but also external factors that could modify its expression, even going so far as to make it heritable. Epigenetic mechanisms challenge this traditional view and demonstrate that environmental influences can have an impact on heredity, offering a more nuanced understanding of the complexities involved in genetic inheritance.

6.4.2.5 Genome Transposons

Linked to the framework also provided by epigenetics to mitigate the unidirectional view of DNA, the history surrounding the discovery, and studies on transposable elements within the genome indicates that this was not an isolated event, restricted to certain plants, as was initially thought. The discovery of transposons dates back to the mid-twentieth century, as previously mentioned in the context of epigenetics. This was made possible through the groundbreaking work of Barbara McClintock, who specifically studied chromosomal breakage in maize plants. From an experiment that aimed to isolate mutants in progeny of self-pollinated maize plants with broken chromosome ends, it soon became evident that modified gene expression patterns were occurring (McClintock, 1942). After observing these results, McClintock concluded that the regulation of the gene expression pattern was associated with an event occurring in mitosis, in which one daughter cell had gained something that the other daughter cell had lost. After that, he turned her attention to determining what one cell had gained, and the other cell had lost. It was then shown that these transposable elements could regulate gene expression precisely (McClintock, 1950, 1951, 1952, 1953). These accidental discoveries, with surprising results for the time, were the origin of much of the subsequent research.

However, her discoveries were not appreciated until the first studies on transposons in bacteria became known, which allowed an understanding of the importance of these segments in the bacterial genome and DNA transfer (Shapiro, 1995; Biémont, 2010). Advances in biology have confirmed that these elements are present in bacteria and plants and a wide range of eukaryotes, including several protozoa. Therefore, we know that DNA transposons no longer represent a rare natural phenomenon, as initially thought. In fact, they must have diversified very early in evolution and have been maintained in all major branches of the eukaryotic tree of life (Feschotte & Pritham, 2007; Pritham, 2009). These transposable DNA segments are an obvious source of mutations, and their activation can affect both genome structure and function, gene regulation and expression, and alter their epigenetic state (Lisch, 2009; Sahebi et al., 2018).

The above has important implications in the combined use of mediated adaptation processes in organisms via epigenetics or transposons, making the issue more interesting, especially in the face of global climate and environmental change (Rey et al., 2016). Transposons play a vital role in evolution by generating new genetic information and regulating stress responses. They contribute to the development of phenotypes that are tolerant to various stresses, including those related to fruit color and shape in plants (Makałowski et al., 2019). Numerous studies have shown how transposons can adapt under different conditions, thus facilitating the survival of organisms in specific environments and, as a consequence, the expression of genes in response to stimuli in different organs (Woodrow et al., 2010; Lu et al., 2011; Ishiguro et al., 2014).

The frequent mobility of transposons can accelerate mutation rates, generating genetic variability. One example we can mention and demonstrate is the activity of the Tol2 element in the medaka fish *Oryzias latipes*, which is responsible for a 1000-fold increase in mutation rate in the pigmentation gene, resulting in different phenotypes of albino-like individuals (Koga et al., 2006). In addition, transposons can also generate genetic variants from which functional evolutionary innovations can arise (Lisch, 2013) and thus also generate genetic diversity. They can affect the expression of particular genes if inserted into gene-associated regions or promoter regions (Rey et al., 2016). However, the most important and studied perspective is related to the epigenetic mechanisms mentioned above, especially DNA methylation (Galindo-González et al., 2018). Epigenetic mechanisms help to respond rapidly to environmental stress through transient or heritable changes. In this context, transposons also respond to stress and are influenced by changes in the epigenetic state of the genome. This process is essential, given that when a transposable element is inserted in or near a coding region, it can modify the epigenetic state of adjacent gene regions (de Souza et al., 2013). In summary, this new perspective discussed in these last two sections is very promising. This finding reinforces the fact that epigenetic mechanisms (not only methylation) and transposons play a fundamental role as "classical" alternatives (e.g., random mutation and subsequent natural selection) in the adaptation of living beings to changing environments. They are no longer viewed as rare mechanisms and processes but contribute to shaping the genome architecture, leading to the emergence of genetic innovations from a wide variety of lineages (Feschotte & Pritham, 2007).

6.4.2.6 Microbiome and Host

Recent advances in microscopy techniques have greatly contributed to our understanding of the intimate association between the intestinal microbiome and the host (Shapira, 2016). This knowledge has been facilitated, in part, by routine use of advanced biochemical techniques for identifying the bacterial flora within the digestive system. Indeed, while it could once have been considered an exception, thanks to the use of new technologies such as genomics, transcriptomics, or metabolomics, it is now accepted that all animals harbor diverse symbionts that aid in the genetic variation of the individual (Gilbert et al., 2012). Bacteriophages, for example, are known to be vehicles that allow horizontal gene exchange between different bacterial species and are, therefore, determinants in bacterial evolution (Brüssow et al., 2004). The ecological and evolutionary significance of these symbioses is of fundamental importance when considering the applications of such knowledge. Furthermore, it has been possible to exponentially increase knowledge about microbial communities and to characterize, for example, human and animal microbiota and describe new genes, proteins and organisms in various ecological niches. Continuous technological progress and the search for novel strategies have become increasingly accessible (Quiroga, 2016).

Symbiosis represents biodiversity in action and is a fundamental life strategy requiring the exchange of information, products, and mutual regulation between individuals to ensure stability and adaptability. The microbiome is the most striking evidence of this (Munzi et al., 2019). We know that the microbiota influences several aspects of the biology of its host animal, including food choice, reproduction, and immune response (Fraune & Bosch, 2010; Ezenwa et al., 2012; McFall-Ngai et al., 2013). As previously discussed in the chapter on rules and earlier sections of this book, the profound implications of understanding the intimate interaction between the intestinal microbiome and the host are evident.

A remarkable example of this interaction is the discovery that the gut microbiome of flies can influence their behavior at the brain level, prompting them to consume specific types of food containing essential amino acids needed by the bacteria (Leitão-Gonçalves et al., 2017). In this case, essential amino acids and gut bacteria are critical modulators of protein appetite in flies. The lack of any essential amino acids in the diet of flies causes them to have a strong appetite for protein-rich foods. However, flies with a good microbiome do not. This occurs because two species of gut bacteria (*Acetobacter pomorum* and *Lactobacillus* sp.) work to suppress the appetite for protein and cause flies not to reduce their reproductive allocation (Leitão-Gonçalves et al., 2017). The impact that bacteria have on the host is due to complex interactions. Therefore, mapping metabolic interactions in the gut microbiome is key to understanding how it influences the host. Advances in metabolomics have helped to understand and demonstrate how relationships between flies and gut bacteria function to enable them to alter the feeding decisions of the host. These metabolic interactions allow bacteria to become resistant to detrimental host diets and ensure the constant flow of metabolites used by the microbiome to alter their reproduction and behavior (Henriques et al., 2020).

The microbial community of each animal species has its own functions. For example, the intestinal microbiota in humans performs metabolic activities that prevent energy losses, help absorb nutrients, and protect the host against pathogenic microorganisms (Guarner & Malagelada, 2003). A healthy intestinal flora is said to be the first barrier to avoiding diseases. It creates a protective barrier that hinders the establishment of pathogenic organisms (Doe, 1989; Ramiro-Puig et al., 2008). To understand these functions, it is necessary to know the structure and behavior of microbial communities (Human Microbiome Project), which is why new DNA sequencing technologies have been advancing rapidly (Escribano, 2016). The same happens with technologies applied to other vertebrates, such as poultry, where for example, the advance of metagenomics stands out, offering an alternative for a better understanding of the bacterial microbiota and the identification of antibiotic resistance genes, identification of mobile genetic elements and the effective design of intervention strategies to break, for example, the transmission of pathogenic microorganisms during the poultry production cycle (Mantilla & Sáez, 2019).

Thus, data collection from different symbiotic systems using standardized research methodologies provides insights that are exploited through comparative analyses and using bioinformatics tools that were previously unknown (Munzi et al., 2019). Multiomics technologies provide detailed information on different microbial activities in the environment. Over the past few years, the cultivation of microorganisms has been overlooked and underestimated. However, its significance has now been recognized, allowing for the combination of omics techniques to analyze microbe–host interactions. This has revealed that we inhabit a symbiotic world, where what was once considered an exception is now acknowledged as the norm (Munzi et al., 2019). Global data analysis is also becoming more critical to achieving a broader view of the studied systems. This led to new forms of research, such as interactomics, to study interactions and transomics to build global biochemical networks using integrated data (Quiroga, 2016). There is still much to be known, and we must have reasonable access to new technologies and regulations for their use. This entails a primarily ethical discussion in terms of research involving humans and animals, which should start soon (Quiroga, 2016).

Holobionts

The knowledge gained in recent years about the existence of an intimate association between the microbiome and its host has allowed the postulation of principles that would have been almost unthinkable before, since this phenomenon was considered a rarity or something exceptional. Animals and plants are no longer announced as autonomous entities but as biomolecular networks composed of the host plus its associated microbes, i.e., "holobionts" (Bordenstein & Theis, 2015). This concept, which we discussed in previous sections (see Sect. 6.2.1.5), was mentioned by Lynn Margulis in her article "*Words as battle cries—symbiogenesis and the new field of endocytobiology*" (Margulis, 1990). She uses this term to refer to symbionts transmitted from one generation to the next. Margulis realized that there were life forms that were linked almost to their entire existence through a relationship that defined them. Following his theory, Margulis referred to the holobiont concept as "the symbiotic complex" (Margulis, 1991) or "the product, temporary or permanent, of the association between its constituent bionts" (Margulis, 1993). Before her, Meyer-Abich used a somewhat different holobiont concept, describing holobionts as "specific bionts within an integrated whole," which he called the "holobiome." However, he addressed the same evolutionary and developmental phenomena as Margulis (Meyer-Abich, 1943).

Subsequently, "holobionts" began to be used to talk about organisms that have a long-lasting symbiotic relationship (Díaz, 2015) and then to refer to the symbiosis between a host and several microorganisms (Mindell, 1992; Jorgensen, 1993; Rohwer et al., 2002). It finally gained popularity when it was associated with the concept of hologenomic evolution in 2008 by Eugène Rosenberg and Illana Zilber-Rosenberg. They postulated that all macroorganisms are holobionts and that the unity between the host and its entire microbiota constituted a biological individual, a unit of selection (Zilber-Rosenberg & Rosenberg, 2008). In this paper, the authors proposed to describe corals as holobionts formed by cnidarians and their algal symbionts.

The concept of the biological individual, as we saw earlier in this book, allowed the study of organisms to be facilitated in several disciplines, such as physiology and genetics, and, added to this, technological advances in these disciplines made it possible to demonstrate the existing associations between hosts and their symbionts (Gilbert et al., 2012). We now know that a holobiont is an entity formed by associating different species that give rise to an ecological unit. This concept has some similarities with the so-called superorganism; however, they should be distinct. Wheeler initially used the latter term to refer to eusocial insects (Wheeler, 1928). It is a term used for organisms that establish colonies of individuals with an organized social structure and division of labor among their members. Superorganisms are formed by individuals of the same species, different from holobionts (Kramer & Bressan, 2015). In turn, due to the intimate associations between the respective genomes of the holobiont individuals, they are all taken as a functional unit in terms of physiological and ecological networks, the term hologenome being used for this purpose (Theis et al., 2016; Rosenberg & Zilber-Rosenberg, 2016).

There is a tendency to argue that it would be incomplete not to consider the holobiont concept in models on evolution, emergence, and biological complexity in living beings, suggesting replacing a more reductionist classical view with a more holistic one. It is in this framework that ten principles (Bordenstein & Theis, 2015) on holobionts and hologenomes have been proposed (Bordenstein & Theis, 2015) to organize this knowledge for future studies in the field, namely, (1) holobionts and hologenomes are units of biological organization; this means that complex eukaryotes are not autonomous organisms but are organizations of several symbionts and their genomes (Rosenberg & Zilber-Rosenberg, 2013). The associations between them are similar to an interaction of genotypes rather than an interaction between genotype and environment (Thompson, 2013). (2) Holobionts and hologenomes are not organ systems, superorganisms, or metagenomes, and this should be distinguished within biology. These terms are used to refer to interactions between symbionts and their hosts but do not imply organization, which is one of the principles when we talk about holobionts (Baquero & Nombela, 2012). (3) The hologenome is an integral genetic system consisting of a nuclear genome, organelles, and microbiome. Thus, variations occurring in any of the genomic subunits cause hologenomic variation (Bordenstein & Theis, 2015). (4) The concept of the hologenome again takes into account some elements of Lamarckian evolution, such as that the microbiome is initially acquired from the environment and can be inherited (Rosenberg & Zilber-Rosenberg, 2018). (5) The variation of hologenomes includes all mutational mechanisms. Each hologenome is composed of multiple mutants, i.e., it encompasses mutations occurring in each of the individual genomes of its parts (nucleus, organelles, and microbiome) (Rosenberg & Zilber-Rosenberg, 2013). (6) Hologenome evolution becomes more comprehensible when we draw an analogy between a gene in the nucleus genome and a microbe in the microbiome. The evolution of genes and symbionts is a change in population frequency observed over successive generations, i.e., the fraction of holobionts carrying that particular gene or microbe. Moreover, this evolution follows a joint inheritance of gene combinations, which ultimately functions like population genetics (Bordenstein & Theis, 2015). (7) The concept of a hologenome is adapted to selection theory. A holobiont is exclusively a host and its associated microbiota; therefore, different holobionts, for example, a pollinator and a flower, interact with each other under specific ecological characteristics (Lloyd, 2007). (8) The hologenome is shaped by selection and neutrality, which means that selection can propagate advantageous nuclear mutations or microbes or eliminate detrimental mutations. When selection does not act, the propagation of hologenome variation occurs exclusively by chance, i.e., as a stochastic process (Hubbell, 2005; Sloan et al., 2006; Nemergut et al., 2013). (9) Hologenomic speciation combines genetics and symbiosis. The study of hologenomic speciation is no longer optional. It is a necessary frontier that must be crossed to fully understand the origin of species (Bordenstein & Theis, 2015). (10) Holobionts and their hologenomes do not change the rules of evolutionary biology. Although the concept of what an animal or a plant is redefined, this does not change evolutionary theory. If the microbiome is a significant component of holobiont DNA, genetic variation can result in new adaptations and speciation (Peiffer et al., 2013; Goodrich et al., 2014).

The structure of the holobiont makes us consider that selection then acts at multiple levels. Within the microbiome itself, all kinds of biological and ecological interactions are present. Therefore, host characteristics are only one of the factors that determine it. Biotic interactions within this holobiont and selection pressures may play an essential role in determining this structure (Cerqueda-García & Falcón, 2016). In conclusion, the discovery of the holobiome and its subsequent in-depth study opens up new opportunities and possibilities for research about the factors influencing the evolution of species, which until now were limited. The concepts of holobionts and the hologenome do not transform evolutionary thinking. Instead, they affirm that hosts and their symbiotic microbes are complex units of organization upon which ecology, evolution, and selection act (Bordenstein & Theis, 2015).

6.4.2.7 About "Cryptic" in Biology

In various fields of biology, researchers have unveiled the existence of mechanisms and cryptic processes that are not immediately evident but are nonetheless present. These processes are more prevalent than previously believed, challenging the notion that they are mere rare occurrences. In a broad sense, by being cryptic, we refer to different characteristics that organisms can present that make it possible for them to go unnoticed by other animals (Allaby, 2015). Among these characteristics, one notable example is camouflage, which can either involve fixed coloration (as seen in classic examples such as striped patterns in zebras and tigers) or changing coloration (as observed in chameleons and cuttlefish that adapt their color based on their environment). In addition to visual crypsis, there are examples of olfactory crypsis, such as the case of squid ink (Conover, 2007) or the case of acoustic crypsis in some nocturnal butterflies to confuse predators (Miller & Surlykke, 2001), to name just a few examples. Indeed, these examples have been nothing new to the reader since they are widely accepted mechanisms in the contexts where they have been described. However, as we acquire more knowledge of the biodiversity surrounding us, these examples are multiplying. Let us also think of new technologies for detecting ultrasonic sounds, such as acoustic crypsis, where we were unaware of their existence because we did not have the necessary equipment for their detection.

In this context, it is crucial for scientists to acknowledge the minor genetic variations that may be involved in cryptic processes. Even though these variations may have a limited or no immediate impact on the phenotype, they can manifest themselves under specific conditions. The "crypsis," the hidden, can also encompass physiology. For example, hidden processes, not "so visible" that have been neglected, as was the communication in plants that we have already discussed earlier in this chapter. The concept of cryptic holds significant potential for analysis, as it transforms from being disregarded or viewed as rare and abnormal to something familiar. It is possible that we had not paid enough attention to it before due to technological, methodological, or theoretical challenges (such as paradigms, perspectives, biases, etc.). As we delve deeper into this concept, we uncover its importance and relevance in various biological phenomena. In this sense, we will focus on two examples of the cryptic biology studied recently, mainly due to advances in schools of thought, technologies, and more excellent knowledge of diversity.

Cryptic Genetic Variation

The variation in the genome that is not expressed in the phenotype observed in a population is called cryptic genetic variation (CGV) (McGuigan & Sgro, 2009; Gibson & Dworkin, 2004; Paaby & Rockman, 2014). The CGV indicates that the genetic diversity observed in a population is only a part of the potential diversity that can be produced and is a long-standing topic of study in evolutionary genetics, driven by the need to explain the adaptive capacity of populations to the constant changes that are occurring in the environment they inhabit (Gibson & Dworkin, 2004). Stress conditions that are generated in the environment where a given population inhabits and new conditions that populations may face (e.g., colonization of new environments, see Sect. 6.2.2.1) are suitable scenarios for CGV to become visible, as they contain mutations that were invisible when they emerged but may become evident under new circumstances (Paaby & Rockman, 2014).

In this sense, CGV studies became more frequent when analyses of the high levels of genetic variation associated with new phenotypes in a given environment, for example, after disturbance processes, began to be carried out (Gibson & Dworkin, 2004). However, what is the importance of having the tools provided, thanks to advances in genomic technologies, that allow us to know the CGV? Here, we can name some critical advantages that their knowledge offers us. For example, being able to know how organisms might respond to specific mutations, both in evolutionary and health contexts (such as diabetes and autoimmune diseases, for example), would undoubtedly be beneficial since we would know in advance how organisms might adapt or not adapt to specific changes in the environment, for example (Gibson & Dworkin, 2004). CGV could also take place in artificial selection processes in animals to contribute to the stabilization of new phenotypes, for example (see Sect. 6.3.4.1) (Gibson & Dworkin, 2004).

CGV can be viewed from a molecular or quantitative point of view (Paaby & Rockman, 2014). From molecular genetics, cryptic genetic variations are polymorphic loci that do not affect phenotype until they are perturbed by some rare condition (Gibson & Dworkin, 2004). From quantitative genetics, cryptic genetic variation is an increase in heritable phenotypic variation that arises when a population is exposed to rare conditions (Hermisson & Wagner, 2004; Paaby & Rockman, 2014). Additionally, CGVs may have a role in disease development. In this sense, alleles that accumulate while hidden may cause the onset of diseases in humans. However, thus far, the existing evidence is insufficient to support this hypothesis. Undoubtedly, studies in this field could be very advantageous when predicting, for example, the onset of certain diseases (Paaby & Rockman, 2014). To date, the models traditionally used for the detection of CGV have been *Drosophila melanogaster* and *Arabidopsis thaliana* (Gibson & Dworkin, 2004). However, these study models will undoubtedly be expanded in the coming years.

Cryptic Female Choice: From Restricted to Rule

A field where the cryptic aspect has been analyzed in increasing depth is sexual selection. We will devote more space to it since it is a rich subject. First, let us put the subject in context. Sexual selection mechanisms can occur in a precopulatory context through competition between males for access to females through ritualized fights or demonstrations (intrasexual selection) and choice of females where the evaluation criteria can be very varied (intersexual selection), obtaining direct benefits such as food or shelter, or indirect benefits through higher quality offspring (Darwin, 1874; Trivers, 1972). These mechanisms have mainly influenced the development of secondary sexual characteristics in males. Think of the classic examples of the antlers of cervids or the striking colors of the peacock's tail (Darwin, 1874; Trivers, 1972; Firman et al., 2017).

Concerning the sexual selection mechanisms associated with cryptic organisms, our focus of interest here is after copulation has begun, i.e., in a postcopulatory context (Andersson & Iwasa, 1996; Eberhard, 1996; Jennions & Kokko, 2010; Hosken & House, 2011). In this sense, if we consider processes that are not directly available to the observer's view as cryptic, we can name sperm competition (Parker, 1970) and cryptic female choice (Thornhill, 1983; Eberhard, 1996), as these processes take place within the female reproductive tract. Sperm competition is a postcopulatory extension of male–male competition, where males compete through their ejaculates within the reproductive tract of females for access to fertilization of their eggs (Parker, 1970). This competition is framed in game theory as males try to do their best to maximize their reproductive success against the interests of other males. The male needs to think about what is best for him, but it depends on what the other males do. Selection pressure favors traits in males that will help them increase their fertilization success, favoring the ability to outcompete the sperm of males that were previously copulated with the female and reducing competition with sperm from subsequent males (Thornhill, 1984; Cordero, 1990; Alcock, 1994; Calbacho-Rosa & Peretti, 2015). Studies of sperm competition that show their evolutionary importance in the processes linked to sexual selection have received particular interest in recent years, showing that they occur in many species of different taxa, no longer as specific exceptions (Andersson, 1994; Birkhead & Møller, 1998). The increase in this type of study has undoubtedly been possible thanks to technological advances, for example, the development of filming equipment that has allowed us to observe the process of intromission of the copulatory organs of males in the reproductive tract of females to be able to associate them, for example, with the placement of sperm in a particular area to favor its use, or the removal of sperm that the female may already have from previous copulations, to name just a few examples.

However, the term cryptic has been explicitly used as a form of female choice not to confuse the reader; when we refer to cryptic, we refer to female choice processes. Therefore, cryptic female choice (CFC) is the process of female choice that continues during and after mating, allowing females to bias (through morphological, behavioral, and physiological mechanisms—Arnqvist, 2014) paternity by "choosing" the sperm from which males will fertilize the majority of their eggs (Thornhill, 1983; Eberhard, 1996; Biswas et al., 2022). Let us keep in mind that until only a few years ago, it was thought that males controlled all processes associated with reproduction in animals. In this sense, as mentioned above, males competed for access to females in a precopulatory sexual selection framework, and their ejaculates continued to compete within the reproductive tract of females for fertilization of eggs in a postcopulatory sexual selection framework; and females … What role were females supposed to have? None of their roles was passive, and they were only considered "simple reservoirs" of sperm. The idea that monogamy was the rule and that polygamous species were instead the exception, mainly in the animal kingdom, was also deeply rooted. However, thanks to changes in scientific interpretations, undoubtedly based on advances in sociocultural thinking, females began to be seen as active participants with the capacity to "choose" or bias in reproductive processes. In the same way, it was seen that the rule in nature was polygamy favoring these processes of sexual selection. In this sense, a female can also select with which male she will copulate, a female choice in a precopulatory framework, and even choose with the sperm of which male she will fertilize her eggs, a cryptic female choice in a postcopulatory framework. So you see, females have an active role just like males. This acceptance has undoubtedly had a significant impact on the current paradigm shifts. Our interpretations have undergone such significant changes that concepts such as female choice no longer seem strange to us. In fact, we have progressed to the point where considering the possibility of cryptic female choice does not appear far-fetched. This openness to new ideas and possibilities reflects the evolving nature of scientific understanding and our willingness to explore and embrace novel concepts in biology. From the beginning of copulations, several processes are controlled by females. For example, staying quiescent to facilitate the spatial location of the male and the insertion of intromittent organs, allowing the placement or transfer of sperm to storage and fertilization sites, sperm nutrition, egg maturation, and implantation in the uterus, among others. These examples almost naturally show that females have an active role in sexual interactions (Eberhard, 1985, 1991, 1996; Shuker & Simmons, 2014; Biswas et al., 2022).

In his book, published in 1996, Eberhard shows that CFC processes are widespread in numerous insect species based on different behaviors that males develop during or after mating. These behaviors exhibited by males, which aim to stimulate females to utilize the sperm of a specific male, are referred to as copulatory courtship. This courtship behavior can be classified into two types: genital or nongenital, depending on which body structure the male uses to stimulate females (Otronen, 1990; Eberhard, 1991, 1994, 1996, 2009; Calbacho-Rosa & Peretti, 2015). Technological advancements, such as the development of video cameras with slow-motion capabilities, played a crucial role in drawing attention to male behaviors associated with copulatory courtship. These cameras enabled the detection of high-speed behaviors that were previously challenging to observe (Dell et al., 2014), as mentioned earlier in Sect. 6.4.2.1. Additionally, detailed descriptions to determine frequencies, relationships, and chronologies of particular behaviors in sexual interactions allowed for fine-grained descriptions of behavioral patterns. However, to demonstrate that a behavior functions as copulatory courtship, it must also be associated with increased fertilization success for a given male. This idea proposes that males stimulate females, which allows the female to bias paternity toward the male she "thinks" did better! In this sense, genomic tools such as multiallelic variable number of tandem repeat (VNTR) polymorphism analyses were also very useful (Pena & Chakraborty, 1994). Additionally, many studies in arthropods have shown that copulations are more prolonged than necessary if the only function involved is sperm transfer; this gives the possibility that males are developing other behaviors, for example, those linked to copulatory courtship (Eberhard, 1996, 2009; Calbacho-Rosa et al., 2013).

On the other hand, and associated with CFC, we can name genital evolution, which has been studied and demonstrated in numerous species in recent years. One of the most striking evolutionary trends is the significant divergence and complexity observed in male genitalic structures across various taxa with internal fertilization (Eberhard, 1985; Leonard & Córdoba-Aguilar, 2010; Peretti & Aisenberg, 2015; Eberhard & Lehmann, 2019). In contrast, females do not exhibit such divergence, and their genitalia typically have a uniform appearance. Like copulatory duration, this great complexity and variation in male genitalia would not make much sense if the only function was sperm transfer to females (Eberhard, 2009). Therefore, these genitalia may serve other functions, such as those associated with CFC, where male genitalia, through copulatory courtship, stimulate females to bias their sperm use during egg fertilization (Eberhard 1985, 1996; Eberhard & Lehmann, 2019). The advancement of equipment, including powerful stereo microscopes and various types of microscopes (e.g., compound microscopy, transmission microscopy, phase-contrast microscopy, confocal microscopy, and scanning electron microscopy), has facilitated the observation of genital structures with exceptional precision. These developments, along with progress in physiology and histology, have been instrumental in investigating and identifying significant variations in genital structures. As Eberhard (1991, 1996) points out, in various species with internal fertilization and where CFC takes place, copulation does not always lead to insemination, nor does insemination guarantee egg fertilization. Consequently, it is reasonable, and perhaps even logical, to consider the occurrence of cryptic mechanisms in such a crucial process as offspring production. These mechanisms are more likely to be a pattern rather than an exception.

Due to the substantial resistance among many researchers to accept the possibility of CFC while acknowledging sperm competition, Eberhard and Lehmann (2019) raise an important question in their paper. They inquire about the amount of evidence needed to definitively establish the existence of the CFC, considering the skepticism surrounding this hypothesis. Crickets of the species *Roeseliana roeselii* and flies of the genus *Glossina* show us significant evidence in favor of CFCs. In the case of these flies, sperm transfer occurs at the end of copulation, and males perform various nongenital behaviors during copulation. The function of these behaviors is copulatory courtship to stimulate females (Jaensen, 1979; Briceño et al., 2007). On the other hand, the males of *R. roeselii*, as occurs in many other cricket species, have titillators. These titillators are paired chitinized structures that do not directly function in sperm transfer (Wulff & Lehmann, 2016). Movement of these structures during copulations is associated with female stimulation under CFC (Wulff & Lehmann, 2016; Eberhard & Lehmann, 2019). According to Eberhard and Lehman, while expecting absolute confirmation of the hypothesis may not be logical, there is a wealth of evidence in the form of morphological and behavioral traits in both males and females. These pieces of evidence strongly support the existence of CFCs in numerous species across different taxa. From our perspective, scientific interpretations are evolving due to advancements in technology and sociocultural changes. Cryptic phenomena are no longer considered mere rarities but are increasingly recognized as widespread occurrences. This shift in understanding is facilitated by improved technological capabilities and changes in societal attitudes.

Beyond the mere behavioral evidence, there are some other cases that favor the CFC. In some species of insects and spiders, for example, males give nuptial gifts to females at mating. These gifts are usually prey that the females may eat during mating. The nuptial gifts may be an honest signal representing the quality of the male, so it is a trait that may be under CFC. Among insects, for example, females regulate oviposition according to the size of prey given as nuptial gifts, resulting in a bias toward those males in fertilization (Thornhill, 1983). Another example is the nursery web spider *Pisaura mirabilis*, where gift-giving males have more significant mating and fertilization success (Stalhandske, 2001; Prokop & Maxwell, 2009).

The CFC hypothesis has also been linked to the evolution of male parasperm, a nonfertile type of sperm (Hayakawa, 2007). While previous work shows that sperm and nonsperm components of the ejaculate can serve in the CFC (Eberhard & Cordero, 1995; Cordero, 1998), it has recently been suggested that the parasperm evolves as a function of CFC (Swallow & Wilkinson, 2002; Oppliger et al., 2003; Holman, & Snook, 2006). Parasperm could lead to increased ejaculate size and movement within the spermatheca (Shepherd & Bonk, 2021), which could stimulate females to use the sperm of a particular male for the fertilization of their eggs (Holman & Snook, 2006). This has been observed in butterfly species such as *Pieris rapae* (Wedell & Cook, 1999) and *Plodia interpunctella* (Gage, 1995). It should be noted that this type of analysis has been possible thanks to more advanced physiological studies that have made it possible to differentiate which components of the ejaculate are spermatozoa and which are not so that different functions can be attributed to them.

Another less studied advantage attributed to CFC is that females could avoid inbreeding depression by avoiding fertilizing their eggs with sperm from related males. CFC would favor utilizing sperm from genetically different males (Løvlie et al., 2013), as seen in the cricket *Gryllus bimaculatus* (Tregenza & Wedell, 2002) and the domestic fowl *Gallus gallus* (Birkhead et al., 2004). Similarly, this preference of females to avoid inbreeding could be detected thanks to technological advances in genetic studies that allow us to know the degree of kinship between individuals. The possibility of CFC in humans is much more controversial and even less studied. Currently, some studies seem to show that follicular fluid may differentially regulate sperm behavior and thus influence fertilization success (Ralt et al., 1991; Fitzpatrick et al., 2020), although many studies are still needed to obtain sufficient evidence to support the occurrence of this process. Undoubtedly, with the advancement of science, this type of study could be developed with much more frequency and accuracy, having precise results on issues that today seem distant.

As we have been saying, CFC is the process of choice by females, which takes place during and after mating, allowing them to "choose" the sperm of which male they will use to fertilize most of their eggs. However, Kekäläinen (2022) suggests that there is evidence to show that females may also select among sperm from the same male. In this sense, CFC acts through prefertilization processes at the cellular and molecular level that facilitate gamete-mediated mate choice (GMMC, Kekäläinen & Evans, 2018), allowing fertilization bias toward genetically compatible sperm haplotypes (Kekäläinen, 2022). This may happen as gametes are genetically unique, and then different sperm can be expected to differ in their compatibility with female eggs, which will ultimately have consequences on offspring fitness (Immler et al., 2014; Alavioon et al., 2017, 2019; Kekäläinen, 2022). This raises the possibility that GMMC evolves to select compatible sperm genotypes within the ejaculate of the same male (Kekäläinen, 2022).

As we have already said, trends in scientific thought have always been associated with the sociocultural context of the moment. The study of female choice processes has received relatively little attention within the field of evolutionary biology and has often been regarded as rare exceptions. In the past, the prevailing paradigm in the context of sexual events focused on the notion that males solely controlled these processes. In the past, it was not widely acknowledged that females could actively participate in these scenarios by exercising choice in selecting their mating partners. Moreover, the idea that mate choice processes could extend beyond the initial stages of mating was not commonly believed, as mentioned earlier in this discussion. However, as research has progressed, we are now beginning to recognize and appreciate the active role that females play in these processes and the potential for mate choice to continue beyond the initial mating stages. The hypothesis of female choice, encompassing both precopulatory and postcopulatory choices, has been steadily gaining acceptance and overcoming previous biases. We now acknowledge that females are generally polyandrous, engaging in multiple mating, and actively contributing to decisions related to reproduction. It has been a significant change in the paradigm that predominated until few years ago. Scientific interpretations no longer only focus on the explanations of the operant processes from the approach of the males, analyzing their costs and benefits but also from the point of view of the advantages of the females.

Therefore, what is certain is that, to a greater or lesser extent, what is happening is a significant change in the views of the roles of each sex and the mechanisms involved in decision-making, among others (Eberhard, 1996; Peretti & Aisenberg 2015). Indeed, the need for a robust theoretical framework shows that cryptic processes are much more frequent than we think. This lack may also have been responsible for these rejections of the increasingly eloquent evidence that it is, for example, the CFC hypothesis that specific mechanisms are operating (Eberhard, 1991).

Finally, the study of sexual cryptic processes requires an understanding of mechanisms associated with insemination, storage, and fertilization that, until a few decades ago, were little explored areas (Birkhead, 1998; Biswas et al., 2022). Consider the technological advances that have allowed differential paternity studies, the key to supporting sperm competition, and CFC hypotheses (Simmons & Wedell, 2020). For example, innovative techniques that allow assessment of sperm storage and fertilization sites within the female genital tract, techniques that consist of sterilization of males by exposure to X-rays (which do not affect the fertilization capacity of sperm but do cause embryo death at an early stage of development—e.g., Kaster & Jakob, 1997 in spiders) and innovative microsatellite techniques to be able to determine, by DNA analysis, the paternity of eggs (Weber et al., 2018; Forsdick et al., 2021). Additionally, in vitro technologies developed in recent years have aided the understanding of cryptic processes, which, together with an increased acceptance of female participation, have favored the study and development of cryptic female choice (Firman et al., 2017).

6.4.2.8 Sexual Selection in Plants: From Simple Curiosities to a Generalized Fact

As we said, the history of sexual selection is complicated from its beginnings, and many factors influenced this. One of the strongest was that Darwin developed this theory in the Victorian era. This was quite a turbulent time for such revolutionary and new theories. Victorian society was entirely dominated by demureness, moralism, and discipline, with rigid prejudices and severe interdictions, which can be roughly summarized as repression (Aksenchuk, 2006). It was a society where men were the ones who ordered and dominated the public space, and women were limited to the private space under a submissive status. Victorian values could currently be classified not only as sexist but also as "puritanical," and the most prominent were the values of thrift, hard work, the extreme importance of morality, the duties of faith, and Sunday rest (Charlot & Marx, 1993). Let us imagine then how difficult it must have been to develop a theory with a taboo subject such as sex in such a puritan society (Vandermassen, 2004). Darwin was struck by the number of characters that are exaggerated and hard to explain by natural selection. He saw that, in general, males develop fighting or attention-seeking characters and that females develop an "*advent for beauty*" (Stephenson & Bertin, 1983; Andersson, 1994). Thus, males fought directly or indirectly among themselves to be chosen by females and to leave offspring.

As mentioned in the previous section, sexual selection arises as an extension of natural selection in which both males and females develop tactics and traits that allow them to reproduce more and better. Just as in natural selection, the organism that does not adapt does not survive; in sexual selection, the organism that is not reproductively successful does not necessarily die but leaves no offspring, something like a "reproductive death." The theory of sexual selection was conceived based on higher vertebrates such as mammals and birds and has a strong influence on human beings. Darwin did not wholly account for lower organisms, not even arthropods, and much fewer plants (Andersson, 1994; Arnold, 1994; Moore & Pannell, 2011). The latter fact is striking, considering the avid studies on plants carried out by Darwin and his son Francis (Darwin, 1875; Darwin, 1899; Darwin & Darwin 1880, 1888). The prevailing paradigm or the way plants were perceived at the time dated back to the time of Aristotle. The view was that plants were considered vegetative organisms lacking a "soul" and a "specific purpose in life." This idea contrasted with animals, organisms that had voluntary movement. Because of this view, the scientific models were primarily animals, especially vertebrates, that showed much more visible behavior to researchers. Added to this strong bias toward animals was the strong anthropocentrism reigning since antiquity. Consequently, plants have received limited scientific attention beyond their practical applications in everyday food and medicine (Jones, 1996; Petrovska, 2012).

On the other hand, we must keep in mind that sexual selection was developed with dioecious organisms. However, it only requires sexual reproduction and not necessarily two sexes (Andersson, 1994). Again, with this premise, some lower organisms and most plants are left out. In addition, male reproductive success in plants is more difficult to quantify than in animals (Andersson, 1994). Finally, the concept of sexual selection was perhaps too strict for plants to fit. A subtle relaxation of the definition allows us to incorporate plants and other organisms, which were not contemplated at the beginning of the theory (Andersson, 1994; Moore & Pannell, 2011). Sexual selection was not a theory embraced by the scientific community quickly (Moore & Pannell, 2011). Indeed, except for a brief lapse in the early twentieth century (e.g., Fisher, 1930; Noble 1934, 1936), it was not until the late 1960s and early 1970s that sexual selection, and especially female choice, was progressively analyzed in depth both theoretically and empirically.

Plants have always seemed to fall outside the framework of sexual selection. Darwin did not consider them and mentioned them vaguely in his theory. He considered that sexual selection did not represent a crucial selective force in them and in lower organisms since they could not "*appreciate the beauty in the other individual*" (Moore & Pannell, 2011). Thus, until relatively recently, sexual selection in plants was mainly ignored (Stephenson & Bertin, 1983). In botany, at most, anecdotal examples were cited that were no more than curiosities or exceptions at the time but could rarely be generalized to the majority of the group. In the early twentieth century, Correns (1917, 1928), Heribert-Nilsson (1923), and Brink (1927) believed that pollen competition could have evolutionary consequences. Similarly, Haldane (1932) and Mulcahy (1979) postulated that competition between pollen grains is essential for the reproductive success of plants. Haldane (1932) also realized that plants produce much more pollen than is necessary to fertilize ovules. Huxley (1942) suggested that competition between pollen grains favors the speed of pollen tube growth. Until that time, these were all anecdotal observations, with a certain amount of hypothesis and speculation. It was not until 1948 that there was an essential change in the conception of sexual selection in plants. Bateman unequivocally suggested the existence of sexual selection in plants (Andersson, 1994; Birkhead & Møller, 1998). Recall that Bateman developed the principle that female reproductive success is limited by the number of eggs the female produces (fecundity), while male reproductive success is limited by the number of matings the male can make (Bateman, 1948; Trivers, 1972).

At present, we can say that there is a break in thinking, going from perceiving sexual selection as something rare in plants to a phenomenon that occurs. However, there still needs to be more discussion of the extent of this phenomenon in plants as a group. Although it was an important milestone to have determined that sexual selection could also operate in plants, there was little interest for almost 30 years. It was not until the 1970s that numerous papers came to light that spoke of pollen benefiting from larger inflorescences (Willson & Rathcke, 1974), reaffirming the idea that there is an increase in both pollen and nectar and linked it to male flowers (Gilbert, 1975), and spoke of mating preference in plants (Janzen, 1977—although he does not mention sexual selection). Willson (1979) suggests that sexual selection may be responsible for different evolutionary phenomena, such as the emergence of pollinia and different mating systems. More recent work in the 1980s considers the influence of sexual selection on the evolution of plants and mating systems, with emphasis on dioecious systems (Lloyd, 1979; Bawa, 1980a, 1980b; Givnish, 1980; Bawa & Beach, 1981; Webb, 1981; Casper & Charnov, 1982). By that time, botanists had already applied the idea of sexual selection with plants in mind, but some difficulties arose (Moore & Pannell, 2011). For example, we already cited that Darwin avoided hermaphroditic organisms, and we know that many systems in plants are hermaphroditic. In addition, conspicuous secondary sexual characters are absent, such as those in animals. Since the 1990s, numerous papers have explored different examples of characters shaped by sexual selection (reviewed in Skogsmyr & Lankinen, 2002); these authors state that although sexual selection in plants is not denied today, there is much controversy in the interpretation of specific mechanisms and the function of certain structures.

Analogies Between Plants and Animals

The rule is that sexual selection operates in plants. Perhaps what is confusing is trying to visualize sexual selection at the prefertilization stage of plants. The exception at this level is the production of enormous amounts of pollen, even far over the number of ovules to be fertilized. This can be directly compared to the production of large quantities of sperm in the males of most animals. However, it will be harder to see if we are looking for something similar to precopulatory competition between animal males. Imagine that the display of males is hardly perceived by "females" in plants, or at least it has not been demonstrated thus far. Thus, the showiest flowers do not target the female flowers or the female part of the plant, but their primary target is the pollinators. Recall that to transfer pollen, pollinators or means of pollination, such as wind, are needed. This fact makes the idea of sexual selection even more distant and complex if we compare it to animals. However, once the pollen reaches the pistil and starts to compete, it becomes a much more familiar selection scenario comparable to sperm competition in animals.

From a precopulatory point of view, physical confrontation was evidence that helped Darwin devise the concept of sexual selection; thus, confrontational behavior between males is commonly known among animals (Andersson, 1994). This was thought to be rare or absent in plants (Lloyd & Webb, 1977; Queller, 1987; Grant, 1995; Murphy, 1998; Skogsmyr & Lankinen, 2002; Delph & Ashman, 2006; Moore & Pannell, 2011), but recent studies in pollinaria strongly suggest that this would not be difficult to find in plants. For example, direct competition has been found to exist between pollinaria of Apocynaceae (milkweeds) (Cocucci et al., 2014). Pollinaria compete for access to specific points on the pollinator's body and can sometimes concatenate to existing pollinaria. This causes a decrease in the reproductive display of the resident pollinium. Thus, from a sexual selection standpoint, there are individuals who better exploit optimal anchorage sites on the pollinator previously occupied by pollinia of rival individuals (but see Coombs et al., 2012 for a noncompetitive explanation). There is evidence that horn development in pollinators is a strategy that avoids concatenation. Here, we see that reproductive success depends on physical interference between individuals in the absence of autonomous movement and sensory perception as present in animals.

From a perspective closer to postcopulatory in animals, in addition to the production of large amounts of pollen, pollen tubes compete with each other once they reach the pistil, similar to how sperm from different males compete to fertilize eggs. This phenomenon can be taken as an extension of the intrasexual competition that we cited as an example previously among pollinia (Cocucci et al., 2014) and similar to what occurs between male–male competition in animals and its extension to the postcopulatory level (Birkhead & Møller, 1998; Simmons 2001, 2014). Gamete aggregation structures also transfer more significant numbers of gametes in a compact form. Examples are pollinia, as mentioned earlier in plants (e.g., orchids and *Asclepias*), and sperm packages, sperm bundles or spermatozeugmata in animals (Pitnick et al., 2009). These would be mechanisms to win the competition between different individuals. However, will there be mechanisms to prevent competition? In animals, structures such as genital plugs prevent the female from mating with a new male, something like a chastity belt (Simmons, 2001; Kunz et al., 2014; Stockley et al., 2020; Fromhage, 2012). In addition, there are guarding behaviors where males generally do not let other males contact the female (mate guarding) (Alcock, 1994; Simmons, 2001). In plants, structures similar to genital plugs are not cited, but there is a similar mechanism: the inhibition of pollen grain germination by the presence of adjacent pollen. Some pollen grains do not germinate in the presence of other grains, either from the same male or from other males (Skogsmyr & Lankinen, 2002). There are also examples where large pollen deposits cover the female stigma, preventing the germination of pollen from other individuals (Skogsmyr & Lankinen, 2002). In plants, pollinators favor pollen transfer. In this sense, if we consider that pollinators allow transport and reach the site of gamete deposition (the pistil), we could say that they fulfill a function partly analogous to the structures that animals use for sperm transfer (e.g., spermatophore, insect aedeagus, penis).

On the other hand, there is room for cryptic female choice in plants (e.g., Janzen, 1977; Charnov, 1979; Willson & Burley, 1983; Galen & Rotenberry, 1988; reviewed in Charlesworth et al., 1987; Queller, 1987; Lyons et al., 1989; Marshall & Folsom, 1991). Cryptic choice appears once pollen reaches female reproductive structures (Bernasconi et al., 2004; Paterno et al., 2020). These structures can select pollen according to their compatibility (self-incompatibility—Haring et al., 1990; Takayama & Isogai, 2005) according to genetic compatibility systems or by distance from which the pollen originates (Outcrossing distance Price & Waser, 1979; Waddington, 1983; reviewed by Bertin, 1988; Doust & Doust, 1988; Waser, 1993). In the first case, this system may have evolved to avoid inbreeding, and this type of choice does not favor any particular phenotypic trait (Waser, 1987; Andersson, 1994). We see that histochemical control is common in pistil receptivity and pollen germination itself (whether she "allows" it or not), but there is also evidence of selective seed abortion, something similar to what can occur in animals (Simmons, 2001; Simmons, 2014). However, some researchers suggest that selective abortion is not a mechanism to select mates but rather one to ensure a fitter offspring (e.g., Queller, 1987).

In conclusion, to summarize the idea of this section on sexual selection in plants, we see that there has been a profound paradigm shift since Darwin. The issue of sexual selection was resisted at first, but then it was taken as a strong evolutionary force in animals. Plants were slow to join this paradigm due to old conceptions that considered plants "vegetables" and "soulless" organisms. The change of conception added to the relaxation of the definitions of sexual selection, and the advent of microscopic molecular studies and laboratory assays allowed the shape of the idea of sexual selection in plants from an exceptional idea to a fairly widespread, massive rule in the plant kingdom.

6.5 Key Concepts and Ideas in This Chapter

- In biology, as occurs at the epistemological level with the advent of new paradigms, there are transitions from rarities or exceptions to patterns or rules. This is part of an intrinsic dynamism linked to our way of conceiving scientific knowledge, but it is also a property of biological systems. We review multiple examples of these transitions in biology.
- The dynamics of E-R can be the result of natural processes, anthropological processes linked to activities of human beings, and finally changes in the perception of science and nature and the development of technological advances.
- Although life is currently the rule on planet Earth, it was originally an exception. In primitive environments where the rule was a lack of oxygen, the first organ-

isms were responsible for modifying the atmosphere so that life could become the rule we know today.

- Other illustrative and key examples in the evolution of life as we know it today were the appearance of eukaryotes, multicellularity, and sex. These rare and uncommon events can be seen as "teratological" phenomena that were successful, linked to changes in gene expression, becoming rules later on.
- Today, the earth is dominated by living beings, but there was a time when the rule was the opposite and there were no terrestrial organisms. The terrestrial conquest was an exceptional milestone, and the organisms that achieved it were also exceptional. Most likely, these organisms were rare, with "extreme" qualities to tolerate transitional terrestrial systems. That is, traits that are rare and a priori useless at one time can become an advantage at another time and make the trait the rule at another time under different conditions.
- In general, when colonizing individuals possess genes that are not representative of the population, the new population descending from them may have an altered gene frequency. This founder effect can lead to the appearance of traits or diseases that were rare in the original population. Thus, there is a change from exception to rule but also from rule to exception since the few individuals from the original population that may be in the new population carry the alleles that will now be rare in this new population.
- Phenotypic plasticity is a phenomenon whereby organisms with the same genotype can express a different phenotypic range depending on the environment. The characters to be expressed may at first be hidden and suddenly, if the environment is favorable, generate a novel character that may initially be a rarity. This change in the environment can also be due to atypical, rare conditions that did not occur before and did not allow the character in question to be revealed.
- Another example of the dynamism of exceptions to rules is observed in introduced exotic species, which in general tend to present a great competitive ability and capacity for propagation, which makes them quickly become common and abundant in new environments.
- Global processes such as diseases that can even become pandemics begin as an exception, which in a certain space and time, with the help of changing human habits and relationships, can become a rule.
- Human action processes such as environmental pollution and climate change have accelerated in recent times, so many patterns that were rare before are very common now.
- Trends in scientific ideas have always been very closely linked to the sociocultural context that is validated or current at the time. Some witness cases that have been changing are, for example, the field of animal learning or personality, the senses in plants, or female cryptic choice in sexual selection. The changes in perspective and increase in the number of studies have led us to know more examples, and although these phenomena are not considered as rules, they are now accepted not only as mere exceptions.

References

Aanen, D. K., & Eggleton, P. (2017). Symbiogenesis: Beyond the endosymbiosis theory? *Journal of Theoretical Biology, 434*, 99–103.

Aberth, J. (2011). *Plagues in world history*. Rowman & Littlefield Publishers.

Abrahamson, W. G. (1989). *Plant-animal interactions*. McGraw-Hill.

Ackerman, J. D., Cuevas, A. A., & Hof, D. (2011). Are deception-pollinated species more variable than those offering a reward? *Plant Systematics and Evolution, 293*, 91–99.

Adler, L. S. (2000). The ecological significance of toxic nectar. *Oikos, 91*(3), 409–420.

Adli, M. (2018). The CRISPR tool kit for genome editing and beyond. *Nature Communications, 9*(1).

Aguado-Aguilar, L. (2001). Aprendizaje y memoria. *Revista de neurología, 32*(4), 373–381.

Ah-King, M. (2023). *The female turn: How evolutionary science shifted perceptions about females*. Springer Nature.

Ahmad, H., Sehgal, S., Mishra, A., & Gupta, R. (2012). *Mimosa pudica* L. (Laajvanti): an overview. *Pharmacognosy Reviews, 6*(12), 115.

Aksenchuk, R. (2006). De la moral victoriana al goce postmoderno: Freud Lacan y Zizek. *Observaciones filosóficas, 2*, 11.

Aktipis, C. A., Boddy, A. M., Jansen, G., Hibner, U., Hochberg, M. E., Maley, C. C., & Wilkinson, G. S. (2015). Cancer across the tree of life: Cooperation and cheating in multicellularity. *Philosophical Transactions of the Royal Society B: Biological Sciences, 370*(1673), 20140219.

Alavioon, G., Hotzy, C., Nakhro, K., Rudolf, S., Scofield, D. G., Zajitschek, S., Maklakov, A. A., & Immler, S. (2017). Haploid selection within a single ejaculate increases offspring fitness. *Proceedings of the National Academy of Sciences, 114*, 8053–8058.

Alavioon, G., Garcia, A. C., LeChatelier, M., Maklakov, A., & Immler, S. (2019). Selection for longer lived sperm within ejaculate reduces reproductive ageing in offspring. *Evolution Letters, 3*, 198–206.

Alcock, J. (1994). Postinsemination associations between males and females in insects: The mate-guarding hypothesis. *Annual Review of Entomology, 39*(1), 1–21.

Alcock, J. (2000). Interactions between the sexually deceptive orchid *Spiculaea ciliata* and its wasp pollinator *Thynnoturneria* sp. (Hymenoptera: Thynninae). *Journal of Natural History, 34*(4), 629–636.

Alegado, R. A., & King, N. (2014). Bacterial influences on animal origins. *CSH Perspect Biol, 6*(11), a016162.

Alekseyenko, O. V., Chan, Y. B., Li, R., & Kravitz, E. A. (2013). Single dopaminergic neurons that modulate aggression in *Drosophila*. *Proceedings of the National Academy of Sciences of the United States of America, 110*, 6151–6156.

Alibardi, L. (2010). *Morphological and cellular aspects of tail and limb regeneration in lizards. A model system with implications for regeneration in mammals*. Springer.

Allaby, M. (2015). Crypsis. In *A dictionary of ecology* (5th ed.). Oxford University Press.

Allen, R. D. (1969). Mechanism of the seismonastic reaction in *Mimosa pudica*. *Plant Physiology, 44*(8), 1101–1107.

Alonso, L. (2013). *Biología sintética: La nueva frontera de la investigación*. Investigación y Ciencia.

Al-Qahtani, A. A. (2020). Severe acute respiratory syndrome coronavirus 2 (SARS-CoV-2): Emergence, history, basic and clinical aspects. *Saudi Journal of Biological Sciences, 27*(10), 2531–2538.

Altieri, M. A., Letourneau, D. K., & Risch, S. J. (1984). Vegetation diversity and insect pest outbreaks. *Critical Reviews in Plant Sciences, 2*(2), 131–169.

Altman, S. (1990). Enzymatic cleavage of RNA by RNA (Nobel lecture). *Angewandte Chemie International Edition in English, 29*(7), 749–758.
Altman, S. (2000). The road to RNase P. *Nature Structural Biology, 7*(10), 827–828.
Alvarez, L. W., Alvarez, W., Asaro, F., & Michel, H. V. (1980). Extraterrestrial cause for the cretaceous-tertiary extinction. *Science, 208*(4448), 1095–1108.
Alvarez-Buylla, A. (1992). Neurogenesis and plasticity in the CNS of adult birds. *Experimental Neurology, 115*, 110–114.
Amasino, R. (2010). Seasonal and developmental timing of flowering. *The Plant Journal, 61*(6), 1001–1013.
Amodio, P., Boeckle, M., Schnell, A. K., Ostojic, L., Fiorito, G., & Clayton, N. S. (2019). Grow smart and die young: Why did cephalopods evolve intelligence? *Trends in Ecology & Evolution, 34*(1), 45–56.
An, Y., Tsang, K. K., & Zhang, H. (2006). Potential of stem cell-based therapy and tissue engineering in the regeneration of the central nervous system. *Biomedical Materials, 1*(2).
Anderson, P. W. (1983). Suggested model for prebiotic evolution: The use of chaos. *Proceedings of the National Academy of Sciences, 80*(11), 3386–3390.
Anderson, D., Glibert, P. M., & Burkholder, J. M. (2002). Harmful algal blooms and eutrophication: Nutrient sources, composition, and consequences. *Estuaries, 25*, 704–726.
Andersson, M. (1994). *Sexual selection*. Princeton University Press.
Andersson, M., & Iwasa, Y. (1996). Sexual selection. *Trends in Ecology & Evolution, 11*(2), 53–58.
Andersson, S. G., Karlberg, O., Canback, B., & Kurland, C. G. (2003). On the origin of mitochondria: A genomics perspective. *Philosophical transactions of the Royal Society of London. Series B: Biological Sciences, 358*, 165–177.
Anthony, K., Kline, D., Diaz-Pulido, G., Dove, S., & Hoegh-Guldberg, O. (2008). Ocean acidification causes bleaching and productivity loss in coral reef builders. *Proceedings of the National Academy of Sciences, 105*(45), 17442–17446.
Apple, J., & Feener, D. (2001). Ant visitation of extrafloral nectaries of *Passiflora*: The effects of nectary attributes and ant behavior on patterns in facultative ant-plant mutualisms. *Oecologia, 127*(3), 409–416.
Applewhite, P. B. (1968). Non-local nature of habituation in a rotifer and protozoan. *Nature, 218*, 287–288.
Apprill, A. (2017). Marine animal microbiomes: Toward understanding host-microbiome interactions in a changing ocean. *Frontiers in Marine Science, 4*, 222.
Arkin, A. (2008). Setting the standard in synthetic biology. *Nature Biotechnology, 26*(7), 771–774.
Arnold, S. J. (1994). Bateman's principles and the measurement of sexual selection in plants and animals. *The American Naturalist, 144*, S126–S149.
Arnold, M. L. (1997). *Natural hybridization and evolution*. Oxford University Press.
Arnqvist, G. (2014). Cryptic female choice. *The evolution of insect mating systems* (pp. 204–220).
Attenborough, D. (1995). *The private life of plants*. BBC Natural History Unit, British Broadcasting Corporation, London. TV production in association with Turner Broadcasting Systems, London.
Aubret, F., Shine, R., & Bonnet, X. (2004). Adaptive developmental plasticity in snakes. *Nature, 431*(7006), 261–262.
Audesirk, T., Audesirk, G., & Byers, B. E. (2008). *Biology: Life on earth with physiology* (8th ed.). Prentice Hall.
Avise, J. (2011). *Hermaphroditism: A primer on the biology, ecology, and evolution of dual sexuality*. Columbia University Press.
Awan, H., Adve, R. S., Wallbridge, N., Plummer, C., & Eckford, A. W. (2019). Information theoretic based comparative analysis of different communication signals in plants. *IEEE Access, 7*, 117075–117087.
Awramik, S. M. (1971). Precambrian columnar stromatolite diversity: Reflection of metazoan appearance. *Science, 174*, 825–827.

Babla, M., Cai, S., Chen, G., Tissue, D. T., Cazzonelli, C. I., & Chen, Z. H. (2019). Molecular evolution and interaction of membrane transport and photoreception in plants. *Frontiers in Genetics, 10*, 956.

Bagley, S. T. (1985). Habitat association of *Klebsiella* species. *Infection Control & Hospital Epidemiology, 6*(2), 52–58.

Baker, H. G. (1976). 'Mistake' pollination as a reproductive system with special reference to the Caricaceae. *Linnean Society Symposium.*

Baker, R. R., & Parker, G. A. (1973). The origin and evolution of sexual reproduction up to the evolution of the male–female phenomenon. *Acta Biotheoretica, 22*(2), 49–77.

Ballari, S. A., & Barrios-García, M. N. (2014). A review of wild boar *Sus scrofa* diet and factors affecting food selection in native and introduced ranges. *Mammal Review, 44*(2), 124–134.

Ballari, S. A., Cuevas, M. F., Cirignoli, S., & Valenzuela, A. E. (2015). Invasive wild boar in Argentina: Using protected areas as a research platform to determine distribution, impacts and management. *Biological Invasions, 17*, 1595–1602.

Baluška, F., & Mancuso, S. (2007). Plant neurobiology as a paradigm shift not only in the plant sciences. *Plant Signaling & Behavior, 2*(4), 205–207.

Baluška, F., & Mancuso, S. (2009). Plant neurobiology: From sensory biology, via plant communication, to social plant behavior. *Cognitive Processing, 10*, 3–7.

Baluška, F., Gagliano, M., & Witzany, G. (Eds.). (2018). *Memory and learning in plants*. Springer.

Baquero, F., & Nombela, C. (2012). The microbiome as a human organ. *Clinical Microbiology and Infection, 18*, 2–4.

Barbour, I. G. (1990). *Religion in an age of science* (p. 208). Harper & Row.

Barlow, S. E., & O'Neill, M. A. (2020). Technological advances in field studies of pollinator ecology and the future of e-ecology. *Current Opinion in Insect Science, 38*, 15–25.

Barrenchea-Mellado, M. J., Ramírez-Rivera, C., & Figueroa-Caro, A. C. (2015). Composición de endosimbiontes facultativos en *Macrosiphum euphorbiae* Thomas y su relación con la planta hospedera [Doctoral dissertation, Universidad de Talca (Chile). Escuela de Agronomía].

Barrière, A., & Félix, M. A. (2005). High local genetic diversity and low outcrossing rate in *Caenorhabditis elegans* natural populations. *Current Biology, 15*(13), 1176–1184.

Barry, C. S., & Giovannoni, J. J. (2007). Ethylene and fruit ripening. *Journal of Plant Growth Regulation, 26*, 143–159.

Bartels, T. (2003). Variations in the morphology, distribution, and arrangement of feathers in domesticated birds. *Journal of Experimental Zoology: Molecular and Developmental Evolution, 298*(1), 91–108.

Barton, N. H. (2001). The role of hybridisation in evolution. *Molecular Ecology, 10*, 551–568.

Bateman, A. J. (1948). Intrasexual selection in *Drosophila. Heredity, 2*(3), 349–368.

Baulcombe, D. C., & Dean, C. (2014). Epigenetic regulation in plant responses to the environment. *Cold Spring Harbor Perspectives in Biology, 6*, a019471.

Bavestrello, G., & Sarà, M. (2009). Porifera, Cnidaria, and Ctenophora. In *Biological science fundamentals and systematics* (Vol. III, p. 129). EOLSS Publications.

Bawa, K. S. (1980a). Mimicry of male by female flowers and intrasexual competition for pollinators in *Jacaratia dolichaula* (D. Smith) Woodson (Caricaceae). *Evolution*, 467–474.

Bawa, K. S. (1980b). Evolution of dioecy in flowering plants. *Annual Review of Ecology and Systematics, 11*(1), 15–39.

Bawa, K. S., & Beach, J. H. (1981). Evolution of sexual systems in flowering plants. *Annals of the Missouri Botanical Garden*, 68–86.

Becerra, J. C., Bildstein, L. S., & Gach, J. S. (2016). Recent insights into the HIV/AIDS pandemic. *Microbial Cell, 3*(9), 451.

Beekman, M., Allsopp, M. H., Lim, J., Goudie, F., & Oldroyd, B. P. (2011). Asexually produced cape honeybee queens (*Apis mellifera capensis*) reproduce sexually. *Journal of Heredity, 102*(5), 562–566.

Belkind-Gerson, J., & Suárez-Rodríguez, R. (2004). Regeneración cerebral. Realidades, posibilidades y esperanzas. *Anales Médicos de la Asociación Médica del Centro Médico. ABC, 49*(4), 201–207.

Bell, G. (1982). *The masterpiece of nature: The evolution and genetics of sexuality*. University of California Press.

Bellver, V. (2002). Razones Para el rechazo de la clonación con fines de investigación biomédica (CIB). *Cuadernos de bioética*, 13(47–49), 75–88. Valencia, .

Bely, A. E., & Nyberg, K. G. (2010). Evolution of animal regeneration: Re-emergence of a field. *Trends in Ecology & Evolution, 25*, 161–170.

Bengtsson, B. O., & Ceplitis, A. (2000). The balance between sexual and asexual reproduction in plants living in variable environments. *Journal of Evolutionary Biology, 13*(3), 415–422.

Benner, S. A., Ricardo, A., & Carrigan, M. A. (2010). Is there a common chemical model for life in the universe? In M. A. Bedau & C. E. Cleland (Eds.), *The nature of life: Classical and contemporary perspectives from philosophy and science* (p. 164). Cambridge University Press.

Benton, M. J. (2010). The origins of modern biodiversity on land. *Philosophical Transactions of the Royal Society B: Biological Sciences, 365*(1558), 3667–3679.

Benton, M. J., & Twitchett, R. J. (2003). How to kill (almost) all life: The end-Permian extinction event. *Trends in Ecology & Evolution, 18*(7), 358–365.

Benzing, D. H. (1990). *Vascular epiphytes: General biology and related biota*. Cambridge University Press.

Berends, M., Goldring, E., Stein, M., & Cravens, X. (2010). Instructional conditions in charter schools and students' mathematics achievement gains. *American Journal of Education, 116*, 303–335.

Berendsen, R. L., Pieterse, C. M., & Bakker, P. A. (2012). The rhizosphere microbiome and plant health. *Trends in Plant Science, 17*(8), 478–486.

Berggren, J. R., Hulver, M. W., & Houmard, J. A. (2005). Fat as an endocrine organ: Influence of exercise. *Journal of Applied Physiology, 99*(2), 757–764.

Bernasconi, G., Ashman, T. L., Birkhead, T. R., Bishop, J. D. D., Grossniklaus, U., Kubli, E., et al. (2004). Evolutionary ecology of the prezygotic stage. *Science, 303*(5660), 971–975.

Bernhardt, H. S. (2012). The RNA world hypothesis: The worst theory of the early evolution of life (except for all the others). *Biology Direct, 7*, 23.

Bernstein, H., & Bernstein, C. (2010). Evolutionary origin of recombination during meiosis. *Bioscience, 60*(7), 498–505.

Bernstein, H., & Bernstein, C. (2013). Evolutionary origin and adaptive function of meiosis. *Meiosis, 1*, 56557.

Bernstein, H., Byerly, H. C., Hopf, F. A., & Michod, R. E. (1984). Origin of sex. *Journal of Theoretical Biology, 110*(3), 323–351.

Bernstein, H., Byerly, H. C., Hopf, F. A., & Michod, R. E. (1985). Genetic damage, mutation, and the evolution of sex. *Science, 229*(4719), 1277–1281.

Bernstein, H., Hopf, F. A., & Michod, R. E. (1987). The molecular basis of the evolution of sex. *Advances in Genetics, 24*, 323–370.

Bertin, R. I. (1988). Paternity in plants. In *Plant reproductive ecology: Patterns and strategies* (pp. 30–59).

Besseling, E., Quik, J. T., Sun, M., & Koelmans, A. A. (2017). Fate of nano- and microplastic in freshwater systems: A modeling study. *Environmental Pollution, 220*, 540–548.

Betlejewski, S. (2007). Social diseases, civilization diseases or lifestyle diseases? *Wiadomosci Lekarskie (Warsaw, Poland), 60*(9–10), 489–492.

Betts, H. C., Puttick, M. N., Clark, J. W., Williams, T. A., Donoghue, P. C. J., & Pisani, D. (2018). Integrated genomic and fossil evidence illuminates life's early evolution and eukaryote origin. *Nature Ecology & Evolution, 2*, 1556–1562.

Biémont, C. (2010). A brief history of the status of transposable elements: From junk DNA to major players in evolution. *Genetics, 186*, 1085–1093.

Bird, A. (2002). DNA methylation patterns and epigenetic memory. *Genes & Development, 16*, 6–21.

Birdsell, J. B. (1957). Some population problems involving Pleistocene man. In *Cold Spring Harbor symposia on quantitative biology* (Vol. 22, pp. 47–69). Cold Spring Harbor Laboratory Press.

Birdsell, J. B. (1958). On population structure in generalized hunting and collecting populations. *Evolution, 12*(2), 189–205.

Biressi, A. C. M., Zou, T., Dupont, S., Dahlberg, C., Di Benedetto, C., Bonasoro, F., et al. (2010). Wound healing and arm regeneration in *Ophioderma longicaudum* and *Amphiura filiformis* (Ophiuroidea, Echinodermata): Comparative morphogenesis and histogenesis. *Zoomorphology, 129*, 1–19.

Birkhead, T. R. (1998). Cryptic female choice: Criteria for establishing female sperm choice. *Evolution, 52*(4), 1212–1218.

Birkhead, T. R., & Møller, A. P. (1998). Sperm competition and sexual selection. Academic Press, pp1–101.

Birkhead, T. R., Chaline, N., Biggins, J. D. T., Burke, T., & Pizzari, T. (2004). Nontransivity of paternity in a bird. *Evolution, 58*, 416–420.

Biswas, P., Chattopadhyay, A., & Ghosh, S. M. (2022). Cryptic female choice: How it changes the narrative of reproduction? In *Reproductive strategies in insects* (pp. 225–241). CRC Press.

Bitar, A. R. (2018). *Diet and the disease of civilization*. Rutgers University Press.

Bitterman, M. E., Menzel, R., Fietz, A., & Schafer, S. (1983). Classical conditioning of proboscis extension in honeybees (*Apis mellifera*). *Journal of Comparative Physiology, 97*, 107–119.

Blackburn, T. M., Lockwood, J. L., & Cassey, P. (2015). The influence of numbers on invasion success. *Molecular Ecology, 24*, 1942–1953.

Blackwell, H. E., & Fuqua, C. (2011). Introduction to bacterial signals and chemical communication. *Chemical Reviews, 111*(1), 1–3.

Blanco, M. A., & Barboza, G. (2005). Pseudocopulatory pollination in *Lepanthes* (Orchidaceae: Pleurothallidinae) by fungus gnats. *Annals of Botany, 95*(5), 763–772.

Blumberg, M. S. (2009). *Freaks of nature: And what they tell us about development and evolution*. Oxford University Press.

Blumer, M. (1971). Scientific aspects of the oil spill problem. *Environmental Affairs, 1*, 54.

Blumler, M. A. (2003). Introgression as a spatial phenomenon. *Physical Geography, 24*, 414–432.

Böhm, J., Scherzer, S., Krol, E., Kreuzer, I., Von Meyer, K., Lorey, C., et al. (2016). The Venus flytrap *Dionaea muscipula* counts prey-induced action potentials to induce sodium uptake. *Current Biology, 26*(3), 286–295.

Bohman, B., Phillips, R. D., Menz, M. H., Berntsson, B. W., Flematti, G. R., Barrow, R. A., et al. (2014). Discovery of pyrazines as pollinator sex pheromones and orchid semiochemicals: Implications for the evolution of sexual deception. *New Phytologist, 203*(3), 939–952.

Bøhn, T., Amundsen, P. A., & Sparrow, A. (2008). Competitive exclusion after invasion? *Biological Invasions, 10*, 359–368.

Bollet, A. J., & Jay, A. B. (2004). *Plagues & poxes: The impact of human history on epidemic disease*. Demos Medical Publishing.

Bolondi, A., Caldarelli, F., Di Felice, F., Durano, D., Germani, G., Michetti, L., et al. (2017). What is a gene? A two sided view. *Evolutionary Biology, 44*, 1–4.

Bondi, M. W., Edmonds, E. C., & Salmon, D. P. (2017). Alzheimer's disease: Past, present, and future. *Journal of the International Neuropsychological Society, 23*(9–10), 818–831.

Bonfante, P., & Genre, A. (2010). Mechanisms underlying beneficial plant–fungus interactions in mycorrhizal symbiosis. *Nature Communications, 1*(1), 48.

Bonita, R., Beaglehole, R., & Kjellström, T. (2006). Basic epidemiology. *World Health*. Organization.

Bonner, J. T. (2003). On the origin of differentiation. *Journal of Biosciences, 28*, 523–528.

Booth, W., & Deimel, P. A. (2019). Parthenogenesis in reptiles. *Biological Reviews, 94*(3), 797–814.

Booth, W., Smith, C. F., Eskridge, P. H., Hoss, S. K., Mendelson, J. R., III, & Schuett, G. W. (2012). Facultative parthenogenesis discovered in wild vertebrates. *Biology Letters, 8*(6), 983–985.
Boraas, M. E., Seale, D. B., & Boxhorn, J. E. (1998). Phagotrophy by a flagellate selects for colonial prey: A possible origin of multicellularity. *Evolutionary Ecology, 12*(2), 153–164.
Bordenstein, S. R., & Theis, K. R. (2015). Host biology in light of the microbiome: Ten principles of holobionts and hologenomes. *PLoS Biology, 13*(8), e1002226.
Bosch, T. C. (2007). Why polyps regenerate and we don't: Towards a cellular and molecular framework for *Hydra* regeneration. *Developmental Biology, 303*, 421–433.
Bose, J. C. (1925). Transmission of stimuli in plants. *Nature, 115*(2891), 457–457.
Bose, J. C., & Das, G. P. (1925). Physiological and anatomical investigations on *Mimosa pudica*. *Proceedings of the Royal Society of London Series B, Containing Papers of a Biological Character, 98*(690), 290–312.
Bouwmeester, H. J., Roux, C., Lopez-Raez, J. A., & Becard, G. (2007). Rhizosphere communication of plants, parasitic plants and AM fungi. *Trends in Plant Science, 12*(5), 224–230.
Bowler, P. J. (1976). Malthus, Darwin, and the concept of struggle. *Journal of the History of Ideas, 37*(4), 631–650.
Bradley, R. S., & Jonest, P. D. (1993). 'Little Ice Age' summer temperature variations: Their nature and relevance to recent global warming trends. *The Holocene, 3*(4), 367–376.
Bradshaw, A. D. (1965). Evolutionary significance of phenotypic plasticity in plants. *Advances in Genetics, 13*, 115–155.
Brain, W. R., & Walton, J. N. (1969). *Brain's diseases of the nervous system* (seventh ed.). Oxford University Press.
Brembs, B., & Heisenberg, M. (2000). The operant and the classical in conditioned orientation of *Drosophila melanogaster* at the flight simulator. *Learning & Memory, 7*, 104–115.
Brenner, E. D., Stahlberg, R., Mancuso, S., Vivanco, J., Baluška, F., & Van Volkenburgh, E. (2006). Plant neurobiology: An integrated view of plant signaling. *Trends in Plant Science, 11*(8), 413–419.
Briceño, R. D., Eberhard, W. G., & Robinson, A. (2007). Copulation behavior of *Glossina pallidipes* (Diptera: Muscidae) outside and inside the female, and genitalic evolution. *Bulletin of Entomological Research, 97*, 1–18.
Bricker, S. B., Clement, C. G., Pirhalla, D. E., Orland, S. P., & Farrow, D. G. (1999). *National estuarine eutrophication assessment: A summary of conditions, historical trends, and future outlook*. National Ocean Service.
Brink, R. A. (1927). The sugary gene in maize as a modifier of the waxy ratio. *Genetics, 12*, 461–491.
Brooks, R., & Kemp, D. J. (2001). Can sexual selection drive the evolution of sperm quality? *Evolution, 55*(1), 1–8.
Brosnan, S. F., & Hopper, L. M. (2014). Psychological limits on animal innovation. *Animal Behaviour, 92*, 325–332.
Brown, A. (2009). Therapeutic cloning: The ethical road to regulation part I: Arguments for and against & regulations. *Human Reproduction & Genetic Ethics, 15*(2), 75–86.
Brown, C. R., & Brown, M. B. (2013). Where has all the road kill gone? *Current Biology, 23*(6), R233–R234.
Brown, G. P., Phillips, B. L., Dubey, S., & Shine, R. (2015). Invader immunology: Invasion history alters immune system function in cane toads (*Rhinella marina*) in tropical Australia. *Ecology Letters, 18*(1), 57–65.
Brunton-Martin, A. L., O'Hanlon, J. C., & Gaskett, A. C. (2020). Orchid sexual deceit affects pollinator sperm transfer. *Functional Ecology, 34*(7), 1336–1344.
Brüssow, H., Canchaya, C., & Hardt, W. D. (2004). Phages and the evolution of bacterial pathogens: From genomic rearrangements to lysogenic conversion. *Microbiology and Molecular Biology Reviews, 68*(3), 560–602.
Bulmer, M. G. (1971). The effect of selection on genetic variability. *The American Naturalist, 105*(940), 201–211.

Bunge, M. (1979). Philosophical inputs and outputs of technology. In G. Bugliarello & D. B. Doner (Eds.), *The history and philosophy of technology* (pp. 262–281). University of Illinois Press.
Burdon-Sanderson, J. S. (1882). I. On the electromotive properties of the leaf of *Dionaea* in the excited and unexcited states. *Philosophical Transactions of the Royal Society of London, 173*, 1–55.
Burdon-Sanderson, J. S., & Page, F. J. (1877). I. On the mechanical effects and on the electrical disturbance consequent on excitation of the leaf of *Dionaea muscipula*. *Proceedings of the Royal Society of London, 25*(171–178), 411–434.
Burger, O., Baudisch, A., & Vaupel, J. W. (2012). Human mortality improvement in evolutionary context. *Proceedings of the National Academy of Sciences, 109*(44), 18210–18214.
Buriticá, S. M. (2014). Los hipopótamos (*Hippopotamus amphibius*) en Colombia. Especie exótica, introducida e invasora. *Revista Lasallista de Investigación, 11*(1), 229–230.
Burkard, C. G., Opriessnig, T., Mileham, A. J., Stadejek, T., Ait-Ali, T., & Lillico, S. G. (2018). Pigs lacking the scavenger receptor cysteine-rich domain 5 of CD163 are resistant to porcine reproductive and respiratory syndrome virus 1 infection. *Journal of Virology, 92*(4), e01996–e01917.
Burke, N. W., & Bonduriansky, R. (2017). Sexual conflict, facultative asexuality, and the true paradox of sex. *Trends in Ecology & Evolution, 32*(9), 646–652.
Burke, G. R., Normark, B. B., Favret, C., & Moran, N. A. (2009). Evolution and diversity of facultative symbionts from the aphid subfamily Lachninae. *Applied and Environmental Microbiology, 75*(16), 5328–5335.
Burke, N. W., Crean, A. J., & Bonduriansky, R. (2015). The role of sexual conflict in the evolution of facultative parthenogenesis: A study on the spiny leaf stick insect. *Animal Behaviour, 101*, 117–127.
Burkholder, J. M. (2001). Eutrophication/oligotrophication. In S. Levin (Ed.), *Encyclopedia of Biodiversity* (Vol. 2, pp. 649–670). Academic Press.
Burkholder, J. M., Tomasko, D. A., & Touchette, B. W. (2007). Seagrasses and eutrophication. *Journal of Experimental Marine Biology and Ecology, 350*(1–2), 46–72.
Burns, C. W. (2013). Predictors of invasion success by *Daphnia* species: Influence of food, temperature and species identity. *Biological Invasions, 15*(4), 859–869.
Butchart, S. H. M., et al. (2010). Global biodiversity: Indicators of recent declines. *Science, 328*, 1164–1168.
Butchart, S. H. M., et al. (2015). Short falls and solutions for meeting national and global conservation area targets. *Conservation Letters, 8*, 329–337.
Butterfield, N. J. (2000). *Bangiomorpha pubescens* n. gen., n. sp.: Implications for the evolution of sex, multicellularity, and the mesoproterozoic/neoproterozoic radiation of eukaryotes. *Paleobiology, 26*(3), 386–404.
Butterfield, N. J. (2015). Early evolution of the Eukaryota. *Palaeontology, 58*(1), 5–17.
Calado, R., Dionísio, G., & Dinis, M. T. (2007). Decapod crustaceans associated with the snakelock anemone *Anemonia sulcata*. Living there or just passing by? *Scientia Marina, 71*(2), 287–292.
Calbacho-Rosa, L., & Peretti, A. V. (2015). Copulatory and post-copulatory sexual selection in haplogyne spiders, with emphasis on Pholcidae and Oonopidae. In *Cryptic female choice in arthropods* (pp. 109–144). Springer.
Calbacho-Rosa, L., Galicia-Mendoza, I., Dutto, M. S., Córdoba-Aguilar, A., & Peretti, A. V. (2013). Copulatory behavior in a pholcid spider: Males use specialized genitalic movements for sperm removal and copulatory courtship. *Naturwissenschaften, 100*(5), 407–416.
Calbacho-Rosa, L., Cargnelutti, F., Córdoba-Aguilar, A., & Peretti, A. V. (2019). Consistency of females' stridulatory behaviour during inter-sexual interactions in spiders. *Ethology, 125*(8), 548–554.
Callahan, M. P., Smith, K. E., Cleaves, H. J., Ruzicka, J., Stern, J. C., Glavin, D. P., et al. (2011). Carbonaceous meteorites contain a wide range of extraterrestrial nucleobases. *Proceedings of the National Academy of Sciences, 108*(34), 13995–13998.

Cameron, D. E., Bashor, C. J., & Collins, J. J. (2014). A brief history of synthetic biology. *Nature Reviews Microbiology, 12*, 381–390.
Campbell, K. H., McWhir, J., Ritchie, W. A., & Wilmut, I. (1996). Sheep cloned by nuclear transfer from a cultured cell line. *Nature, 380*(6569), 64–66.
Campbell, N. A., & Reece, J. B. (2005). *Biology* (7th ed.). Pearson Benjamin Cummings.
Cañas, J. (2017). Chumbera: la invasora amenazada [Article]. El País.. https://elpais.com/elpais/2017/01/24/ciencia/1485271641_736556.html#:~:text=Fue%20una%20de%20las%20primeras,carm%C3%ADn%2C%20empleada%20para%20producir%20tintes
Canestrelli, D., Porretta, D., Lowe, W. H., Bisconti, R., Carere, C., & Nascetti, G. (2016). The tangled evolutionary legacies of range expansion and hybridization. *Trends in Ecology & Evolution, 31*(9), 677–688.
Cardoso, G. C., Batalha, H. R., Reis, S., & Lopes, R. J. (2014). Increasing sexual ornamentation during a biological invasion. *Behavioral Ecology, 25*(4), 916–923.
Carlin, R. K., & Siekevitz, P. (1983). Plasticity in the central nervous system: Do synapses divide? *Proceedings of the National Academy of Sciences, 80*(11), 3517–3521.
Carlson, R. H. (2016). Estimating the biotech sector's contribution to the US economy. *Nature Biotechnology, 34*, 247–255.
Carny, O., & Gazit, E. (2005). A model for the role of short self-assembled peptides in the very early stages of the origin of life. *The FASEB Journal, 19*(9), 1051–1055.
Carrera-Bastos, P., Fontes-Villalba, M., O'Keefe, J. H., Lindeberg, S., & Cordain, L. (2011). The western diet and lifestyle and diseases of civilization. *Research Reports in Clinical Cardiology*, 15–35.
Carson, H. L., & Johnson, W. E. (1975). Genetic variation in Hawaiian *Drosophila* I. Chromosome and allozyme polymorphism in *D. setosimentum* and *D. ochrobasis* from the island of Hawaii. *Evolution*, 11–23.
Carson, H. L., & Kaneshiro, K. Y. (1976). *Drosophila* of Hawaii: Systematics and ecological genetics. *Annual Review of Ecology and Systematics*, 311–345.
Casper, B. B., & Charnov, E. L. (1982). Sex allocation in heterostylous plants. *Journal of Theoretical Biology, 96*(2), 143–149.
Castaneda, L. E., Sandrock, C., & Vorburger, C. (2010). Variation and covariation of life history traits in aphids are related to infection with the facultative bacterial endosymbiont *Hamiltonella defensa*. *Biological Journal of the Linnean Society, 100*(1), 237–247.
Castro, L., & Wasserman, E. A. (2010). Animal learning. *Wiley Interdisciplinary Reviews: Cognitive Science, 1*(1), 89–98.
Castro-Delgado, A., Elizondo-Mesén, S., Valladares-Cruz, Y., & Rivera-Méndez, W. (2020). Wood wide web: Communication through the mycorrhizal network. *Tecnología en Marcha, 33–4*, 114–125.
Cavalier-Smith, T. (2002a). The phagotrophic origin of eukaryotes and phylogenetic classification of Protozoa. *International Journal of Systematic and Evolutionary Microbiology, 52*, 297–354.
Cavalier-Smith, T. (2002b). The neomuran origin of archaebacteria, the negibacterial root of the universal tree and bacterial megaclassification. *International Journal of Systematic and Evolutionary Microbiology, 52*, 7–76.
Cavalier-Smith, T. (2006). Cell evolution and earth history: Stasis and revolution. *Philosophical Transactions of the Royal Society B: Biological Sciences, 361*(1470), 969–1006.
Centers for Disease Control and Prevention. (2004). Prevalence of overweight and obesity among adults with diagnosed diabetes —United States, 1988–1994 and 1999–2002. *MMWR. Morbidity and Mortality Weekly Report, 53*(45), 1066–1068.
Cech, T. R. (1990). Self-splicing and enzymatic activity of an intervening sequence RNA from *Tetrahymena* (Nobel lecture). *Angewandte Chemie International Edition in English, 29*(7), 759–768.
Ceja, J., Espejo-Serna, A., López-Ferarri, A., García, C. J., Mendoza, R. A., & Pérez, G. B. (2008). Las plantas epífitas, su diversidad e importancia. *Ciencias, 91*, 35–42.

Cerqueda-García, D., & Falcón, L. I. (2016). La construcción del nicho y el concepto de holobionte, hacia la reestructuración de un paradigma. *Revista Mexicana de Biodiversidad, 87*(1), 239–241.

Chaix, R., Alvarez-López, M. J., Fagny, M., Lemee, L., Regnault, B., Davidson, R. J., et al. (2017). Epigenetic clock analysis in long-term meditators. *Psychoneuroendocrinology, 85*, 210–214.

Chamovitz, D. (2020). *What a plant knows: A field guide to the senses: Updated and expanded edition*. Scientific American/Farrar, Straus and Giroux.

Champagne, F. A. (2014). Epigenetics of mammalian parenting. In D. Narvaez, K. Valentino, A. Fuentes, J. J. McKenna, & P. Gray (Eds.), *Ancestral landscapes in human evolution: Culture, childrearing and social wellbeing* (pp. 18–37). Oxford University Press.

Chandran, S. S., & Klebanoff, C. A. (2019). T cell receptor-based cancer immunotherapy: Emerging efficacy and pathways of resistance. *Immunological Reviews, 290*(1), 127–147.

Chang, Y. J., Stephen, J. R., Richter, A. P., Venosa, A. D., Brüggemann, J., Macnaughton, S. J., et al. (2000). Phylogenetic analysis of aerobic freshwater and marine enrichment cultures efficient in hydrocarbon degradation: Effect of profiling method. *Journal of Microbiological Methods, 40*(1), 19–31.

Chang, E. S., Neuhof, M., Rubinstein, N. D., Diamant, A., Philippe, H., Huchon, D., & Cartwright, P. (2015). Genomic insights into the evolutionary origin of Myxozoa within Cnidaria. *Proceedings of the National Academy of Sciences, 112*(48), 14912–14917.

Charlesworth, D. (1993). Why are unisexual flowers associated with wind pollination and unspecialized pollinators? *The American Naturalist, 141*(3), 481–490.

Charlesworth, B. (2003). Sex determination in the honeybee. *Cell, 114*(4), 397–398.

Charlesworth, D., & Charlesworth, B. (1987). Inbreeding depression and its evolutionary consequences. *Annual Review of Ecology and Systematics, 18*, 237–268.

Charlesworth, D., Schemske, D. W., & Sork, V. L. (1987). The evolution of plant reproductive characters; sexual versus natural selection. In *The evolution of sex and its consequences* (pp. 317–335). Elsevier.

Charlesworth, C. T., Deshpande, P. S., Dever, D. P., Camarena, J., Lemgart, V. T., Cromer, M. K., & Porteus, M. H. (2019). Identification of preexisting adaptive immunity to Cas9 proteins in humans. *Nature Medicine, 25*(2), 249–254.

Charlot, M., & Marx, R. (1993). *Londres 1851–1901. La era victoriana o el Triunfo de las desigualdades*. Alianza Editorial.

Charnov, E. L. (1979). The genetical evolution of patterns of sexuality: Darwinian fitness. *The American Naturalist, 113*(4), 465–480.

Chaves, I., Pokorny, R., Byrdin, M., Hoang, N., Ritz, T., Brettel, K., Essen, L. O., van der Horst, G. T., Batschauer, A., & Ahmad, M. (2011). The cryptochromes: Blue light photoreceptors in plants and animals. *Annual Review of Plant Biology, 62*, 335–364.

Chen, I. A., Roberts, R. W., & Szostak, J. W. (2004). The emergence of competition between model protocells. *Science, 305*, 1474–1476.

Cheng, X. Y., Xie, P. Z., Cheng, F. X., Xu, R. M., & Xie, B. Y. (2009). Competitive displacement of the native species *Bursaphelenchus mucronatus* by an alien species *Bursaphelenchus xylophilus* (Nematoda: Aphelenchida: Aphelenchoididae): A case of successful invasion. *Biological Invasions, 11*(2), 205–213.

Chodasewicz, K. (2014). Evolution, reproduction and definition of life. *Theory in Biosciences, 133*, 39–45.

Chrousos, G. P. (2009). Stress and disorders of the stress system. *Nature Reviews Endocrinology, 5*(7), 374–381.

Chuaire, L., Sánchez, M. C., & Franco, M. L. (2004). Clonación animal: avances y perspectivas. *Colombia Médica, 35*(2), 101–111.

Chyba, C. F., Thomas, P. J., Brookshaw, L., & Sagan, C. (1990). Cometary delivery of organic molecules to the early. *Earth Science, 249*(4967), 366–373.

Cibelli, J. B., Kiessling, A. A., Cunnif, K., Richards, C., Lanza, R. P., & West, M. D. (2001). Somatic cell nuclear transfer in humans: Pronuclear and early embryonic development. *Journal of Regenerative Medicine, 2*, 25–31.

Cibelli, J. B., Campbell, K. H., Seidel, G. E., West, M. D., & Lanza, R. P. (2002). The health profile of cloned animals. *Nature Biotechnology, 20*, 13–14.

Claessen, D., Rozen, D. E., Kuipers, O. P., Søgaard-Andersen, L., & Van Wezel, G. P. (2014). Bacterial solutions to multicellularity: A tale of biofilms, filaments and fruiting bodies. *Nature Reviews Microbiology, 12*(2), 115–124.

Clegg, S. M., Degnan, S. M., Kikkawa, J., Moritz, C., Estoup, A., & Owens, I. P. (2002). Genetic consequences of sequential founder events by an Island-colonizing bird. *Proceedings of the National Academy of Sciences, 99*(12), 8127–8132.

Coblentz, B. E., & Baber, D. W. (1987). Biology and control of feral pigson Isla Santiago, Galapagos. *Ecuador Journal of Applied Ecology, 24*, 403–418.

Cockburn, A. (1963). *The evolution and eradication of infectious diseases*. Johns Hopkins University Press.

Cockell, C. S. (2016). The similarity of life across the universe. *Molecular Biology of the Cell, 27*(10), 1553–1555.

Cockell, C. S. (2020). *Astrobiology: Understanding life in the universe*. John Wiley & Sons.

Cocucci, A. A., Marino, S., Baranzelli, M., Wiemer, A. P., & Sérsic, A. (2014). The buck in the milkweed: Evidence of male-male interference among pollinaria on pollinators. *New Phytologist, 203*(1), 280–286.

Colautti, R. I., Grigorovich, I. A., & MacIsaac, H. J. (2006). Propagule pressure: A null model for biological invasions. *Biological Invasions, 8*, 1023–1037.

Colciago, A., Casati, L., Negri-Cesi, P., & Celotti, F. (2015). Learning and memory: Steroids and epigenetics. *The Journal of Steroid Biochemistry and Molecular Biology, 150*, 64–85.

Cole, P. D., & Adamo, S. A. (2005). Cuttlefish (*Sepia officinalis*: Cephalopoda) hunting behavior and associative learning. *Animal Cognition, 8*, 27–30.

Coleman, E. (1929). Pollination of an Australian orchid by the male ichneumonid *Lissopimpla semipunctata*, Kirby. *Transactions of the Royal Entomological Society of London, 76*(2), 533–539.

Collins English Dictionary. (2003). *Collins English Dictionary*. HarperCollins.

Colman, A., & Kind, A. (2000). Therapeutic cloning: Concepts and practicalities. *Trends in Biotechnology, 18*(5), 192–196.

Coman, B. (2010). *Tooth and nail: The story of the rabbit in Australia*. Text Publishing.

Comparatore, V., & Mazzolari, A. (2009). *Invasión de zarzamora en Laguna de los padres*. Argentina investiga. *Divulgación científica y noticias universitarias*. Universidad Nacional de Mar del Plata - Facultad de Ciencias Exactas y Naturales.

Connell, J. H., & Slatyer, R. O. (1977). Mechanisms of succession in natural communities and their role in community stability and organization. *The American Naturalist, 111*(982), 1119–1144.

Conner, J. K. (2003). Artificial selection: A powerful tool for ecologists. *Ecology, 84*(7), 1650–1660.

Conover, M. R. (2007). *Predator-prey dynamics: The role of olfaction*. CRC Press.

Conrad, L. I. (1998). Umar at Sargh: The evolution of an Umayyad tradition on flight from the plague. In S. Leder (Ed.), *Story-telling in the framework of non-fictional Arabic literature* (pp. 488–528). Harrassowitz.

Cook, R. E. (1979). Asexual reproduction: A further consideration. *The American Naturalist, 113*(5), 769–772.

Cook, J. M., & Rasplus, J. Y. (2003). Mutualists with attitude: Coevolving fig wasps and figs. *Trends in Ecology & Evolution, 18*(5), 241–248.

Cook, L. M., Mani, G. S., & Varley, M. E. (1986). Postindustrial melanism in the peppered moth. *Science, 231*(4738), 611–613.

Coombs, G., Dold, A. P., Brassine, E. I., & Peter, C. I. (2012). Large pollen loads of a south African asclepiad do not interfere with the foraging behaviour or efficiency of pollinating honey bees. *Naturwissenschaften, 99*, 545–552.

Cooper, J. P. (1960). Selection and population structure in Lolium, IV. Correlated response to selection. *Heredity, 14*, 229–246.
Cordero, A. (1990). The adaptive significance of the prolonged copulations of the damselfly, *Ischnura graellsii* (Odonata: Coenagrionidae). *Animal Behaviour, 40*, 43–48.
Cordero, C. (1998). Chemical ornaments of semen. *Journal of Theoretical Biology, 192*, 581–584.
Corlett, R. T. (2017). A bigger toolbox: Biotechnology in biodiversity conservation. *Trends in Biotechnology, 35*(1), 55–65.
Cornuet, J. M., & Luikart, G. (1996). Description and power analysis of two tests for detecting recent population bottlenecks from allele frequency data. *Genetics, 144*(4), 2001–2014.
Correns, C. (1917). *Ein Fall experimenteller Verschiebung des Geschlechtsverhaltnisses.* Sitzungsberichte der Koeniglicheen Preussichen Akademie der Wissenschaften 685–717.
Correns, C. (1928). Bestimmung, Vererbung und Verteilung des Geschlechtes bei den hoheren Pflanzen. *Handbuch der Vererbungswissenschaft, 2*, 1–38.
Cotman, C. W., & Nieto-Sampedro, M. (1984). Cell biology of synaptic plasticity. *Science, 225*, 1287–1294.
Crews, D., & Fitzgerald, K. T. (1980). Sexual behavior in parthenogenetic lizards (*Cnemidophorus*). *Proceedings of the National Academy of Sciences, 77*(1), 499–502.
Crick, F. H. C. (1958). On protein synthesis. In *Symposia of the Society for Experimental Biology; number XII: The biological replication of macromolecules* (pp. 138–163). Cambridge University Press.
Crick, F. (1981). *Life itself: Its origin and nature*. Simon and Schuster.
Crispo, E. (2008). Modifying effects of phenotypic plasticity on interactions among natural selection, adaptation and gene flow. *Journal of Evolutionary Biology, 21*(6), 1460–1469.
Cronin, T. W., Caldwell, R. L., & Marshall, J. (2006). Learning in stomatopod crustaceans. *International Journal of Comparative Psychology, 19*(3).
Crow, J. F. (1994). Advantages of sexual reproduction. *Developmental Genetics, 15*(3), 205–213.
Crowley, T. J., & North, G. R. (1988). Abrupt climate change and extinction events in earth history. *Science, 240*(4855), 996–1002.
Cubas, P., Vincent, C., & Coen, E. (1999). An epigenetic mutation responsible for natural variation in floral symmetry. *Nature, 401*, 157–161.
Curtis, M. A., Kam, M., Nannmark, U., Anderson, M. F., Axell, M. Z., Wikkelso, C., Holtas, S., van Roon-Mom, W. M. C., Björk-Eriksson, T., & Nordborg, C. (2007). Human neuroblasts migrate to the olfactory bulb via a lateral ventricular extension. *Science, 315*, 1243–1249.
Curtis, H., Schnek, A., Barnes, S., & Massarini, A. (2008). Biología de Curtis. Editorial Médica Panamericana. (7ma Ed.).
Dafni, A. (1983). Pollination of *Orchis caspia* –a nectarless plant which deceives the pollinators of nectariferous species from other plant families. *The Journal of Ecology, 71*(2), 467–474.
Dall, S. R. X., Houston, A. I., & McNamara, J. M. (2004). The behavioural ecology of personality: Consistent individual differences from an adaptive perspective. *Ecology Letters, 7*, 734–739.
Darmaillacq, A. S., Dickel, L., Chichery, M., Agin, V., & Chichery, R. (2004). Rapid taste aversion learning in adult cuttlefish, *Sepia officinalis*. *Animal Behaviour, 68*, 1291–1298.
Darmaillacq, A. S., Dickel, L., & Mather, J. (2014). *Cephalopod cognition*. Cambridge University Press.
Darnaudery, M., & Maccari, S. (2008). Epigenetic programming of the stress response in male and female rats by prenatal restraint stress. *Brain Research Reviews, 57*(2), 571–585.
Darwin, C. (1859). *On the origins of species by means of natural selection* (Vol. 247, p. 1859). Murray.
Darwin, C. (1862). *On the various contrivances by which British and foreign orchids are fertilised by insects, and on the good effect of intercrossing*. John Murray.
Darwin, C. (1869). XVI.—Notes on the fertilization of orchids: To the editors of Annals and Magazine of Natural History. *Journal of Natural History, 4*(21), 141–159.
Darwin, C. (1871). *The Descent of man, and selection in relation to sex*. Princeton Univ. Press.
Darwin, C. (1874). *The descent of man and selection in relation to sex*. A. L.

Darwin, C. (1875). *Insectivorous plants*. John Murray.
Darwin, C. (1877). *The various contrivances by which orchids are fertilised by insects*. John Murray.
Darwin, F. (1899). The botanical work of Darwin. *Annals of Botany, 13*(52), ix–xix.
Darwin, C., & Darwin, F. (1880). *The power of movement in plants*. John Murray.
Darwin, C., & Darwin, F. (1888). *Insectivorous plants*. J. Murray.
Davies, P. C., & Lineweaver, C. H. (2011). Cancer tumors as Metazoa 1.0: Tapping genes of ancient ancestors. *Physical Biology, 8*(1), 015001.
Daxinger, L., & Whitelaw, E. (2012). Understanding transgenerational epigenetic inheritance via the gametes in mammals. *Nature Reviews Genetics, 13*, 153–162.
de la Cruz, R. R., Pastor, A. M., & Delgado-García, J. M. (1996). Influence of the postsynaptic target on the functional properties of neurons in the adult mammalian central nervous system. *Reviews in the Neurosciences, 7*, 115–149.
de la Cruz, R. R., Delgado-García, J. M., & Pastor, A. M. (2000). Discharge characteristics of axotomized abducens internuclear neurons in the adult cat. *Journal of Comparative Neurology, 427*, 391–404.
de Lorenzo, V. (2014). Biología sintética: la ingeniería al asalto de la complejidad biológica. *Arbor, 190*(768), 149–149.
de Lorenzo, V. (2020). Y la biología se convirtió en ingeniería: la adopción de estándares para sistemas vivos. *Mètode: Revista de difusión de la Investigación, 2*(105), 44–51.
de Lorenzo, V., & Danchin, A. (2008). Synthetic biology: Discovering new worlds and new words: The new and not so new aspects of this emerging research field. *EMBO Reports, 9*(9), 822–827.
de Lorenzo, V., & Schmidt, M. (2017). Biological standards for the knowledge-based BioEconomy: What is at stake. *New Biotechnology, 40*, 170–180.
de Sousa, P. A., Dobrinsky, J. R., & Zhu, J. (2002). Somatic cell nuclear transfer in the pig: Control of pronuclear formation and integration with improved methods for activation and maintenance of pregnancy. *Biology of Reproduction, 66*, 642–650.
de Souza, F. S. J., Franchini, L. F., & Rubinstein, M. (2013). Exaptation of transposable elements into novel cis-regulatory elements: Is the evidence always strong? *Molecular Biology and Evolution, 30*(6), 1239–1251.
Dean, G. (1982). Porphyria variegata. *Acta Dermato-Venereologica, 100*, 81–85.
Dell, A. I., Bender, J., Branson, K., Couzin, I. D., de Polavieja, G. G., & Noldus, L. (2014). Automated image-based tracking and its application in ecology. *Trends in Ecology & Evolution, 29*, 417–428.
DeLong, E. F., & Pace, N. R. (2001). Environmental diversity of bacteria and archaea. *Systematic Biology, 50*(4), 470–478.
Delph, L. F. (1999). Sexual dimorphism in life history. In *Gender and sexual dimorphism in flowering plants* (pp. 149–173). Springer Berlin Heidelberg.
Delph, L. F., & Ashman, T. L. (2006). Trait selection in flowering plants: How does sexual selection contribute? *Integrative and Comparative Biology, 46*(4), 465–472.
Des-Marais, D. J. (2000). When did photosynthesis emerge on earth? *Science, 289*(5485), 1703–1705.
Díaz, J. S. (2015). El mecanismo evolutivo de Margulis y los niveles de selección. *Contrastes: revista internacional de filosofía, 20*(1), 101–118.
Dingemanse, N. J., Kazem, A. J., Réale, D., & Wright, J. (2010). Behavioural reaction norms: Animal personality meets individual plasticity. *Trends in Ecology & Evolution, 25*(2), 81–89.
Dinnyes, A., & Szmolenszky, A. (2005). Animal cloning by nuclear transfer: State-of-the-art and future perspectives. *Acta Biochimica Polonica, 52*(3), 585–588.
Dismukes, G. C., Klimov, V. V., Baranov, S. V., Kozlov, Y. N., DasGupta, J., & Tyryshkin, A. (2001). The origin of atmospheric oxygen on earth: The innovation of oxygenic photosynthesis. *Proceedings of the National Academy of Sciences, 98*(5), 2170–2175.
Dodd, M. S., Papineau, D., Grenne, T., Slack, J. F., Rittner, M., Pirajno, F., O'Neil, J., & Little, C. T. (2017). Evidence for early life in Earth's oldest hydrothermal vent precipitates. *Nature, 543*(7643), 60–64.

Doe, W. F. (1989). The intestinal immune system. *Gut, 30*(12), 1679.
Doetsch, F., & Scharff, C. (2001). Challenges for brain repair: Insight from adult neurogenesis in birds and mammals. *Brain Behavior and Evolution, 58*, 306–322.
Dols, M. W. (1974). Plague in early Islamic history. *Journal of the American Oriental Society*, 371–383.
Dorey, N. R. (2020). Learning theory. In *Zoo animal learning and training* (pp. 3–13).
Dörgeloh, W. G. (2001). Relationship between distribution of kudu and woody plant structure. *South African Journal of Wildlife Research, 31*(3), 85–91.
Doron-Mandel, E., Fainzilber, M., & Terenzio, M. (2015). Growth control mechanisms in neuronal regeneration. *FEBS Letters, 589*(14), 1669–1677.
Doudna, J. A., & Charpentier, E. (2014). The new frontier of genome engineering with CRISPR-Cas9. *Science, 346*, 1258096.
Doust, J. L., & Doust, L. L. (1988). Sociobiology of plants: An emerging synthesis. In *Plant reproductive ecology: Patterns and strategies* (pp. 5–29). Oxford University Press.
Drent, P. J., van Oers, K., & van Noordwijk, A. J. (2003). Realized heritability of personalities in the great tit (*Parus major*). *Proceedings of the Royal Society B: Biological Sciences, 270*(1510), 45–51.
Dubinsky, Z., & Jokiel, P. L. (1994). Ratio of energy and nutrient fluxes regulates symbiosis between zooxanthellae and corals. *Pacific Science, 48*(3), 313–324.
Dudgeon, C. L., Coulton, L., Bone, R., Ovenden, J. R., & Thomas, S. (2017). Switch from sexual to parthenogenetic reproduction in a zebra shark. *Scientific Reports, 7*(1), 1–8.
Dukas, R. (1998). *Cognitive ecology: The evolutionary ecology of information processing and decision making*. University of Chicago Press.
Dukas, R. (2008). Evolutionary biology of insect learning. *Annual Review of Entomology, 53*, 145–160.
Dukas, R., Clark, C. W., & Abbott, K. (2006). Courtship strategies of male insects: When is learning advantageous? *Animal Behaviour, 72*(6), 1395–1404.
Dupont, S., & Thorndyke, M. C. (2006). Growth or differentiation? Adaptive regeneration in the brittlestar *Amphiura filiformis*. *Journal of Experimental Biology, 209*, 3873–3881.
Dupraz, S., Grassi, D., Karnas, D., Nieto Guil, A. F., Hicks, D., & Quiroga, S. (2013). The insulin-like growth factor 1 receptor is essential for axonal regeneration in adult central nervous system neurons. *Plos One, 8*(1), e54462.
Duron, O., Bouchon, D., Boutin, S., Bellamy, L., Zhou, L., Engelstadter, J., & Hurst, G. D. D. (2008). The diversity of reproductive parasites among arthropods: *Wolbachia* do not walk alone. *BMC Biology, 6*, 27.
Dyson, F. (1999). *Origins of life*. Cambridge University Press.
Dziubinska, H. A. (2003). Ways of signal transmission and physiological role of electrical potentials in plants. *Acta Societatis Botanicorum Poloniae, 72*(4).
Eberhard, W. G. (1985). *Sexual selection and animal genitalia*. Harvard University Press.
Eberhard, W. G. (1991). Copulatory courtship and cryptic female choice in insects. *Biological Reviews, 66*(1), 1–31.
Eberhard, W. G. (1994). Evidence for widespread courtship during copulation in 131 species of insects and spiders, and implications for cryptic female choice. *Evolution, 48*, 711–733.
Eberhard, W. G. (1996). *Female control: Sexual selection by cryptic female choice*. Princeton University Press.
Eberhard, W. G. (2009). Post-copulatory sexual selection: Darwin's omission and its consequences. *Proceedings of the National Academy of Sciences of the United States of America, 106*, 10025–10032.
Eberhard, W. G., & Cordero, C. (1995). Sexual selection by cryptic female choice on male seminal products-a new bridge between sexual selection and reproductive physiology. *Trends in Ecology & Evolution, 10*(12), 493–496.
Eberhard, W. G., & Lehmann, G. U. (2019). Demonstrating sexual selection by cryptic female choice on male genitalia: What is enough? *Evolution, 73*(12), 2415–2435.

Egnor, S. R., & Branson, K. (2016). Computational analysis of behavior. *Annual Review of Neuroscience, 39*, 217–236.
Eisen, J. A. (2010). Genome sequence of the pea aphid *Acyrthosiphon pisum*. *PLoS Biology, 8*(2).
El Karoui, M., Hoyos-Flight, M., & Fletcher, L. (2019). Future trends in synthetic biology—A report. *Frontiers in Bioengineering and Biotechnology, 7*, 175.
Elliott, C. G., & Maheshwari, R. (1994). Reproduction in fungi: Genetical and physiological aspects. *Journal of Genetics, 73*(1), 55.
Ellis, A. G., & Johnson, S. D. (2010). Floral mimicry enhances pollen export: The evolution of pollination by sexual deceit outside of the Orchidaceae. *The American Naturalist, 176*(5), E143–E151.
Ellstrand, N. C., & Schierenbeck, K. A. (2000). Hybridization as a stimulus for the evolution of invasiveness in plants? *Proceedings of the National Academy of Sciences, 97*, 7043–7050.
Elsdale, T. R., Gurdon, J. B., & Fischberg, M. (1960). A description of the technique for nuclear transplantation in *Xenopus laevis*. *Development, 8*(4), 437–444.
Elton, C. S. (2020). *The ecology of invasions by animals and plants*. Springer Nature.
Endersby, J. (2016). Deceived by orchids: Sex, science, fiction and Darwin. *The British Journal for the History of Science, 49*(2), 205–229.
Endler, J. A. (1980). Natural selection on color patterns in *Poecilia reticulata*. *Evolution, 34*(1), 76–91.
Endy, D. (2005). Foundations for engineering biology. *Nature, 438*(7067), 449–453.
Endy, D., Knight, T., & Ha, L. (2010). The BioBricks foundation Inc. http://bbf.openwetware.org
Epstein, P. (1997). The threatened plague. *People & The planet, 6*(3), 14–17.
Ereskovsky, A. V., & Tokina, D. B. (2007). Asexual reproduction in homoscleromorph sponges (Porifera; Homoscleromorpha). *Marine Biology, 151*(2), 425–434.
Eriksson, P. S., Perfilieva, E., Bjork-Eriksson, T., Alborn, A. M., Nordborg, C., Peterson, D. A., & Gage, F. H. (1998). Neurogenesis in the adult human hippocampus. *Nature Medicine, 4*, 1313–1317.
Escribano, A. V. (2016). *Técnicas de secuenciación de nueva generación Para el estudio del microbioma humano*. [Doctoral thesis]. Universidad Complutense.
Evans, J. H., & Evans, M. S. (2008). Religion and science: Beyond the epistemological conflict narrative. *Annual Review of Sociology, 34*, 87–105.
Ezenwa, V. O., Gerardo, N. M., Inouye, D. W., Medina, M., & Xavier, J. B. (2012). Animal behavior and the microbiome. *Science, 338*(6104), 198–199.
Faber, N. R., McFarlane, G. R., Gaynor, R. C., Pocrnic, I., Whitelaw, C. B. A., & Gorjanc, G. (2021). Novel combination of CRISPR-based gene drives eliminates resistance and localises spread. *Scientific Reports, 11*(1), 1–15.
Fairclough, S. R., Chen, Z., Kramer, E., Zeng, Q., Young, S., Robertson, H. M., et al. (2013). Premetazoan genome evolution and the regulation of cell differentiation in the choanoflagellate *Salpingoeca rosetta*. *Genome Biology, 14*(2), 1–15.
Faisst, J., Keenan, C. L., & Koopowitz, H. (1980). Neuronal repair and avoidance behavior in the flatworm, *Notoplana acticola*. *Journal of Neurobiology, 11*(5), 483–496.
Farmer, A. (2002). *Managing environmental pollution*. Routledge.
Fay, M. F., & Chase, M. W. (2009). Orchid biology: From Linnaeus via Darwin to the 21st century. *Annals of Botany, 104*(3), 359–364.
Federle, M. J., & Bassler, B. L. (2003). Interspecies communication in bacteria. *The Journal of Clinical Investigation, 112*(9), 1291–1299.
Feil, R., & Fraga, M. F. (2012). Epigenetics and the environment: Emerging patterns and implications. *Nature Reviews Genetics, 13*, 97–109.
Feldheim, K. A., Chapman, D. D., Sweet, D., Fitzpatrick, S., Prodöhl, P. A., Shivji, M. S., & Snowden, B. (2010). Shark virgin birth produces multiple, viable offspring. *Journal of Heredity, 101*(3), 374–377.
Fernandez, S., & Alonso, G. (2009). Colera y *Vibrio cholerae*. *Revista del Instituto Nacional de Higiene Rafael Rangel, 40*(2), 50–69.

Ferveur, J. F. (2005). Cuticular hydrocarbons: Their evolution and roles in *Drosophila* pheromonal communication. *Behavior Genetics, 35*, 279–295.
Feschotte, C., & Pritham, E. J. (2007). DNA transposons and the evolution of eukaryotic genomes. *Annual Review of Genetics, 41*, 331–368.
Fields, A. T., Feldheim, K. A., Poulakis, G. R., & Chapman, D. D. (2015). Facultative parthenogenesis in a critically endangered wild vertebrate. *Current Biology, 25*(11), R446–R447.
Finegan, B. (1984). Forest succession. *Nature, 312*(5990), 109–114.
Finn, J. K., Tregenza, T., & Norman, M. D. (2009). Defensive tool use in a coconut-carrying octopus. *Current Biology, 19*(23), R1069–R1070.
Firman, R. C., Gasparini, C., Manier, M. K., & Pizzari, T. (2017). Postmating female control: 20 years of cryptic female choice. *Trends in Ecology & Evolution, 32*(5), 368–382.
Fisher, R. A. (1930). *The genetical theory of natural selection*. Clarendon Press.
Fisher, M. C., Henk, D. A., Briggs, C. J., Brownstein, J. S., Madoff, L. C., McCraw, S. L., & Gurr, S. J. (2012). Emerging fungal threats to animal, plant and ecosystem health. *Nature, 484*(7393), 186–194.
Fitz, D., Reiner, H., & Rode, B. M. (2007). Chemical evolution toward the origin of life. *Pure and Applied Chemistry, 79*(12), 2101–2117.
Fitzpatrick, J. L., Willis, C., Devigili, A., Young, A., Carroll, M., Hunter, H. R., & Brison, D. R. (2020). Chemical signals from eggs facilitate cryptic female choice in humans. *Proceedings of the Royal Society B, 287*(1928), 20200805.
Fix, A. G. (1997). Gene frequency clines produced by kin-structured founder effects. *Human Biology, 69*(5), 663–673.
Flecknoe, D., Charles Wakefield, B., & Simmons, A. (2018). Plagues & wars: The 'Spanish Flu' pandemic as a lesson from history. *Medicine, Conflict and Survival, 34*(2), 61–68.
Flori, L., Fritz, S., Jaffrézic, F., Boussaha, M., Gut, I., Heath, S., & Gautier, M. (2009). The genome response to artificial selection: A case study in dairy cattle. *PLoS One, 4*(8), e6595.
Fogden, M., & Fogden, P. (2018). *The natural history of flowers*. Texas A&M University Press.
Font, E., Desfilis, E., Pérez-Cañellas, M. M., & García-Verdugo, J. M. (2001). Neurogenesis and neuronal regeneration in the adult reptilian brain. *Brain, Behavior and Evolution, 58*(5), 276–295.
Forsdick, N. J., Cubrinovska, I., Massaro, M., & Hale, M. L. (2021). Microsatellite genotyping detects extra-pair paternity in the Chatham Island Black Robin, a highly inbred, socially monogamous passerine. *Emu-Austral Ornithology, 121*(1–2), 68–74.
Forsman, A. (2014). Effects of genotypic and phenotypic variation on establishment are important for conservation, invasion, and infection biology. *Proceedings of the National Academy of Sciences, 111*, 302–307.
Forterre, Y. (2013). Slow, fast and furious: Understanding the physics of plant movements. *Journal of Experimental Botany, 64*(15), 4745–4760.
Forterre, P., Gribaldo, S., & Brochier, C. (2005). Luca: The last universal common ancestor. *Médecine Sciences, 21*(10), 860–865.
Foss, K. A. (2020). Kansas memory: Maintained by the Kansas Historical Society. *American Journalism, 37*(2), 271–273.
Fournier, D., Estoup, A., Orivel, J., Foucaud, J., Jourdan, H., Le Breton, J., & Keller, L. (2005). Clonal reproduction by males and females in the little fire ant. *Nature, 435*, 1230–1234.
Francis, D. D., Szegda, K., Campbell, G., Martin, W. D., & Insel, T. R. (2003). Epigenetic sources of behavioral differences in mice. *Nature Neuroscience, 6*(5), 445–446.
Frank, H. (1980). Evolution of canine information processing under conditions of natural and artificial selection. *Zeitschrift für Tierpsychologie, 53*(4), 389–399.
Frankham, R. (2010). Challenges and opportunities of genetic approaches to biological conservation. *Biological Conservation, 143*(9), 1919–1927.
Frankham, R. (2015). Genetic rescue of small inbred populations: Meta-analysis reveals large and consistent benefits of gene flow. *Molecular Ecology, 24*(11), 2610–2618.

Fraune, S., & Bosch, T. C. G. (2010). Why bacteria matter in animal development and evolution. *BioEssays, 32*(7), 571–580.
Freeman, S., & Herron, J. C. (2007). *Evolutionary analysis*. Pearson Prentice Hall.
Freeman, S., Quilin, K., Allison, L., Black, M., Podgorski, G., Taylor, E., & Carmichael, J. (2017). *Biological sciences* (6th ed., Global ed.). Pearson Education Limited.
Freidenfelds, N. A., Robbins, T. R., & Langkilde, T. (2012). Evading invaders: The effectiveness of a behavioral response acquired through lifetime exposure. *Behavioral Ecology, 23*, 659–664.
Frentiu, F. D., & Chenoweth, S. F. (2010). Clines in cuticular hydrocarbons in two *Drosophila* species with independent population histories. *Evolution: International Journal of Organic Evolution, 64*(6), 1784–1794.
Fresquet, N. (1999). Effects of aging on the acquisition and extinction of excitatory conditioning in *Drosophila melanogaster*. *Physiology & Behavior, 67*, 205–211.
Fromhage, L. (2012). Mating unplugged: A model for the evolution of mating plug (dis-) placement. *Evolution, 66*(1), 31–39.
Furusawa, C., & Kaneko, K. (2002). Origin of multicellular organisms as an inevitable consequence of dynamical systems. *The Anatomical Record: An Official Publication of the American Association of Anatomists, 268*(3), 327–342.
Fusco, G., & Minelli, A. (2019). *The biology of reproduction*. Cambridge University Press.
Futuyma, D. J. (2013). Natural selection and adaptation. In the Princeton guide to evolution (p. 189).
Gabrys, J. (2017). Monitoring and remediating a garbage patch. In *Research objects in their technological setting* (pp. 201–214). Routledge.
Gage, M. J. (1995). Continuous variation in reproductive strategy as an adaptive response to population density in the moth *Plodia interpunctella*. *Proceedings of the Royal Society of London B Biological Sciences, 261*, 25–30.
Gage, F. H., & Temple, S. (2013). Neural stem cells: Generating and regenerating the brain. *Neuron, 80*(3), 588–601.
Gage, J. L., Jarquin, D., Romay, C., Lorenz, A., Buckler, E. S., Kaeppler, S., & De Leon, N. (2017). The effect of artificial selection on phenotypic plasticity in maize. *Nature Communications, 8*(1), 1–11.
Gagliano, M. (2013). Seeing green: The re-discovery of plants and nature's wisdom. *Societies, 3*(1), 147–157.
Gagliano, M. (2015). In a green frame of mind: Perspectives on the behavioral ecology and cognitive nature of plants. *AoB Plants, 7*.
Gagliano, M., Renton, M., Duvdevani, N., Timmins, M., & Mancuso, S. (2012). Out of sight but not out of mind: Alternative means of communication in plants. *PLoS One, 7*(5), e37382.
Galen, C., & Rotenberry, J. T. (1988). Variance in pollen carryover in animal-pollinated plants: Implications for mate choice. *Journal of Theoretical Biology, 135*(3), 419–429.
Galindo-González, L., Sarmiento, F., & Quimbaya, M. A. (2018). Shaping plant adaptability, genome structure and gene expression through transposable element epigenetic control: Focus on methylation. *Agronomy, 8*(9), 180.
Gao, L., Guan, W., Wang, M., Wang, H., Yu, J., Liu, Q., & Pei, G. (2017). Direct generation of human neuronal cells from adult astrocytes by small molecules. *Stem Cell Reports, 8*(3), 538–547.
García, J. M. D. (2003). Plasticidad y regeneración neuronal: mitos y expectativas. *MIO, 3*(1), 289.
Garrett, P. (1970). Phanerozoic stromatolites: Noncompetitive ecologic restriction by grazing and burrowing animals. *Science, 169*, 171–173.
Garro, G. (2016). Biología sintética: Una nueva era genómica más allá de los transgénicos. *Investigación y Ciencia, 27*, 2–2.
Gascuel, F., Choisy, M., Duplantier, J. M., Débarre, F., & Brouat, C. (2013). Host resistance, population structure and the long-term persistence of bubonic plague: Contributions of a modelling approach in the Malagasy focus. *PLoS Computational Biology, 9*(5), e1003039.
Gaskett, A. C., Winnick, C. G., & Herberstein, M. E. (2008). Orchid sexual deceit provokes ejaculation. *The American Naturalist, 171*(6), E206–E212.

Gazave, E., Lapébie, P., Ereskovsky, A. V., Vacelet, J., Renard, E., Cárdenas, P., & Borchiellini, C. (2012). No longer Demospongiae: Homoscleromorph sponges revisited by molecular phylogeny, Linnaean classification and the PhyloCode. *Hydrobiologia, 687*(1), 3–10.

Gee, H. (1999). Size and the single sex cell. *Nature*.

Gehrke, A. R., Neverett, E., Luo, Y. J., Brandt, A., Ricci, L., Hulett, R. E., & Srivastava, M. (2019). Acoel genome reveals the regulatory landscape of whole-body regeneration. *Science, 363*(6432).

Geyer, R. (2020). A brief history of plastics. In *Mare Plasticum-the Plastic Sea: Combatting plastic pollution through science and art* (pp. 31–47).

Ghalambor, C. K., McKay, J. K., Carroll, S. P., & Reznick, D. N. (2007). Adaptive versus non-adaptive phenotypic plasticity and the potential for contemporary adaptation in new environments. *Functional Ecology, 21*(3), 394–407.

Ghosh-Roy, A., Wu, Z., Goncharov, A., Jin, Y., & Chisholm, A. D. (2010). Calcium and cyclic AMP promote axonal regeneration in *Caenorhabditis elegans* and require DLK-1 kinase. *Journal of Neuroscience, 30*, 3175–3183.

Giacometti, D. C. (1990). Impacto Atual da Cultura de Tecidos de Plantas. In A. C. Torres (Ed.), *Técnicas e aplicações de cultura de tecidos de plantas*. Brasília.

Gibson, G., & Dworkin, I. (2004). Uncovering cryptic genetic variation. *Nature Reviews Genetics, 5*(9), 681–690.

Gil, R., & Latorre, A. (2019). Unity makes strength: A review on mutualistic symbiosis in representative insect clades. *Life, 9*(1), 21.

Gilbert, L. E. (1975). Ecological consequences of a coevolved mutualism between butterflies and plants. In *Coevolution of animals and plants* (pp. 210–240). University of Texas Press.

Gilbert, W. (1986). Origin of life: The RNA world. *Nature, 319*(6055), 618.

Gilbert, S. F. (2003). *Developmental biology*. Sinauer Associates, Inc..

Gilbert, S. F. (2011). Symbionts as an epigenetic source of heritable variation. In *Transformations of Lamarckism: From subtle fluids to molecular biology* (pp. 283–293).

Gilbert, S. F. (2014). Symbiosis as the way of eukaryotic life: The dependent co-origination of the body. *Journal of Biosciences, 39*, 201–209.

Gilbert, S. F., Sapp, J., & Tauber, A. I. (2012). A symbiotic view of life: We have never been individuals. *The Quarterly Review of Biology, 87*(4), 325–341.

Giorgis, M. A., & Tecco, P. A. (2014). Árboles y arbustos invasores de la Provincia de Córdoba (Argentina): una contribución a la sistematización de bases de datos globales. *Boletín de la Sociedad Argentina de Botánica, 49*(4), 581–603.

Giorgis, M. A., Tecco, P. A., Marcora, P., Cingolani, M., Paiaro, V., & Renison, D. (2006). Las invasiones biológicas y su problemática ambiental–manual Para docentes.

Giribet, G., & Edgecombe, G. D. (2020). *The invertebrate tree of life*. Princeton University Press.

Giurfa, M. (2007). Behavioral and neural analysis of associative learning in the honeybee: A taste from the magic well. *Journal of Comparative Physiology, 193*, 801–824.

Giurfa, M. (2013). Cognition with few neurons: Higher order learning in insects. *Trends in Neurosciences, 36*, 285–294.

Giurfa, M. (2015). Learning and cognition in insects. *Wiley Interdisciplinary Reviews: Cognitive Science, 6*(4), 383–395.

Giurfa, M., & Sandoz, J. C. (2012). Invertebrate learning and memory: Fifty years of olfactory conditioning of the proboscis extension response in honeybees. *Learning & Memory, 19*(2), 54–66.

Givnish, T. J. (1980). Ecological constraints on the evolution of breeding systems in seed plants: Dioecy and dispersal in gymnosperms. *Evolution*, 959–972.

Glansdorff, N., Xu, Y., & Labedan, B. (2008). The last universal common ancestor: Emergence, constitution and genetic legacy of an elusive forerunner. *Biology Direct, 3*(1), 1–35.

Glasier, J. R. N., Poore, A. G. B., & Eldridge, D. J. (2018). Do mutualistic associations have broader host ranges than neutral or antagonistic associations? A test using myrmecophiles as model organisms. *Insectes Sociaux, 65*(4), 639–648.

Glibert, P. M., Seitzinger, S., Heil, C. A., Burkholder, J. M., Parrow, M. W., Codispoti, L. A., & Kelly, V. (2005). Eutrophication. *Oceanography, 18*(2), 198.

Godin, J., & Dugatkin, L. A. (1996). Female mating preference for bold males in the guppy, *Poecilia reticulata*. *Proceedings of the National Academy of Science USA, 93*, 10262–10267.
Gómez, M. C., Pope, C. E., Ricks, D. M., Lyons, J., Dumas, C., & Dresser, B. L. (2008). Cloning endangered felids using heterospecific donor oocytes and interspecies embryo transfer. *Reproduction, Fertility and Development, 21*(1), 76–82.
González Jiménez, A. E. (1998). Generalidades del cultivo in vitro. In J. N. Pérez Ponce (Ed.), *Propagación y Mejora de Plantas por Biotecnología* (pp. 397–398). Instituto de Biotecnología de las Plantas.
Goodrich, J. K., Waters, J. L., Poole, A. C., Sutter, J. L., Koren, O., Blekhman, R., et al. (2014). Human genetics shape the gut microbiome. *Cell, 159*(4), 789–799.
Gordis, L. (2013). *Epidemiology e-book*. Elsevier Health Sciences.
Gosling, S. (2001). From mice to men: What can we learn about personality from animal research. *Psychology Bulletin, 127*, 45–86.
Govindarajulu, M., Pfeffer, P. E., Jin, H. R., Abubaker, J., Douds, D. D., Allen, J. W., et al. (2005). Nitrogen transfer in the arbuscular mycorrhizal symbiosis. *Nature, 435*, 819–823.
Gracía, L. M. P. (1999). Consideraciones bioéticas sobre la clonación humana y animal. *Persona y Bioética, 5*, 2–17.
Gradstein, S. R., & Kerp, H. (2012). A brief history of plants on earth. *The geologic time scale, 29*, 233–237.
Graham, L. E., Graham, J. M., & Wilcox, L. W. (2006). *Plant biology* (2nd ed.).
Graham, S. P., Freidenfelds, N. A., McGormick, G. L., & Langkilde, T. (2012). The impacts of invaders: Basal and acute stress glucocorticoid profiles and immune function in native lizards threatened by invasive ants. *General and Comparative Endocrinology, 176*, 400–408.
Grant, V. (1995). Sexual selection in plants: Pros and cons. *Proceedings of the National Academy of Sciences, 92*(5), 1247–1250.
Gray, R. H. (2015). The Fermi paradox is neither Fermi's nor a paradox. *Astrobiology, 15*(3), 195–199.
Graziadei, P. P. C., & DeHan, R. S. (1973). Neuronal regeneration in frog olfactory system. *The Journal of Cell Biology, 59*(2), 525–530.
Greenberg, E. P. (2003). Bacterial communication and group behavior. *The Journal of Clinical Investigation, 112*(9), 1288–1290.
Gregory, T. R. (2009). Artificial selection and domestication: Modern lessons from Darwin's enduring analogy. *Evolution: Education and Outreach, 2*(1), 5–27.
Gribaldo, S., Poole, A. M., Daubin, V., Forterre, P., & Brochier-Armanet, C. (2010). The origin of eukaryotes and their relationship with the archaea: Are we at a phylogenomic impasse? *Nature Reviews Microbiology, 8*(10), 743–752.
Grierson, D. (2013). Ethylene and the control of fruit ripening. *The molecular biology and biochemistry of fruit ripening*, 43–73.
Griffin, J. P. (2008). Changing life expectancy throughout history. *Journal of the Royal Society of Medicine, 101*(12), 577.
Griffiths, A. J. F., Gelbart, W. M., Miller, J. H., & Lewontin, R. C. (1999). *The molecular basis of mutation*. Mod Genet Anal.
Groothuis, T., & Trillmich, F. (2011). Unfolding personalities: The importance of studying ontogeny. *Developmental Psychobiology, 53*, 641–655.
Grosberg, R. K., & Strathmann, R. R. (2007). The evolution of multicellularity: A minor major transition? *Annual Review of Ecology, Evolution, and Systematics, 38*, 621–654.
Gruart, A., & Delgado-García, J. M. (2004). Regeneración neuronal y recuperación funcional tras la lesión del sistema nervioso periférico. *Reviews in the Neurosciences, 38*(8), 746–756.
Grützner, F., Rens, W., Tsend-Ayush, E., El-Mogharbel, N., O'Brien, P. C., Jones, R. C., & Graves, J. (2004). In the platypus a meiotic chain of ten sex chromosomes shares genes with the bird Z and mammal X chromosomes. *Nature, 432*(7019), 913–917.
Guardo-Gómez, V., Fajardo-Perdomo, M. A., Muñoz, A. L., & Gómez, L. A. (2020). Ingeniería Tisular Para Regeneración Nerviosa: Una Revisión. *Revista Chilena de Neuro-Psiquiatría, 58*(1), 50–60.

Guarner, F., & Malagelada, J. R. (2003). Gut flora in health and disease. *The Lancet, 361*(9356), 512–519.
Gunn-Moore, D., Kaidanovich-Beilin, O., Iradi, M. C., Gunn-Moore, F., & Lovestone, S. (2018). Alzheimer's disease in humans and other animals: A consequence of postreproductive life span and longevity rather than aging. *Alzheimer's & Dementia, 14*(2), 195–204.
Guo, G. L., Tseng, D. H., & Huang, S. L. (2001). Co-metabolic degradation of trichloroethylene by *Pseudomonas putida* in a fibrous bed bioreactor. *Biotechnology Letters, 23*, 1653–1657.
Gurdon, J. B. (1996). Introductory comments: *Xenopus* as a laboratory animal. In R. C. Tinsley & H. R. Kobel (Eds.), *The biology of Xenopus*. Oxford University Press.
Gurdon, J. B. (2006). Nuclear transplantation in *Xenopus*. In *Nuclear reprogramming* (pp. 1–9). Humana Press.
Gurdon, J. B., & Byrne, J. A. (2003). The first half-century of nuclear transplantation. *Proceedings of the National Academy of Sciences, 100*, 8048–8052.
Gurevitch, J., & Padilla, D. K. (2004). Are invasive species a major cause of extinctions? *Trends in Ecology & Evolution, 19*(9), 470–474.
Haase, A. T. (1986). Pathogenesis of lentivirus infections. *Nature, 322*(6075), 130–136.
Hagerman, A. E., & Butler, L. G. (1981). The specificity of proanthocyanidin-protein interactions. *Journal of Biological Chemistry, 256*(9), 4494–4497.
Hahn, B. H., Shaw, G. M., De Cock, K. M., & Sharp, P. M. (2000). AIDS as a zoonosis: Scientific and public health implications. *Science, 287*(5453), 607–614.
Haldane, J. B. S. (1929). Natural selection. *Nature, 124*(3125), 444–444.
Haldane, J. B. S. (1932). *The causes of evolution*. Longman, Green and Co. Princeton University Press.
Hallatschek, O., Hersen, P., Ramanathan, S., & Nelson, D. R. (2007). Genetic drift at expanding frontiers promotes gene segregation. *Proceedings of the National Academy of Sciences, 104*(50), 19926–19930.
Hamilton, G. A. (1953). Thelytokous parthenogenesis for four generations in the desert locust (*Schistocerca gregaria* Forsk). *Nature, 173*, 1153–1154.
Hammer, M. (1997). The neural basis of associative reward learning in honeybees. *Trends in Neurosciences, 20*, 245–252.
Hammer, T. J., Hata, C., & Nieh, J. C. (2009). Thermal learning in the honeybee, *Apis mellifera*. *Journal of Experimental Biology, 212*, 3928–3934.
Hammer, J., Kraak, M. H., & Parsons, J. R. (2012). Plastics in the marine environment: The dark side of a modern gift. *Reviews of Environmental Contamination and Toxicology*, 1–44.
Hammerschmidt, K., Rose, C. J., Kerr, B., & Rainey, P. B. (2014). Life cycles, fitness decoupling and the evolution of multicellularity. *Nature, 515*(7525), 75–79.
Hanke, F. D., & Osorio, D. C. (2018). Vision in cephalopods. *Frontiers in Physiology, 9*, 18.
Hanot, P., Herrel, A., Guintard, C., & Cornette, R. (2018). The impact of artificial selection on morphological integration in the appendicular skeleton of domestic horses. *Journal of Anatomy, 232*(4), 657–673.
Hanschen, E. R., Marriage, T. N., Ferris, P. J., Hamaji, T., Toyoda, A., Fujiyama, A., et al. (2016). The *Gonium pectorale* genome demonstrates co-option of cell cycle regulation during the evolution of multicellularity. *Nature Communications, 7*, 11370.
Haring, V., Gray, J. E., McClure, B. A., Anderson, M. A., & Clarke, A. E. (1990). Self-incompatibility: A self-recognition system in plants. *Science, 250*(4983), 937–941.
Harold, F. M. (2001). *The way of the cell: Molecules, organisms and the order of life*. Oxford University Press.
Harrison, M. J., Dewbre, G. R., & Liu, J. (2002). A phosphate transporter from Medicago truncatula involved in the acquisition of phosphate released by arbuscular mycorrhizal fungi. *The Plant Cell, 14*(10), 2413–2429.
Harrison, M. J. (2005). Signaling in the arbuscular mycorrhizal symbiosis. *Annual Review of Microbiology, 59*, 19–42.

Hartl, M., Mitterstiller, A. M., Valovka, T., Breuker, K., Hobmayer, B., & Bister, K. (2010). Stem cell-specific activation of an ancestral myc protooncogene with conserved basic functions in the early metazoan *Hydra*. *Proceedings of the National Academy of Sciences, 107*(9), 4051–4056.

Hassanpour, S., MaheriSis, N., & Eshratkhah, B. (2011). Plants and secondary metabolites (Tannins): A review. *International Journal of Forest, Soil and Erosion, 1*, 47–53.

Hawksworth, D. L. (1991). The fungal dimension of biodiversity: Magnitude, significance, and conservation. *Mycological Research, 95*, 641–655.

Hawksworth, D. L., Kirk, P. M., Sutton, B. C., & Pegler, D. N. (1995). *Dictionary of the fungi*. CAB.

Hayakawa, Y. (2007). Parasperm: Morphological and functional studies on nonfertile sperm. *Ichthyological Research, 54*, 111–130.

Haynes, V., & Bojcun, M. (1988). *The Chernobyl disaster*. Hogarth Press.

He, W., Yuan, S., Zhong, W. H., Siddikee, M. A., & Dai, C. C. (2016). Application of genetically engineered microbial whole-cell biosensors for combined chemosensing. *Applied Microbiology and Biotechnology, 100*(3), 1109–1119.

Heckman, D. S., Geiser, D. M., Eidell, B. R., Stauffer, R. L., Kardos, N. L., & Hedges, S. B. (2001). Molecular evidence for the early colonization of land by fungi and plants. *Science, 293*(5532), 1129–1133.

Hedhammar, A., Olsson, S. E., Andersson, S. A., Persson, L., Pettersson, L., Olausson, A., & Sundgren, P. E. (1979). Canine hip dysplasia: Study of heritability in 401 litters of German Shepherd dogs. *Journal of the American Veterinary Medical Association, 174*(9), 1012–1016.

Hedrick, P. W. (2005). *Genetics of populations*. Jones and Bartlett Publishers.

Hedrick, P. W. (2015). Heterozygote advantage: the effect of artificial selection in livestock and pets. *Journal of Heredity, 106*(2), 141–154.

Heijmans, B. T., et al. (2008). Persistent epigenetic differences associated with prenatal exposure to famine in humans. *Proceedings of the National Academy of Sciences, 105*, 17046–17049.

Heil, M., Greiner, S., Meimberg, H., Krüger, R., Noyer, J. L., Heubl, G., et al. (2004). Evolutionary change from induced to constitutive expression of an indirect plant resistance. *Nature, 430*(6996), 205–208.

Henriques, S. F., Dhakan, D. B., Serra, L., Francisco, A. P., Carvalho-Santos, Z., Baltazar, C., et al. (2020). Metabolic cross-feeding in imbalanced diets allows gut microbes to improve reproduction and alter host behaviour. *Nature Communications, 11*(1), 1–15.

Henry, L. M., Maiden, M. C., Ferrari, J., & Godfray, H. C. J. (2015). Insect life history and the evolution of bacterial mutualism. *Ecology Letters, 18*, 516–525.

Herbert, S. (1971). Darwin, Malthus, and selection. *Journal of the History of Biology*, 209–217.

Heribert-Nilsson, N. (1923). Zertationsversuche mit durchtrennung des griffes bei *Oenothera lamarckiana*. *Hereditas, 4*, 177–190.

Hermisson, J., & Wagner, G. P. (2004). The population genetic theory of hidden variation and genetic robustness. *Genetics, 168*, 2271–2284.

Herrel, A., Huyghe, K., Vanhooydonck, B., Backeljau, T., Breugelmans, K., Grbac, I., & Irschick, D. (2008). Rapid large-scale evolutionary divergence in morphology and performance associated with exploitation of a different dietary resource. *Proceedings of the National Academy of Sciences, 105*(12), 4792–4795.

Herrera, C. M. (1996). Floral traits and plant adaptation to insect. In D. G. Lloyd & S. C. H. Barret (Eds.), *Floral biology: Studies on floral evolution in animal-pollinated plants* (p. 65). Chapman & Hall.

Herrera, C. M., & Bazaga, P. (2010). Epigenetic differentiation and relationship to adaptive genetic divergence in discrete populations of the violet *Viola cazorlensis*. *New Phytologist, 187*, 867–876.

Herrero, L. B., & Ferradaz, M. A. L. (2011). Aportes del ejercicio físico a la actividad cerebral. *Lecturas: Educación física y deportes, 160*, 1–7.

Hetzer, H. O., & Harvey, W. R. (1967). Selection for high and low fatness in swine. *Journal of Animal Science, 26*(6), 1244–1251.

Higgs, P. G., & Lehman, N. (2015). The RNA world: Molecular cooperation at the origins of life. *Nature Reviews Genetics, 16*(1), 7–17.

Higham, J. P., & Hebets, E. A. (2013). An introduction to multimodal communication. *Behavioral Ecology and Sociobiology, 67*, 1381–1388.

Hill, W. G., & Caballero, A. (1992). Artificial selection experiments. *Annual Review of Ecology and Systematics, 23*(1), 287–310.

Hill, J. R., Burghardt, R. C., & Jones, K. (2000). Evidence for placental abnormality as the major cause of mortality in first trimester somatic cell cloned bovine fetuses. *Biology of Reproduction, 63*, 1787–1794.

Hochedlinger, K., & Jaenisch, R. (2002). Nuclear transplantation: Lessons from frogs and mice. *Current Opinion in Cell Biology, 14*, 741–748.

Hochedlinger, K., & Jaenisch, R. (2003). Nuclear transplantation, embryonic stem cells, and the potential for cell therapy. *New England Journal of Medicine, 349*, 275–286.

Hodick, D., & Sievers, A. (1988). The action potential of *Dionaea muscipula* Ellis. *Planta, 174*, 8–18.

Hoegh-Guldberg, O., Hughes, L., McIntyre, S., Lindenmayer, D. B., Parmesan, C., Possingham, H. P., & Thomas, C. D. (2008). Assisted colonization and rapid climate change. *Science, 321*, 345–346.

Hoekstra, B. (2000). Plant blindness—The ultimate challenge to botanists. *The American Biology Teacher, 62*(2), 82–83.

Hoezel, A. R. (1999). Impact of population bottlenecks on genetic variation and the importance of life-history; a case study of the northern elephant seal. *Biological Journal of the Linnean Society, 68*(1–2), 23–39.

Holland, J., Roberts, D., & Liscum, E. (2009). Understanding phototropism: From Darwin to today. *Journal of Experimental Botany, 60*(7), 1969–1978.

Holman, L., & Snook, R. R. (2006). Spermicide, cryptic female choice and the evolution of sperm form and function. *Journal of Evolutionary Biology, 19*(5), 1660–1670.

Holmes, E. C. (2001). On the origin and evolution of the human immunodeficiency virus (HIV). *Biological Reviews, 76*(2), 239–254.

Holoch, D., & Moazed, D. (2015). RNA-mediated epigenetic regulation of gene expression. *Nature Reviews Genetics, 16*, 71–84.

Honey, R. C., Iordanova, M. D., & Good, M. (2014). Associative structures in animal learning: Dissociating elemental and configural processes. *Neurobiology of Learning and Memory, 108*, 96–103.

Hooper, R. (2021). The wisdom of the woods. *New Scientist, 250*(3332), 39–43.

Hooper, D. U., Adair, E. C., Cardinale, B. J., Byrnes, J. E., Hungate, B. A., Matulich, K. L., et al. (2012). A global synthesis reveals biodiversity loss as a major driver of ecosystem change. *Nature, 486*(7401), 105–108.

Horiike, T., Hamada, K., Kanaya, S., & Shinozawa, T. (2001). Origin of eukaryotic cell nuclei by symbiosis of archaea in bacteria is revealed by homology-hit analysis. *Nature Cell Biology, 3*, 210–214.

Horodyski, R. J., & Knauth, L. P. (1994). Life on land in the Precambrian. *Science, 263*, 494–498.

Hosken, D. J., & House, C. M. (2011). Sexual selection. *Current Biology, 21*(2), R62–R65.

Hough, A. R., Case, J., & Boal, J. G. (2016). Learned control of body patterning in cuttlefish *Sepia officinalis* (Cephalopoda). *Journal of Molluscan Studies, 82*, 427–431.

Howard, W. E. (2018). Rodent pest management: The principles. In *Rodent pest management* (pp. 285–293). CRC Press.

Howard, R. S., & Lively, C. M. (1998). The maintenance of sex by parasitism and mutation accumulation under epistatic fitness functions. *Evolution, 52*(2), 604–610.

Howarth, R. W., Sharpley, A., & Walker, D. (2002). Sources of nutrient pollution to coastal waters in the United States: Implications for achieving coastal water quality goals. *Estuaries, 25*, 656–676.

Hubbell, S. P. (2005). Neutral theory in community ecology and the hypothesis of functional equivalence. *Functional Ecology, 19*(1), 166–172.
Huber, B. T., MacLeod, K. G., & Wing, S. L. (2000). *Warm climates in earth history*. Cambridge University Press.
Hudson, C. M., Phillips, B. L., Brown, G. P., & Shine, R. (2015). Virgins in the vanguard: Low reproductive frequency in invasion-front cane toads. *Biological Journal of the Linnean Society, 116*(4), 743–747.
Hughes, A. R., Inouye, B. D., Johnson, M. T., Underwood, N., & Vellend, M. (2008). Ecological consequences of genetic diversity. *Ecology Letters, 11*(6), 609–623.
Hulsebosch, C. E., & Bittner, G. D. (1980). Evolution of abilities to regenerate neurons in central nervous systems. *The American Naturalist, 115*(2), 276–284.
Humpherys, D., Eggan, K., & Akutsu, H. (2001). Epigenetic instability in ES cells and cloned mice. *Science, 293*, 95–97.
Huston, M., & Smith, T. (1987). Plant succession: Life history and competition. *The American Naturalist, 130*(2), 168–198.
Huxley, J. S. (1942). *Evolution, the modern synthesis*. Allen and Unwin.
Immler, S., Hotzy, C., Alavioon, G., Petersson, E., & Arnqvist, G. (2014). Sperm variation within a single ejaculate affects offspring development in Atlantic salmon. *Biology Letters, 10*, 20131040.
Inoue, M., Nakamura, T., Tanaka, Y., Suzuki, A., Yokoyama, Y., Kawahata, H., et al. (2018). A simple role of coral-algal symbiosis in coral calcification based on multiple geochemical tracers. *Geochimica et Cosmochimica Acta, 235*, 76–88.
Inouye, D. W. (1980). The terminology of floral larceny. *Ecology, 61*, 1251–1253.
Inouye, D. W., & Ogilvie, J. E. (2001). Role of pollinators. *Encyclopedia of Biodiversity*, 723–730.
Iqbal, N., Khan, N. A., Ferrante, A., Trivellini, A., Francini, A., & Khan, M. I. (2017). Ethylene role in plant growth, development and senescence: Interaction with other phytohormones. *Frontiers in Plant Science, 8*, 475.
Irwin, R. E., Bronstein, J. L., Manson, J. S., & Richardson, L. (2010). Nectar robbing: Ecological and evolutionary perspectives. *Annual Review of Ecology, Evolution, and Systematics, 41*, 271–292.
Ishiguro, S., Ogasawara, K., Fujino, K., Sato, Y., & Kishima, Y. (2014). Low temperature-responsive changes in the anther transcriptome's repeat sequences are indicative of stress sensitivity and pollen sterility in rice strains. *Plant Physiology, 164*(2), 671–682.
Iwasaki, M., & Paszkowski, J. (2014). Epigenetic memory in plants. *The EMBO Journal, 33*(18), 1987–1998.
Izawa, T. (2007). Adaptation of flowering-time by natural and artificial selection in *Arabidopsis* and rice. *Journal of Experimental Botany, 58*(12), 3091–3097.
Izawa, T. (2008). Photoperiodic flowering in rice. In *Rice biology in the genomics era* (pp. 163–176).
Jablonka, E., & Lamb, M. J. (2005). *Evolution in four dimensions: Genetic, epigenetic, behavioral, and symbolic variation in the history of life*. MIT Press.
Jaenike, J. (1987). A hypothesis to account for the maintenance of sex within populations. *Evolutionary Theory, 3*, 191–194.
Jaensen, T. (1979). Mating of males of *Glossina pallidipes* Austen (Diptera: Glossinidae). *Bulletin of Entomological Research, 69*, 573–588.
Jaffe, M. J., & Forbes, S. (1993). Thigmomorphogenesis: The effect of mechanical perturbation on plants. *Plant Growth Regulation, 12*, 313–324.
Jakob, E., Skow, C. D., & Long, S. M. (2011). Plasticity, learning and cognition. In *Spider behavior: Flexibility and versatility*.
Jaksic, F. M., Iriarte, J. A., Jiménez, J. E., & Martínez, D. R. (2002). Invaders without frontiers: Cross-border invasions of exotic mammals. *Biological Invasions, 4*, 157–173.
James, W. P. (2008). The epidemiology of obesity: The size of the problem. *Journal of Internal Medicine, 263*(4), 336–352.

Janzen, D. H. (1977). A note on optimal mate selection by plants. *The American Naturalist, 111*(978), 365–371.
Jarne, P., & Auld, J. R. (2006). Animals mix it up too: The distribution of self-fertilization among hermaphroditic animals. *Evolution, 60*(9), 1816–1824.
Jennions, M. D., & Kokko, H. (2010). Sexual selection. *Evolutionary Behavioral Ecology*, 343–364.
Jersáková, J., Johnson, S. D., & Kindlmann, P. (2006). Mechanisms and evolution of deceptive pollination in orchids. *Biological Reviews, 81*(2), 219–235.
Johnson, J. A., Altwegg, R., Evans, D. M., Ewen, J. G., Gordon, I. J., Pettorelli, N., & Young, J. K. (2016). Is there a future for genome-editing technologies in conservation? *Animal Conservation, 19*, 97–101.
Jones, F. A. (1996). Herbs–useful plants. Their role in history and today. *European Journal of Gastroenterology & Hepatology, 8*(12), 1227–1231.
Jones, L. K., Jennings, B. M., Goelz, R., Haythorn, K. W., Zivot, J. B., & de Waal, F. B. (2016). An ethogram to quantify operating room behavior. *Annals of Behavioral Medicine, 50*, 487–496.
Jorgensen, R. (1993). The origin of land plants: A union of alga and fungus advanced by flavonoids? *Biosystems, 31*(2–3), 193–207.
Jose, S. B., Wu, C. H., & Kamoun, S. (2019). Overcoming plant blindness in science, education, and society. *Plants, People, Planet, 1*(3), 169–172.
Jozet-Alves, C., Boal, J. G., & Dickel, L. (2008). Short-distance navigation in cephalopods: A review and synthesis. *Cognitive Processes, 9*, 239–247.
Jozet-Alves, C., Bertin, M., & Clayton, N. S. (2013). Evidence of episodic-like memory in cuttlefish. *Current Biology, 23*, R1033–R1035.
Judson, O. P. (2017). The energy expansions of evolution. *Nature Ecology & Evolution, 0138.*
Jullesson, D., David, F., Pfleger, B., & Nielsen, J. (2015). Impact of synthetic biology and metabolic engineering on industrial production of fine chemicals. *Biotechnology Advances, 33*(7), 1395–1402.
June, C. H., O'Connor, R. S., Kawalekar, O. U., Ghassemi, S., & Milone, M. C. (2018). CAR T cell immunotherapy for human cancer. *Science, 359*, 1361–1365.
Juster, R. P., McEwen, B. S., & Lupien, S. J. (2010). Allostatic load biomarkers of chronic stress and impact on health and cognition. *Neuroscience & Biobehavioral Reviews, 35*(1), 2–16.
Kahn, A. (1997). Clone mammals... Clone man? *Nature, 386*(6621), 119.
Kaidar-Person, O., Person, B., Szomstein, S., & Rosenthal, R. J. (2008). Nutritional deficiencies in morbidly obese patients: A new form of malnutrition? Part A: Vitamins. *Obesity Surgery, 18*, 870–876.
Kaiser, D. (2001). Building a multicellular organism. *Annual Review of Genetics, 35*(1), 103–123.
Kaiser, J. (2010). The dirt on ocean garbage patches. *Science, 328*, 1506–1506.
Kaiser, V. B., & Bachtrog, D. (2010). Evolution of sex chromosomes in insects. *Annual Review of Genetics, 44*, 91–112.
Kandel, E. R., Schwartz, J. H., & Jessell, T. M. (2000). *Principles of neural science*. McGraw-Hill.
Karban, R. (2015). Plant sensing and communication. In *Plant sensing and communication*. University of Chicago Press.
Karban, R. (2021). Plant communication. *Annual Review of Ecology, Evolution, and Systematics, 52*, 1–24.
Karlberg, O., Canback, B., Kurland, C. G., & Andersson, S. G. (2000). The dual origin of the yeast mitochondrial proteome. *Yeast, 17*, 170–187.
Karson, M. A., Boal, J. G., & Hanlon, R. T. (2003). Experimental evidence for spatial learning in cuttlefish (*Sepia officinalis*). *Journal of Comparative Psychology, 117*, 149–155.
Kaster, J. L., & Jakob, E. M. (1997). Last-male sperm priority in a haplogyne spider (Araneae: Pholcidae): Correlations between female morphology and patterns of sperm usage. *Annals of the Entomological Society of America, 90*(2), 254–259.
Kauffman, S. A. (2011). Approaches to the origin of life on earth. *Life, 1*(1), 34–48.

Kay, K. M., & Sargent, R. D. (2009). The role of animal pollination in plant speciation: Integrating ecology, geography, and genetics. *Annual Review of Ecology, Evolution, and Systematics, 40*, 637–656.

Keasling, J. (2007). Setting the standard in synthetic biology. Synthetic biology for synthetic chemistry. *ACS Chemical Biology, 3*(1), 64–76.

Keefer, C. L. (2015). Artificial cloning of domestic animals. *Proceedings of the National Academy of Sciences, 112*(29), 8874–8878.

Keeling, P. J., & Doolittle, W. F. (1995). Archaea: Narrowing the gap between prokaryotes and eukaryotes. *Proceedings of the National Academy of Sciences, 92*(13), 5761–5764.

Keeling, M. J., & Gilligan, C. A. (2000). Metapopulation dynamics of bubonic plague. *Nature, 407*(6806), 903–906.

Keiser, C. N., Lichtenstein, J. L., Wright, C. M., Chism, G. T., & Pruitt, J. N. (2018). Personality and behavioral syndromes in insects and spiders. In A. Córdoba-Aguilar, D. González-Tokman, & I. González-Santoyo (Eds.), *Insect behavior: From mechanisms to ecological and evolutionary consequences* (pp. 236–256). Oxford University Press.

Kekäläinen, J. (2022). Cryptic female choice within individual males—A neglected component of the postmating sexual selection? *Journal of Evolutionary Biology, 35*(11), 1407–1413.

Kekäläinen, J., & Evans, J. P. (2018). Gamete-mediated mate choice: Towards a more inclusive view of sexual selection. *Proceedings of the Royal Society B, 285*, 20180836.

Keller, G., & Snodgrass, H. R. (1999). Human embryonic stem cells: The future is now. *Nature Medicine, 5*, 151–152.

Keller, L., & Surette, M. G. (2006). Communication in bacteria: An ecological and evolutionary perspective. *Nature Reviews Microbiology, 4*(4), 249–258.

Kelley, D. B. (1988). Sexually dimorphic behaviors. *Annual Review of Neuroscience, 11*(1), 225–251.

Kempermann, G. (2002). Neuronal stem cells and adult neurogenesis. *Ernst Schering Research Foundation Workshop, 35*, 17–28.

Kenny, C. (2021). *The plague cycle: The unending war between humanity and infectious disease*. Simon and Schuster.

Kernie, S. G., & Parent, J. M. (2010). Forebrain neurogenesis after focal ischemic and traumatic brain injury. *Neurological Disease, 37*, 267–274.

Kessels, R. P. C., de Haan, E. H. F., Kappele, L. J., & Postma, A. (2001). Varieties of human spatial memory: A meta-analysis on the effects of hippocampal lesions. *Brain Research Reviews, 35*, 295–303.

Kettlewell, H. B. (1959). Darwin's missing evidence. *Scientific American, 200*(3), 48–53.

Khanna, P., Kaushik, R., & Kaur, G. (2012). Changing dietary pattern and lifestyle on diseases. *Asian Journal of Multidimensional Research (AJMR), 1*(6), 49–54.

Kholosha, V. I., Tabachnyi, L. Y., Davydchuk, S. V., & Lytvynenko, O. Y. (2008). *Atlas*. Ukraine Radioactive contamination. Ministry of Ukraine of Emergencies and Affairs of Population Protection from the Consequences of Chornobyl Catastrophe.

Kilpatrick, T. J., & Bartlett, P. F. (1993). Cloning and growth of multipotential neural precursors: Requirements for proliferation and differentiation. *Neuron, 10*, 255–265.

Kim, B. H., Gadd, G. M., & Chang, P. H. (2019). *Bacterial physiology and metabolism*. Cambridge University Press.

King, N. (2005). Choanoflagellates. *Current Biology, 15*(4), R113–R114.

King, N., Westbrook, M. J., Young, S. L., Kuo, A., Abedin, M., Chapman, J., et al. (2008). The genome of the choanoflagellate *Monosiga brevicollis* and the origin of metazoans. *Nature, 451*(7180), 783–788.

Kinoshita, T., & Seki, M. (2014). Epigenetic memory for stress response and adaptation in plants. *Plant and Cell Physiology, 55*(11), 1859–1863.

Kishimoto, N., Shimizu, K., & Sawamoto, K. (2012). Neuronal regeneration in a zebrafish model of adult brain injury. *Disease Models & Mechanisms, 5*(2), 200–209.

Klein, S. B. (2015). What memory is. *Wiley Interdisciplinary Reviews: Cognitive Science, 6*(1), 1–38.
Klimont, Z., Smith, S. J., & Cofala, J. (2013). The last decade of global anthropogenic sulfur dioxide: 2000–2011 emissions. *Environmental Research Letters, 8*(1), 014003.
Klopfstein, S., Currat, M., & Excoffier, L. (2006). The fate of mutations surfing on the wave of a range expansion. *Molecular Biology and Evolution, 23*(3), 482–490.
Knoll, A. H. (2011). The multiple origins of complex multicellularity. *Annual Review of Earth and Planetary Sciences, 39*, 217–239.
Knoll, A. H., Javaux, E. J., Hewitt, D., & Cohen, P. (2006). Eukaryotic organisms in Proterozoic oceans. *Philosophical Transactions of the Royal Society B: Biological Sciences, 361*(1470), 1023–1038.
Koffel, T., Boudsocq, S., Loeuille, N., & Daufresne, T. (2018). Facilitation- vs. competition-driven succession: The key role of resource-ratio. *Ecology Letters, 21*(7), 1010–1021.
Koga, A., Iida, A., Hori, H., Shimada, A., & Shima, A. (2006). Vertebrate DNA transposon as a natural mutator: The medaka fish Tol2 element contributes to genetic variation without recognizable traces. *Molecular Biology and Evolution, 23*, 1414–1419.
Kokko, H. (2020). When synchrony makes the best of both worlds even better: How well do we really understand facultative sex? *The American Naturalist, 195*(2), 380–392.
Kokko, H., & Jennions, M. D. (2008). Parental investment, sexual selection and sex ratios. *Journal of Evolutionary Biology, 21*(4), 919–948.
Kolb, H. (2012). *Photoreceptors*. The Organization of the Retina and Visual System. Retrieved from https://www.ncbi.nlm.nih.gov/books/NBK11510/
Kolb, B., & Whishaw, I. Q. (1998). Brain plasticity and behavior. *Annual Review of Psychology, 49*, 43–64.
Komiyama, T., Kobayashi, H., Tateno, Y., Inoko, H., Gojobori, T., & Ikeo, K. (2009). An evolutionary origin and selection process of goldfish. *Gene, 430*(1–2), 5–11.
Kondrashov, A. S. (1982). Selection against harmful mutations in large sexual and asexual populations. *Genetics Research, 40*(3), 325–332.
Kondrashov, A. S. (1984). Deleterious mutations as an evolutionary factor: 1. The advantage of recombination. *Genetical Research, 44*(2), 199–217.
Kondrashov, A. S. (1988). Deleterious mutations and the evolution of sexual reproduction. *Nature, 336*(6198), 435–440.
Kondrashov, A. S. (1993). Classification of hypotheses on the advantage of amphimixis. *Journal of Heredity, 84*(5), 372–387.
Koolhaas, J. M., Korte, S. M., De Boer, S. F., Van Dervegt, B. J., Van Reenen, C. G., Hopster, H., De Jong, I. C., Ruis, M. A., & Blokhuis, H. J. (1999). Coping style in animals: Current status in behavior and stress-physiology. *Neuroscience and Biobehavioral Reviews, 23*, 925–935.
Kopp, W. (2019). How western diet and lifestyle drive the pandemic of obesity and civilization diseases. *Diabetes, Metabolic Syndrome and Obesity: Targets and Therapy, 12*, 2221–2236.
Kostyuchenko, R. P., Kozin, V. V., & Kupriashova, E. E. (2016). Regeneration and asexual reproduction in annelids: Cells, genes, and evolution. *Biology Bulletin, 43*(3), 185–194.
Kotradyová, V., Petelen, I., & Žúžiová, Z. (2013). Civilization diseases and their relation to interior design of working spaces. *Human Resources Management & Ergonomics, 7*(2).
Kotrschal, A., Lievens, E. J., Dahlbom, J., Bundsen, A., Semenova, S., Sundvik, M., Maklakov, A., Winberg, S., Panula, P., & Kolm, N. (2014). Artificial selection on relative brain size reveals a positive genetic correlation between brain size and proactive personality in the guppy. *Evolution, 68*(4), 1139–1149.
Kralj-Fišer, S., & Schuett, W. (2014). Studying personality variation in invertebrates: Why bother? *Animal Behaviour, 91*, 41–52.
Kramer, P., & Bressan, P. (2015). Humans as superorganisms: How microbes, viruses, imprinted genes, and other selfish entities shape our behavior. *Perspectives on Psychological Science, 10*(4), 464–481.
Krasne, F. B., & Teshiba, T. M. (1994). Habituation of an invertebrate escape reflex due to modulation by higher centers rather than local events. *Proceedings of the National Academy of Sciences, 92*(8), 3362–3366.

Krasne, F. B., & Wine, J. J. (1987). Evasion response of the crayfish. In D. M. Guthrie (Ed.), *Aims and methods in neuroethology* (pp. 10–45). Springer.
Kua, E. H., Ho, E., Tan, H. H., Tsoi, C., Thng, C., & Mahendran, R. (2014). The natural history of dementia. *Psychogeriatrics, 14*(3), 196–201.
Kühhölzer, B., Hawley, R. J., Lai, L., Kolber-Simonds, D., & Prather, R. S. (2001). Clonal lines of transgenic fibroblast cells derived from the same fetus result in different development when used for nuclear transfer in pigs. *Biology of Reproduction, 64*, 1695–1698.
Kuhn, T. S. (1976). Scientific revolutions as changes of world view. In S. G. Harding (Ed.), *Can theories be refuted?* (pp. 81–97). Synthese Library. Springer.
Kuhn, H. G., Dickinson-Anson, H., & Gage, F. H. (1996). Neurogenesis in the dentate gyrus of the adult rat: Age-related decrease of neuronal progenitor proliferation. *Journal of Neuroscience, 16*, 2027–2033.
Kunz, K., Witthuhn, M., & Uhl, G. (2014). Do the size and age of mating plugs alter their efficacy in protecting paternity? *Behavioral Ecology and Sociobiology, 68*, 1321–1328.
Kutschera, U., & Briggs, W. R. (2009). From Charles Darwin's botanical country-house studies to modern plant biology. *Plant Biology, 11*(6), 785–795.
Kweku, D. W., Bismark, O., Maxwell, A., Desmond, K. A., Danso, K. B., Oti-Mensah, E. A., Quachie, A. T., & Adormaa, B. B. (2018). Greenhouse effect: Greenhouse gases and their impact on global warming. *Journal of Scientific Research and Reports, 17*(6), 1–9.
Kyselý, R. (2008). Aurochs and potential crossbreeding with domestic cattle in Central Europe in the Eneolithic period. A metric analysis of bones from the archaeological site of Kutná Hora-Denemark (Czech Republic). *Anthropozoologica, 43*(2), 7–37.
Kyzar, E. J., Pham, M., Roth, A., Cachat, J., Green, J., Gaikwad, S., & Kalueff, A. V. (2012). Alterations in grooming activity and syntax in heterozygous SERT and BDNF knockout mice: The utility of behavior-recognition tools to characterize mutant mouse phenotypes. *Brain Research Bulletin, 89*(5–6), 168–176.
Lake, J. A., & Rivera, M. C. (1994). Was the nucleus the first endosymbiont? *Proceedings of the National Academy of Sciences USA, 91*, 2880–2881.
Laloi, D., Sandoz, J., Picard-Nizou, A., Marchesi, A., Pouvreau, A., Tasei, J., Poppy, G., & Pham-Delegue, M. (1999). Olfactory conditioning of the proboscis extension in bumble bees. *Entomologia Experimentalis et Applicata, 90*, 123–129.
Lamarck, J. B. (1809). *Philosophie zoologique* (translated, with an introduction, by H. Elliot). Hafner Publishing Company, 1963.
Lammoglia, M., Lozano, R., García, C., Avilez, C., Trejo, V., Muñoz, R., & López, C. (2016). La revolución en ingeniería genética: Sistema CRISPR/Cas. *Investigación en Discapacidad, 5*(2), 116–128.
Lande, R. (1976). Natural selection and random genetic drift in phenotypic evolution. *Evolution*, 314–334.
Langkilde, T. (2009). Invasive fire ants alter behavior and morphology of native lizards. *Ecology, 90*, 208e217.
Langkilde, T., & Freidenfelds, N. A. (2010). Consequences of envenomation: Red imported fire ants have delayed effects on survival but not growth of native fence lizards. *Wildlife Research, 37*, 566e573.
Lango, M. N. (2020). How did we get here? Short history of COVID-19 and other coronavirus-related epidemics. *Head & Neck, 42*(7), 1535–1538.
Lanteri, A., Confalonieri, V., & Rodriguero, M. (2010). Formas curiosas de reproducción animal, la partenogénesis. *Ciencia Hoy, 20*(119), 14–22.
Lavrov, A. I., & Kosevich, I. A. (2014). Sponge cell reaggregation: Mechanisms and dynamics of the process. *Russian Journal of Developmental Biology, 45*(4), 205–223.
Lazcano, A., & Miller, S. L. (1996). The origin and early evolution of life: Prebiotic chemistry, the pre-RNA world, and time. *Cell, 85*(6), 793–798.

Lazcano-Araujo, A. R., & Oró, J. (1981). Cometary material and the origins of life on earth. In C. Ponnamperuma (Ed.), *Comets and the origin of life proceedings of the College Park colloquia, vol 5*. Springer.

Le Feuvre, R., & Scrutton, N. (2018). A living foundry for synthetic biological materials: A synthetic biology roadmap to new advanced materials. *Synthetic and Systems Biotechnology, 3*, 105–112.

Le Trionnaire, G., Hardie, J., Jaubert-Possamai, S., Simon, J. C., & Tagu, D. (2008). Shifting from clonal to sexual reproduction in aphids: Physiological and developmental aspects. *Biology of the Cell, 100*(8), 441–451.

Lebreton, L., Slat, B., Ferrari, F., Sainte-Rose, B., Aitken, J., Marthouse, R., Hajbane, S., Cunsolo, S., Schwarz, A., Levivier, A., & Noble, K. (2018). Evidence that the great Pacific garbage patch is rapidly accumulating plastic. *Scientific Reports, 8*(1), 1–5.

Ledón-Rettig, C. C., Pfennig, D. W., Chunco, A. J., & Dworkin, I. (2014). Cryptic genetic variation in natural populations: A predictive framework. *Evolution, 68*(3), 783–793.

Lee, B. R., Cho, S., Song, Y., Kim, S. C., & Cho, B. K. (2013). Emerging tools for synthetic genome design. *Molecules and Cells, 35*(5), 359–370.

Lehnebach, C. A., Robertson, A. W., & Hedderley, D. (2005). Pollination studies of four New Zealand terrestrial orchids and the implication for their conservation. *New Zealand Journal of Botany, 43*(2), 467–477.

Lei, L. (2017). Lignin evolution: Invasion of land. *Nature Plants, 3*(4), 1–1.

Leitão-Gonçalves, R., Carvalho-Santos, Z., Francisco, A. P., Tondolo Fioreze, G., Anjos, M., Baltazar, C., Elias, A. P., Itskov, P. M., Piper, M. D. W., & Ribeiro, C. (2017). Commensal bacteria and essential amino acids control food choice behavior and reproduction. *PLoS Biology, 15*(4), e2000862.

Lemoine, N. P., Burkepile, D. E., & Parker, J. D. (2016). Quantifying differences between native and introduced species. *Trends in Ecology & Evolution, 31*(5), 372–381.

Lendon, H., & Martin, B. (2007). Environmental disasters. In B. Martin (Ed.), *Justice ignited: The dynamics of backfire* (pp. 99–112). Rowman and Littlefield.

Lenormand, T. (2003). The evolution of sex in a changing environment. *Journal of Evolutionary Biology, 16*(6), 1146–1153.

Leonard, J., & Córdoba-Aguilar, A. (2010). *The evolution of primary characters in animals*. Oxford University Press.

Levernier, N., Pouliquen, O., & Forterre, Y. (2021). An integrative model of plant gravitropism linking statoliths position and auxin transport. *Frontiers in Plant Science, 12*, 651928.

Levis, N. A., & Pfennig, D. W. (2016). Evaluating "plasticity-first" evolution in nature: Key criteria and empirical approaches. *Trends in Ecology & Evolution, 31*, 563–574.

Levis, N. A., de la Serna Buzón, S., & Pfennig, D. W. (2015). An inducible offense: Carnivore morph tadpoles induced by tadpole carnivory. *Ecology and Evolution, 5*, 1405–1411.

Levis, N. A., Isdaner, A. J., & Pfennig, D. W. (2018). Morphological novelty emerges from pre-existing phenotypic plasticity. *Nature Ecology & Evolution, 2*, 1289–1297.

Lewis, D. H. (1987). Evolutionary aspects of mutualistic associations between fungi and photosynthetic organisms. *Evolutionary Biology of the Fungi*, 161–178.

Li, Y. (2021). Modern epigenetics methods in biological research. *Methods, 187*, 104–113.

Li, L. F., & Ge, X. J. (2017). Origin and domestication of cultivated banana. *Ecological Genetics and Genomics, 2*, 1–2.

Li, X., Li, Z., Jouneau, A., Zhou, Q., & Renard, J. P. (2003). Nuclear transfer: Progress and quandaries. *Reproductive Biology and Endocrinology, 1*, 84–89.

Li, Y., Jiang, Y., Chen, H., Liao, W., Li, Z., Weiss, R., & Xie, Z. (2015). Modular construction of mammalian gene circuits using TALE transcriptional repressors. *Nature Chemical Biology, 11*(3), 207–213.

Li, Y., Wang, X., Fu, W., Xia, X., Liu, C., Min, J., Zhang, W., & Crittenden, J. C. (2019). Interactions between nano/micro plastics and suspended sediment in water: Implications on aggregation and settling. *Water Research, 161*, 486–495.

Lichtenberg, E. M., Irwin, R. E., & Bronstein, J. L. (2018). Costs and benefits of alternative food handling tactics help explain facultative exploitation of pollination mutualisms. *Ecology, 99*(8), 1815–1824.
Liedvogel, M., Chapman, B. B., Muheim, R., & Åkesson, S. (2013). The behavioural ecology of animal movement: reflections upon potential synergies. *Animal Migration, 1*(1), 39–46.
Lieskovská, V., & Megyesiová, S. (2015). Managing activities supporting the elimination of civilization diseases. In *XIV international business and economy conference* (pp. 5–15).
Lima, S. L., & Dill, L. M. (1990). Behavioral decisions made under the risk of predation: A review and prospectus. *Canadian Journal of Zoology, 68*, 619–640.
Linares, C., & Maciel, L. (2017). Biotecnología animal en Argentina: Panorama interno y desafíos jurídico-comerciales internacionales. *RECORDIP, 1*.
Lind, E. M., & Parker, J. D. (2010). Novel weapons testing: Are invasive plants more chemically defended than native plants? *PLoS One, 5*(5), e10429.
Lira-Medeiros, C. F., Parisod, C., Fernandes, R. A., Mata, C. S., Cardoso, M. A., & Ferreira, P. C. G. (2010). Epigenetic variation in mangrove plants occurring in contrasting natural environments. *PLoS One, 5*, e10326.
Lisch, D. (2009). Epigenetic regulation of transposable elements in plants. *Annual Review of Plant Biology, 60*, 43–66.
Lisch, D. (2013). How important are transposons for plant evolution? *Nature Reviews Genetics, 14*, 49–61.
Little, R. J. (1983). A review of floral food deception mimicries with comments on floral mutualism. In *Handbook of experimental pollination biology* (pp. 294–309).
Liu, K., Tedeschi, A., Park, K. K., & He, Z. (2011). Neuronal intrinsic mechanisms of axon regeneration. *Annual Review of Neuroscience, 34*, 131–152.
Liu, D., Hoynes-O'Connor, A., & Zhang, F. (2013). Bridging the gap between systems biology and synthetic biology. *Frontiers in Microbiology, 4*, 1–8.
Liu, Z., Cai, Y., Wang, Y., Wang, Z., Poo, M., & Sun, Q. (2018). Cloning of macaque monkeys by somatic cell nuclear transfer. *Cell, 172*, 881–887.
Lloyd, J. E. (1979). Sexual selection in luminescent beetles. In M. S. Blum & N. A. Bluym (Eds.), *Sexual selection and reproductive competition in insects* (pp. 293–342). Academic Press.
Lloyd, E. A. (2007). Units and levels of selection. In D. L. Hull & M. Ruse (Eds.), *Cambridge companion to the philosophy of biology* (pp. 44–65). Cambridge University Press.
Lloyd, D. G., & Webb, C. J. (1977). Secondary sex characters in plants. *The Botanical Review, 43*(2), 177–216.
Lodé, T. (2012). Have sex or not? Lessons from bacteria. *Sex Development, 6*(6), 325–328.
Lodge, D. J. (2000). Ecto-or arbuscular mycorrhizas–which are best? *The New Phytologist, 146*(3), 353–354.
Logares, R., Boltovskoy, A., Bensch, S., Laybourn-Parry, J., & Rengefors, K. (2009). Genetic diversity patterns in five protist species occurring in lakes. *Protist, 160*(2), 301–317.
Lom, J., & Dyková, I. (2006). Myxozoan genera: Definition and notes on taxonomy, life-cycle terminology and pathogenic species. *Folia Parasitologica, 53*(1), 1–36.
Lonsdale, W. M., Miller, I. L., & Forno, I. W. (1995). Mimosa pigra L. In R. H. Groves, R. C. H. Sheppard, & R. G. Richardson (Eds.), *The Biology of Australian Weeds* (pp. 169–188). R.G. and F.J. Richardson Publishers.
López-García, P., Eme, L., & Moreira, D. (2017). Symbiosis in eukaryotic evolution. *Journal of Theoretical Biology, 434*, 20–33.
Løvlie, H., Gillingham, M. A., Worley, K., Pizzari, T., & Richardson, D. S. (2013). Cryptic female choice favours sperm from major histocompatibility complex-dissimilar males. *Proceedings of the Royal Society B: Biological Sciences, 280*(1769), 20131296.
Lowe, S., Browne, M., Boudjelas, S., & De Poorter, M. (2000). *100 of the World's worst invasive alien species: A selection from the global invasive species database.* The invasive species specialist group (ISSG) a specialist group of the species survival commission (SSC) of the

world conservation union (IUCN). Retrieved from https://s3.amazonaws.com/academia.edu.documents/33655728/100_world_worst_invasive_alien_species_English.pdf.

Lu, H.-F., Dong, H.-T., Sun, C.-B., Qing, D.-J., Li, N., Wu, Z.-K., et al. (2011). The panorama of physiological responses and gene expression of whole plant of maize inbred line YQ7-96 at the three-leaf stage under water deficit and re-watering. *Theoretical and Applied Genetics, 123*(6), 943.

Lüdke, J., & Lakes-Harlan, R. (2008). Regeneration of the tibia and somatotopy of regenerated hair sensilla in *Schistocerca gregaria* (Forskål). *Arthropod Structure & Development, 37*(3), 210–220.

Luisi, P. L. (1996). Self-reproduction of micelles and vesicles: Models for the mechanisms of life from the perspective of compartmented chemistry. *Advances in Chemistry and Physics, 92*, 425–438.

Luong, L. T., & Mathot, K. J. (2019). Facultative parasites as evolutionary stepping-stones towards parasitic lifestyles. *Biology Letters, 15*(4), 20190058.

Luong, L. T., Brophy, T., Stolz, E., & Chan, S. J. (2017). State-dependent parasitism by a facultative parasite of fruit flies. *Parasitology, 144*, 1468–1475.

Lusher, A. L., Welden, N. A., Sobral, P., & Cole, M. (2020). Sampling, isolating and identifying microplastics ingested by fish and invertebrates. In *Analysis of Nanoplastics and microplastics in food* (pp. 119–148).

Luteijn, M. J., & Ketting, R. F. (2013). PIWI-interacting RNAs: From generation to transgenerational epigenetics. *Nature Reviews Genetics, 14*, 523–534.

Lutzoni, F., Pagel, M., & Reeb, V. (2001). Major fungal lineages are derived from lichen symbiotic ancestors. *Nature, 411*(6840), 937–940.

Lynch, M., & Conery, J. S. (2003). The origins of genome complexity. *Science, 302*, 1401–1404.

Lyons, N. A., & Kolter, R. (2015). On the evolution of bacterial multicellularity. *Current Opinion in Microbiology, 24*, 21–28.

Lyons, E. E., Waser, N. M., Price, M. V., Antonovics, J., & Motten, A. F. (1989). Sources of variation in plant reproductive success and implications for concepts of sexual selection. *The American Naturalist, 134*(3), 409–433.

MacIver, M. A., Schmitz, L., Mugan, U., Murphey, T. D., & Mobley, C. D. (2017). Massive increase in visual range preceded the origin of terrestrial vertebrates. *Proceedings of the National Academy of Sciences, 114*(12), E2375–E2384.

Mackie, G. O. (1970). Neuroid conduction and the evolution of conducting tissues. *The Quarterly Review of Biology, 45*(4), 319–332.

Mahandra, B. (1984). *Dementia: A survey of the syndrome of dementia*. MTP.

Mahgoub, M., & Monteggia, L. M. (2013). Epigenetics and psychiatry. *Neurotherapeutics, 10*(4), 734–741.

Majerus, M. E. (2009). Industrial melanism in the peppered moth, *Biston betularia*: An excellent teaching example of Darwinian evolution in action. *Evolution: Education and Outreach, 2*(1), 63–74.

Makałowski, W., Gotea, V., Pande, A., & Makałowska, I. (2019). Transposable elements: Classification, identification, and their use as a tool for comparative genomics. In M. Anisimova (Ed.), *Evolutionary genomics: Statistical and computational methods* (pp. 177–207). Springer.

Malenka, R. C. (1995). LTP and LTD: Dynamic and interactive processes of synaptic plasticity. *The Neuroscientist, 1*, 35–42.

Mancuso, S., & Shabala, S. (2006). *Rhythms in plants*. Springer.

Mandrycky, C., Wang, Z., Kim, K., & Kim, D. H. (2016). 3D bioprinting for engineering complex tissues. *Biotechnology Advances, 34*(4), 422–434.

Manetas, Y. (2012). Are plants intelligent organisms after all? In *Alice in the land of plants: Biology of plants and their importance for planet earth* (pp. 323–342).

Mann, M. E. (2002). Little ice age. In *Encyclopedia of global environmental change* (p. e509).

Mann, M. E., Bradley, R. S., & Hughes, M. K. (1999). Northern hemisphere temperatures during the past millennium: Inferences, uncertainties, and limitations. *Geophysical Research Letters, 26*(6), 759–762.
Mansy, S. S., Schrum, J. P., Krishnamurthy, M., Tobé, S., Treco, D. A., & Szostak, J. W. (2008). Template directed synthesis of a genetic polymer in a model protocell. *Nature, 454*, 122–126.
Mantilla, M. J., & Sáez, R. G. T. (2019). Enfoque metagenómico para la caracterización del microbioma de aves corral. *Revisión. Revista Colombiana de Biotecnología, 21*(2), 77–97.
Margulis, L. (1981). Symbiosis in cell evolution: Life and its environment on the early earth.
Margulis, L. (1990). Words as battle cries: Symbiogenesis and the new field of endocytobiology. *Bioscience, 40*(9), 673–677.
Margulis, L. (1991). Symbiogenesis and symbionticism. In L. Margulis & R. Fester (Eds.), *Symbiosis as a source of evolutionary innovation: Speciation and morphogenesis* (pp. 1–14). MIT Press.
Margulis, L. (1993). *Symbiosis in cell evolution: Microbial communities in the Archean and Proterozoic eons*. Freeman.
Margulis, L. (2010). Symbiogenesis. A new principle of evolution rediscovery of Boris Mikhaylovich Kozo-Polyansky (1890–1957). *Journal of Paleontology, 44*, 1525–1539.
Margulis, L., & Chapman, M. J. (1998). Endosymbioses: Cyclical and permanent in evolution. *Trends in Microbiology, 6*(9), 342–345.
Margulis, L., & Sagan, D. (1986). *Origins of sex: Three billion years of genetic recombination*. Yale University Press.
Margulis, L., & Sagan, D. (2002). *Acquiring genomes: A theory of the origin of species*. Basic Books.
Mariette, J., Carcaud, J., & Sandoz, J. C. (2021). The neuroethology of olfactory sex communication in the honeybee *Apis mellifera L. Cell and Tissue Research, 383*, 177–194.
Markham, A. C. (2019). *A brief history of pollution*. Routledge.
Marks, P. L. (1974). The role of pin cherry (*Prunus pensylvanica* L.) in the maintenance of stability in northern hardwood ecosystems. *Ecological Monographs, 44*(1), 73–88.
Marshall, D. L., & Folsom, M. W. (1991). Mate choice in plants: An anatomical to population perspective. *Annual Review of Ecology and Systematics, 22*(1), 37–63.
Martin, W. (1999). A briefly argued case that mitochondria and plastids are descendants of endosymbionts, but that the nuclear compartment is not. *Proceedings of the Royal Society B: Biological Sciences, 266*, 1387–1395.
Martin, W., & Russell, M. J. (2003). On the origins of cells: A hypothesis for the evolutionary transitions from abiotic geochemistry to chemoautotrophic prokaryotes, and from prokaryotes to nucleated cells. *Philosophical Transactions of the Royal Society B: Biological Sciences, 358*, 59–85.
Martin, W. F., & Sousa, F. L. (2016). Early microbial evolution: The age of anaerobes. *Cold Spring Harbor Perspectives in Biology, 8*(2), a018127.
Martin, W., Hoffmeister, M., Rotte, C., & Henze, K. (2001). An overview of endosymbiotic models for the origins of eukaryotes, their ATP-producing organelles (mitochondria and hydrogenosomes), and their heterotrophic lifestyle. *Biological Chemistry, 382*, 1521–1539.
Martin, W. F., Bryant, D. A., & Beatty, J. T. (2018). A physiological perspective on the origin and evolution of photosynthesis. *FEMS Microbiology Reviews, 42*(2), 205–231.
Martínez, V., Reddy, P., & Zoran, M. (2006). Asexual reproduction and segmental regeneration, but not Morphallaxis, are inhibited by boric acid in *Lumbriculus variegatus* (Annelida: Clitellata: Lumbriculidae). *Hydrobiologia, 564*, 73–86.
Marx, P. A., Apetrei, C., & Drucker, E. (2004). AIDS as a zoonosis? Confusion over the origin of the virus and the origin of the epidemics. *Journal of Medical Primatology, 33*(5–6), 220–226.
Mather, J. A., & Anderson, R. C. (1993). Personalities of octopuses (*Octopus rubescens*). *Journal of Comparative Psychology, 107*, 336–340.
Mathger, L. M., Shashar, N., & Hanlon, R. T. (2009). Do cephalopods communicate using polarized light reflections from their skin? *Journal of Experimental Biology, 212*(14), 2133–2140.

Matossian, M. A. (1989). *Poisons of the past: Molds, epidemics, and history*. Yale University Press.
Matsuura, K., Vargo, E. L., Kawatsu, K., Labadie, P. E., Nakano, H., Yashiro, T., & Tsuji, K. (2009). Queen succession through asexual reproduction in termites. *Science, 323*, 1687.
Matthews, J. A., & Briffa, K. R. (2005). The 'Little ice Age': Re-evaluation of an evolving concept. *Geografiska Annaler: Series A, Physical Geography, 87*(1), 17–36.
Mayhew, P. J. (2007). Why are there so many insect species? Perspectives from fossils and phylogenies. *Biological Reviews, 82*(3), 425–454.
Mayr, E. (1963). *Animal species and evolution*. Harvard University Press.
McClintock, B. (1942). The fusion of broken ends of chromosomes following nuclear fusion. *Proceedings of the National Academy of Sciences of the United States of America, 28*(11), 458–463.
McClintock, B. (1950). The origin and behavior of mutable loci in maize. *Proceedings of the National Academy of Sciences, 36*, 344–355.
McClintock, B. (1951). Chromosome organization and genic expression. In *Cold Spring Harbor symposia on quantitative biology* (Vol. 16, pp. 13–47). Cold Spring Harbor Laboratory Press.
McClintock, B. (1952). Mutable loci in maize: Origins of instability at the A1 and A2 loci. Instability of Sh1 action induced by Ds. Summary. *Carnegie Institution of Washington Year Book, 51*, 212–219.
McClintock, B. (1953). Induction of instability at selected loci in maize. *Genetics, 38*, 579–599.
McClintock, B. (1984). Nobel lecture. The significance of response of the genome to challenge. *Science, 226*, 792–801.
McCormick, S. D., Hansen, L. P., Quinn, T. P., & Saunders, R. L. (1998). Movement, migration, and smolting of Atlantic salmon (*Salmo salar*). *Canadian Journal of Fisheries and Aquatic Sciences, 55*(S1), 77–92.
McFadden, G. I., & van Dooren, G. G. (2004). Evolution: Red algal genome affirms a common origin of all plastids. *Current Biology, 14*, R514–R516.
McFall-Ngai, M., Hadfield, M. G., Bosch, T. C. G., Carey, H. V., Domazet-Lošo, T., Douglas, A. E., et al. (2013). Animals in a bacterial world, a new imperative for the life sciences. *Proceedings of the National Academy of Sciences, 110*(9), 3229–3236.
McGuigan, K., & Sgro, C. M. (2009). Evolutionary consequences of cryptic genetic variation. *Trends in Ecology & Evolution, 24*(6), 305–311.
McGuire, T. R. (1984). Learning in three species of Diptera: The blow fly *Phormia regina*, the fruit fly *Drosophila melanogaster*, and the house fly *Musca domestica*. *Behavior Genetics, 14*, 479–526.
McKusick, V. A. (2000). Ellis-van Creveld syndrome and the Amish. *Nature Genetics, 24*(3), 203–204.
McNeill, W. H. (1998). *Plagues and peoples*. Anchor P.
Meiners, S. J. (2007). Native and exotic plant species exhibit similar population dynamics during succession. *Ecology, 88*(5), 1098–1104.
Meinertzhagen, I. A., & Okamura, Y. (2001). The larval ascidian nervous system: The chordate brain from its small beginnings. *Trends in Neurosciences, 24*, 401–410.
Meng, L., Ely, J. J., Stouffer, R. L., & Wolf, D. P. (1997). Rhesus monkeys produced by nuclear transfer. *Biology of Reproduction, 57*, 454–459.
Menzel, R. (1999). Memory dynamics in the honeybee. *Journal of Comparative Physiology, 185*, 323–340.
Menzel, R. (2001). Searching for the memory trace in a mini-brain, the honeybee. *Learning & Memory, 8*, 53–62.
Menzel, R., Manz, G., Menzel, R. M., & Greggers, U. (2001). Massed and spaced learning in honeybees: The role of CS, US, the inter-trial interval, and the test interval. *Learning & Memory, 8*, 198–208.
Mereschkowsky, C. (1910). Theorie der zwei Plasmaarten als Grundlage der Symbiogenesis, einer neuen Lehre von der Entstehung der Organismen. *Biologisches Zentralblatt, 30*, 278–303.
Merriam-Webster. (2010). *Merriam-Webster online dictionary*. https://www.merriam-webster.com/

Meteosim. (2021). World most polluted cities. https://meteosim.com/en/world-most-polluted-citie/
Méthot, P. O. (2012). Why do parasites harm their host? On the origin and legacy of Theobald Smith's "law of declining virulence" 1900–1980. *History and Philosophy of the Life Sciences, 34*, 561–601.
Meyer-Abich, A. (1943). Beiträge zur Theorie der Evolution der Organismen. I. Das typologische Grundgesetz und seine Folgerungen für Phylogenie und Entwicklungsphysiologie. *Acta Biotheoretica, 7*, 1–80.
Mian, I. S., & Rose, C. (2011). Communication theory and multicellular biology. *Integrative Biology, 3*(4), 350–367.
Michaloud, G., Carriere, S., & Kobbi, M. (1996). Exceptions to the one: One relationship between African fig trees and their fig wasp pollinators: Possible evolutionary scenarios. *Journal of Biogeography, 23*(4), 513–520.
Micheneau, C., Johnson, S. D., & Fay, M. F. (2009). Orchid pollination: From Darwin to the present day. *Botanical Journal of the Linnean Society, 161*(1), 1–19.
Michod, R. E., Viossat, Y., Solari, C. A., Hurand, M., & Nedelcu, A. M. (2006). Life-history evolution and the origin of multicellularity. *Journal of Theoretical Biology, 239*(2), 257–272.
Miller, S. L. (1953). A production of amino acids under possible primitive earth conditions. *Science, 117*(3046), 528–529.
Miller, I. L., & Lonsdale, W. M. (1987). Early records of '*Mimosa pigra*' in the Northern Territory. *Plant Protection Quarterly, 2*, 140–142.
Miller, L. A., & Surlykke, A. (2001). How some insects detect and avoid being eaten by bats: Tactics and countertactics of prey and predator. *Bioscience, 51*(7), 570–581.
Mindell, D. P. (1992). Phylogenetic consequences of symbioses: Eukarya and eubacteria are not monophyletic taxa. *Biosystems, 27*(1), 53–62.
Mitchell, R. (2014, July 11). *6 animals shaped by artificial selection into beautiful monsters*. Gizmodo. https://gizmodo.com/6-animals-shaped-by-artificial-selection-into-beautiful-1603184303
Mitchell, R. J., Irwin, R. E., Flanagan, R. J., & Karron, J. D. (2009). Ecology and evolution of plant-pollinator interactions. *Annals of Botany, 103*(9), 1355–1363.
Mitchell, N. S., Catenacci, V. A., Wyatt, H. R., & Hill, J. O. (2011). Obesity: Overview of an epidemic. *Psychiatric Clinics, 34*(4), 717–732.
Mohammadi, K., Khalesro, S., Sohrabi, Y., & Heidari, G. (2011). A review: Beneficial effects of the mycorrhizal fungi for plant growth. *Journal of Applied Environmental and Biological Sciences, 1*(9), 310–319.
Molinier, J., Ries, G., Zipfel, C., & Hohn, B. (2006). Transgeneration memory of stress in plants. *Nature, 442*(7106), 1046–1049.
Møller, A. P., & Jennions, M. D. (2001). How important are direct fitness benefits of sexual selection? *Naturwissenschaften, 88*(8), 401–415.
Møller, A. P., Barnier, F., & Mousseau, T. A. (2012). Ecosystems effects 25 years after Chernobyl: Pollinators, fruit set and recruitment. *Oecologia, 170*, 1155–1165.
Mondal, A. K., & Parui, S. M. (2013). What makes plants 'Sleep'? *Science Reporter, 50*(7), 45–48.
Monsalve, S. (2014). Los hipopótamos (*Hippopotamus amphibius*) en Colombia. Especie exótica, introducida e invasora. *Revista Lasallista de investigación, 11*(1), 32.
Montagu, J. D. (1994). Length of life in the ancient world: A controlled study. *Journal of the Royal Society of Medicine, 87*(1), 25–26.
Monteith, L. G. (1963). Habituation and associative learning in *Drino bohemica*. Men. (Diptera: Tachinidae). *The Canadian Entomologist, 95*, 418–426.
Moore, J. C., & Pannell, J. R. (2011). Sexual selection in plants. *Current Biology, 21*(5), R176–R182.
Moran, N. A., Munson, M. A., Baumann, P., & Ishikawa, H. (1993). A molecular clock in endosymbiotic bacteria is calibrated using the insect hosts. *Proceedings of the Royal Society of London. Series B: Biological Sciences, 253*, 167–171.

Moran, N. A., Russell, J. A., Koga, R., & Fukatsu, T. (2005). Evolutionary relationships of three new species of Enterobacteriaceae living as symbionts of aphids and other insects. *Applied and Environmental Microbiology, 71*(6), 3302–3310.

Moreno, A. (1994). Aprendizaje automático. *Universitat politécnica de Catalunya.*. https://upcommons.upc.edu/handle/2099.3/36157

Mori, I. (1999). Genetics of chemotaxis and thermotaxis in the nematode *Caenorhabditis elegans*. *Annual Review of Genetics, 33*(1), 399–422.

Mori, E., Iacucci, A., Castiglia, R., & Santini, L. (2017). Sexual-size dimorphism in two synanthropic rat species: Comparison and eco-evolutionary perspectives. *Mammalian Biology, 83*, 78–80.

Morran, L. T., Schmidt, O. G., Gelarden, I. A., Parrish, R. C., & Lively, C. M. (2011). Running with the red queen: Host-parasite coevolution selects for biparental sex. *Science, 333*(6039), 216–218.

Moulia, B., Bastien, R., Chauvet-Thiry, H., & Leblanc-Fournier, N. (2019). Posture control in land plants: Growth, position sensing, proprioception, balance, and elasticity. *Journal of Experimental Botany, 70*(14), 3467–3494.

Moulia, B., Douady, S., & Hamant, O. (2021). Fluctuations shape plants through proprioception. *Science, 372*(6540), eabc6868.

Mousseau, T. A. (2021). The biology of Chernobyl. *Annual Review of Ecology, Evolution, and Systematics, 52*, 87–109.

Mousseau, T. A., & Møller, A. P. (2012). Chernobyl and Fukushima: Differences and similarities a biological perspective. *Transactions of the American Nuclear Society, 107*, 200.

Moya, A., Galiana, A., & Ayala, F. J. (1995). Founder-effect speciation theory: Failure of experimental corroboration. *Proceedings of the National Academy of Sciences, 92*(9), 3983–3986.

Mulcahy, D. L. (1979). The rise of the angiosperms: A genecological factor. *Science, 206*, 20–23.

Muller, H. J. (1964). The relation of recombination to mutational advance. *Mutation Research/Fundamental and Molecular Mechanisms of Mutagenesis, 1*(1), 2–9.

Muñoz-Miranda, L. A., Higuera-Ciapara, I., Gschaedler-Mathis, A. C., Rodríguez-Zapata, L. C., Pereira-Santana, A., & Figueroa-Yáñez, L. J. (2019). Breve Descripción de la Biología Sintética y la Importancia de su Relación con otras Disciplinas. *Revista mexicana de ingeniería biomédica, 40*(1).

Munzi, S., Cruz, C., & Corrêa, A. (2019). When the exception becomes the rule: An integrative approach to symbiosis. *Science of the Total Environment, 672*, 855–861.

Murashko, O. N., & Lin-Chao, S. (2017). *Escherichia coli* responds to environmental changes using enolasic degradosomes and stabilized DicF sRNA to alter cellular morphology. *Proceedings of the National Academy of Sciences, 114*(38), E8025–E8034.

Murphey, R. K., Bacon, J. P., Sakaguchi, D. S., & Johnson, S. E. (1983). Transplantation of cricket sensory neurons to ectopic locations: Arborizations and synaptic connections. *Journal of Neuroscience, 3*, 659–672.

Murphy, C. G. (1998). Interaction-independent sexual selection and the mechanisms of sexual selection. *Evolution, 52*(1), 8–18.

Murphy, S. V., & Atala, A. (2014). 3D bioprinting of tissues and organs. *Nature Biotechnology, 32*(8), 773–785.

Nabavizadeh, S. L., Mehrabani, D., Vahedi, Z., & Manafi, F. (2016). Cloning: A review on bioethics, legal, jurisprudence and regenerative issues in Iran. *World Journal of Plastic Surgery, 5*(3), 213–225.

Navas, J. R. (1987). Los vertebrados exóticos introducidos en la Argentina. *Revista del Museo Argentino de Ciencias Naturales Bernardino Rivadavia Zool, 14*(2), 7–38.

Naylor, R. L., Williams, S. L., & Strong, D. R. (2001). Aquaculture—A gateway for exotic species. *Science, 294*, 1655–1656.

Nedelcu, A. M. (2020). The evolution of multicellularity and cancer: Views and paradigms. *Biochemical Society Transactions, 48*(4), 1505–1518.

Nei, M., Maruyama, T., & Chakraborty, R. (1975). The bottleneck effect and genetic variability in populations. *Evolution, 29*(1), 1–10.

Neiman, M., Lively, C. M., Meirmans, S., et al. (2017). Why sex? A pluralist approach revisited. *Trends in Ecology & Evolution, 32*(8), 589–600.

Nelson-Smith, A. (1971). The problem of oil pollution of the sea. In *Advances in marine biology* (Vol. 8, pp. 215–306). Academic Press.

Nemergut, D. R., Schmidt, S. K., Fukami, T., O'Neill, S. P., Bilinski, T. M., Stanish, L. F., et al. (2013). Patterns and processes of microbial community assembly. *Microbiology and Molecular Biology Reviews, 77*(3), 342–356.

Nesterenko, V. B., & Yablokov, A. V. (2009). Chapter I. Chernobyl contamination: An overview. Annals of the New York Academy of Sciences, 1181(1), 4–30.

Niemela, P. T., DiRienzo, N., & Hedrick, A. V. (2012). Predator-induced changes in the boldness of naive field crickets, *Gryllus integer*, depend on behavioral type. *Animal Behaviour, 84*, 129–135.

Niklas, K. J., & Newman, S. A. (2013). The origins of multicellular organisms. *Evolution & Development, 15*(1), 41–52.

Niklas, K. J., & Newman, S. A. (2016). *Multicellularity: Origins and evolution* (Vol. 18). MIT Press.

Niklas, K. J., & Newman, S. A. (2020). The many roads to and from multicellularity. *Journal of Experimental Botany, 71*(11), 3247–3253.

Niklas, K. J., Wayne, R., Benítez, M., & Newman, S. A. (2019). Polarity, planes of cell division, and the evolution of plant multicellularity. *Protoplasma, 256*, 585–599.

Nilsson, D. E. (2017). Evolution: An irresistibly clear view of land. *Current Biology, 27*(14), R715–R717.

Nilsson, D. E., Warrant, E. J., Johnsen, S., Hanlon, R., & Shashar, N. (2012). A unique advantage for giant eyes in giant squid. *Current Biology, 22*(8), 683–688.

Noble, G. K. (1934). Experimenting with the courtship of lizards. *Natural History, 34*, 5–15.

Noble, G. K. (1936). Courtship and sexual selection of the flicker (*Colaptes auratus luteus*). *The Auk, 53*, 269–282.

Novillo, A., & Ojeda, R. A. (2008). The exotic mammals of Argentina. *Biological Invasions, 10*, 1333–1344.

O'Malley, M., Powell, A., Davies, J., & Calvert, J. (2008). Knowledge-making distinctions in synthetic biology. *BioEssays, 30*(1), 57–65.

O'Malley, M. A., Simpson, A. G., & Roger, A. J. (2013). The other eukaryotes in light of evolutionary protistology. *Biology & Philosophy, 28*(2), 299–330.

Officer, C. B., & Drake, C. L. (1983). The cretaceous-tertiary transition. *Science, 219*(4591), 1383–1390.

Ollerton, J. (2006). "Biological Barter": Patterns of specialization compared across different mutualisms. In N. M. Waser & J. Ollerton (Eds.), *Plant-pollinator interactions: From specialization to generalization* (pp. 411–435). University of Chicago Press.

Ollivier, M., Gabaldón, T., Poulain, J., Gavory, F., Leterme, N., Gauthier, J. P., et al. (2012). Comparison of gene repertoires and patterns of evolutionary rates in eight aphid species that differ by reproductive mode. *Genome Biology and Evolution, 4*, 155–167.

Oparin, A. I. (1924). *Proiskhozhdenie zhizni*. Voennoe Izd.

Oppliger, A., Naciri-Graven, Y., Ribi, G., & Hosken, D. J. (2003). Sperm length influences fertilization success during sperm competition in the snail *Viviparus ater*. *Molecular Ecology, 12*, 485–492.

Orgel, L. E. (1998a). The origin of life—a review of facts and speculations. *Trends in Biochemical Sciences, 23*(12), 491–495.

Orgel, L. E. (1998b). The origin of life–how long did it take? *Origins of Life and Evolution of the Biosphere, 28*(1), 91–96.

Ortega, A., & Olivares-Bañuelos, T. N. (2020). Neurons and glia cells in marine invertebrates: An update. *Frontiers in Neuroscience, 14*, 121.

Ortiz, H. (2007). *Zarzamora, la berrie rústica que llegó para quedarse*. La Nación. https://www.lanacion.com.ar/economia/campo/zarzamora-la-berrie-rustica-que-llego-para-quedarse-nid974624/.
Otronen, M. (1990). Mating behavior and sperm competition in the fly, *Dryomyza anilis*. *Behavioral Ecology and Sociobiology, 26*, 349–356.
Otto, S. (2008). Sexual reproduction and the evolution of sex. *Nature Education, 1*(1).
Otto, S. P. (2009). The evolutionary enigma of sex. *The American Naturalist, 174*(S1), S1–S14.
Otto, S. P., & Lenormand, T. (2002). Resolving the paradox of sex and recombination. *Nature Reviews Genetics, 3*(4), 252–261.
Oxford, J. S., Lambkin, R., Sefton, A., Daniels, R., Elliot, A., Brown, R., & Gill, D. (2005). A hypothesis: The conjunction of soldiers, gas, pigs, ducks, geese and horses in northern France during the great war provided the conditions for the emergence of the "Spanish" influenza pandemic of 1918–1919. *Vaccine, 23*(7), 940–945.
Ozawa, M., & Kawaoka, Y. (2013). Cross talk between animal and human influenza viruses. *Annual Review of Animal Biosciences, 1*(1), 21–42.
Paaby, A. B., & Rockman, M. V. (2014). Cryptic genetic variation: Evolution's hidden substrate. *Nature Reviews Genetics, 15*(4), 247–258.
Pagani-Balletti, R. (2020). Enfermedades epidémicas y pandémicas: Causas, cronología e implicaciones socioculturales. *Anales Real Academia de Farmacia*, 189–214.
Pank, F., Matzk, F., Kästner, U., Blüthner, W. D., Foltys de Garcia, E., Meister, A., et al. (2003). Reproductive diversity and strategies for breeding in St. John's wort (*Hypericum perforatum* L.). *Euphytica, 134*, 77–84.
Papini, M. R., & Bitterman, M. E. (1991). Appetitive conditioning in *Octopus cyanea*. *Journal of Comparative Physiology, 105*, 107–114.
Paps, J., & Holland, P. W. (2018). Reconstruction of the ancestral metazoan genome reveals an increase in genomic novelty. *Nature Communications, 9*(1), 1–8.
Pardo, A. (1997). La clonación humana. *Persona y Bioética, 1*, 44–56.
Parker, G. A. (1970). Sperm competition and its evolutionary consequences in the insects. *Biological Reviews, 45*, 525–567.
Parrott-Holden, J. (1987). The Budapest Short-faced Tumbler. *AFA Watchbird, 13*(6), 52–56.
Paterno, G. B., Silveira, C. L., Kollmann, J., Westoby, M., & Fonseca, C. R. (2020). The maleness of larger angiosperm flowers. *Proceedings of the National Academy of Sciences, 117*(20), 10921–10926.
Patiny, S. (Ed.). (2011). *Evolution of plant-pollinator relationships*. Cambridge University Press.
Pätschke, A., Bicker, G., & Stern, M. (2004). Axonal regeneration of proctolinergic neurons in the central nervous system of the locust. *Developmental Brain Research, 150*(1), 73–76.
Patterson, K. D., & Pyle, G. F. (1991). The geography and mortality of the 1918 influenza pandemic. *Bulletin of the History of Medicine, 65*(1), 4–21.
Pavlov, I. P. (1927). *Conditioned reflexes*. Oxford University Press.
Paxton, H., Anthony, N. B., Corr, S. A., & Hutchinson, J. R. (2010). The effects of selective breeding on the architectural properties of the pelvic limb in broiler chickens: A comparative study across modern and ancestral populations. *Journal of Anatomy, 217*(2), 153–166.
Peakall, R. (1990). Responses of male *Zaspilothynnus trilobatus* turner wasps to females and the sexually deceptive orchid it pollinates. *Functional Ecology, 4*(1), 159–167.
Pearce, F. (2003). Going bananas. *New Scientist, 177*(2378), 26–27.
Pearce, F. (2008). The sterile banana. *Conservation Magazine, 9*(4).
Pearcy, M., Aron, S., Doums, C., & Keller, L. (2004). Conditional use of sex and parthenogenesis for worker and queen production in ants. *Science, 306*(5702), 1780–1783.
Pechenik, J. A. (2015). *Biology of the invertebrates*. McGraw-Hill Education.
Peck, J. R. (1993). Frequency-dependent selection, beneficial mutations, and the evolution of sex. *Proceedings of the Royal Society of London Series B: Biological Sciences, 254*(1340), 87–92.
Peeters, M., Jung, M., & Ayouba, A. (2013). The origin and molecular epidemiology of HIV. *Expert Review of Anti-Infective Therapy, 11*(9), 885–896.

Peiffer, J. A., Spor, A., Koren, O., Jin, Z., Tringe, S. G., Dangl, J. L., et al. (2013). Diversity and heritability of the maize rhizosphere microbiome under field conditions. *Proceedings of the National Academy of Sciences, 110*(16), 6548–6553.
Pena, S. D., & Chakraborty, R. (1994). Paternity testing in the DNA era. *Trends in Genetics, 10*(6), 204–209.
Peng, X., Qiao, X., & Chen, M. (2017). Responses of holocyclic and anholocyclic *Rhopalosiphum padi* populations to low temperature and short-photoperiod induction. *Ecology and Evolution, 7*, 1030–1042.
Peretti, A. V., & Aisenberg, A. (2015). *Cryptic female choice in arthropods. Patterns, mechanisms and prospects*. Springer.
Pernet, V., & Schwab, M. E. (2014). Lost in the jungle: New hurdles for optic nerve axon regeneration. *Trends in Neurosciences, 37*, 381–387.
Perotti, M. A., Young, D. K., & Braig, H. R. (2016). The ghost sex-life of the paedogenetic beetle *Micromalthus debilis*. *Scientific Reports, 6*(1), 27364.
Perry, C. J., Barron, A. B., & Cheng, K. (2013). Invertebrate learning and cognition: Relating phenomena to neural substrate. *Wiley Interdisciplinary Reviews: Cognitive Science, 4*(5), 561–582.
Pershin, Y. V., La Fontaine, S., & Di Ventra, M. (2009). Memristive model of amoeba learning. *Physical Review E, 80*(2), 021926.
Peter, B. M., & Slatkin, M. (2015). The effective founder effect in a spatially expanding population. *Evolution, 69*(3), 721–734.
Petrovska, B. B. (2012). Historical review of medicinal plants' usage. *Pharmacognosy Reviews, 6*(11), 1.
Petsko, G. A. (2006). The next epidemic. *Genome Biology, 7*, 1–2.
Pfennig, D. W., Wund, M. A., Snell-Rood, E. C., Cruickshank, T., Schlichting, C. D., & Moczek, A. P. (2010). Phenotypic plasticity's impacts on diversification and speciation. *Trends in Ecology & Evolution, 25*(8), 459–467.
Phillips, B. L., & Shine, R. (2006). An invasive species induces rapid adaptive change in a native predator: Cane toads and black snakes in Australia. *Proceedings of the Royal Society B: Biological Sciences, 273*(1593), 1545–1550.
Phillips, B. L., Brown, G. P., & Shine, R. (2003). Assessing the potential impact of cane toads on Australian snakes. *Conservation Biology, 17*(6), 1738–1747.
Phillips, B. L., Brown, G. P., & Shine, R. (2010). Life-history evolution in range-shifting populations. *Ecology, 91*, 1617–1627.
Phillips, R. D., Scaccabarozzi, D., Retter, B. A., Hayes, C., Brown, G. R., Dixon, K. W., & Peakall, R. (2014). Caught in the act: Pollination of sexually deceptive trap-flowers by fungus gnats in *Pterostylis* (Orchidaceae). *Annals of Botany, 113*(4), 629–641.
Piaggio, A. J., Segelbacher, G., Seddon, P. J., Alphey, L., Bennett, E. L., Carlson, R. H., et al. (2017). Is it time for synthetic biodiversity conservation? *Trends in Ecology & Evolution, 32*(2), 97–107.
Pickett, S. T., Collins, S. L., & Armesto, J. J. (1987). Models, mechanisms and pathways of succession. *Botanical Review, 53*(3), 335–371.
Pigna, F. (2020). Historia de las pandemias [audio podcast episode]. H*istorias de nuestra historia*. Radio Nacional Argentina. https://www.radionacional.com.ar/la-historia-de-las-pandemias/
Pimentel, D., Zuniga, R., & Morrison, D. (2005). Update on the environment and economic costs associated with alien-invasive species in the United States. *Ecological Economics, 52*, 273–288.
Piprek, R. P. (2009). Genetic mechanisms underlying male sex determination in mammals. *Journal of Applied Genetics, 50*(4), 347–360.
Pisani, D., Poling, L. L., Lyons-Weiler, M., & Hedges, S. B. (2004). The colonization of land by animals: Molecular phylogeny and divergence times among arthropods. *BMC Biology, 2*(1), 1–10.
Pitnick, S., Hosken, D. J., & Birkhead, T. R. (2009). Sperm morphological diversity. In *Sperm biology* (pp. 69–149). Academic Press.

Plackett, A. R., & Coates, J. C. (2016). Life's a beach–the colonization of the terrestrial environment. *New Phytologist, 212*(4), 831–835.

Podgórski, T., Baś, G., Jędrzejewska, B., Sönnichsen, L., Śnieżko, S., Jędrzejewski, W., & Okarma, H. (2013). Spatiotemporal behavioral plasticity of wild boar (*Sus scrofa*) under contrasting conditions of human pressure: Primeval forest and metropolitan area. *Journal of Mammalogy, 94*(1), 109–119.

Ponnamperuma, C. (1978). Prebiotic molecular evolution. In *Origin of life* (pp. 67–81). Center for Academic Publications Japan.

Poole, A., & Penny, D. (2001). Does endosymbiosis explain the origin of the nucleus? *Natural Cell Biology, 3*, E173–E174.

Popham, D. L. (2013). Binary fission in bacteria. *Microbiology and Molecular Biology Reviews, 77*(3), 200–216.

Poss, K. D. (2010). Advances in understanding tissue regenerative capacity and mechanisms in animals. *Nature Reviews Genetics, 11*(10), 710–722.

Poss, K. D., Keating, M. T., & Nechiporuk, A. (2003). Tales of regeneration in zebrafish. *Developmental Dynamics, 226*(2), 202–210.

Powell, J. R. (1978). The founder-flush speciation theory: An experimental approach. *Evolution*, 465–474.

Power, J. F., & Follett, R. F. (1987). Monoculture. *Scientific American, 256*(3), 78–87.

Prave, A. R. (2002). Life on land in the Proterozoic: Evidence from the Torridonian rocks of Northwest Scotland. *Geology, 30*, 811–814.

Press, M. C., Graves, J. D., & Stewart, G. R. (1990). Physiology of the interaction of angiosperm parasites and their higher plant hosts. *Plant Cell and Environment, 13*, 91–104.

Pressman, A., Blanco, C., & Chen, I. A. (2015). The RNA world as a model system to study the origin of life. *Current Biology, 25*(19), R953–R963.

Price, M. V., & Waser, N. M. (1979). Pollen dispersal and optimal outcrossing in *Delphinium nelsoni*. *Nature, 277*(5694), 294–297.

Pritham, E. J. (2009). Transposable elements and factors influencing their success in eukaryotes. *The Journal of Heredity, 100*(5), 648–655.

Proctor, M., Yeo, P., & Lack, A. (1996). *The natural history of pollination*. HarperCollins Publishers.

Prokop, P., & Maxwell, M. R. (2009). Female feeding and polyandry in the nuptially feeding nursery web spider, *Pisaura mirabilis*. *Naturwissenschaften, 96*, 259–265.

Prosdocimi, F., José, M. V., & Farias, S. T. D. (2018). *Be introduced to the first universal common ancestor (FUCA): The great-grandmother of LUCA (Last Universal Common Ancestor)*. Europe PMC.

Prosdocimi, F., José, M. V., & de Farias, S. T. (2019). The first universal common ancestor (FUCA) as the earliest ancestor of LUCA's (last UCA) lineage. In *Evolution, origin of life, concepts and methods* (pp. 43–54). Springer.

Provine, W. B. (2004). Ernst Mayr: Genetics and speciation. *Genetics, 167*(3), 1041–1046.

Qiu, Y.-L., Li, L., Bin, W., Chen, Z., Knoop, V., Groth-Malonek, M., et al. (2006). The deepest divergences in land plants inferred from phylogenomic evidence. *Proceedings of the National Academy of Sciences, 103*, 15511–15516.

Queitsch, C., Sangster, T. A., & Lindquist, S. (2002). Hsp90 as a capacitor of phenotypic variation. *Nature, 417*, 618–624.

Queller, D. (1987). Sexual selection in flowering plants. In J. Bradbury & M. Andersson (Eds.), *Sexual selection* (pp. 165–179). Wiley.

Quinn, S. D. P., & De Boni, U. (1991). Enhanced neuronal regeneration by retinoic acid of murine dorsal root ganglia and of fetal murine and human spinal cord in vitro. *In Vitro Cellular & Developmental Biology-Animal, 27*(1), 55–62.

Quiroga, C. (2016). Las tecnologías «ómicas»: situación actual y desafíos futuros. *Revista Argentina de Microbiología, 48*(4), 265–266.

Ralt, D., Goldenberg, M., Fetterolf, P., Thompson, D., Dor, J., Mashiach, S., Barbers, D. L., & Eisenbach, M. (1991). Sperm attraction to a follicular factor(s) correlates with human egg fertilizability. *Proceedings of the National Academy of Sciences of the United States of America, 88*, 2840–2844.

Ramachandran, S., Deshpande, O., Roseman, C. C., Rosenberg, N. A., Feldman, M. W., & Cavalli-Sforza, L. (2005). Support from the relationship of genetic and geographic distance in human populations for a serial founder effect originating in Africa. *Proceedings of the National Academy of Sciences, 102*(44), 15942–15947.

Ramiro-Puig, E., Pérez-Cano, F. J., Castellote, C., Franch, A., & Castell, M. (2008). El intestino: pieza clave del sistema inmunitario. *Revista Española de Enfermedades Digestivas, 100*(1), 29–34.

Ramm, S. A. (2017). Exploring the sexual diversity of flatworms: Ecology, evolution, and the molecular biology of reproduction. *Molecular Reproduction and Development, 84*(2), 120–131.

Rampelotto, P. H. (2013). Extremophiles and extreme environments. *Life, 3*(3), 482–485.

Rana, J. S., Nieuwdorp, M., Jukema, J. W., & Kastelein, J. J. (2007). Cardiovascular metabolic syndrome–an interplay of obesity, inflammation, diabetes and coronary heart disease. *Diabetes, Obesity and Metabolism, 9*(3), 218–232.

Rankin, C. H. (2004). Invertebrate learning: What can't a worm learn? *Current Biology, 14*(15), R617–R618.

Rasplus, J. Y. (1996). The one-to-one species specificity of the *Ficus*-Agaoninae mutualism: how casual?. In *The Biodiversity of African Plants: Proceedings XIVth AETFAT Congress 22–27 August 1994, Wageningen, The Netherlands* (pp. 639–649). Springer .

Raup, D. M., & Valentine, J. W. (1983). Multiple origins of life. *Proceedings of the National Academy of Sciences, 80*(10), 2981–2984.

Raven, P. H., Johnson, G. B., Losos, J. B., & Singer, S. R. (2002). *Biology*. McGraw Hill.

Ray, R. P., Nakata, T., Henningsson, P., & Bomphrey, R. J. (2016). Enhanced flight performance by genetic manipulation of wing shape in *Drosophila*. *Nature Communications, 7*(1), 1–8.

Razmjou, J., Vorburger, C., Moharramipour, S., Mirhoseini, S. Z., & Fathipour, Y. (2010). Host-associated differentiation and evidence for sexual reproduction in Iranian populations of the cotton aphid, *Aphis gossypii*. *Entomologia Experimentalis et Applicata, 134*, 191–199.

Read, C. F., Elith, J., & Vesk, P. A. (2016). Testing a model of biological soil crust succession. *Journal of Vegetation Science, 27*(1), 176–186.

Reaka, M. L. (1980). On learning and living in holes by mantis shrimp. *Animal Behaviour, 28*(1), 111–115.

Réale, D., Reader, S. M., Sol, D., McDougall, P. T., & Dingemanse, N. J. (2007). Integrating animal temperament within ecology and evolution. *Biological Reviews, 82*(2), 291–318.

Reddien, P., & Sánchez Alvarado, A. (2004). Fundamentals of planarian regeneration. *Annual Review of Cell and Developmental Biology, 20*, 725–757.

Redecker, D., Kodner, R., & Graham, L. E. (2000). Glomalean fungi from the Ordovician. *Science, 289*, 1920–1921.

Redford, K. H., Adams, W., & Mace, G. M. (2013). Synthetic biology and conservation of nature: Wicked problems and wicked solutions. *PLoS Biology, 11*(4), e1001530.

Redford, K. H., Adams, W. M., Carlson, R., Mace, G. M., & Ceccarelli, B. (2021). *Strange natures: Conservation in the era of synthetic biology*. Yale University Press.

Reichert, B. E., Sovie, A. R., Udell, B. J., Hart, K. M., Borkhataria, R. R., Bonneau, M., et al. (2017). Urbanization may limit impacts of an invasive predator on native mammal diversity. *Diversity and Distributions, 23*, 355–367.

Reid, R. G. (1985). *Evolutionary theory: The unfinished synthesis*. Cornell University Press.

Reinhard, J., Sinclair, M., Srinivasan, M. V., & Claudianos, C. (2010). Honeybees learn odour mixtures via a selection of key odorants. *PLoS One, 5*(2), e9110.

Rejmánek, M., & Richardson, D. M. (2013). Trees and shrubs as invasive alien species—2013 update of the global database. *Diversity and Distributions, 19*, 1–2.

Rendel, J. M. (1970). The time scale of genetic change. In *The impact of civilization on the biology of man* (pp. 27–47).
Renner, S. S., & Ricklefs, R. E. (1995). Dioecy and its correlates in the flowering plants. *American Journal of Botany, 82*(5), 596–606.
Rey, O., Danchin, E., Mirouze, M., Loot, C., & Blanchet, S. (2016). Adaptation to global change: A transposable element-epigenetics perspective. *Trends in Ecology & Evolution, 31*(7), 514–526.
Reynolds, B. A., & Weiss, S. (1992). Generation of neurons and astrocytes from isolated cells of the adult mammalian central nervous system. *Science, 255*, 1707–1710.
Rice, W. R. (2002). Experimental tests of the adaptive significance of sexual reproduction. *Nature Reviews Genetics, 3*(4), 241–251.
Richards, A. J. (1997). *Plant breeding systems* (2nd ed.). Chapman & Hall.
Richardson, K., & Jørgensen, B. B. (1996). Eutrophication: Definition, history and effects. *Eutrophication in Coastal Marine Ecosystems, 52*, 1–9.
Richter, D. J., Fozouni, P., Eisen, M. B., & King, N. (2018). Gene family innovation, conservation and loss on the animal stem lineage. *eLife, 7*, e34226.
Rideout, W. M., III, Eggan, K., & Jaenisch, R. (2001). Nuclear cloning and epigenetic reprogramming of the genome. *Science, 293*, 1093–1098.
Rideout, W. M., Hochedlinger, K., Kyba, M., Daley, G. Q., & Jaenisch, R. (2002). Correction of a genetic defect by nuclear transplantation and combined cell and gene therapy. *Cell, 109*, 17–27.
Ridley, M. (2004). *Evolution* (3rd ed.). Blackwell Publishing.
Riechert, S. E., & Hedrick, A. V. (1993). A test of correlations among fitness-related behavioral traits in the spider, *Agelenopsis aperta* (Araneae, Agelinadae). *Animal Behaviour, 46*, 669–675.
Rieseberg, L. H., Kim, S. C., Randell, R. A., Whitney, K. D., Gross, B. L., Lexer, C., & Clay, K. (2007). Hybridization and the colonization of novel habitats by annual sunflowers. *Genetica, 129*, 149–165.
Riley, S. P. D., Shaffer, H. B., Voss, S. R., & Fitzpatrick, B. M. (2003). Hybridization between a rare, native tiger salamander and its introduced congener. *Ecological Applications, 13*, 1263–1275.
Rivera, M. C., & Lake, J. A. (2004). The ring of life provides evidence for a genome fusion origin of eukaryotes. *Nature, 431*(7005), 152–155.
Rivest, S., & Forrest, J. R. (2020). Defence compounds in pollen: Why do they occur and how do they affect the ecology and evolution of bees? *New Phytologist, 225*(3), 1053–1064.
Robertson, A. (1961). Inbreeding in artificial selection programmes. *Genetics Research, 2*(2), 189–194.
Robertson, A. L., & Wolf, D. E. (2012). The role of epigenetics in plant adaptation. *Trends in Evolutionary Biology, 4*(1), e4.
Robinson, G. E., & Huang, Z. Y. (1998). Colony integration in honey bees: Genetic, endocrine and social control of division of labor. *Apidologie, 29*(1–2), 159–170.
Robson, A. J., & Kaplan, H. S. (2003). The evolution of human life expectancy and intelligence in hunter-gatherer economies. *American Economic Review, 93*(1), 150–169.
Rodrigues, D., Kaminski, L. A., Freitas, A. V. L., & Oliveira, P. S. (2010). Tradeoffs underlying polyphagy in a facultative ant-tended florivorous butterfly: The role of host plant quality and enemy-free space. *Oecologia, 163*, 719–728.
Rodríguez Lozano, A. (2009). Aventuras y desventuras del virus de la gripe(I). [audio podcast episode]. In *Hablando con Científicos*. URL: https://cienciaes.com/entrevistas/2009/05/23/aventuras-y-desventuras-del-virus-de-la-gripei-hablamos-con-enrique-tabar-s/
Rohwer, F., Seguritan, V., Azam, F., & Knowlton, N. (2002). Diversity and distribution of coral-associated bacteria. *Marine Ecology Progress Series, 243*, 1–10.
Rokas, A. (2008). The molecular origins of multicellular transitions. *Current Opinion in Genetics & Development, 18*(6), 472–478.
Rose, J. K., & Rankin, C. H. (2001). Analyses of habituation in *Caenorhabditis elegans*. *Learning & Memory, 8*, 63–69.

Rosenberg, E., & Zilber-Rosenberg, I. (2013). *The hologenome concept: Human, animal and plant microbiota*. Springer.
Rosenberg, E., & Zilber-Rosenberg, I. (2016). Microbes drive evolution of animals and plants: The hologenome concept. *MBio, 7*(2), e01395–e01315.
Rosenberg, E., & Zilber-Rosenberg, I. (2018). The hologenome concept of evolution after 10 years. *Microbiome, 6*(1), 1–14.
Rosenfeld, C. S. (2010). Animal models to study environmental epigenetics. *Biology of Reproduction, 82*, 473–488.
Rossi, V. (2018). *El pastor alemán*. Parkstone International.
Roth, J., Qiang, X., Marbán, S. L., Redelt, H., & Lowell, B. C. (2004). The obesity pandemic: Where have we been and where are we going? *Obesity Research, 12*(S11), 88S–101S.
Rothman, K. J. (1976). Causes. *American Journal of Epidemiology, 104*(6), 587–592.
Rothman, K. J. (2012). *Epidemiology: An introduction*. Oxford University Press.
Rowbury, R. (2009). Darwin, his publisher and science progress. *Science Progress, 92*(3–4), 205–210.
Rubilar, T., de Ward, C. T. P., & de Vivar, M. E. D. (2005). Sexual and asexual reproduction of *Allostichaster capensis* (Echinodermata: Asteroidea) in Golfo Nuevo. *Marine Biology, 146*(6), 1083–1090.
Rubinstein, C. V., Gerrienne, P., de la Puente, G. S., Astini, R. A., & Steemans, P. (2010). Early middle Ordovician evidence for land plants in Argentina (eastern Gondwana). *New Phytologist, 188*, 365–369.
Ruder, W. C., Lu, T., & Collins, J. J. (2011). Synthetic biology moving into the clinic. *Science, 333*(6047), 1248–1252.
Rudin, F. S., & Biffa, M. (2012). Is boldness a resource-holding potential trait? Fighting prowess and changes in startle response in the sea anemone, *Actinia equina*. *Proceedings of the Royal Society B: Biological Sciences, 279*, 1904–1910.
Ruíz, R., & Ayala, F. J. (1998). *El método en las ciencias. Epistemología y darwinismo*. S. L. Fondo de Cultura Económica de España (Colección Ciencia y Tecnología), México DF.
Ruiz Mirazo, P., & Moreno Bergareche, Á. J. (2012). Biología sintética: enfrentándose a la vida para comprenderla, utilizarla o extenderla. *Revista de Pensamiento Contemporáneo, 38*, 28–37.
Ruiz-Trillo, I., & Nedelcu, A. M. (Eds.). (2015). *Evolutionary transitions to multicellular life: Principles and mechanisms* (Vol. 2). Springer.
Ruppert, E. E., & Barnes, R. D. (1996). *Invertebrate zoology* (6th ed.). Saunders College Publishing, Ft.
Ruppert, E. E., Fox, R. S., & Barnes, R. D. (2004). *Invertebrate zoology*. Brooks/Cole.
Sadava, D., Hillis, D. M., Heller, H. C., & Berenbauqm, M. R. (2011). *Life, the science of biology* (Evolution, diversity and ecology) (Vol. 2, 2nd ed., p. 615). Sinauer, 618–619.
Sahebi, M., Hanafi, M. M., van Wijnen, A. J., Rice, D., Rafii, M. Y., Azizi, P., Osman, M., Taheri, S., Bakar, M. F. A., Isa, M. N. M., & Noor, Y. M. (2018). Contribution of transposable elements in the plant's genome. *Gene, 665*, 155–166.
Sakai, A. K., Allendorf, F. W., Holt, J. S., Lodge, D. M., Molofsky, J., With, K. A., et al. (2001). The population biology of invasive species. *Annual Review of Ecology and Systematics, 32*(1), 305–332.
Sánchez, O. (2007). Ecosistemas acuáticos: diversidad, procesos, problemática y conservación. In *Perspectivas sobre conservación de ecosistemas acuáticos en México* (Vol. 11).
Sánchez Alvarado, A. (2006). Fragmentation in planarians. *Current Opinion in Genetics & Development, 16*(4), 384–388.
Sandler, R. (2020). The ethics of genetic engineering and gene drives in conservation. *Conservation Biology, 34*(2), 378–385.
Sanford, R. A., Wagner, D. D., Wu, Q., Chee-Sanford, J. C., Thomas, S. H., Cruz-García, C., et al. (2012). Unexpected nondenitrifier nitrous oxide reductase gene diversity and abundance in soils. *Proceedings of the National Academy of Sciences, 109*(48), 19709–19714.

Santín, L. J., Rubio, S., Begega, A., Miranda, R., & Arias, J. L. (2000). Aprendizaje espacial e hipocampo. *Revista de Neurologia, 31*, 455–462.

Santos, B. S., Friedrichs, M. A., Rose, S. A., Barco, S. G., & Kaplan, D. M. (2018). Likely locations of sea turtle stranding mortality using experimentally-calibrated, time and space-specific drift models. *Biological Conservation, 226*, 127–143.

Saragusty, J., Diecke, S., Drukker, M., Durrant, B., Friedrich Ben-Nun, I., Galli, C., & Hildebrandt, T. B. (2016). Rewinding the process of mammalian extinction. *Zoo Biology, 35*(4), 280–292.

Sartí, J. E. (2012). ¿Cuáles Estrategias De Vida Presentan Las Plantas Epífitas? *Desde El Herbario CICY, 4*(30–31), 2012.

Satoshi, T., Blauch, L. R., Tang, S. K. Y., Morsut, L., & Lim, W. (2018). Programming self-organizing multicellular structures with synthetic cell-cell signaling. *Science, 361*, 156–162.

Scala, J., Iott, K., Schwab, D. W., & Semersky, F. E. (1969). Digestive secretion of *Dionaea muscipula* (Venus's flytrap). *Plant Physiology, 44*(3), 367–371.

Schiestl, F. P. (2005). On the success of a swindle: Pollination by deception in orchids. *Naturwissenschaften, 92*, 255–264.

Schlichting, C. D. (2003). Origins of differentiation via phenotypic plasticity. *Evolution & Development, 5*(1), 98–105.

Schne, A., Curtis, H., & Barnes, S. (2008). *Curtis biología*. Panamericana.

Schnell, A. K., & Clayton, N. S. (2019). Cephalopod cognition. *Current Biology, 29*(15), R726–R732.

Schnell, A. K., Amodio, P., Boeckle, M., & Clayton, N. S. (2021). How intelligent is a cephalopod? Lessons from comparative cognition. *Biological Reviews, 96*(1), 162–178.

Schöneich, S. (2020). Neuroethology of acoustic communication in field crickets—From signal generation to song recognition in an insect brain. *Progress in Neurobiology, 194*, 101882.

Schoonhoven, L. M., Van Loon, J. J., & Dicke, M. (2005). *Insect-plant biology*. Oxford University Press.

Schopf, J. W. (2006). Fossil evidence of Archaean life. *Philosophical Transactions of the Royal Society B: Biological Sciences, 361*(1470), 869–885.

Schopf, J. W., Kudryavtsev, A. B., Czaja, A. D., & Tripathi, A. B. (2007). Evidence of Archean life: Stromatolites and microfossils. *Precambrian Research, 158*(3–4), 141–155.

Schrey, A. W., Coon, C. A., Grispo, M. T., Awad, M., Imboma, T., McCoy, E. D., et al. (2012). Epigenetic variation may compensate for decreased genetic variation with introductions: A case study using house sparrows (*Passer domesticus*) on two continents. *Genetics Research International*, 979751.

Schuett, W., Tregenza, T., & Dall, S. R. (2010). Sexual selection and animal personality. *Biological Reviews, 85*, 217–224.

Schultheis, E. H., Berardi, A. E., & Lau, J. A. (2015). No release for the wicked: Enemy release is dynamic and not associated with invasiveness. *Ecology, 96*(9), 2446–2457.

Schulze-Makuch, D., & Irwin, L. N. (2018). Definition of life. In *Life in the universe* (Springer Praxis Books). Springer.

Scorza, L. C., & Dornelas, M. C. (2011). Plants on the move: Towards common mechanisms governing mechanically-induced plant movements. *Plant Signaling & Behavior, 6*(12), 1979–1986.

Scott, B. I. (1962). Electricity in plants. *Scientific American, 207*(4), 107–118.

Scoville, A. G., & Pfrender, M. E. (2010). Phenotypic plasticity facilitates recurrent rapid adaptation to introduced predators. *Proceedings of the National Academy of Sciences, 107*(9), 4260–4263.

Sebé-Pedrós, A., & de Mendoza, A. (2015). Transcription factors and the origin of animal multicellularity. In *Evolutionary transitions to multicellular life: Principles and mechanisms* (pp. 379–394).

Sebé-Pedrós, A., Degnan, B. M., & Ruiz-Trillo, I. (2017). The origin of Metazoa: A unicellular perspective. *Nature Reviews Genetics, 18*(8), 498–512.

Seddon, P. J., Moehrenschlager, A., & Ewen, J. (2014). Reintroducing resurrected species: Selecting extinction candidates. *Trends in Ecology & Evolution, 29*(3), 140–147.

Seiter, M., & Stockmann, M. (2017). The life history of the parthenogenetic scorpion *Lychas tricarinatus* (Simon, 1884) from Odisha province, India and supplementary notes on *Tityus trivittatus* Kraepelin, 1898 (Scorpiones, Buthidae). *Zoological Anzeiger, 270*, 155–165.

Seitzinger, S. P., Kroeze, C., Bouwman, A. F., Caraco, N., Dentener, F., & Styles, R. V. (2002). Global patterns of dissolved inorganic and particulate nitrogen inputs to coastal systems: Recent conditions and future projections. *Estuaries, 25*, 640–655.

Selosse, M. A., & Le Tacon, F. (1998). The land flora: A phototroph-fungus partnership? *Trends in Ecology & Evolution, 13*(1), 15–20.

Selosse, M. A., Richard, F., He, X., & Simard, S. W. (2006). Mycorrhizal networks: Des liaisons dangereuses? *Trends in Ecology & Evolution, 21*(11), 621–628.

Serrano, L. (2007). Synthetic biology: Promises and challenges. *Molecular Systems Biology, 3*(1), 158.

Sha, H. H., Wang, D. D., Yan, D. L., Hu, Y., Yang, S. J., Liu, S. W., & Feng, J. F. (2017). Chimaeric antigen receptor T-cell therapy for tumour immunotherapy. *Bioscience Reports, 37*(1).

Shapira, M. (2016). Gut microbiotas and host evolution: Scaling up Symbiosis. *Trends in Ecology & Evolution, 31*(7), 539–549.

Shapiro, R. (1988). Prebiotic ribose synthesis: A critical analysis. *Origins of Life and Evolution of the Biosphere, 18*(1), 71–85.

Shapiro, J. (1995). The discovery and significance of mobile genetic elements. In D. J. Sherratt (Ed.), *Mobile genetic elements* (pp. 1–13). IRL Press—Oxford Univ. Press.

Shapiro, H. T. (1997). Ethical and policy issues of human cloning. *Science, 277*(5323), 195–196.

Shapiro, R. (2007). A simpler origin for life. *Scientific American, 296*(6), 46–53.

Sharp, P. M., & Hahn, B. H. (2010). The evolution of HIV-1 and the origin of AIDS. *Philosophical Transactions of the Royal Society B: Biological Sciences, 365*(1552), 2487–2494.

Sharp, P. M., & Hahn, B. H. (2011). Origins of HIV and the AIDS pandemic. *Cold Spring Harbor Perspectives in Medicine, 1*(1), a006841.

Shaw, G. H. (2018). End of cretaceous extinction: The end of the dinosaurs. In *Great moments in the history of life* (pp. 65–68).

Shepherd, J. G., & Bonk, K. S. (2021). Activation of parasperm and eusperm upon ejaculation in Lepidoptera. *Journal of Insect Physiology, 130*, 104201.

Shi, K., Xue, J., Xiao, X., Qiao, Y., Wu, Y., & Gao, Y. (2019). Mechanism of degrading petroleum hydrocarbons by compound marine petroleum-degrading bacteria: Surface adsorption, cell uptake, and biodegradation. *Energy & Fuels, 33*(11), 11373–11379.

Shigenobu, S., Watanabe, H., Hattori, M., Sakaki, Y., & Ishikawa, H. (2000). Genome sequence of the endocellular bacterial symbiont of aphids Buchnera sp. APS. *Nature, 407*, 81–86.

Shin, T., Kraemer, D., & Pryor, J. (2002). A cat cloned by nuclear transplantation. *Nature, 415*, 859.

Shine, R., Brown, G. P., & Phillips, B. L. (2011). An evolutionary process that assembles phenotypes through space rather than through time. *Proceedings of the National Academy of Sciences of the United States of America, 108*, 5708–5711.

Shrestha, M., Dyer, A. G., Dorin, A., Ren, Z. X., & Burd, M. (2020). Rewardlessness in orchids: How frequent and how rewardless? *Plant Biology, 22*(4), 555–561.

Shuker, D. M., & Simmons, L. W. (Eds.). (2014). *The evolution of insect mating systems* (Vol. No. 27). Oxford University Press.

Shuo, Y. A. N., Wang, W. X., & Jie, S. H. E. N. (2020). Reproductive polyphenism and its advantages in aphids: Switching between sexual and asexual reproduction. *Journal of Integrative Agriculture, 19*(6), 1447–1457.

Sih, A., & Watters, J. V. (2005). The mix matters: Behavioral types and group dynamics in water striders. *Behaviour, 141*, 1417–1431.

Sih, A., Lauer, M., & Krupa, J. J. (2002). Path analysis and the relative importance of male–female conflict, female choice and male–male competition in water striders. *Animal Behaviour, 63*(6), 1079–1089.

Sih, A., Kats, L. B., & Maurer, E. F. (2003). Behavioral correlations across situations and the evolution of antipredator behaviour in a sunfish–salamander system. *Animal Behaviour, 65*, 29–44.

Sih, A., Bell, A., & Johnson, J. C. (2004a). Behavioral syndromes: An ecological and evolutionary overview. *Trends in Ecology & Evolution, 19*(7), 372–378.
Sih, A., Bell, A. M., Johnson, J. C., & Ziemba, R. E. (2004b). Behavioral syndromes: An integrative overview. *Quarterly Review of Biology, 79*, 241–277.
Simard, S. W., Perry, D. A., Jones, M. D., Myrold, D. D., Durall, D. M., & Molina, R. (1997). Net transfer of carbon between ectomycorrhizal tree species in the field. *Nature, 388*(6642), 579–582.
Simberloff, D. (2001). Biological invasions—How are they affecting us, and what can we do about them? *Western North American Naturalist, 61*, 308–315.
Simmons, L. G. (2001). *Sperm competition and its evolutionary consequences in insects*. Princeton University Press.
Simmons, L. W. (2014). Sperm competition. In *The evolution of insect mating systems* (pp. 181–203). Cambridge University Press.
Simmons, L. W., & Wedell, N. (2020). Fifty years of sperm competition: The structure of a scientific revolution. *Philosophical Transactions of the Royal Society B, 375*(1813), 20200060.
Simon, J. C., Stoeckel, S., & Tagu, D. (2010). Evolutionary and functional insights into reproductive strategies of aphids. *Comptes Rendus Biologies, 333*(6–7), 488–496.
Singec, I., Jandial, R., Crain, A., Nikkhah, G., & Snyder, E. Y. (2007). The leading edge of stem cell therapeutics. *Annual Review of Medicine, 58*, 313–328.
Singhal, T. (2020). A review of coronavirus disease-2019 (COVID-19). *The Indian Journal of Pediatrics, 87*(4), 281–286.
Sinn, D. L., & Moltschaniwskyj, N. A. (2005). Personality trait in dumpling squid (*Euprymna tasmanica*): Context-specific traits and their correlation with biological characteristics. *Journal of Comparative Psychology, 119*, 99–110.
Sinn, D. L., Apiolaza, L. A., & Moltschaniwskyj, N. A. (2006). Heritability and fitness-related consequences of squid personality traits. *Journal of Evolutionary Biology, 19*, 1437–1447.
Skogsmyr, I. O., & Lankinen, Å. (2002). Sexual selection: An evolutionary force in plants? *Biological Reviews, 77*(4), 537–562.
Slabbekoorn, H. (2013). Songs of the city: noise-dependent spectral plasticity in the acoustic phenotype of urban birds. *Animal Behaviour, 85*(5), 1089–1099.
Slatkin, M., & Excoffier, L. (2012). Serial founder effects during range expansion: A spatial analog of genetic drift. *Genetics, 191*(1), 171–181.
Sloan, W. T., Lunn, M., Woodcock, S., Head, I. M., Nee, S., & Curtis, T. P. (2006). Quantifying the roles of immigration and chance in shaping prokaryote community structure. *Environmental Microbiology, 8*(4), 732–740.
Smayda, T. J. (1990). Novel and nuisance phytoplankton blooms in the sea: Evidence for a global epidemic. In *Toxic marine phytoplankton* (pp. 29–40). Elsevier.
Smith, T. (1934). Parasitism and disease. .
Smith, H. (1982). Light quality, photoperception, and plant strategy. *Annual Review of Plant Physiology, 33*(1), 481–518.
Smith, J. M. (1991). Theories of sexual selection. *Trends in Ecology & Evolution, 6*(5), 146–151.
Smith, L. C., Bordignon, V., Babkine, M., Fecteau, G., & Keefer, C. (2000). Benefits and problems with cloning animals. *The Canadian Veterinary Journal, 41*(12), 919.
Smith, C. A., Katz, M., & Sinclair, A. H. (2003). DMRT1 is upregulated in the gonads during female-to-male sex reversal in ZW chicken embryos. *Biology of Reproduction, 68*(2), 560–570.
Snell, T. W., Kubanek, J., Carter, W., Payne, A. B., Kim, J., Hicks, M. K., & Stelzer, C. P. (2006). A protein signal triggers sexual reproduction in *Brachionus plicatilis* (Rotifera). *Marine Biology, 149*(4), 763–773.
Sol, D. (2008). Artificial selection, naturalization, and fitness: Darwin's pigeons revisited. *Biological Journal of the Linnean Society, 93*(4), 657–665.
Solomon, E. C., Martin, D. W., Martin, L. R., & Berg, L. R. (2018). *Biology* (11th ed.). Cengage Learning.

Sosa, M. M., & Tressen, S. G. (2002). Las plantas parásitas. In M. M. Arbo & S. G. Tressen (Eds.), *Flora del Iberá* (pp. 167–178). Editorial Eudene. Corrientes.
Sovie, A., McCleery, R., Fletcher, R., & Hart, K. (2016). Invasive pythons, not anthropogenic stressors, explain the distribution of a keystone species. *Biological Invasions, 18*, 3309–3318.
Spang, A., Saw, J. H., Jorgensen, S. L., Zaremba-Niedzwiedzka, K., Martijn, J., Lind, A. E., Van, E. R., Schleper, C., Guy, L., & Ettema, T. J. (2015). Complex archaea that bridge the gap between prokaryotes and eukaryotes. *Nature, 521*, 173–179.
Speidel, M. (2000). The parasitic host: Symbiosis contra neo-Darwinism. *Pli, 9*, 119–138.
Speziale, K. L., Lambertucci, S. A., Carrete, M., & Tella, J. L. (2012). Dealing with non-native species: What makes the difference in South America? *Biological Invasions, 14*, 1609–1621.
Srivastava, M., Simakov, O., Chapman, J., Fahey, B., Gauthier, M. E., Mitros, T., et al. (2010). The *Amphimedon queenslandica* genome and the evolution of animal complexity. *Nature, 466*(7307), 720–726.
Stalhandske, P. (2001). Nuptial gift in the spider *Pisaura mirabilis* maintained by sexual selection. *Behavioral Ecology, 12*(6), 691–697.
Stamps, J. A. (2007). Growth-mortality tradeoffs and 'personality' traits in animals. *Ecology Letters, 10*, 355–363.
Stamps, J. A., Briffa, M., & Biro, P. A. (2012). Unpredictable animals: Individual differences in intraindividual variability (IIV). *Animal Behaviour, 83*, 1325–1334.
Stanier, R. Y., & Villanueva, J. R. (1996). Microbiología. *Reverté.*
Stapley, J. (2006). Individual variation in preferred body temperature covaries with social behaviours and colour in male lizards. *Journal of Thermal Biology, 31*(4), 362–369.
Stasiuk, S. J., Scott, M. J., & Grant, W. N. (2012). Developmental plasticity and the evolution of parasitism in an unusual nematode. *Parastrongyloides trichosuri. EvoDevo, 3*, 14.
Stearns, S. C. (Ed.). (1987). *The evolution of sex and its consequences*. Springer Basel AG.
Stearns, S. C., & Hoekstra, R. F. (2005). *Evolution: An introduction*. Oxford University Press.
Steele, R. E. (2002). Budding in *Hydra*. *Developmental Biology, 248*(1), 24–43.
Steer, M., & Thompson, R. C. (2020). Plastics and microplastics: Impacts in the marine environment. In M. Streit-Bianchi, M. Cimadevila, & W. Trettnak (Eds.), *Mare Plasticum-the Plastic Sea: Combatting plastic pollution through science and art* (pp. 49–72). Springer.
Stenseth, N. C., Atshabar, B. B., Begon, M., Belmain, S. R., Bertherat, E., Carniel, E., Gage, K. L., Leirs, H., & Rahalison, L. (2008). Plague: past, present, and future. *PLoS Medicine, 5*(1), e3.
Stephenson, A. G., & Bertin, R. I. (1983). Male competition, female choice, and sexual selection in plants. *Pollination Biology, 109*, 149.
Stern, D. L. (2008). Aphids. *Current Biology, 18*(12), R504–R505.
Stern, M., Ediger, V. L., Gibbon, C. R., Blagburn, J. M., & Bacon, J. P. (1997). Regeneration of cercal filiform hair sensory neurons in the first instar cockroach restores escape behavior. *Journal of Neurobiology, 33*, 439–458.
Stevenson, P. C. (2020). For antagonists and mutualists: The paradox of insect toxic secondary metabolites in nectar and pollen. *Phytochemistry Reviews, 19*(3), 603–614.
Stockley, P., Franco, C., Claydon, A. J., Davidson, A., Hammond, D. E., Brownridge, P. J., et al. (2020). Revealing mechanisms of mating plug function under sexual selection. *Proceedings of the National Academy of Sciences, 117*(44), 27465–27473.
Strahl, B. D., & Allis, C. D. (2000). The language of covalent histone modifications. *Nature, 403*, 41–45.
Strong, P. J., Xie, S., & Clarke, W. P. (2015). Methane as a resource: Can the methanotrophs add value? *Environmental Science & Technology, 49*(7), 4001–4018.
Subramanian, H., Brown, J., & Gatenby, R. (2020). Prebiotic competition and evolution in self-replicating polynucleotides can explain the properties of DNA/RNA in modern living systems. *BMC Evolutionary Biology, 20*(1), 1–9.
Suga, H., Chen, Z., De Mendoza, A., Sebé-Pedrós, A., Brown, M. W., Kramer, E., et al. (2013). The *Capsaspora* genome reveals a complex unicellular prehistory of animals. *Nature Communications, 4*(1), 1–9.

Sung, S., & Amasino, R. M. (2005). Remembering winter: Toward a molecular understanding of vernalization. *Annual Review of Plant Biology, 56*, 491–508.
Susoy, V., Ragsdale, E. J., Kanzaki, N., & Sommer, R. J. (2015). Rapid diversification associated with a macroevolutionary pulse of developmental plasticity. *eLife, 4*, e05463.
Swallow, J. G., & Wilkinson, G. S. (2002). The long and the short of sperm polymorphisms in insects. *Biological Reviews, 77*, 153–182.
Swallow, J. G., Carter, P. A., & Garland, T. (1998). Artificial selection for increased wheel-running behavior in house mice. *Behavior Genetics, 28*(3), 227–237.
Szűcs, M., Vahsen, M. L., Melbourne, B. A., Hoover, C., Weiss-Lehman, C., & Hufbauer, R. A. (2017a). Rapid adaptive evolution in novel environments acts as an architect of population range expansion. *Proceedings of the National Academy of Sciences, 114*(51), 13501–13506.
Szűcs, M., Melbourne, B. A., Tuff, T., Weiss-Lehman, C., & Hufbauer, R. A. (2017b). Genetic and demographic founder effects have long-term fitness consequences for colonizing populations. *Ecology Letters, 20*(4), 436–444.
Tabar, V., & Studer, L. (2014). Pluripotent stem cells in regenerative medicine: Challenges and recent progress. *Nature Reviews Genetics, 15*(2), 82–92.
Taff, C. C., & Vitousek, M. N. (2016). Endocrine flexibility: Optimizing phenotypes in a dynamic world? *Trends in Ecology & Evolution, 31*(6), 476–488.
Taga, M. E., & Bassler, B. L. (2003). Chemical communication among bacteria. *Proceedings of the National Academy of Sciences, 100*(2), 14549–14554.
Takayama, S., & Isogai, A. (2005). Self-incompatibility in plants. *Annual Review of Plant Biology, 56*, 467–489.
Takeda, K. (1961). Classical conditioned response in the honey bee. *Journal of Insect Physiology, 6*, 168–179.
Tamashiro, K., Wakayama, T., & Yamazaki, Y. (2003). Phenotype of cloned mice: Development, behavior and physiology. *Experimental Biology and Medicine, 228*, 1193–1200.
Taylor, J. W., Hann-Soden, C., Branco, S., Sylvain, I., & Ellison, C. E. (2015). Clonal reproduction in fungi. *Proceedings of the National Academy of Sciences, 112*(29), 8901–8908.
Templeton, A. R. (1979a). Once again, why 300 species of Hawaiian *Drosophila*? *Evolution, 33*(1), 513–517.
Templeton, A. R. (1979b). The unit of selection in *Drosophila mercatorum*. II. Genetic revolution and the origin of coadapted genomes in parthenogenetic strains. *Genetics, 92*(4), 1265–1282.
Templeton, A. R. (1980). The theory of speciation via the founder principle. *Genetics, 94*(4), 1011–1038.
Templeton, A. R. (2008). The reality and importance of founder speciation in evolution. *BioEssays, 30*(5), 470–479.
Templeton, A. R., & Rankin, M. A. (1978). Genetic revolutions and control of insect populations. *The screwworm problem*, 83–112.
Tepfer, D. (2008). The origin of life, panspermia and a proposal to seed the universe. *Plant Science, 175*(6), 756–760.
The American Heritage® dictionary of the English language. (2003). *The American Heritage dictionary of the English language*. Houghton Mifflin.
Theis, K. R., Dheilly, N. M., Klassen, J. L., Brucker, R. M., Baines, J. F., Bosch, T. C., et al. (2016). Getting the hologenome concept right: An eco-evolutionary framework for hosts and their microbiomes. *mSystems, 1*(2), e00028–e00016.
Thompson, J. N. (2013). *Relentless evolution*. University of Chicago Press.
Thompson, R., & McConnell, J. (1955). Classical conditioning in the planarian, *Dugesia dorotocephala*. *Journal of Insect Physiology, 48*, 65–68.
Thornhill, R. (1983). Cryptic female choice and its implications in the scorpionfly *Harpobittacus nigriceps*. *The American Naturalist, 122*(6), 765–788.
Thornhill, R. (1984). Alternative hypotheses for traits believed to have evolved by sperm competition. In R. L. Smith (Ed.), *Sperm competition and the evolution of animal mating systems* (pp. 151–178). Academic Press.

Thorpe, W. H. (1943). Types of learning in insects and other arthropods. Part II. *British Journal of Psychology, 34*(1), 20.
Thorpe, W. H. (1956). *Learning and instinct in animals*. Methuen.
Tittensor, D. P., Walpole, M., Hill, S. L., Boyce, D. G., Britten, G. L., Burgess, N. D., & Ye, Y. (2014). A mid-term analysis of progress toward international biodiversity targets. *Science, 346*(6206), 241–244.
Tollefsbol, T. O. (2011). Advances in epigenetic technology. In T. Tollefsbol (Ed.), *Epigenetics protocols* (pp. 1–10). Humana Press.
Torres, A. C., Ferreira, A. T., Sá, F. G., Buso, J. A., Caldas, L. S., Nascimento, A. S., et al. (2000). Glossário de Biotecnologia Vegetal. *Embrapa Hortaliças.*
Tortora, G. J., Case, C. L., & Funke, B. R. (2016). Microbiologia-12ª Edição.. Artmed Editora.
Toussaint, B., Raffael, B., Angers-Loustau, A., Gilliland, D., Kestens, V., Petrillo, M., et al. (2019). Review of micro-and nanoplastic contamination in the food chain. *Food Additives & Contaminants: Part A, 36*(5), 639–673.
Towe, K. M. (1981). Environmental conditions surrounding the origin and early Archean evolution of life: A hypothesis. *Precambrian Research, 16*(1–2), 1–10.
Tramontin, A. D., & Brenowitz, E. A. (2000). Seasonal plasticity in the adult brain. *Trends in Neurosciences, 23*, 251–258.
Travis, J. M., & Dytham, C. (2002). Dispersal evolution during invasions. *Evolutionary Ecology Research, 4*(8), 1119–1129.
Tregenza, T., & Wedell, N. (2000). Genetic compatibility, mate choice and patterns of parentage: Invited review. *Molecular Ecology, 9*(8), 1013–1027.
Tregenza, T., & Wedell, N. (2002). Polyandrous females avoid costs of inbreeding. *Nature, 415*, 71–73.
Trewavas, A. (2003). Aspects of plant intelligence. *Annals of Botany, 92*(1), 1–20.
Trewavas, A. (2005). Plant intelligence. *Naturwissenschaften, 92*, 401–413.
Trewavas, A. (2014). *Plant behaviour and intelligence*. Oxford University Press.
Trigos, A. S., Pearson, R. B., Papenfuss, A. T., & Goode, D. L. (2018). How the evolution of multicellularity set the stage for cancer. *British Journal of Cancer, 118*(2), 145–152.
Tripp, E. A., Zhang, N., Schneider, H., Huang, Y., Mueller, G. M., Hu, Z., Häggblom, M., & Bhattacharya, D. (2017). Reshaping Darwin's tree: Impact of the symbiome. *Trends in Ecology & Evolution, 32*(8), 552–555.
Trivers, R. L. (1972). Parental investment and sexual selection. In B. Campbell (Ed.), *Sexual selection and the descent of man 1871–1971* (pp. 136–179). Aldine-Atherton.
Tsukahara, N. (1981). Synaptic plasticity in the mammalian central nervous system. *Annual Review of Neuroscience, 4*, 351–379.
Tsunoda, Y., & Kato, Y. (2002). Recent progress and problems in animal cloning. *Differentiation, 69*(4–5), 158–161.
Tulving, E. (1985). How many memory systems are there? *American Psychologist, 40*(4), 385.
Tyagi, A., Kumar, A., Aparna, S., Mallappa, R., Grover, S., & Batish, V. (2014). Synthetic biology: Applications in the food sector. *Critical Reviews in Food Science and Nutrition, 56*(11), 1777–1789.
Tyler, M. J. (1994). *Australian frogs A natural history*. Reed Books.
Tyre, A. J., & Addicott, J. F. (1993). Facultative non-mutualistic behaviour by an "obligate" mutualist: "cheating" by yucca moths. *Oecologia, 94*, 173–175.
Uetz, P., Freed, P. A., & Hošek, J. (2019). *The reptile database*. http://www.reptile-database.org
Umen, J. G. (2014). Green algae and the origins of multicellularity in the plant kingdom. *Cold Spring Harbor Perspectives in Biology, 6*(11), a016170.
Unden, G., Becker, S., Bongaerts, J., Schirawski, J., & Six, S. (1994). Oxygen regulated gene expression in facultatively anaerobic bacteria. *Antonie Van Leeuwenhoek, 66*(1), 3–22.
Unden, G., Becker, S., Bongaerts, J., Holighaus, G., Schirawski, J., & Six, S. (1995). O2-sensing and O2-dependent gene regulation in facultatively anaerobic bacteria. *Archives of Microbiology, 164*(2), 81–90.

Urru, I., Stensmyr, M. C., & Hansson, B. S. (2011). Pollination by brood-site deception. *Phytochemistry, 72*(13), 1655–1666.
Vaccari, N. E., Edgecombe, G. D., & Escudero, C. (2004). Cambrian origins and affinities of an enigmatic fossil group of arthropods. *Nature, 430*(6999), 554–557.
Vamosi, J. C., & Vamosi, S. M. (2004). The role of diversification in causing the correlates of dioecy. *Evolution, 58*(4), 723–731.
Van Der Heijden, M. G., Streitwolf-Engel, R., Riedl, R., Siegrist, S., Neudecker, A., Ineichen, K., et al. (2006). The mycorrhizal contribution to plant productivity, plant nutrition, and soil structure in experimental grassland. *New Phytologist, 172*(4), 739–752.
Van der Waal, C. (2005). Kudu foraging behavior: Influenced by animal density? *African Journal of Range & Forage Science, 22*(1), 11–16.
van Doren, D., Koenigstein, S., & Reiss, T. (2013). The development of synthetic biology: A patent analysis. *Systems and Synthetic Biology, 7*, 209–220.
Van Hoven, W. T. (1984). Tannins and digestibility in greater kudu. *Canadian Journal of Animal Science, 64*(5), 177–178.
Van Oers, K., & Sinn, D. L. (2011). Toward a basis for the phenotypic gambit: Advances in the evolutionary genetics of animal personality. In M. Inoue-Murayama, S. Kawamura, & A. Weiss (Eds.), *From genes to animal behavior* (pp. 165–183). Springer.
Van Oers, K., De Jong, G., Drent, P. J., & van Noordwijk, A. J. (2004). Genetic correlations of avian personality traits: Correlated response to artificial selection. *Behavior Genetics, 34*, 611–619.
VandenBosch, L. S., & Reh, T. A. (2020). Epigenetics in neuronal regeneration. In *Seminars in cell & developmental biology* (Vol. 97, pp. 63–73). Academic Press.
Vandenbrink, J. P., & Kiss, J. Z. (2019). Plant responses to gravity. In *Seminars in cell & developmental biology* (Vol. 92, pp. 122–125). Academic Press.
Vandermassen, G. (2004). Sexual selection: A tale of male bias and feminist denial. *European Journal of Women's Studies, 11*(1), 9–26.
Velasco, J. (2018). Universal common ancestry, LUCA, and the tree of life: Three distinct hypotheses about the evolution of life. *Biology & Philosophy, 33*, 31.
Vennerstrom, J. L., Arbe-Barnes, S., Brun, R., Charman, S. A., Chiu, F. C., Chollet, J., et al. (2004). Identification of an antimalarial synthetic trioxolane drug development candidate. *Nature, 430*, 900–904.
Ventura González, I., Sanmartin Santos, I. S., Lloret Alcañiz, A., Revert Ros, F., & Prieto Ruiz, J. (2020). La Biología Sintética; reto biotecnológico y bioético en las Ciencias de la Vida. In *IN-RED 2020: VI Congreso de Innovación Educativa y Docencia en red* (pp. 264–269). Editorial Universitat Politècnica de València.
Vereecken, N. J., Wilson, C. A., Hötling, S., Schulz, S., Banketov, S. A., & Mardulyn, P. (2012). Pre-adaptations and the evolution of pollination by sexual deception: Cope's rule of specialization revisited. *Proceedings of the Royal Society B: Biological Sciences, 279*(1748), 4786–4794.
Verghese, M. W., & Nordskog, A. W. (1968). Correlated responses in reproductive fitness to selection in chickens. *Genetics Research, 11*(3), 221–238.
Viano, L. (2008). *¡Invaden Córdoba!: Grateus o tomatito*. Aquí hay cienciea. http://aquihayciencia.blogspot.com/2008/05/invaden-crdoba-grateus-o-tomatito.html.
Viano, L. (2014). La proliferación de plantas exóticas es la gran amenaza del bosque nativo. *La Voz del Interior.*. https://www.lavoz.com.ar/ciudadanos/la-proliferacion-de-plantas-exoticas-es-la-gran-amenaza-del-bosque-nativo/
Vicens, P., Redolat, R., & del Carmen Carrasco, M. (2003). Aprendizaje espacial y laberinto de agua: metodología y aplicaciones. *Psicothema, 15*(4), 539–544.
Viney, M. E. (1996). Developmental switching in the parasitic nematode *Strongyloides ratti*. *Proceedings of the Royal Society of London B, 263*, 201–208.
Völkle, H. (2015). A brief history of nuclear disasters: Prevention, consequences and re-coverage. *GRF Davos Planet@Risk, 3*(2), 271–280.
Volkov, A. G., Adesina, T., Markin, V. S., & Jovanov, E. (2008). Kinetics and mechanism of *Dionaea muscipula* trap closing. *Plant Physiology, 146*(2), 694.

Volkov, A. G., Foster, J. C., & Markin, V. S. (2010). Signal transduction in *Mimosa pudica*: Biologically closed electrical circuits. *Plant, Cell & Environment, 33*(5), 816–827.

Waddington, K. D. (1983). Pollen flow and optimal outcrossing distance. *The American Naturalist, 122*(1), 147–151.

Waelti, P., Dickinson, A., & Schultz, W. (2001). Dopamine responses comply with basic assumptions of formal learning theory. *Nature, 412*(6842), 43–48.

Waller, J. C. (2008). In a spin: The mysterious dancing epidemic of 1518. *Endeavour, 32*(3), 117–121.

Waller, J. (2009). Looking back: Dancing plagues and mass hysteria. *Psychologist, 22*(7), 644–647.

Walsh, M. R. (2013). The link between environmental variation and evolutionary shifts in dormancy in zooplankton. *Integrative and Comparative Biology, 53*(4), 713–722.

Walter, M. R., & Heys, G. R. (1985). Links between the rise of the metazoa and the decline of stromatolites. *Precambrian Research, 29*, 149–174.

Wang, S., Hu, M., Zheng, J., Huang, W., Shang, Y., Fang, J. K., Shi, H., & Wang, Y. (2021). Ingestion of nano/micro plastic particles by the mussel *Mytilus coruscus* is size dependent. *Chemosphere, 263*, 127957.

Ward, P. (2018). *Lamarck's revenge: How epigenetics is revolutionizing our understanding of Evolution's past and present*. Bloomsbury.

Ward, P., & Kirschvink, J. (2015). *A new history of life: The radical new discoveries about the origins and evolution of life on earth*. Bloomsbury.

Waser, N. M. (1987). Spatial genetic heterogeneity in a population of the montane perennial plant *Delphinium nelsonii*. *Heredity, 58*(2), 249–256.

Waser, N. M. (1993). Population structure, optimal outbreeding, and assortative mating in angiosperms. In N. W. Thornhill (Ed.), *The natural history of inbreeding and outbreeding: Theoretical and empirical perspectives* (pp. 173–199). University of Chicago Press.

Watanabe, K. (2001). Microorganisms relevant to bioremediation. *Current Opinion in Biotechnology, 12*(3), 237–241.

Waterland, R. A., Kellermayer, R., Laritsky, E., Rayco-Solon, P., Harris, R. A., Travisano, M., et al. (2010). Season of conception in rural *Gambia* affects DNA methylation at putative human metastable epialleles. *PLoS Genetics, 6*(12), e1001252.

Waters, E. R. (2003). Molecular adaptation and the origin of land plants. *Molecular Phylogenetics and Evolution, 29*(3), 456–463.

Watts, P. C., et al. (2006). Parthenogenesis in komodo dragons. *Nature, 444*(7122), 1021–1022.

Webb, C. J. (1981). Andromonoecism, protandry, and sexual selection in Umbelliferae. *New Zealand Journal of Botany, 19*(4), 335–338.

Webber, B. L., Raghu, S., & Edwards, O. R. (2015). Opinion: Is CRISPR-based gene drive a biocontrol silver bullet or global conservation threat? *Proceedings of the National Academy of Sciences, 112*(34), 10565–10567.

Weber, K., Hoelzl, F., Cornils, J. S., Smith, S., Bieber, C., Balint, B., & Ruf, T. (2018). Multiple paternity in a population of free-living edible dormice (Glis glis). *Mammalian Biology, 93*, 45–50.

Webster, R. G. (2002). The importance of animal influenza for human disease. *Vaccine, 20*(5), S16–S20.

Wedell, N., & Cook, P. A. (1999). Butterflies tailor their ejaculate in response to sperm competition risk and intensity. *Proceedings of the Royal Society of London B Biological Sciences, 266*, 1033–1039.

Wehr, N. H. (2021). Historical range expansion and biological changes of *Sus scrofa* corresponding to domestication and feralization. *Mammal Research, 66*(1), 1–12.

Weinberg, R. A. (1983). Oncogenes and the molecular biology of cancer. *The Journal of Cell Biology, 97*(6), 1661–1662.

Weinberg, R. A. (2007). *The biology of cancer*. Garland Science. Taylor & Francis Group, LLC.

Weiss, R. A. (1996). Retrovirus classification and cell interactions. *Journal of Antimicrobial Chemotherapy, 37*(B), 1.

Weiss, M. C., Preiner, M., Xavier, J. C., Zimorski, V., & Martin, W. F. (2018). The last universal common ancestor between ancient earth chemistry and the onset of genetics. *PLoS Genetics, 14*(8), e1007518.
Weiss-Lehman, C., Hufbauer, R. A., & Melbourne, B. A. (2017). Rapid trait evolution drives increased speed and variance in experimental range expansions. *Nature Communications, 8*(1), 1–7.
Wellman, C. H., Oterhoff, P. L., & Mohiuddin, U. (2003). Fragments of the earliest land plants. *Nature, 425*(6955), 282–285.
Wells, D. N. (2005). Animal cloning: Problems and prospects. *Revue Scientifique Et Technique-Office International Des Epizooties, 24*(1), 251.
Wells, C. L., & Pigliucci, M. (2000). Adaptive phenotypic plasticity: The case of heterophylly in aquatic plants. *Perspectives in Plant Ecology, Evolution and Systematics, 3*(1), 1–18.
West, S. A., Lively, C. M., & Read, A. F. (1999). A pluralist approach to sex and recombination. *Journal of Evolutionary Biology, 12*(6), 1003–1012.
West-Eberhard, M. J. (2003). *Developmental plasticity and evolution*. Oxford University Press.
West-Eberhard, M. J. (2005). Developmental plasticity and the origin of species differences. *Proceedings of the National Academy of Sciences, 102*(1), 6543–6549.
Westwood, J. H., Yoder, J. I., Timko, M. P., & Depamphilis, C. W. (2010). The evolution of parasitism in plants. *Trends in Plant Science, 15*(4), 227–235.
Wheeler, W. M. (1928). *The social insects: Their origin and evolution*. Kegan Paul, Trench, Trubner and, Ltd..
Whitman, W. B., Coleman, D. C., & Wiebe, W. J. (1998). Prokaryotes: The unseen majority. *Proceedings of the National Academy of Sciences, 95*(12), 6578–6583.
Whittaker, R. H. (1969). New concepts of kingdoms of organisms: Evolutionary relations are better represented by new classifications than by the traditional two kingdoms. *Science, 163*(3863), 150–160.
Whittet, D. C. (1997). Is extraterrestrial organic matter relevant to the origin of life on earth? In D. C. Whittet (Ed.), *Planetary and interstellar processes relevant to the origins of life* (pp. 249–262). Springer Science & Business Media.
Wickramasinghe, C. (2012). Origin of life and panspermia. In: J. Seckbach (Ed.), Genesis-in the beginning: Precursors of life, chemical models and early biological evolution (pp. 621–649). : Springer Netherlands.
Wiese, L., Wiese, W., & Edwards, D. A. (1979). Inducible anisogamy and the evolution of oogamy from isogamy. *Annals of Botany, 44*(2), 131–139.
Wignall, A. E., Kemp, D. J., & Herberstein, M. E. (2014). Extreme short-term repeatability of male courtship performance in a tropical orb-web spider. *Behavioral Ecology, 25*, 1083–1088.
Williams, G. C. (1975). *Sex and evolution*. Princeton University Press.
Williams, J. L., Hufbauer, R. A., & Miller, T. E. (2019). How evolution modifies the variability of range expansion. *Trends in Ecology & Evolution, 34*(10), 903–913.
Willmer, P. (2011). *Pollination and floral ecology*. Princeton University Press.
Willson, M. F. (1979). Sexual selection in plants. *American Naturalist, 113*, 777–790.
Willson, M. F., & Ågren, J. (1989). Differential floral rewards and pollination by deceit in unisexual flowers. *Oikos, 56*, 23–29.
Willson, M. F., & Burley, N. (1983). *Mate choice in plants: Tactics, mechanisms, and consequences*. Princeton University Press.
Willson, M. F., & Rathcke, B. J. (1974). Adaptive design of the floral display in *Asclepias syriaca* L. *American Midland Naturalist*, 47–57.
Wilson, D. S. (1994). Adaptive genetic-variation and human evolutionary psychology. *Ethology and Sociobiology, 15*, 219–235.
Wilson, A. C. C., & Duncan, R. P. (2015). Signatures of host/symbiont genome coevolution in insect nutritional endosymbioses. *Proceedings of the National Academy of Sciences, 112*, 10255–10261.

Winfield, I. J., & Hoare, G. (2015). *Parthenogenesis: The development of unfertilized eggs*. Wiley-Blackwell.
Woese, C. R. (1987). Bacterial evolution. *Microbiological Reviews, 51*(2), 221.
Wolf, H. (2017). Scorpions pectines–idiosyncratic chemo-and mechanosensory organs. *Arthropod Structure & Development, 46*(6), 753–764.
Wong, B. B., & Schiestl, F. P. (2002). How an orchid harms its pollinator. *Proceedings of the Royal Society of London Series B: Biological Sciences, 269*(1500), 1529–1532.
Woodrow, P., Pontecorvo, G., Fantaccione, S., Fuggi, A., Kafantaris, I., Parisi, D., & Carillo, P. (2010). Polymorphism of a new Ty1-copia retrotransposon in durum wheat under salt and light stresses. *Theoretical and Applied Genetics, 121*, 311–322.
Work, H. P., & Flowers, H. P. M. (2011). *The wood-wide-web: Are plants inter-connected by a subterranean fungal network?* https://www.howplantswork.com/2011/07/26/the-wood-wide-web-are-plants-inter-connected-by-a-subterranean-fungal-network/
Wray, M. K., Mattila, H. R., & Seeley, T. D. (2011). Collective personalities in honeybee colonies are linked to colony fitness. *Animal Behaviour, 81*(3), 559–568.
Wright, J. W., & Lowe, C. H. (1968). Weeds, polyploids, parthenogenesis and the geographical and ecological distribution of all-female species of *Cnernidophorus*. *Copeia, 1968*, 128–138.
Wright, G. A., & Schiestl, F. P. (2009). The evolution of floral scent: The influence of olfactory learning by insect pollinators on the honest signaling of floral rewards. *Functional Ecology, 23*(4), 841–851.
Wright, J. P., Jones, C. G., & Flecker, A. S. (2002). An ecosystem engineer, the beaver, increases species richness at the landscape scale. *Oecologia, 132*, 96–101.
Wright, S. I., Bi, I. V., Schroeder, S. G., Yamasaki, M., Doebley, J. F., McMullen, M. D., & Gaut, B. S. (2005). The effects of artificial selection on the maize genome. *Science, 308*(5726), 1310–1314.
Wu, H., Fang, Y., Yu, J., & Zhang, Z. (2014). The quest for a unified view of bacterial land colonization. *The ISME Journal, 8*(7), 1358–1369.
Wulff, N. C., & Lehmann, G. U. (2016). Function of male genital titillators in mating and spermatophore transfer in the tettigoniid bushcricket *Metrioptera roeselii*. *Biological Journal of the Linnean Society, 117*(2), 206–216.
Wund, M. A. (2012). Assessing the impacts of phenotypic plasticity on evolution. *Integrative and Comparative Biology, 52*, 5–15.
Xu, S., & Chong, K. (2018). Remembering winter through vernalisation. *Nature Plants, 4*(12), 997–1009.
Yashiro, T., & Matsuura, K. (2014). Termite queens close the sperm gates of eggs switch from sexual to asexual reproduction. *Proceedings of the National Academy of Sciences of the United States of America, 111*, 17212–17217.
Zaffiri, L., Gardner, J., & Toledo-Pereyra, L. H. (2012). History of antibiotics. From salvarsan to cephalosporins. *Journal of Investigative Surgery, 25*(2), 67–77.
Zalasiewicz, J., Waters, C. N., Do Sul, J. A. I., Corcoran, P. L., Barnosky, A. D., Cearreta, A., et al. (2016). The geological cycle of plastics and their use as a stratigraphic indicator of the Anthropocene. *Anthropocene, 13*, 4–17.
Zannas, A. S., Arloth, J., Carrillo-Roa, T., Iurato, S., Röh, S., Ressler, K. J., et al. (2015). Lifetime stress accelerates epigenetic aging in an urban, African American cohort: Relevance of glucocorticoid signaling. *Genome Biology, 16*(1), 1–12.
Zeegers, M., Vlietinck, R., van Poppel, F., Spruijt, L., & Ostrer, H. (2003). Founder mutations among the Dutch. *American Journal of Human Genetics, 73*(5).
Zepeda, E. A., Veline, R. J., & Crook, R. J. (2017). Rapid associative learning and stable long-term memory in the squid *Euprymna scolopes*. *Biological Bulletin, 232*, 212–218.
Zhang, H., & Liu, L. (1998). The genetic diversity of *Castanea mollissima* and the effect of artificial selection. *Acta Botanica Yunnanica, 20*(1), 81–88.
Zhang, X., & Zhang, Y. (2015). Tissue engineering applications of three-dimensional bioprinting. *Cell Biochemistry and Biophysics, 72*(3), 777–782.

Zhao, J., Hussain, M., Wang, M., Li, Z., & He, N. (2020). Embedded 3D printing of multi-internal surfaces of hydrogels. *Additive Manufacturing, 32*, 101097.

Zheng, Y. G., & Bernstein, A. I. (2015). *Epigenetic technological applications*. Elsevier Science & Technology.

Zheng, J. Q., Kelly, T. K., Chang, B., Ryazantsev, S., Rajasekaran, A. K., Martin, K. C., & Twiss, J. L. (2001). A functional role for intra-axonal protein synthesis during axonal regeneration from adult sensory neurons. *Journal of Neuroscience, 21*(23), 9291–9303.

Zilber-Rosenberg, I., & Rosenberg, E. (2008). Role of microorganisms in the evolution of animals and plants: The hologenome theory of evolution. *FEMS Microbiology Reviews, 32*(5), 723–735.

Zillig, W., Klenk, H. P., Palm, P., Leffers, H., Puhler, G., Gropp, F., & Garrett, R. A. (1989). Did eukaryotes originate by a fusion event. *Endocytobiosis and Cell Research, 6*(1), 1–25.

Zimorski, V., Ku, C., Martin, W. F., & Gould, S. B. (2014). Endosymbiotic theory for organelle origins. *Current Opinion in Microbiology, 22*, 38–48.

Zou, Y., Chiu, H., Zinovyeva, A., Ambros, V., Chuang, C. F., & Chang, C. (2013). Developmental decline in neuronal regeneration by the progressive change of two intrinsic timers. *Science, 340*(6130), 372–376.

Zuk, M., Bastiaans, E., Langkilde, T., & Swanger, E. (2014). The role of behaviour in the establishment of novel traits. *Animal Behaviour, 92*, 333–344.

Zulet, M. I., Fontes, L. P., Blanco, T. A., Bescos, F. L., & Iriarte, M. M. (2017). Modificaciones epigenéticas en neurología: alteraciones en la metilación del ADN en la esclerosis múltiple. *Neurología, 32*(7), 463–468.

Zupanc, G. K. (2001). Adult neurogenesis and neuronal regeneration in the central nervous system of teleost fish. *Brain, Behavior and Evolution, 58*, 250–275.

Zupanc, G. K. H. (2006). Neurogenesis and neuronal regeneration in the adult fish brain. *Journal of Comparative Physiology A, 192*(6), 649–670.

Zupanc, G. K. (2008). Adult neurogenesis and neuronal regeneration in the brain of teleost fish. *Journal of Physiology-Paris, 102*(4–6), 357–373.

Zupanc, G., & Sîrbulescu, R. F. (2011). Adult neurogenesis and neuronal regeneration in the central nervous system of teleost fish. *European Journal of Neuroscience, 34*(6), 917–929.

Zwarts, L., Versteven, M., & Callaerts, P. (2012). Genetics and neurobiology of aggression in *Drosophila*. *Fly, 6*, 35–48.

Chapter 7
Focusing on Dynamics: When a Rule Becomes an Exception

In science it is important to modify and change own ideas as science advances

—Herbert Spencer

Contents

A. V. Peretti et al., *Rules and Exceptions in Biology: from Fundamental Concepts to Applications*, https://doi.org/10.1007/978-3-031-55382-0_7

Abstract In this chapter, we look at the other side, the cases where the rules have become exceptions. The factors responsible for these changes are diverse. Let us consider, for example, those changes produced by nature itself, where certain characters, behaviors, and even interactions that were previously very common became rare or even disappeared; changes associated with climatic relics, relict species, up to major extinction events. Let us also think of the changes that humans have made in nature that are responsible for certain rules becoming exceptions, the effects of artificial selection, deforestation, species introduced into environments that are not natural, and of course the climate change in which we are major participants, to name just a few examples. Additionally, the changes from rules to exceptions can result from changes in scientific interpretations, such as biases in study approaches, biases in the choice of model species for research, and their general extrapolation of results without, in many cases, the necessary precautions, in addition to the biases in interpretations associated with the use of certain current equipment and methodologies.

7.1 Introduction

As we explored in the previous chapter, on exceptions that became rules, in many cases, this includes the reverse situation, i.e., a rule that became an exception. For example, an exotic organism introduced into a given territory was initially a rarity numerically speaking, but over the years, it became a rule compared to other organisms inhabiting that territory. This, therefore, implied that other native organisms that, for example, were displaced from their niches by the intruder gradually became rarer and scarcer, becoming the exception. In this example, they shifted from being considered a rule at the group level to becoming an exception.

In the following, we will expose some examples that imply a passage from rules to exceptions, sometimes following this idea of opposites, implicit twins, as we have already mentioned. Other examples may not have been discussed in previous chapters. The idea is that we observe specific examples of rules falling, not through a complete replacement by another rule leading to their disappearance but rather through their "restriction" to exceptions within certain contexts of time, space, and group.

7.2 R-E Changes Due to Nature's Own Action

7.2.1 Reductions and Disappearances of Characters and Interactions

When it is generally mentioned in the biological evolution literature about the reduction of a trait (e.g., Darwin, 1859; Ha & Nehm, 2014; Kraaijeveld et al., 2016), the topic itself is closely connected to what we have discussed about how a rule could vary over time from the condition of being a "norm" to becoming a rarity later on. In essence, we are discussing rules (represented by the mean phenotype, the most common one) from an adaptive standpoint. Over time, due to directional selection favoring another phenotype (which might have been previously at the extreme end of the distribution), it becomes the less common pattern. This shift can cause the altered structure to remain within the range of characteristics found in the standard distribution but now extend toward the extremes. This process could be attributed to the influence of natural selection and sexual selection acting on that specific trait. For example, by changing predation pressure, the environment, etc., it becomes the exception now (e.g., in a specific environment, with x predator in place). We are then talking about a "mix" of change in one or more traits of a given group, in a given space, and in a given time. It would be reductionist to limit this situation to only one of these contexts. At this point, then, when we speak of patterns that became rarities, we can refer to those traits that we observe in nature, whether morphological or behavioral, for example, that were once very frequent and are no longer so (Darwin, 1859; Porter & Crandall, 2003). Indeed, these traits once had a specific function. Nonetheless, certain changes, as mentioned earlier, such as environmental shifts or interactions with other species, rendered their function unnecessary for individuals or, in extreme cases, even harmful to them (Darwin, 1859; Lahti et al., 2009; Maughan et al., 2006; Walker-Larsen & Harder, 2001).

One of the research goals of evolutionary biology is to determine why new traits arise and how they influence the better fitness of individuals (Ellers et al., 2012). However, less attention has been given to the evolutionary loss or decline of particular traits (Darwin, 1859; Royer et al., 2016). Only in recent years have researchers paid more attention to the study of the processes that trigger the loss of traits, taking them from being the most frequent to just rarities (Ellers et al., 2012; Lehmann et al., 2007). As with acquiring new traits, trait loss can occur at different levels ranging from total loss in extreme cases to vestigialization (see below, subitem 7.2.1.1). In this sense, the evolution of morphological traits, which includes the acquisition and loss of traits, is made possible by changes in embryonic development (Gould, 1977; Prud'homme et al., 2007). In the 1980s, significant technological advancements, particularly in histochemistry, biochemistry, and DNA and RNA analyses, allowed for a substantial increase in the number of analyses and interpretations of previously unknown processes. As a result, the identification of genes involved in morphological evolution began in specific animal models (Prud'homme et al., 2007). The development of morphological traits, involving their acquisition and loss, is based on the expression of particular genes encoding transcription factors and molecules for cell signaling (Carroll, 1995; King & Wilson, 1975; Prud'homme et al., 2007).

Different evolutionary causes can lead to the loss of traits (Ha & Nehm, 2014; Kraaijeveld et al., 2016), in this sense going from being the most frequent and standard to being considered rare when observed in nature. Some of these causes include natural selection (such as changes in the environment and predation pressure, as mentioned earlier), as well as intraspecific and interspecific competition, external biotic factors (e.g., interspecific interactions, as discussed below), and abiotic factors (such as food and resource availability and colonization of new environments) (Darwin, 1859; Royer et al., 2016). These factors ultimately lead to a particular trait or structure becoming obsolete in its function. Making an analogy with item 6.2.2.1, where individuals who colonize new environments may not possess the most common allele frequency in the population, and due to gene drift processes, such as the founder effect and bottleneck, these allele frequencies may become common in the new population. The opposite may also occur in a nonexclusive manner. Specific allele frequencies that were common or the rule may now become rare or may not even remain in the population once it has settled in the new location. In addition, the loss of certain traits may be the product of molecular causes, such as deleterious changes and mutations in specific genes that alter their function (Drouin et al., 2011; Hiller et al., 2012; Kraaijeveld et al., 2016). In some cases, trait loss can occur rapidly over time, which is an exception to the general rule that evolution occurs through small changes over long periods (Johnson et al., 2012). In the following, we will analyze some classic and not-so-classic cases of traits that are no longer common.

7.2.1.1 From Common to Vestigial and Rare Structures

When we discuss vestigial organs, we refer to those structures that have lost, during evolution, their original function, which contrasts with the functionality they may have had in ancestral species and related species (Smith & Wright, 2018). If the presence of the trait does not cause any disadvantage in individuals, it might not be eliminated or take a long time to do so. However, all structures need energy for their development and maintenance, which also means that there is selective pressure to eliminate structures that do not increase the fitness of individuals. At one extreme, if the trait becomes detrimental, it should be eliminated more quickly.

Although vestigial structures have lost their original function, they may still have other minor functions (Muller, 2002). Darwin (1859) said, "*An organ may become rudimentary for its primary purpose and be used for a different purpose.*" For example, penguins' wings are vestigial, as they have lost their original flight function but have a new function associated with locomotion in water. In ostriches, wings have also lost their original flight function but now have functions associated with temperature control and display in males in reproductive contexts. Boas and python snakes show traces of the pelvis, and the toes in some species, such as horses, are examples of vestigial structures. In cetaceans (whales, dolphins, and porpoises), limbs and fur have been reduced or even disappeared. The loss of limbs was favored by natural selection for greater swimming efficiency (Johnson et al., 2012; Xiong et al., 2009). However, one of the most notorious examples of vestigial structures is those observed in cave-dwelling species. In these species, it is common for the eyes to have ceased to be functional (in addition to a reduction in pigment and metabolic rate and hypertrophy in nonoptical sensory organs) (Borowsky & Wilkens, 2002; Porter & Crandall, 2003). For example, in the fish *Astyanax mexicanus*, although embryos develop optic primordia, these degenerate during ontogeny so that adults lack eyes (Jeffery & Martasian, 1998; Yamamoto & Jeffery, 2000). Other traits go from being a pattern to just a rarity in terms of their frequency in nature; i.e., the loss of a particular trait occurs in cases where sexually reproducing organisms switch to an asexual mode of reproduction (Kraaijeveld et al., 2016). For example, among the traits lost by these organisms, we can name the ability to attract mates and fertilize eggs (van der Kooi & Schwander, 2014).

Plants may also show vestigial structures, such as stipules and carpels, which no longer have a clear function (Knobloch, 1951), as well as a reduction in flower size in those that reproduce without crossing (Eckert, 2002). The aquatic plant *Decodon verticillatus* shows a sexual or asexual reproductive system depending on its distribution area. The loss of the ability to reproduce sexually in this species may have occurred several times in evolution, where populations with asexual reproduction show lower values of genetic and morphological variability (Dorken & Eckert, 2001; Johnson et al., 2012). Among the hypotheses given by the authors to explain why the loss of sex could have occurred, we can name, for example, mutations that lead to the loss of sexuality and advantages of asexual reproduction in the fact that it frees resources previously destined for sexual reproduction (Eckert, 2002; Johnson et al., 2012).

In humans, since Darwin (1859), organs or structures that would be in the vestigial category have been recognized; here, we will give just a few examples. The appendix, for example, is considered vestigial because it does not maintain its digestive function. However, the appendix maintains functions associated with immunology and maintains the intestinal flora (Muller, 2002). In turn, the wisdom teeth are vestigial from human ancestors, where their function was associated with grinding food, but this function was gradually lost as the human diet shifted toward more prepared and softer foods; in some people, these teeth are no longer present (Bollinger et al., 2007; Rozkovcová et al., 1999). Some muscles of the human body have also become vestigial. The musculature in monkeys' ears is significantly more developed than that in humans, granting them the ability to move their ears for enhanced hearing, especially in response to potential threats. In contrast, as mentioned before, humans have less developed ear muscles and compensate for the inability to move their ears by having the capacity to turn their heads. The muscle on the dorsal part of the head that attaches to the ear (auricular superior muscle) varies in its frequency of occurrence in the current population; for example, it is present in 36% of the European population, 56% of the African population, and 50% of the Asian population (Macalister, 1871). The palmaris muscle is absent in 14% of the population, with no function today, but is believed to have been involved in locomotion on trees in primates (Tamatsu et al., 2007). The vomeronasal organ, an organ associated with olfaction, whose primary function is the detection of pheromones in sociosexual contexts, is considered a vestigial structure in humans (Meredith, 1991).

Among the vestigial structures in humans, we can also include embryological remnants that serve a function during development, such as the belly button, and analogous structures between the biological sexes. For example, males are born with two nipples, and we do not know what function they serve compared to females. The tailbone or coccyx, despite being a vestige of the primate tail, serves a functional purpose for us, such as providing the insertion point for pelvic muscles and muscles that aid in bowel movements (Saraga-Babić et al., 1994). Human embryos develop a tail formed by the last vertebrae in the early stages of development, but after approximately 8 weeks, these disappear completely (Smith & Wright, 2018) (see below how, in some cases where this disappearance does not occur, they are considered teratologies). The loss of the tail, a pattern in our ancestors, is thought to have occurred as part of the evolution of humans to favor our bipedal stance. Determining how the loss of this conspicuous morphological trait occurred in our ancestors has been a source of research for scientists (Smith & Wright, 2018). In recent years, thanks to significant technological advances, it has been possible to determine that the loss of this trait would have its basis, similar to most evolutionary changes, in genetics. The cause could have been the alteration in the location of the gene related to the presence of the tail or a different site in the genome in a primate species, which generated a mutation that ultimately resulted in the disappearance of the trait (Smith & Wright, 2018). Vestigial molecular structures also exist in humans. For instance, the L-gulonolactone oxidase gene, responsible for producing an enzyme that synthesizes vitamin C, is functional in most mammals. However, a mutation in this gene in humans has rendered it incapable of performing such a function (Ohta & Nishikimi, 1999).

However, with many of these vestigial cases becoming extreme or exceptional cases, this change from being the pattern to rarity could result in such a modification of the trait in question that it becomes something "quasi-teratological" from a biomedical point of view, that is, if such a structure is no longer used and is strikingly odd when it appears. This implies that rather than on a continuous trait, something discrete, qualitative, quite distinct, does or does not appear, and when it does, it is very striking to the rest of the population. This may occur in ancient structures, without current use and due to a "long-term time" action, which means that what was the norm is now no longer even an extreme in the original distribution. However, if it appears (i.e., is rarer, outside the new distribution limit of the trait), it is teratological in the classical concept, already discussed in the chapter focused on exceptions.

Is this possible? We will mention some examples below, which do not make it thus far-fetched, mainly when focusing on humans. As we said before, we have the example of tail loss in humans. There are exceptional cases where these tails do not disappear. Babies can be born with tails, which can be considered teratology. These are vestigial protrusions and are considered pseudotails, not true tails, because although they have muscles and nerves and are covered by skin, they do not possess bone and are not connected to the spinal cord (Mukhopadhyay et al., 2012; Smith & Wright, 2018; Xia et al., 2021). A congenital disability results in babies being born with this pseudotail (Dao & Netsky, 1984; Dubrow et al., 1998). Again, technological advances allow these pseudotails to be surgically removed and the individual to live an everyday life (Spiegelmann et al., 1985). Another trait that can be considered teratology in humans is excessive hairiness. In the 1860s, a woman of Mexican nationality showed some features that made her "attractive" at that time. Features such as short stature, multiple teeth, protrusions with excrescences, and a significant presence of hairiness all over her face and body made her an "*ape-like freak*" to use the terminology of the time. Worthy of display in circuses, this woman was presented as a hybrid between a human and an orangutan (Moros, 2004).

7.2.1.2 Behaviors that "No Longer Fit"

In the previous section, we discussed and provided examples of morphological traits that can be considered vestigial. Due to the loss of their function, these traits have transitioned from being predominant patterns to becoming rare occurrences, and in the most extreme cases, they are considered teratologies. However, the traits we can refer to are not only morphological, but we can also include certain behaviors that, by loss of their function, cease to be the predominant pattern (Rognini, 2018). For example, let us think about the classic cricket song; in most species, males perform a chirp, a function mainly associated with mate attraction. However, the males of a population of *Teleogryllus oceanicus* do not perform this characteristic behavior. The reason is linked to the fact that this song attracts, in addition to females, the parasitoid fly *Ormia ochracea*. Thus, when males sing, they have a higher risk of parasitism (Johnson et al., 2012; Schneider et al., 2018). Another case to mention is that of some species of frugivorous birds, where males have changed their usual pattern of showy coloration to a more monotone hue, a trait directly associated with attracting females (Johnson et al., 2012). This loss of sexual traits may be linked to the presence of a new parasite or predator, which is also attracted to this trait (Johnson et al., 2012). These examples illustrate how selection pressures, in this instance, the presence of a parasitoid species, can influence the modification of traits from being the most common behavioral pattern (Tinghitella & Zuk, 2009; Zuk et al., 2006). Another factor that can lead to a sexually selected trait becoming rare is if the communication channel between individuals is disrupted. For example, the water in some lakes has become polluted to the point of becoming so turbid that it makes it impossible for females of some fish species to see underwater, which reduces the selection pressure for conspicuous coloration of males, leading to a shift from conspicuous coloration, which was the predominant pattern, to a duller coloration (Johnson et al., 2012).

In humans, we can also name some behaviors that no longer "fit" today. The formation of goosebumps in humans under stress is a vestigial reflex; its function in human ancestors was to raise the hair from the body to create a larger appearance, aiding in scaring away predators and maintaining a higher body temperature through the creation of an extra layer of air (Foltán & Šedý, 2009). Another behavior linked to vestigial reflexes is hiccups; in this case, researchers have proposed that hiccups are a remnant of amphibious respiration (Straus et al., 2003). Children's and some adults' fears of the unknown, darkness, and loud noises have their basis in ancestral self-protective behaviors in the face of potential danger (Carleton, 2016; Craske, 1999; Rognini, 2018). Additionally, infants' reflex behaviors, such as finger-prehensile behavior, are considered vestigial. It is postulated that an ancestral primate would have had sufficient body hair for the offspring to grasp tightly, unlike in modern humans, where the prehensile response, while evident, is comparatively of lower strength (Futagi et al., 2012; Rognini, 2018). The sudden reflex extension of the arms and legs in infants in response to sudden stimuli, which may have served the purpose of staying close to the mother, could also be considered vestigial behavior (Ropper & Brown, 2005).

7.2.1.3 Altered Interspecific Interactions

Several factors can lead to changes in interspecific interactions or even specific traits in interacting species, resulting in the most common pattern (rule) becoming rare. One example arises when one of the species experiences a decline or complete extinction in its population due to natural or artificial causes, thereby significantly impacting the other species' way of life. Here, we will focus on changes in the interactions between species produced by natural causes and, later in this chapter, on those interactions that have been altered by human action.

Numerous examples of changes in interspecific interactions resulting from the disappearance of one species can be traced back to the great extinction of megafauna at the end of the Pleistocene, approximately 12,000 years ago. This extinction was primarily caused by the direct or indirect effects of modern humans (Araujo et al., 2015; Galetti et al., 2018; Guimarães Jr. et al., 2008) and climate change (Alroy, 2001; Guimarães Jr. et al., 2008). For instance, in South America, at least seven genera of large mammals became extinct (Guimarães Jr. et al., 2008).

While the causes of this mass extinction have been subject to extensive debate, the current focus is on its consequences (Galetti et al., 2018; Johnson, 2009; Malhi et al., 2016; Smith et al., 2016). Among the consequences of this great extinction, we can expect the loss of other species that were dependent on extinct species, a phenomenon known as coextinction, especially in closely intertwined interactions such as parasitism and mutualism (Colwell et al., 2012; Galetti et al., 2018; Pires & Guimarães Jr., 2013). As a result, interspecific relationships that were once prevalent are now almost nonexistent. However, some species associated with extinct megafauna managed to survive by adapting and establishing new interactions with their environment. For instance, they adopted asexual reproduction strategies and even formed associations with secondary dispersers such as humans, though often displaying examples of evolutionary anachronisms (Galetti et al., 2018; Janzen & Martin, 1982).

Moreover, what do we mean by evolutionary anachronism? This concept in evolutionary biology refers to traits of living species that are best explained by coevolution with species that have become extinct. Evolutionary anachronism has been discussed mainly considering the contexts of seed dispersal (fruit morphology) and defense strategies shown by numerous plant species associated with interactions with extinct animal species (Guimarães Jr. et al., 2008; Janzen & Martin, 1982). Large herbivores disappeared, but the plants that evolved along with them did not, remaining without their primary natural disperser and forming part of what has been called "evolutionary ghosts" (Barlow, 2001; Guimarães Jr. et al., 2008). Let us keep in mind that a trait that is considered an evolutionary anachronism is not the same as a trait that is considered vestigial (see item 7.2.1.1) since the former retains its original function while the vestigial trait does not.

Now, let us explore examples of species with anachronistic evolutionary traits resulting from changes in species associations over time. One well-studied case involves the morphological characteristics of large, fleshy fruits, which once relied on large mammals for seed dispersal (Galetti et al., 2018; Janzen & Martin, 1982). These large mammals were extinct 12,000 years ago. We have the case of avocado *Persea americana*, a medium-sized tree with a tasty fleshy fruit containing an enormous seed. The animals that consumed this fruit were large animals, such as giant elephant species of the Gomphotheriidae family, giant sloths, and toxodonts, now extinct. The size of its seed makes it almost impossible for today's animals to consume and disperse it; however, the avocado survived because it is a species cultivated by humans, first in Mesoamerica and now worldwide. A similar case of a plant surviving through human intervention despite the extinction of its consumers and main dispersers is the mango (*Mangifera indica*) (Agnolin et al., 2001).

Other illustrative cases are the Louisiana orange tree (*Maclura pomifera*) with ample fruits and the Kentucky coffee tree (*Gymnocladus dioicus*) with thick, sweet-smelling seed pods. Additionally, the large, fleshy fruits of the lucumo (*Pouteria splendens*) and lucumillo (*Myrcianthes coquimbensis*) found in northern Chile may have been eaten by the mylodon (*Mylodon darwini*). Additionally, the classic example of the pink shower tree *Cassia grandis*, which Janzen and Martin (1982) noted that the local fauna did not consume its large fruit, on which they based their argument of evolutionary anachronism. Guimarães Jr. et al., (2008) formulated an extensive list of characters that show the fruits of the plants cataloged under evolutionary anachronism; for example, the large, mature fruits germinate where they fall, they are not attractive to arboreal mammals or current fruit-eating birds, and their distribution is fragmentary since their main dispersers are extinct (Agnolin et al., 2001).

Other traits observed in some plants and considered evolutionary anachronisms are spines and toxins that protect them against large herbivores that no longer exist (Bond & Silander Jr, 2007; Galetti et al., 2018; Greenwood & Atkinson, 1977). In the first case, Janzen & Martin (1982) also included among plants with anachronisms several species of Bromeliaceae, such as pineapple (*Ananas bracteatus*), because of its complex, spiny leaves around the fruit (Agnolin et al., 2001). Another example is *Ibicella lutea*, whose fruits have long spines and sticky hairs, which would have allowed them to adhere to the legs of extinct megafauna for dispersal

(Agnolin et al., 2001; Barlow, 2001). Regarding the presence of toxins, Janzen and Martin (1982) also included among the plants with fruits with traits associated with evolutionary anachronism the *Genipa americana* tree with fruits with a thick and toxic shell (Agnolin et al., 2001). Additionally, the Kentucky coffeetree named above is an example since its peel and pulp are poisonous (Zaya & Howe, 2009). Other examples are several species of the Cucurbitaceae family, such as *Cucurbita foetidissima*, which produce bitter-tasting toxins and are harmful to humans and livestock (Agnolin et al., 2001; Barlow, 2001).

The study of the consequences of the Pleistocene megafaunal extinction on plant-frugivore interactions has been deepened in recent decades (e.g., Guimarães Jr. et al., 2008; Loiselle & Blake, 2002). From the perspective of our book, what is certain is that numerous processes that were the norm, the rule in interspecific interactions, became rarities or even disappeared altogether, as we have seen. For example, we can think of the dispersal of a smaller number of seeds over much shorter distances than usual, as well as alterations in the genetic structure of populations due to a decrease in gene flow (Guimarães Jr. et al., 2008; Jordano et al., 2007). The extinction of megafauna also produced alterations in other animal species with which they interacted. For example, in New Zealand, there are bird species with specific coloration patterns and nocturnal habits, which could reflect the existence of predatory species that influence these behaviors and are now extinct (Johnson, 2009). Another behavior considered an evolutionary anachronism is shown by the American pronghorn, *Antilocapra americana*, which develops a high escape speed from its current predators, much faster than necessary for a successful escape. In this sense, this incredible speed was necessary to escape *Miracinonyx*, which is now an extinct cheetah-like predator that inhabited North America. The extinction of numerous herbivores also produced alterations in their predators that managed to survive. For example, the jaguar *Panthera onca* prefers smaller prey than other felids. It is believed that the species underwent a dietary shift toward smaller prey, as larger prey became extinct. This adaptability allowed the species to survive (Galetti et al., 2018; Hayward et al., 2016).

7.2.2 From Dominant to Rarities and Major Extinctions

The history of life on our planet is a continuous process of the creation of new species and the extinction of others, which made it possible to form an enormous variety of natural branches in the tree of life. From the beginning, as we have already mentioned earlier in this book, from the appearance of eukaryotic cells and their organelles, the patterns began to change, going from organisms that were the only known or standard form to slightly more complex organisms. The same is true for the transition from unicellular to multicellular life forms, which has occurred more than once throughout evolution. Organisms have evolved, and while some remain unicellular, others have become multicellular organisms, such as animals and plants. This shows how patterns change evolutionarily, leaving some organisms that were once the rule as rare or less common forms.

Suppose we continue this line of thought at another level of organization. In that case, the same thing happens with the appearance of new species and the reduction or spatial disappearance/extinction of others. Concerning this, in the history of life, disturbances caused by geological, astronomical, or climatic agents have led, at some points in history, to significant extinctions, in which many species have disappeared in a short time (Keller & Kerr, 2014). When a species disappears, it ends its entire evolutionary history. Additionally, it affects many other species that have evolved in an intimate relationship with it and, sometimes, are even dependent on it. Some species are forced to change and sometimes even disappear along with the species to which they are related. Countless extinct organisms, populations, and species were also part of evolution since they left offspring that survived to reproduce with the same or different characteristics (Emer et al., 2019). Therefore, "evolution" refers to the "changes" that were transmitted from one generation to the next, comprising all those that took place at the morphological, physiological, behavioral, biochemical, genetic, molecular, and ecosystemic levels (Madrigal & González-José, 2016). These changes mean that the species and populations that we observe today are not the same as those that were dominant in the past, and this causes the rule to change, and the pattern species become rare species or directly disappear. In the history of the Earth, mass extinctions, for example, repeatedly changed the course of evolution: It was probably no accident that reptiles radiated in the Triassic after the Permian mass extinction wiped out other groups, just as mammals radiated in the Paleocene after the tremendous Cretaceous extinction wiped out most reptiles (O'Leary et al., 2013; Sue, 2019).

Patterns in the abundance or existence of species change over time due to different natural processes. In addition to significant historical events, such as the extinctions already mentioned, these patterns can change due to comparatively "simpler" natural events. An example of this may be a change in the abundance of species in different areas of their distribution due to interactions between them or environmental factors such as climate or altitude. Then, species that are the exception in some places could be the rule in others, sometimes even remaining as relicts in a geographical space. In the following sections, we discuss all these cases, focusing on how nature can shape patterns over time and the evolutionary history of species.

7.2.2.1 Changes in Population According to Inhabited Space

The areas of distribution of species on the planet are not random. They evolve as different environmental factors, such as relief, soil, climate, relationships between organisms, and even human activity, and as a function of species evolution (Bravo, 2003). Some species have a large area of distribution on the planet, but within it, they have a different abundance everywhere. Climate change, for example, affects species interactions, community structure, and biodiversity and can induce significant shifts in geographic ranges (Post, 2013; Visser, 2016). However, not all species will respond to these changes in a similar way (Springate & Kover, 2014). When the environment is variable and changing, species with broad global distributions will have a more significant genetic variation that allows them to adapt to climatic and environmental gradients (Jump & Peñuelas, 2005). Cosmopolitan species, for example, have a high rate of genetic diversity and high phenotypic variation, which gives them greater flexibility to evolve in response to change, unlike species with a more restricted geographic distribution (Lavergne & Molofsky, 2007). For example, the common reed (*Phragmites australis*) has a global distribution, has high genetic and phenotypic variation, and is found in many environments (Eller et al., 2017). However, some populations may experience a decrease in genetic diversity during acclimation and adaptation processes (Almeida et al., 2013). While it is unlikely for species with high intraspecific diversity to face total extinction, they may still undergo changes in genetic composition, leading to genetic impoverishment within a population (Franks et al., 2014; Valladares et al., 2014). Thus, while some species may inhabit extensive areas and adapt to heterogeneous conditions, they will have a different abundance everywhere, so they may be expected in some places. In contrast, they will be rarer species in others.

In this same context, there are species that, having adapted to specific climates better than others, are limited to regional distribution, being common only in some environments. An example of these species is the coastal penguin *Pygoscelis papua*, which inhabits the sub-Antarctic region but is not found elsewhere (BirdLife International, 2024a). The same is true, for example, of the Andean condor (*Vultur gryphus*), which inhabits South America and ranges along the Andes Mountains and coasts of the Pacific and Atlantic Oceans (BirdLife International, 2024b). These species, like many others, are the rule in their regional area of distribution but are the exception in other parts of the world. Within plants, the same can be observed for Coihue (*Nothofagus dombeyi*) growing in the Patagonian forests of South America (Barstow et al., 2017) or araucaria (*Araucaria araucana*) of the sub-Antarctic forests (Premoli et al., 2013), among many other species of regional distribution in the world.

Similarly, species adapt to various environments based on the factors mentioned above, with some being more successful in adapting to urban areas. In cities, natural habitats undergo significant modification and alteration due to construction and the replacement of native species with nonnative vegetation (McDonnell et al., 1993; McKinney, 2002). This means that species with more significant adaptive capacity benefit and can colonize urban environments or remain in them. Cities create their microclimate, increase the average temperature concerning their surroundings, and offer shelter in their buildings and various feeding possibilities. However, they are an inhospitable habitat for most species. A small number of animals make use of urban settlements. They are cosmopolitan and opportunistic organisms that are generally prolific and omnivorous. These species include pigeons (*Columba livia*), sparrows (*Passer domesticus*), rats (*Rattus norvegicus*), and mice (*Mus musculus*), for example (Giménez, 2009). In Argentina, certain species of scorpions, such as the medically significant *Tityus carrilloi*. Populations living in urban areas of the center of the country reproduce through parthenogenesis and demonstrate remarkable adaptability to urbanized environments. Consequently, their populations in cities are steadily increasing, leading to frequent encounters with them (de Roodt, 2014). Due to their ability to adapt and take advantage of resources, all these species and many more can become a rule in urban settlements and large cities. In contrast, many others that do not possess these capabilities eventually disappear.

The variation in species abundance across different regions on the planet illustrates how climate, biotic interactions, environmental factors, and community dynamics can influence the patterns and processes of species. Consequently, we can observe that species we sometimes see as standard become rare or exceptions according to changes in their habitat, resulting in dynamic communities throughout their life history. In the following, we will mention two situations linked to the reduction of something generalized to something circumscribed or restricted, mainly by the action of nature and at the level of space and group, respectively.

7.2.2.2 Climate Relics

Closely linked to the previous point, there are cases where climate change (natural or not) results in isolated populations called climatic relicts, which have been fragmented by environmental changes and habitat loss caused by climate (Hampe & Jump, 2011; Woolbright et al., 2014). These climatic relicts form natural laboratories regarding the dynamics of the populations involved (ecology) and their evolution. They preserve ecological and evolutionary histories that can span millennia. They make it possible to study how environmental change affects species and populations over the long term. However, they can also be beneficial for studying interactions between species and how they can affect communities and the ecosystem, for example. The main characteristic of these relicts is the lack of gene flow between them, which causes their low genetic differentiation. In turn, within relicts, natural selection and genetic drift also act by decreasing genetic variation within populations (Woolbright et al., 2014).

In these relicts, the abiotic environment governs population patterns to a greater extent. However, there may also be an influence on biotic interactions. Competition (Ejankowski, 2010), predation (Castro et al., 1999; Hampe & Bairlein, 2000), pathogen infection (Linares et al., 2010), herbivore action, and invasive species can have negative or positive consequences at small population sizes. For example, it is known that aspens (*Populus* spp.) adjacent to an isolated spring, all trees may be clones of the same genotype, having survived through asexual reproduction (Woolbright et al., 2014). This low genetic variation means they have little resistance, for example, to pathogens or herbivores (Busby et al., 2013), and the population may decline rapidly in the event of an attack. As a result of this characteristic, species found in climatic relics can sometimes be common in some sites but rare or even absent in others. In addition, the lack of genetic variation may cause species to disappear due to biological interactions in relicts and thus become rare species even within the same relict.

Climate change is one reason that biotic interactions are changing and may influence the ability of species to adapt to new habitats and persist (Gilman et al., 2010). For example, glacial relict populations of the shrub *Dryas octopetala* are the only hosts of ectomycorrhizal fungi in the Burren steppes of western Ireland (Harrington & Mitchell, 2002; Liston & Harrington, 2012). These fungi are typically associated with forest trees (e.g., Scots pine *Pinus sylvestris*), which are not found in that climatic relict (Roche et al., 2009). The novel associations of the fungus with shrubs in glacial relicts may explain the persistence of the fungus, which would otherwise have become extinct. Variations in this community may be explained by soil and vegetation characteristics that allowed them to live in atypical habitats (Schechter & Bruns, 2008). However, it could also be influenced by genetic variation in the host plant. If this is the case, the population could be a geographic mosaic (Smith et al., 2011; Thompson, 2005), creating natural laboratories that could reveal how climate affects genetically controlled interactions among species, which not only shape communities and ecosystems in changing environments but also drive evolution, diversification, and speciation (Woolbright et al., 2014). Indeed, climate relict species and communities exhibit unique patterns that are not observed in other areas, enabling them to thrive and dominate in their specific habitats, even though they might be considered exceptions in other regions.

7.2.2.3 Relictual Species

The concept of "relictual" species refers to those species that were once abundant in a more extensive distribution area and became confined to smaller areas (Fig. 7.1). They are generally species that are in a state of regression. This regression may be caused by climatic changes, competition, or interactions with other species and is called biogeographic relicts (Lomolino et al., 2006). An example is the relict forests, which remain vestiges of more extensive areas than they were previously occupied. However, there are also taxonomic relicts, species with a few survivors of a once very diverse taxonomic assemblage. The two forms sometimes coincide, for example, in the case of misnamed "living fossils," which we will discuss below.

Relict species, in general, are often valuable for evolutionary biology and biogeography to help address extinction issues. Within molecular phylogeny, they allow, for example, the estimation of extinction rates (Crisp & Cook, 2009; Nee et al., 1994). In addition, these species allow us to identify the characteristics that make them survive extinction and model niches according to environmental change (Graham et al., 2006; Grandcolas et al., 2011). However, in tune with the theme of this book, they are representatives of larger groups that have partially disappeared, so they give us information on the diversity of the past of existing groups in terms of ancestral species (Assmann et al., 2010; Fjeldså & Bowie, 2008; Pellens & Grandcolas, 2015). This characteristic will become increasingly important in the coming decades as the current major extinction crisis will make it necessary to conserve species for ecological services and as representatives in the tree of life (Grandcolas et al., 2014).

One example of a relictual species is termites of *Mastotermes darwiniensis*, the only living representative of the family Mastotermitidae (Legendre et al., 2008). The family is now restricted to northern Australia. They are closer to cockroaches

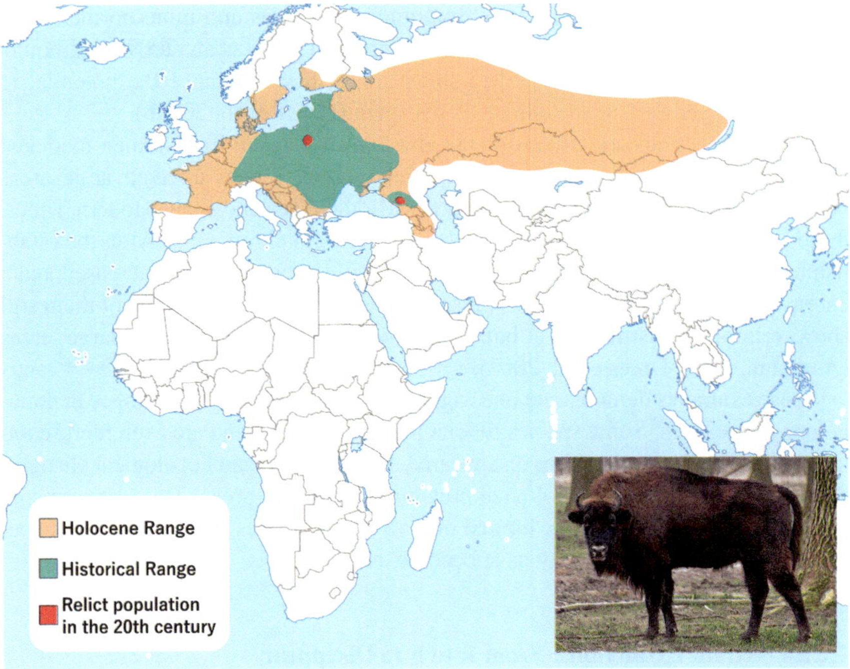

Fig. 7.1 Distribution of two European bison species. (**a**) Holocene (*Bison priscus* only) range map in orange. (**b**) Historical range (*B. bonasus* range in the high middle ages in green). (**c**) Relict *B. bonasus* populations in the twentieth century in red

and share interesting morphological and social characteristics. For example, they care for their young and feed on dead wood (Donovan et al., 2000; Legendre et al., 2008). In addition, they lay an egg mass similar to the ootheca of cockroaches, their reproductive individuals are winged, and their symbiont bacteria are similar to those found in cockroaches. The current distribution of these termites can be interpreted as a worldwide extinction of the family throughout its former distribution, except in Australia. Fossils have been found dating back to the Eocene, indicating that they were present on almost all continents (Grandcolas et al., 2014). This is a clear example of a species that went from being the rule, distributed almost everywhere in the world, to being the exception in only one restricted area.

Many thermophilic organisms are also relevant examples of relictual species. They survived the last ice age in relatively small refugia, such as, for example, on the peninsulas of southern Europe. During interglacial periods, these species were able to colonize large areas. Thus, many temperate species that are widespread today existed as small relict populations only a few 1000 years ago (Schmitt, 2007). In contrast, many cold-adapted species show evidence of formerly broad distributions followed by severe range restrictions during postglacial warming. Butterflies of the species *Proclossiana eunomia* and *Lycaena helle* are examples of the latter.

They are currently found only in high-altitude mountains and more northern latitudes, coinciding with distinctive genetic groupings (Habel et al., 2009a). Although these species today are restricted to small areas, they may have the potential to recolonize larger areas if the climate cools again (Habel et al., 2009b).

Range restrictions and expansion are also known to result from human-mediated landscape changes. For example, in central Europe, forests covered large areas before deforestation. Especially during the Middle Ages and early Modern Times, these extensive forests were degraded, leaving only small remnants. After the establishment of modern forestry and changes in land use, forest cover increased again enormously. The old forest retains preserved plants and animals. Some of them still show relict-like distributions, but others were able to recolonize large areas (Assmann, 1999; Drees et al., 2008).

These examples demonstrate once again the historical dynamic changes in populations and species. Some species restrict their distribution and are even representatives of larger groups that have disappeared due to climatic and ecological changes. Some of them may even recolonize previously inhabited areas. Thus, pattern species soon become exceptions, limited only to certain places or only to certain species of a larger group that may have become extinct.

7.2.2.4 Great Extinctions: From R to E to Disappear

As we outlined at the beginning of the chapter, mass extinctions are another case of the rule-exception dynamic. At a specific moment, some species were the pattern or the rule and became the exception or even ceased to exist. Throughout history, the planet has experienced six major mass extinctions caused by various contingencies, including volcanism, epidemics, interspecies competition, and climate change. These catastrophic events have led to significant losses of biodiversity and have played a crucial role in shaping the course of evolution on Earth. Mass extinctions, in general, were terminal extinctions in which more than half of the species that were present at the time disappeared without leaving descendants (Keller & Kerr, 2014). These extinctions marked a change in the composition of the communities of species that inhabited the Earth, either by the disappearance of some or the evolution of others.

In the Ordovician-Silurian period, two major extinctions occurred, resulting in the second-largest extinction of all occurring on Earth (Bond & Grasby, 2020). At that time, the most complex organisms were marine, and more than 100 biological families became extinct, representing 85% of the species. The groups that suffered the most significant consequences were brachiopods, bryozoans, trilobites, conodonts, and graptolites. Due to the disappearance of large invertebrates, the first fishes occupied empty biological niches until reaching more evolved life forms, such as amphibians, reptiles, or even mammals (Sheehan, 2001).

After this great extinction, during the Devonian, there was a significant diversification of fish, a remarkable growth of coral reefs, the first trees appeared, and the first tetrapods conquered the land. The next mass extinction had more significant effects on the seas at tropical latitudes. The organisms that inhabited the planet's temperate zones suffered the most significant effects (Hallam & Wignall, 1997). There was biodiversity loss because speciation rates were low, and many species almost disappeared. Coral populations declined considerably, with almost 85% of their species becoming extinct (Hallam & Wignall, 1997). Ninety-seven percent of shallow-water rugose corals disappeared, as did 60% of deeper-water corals and 80% of tabulate corals. Calcareous algae also disappeared, as did different species of jawless fish, placoderms, cephalopods, ostracods, and sponges, among others. The species that remained after this extinction began to repopulate the planet until the Permian–Triassic period, when what is commonly known as "The Great Permian Extinction" or "The Great Mortality" occurred. This was the most extensive extinction ever to happen on Earth. Ninety-five percent of marine species and 70% of terrestrial vertebrates have disappeared (Benton, 2003; Sahney & Benton, 2008). After this event, life took a long time to recover. There is no certainty on the possible causes of this extinction. The hypotheses suggest extreme volcanism, the impact of a giant asteroid, or the release of gases from the bottom of the ocean. Nevertheless, it is believed that there must have been a convergence of several factors that caused this level of species disappearance, making this event the most devastating and significant destruction of the biosphere ever known (Song et al., 2013).

After this significant mortality, reptiles and amphibians began to populate the continent. The first dinosaurs, pterosaurs, and crocodylomorphs appeared. The first small mammals and different terrestrial arthropods also appeared. Pteridophytes, cycads, and conifers predominated among the vegetation. In the aquatic environment, ichthyosaurs, plesiosaurs, nothosauroids, and placodonts diversified, as did many groups of fish. Many invertebrate groups also radiated in this period, such as corals, dinoflagellates, brachiopods, and mollusks (Abdala & Ribeiro, 2010; Ogg et al., 2004; Sue & Fraser, 2010).

During the Triassic-Jurassic period, there were substantial volcanic eruptions coupled with climatic fluctuations, sea level changes, and anoxia (Hautmann, 2012). Once again, these factors significantly impacted life on land and in the oceans. Approximately 20% of marine families, archosaurs, therapsids, and large amphibians disappeared. Many biological niches were then released, and dinosaurs took a dominant role in the following period (Hautmann, 2012).

In the Cretaceous, a sudden mass extinction caused three-quarters of the Earth's animal and plant species to disappear. Except for sea turtles, crocodiles, and other ectotherms, no large tetrapods survived (Muench et al., 2000; Renne et al., 2013). The best-known species that did not survive this great extinction were nonavian dinosaurs and many other terrestrial organisms, such as mammals, birds, lizards, insects, and plants. For marine species, plesiosaurs and mosasaurs, mollusks such as ammonites, many teleost fishes, and sharks disappeared. Many plankton species also vanished (Jablonski & Chaloner, 1994; Renne et al., 2013).

However, this event provided new empty niches and evolutionary opportunities for groups that could adaptively radiate and diversify into new species, particularly mammals. The surviving dinosaurs evolved into modern birds. Teleost fish and lizards also diversified (Cracraft, 2001; Longrich et al., 2015; Wilson et al., 2012).

Amid these mass extinctions, other events also occurred that caused a change in the patterns of existing species on the planet. One of them was the Messinian salt crisis, an event characterized by the accumulation of salt at the bottom of the Mediterranean, which also included its almost complete desiccation (Sternai et al., 2017). During the Messinian age, at the end of the Miocene, the Mediterranean Sea was isolated from the Atlantic due to the tectonic uplift of the Betic (south of the Iberian Peninsula) and Rifian (north of Morocco) mountain ranges. This process led to the drying and transformation of this sea into a large salt marsh. Later, the flooding of the Mediterranean by Atlantic waters through the Strait of Gibraltar ended this crisis (Garcia-Castellanos et al., 2009). However, in response to these changes in the natural environment, it is accepted that the Mediterranean flora and fauna come from Atlantic species that have populated it during the last five million years (Hsü et al., 1977; Naveh & Vernet, 1991). This invasion resulted in a change in species patterns, where those common in the Mediterranean Sea mostly disappeared, while Atlantic species became the new pattern in this environment.

The sixth mass extinction or Holocene mass extinction is the last extinction event in the current Holocene period and resulted mainly from human activity. This is the period of large megafaunal disappearance. As we saw, this extinction affected many species, including the mammoth, the dodo, and countless others that continue to become extinct yearly. After the last glaciation, the continents were populated by *Homo sapiens*, whose activity, directly or indirectly, contributed to the extinction of all these species. Activities such as the hunting of large animals, the transformation of the environment by fire, cultivation, erosion, and the transport of exotic species to new environments contribute to the extinction of species (Corlett, 2007; Fernandez, 2015). The uncontrolled action of human beings continues and exacerbates at present, which is why many scientists discuss the possibility of being immersed in the seventh mass extinction, caused in this case by uncontrolled human action (Turvey, 2009). The planet's current biodiversity may be reaching a turning point in which human activity has accelerated the end of several species by destroying their habitats. More than 300 vertebrate species have become extinct since 1500, and invertebrates have decreased their populations by approximately 45%, while the human population has doubled (Ceballos et al., 2015). Globally, for example, pollinators appear to be declining sharply in abundance and diversity. The significant problem is that the decline of affected species will have a cascading effect on ecosystem functioning, ultimately affecting human well-being (Cardoso et al., 2020).

Extinctions occurring across various evolutionary periods up to the present day indicate the diversification of species. Some species inhabit different regions of the planet, while others face extinction. Abundant species, at times, diminished or disappeared, making room for other diversified groups. As a result, both terrestrial and marine ecosystems underwent changes in their community composition, highlighting the dynamic nature of species diversity. This exemplifies the topic discussed in this section, where patterns are not static. Throughout history, patterns change, allowing new species to emerge, while others vanish or decrease in number, creating opportunities in ecological niches. Hence, the evolution and interactions among organisms, the abiotic environment, and human influence maintain the possibility for rules to become exceptions in the history of life on Earth.

7.2.2.5 Living Fossils

Regarding the previous point, certain species have survived mass extinctions and exhibited gradual evolution, closely resembling their ancestors in overall phenotype (Beierkuhnlein, 2007; Lomolino et al., 2006). These are the so-called living fossils, an incorrect designation since, by definition, a fossil is not alive (Fig. 7.2). However, to simplify things, we will take this name. They are organisms documented in the fossil record and are now represented by a few extant species in which the characters have been "fossilized"; i.e., they have remained without much change over time (Grandcolas et al., 2014). They are not identical, as many of these organisms change in some characteristics from their ancient versions (Cavin & Guinot, 2014). Living fossils are unique lineages characterized by stable morphology, low taxonomic diversity, and some rarity (Werth & Shear, 2014). Their apparent lack of diversification and stability suggests that they have adapted so effectively that they reduced the need to exhibit phenotypic changes in response to environmental change (Kin & Błażejowski, 2014). Consequently, these species can serve as valuable long-term evolutionary witnesses. There is no universal consensus on which lineages constitute genuine living fossils, but some universally recognized cases include Ginkgo tree, Tuatara, lungfishes, coelacanths, and nautiluses (Ward, 1984; Werth & Shear, 2014).

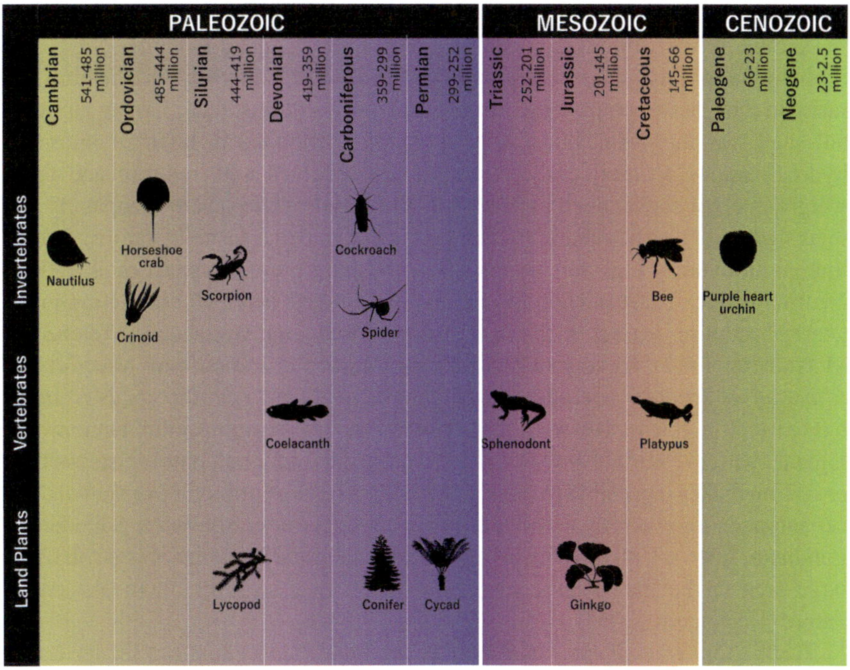

Fig. 7.2 Ancient beginnings of living fossils. The silhouettes show when each "living fossil" first appeared. Every group seen on this tree is still alive today

License: own creation. Free images from https://www.phylopic.org

Ginkgo (*Ginkgo biloba*) is known to have existed since the Jurassic period, 170 million years ago. However, there is no complete understanding of its evolution because of a gap in the fossil record of more than 100 million years. Ginkgo trees grow in China, Korea, Japan, Europe, and the United States. They can be up to 30 meters tall, 15 meters in circumference, and can live up to 1000 years. They were the first plants to grow back after the nuclear bomb was detonated in Hiroshima, with no signs of genetic mutations. Because of this resilience, these trees are frequently planted in large cities such as New York and Tokyo (Jacobs & Browner, 2000).

Another example mentioned above is the coelacanth (*Latimeria* spp.), which inhabits specific sectors of the Indian Ocean. This lobe-finned fish represents an early evolutionary deviation from the lineage that led to terrestrial vertebrates and is of great scientific interest (Amemiya et al., 2013). It has changed very little in appearance from the fossilized coelacanths of the Cretaceous (aprox. 145.5 to 65.5 million years ago) (Cavin & Guinot, 2014). The coelacanth's ancestors evolved more than 400 million years ago, surviving the movement of continents and the great extinction that ended with the dinosaurs. They are very slow-evolving fishes but may be the key to elucidating the rate and mode of evolution from fish to tetrapods. The knowledge of their genome is essential to elucidate how ancestral aquatic vertebrates evolved into terrestrial animals (Nikaido et al., 2013). Additionally, in the aquatic environment, we still find sharks and rays, whose anatomy and life forms appeared in the Devonian (Peristeraki & Megalofonou, 2007). Like the coelacanth, these fish have retained their morphology for approximately 350 million years.

Another remarkable living fossil is the nautilus (*Nautilus pompilius*), the only cephalopod mollusk that retains a true external shell. Currently, this feature is a rarity in the class Cephalopoda. In the basal groups of this class, having an external shell was the norm, but it was gradually abandoned in favor of improved hydrodynamics. Consequently, certain groups evolved with reduced and lightweight internal shells, such as cuttlefish and squid, while others completely lost their shells, such as octopuses (which granted them better maneuverability in benthic environments, e.g., for hiding and ambushing prey) (Ruppert et al., 2003). Nautiloids once constituted a diverse and widespread group within the mollusks, nearly becoming extinct at the end of the Permian and Cretaceous (Teichert & Matsumoto, 1987), leaving only a lineage comprising some living *Nautilus* and *Allonautilus* species. According to paleontological evidence, the genus *Nautilus* has existed for at least 100 million years without any apparent significant morphological changes (Ward, 1984). Nautiloids are a model for paleontologists, as their condition makes it possible to infer the biology of a large group of external-shelled cephalopods known only from fossil records (Harvey et al., 1999; Saunders & Landman, 2010). Unlike other living fossils, nautiloids now inhabit a relatively large area in the Indo-Pacific region, although they are difficult to observe in nature (De Angelis, 2012).

Tuatara (*Sphenodon* spp.) is also an example of a living fossil in the terrestrial environment. It is the only representative of the early-branching sphenodont-lepidosaur group and may be evidence for the antiquity of the New Zealand biota

(Jones et al., 2009). However, this genus shows a high rate of molecular change, inferred from comparisons between subfossils and modern individuals (Hay et al., 2008). Another example of an animal that has changed little or not over millions of years is the platypus (*Ornithorhynchus anatinus*). This monotreme mammal is one of the few descendants of a common ancestor that diverged from mammals 150 million years ago. It retains primitive features such as its leathery eggs and the absence of nipples, unlike more evolved mammals. Skulls of ancestors similar to platypuses have been discovered dating from the Cretaceous period. Like those mentioned above, these animals have a prolonged evolution, causing them to retain their appearance for millions of years, making them almost an exception (Chimento et al., 2013; Komblihtt, 2009).

In arthropods, many insects are also ancient inhabitants of our planet, with cockroaches standing out among the best-known. They appeared more than 300 million years ago and show morphological characteristics that have not changed much since. These insects have adapted to all environmental changes and were even the first animals found undamaged after nuclear explosions (Copeland, 2003). Just as hardy and ancient are scorpions, which emerged along with other arachnids and mites 350 million years ago (Gomes & Gomes, 2015).

The list of "living fossils" is extensive, but the aforementioned cases illustrate how species that were once widespread in their habitats millions of years ago are now apparent exceptions, rarities, and, in some cases (e.g., the coelacanth), even considered previously extinct. All these examples of once dominant groups become rarities, as discussed in this section, provide valuable insights into the patterns of species diversification on our planet and the changes that various groups may undergo due to interactions between species and their environment.

7.3 Human-Induced R-E Changes over Nature

7.3.1 Reduction or Loss of Changing Capacity in Living Organisms

7.3.1.1 In Animals and Plants: The Effect of Artificial Selection and Associates

Human beings are indeed impeding the normal course of evolution in numerous living organisms, particularly in groups inhabiting "hot spots" or facing endangerment. This interference stems from various actions, such as artificial selection, genetic manipulation, breed manipulation, direct or indirect impacts on extinction, climate change, and the introduction or control of species. Specific cases presented below will show that humans affect living organisms through these activities and remove their chances of evolving in time and form in the face of these scenarios imposed by them. It is interesting to discuss this point because it may be central and globalizing for many of the ideas in the book.

Darwin, in The Origin of Species, already saw a parallelism between natural evolution and artificial evolution carried out by humans. Thus, examples of different breeds of pigeons with different attributes are sketched (Darwin, 1859). Since the beginning of time, humans have selected organisms for their benefit. As mentioned earlier in Chap. 6, first in a more natural way, with humans functioning as any other animal in the ecosystem, but then at the time of a more stable settlement, they began to exert a much higher selective pressure, to the point of totally modifying some organisms. The most significant pressure was on organisms linked to the basic requirements of food and shelter. Humans selected the crops they preferred and cultivated the varieties that gave them the highest yields. The original species that were the rule at the beginning of nature declined in abundance or ceased to exist to give way to new species or varieties. A unique example is that of maize, where the original species of maize called perennial teosinte (*Zea perennis*) presented a fruit with a tiny corn cob with four rows of kernels. This original variety, which was the rule in the past, was modified over time until it reached the current corn cob. As we said in Chap. 6, the current crop is a hybrid (*Zea mays*) with a much larger corncob and multiple rows of larger kernels. A similar situation occurred with cattle. It is impossible to compare the musculature, milk-producing capacity, or wool production of the current varieties to the original from which they descended. Essential species that would not naturally exist, such as mules, were also created for work. Beneficial associations were also generated, for example, with wolves, which, when selected, gave rise to modern dogs.

Certainly, while certain organisms experience favorable selection pressures, others have been significantly threatened for various reasons. Some of these reasons include fear factors (e.g., snakes and spiders), associations with diseases or ailments (e.g., rats), superstitions (e.g., cats during medieval times associated with witchcraft), environmental modifications (e.g., deforestation affecting forest species, as discussed in item 7.2.2.3), or indiscriminate hunting (e.g., American bison). Precisely, as already mentioned, a large-scale example of indiscriminate hunting happened in the Quaternary with the mammalian megafauna. Hunting selection pressure by humans and other factors favored the extinction of this large group of animals (Diamond, 1989; Brook & Bowman, 2004; Bulte et al., 2006) (see item 7.2.2.4). Indeed, some Australian species face the detrimental effects of human neglect when placental mammals are introduced through expeditions and colonization voyages (Woinarski et al., 2015). This introduction of foreign species had significant impacts on the local ecosystems, leading to the decline and endangerment of certain native species in Australia. As mentioned elsewhere in the book, the introduction of exotic species is a phenomenon that has shaped the population dynamics and environment of certain regions quite a bit and is generally directly related to human activity. The introduction of exotic species is one of the greatest threats to biodiversity in the world. It has been noted that a small number of species are slowly replacing species that are threatened by human activities. Historical empirical data show what happens when there is a widespread invasion of organisms. South America was a relatively isolated place until the Pleistocene. After the formation of Central America, massive migration and invasion of many organisms began, replacing the South American natives. In fact, approximately half of the current South American mammals originated in North America (Goin et al., 2012; Marshall et al., 1982).

Some groups of organisms are common and declining due to human activities. In many cases, the dominant or common groups are threatened by specific changes or pressures that can reduce their numbers or completely extinguish their population. We have seen examples throughout the book where fish replaced groups such as Paleozoic cephalopods in the marine ecosystem. If we analyzed it in sequence, cephalopods were dominant in the environment and were the rule at that time. Gradually, fish appeared, which were evolutionarily better, and displaced them from the environment until certain groups became extinct, and the number of the group in general decreased notably. Today, we can see something similar where the human being is a crucial participant. With their activities, humans shape the environment, and many organisms are affected and cannot adapt well or fast enough. Hence, they decrease their numbers, become extinct, or have the potential to do so. As we recognize the significant role of human activities in causing extinction, the concept of species conservation has become increasingly prevalent (Akcakaya et al., 2004). Unlike in the past, there is now a greater emphasis on the importance of conservation efforts. However, while striving for a more comprehensive approach to conservation by encompassing numerous groups and species, humans tend to prioritize certain species known as "charismatic" species, which are more beloved (Colléony et al., 2017; Ducarme et al., 2013). Indeed, the selective pressure exerted by humans becomes favorable for certain species that would otherwise face natural extinction (Courchamp et al., 2018). The giant panda (*Ailuropoda melanoleuca*) serves as a clear example of this phenomenon. This species, which might have already disappeared in the wild for various reasons, continues to exist thanks to extensive human efforts, economic contributions, and technological advancements dedicated to its conservation.

Another example of species being selected by humans and linked to scientific and technological advances is the controversial GMOs or genetically modified organisms. These organisms have alterations in their genome, where genetic material from another organism has been introduced (Husby, 2007). There are currently many examples linked to consumption by the human population itself. In nature, for example, if we consider crops or livestock, we find that modified organisms are becoming the rule, and 100% "natural" organisms are slowly becoming the exception. GMOs have become organisms with attractive attributes (Hug, 2008). They present, for example, resistance to insects (Edgerton et al., 2012; Jouanin et al., 1998) and herbicides (De Greef et al., 1989; Duke, 1998), tolerance to different types of stress (Bhatnagar-Mathur et al., 2008), and increased yield potential (Edgerton et al., 2012) and can be used as bioreactors to produce various therapeutic agents (Houdebine, 2000; Jänne et al., 1998), among other outstanding characteristics. However, with these potential benefits come many concerns associated with the properties of the modified organisms, the possibility of spreading the transgenes to other organisms, and the potential problems of their release into nature (Rastogi-Verma, 2013). A case that draws attention and is directly related to genetically modified organisms is the monopoly of seeds by large multinationals. This phenomenon can be cited as a change from rule to exception. In the old days, the rule was to be able to obtain and sow the seeds that the producer wanted. However, now, the rule in many places is that the producer must buy or "ask permission" from the multinational to produce a crop. We must try to make this type of bio-monopoly the exception rather than the rule (see Kloppenburg, 2014).

We will now look at some key and well-known examples of groups that are affected by human activity and, as a result, are seeing their numbers decimated with the potential risk of extinction. This is what we call in this book a change from rule to exception due to human activity.

Global Pollinator Crisis

Pollinating insects play a crucial role in providing essential ecosystem services that enhance and stabilize global crop production (Klein et al., 2007 in Vanbergen & Initiative, 2013; Zayed & Packer, 2005). As discussed in Chap. 6, pollination is the primary process through which plants achieve reproduction. This vital ecosystem service is crucial because plants, unlike animals, cannot move to seek mates and thus rely on pollinators to facilitate their reproduction. Among these pollinators, animal pollinators, particularly arthropods, play a critical role in ensuring successful pollination. There are now many studies showing how various types of pollinators have been declining in number in recent years, both locally and globally (Steffan-Dewenter et al., 2005; Potts et al., 2010; but see Ghazoul, 2005). Much has been discussed about this. The facts confirm this pattern, a decrease in the diversity and abundance of pollinators, mainly arthropods. If this process accelerates, we will discuss a change from rule to exception with potentially catastrophic results (Steffan-Dewenter et al., 2005). This decline has a substantial impact on plant-animal interactions. All pollinators are affected, especially butterflies and bees. Bees are particularly haplodiploid and have genetic conditioning that predisposes them to extinction (Zayed & Packer, 2005).

Human activities are a significant driver of change, leading to accelerated rates of evolution and extinction, resulting in the loss of biodiversity (Cane & Tepedino, 2001; Vanbergen & Initiative, 2013). One of the most cited causes of these changes is the intensification of land use for agricultural activities, often associated with the indiscriminate use of pesticides that can harm both crop pests and vital pollinator species. Climate change also contributes to environmental issues, causing habitat alterations and species interactions, potentially leading to the decline of some species (see below). Additionally, the decline of pollinators is associated with the spread of exotic species and the emergence of pollinator diseases, pathogens, and parasites.

The complex interplay between the mentioned pressures, biological processes, and scales contributes to an overall decline in arthropod pollinators in natural ecosystems (Vanbergen & Initiative, 2013). This decline can affect various animal species that depend on plant production, including humans. For instance, reduced pollination can lead to increased inbreeding and inbreeding depression, which plants typically strive to avoid. Self-incompatible plants may suffer from blocked stigmas due to pollen from neighboring related flowers (Waser, 1978), which can accelerate pistil senescence (Scribailo & Barrett, 1994). These effects ultimately impact fruit and seed production.

7.3.2 In Humans: Reduction of Human Integration to Natural Dynamics

Following the reasoning presented in the previous point, we can observe that, with the exception of certain tribes or Western groups that embrace a more ecological lifestyle, humans, as a species, have disrupted the natural "continuum" to which they once belonged. Human activities now exert control and modification over rhythms and natural processes, resulting in humans no longer being merely a part of the ecosystem and nature. Instead, they have become a dominant and invasive species that poses a threat to nature itself.

Before the establishment of towns and cities, humans functioned as animals within the natural framework of the ecosystem. However, as humans settled, they began to distance themselves from this natural setting and embarked on modifying the environment to suit their needs. Over time, human activities evolved toward dominating and significantly altering nature, particularly with the advent of the Industrial Revolution. As evident from stratigraphic markers in sediments, humans have left a distinct mark that differentiates this new era from the Holocene (Waters et al., 2016). This epoch's inception can be traced back to the emergence of agriculture and deforestation. Other pivotal milestones in human history include the exchange of species following the conquest and colonization of America, the onset of the Industrial Revolution, and, presently, the exponential growth in the human population.

Each of these milestones represents a significant departure from stability, often transforming exceptions into the norm. Globally, this process can be understood as the breaking of the rule that humans are just one of many regular organisms. Humans have become the sole species capable of generating changes of such magnitude on the planet, fundamentally altering ecosystems and the global environment.

7.3.2.1 Toward a More Artificial Human Species?

It has become increasingly uncommon to find individuals who are completely "natural"—that is, without any prostheses, surgeries, repairs, grafts, implants, or foreign bodies within their bodies. This characteristic extends even to various animal groups, including those of economic importance or those under conservation efforts, which might lead to a situation described as "synthetic biodiversity conservation" (Piaggio et al., 2017). In this scenario, future instances of untreated or nongenetically manipulated specimens may become rare.

The advent of technology has played a significant role in advancing practices and manipulations that were once only seen in science fiction novels. Especially in the field of medicine, these technological advancements have allowed humans to live longer, which, in some ways, puts additional pressure on nature. This is because a larger population of longer-lived individuals may necessitate increased resource consumption and pose new challenges for the environment.

Indeed, the trajectory of medical advancements and genetic technologies suggests that in the not-too-distant future, virtually every human being may experience some form of disease or covert ailment that has been addressed from birth or at an early age. This means that not only individuals with well-known conditions such as diabetes, allergies, or immunodeficiencies will be affected, but there will likely be no lineages of individuals who can be considered 100% healthy in a strictly pure sense. The growing utilization of organ donation and "mixed" tissues (e.g., pig + human) due to the scarcity of donors may lead to a rarity of individuals who have never received a "mixed" organ in certain regions (e.g., large cities in developed countries) during their middle to third age. Consequently, in the future, finding a naturally born individual, including newborns, might become uncommon.

However, in preventive medicine, we can say that it is rare to find humans who have never been vaccinated with inactivated viruses or bacteria or their inactivated antigens for certain mass diseases, such as poliomyelitis, tuberculosis, and smallpox. Until a couple of decades ago, this practice was nonexistent and later became rare (e.g., during the early days of health plan application). Over time, it has progressively become the norm, even explicitly mandatory in the childhood immunization schedule. If it continues to remain rare, it may be limited to countries and regions with a low socioeconomic level. This persistence suggests that a problem that should have been eradicated is still being maintained due to socioeconomic issues. However, in developed countries, anti-vaccine movements profess the often conspiratorial ideology that vaccines are harmful. These movements often rely on erroneous premises and sometimes lack a thorough understanding of how vaccines actually work. However, there is also some secrecy and misinformation from different actors directly linked to health, which weave a curtain that facilitates these conspiratorial beliefs.

These ideas can generate the reappearance of many diseases known to have been eradicated or targeted for eradication thanks to vaccines (see Hinman, 1999; Kayser & Ramzan, 2021). This movement can favor diseases thought to have been diminished or eradicated or even the infamous COVID-19 to reappear or perpetuate, becoming a rule again. Currently, there is a unique phenomenon where the rule is to read the news that appears on social networks or comments from influencers or the press and leave aside data and scientific facts consensual by most scientists. This change from rule to exception is hazardous to cultivate. This fact is superbly captured by the movie "Don't Look Up" from 2021. Returning to diseases, this disbelief is a phenomenon that indirectly generates a change in disease dynamics. The incidence of diseases can be higher, as we said, because of beliefs that go against scientific facts that have been studied and proven but that, as with almost everything in science, we cannot assure 100%. Indeed, scientific studies should also be rigorous and have no background interest, which honestly often are utopies and favor even more conspiracy scenarios. These practices leave the substrate ready for disease variants to occur. Any of the mutated strains of the virus can transform from local exceptions to potential pandemics with the capacity to reinfect the population worldwide and generate isolation, psychosis, and general discomfort. We have the tools to avoid this; vaccination is the only and most powerful. Suppose we add antibiotic resistance and the appearance of mutant agents of these diseases. In that case, the conditions could turn from exceptions to potential rules, and the situation would become even more complicated.

The phenomenon of increasing resistance to antibiotics by bacteria is of great concern. This characteristic, among other things, is generated by self-medication and the abuse of antibiotic substances, even for conditions that do not require their use, such as a viral infection. The rule is that antibiotics are potent agents for treating bacterial infections. However, developing effective antibiotics is becoming increasingly challenging, and their action may become exceptional in the coming years. Imagine a person falling off a bicycle, scraping his knee, becoming infected, and after a while dying from a bacterial infection that would be easily treated today. This situation shows a complicated future we would want to avoid reaching. Fortunately, RNA therapies are being developed that will probably replace antibiotic-based therapies (Mourenza et al., 2022; Vogel, 2020).

Additionally, currently, with the demands of everyday life and added to a large number of activities to maintain and strengthen the body, there is an increase in the consumption of vitamin supplements and steroids that improve people's performance. This practice was perhaps uncommon in the past, and all these nutrients were incorporated in smaller amounts through food (Kaebnick, 2013). Today, it is a gigantic industry that also helps to generate artificial humans that are starting to become the rule. Another stepping stone related to superpeople generation is the possibility of modifying genes through gene therapies to generate a desirable physical attribute.

In summary, the examples presented in this discussion highlight how human activity directly or indirectly leads to environmental changes. These changes, in turn, cause certain phenomena or organisms that were once common to alter their structure or abundance, becoming rare and exceptional. This dynamic of change and species replacement has been a natural occurrence on the planet for millions of years, but in this particular case, it is driven by human actions rather than natural processes, often occurring at an unprecedented speed.

7.3.3 Toward the Extinction of Organisms as a Massive Phenomenon ("Folding" Life Diversity)

We have already seen that extinction processes have occurred due to abiotic natural causes (e.g., various cataclysms, gradual modifications of the atmosphere, and terrain) and biotic causes (e.g., interspecific competition) throughout the evolutionary history of organisms. However, we have already addressed the terrible role that human activity has played, directly (hunting, fishing, or uncontrolled deforestation, for example) or indirectly (environmental pollution and climate change), on many of the world's ecosystems. Throughout the history of mankind and more strongly marked after the Industrial Revolution, we know how various human events have implied a radical alteration for local communities or even the destruction of the environment (Fig. 7.3a, b). Since ancient times, the environment has been modified for forestry, livestock, and agriculture. Indeed, humans have had a significant impact on the environment, landscapes, and diverse range of species inhabiting our planet. This influence can be observed across the entire animal kingdom. Among vertebrates, birds and mammals have been particularly affected by human activities, often experiencing near "annihilation" levels in numerous cases. For example, since human colonization, it is estimated that approximately 1800 species of birds have become extinct (Steadman, 2006). In addition, records on extinctions of large mammals, birds, and reptiles date back to the seventeenth century, with species such as the dodo (*Raphus cucullatus*), the Steller's sea cow (*Hydrodamalis gigas*) and the giant tortoise (*Cylindraspis peltastes*) (Ceballos et al., 2015). Reptiles, amphibians, and fish have been documented as extinct since the twentieth century (Baillie & Cokeliss, 2004). Even some species that are not currently threatened face the widespread disappearance of certain populations, which may be a prelude to extinction at the species level (Ceballos & Ehrlich, 2009).

From the perspective of the book, in the case of the extinction of a living species by direct (e.g., hunting) or indirect human action (e.g., pollution, introduction of exotics), we would be talking about a change from Rule to Rule, i.e., from "the species being present" to "the species no longer being present." This transition from Rule to Rule is the extreme case. However, an intermediate case would be species

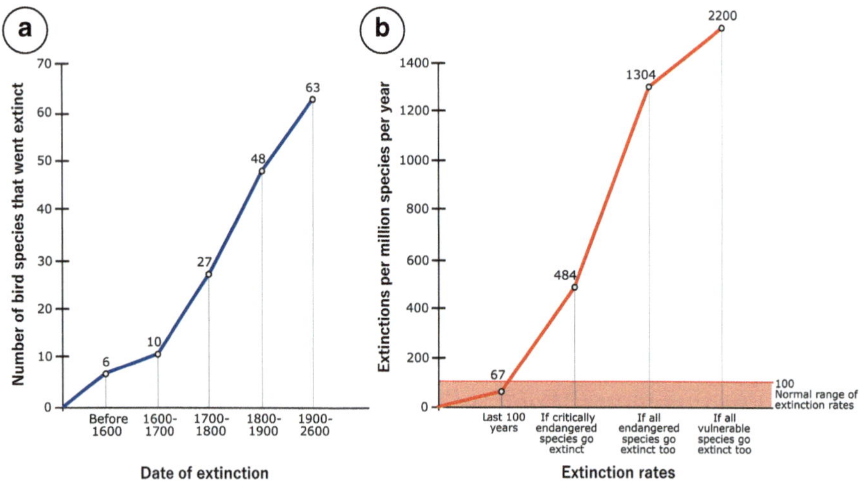

Fig. 7.3 Rate of extinction in bird species produced by man. (**a**) Number of species extinct over time. (**b**) Extinct species per year and extinction rates. (Based on https://evolution.berkeley.edu/mass-extinction/the-earths-sixth-mass-extinction/; from Pimm et al., 2006; Barnosky et al., 2011)

"in danger" of extinction, where their numbers are reduced to values of simple exceptions or rarities, either worldwide or in certain regions. In any case, it is because they are interrelated that they are presented here and not in the Rule-Rule chapter since it is a transition from Rule to Exception and then from Rule to Rule. We must not lose sight of the fact that everything is ultimately closely linked (e.g., if there was an intruder species that now occupies the niche of the native species, as we saw in the previous chapter, essentially, it went from Exception to Rule).

We will now review some of these witness cases, many of which are well-known. The objective is to highlight what we mean when we say that humankind is rapidly and systematically turning the drastic reduction of entire groups of living organisms into a rule, even up to the total disappearance not only of one species but of several.

7.3.3.1 Effect of Deforestation, Hunting, and Fishing

When the balance ruling in ancient human communities for these activities was abandoned, human occupation of natural environments and land-use inevitably changed and led to a reduction in the number of species (Ceballos & Ehrlich, 2002; Dirzo et al., 2014). Even pristine environments have not been immune to species loss, primarily driven by various human exploitation activities, be it for subsistence or commercial purposes (Peres & Palacios, 2007; Young et al., 2016). For instance, the Amazon rainforest and the tropical rainforests of Southeast Asia have experienced significant species loss due to activities such as deforestation and habitat destruction (Donald, 2004; Koh & Wilcove, 2007; Rudel et al., 2009). Similar scenarios can be observed in Madagascar, Africa, and other regions, where large areas are rapidly disappearing as they are converted for commercial agriculture (Vieilledent et al., 2018). Agriculture is a major cause of extinction and threatens wild biodiversity (Norris 2008; Butler et al., 2007). Millions of forests and natural habitats have been cleared for agricultural use, resulting in landscape fragmentation and soil and water degradation, leading to the extinction of many species (Benton & Twitchett, 2003; Foley et al., 2005; Tscharntke, 2002). Sometimes, farmers even eliminate wild species to reduce the effects of pests, predators, and weeds, which affects biodiversity, often leading to the eventual disappearance of plants and animals (McNeely & Scherr, 2003).

However, certain ecosystems have remained relatively unaltered, but they have not been immune to the impacts of activities that exploit natural resources and endanger numerous species. In some cases, exotic species are introduced for recreational activities such as hunting and fishing, leading to competition with native species for resources and potentially displacing them. Additionally, these practices can directly target native species, and if not properly controlled, they may contribute to the disappearance of those species.

An example of indiscriminate hunting of a species is that of the American bison (*Bison bison*), which, before the arrival of Europeans, inhabited North America. Bisons began to be hunted for their meat, then for their skin and finally for sport. The creation of reserves during the nineteenth century favored the survival of only two species currently living (Navarro, 2011). Another well-represented example is black rhinoceros (*Diceros bicornis*), whose constant slaughter by hunters led to the disappearance of at least two of its subspecies. Its horns are heavily traded in the Near Eastern and East Asian markets, either as ornamentals or for medicinal purposes (Martin, 1983; Penny, 1988). The problem becomes even worse, as the decreasing availability of these animals by hunting greatly increases the value of their horns. This supply and demand phenomenon creates truly catastrophic repercussions for the species because all ages and sizes of the animals are now susceptible (Ashley et al., 1990; Leader-Williams, 1988). Additionally, in the northern hemisphere, anthropogenic activity has caused the decline of up to 95% of the extension of gray wolf (*Canis lupus*) populations, which were found in almost all territories of the Iberian Peninsula (Blanco & Cortés, 2003). During the 1950s and 1960s, these populations declined due to human persecution (Valverde, 1971; Vila et al., 1993). In 1970, the species was protected by hunting law, which was key to its conservation (Arija, 2010). In the Southern Hemisphere, we find the same problem with many species threatened by poaching. One of these species is the jaguar (*Panthera onca*) in northern Argentina. This species is becoming very depleted, and a medium-term extinction is highly probable if the threat persists (approximately by the year 2050). Most of these animals are killed by hunters looking for trophies or hides (Paviolo, 2010). The same occurs with the South American tapir (*Tapirus terrestris*), which in Argentina is considered an endangered species (García Fernández et al., 1997) due to hunting and habitat destruction and currently survives only in some populations in the north of the country (García, 1972; Richard & Juliá, 2000).

In the aquatic environment, the same pattern can be observed due to the exploitation of fisheries, which affect the abundance of fish species due to the introduction of other fish or exotic crabs, habitat destruction, and pollution, among others (Doadrio, 2001; Doadrio & Elvira, 2002; Nicola et al., 2013). Fish are one of the most threatened vertebrate groups on the planet (Baillie & Cokeliss, 2004; Burkhead, 2012), and the extinction of many species occurs in a very short time span and in an accelerated manner (Rodríguez et al., 2020). One of the main causes of this decline is excessive and indiscriminate fishing and the lack of compliance with regulations prohibiting capture.

The common hammerhead shark (*Sphyrna lewini*) is one of the most exploited species in fisheries, sought after for its meat and even for its unique shape, as its head is used for ornamental purposes. To protect this endangered species, specific laws have been implemented in certain countries, such as Costa Rica and Ecuador. Additionally, the common hammerhead shark is listed on the red list of threatened species by the International Union for Conservation of Nature and Natural Resources (IUCN, 2023). Other groups of cartilaginous fish, such as rays and sawfishes, are in the same situation, and there is evidence that their extinction is a real possibility. The golden cownose ray (*Rhinoptera steindachneri*), for example, is a very susceptible species, as it produces one young per year and is highly affected by overexploitation (Bizzarro et al., 2007; Burgos-Vázquez et al., 2019). Similarly, the sawfish (*Pristis* and *Anoxypristis* spp.) is being driven to extinction by this cause. The sustainability of the populations of these fish is only possible if mortality and the catch rate per fishery are reduced due to their capacity for recovery, given their reproductive potential (Acevedo et al., 2007).

In the Southern Hemisphere, many fish species are also strongly affected by overfishing. One of these is the Argentine hake (*Merluccius hubbsi*), whose decline occurred in a very short period of time because the current regulations did not prohibit its capture (de Villalobos, 2002). Additionally, the catch of mollusks and crustaceans has also grown in volume in that area, much more than the catch of fish. The Argentine squid (*Illex argentinus*), for example, represents 95% of the catch. This rate of exploitation shows immediate consequences, and the estimates are for a continuous decline in fish and other organism biomass (Casal & Prenski, 2000).

All these examples show that the abundance of species is not only affected by nature. Humankind has a crucial role in these processes. Due to exploitation, species that were the rule in different areas of the planet have become the exception or even disappeared. This reduction is accomplished through activities such as deforestation, hunting, fishing, and the degradation of natural environments. The leading cause of the mismanagement of natural resources and the change in species from being patterns to exceptions due to overexploitation is the lack of management policies. The measures taken concerning these exploitations can help reduce the disappearance of species already in danger of extinction and thus recover some populations. Indeed, while these measures may not fully address irreversible extinctions, they can still be instrumental in mitigating the damage to species that are experiencing declining populations. Implementing such measures can help prevent these species from reaching the point of extinction and provide a chance for their populations to recover and stabilize.

7.3.3.2 Effect of Introduced Species

Another consequence of human interference in nature is the introduction of invasive alien species, which, as mentioned in the previous chapter, pose a threat to native species and the ecosystems they inhabit. These invasions often result in extinction or a significant reduction in the abundance of various local plant and animal species. The detrimental effects of invasive species can include hybridization, transmission of diseases or parasites, competition for resources, alteration of the behavior and reproduction of native species, and other impacts (Allison & Vitousek, 2004; Lambrinos, 2004). As these adverse effects spread, they can transform the structure and functioning of the environment.

Initially, exotic species in relatively low numbers may coexist with native species over an extended period. However, when their reproductive capabilities are exploited, they can become more aggressive and escalate their impact on native species, exacerbating the problem further. The introduction of certain species into new environments is often accidental. This can occur, for example, when species inadvertently hitchhike on ships or other types of transportation, unintentionally being transported to regions far from their native habitats. However, in other cases, exotic species are intentionally introduced, generally for economic purposes. For example, many species are introduced for agriculture or livestock or, as mentioned in the previous item, for recreation (e.g., hunting and fishing). Many plants are introduced for gardening, landscaping, or forestation. All these situations imply the possibility that these species become invasive and go from being the exception to being the rule, as already pointed out in Chap. 6. In this case, we focus inversely on the process that now leaves native species as exceptions in their environment due to the introduction of exotics.

Many times, the native flora and fauna are not prepared to adapt to changes in a short time. The alterations that have occurred in the climate due to global warming favor species that have been introduced voluntarily or involuntarily by human action and can expand into new areas. Due to this expansion, invasive species are responsible for approximately 15% of extinctions worldwide, making them a significant threat. In Europe, as elsewhere in the world, there are a large number of native species that are competing with European flora and fauna. Furthermore, in this competition for survival, native species are losing their place. One example is the European mink (*Mustela lutreola*) in Spain. These animals have always been heavily persecuted for their fur, and in the 1990s, their population declined significantly. Currently, the introduction of the American mink (*Neovison vison*) has led the European mink to be included in the list of endangered species in Europe, as it is suffering rapid decline and is prone to extinction. These animals were introduced for fur production, but accidental escapes or intentional releases of individuals caused them to affect the native fauna (Fasola et al., 2010; Jaksic et al., 2002; Lizarralde & Escobar, 2000). The introduced mink is a very opportunistic animal. In addition, coupled with the ability to compete for food and territory, the American mink is much more resistant to some diseases and has therefore been able to invade European mink habitats easily (Pavés., 2021). Mink constitutes a significant threat to native rodents and birds. This species of mink has also been introduced in the Southern Hemisphere for the same purpose. In Argentina, it has significantly reduced native avifauna, mainly the hooded grebe (*Podiceps gallardoi*) (Roesler et al., 2012).

Next, we will refer to other specific cases of native animals and plants that were displaced by introduced exotics and, in many cases, led to their extinction.

Animals

Cats and dogs have been domesticated for thousands of years, and they now play an important role in human society. In areas where predation is not a major threat and adequate prey is available, some abandoned cats and dogs manage to survive and form feral populations. This phenomenon is common on oceanic islands in Australia and New Zealand (Van't Woudt, 1990). However, outside of oceanic islands and desert regions, where their impact on native animals can be devastating (Medina et al., 2011), feral cats can play a useful role in controlling small vermin populations. Nevertheless, their presence can also harm wildlife, particularly lizards, small marsupials, and certain bird species (Kauhala et al., 2015). The need for control of roaming, stray, or feral cats in rural areas is not entirely clear-cut, as their potential benefits in managing vermin populations need to be weighed against the damage they may cause to native wildlife.

In contrast, dogs, whether roaming, stray, or feral, can be potentially destructive animals and demand strict control measures to mitigate their impact on both wildlife and human communities (Banks & Bryant, 2007; Zapata-Ríos & Branch, 2016). We have already discussed the introduction of rabbits in Australia and their effects on this new environment. Since the end of the nineteenth century, when rabbits began to cause damage to vegetation in Australia, it became clear that introduced species were potentially harmful (Enserink, 1999). Rabbits destroy natural vegetation, grasses, and small shrubs and are responsible for erosion problems as they expose topsoil. This damage is devastating, as this layer of soil takes hundreds of years to regenerate (Cooke, 2012).

One of the most critical environmental risks is introducing exotic aquatic species due to the modification they cause to aquatic habitats and biodiversity (Hopkins, 2001). The introduction of exotic species has been associated with the extinction of 54% of the world's native aquatic fauna (Harrison & Stiassny, 1999), 70% of North American fish (Lassuy, 1995), and 60% of Mexican fish (Contreras-Balderas, 1999). An example of this phenomenon is the common carp (*Cyprinus carpio*), which wreaks havoc on populations and natural environments. It was introduced from Asia, and its method of feeding removes sediment from the river bottom, reducing water quality, generating turbidity, and destroying aquatic plants. It also reduces the abundance of aquatic invertebrates, alters the spawning habitats of other species, competes ecologically with other species, and reduces native biodiversity. It has been responsible for the decline of native fish species, amphibians, reptiles, and birds in various aquatic ecosystems worldwide (Rauque Perez et al., 2017). The painted turtle (*Trachemys scripta*) is also one of the 100 most detrimental invasive alien species (Seburn et al., 2023). It was introduced into Europe from the United States and competes with Iberian turtles, as it is better adapted to polluted environments and human presence.

Many animals were also introduced for petting, as is the case of the monk parakeet (*Myiopsitta monachus*) in Spain. This species competes with and displaces native species, such as the spring blackbird (*Turdus migratorius*) or the magpie (*Pica pica*). In addition, it is a dominant and aggressive species in artificial feeders (Muñoz-Jiménez & Alcántara-Carbajal, 2017) and affects the flora through its nest-building activity (Gallego, 2019). Additionally, as a pet, the raccoon (*Procyon lotor*) has been introduced to Europe from North America, affecting many native species through competition and predation of fish, amphibians, mollusks, and birds, among others (García et al., 2012). Raccoons are voracious and opportunistic animals that can typically hunt and prey on bird nests. They compete intensely with aquatic medium carnivores, such as otters (*Lutra lutra*), European mink (*Mustela lutreola*), or polecats (*Mustela putorius*), for diet or territory overlap (Salgado, 2015).

All these cases demonstrate that species introduced for a particular purpose that favors human activities can end up harming native species. Thus, native species that were the rule in their distribution can become mere exceptions when displaced by exotic species and all as a consequence of human influence on nature.

Plants

The plants introduced for forestation, landscaping, and cultivation may displace other native plants until they are significantly reduced. Many species colonized new sites through agriculture. Some of these species, such as maize (*Zea mays*), wheat (*Triticum* spp.), soy (*Glycine max*), or rice (*Oryza sativa*), have been beneficial because they facilitate food supply. However, they also cause economic losses and ecological impacts (Pimentel et al., 2001). Indeed, when plants are introduced for agricultural production, the soil's physical, chemical, and biological characteristics are modified, which can take years to recover and stabilize (Okada & Harada, 2007; Villenave et al., 2001), but this modification helps the establishment of invasive species. Sorghum (*Sorghum halepense*), an African crop with vegetative reproduction facilitated by soil tillage, was introduced in Argentina. Spreading the rhizomes favors their establishment, proliferation, and dominance (Radosevich et al., 2007).

For landscaping and gardening purposes, the ash-leaved maple (*Acer negundo*), native to North America, has been introduced to many continents. This species is highly invasive and forms a dense layer of foliage that hinders or prevents the growth of other native species. The behavior of foliage allocation in response to the availability of light and nutrients allows this species to outcompete native tree species. This characteristic threatens biodiversity and ecosystem functioning (Porté et al., 2011). Another case is that of dandelion (*Taraxacum officinale*), a perennial herbaceous plant native to the temperate zones of the northern hemisphere, which is now widespread in the temperate zones of almost the entire world, except Africa (Holm et al., 1997). This plant species has apomictic reproduction and can grow in contrasting climates and disturbance regimes. In some parts of the world, these plants are considered weeds and can reduce the pollination and reproduction of native grasses (Kandori et al., 2009; Molina-Montenegro et al., 2010; Muñoz & Cavieres, 2008). Another example of an introduced species is *Banksia ericifolia*, which produces thousands of seeds at an earlier age in southern Africa, where it is exotic. At the same time, in its native environment, Australia, it only begins its reproductive cycle much later, at 25 years of age (Crawley, 1997). This characteristic causes its proliferation to be much more significant and can displace native African species, such as the proteaceous shrub *Leucadendron laureolum*. Both species are killed by fire and rely on stored seeds for recovery. *Banksia ericifolia* produces an average of 16,500 seeds per plant, approximately 30 times more than *L. laureolum*, giving the latter a considerable disadvantage in recovery compared to the invasive potential of the exotic *B. ericifolia* (Honig et al., 1992).

An invasive species spreading uncontrollably in Argentina is the broad-leaf privet (*Ligustrum lucidum*). Also introduced for landscaping, this tree, originally from China, is one of the 100 most invasive species in the world. It grows by taking water away from other native species, occupies their space, and generates biodiversity loss, creating a less conducive habitat for fauna (Ayup et al., 2014). These trees limit the diversity of native birds, such as the black-backed grosbeak (*Pheucticus aureoventris*), the dark-billed cuckoo (*Coccyzus melacoryphus*), and the scimitar-billed woodcreeper (*Drymornis bridgesii*). In addition, they favor the proliferation of exotic generalist birds such as pigeons (*Leptotila verreauxi*,

Patagioenas picazuro, *P. maculosa*) or parrots (*Myiopsitta monachus*) (Bellis et al., 2021). However, due to high growth and reproduction rates, this tree also represents an essential change agent in native Tala (*Celtis tala*) forests (Delucchi & Torres Robles, 2009; Fernandez et al., 2020; Morello et al., 2012). The introduction of all these plants by man, for different purposes, and without planning, causes uncontrolled growth to damage the native flora in the same way that happens with animals. Thus, native species that were common are diminished in their growth, reproduction, and distribution until remaining as exceptions in their original environment.

7.3.3.3 On Secondary Extinctions of Biodiversity

In the context of transitioning from rule to exception, it is intriguing to examine extinctions caused by human activity on a global scale. These extinctions demonstrate how "indirect" actions, such as hunting, can be as impactful as direct actions in causing the extinction of a species. We use the term "secondary extinctions" to describe situations where a species becomes extinct indirectly due to human actions that lead to the elimination of another species with which it had interactions (please refer to Fig. 1 of Brodie et al., 2014). This dynamic is highly intriguing, as it goes beyond solely considering the direct impact of human actions and introduces the additional factor of eliminating one or more interacting species.

Some relationships between species are highly dependent on each other, highly specialized, and have a higher probability of ending in a secondary extinction, as well as between species that evolve more slowly. These are more vulnerable interactions. An example is black-footed ferrets (*Mustela nigripes*), which came to the brink of extinction when the abundance of their only prey, the prairie dog (*Cynomys* spp.), declined alarmingly (Miller et al., 1994). This reduction occurred due to eradication programs by the U.S. government to favor the cattle industry. Actions are now being taken to recover these populations, which are essential to maintaining diversity (Miller et al., 1990). Here, we can see that the indirect action of man to favor the economy puts biodiversity at risk, not only by eliminating one species but also a second species with which it necessarily interacts.

Sometimes, species vulnerability also depends on their resistance to human impacts. If a species is a generalist but all interacting species are affected by the impact of anthropization, that species may also be at risk of extinction. Conversely, specialist species that depend on others very resistant to human-induced change may not be affected (Rezende et al., 2007). This is the case, for example, for the passenger pigeon chewing louse (*Columbicola extinctus*), for which its host species (*Patagioenas fasciata*) is not vulnerable to human impacts (Colwell et al., 2012). Therefore, this specialist species is not affected, thanks to the resistance offered by the primary interacting species, in this case the host.

In the framework of this book, secondary extinctions could be seen as the passage from rule to exception, not only of a species affected by human actions but also as a chain that begins with these actions, turning more than one species into an exception and affecting the natural regulation of populations.

However, the type of dependence between species will also be a determining factor in secondary extinctions. Most of the time, dependence is linked to food. However, there is also dependence on mutualistic species, for example, in the case of pollination. However, in some cases, such dependence does not usually determine the species' survival, so the reduction of the pollinator by man does not always have consequences on the affected species (Brodie et al., 2014). For example, as we already saw, sometimes the elimination of pollinators due to human action causes some pollinator-dependent populations to be reduced. However, it can also cause self-pollination capacity to develop within a few generations, such as in monkeyflower (*Erythranthe guttata*). In this species, populations without pollinators had significantly reduced fitness in the first few generations but recovered as they developed an improved ability to self-fertilize (Roels & Kelly, 2011). As we saw in the previous chapter, evolution toward self-pollination is a typical response to pollinator removal, as is the shift toward pollination by other species or by the wind in other cases (Costa & Machado, 2012; Kalisz et al., 2004; Yamasaki & Sakai, 2013). In any case, humans negatively affect species by interfering with the interaction among them, which could eventually lead them to face a high extinction risk.

However, external factors can also lead to secondary extinction. For example, this occurs when multiple anthropogenic stressors act on species interactions, exacerbating the risk of secondary extinction (Didham et al., 2007). If humans disturb the habitat, this can increase the impact of exotic species on native species, generating an unfavorable chain effect (Didham et al., 2007). However, related to the loss of pollinators, species that respond to climate change may also become incompatible with their pollinators (Memmott et al., 2007), and the loss of native pollinators could be severe for some species (Cox, 1983). If new associations between species are generated, in some cases, the probability of secondary extinctions may be reduced (Brodie et al., 2014).

All these possibilities highlight the significant impact of human actions on both primary extinctions, as discussed in the previous sections, and secondary extinctions of numerous species, as presented here. These actions not only lead to a loss of biodiversity but also trigger shifts in the abundance of species, both directly and indirectly affected. The increasing trend of common species becoming rare or disappearing due to human activities is altering abundance patterns. By addressing the loss of species interactions, we can potentially reduce the risk of secondary extinctions. Therefore, it becomes imperative to consider the number of human-induced secondary extinctions in conservation planning (Brodie et al., 2014).

7.3.4 Climate Change and its Effect on the Dynamics of R to E

We might ask, Climate change again? It is a recurring theme, but it is true that the change in climatic conditions linked to human activity clearly shows constant transitions from rules to exceptions and vice versa. Indeed, in the previous chapter, we discussed how the phenomenon went from being exceptional to escalating and becoming commonplace and dangerous. In the present chapter, we will analyze the phenomenon from another point of view. We will analyze in case of studies the effects of climate change on different systems and how this influence results in a transition from rule to exception.

7.3.4.1 Effect of Global Warming on Organisms

Changes in the Distribution, Hybridization Processes, and Interactions of Animal Species

Temperature strongly influences numerous characteristics of living organisms, ranging from the maturation of plants and insects to the dynamics of prey appearance in ecosystems. Even slight changes in the planet's temperature can significantly impact various biological relationships and timings. Considering the accelerated rates of current climate change compared to natural occurrences (Solomon et al., 2007), the effects become even more pronounced. For example, severe climate change can modify organisms' natural spring and winter (Peñuelas & Filella, 2009). As we will see below, this phenomenon can have adverse effects on organisms and populations. Climate change causes phenology variations, which are living organisms' reactions to seasonal and climatic changes in their environment. This phenomenon is widespread in many species distributed across different taxa (Visser & Both, 2005). For instance, a temperature increase of just 1 °C can advance the initial emergence and peak emergence of certain butterflies by 2–10 days. Although climate change has been shown to alter insect phenology, the relationships can sometimes be complex and even counterintuitive (Forrest, 2016).

Future climate scenarios show that we should expect species range to vary in size, shrink or increase, or cause species shifts in altitude, longitude, or both (see Gómez-Ruiz & Lacher Jr, 2019). One consequence that the planet has experienced in recent decades due to climate change is expanding species ranges into new areas (Crowley & North, 1988). When an invasive species expands its range and overlaps with the distribution of native species, it can lead to mating between different species and even hybridization (Coyne & Orr, 2004; Rhymer & Simberloff, 1996). The explanation for these phenomena is that when two sister species evolve in the exact location, they may partially or completely lose the ability to discriminate against the other species. Thus, interspecies mating and hybridization occur when sister species come into secondary contact (reviewed by Coyne & Orr, 2004; Sánchez-Guillén et al., 2016). One example is posited by Wellenreuther et al. (2010) with damselflies, *Calopteryx splendens*, and *C. virgo*. These authors documented a loss in the ability to discern the same species' mating partner from the sister species in allopatric populations in northern Europe. However, how frequently hybridization occurs is an underexplored topic (see Sánchez-Guillén et al., 2016).

In four major taxa, including fruit flies of the genus *Drosophila* (Coyne & Orr, 1989, 1997), salamanders (Tilley et al., 1990), frogs and toads (Sasa et al., 1998), butterflies (Presgraves, 2002), and gymnosperm plants (Lowry et al., 2008), the degree of reproductive isolation has been shown to correlate directly with Nei's genetic distance (Nei, 1972) between species pairs. Thus, those species pairs at different genetic distances show different levels of speciation. Based on this principle, we can predict which pairs of species may hybridize with invasive species when they come into contact within their range in response to climate change. In this case, climate change affects the distribution of species, modifying their natural or normal range, and this change allows species to come into contact. Encounters between closely related species do not typically occur in the absence of climate change. However, with climate change-induced range expansions, these encounters can lead to hybridizations and cause confusion between mating pairs, resulting in lost copulation opportunities (pseudocopulation). In addition, this encounter can even generate more significant investment in reproduction to compete with an individual of another species (e.g., Naretto et al., 2016; Vrech et al., 2018). All of these factors are detrimental to the maintenance of the species. These disruptions are rare and may become common in the future with potential adverse outcomes for populations.

Modifications in Predator–Prey Interaction: A Witness Case. Climate change-induced changes in phenology or in migration patterns can disrupt the interactions between predators and prey. There is a possibility that the predator and the prey (or its food source) may no longer synchronize their encounters due to these shifts. We will leave aside predator-plant interactions, which are also altered by changes in the distribution of one of the organisms or the phenology of plants. Instead, we will analyze a slightly more complex example of interactions. It is the one described by Lee Cooper (see Lovejoy & Hannah, 2019) in the Bering Sea. In this sea, the shallow shelf can be divided into two well-defined ecosystems. To the south, a pelagic system (in the water column) is dominated by fishes that partition the water column. To the northwest, waters with freezing temperatures do not allow the development of fish. However, they do allow the development of marine macroinvertebrates such as mollusks, annelids, or crustaceans that feed birds and large mammals such as walruses, seals, or some whales. In the northern system, the ice cover has been significantly reduced over the last 30 years. In this area, the increase in temperature caused a noticeable change to a more pelagic ecosystem (Grebmeier et al., 2006). Thus, mammals such as walruses, which depend on benthic organisms (seabed), are severely affected. Ice sheet retreat means that they have to swim a long distance, beyond their standard swimming capabilities, to find food. In turn, food is also diminishing, causing a double ecological problem for these organisms. In typical situations, such as those at the beginning of the twentieth century, the ice sheet was present from November to June. Above it, ice-dependent species such as large terrestrial mammals and birds dominated. Below this point, bottom-associated animal (benthic) production was more significant than water column production. In contrast, the change brought about by the increase in temperature causes less ice to exist. The ice retreats earlier and appears after the periods where it usually did. However, here, the dominant organisms on the ice sheet decreased significantly (e.g., walruses and polar bears), and there was a significant increase in seasonal migrants, such as cetaceans. Finally, there is an increased production of organisms in the water column. This example demonstrates how the change brought about by global warming is detrimental to the feeding of some species, producing situations that were previously exceptional or appeared in another area of the ecosystem and became the norm at the expense of the organisms that make up the ecosystem.

Pollinators and Plants: Phenological and Distributional Changes That Result in Mutualistic Interaction Interference

As we saw in the previous chapter, the relationship between plants and pollinators generates globally essential ecosystem services. This balance between the two groups can be disturbed by global warming, which can modify the phenology of plants, pollinators, or both, generating interaction mismatches (Bartomeus et al., 2013). Spatial mismatches can cause associated problems due to changes in the distribution of involved species or by changes in phenology, as we have seen in the previous paragraph. Displacements can also occur even when the species are in contact. In this case, morphological changes may appear or changes in interaction behaviors. These changes can generate new scaffolds in the community (Gérard et al., 2020). Perhaps the species most likely to be affected are plants associated with specialist pollinators and those that flower in early spring (Gérard et al., 2020; Willmer, 2014).

An Example of Phenological Changes in Pollinator-Plant Interactions. Temperature can differentially affect the flight period of pollinators and plant flowering times (Willmer, 2014). Let us take as an example the interaction between orchids and bees, something we already saw in the previous chapter. Combined studies between herbaria, museums, and field data indicate that bee flight initiation dates are moving ahead of orchid flowering time. This behavior may lead to a future mismatch in the interaction (Willmer, 2014). Specifically, in the case of the orchid *Ophrys sphegodes*, flowering is delayed relative to the emergence of the bee species *Andrena nigroaenea*, and the misalignment is predicted to be total if current high rates of warming continue. This relationship is under a deceive system (see Chap. 6) where flowers are not typical, and male pollinators appear early in the season, before females. These males are easily cheated because they have no females to interact with and take for reference. Instead, they perform false copulations with flowers that mimic the females of the species, as detailed in Chap. 6.

With the warming effect, an increase of only 2 °C makes males appear much earlier. However, this is not the only thing; female bees also appear earlier in the population, favored by the increase in temperature. For example, in a warm spring, we will find male and female bees flying in the population, but the flowers would not be open yet. The orchid's deception system is conditioned in this situation by having male and female pollinators interact. The chances of deceiving male pollinators are severely diminished as temperature increases (Willmer, 2014). Here, again, the shift from rule to exception due to global warming is clear and is reflected in that the orchid's deception system would not function properly due to the subtle change in the phenology of the pollinator and its interaction with the female of its species.

An Example of Range Shift and Change in Pollinator-Plant Relationship. Agaves (*Agave* spp.) are plants that develop flowers pollinated by bats (Fenster et al., 2004). The most important are the long-nosed bats of the genus *Leptonycteris* spp. (Gómez-Ruiz & Lacher Jr, 2019). These bats migrate, and their migration is considered endangered due to their complexity (Arita & Santos-del-Prado, 1999). The loss of this interaction would have catastrophic results ecologically but also economically, socially, and culturally, since, among other things, agave is the source of tequila and mezcal (Colunga-García et al., 2007; Gentry, 1982). The study by Gómez-Ruiz & Lacher Jr. (2019) shows that the overlap of agave and pollinating bat areas will be reduced by approximately 75%. The lack of pollinator foraging zones generates a threat to its survival and, in return, a drop in agave pollination service. Thus, the extinction of the pollinator would generate problems at the genetic and reproductive levels that would make it more vulnerable to future environmental changes.

Something similar has been observed in invertebrates. Evidence shows that some invertebrate species cannot cope with phenological changes in the plant species they pollinate (Hegland et al., 2009; Root et al., 2003). Hence, if pollinators rely on plants for their development and the occurrence of plants is delayed due to temperature increases, it can pose significant problems for pollinators (Durant et al., 2007). This effect is accentuated in specialized mutualistic relationships (Schweiger et al., 2012). Nevertheless, it is essential to note that not all outcomes are negative. As ongoing studies reveal, in certain cases, phenological changes in plants may not have a significant impact on the insects that rely on them. For example, high levels of biodiversity can balance the adverse effects of dyssynchrony at the community level (Bartomeus et al., 2013). However, at the same time, we must remember that overall biodiversity levels in many parts of the planet are being seriously decimated by climate change and environmental pollution. This is a controversial issue, but with a solid scientific base. In this context, a recent study shows how some 700 species are affected by climate change (Wessel, 2017). Therefore, we should not take for granted that if plants lose their pollinators, there will be another pollinator to fulfill the task (Burkle & Alarcón, 2011; Hegland et al., 2009).

7.3.4.2 Modification or Disappearance of Water Sources and Watercourses

Lakes, Rivers, and Permanent or Seasonal Ice

It is not only living organisms that suffer from temperature increases. Geographical agents, such as watercourses and hydrology, may also be affected (Solomon et al., 2007). Climate change will impact river courses and directly affect associated ecosystems (Thompson et al., 2021). Many changes in hydrological systems are linked to the escalating global temperatures caused by human activities. One noticeable effect is the alteration in precipitation patterns and their seasonal distribution. While average precipitation is increasing due to climate change, a substantial portion of it falls over the sea. Consequently, numerous continental regions are projected to experience a reduction in precipitation (Arnell, 1999).

There are also changes in the balance between snow and rain. These changes cause ice to retreat because of an imbalance between snow flow and melting. This imbalance contributes to the retreat of diverse ice masses such as glaciers, snow, or permanent ice (large ice masses). In fact, mountainous permafrost, glaciers, and large continental masses are constantly fighting against their disappearance due to the increase in temperature and the change in the type of precipitation. Water in the area can change from snow to rain or a mixed situation, resulting in the loss or retreat of the ice sheet. This mismatch in precipitation and the rate of ice generation and retreat has a profound impact on the flow of watercourses, including rivers, lakes, and even the ocean itself.

In addition, the increase in evapotranspiration and loss of soil moisture generates desertification, large-scale droughts, and vegetation cover changes linked to modifications in temperature and precipitation. In addition, this moisture loss increases the risk of flooding by losing a natural shield against the advance of water and increases the risk of fires as droughts become more extreme. Temperatures rise in areas with high plant fuel, and the drop in precipitation means that water does not help to mitigate fires (Goudie, 2006). Regional precipitation and temperature patterns are projected to change, affecting natural stream flows (Palmer et al., 2009). In fact, due to projected changes in CO_2 concentration and their resulting change in temperature and precipitation, river discharges are believed to continue to change in many regions much faster than they have historically (Alcamo et al., 2003; Lettenmaier et al., 1994; Vörösmarty et al., 2000).

Thus, climate change is a process that generates many geological changes from rule to exception. This dynamic may be caused by disappearing systems such as the retreat and melting of ice sheets or the decline of water resources such as rivers or lakes. Additionally, by generating a change in the functioning of some systems, such as a change in the evapotranspiration system that triggers problems previously less frequent or less "apocalyptic" (e.g., floods, fires, and other natural disasters). However, we should also know that climate change is not the only risk for many watercourses, as they are also affected by pollution generated by human activities, as we saw in the previous chapter (see also Palmer et al., 2009).

As an illustrative example, let us consider a study conducted on the rivers of the Himalayas, one of the most significant mountain ranges in the world. The research focused on the dynamics of four rivers in the northwestern Himalayas over various periods of the twentieth century (Bhutiyani et al., 2008). In the last decades of the century, a decline in discharge was observed in the Beas and Ravi rivers, with a less pronounced trend in the latter. Both watercourses share similarities in having limited glaciers within their basins and receive minimal contributions from snowmelt. Instead, they primarily depend on monsoon winds for their water supply. However, the Satluj and Chenab rivers are primarily sustained by glaciers, which have acted as buffers, regulating discharge. However, the accelerated retreat of the glacier associated with the Satluj River accounts for the decline in discharge observed during the period from 1991 to 2004. This illustrates the substantial impact on the watersheds, leading to a transition from rule to exception. In this example, long-established watercourses may face the risk of disappearing or undergoing significant modifications due to the influence of climate change on underlying processes.

Oceans and Their Communities

The oceans, along with lentic and lotic watercourses, are experiencing significant modifications due to climate change. The melting of polar ice caps results in the introduction of large amounts of freshwater into marine systems, altering the water's salinity and causing sea levels to rise. This has well-established direct and indirect impacts on the ecosystem. Furthermore, the rise in temperature and changes in pH significantly affect marine ecosystems, leading to alterations and even the disappearance of many organisms. In the following sections, we will examine two cases related to two crucial components of the oceans: plankton and coral reefs.

Plankton: Change in Ecological Function. Plankton is an essential requirement for maintaining the ocean's biological network, being the axis of primary production and the basis of many aquatic trophic chains, transferring biomass to other trophic levels (Murphy et al., 2020). They feed many other organisms and are the basis of critical biogeochemical processes such as carbon cycling in the ecosystem (Doney, 2006). The rule up to this point is that plankton biomass varies within acceptable ranges. These variations are caused by interannual changes in climate and the upwelling of nutrient-filled cold ocean waters (Dugdale, 1985; Kämpf & Chapman, 2016). Human-induced climate change generates numerous impacts on the plankton community by decreasing productivity (Boyce et al., 2010; Murphy et al., 2020). Climate change exerts more direct pressure because ocean temperatures are increasing at a rate never seen before, generating more extended warm periods. This disturbance directly impacts the production and recovery capacity of plankton-forming organisms. If this is exacerbated and sustained over time, the decline in plankton biomass could jeopardize the rest of the trophic levels that depend on them.

The change from rule to exception in this example is that plankton may no longer be the base of the food chain, nor can they cycle carbon and other nutrients efficiently as they have been doing until now. At the same time, and more locally, we will see a change in the diversity of organisms that make up the plankton (Ibarbalz et al., 2019), generating the disappearance of some organisms that are now classified as common, as well as changes in their phenology.

A study by Guinder et al. (2010) shows how the base of the food chain can be affected in the phytoplankton community of estuaries of Bahía Blanca in Argentina. In this study, spanning many years, the researchers observed that the rule was a single explosive appearance (blooming) of organisms between winter and early spring. This blooming was the main pattern from 1978 to 2002. From 2006 onward, some changes were detected. For example, there was an absence of some organisms, changes in their phenology, replacement of dominant species in the blooms, and new species that were previously absent. Nonseasonal blooms were also observed, with short-lived diatoms throughout the year. In addition, large species were replaced by comparatively smaller species. All these changes are due to the region's progressively warmer winters and drier climate.

In this example, the change from rule to the exception is evident, with changes in phenological patterns, abundance, presence, and replacement of organisms in a population, a change directly attributable to the increase in temperature due to human activity.

Coral Reef: Bleaching, Pale Corals. Corals are beautiful organisms. They represent the most beautiful postcards of any trip to the beach. The so-called stony corals (Anthozoa, Scleractinia) are generally colonial organisms, relatives of jellyfish, which form calcium carbonate external skeletons and are capable of generating coral reefs. The most extensive reefs are found in the Great Barrier Reef of Australia, followed by the Belize coral reefs in Central America, in areas close to the equator. A small polyp with tentacles forms the living part of the coral, similar to a small anemone. Inside it forms a symbiotic association with microorganisms called zooxanthellae (Dinoflagellata) that provide food. Coral reefs are currently experiencing a high burden of human-generated stress (Carpenter et al., 2008; Hoegh-Guldberg, 2011; Hoegh-Guldberg et al., 2007). Corals are affected by several factors, some of which are more local, such as eutrophication, offshore development, resource exploitation, or overfishing (Hughes, 1994; Hughes & Connell, 1999). However, one of the significant global culprits responsible for the decline in coral health is human-generated climate change (Eakin et al., 2009; Hoegh-Guldberg, 1999).

There is a phenomenon of great concern to researchers called coral bleaching (Fig. 7.4a-d). This phenomenon causes corals to expel the symbiont zooxanthellae and lose their coloration, becoming pale and eventually dying. It is a reaction to environmental stress. Susceptibility to bleaching varies according to species, depths,

Fig. 7.4 Real-life examples of coral bleaching. (**a**) Stag coral bleaching from the center to the sides (Credits: Gavin Bain). (**b**) Bleached branching coral (foreground) and normal branching coral (background) (Credits: Acropora). (**c**) Bleached brain coral (Credits: NOAA National Ocean Service, 2005). (**d**) A partially bleached coral colony (Credits: National Marine Sanctuaries) https://commons.wikimedia.org/wiki/File:NMSAS_coral_bleaching_(27416199680).jpg

and location (Grottoli et al., 2014) and is influenced by the endosymbiont (Howells et al., 2016; Warner et al., 1999). Corals can recover from a bleaching event, but recurrent annual bleaching and recovery events will generate intense selection and a decrease in coral diversity in the future, as well as an overall decline in reefs in the Caribbean (Grottoli et al., 2014). If bleaching becomes a recurring annual event, more than 90% of coral reefs will be at risk of degradation (Grottoli et al., 2014). Supporting our discussion of the transition from rule to exception, some models project that reefs, as we know them today, will exist as relics in isolated locations (Hoegh-Guldberg et al., 2007). In retrospect, it would be regrettable to lose coral reefs due to our actions and management of ecosystems, becoming an exception, something very rare.

7.4 R-E Changes Due to Modifications in Scientific Interpretations

As seen in science history, there has been a constant evolution in the view of the natural phenomena investigated. These changes may be due to several reasons, such as insufficient knowledge, over-reliance on a single animal model and the assumption of uniform behavior across species, or the use of inadequate techniques, equipment, technology, and the influence of paradigm biases. Inadequate technology, limited information, and geographic, taxonomic, or theoretical bias are variables that can act alone or in combination to erroneously transform an observation into a generalized rule or norm when it should be more restricted than initially assumed. These three scenarios are not as sharp or delimited from each other. There has been a combination of at least two for changing a rule into something admitted only as an exception. For example, a bias in the school of thought and using a model organism, from which data were then extrapolated to many other organisms. Therefore, we can mention some scenarios or areas, as we did in the previous chapter, fully applicable to the present chapter. We will begin with the one referring to theoretical or methodological biases in research.

7.4.1 Key Factors Linked to Change

7.4.1.1 Biases in Study Approaches

Strong theoretical biases during experimental design and interpretation of results have significant consequences on the robustness and future liability to the permanence of a rule for any aspect of biology. In other words, rules may include biases imposed by a prevailing paradigm or idea. Interpreting data beyond the confines of our theoretical framework, in which we are directly or indirectly immersed, can be challenging. Scientific "inertia" may set in if we solely pursue established ideas without engaging in critical thinking. As we discussed in Chap. 1 on rules, distancing ourselves partially or entirely from traditional explanations has always demanded courage and creativity. Within this context, let us consider two situations as examples. The first involves focusing on specific organisms to make progress in experimental topics while neglecting diversity. The second situation pertains to evolutionary biology, particularly sexual selection, where significant changes have occurred in recent years, especially regarding the use of model species and extrapolations.

In the subsequent chapters, we will delve deeper into these topics to explore the challenges and opportunities they present in scientific research.

Use of "Model Species" and Generalized Extrapolation

First, let us clarify what we mean by model species. Model species are those that are studied and analyzed in depth to know and understand different biological processes and apply them to other species (Ankeny & Leonelli, 2011). The projection of knowledge obtained from a particular model species can go to just one species of interest, a family, or even all organisms (Ankeny & Leonelli, 2011). Most, if not all fields of biology use model species to test hypotheses since it is unlikely that we can know everything about every nonmodel species (Kunkel, 2006; Zuk et al., 2014). Some of the most notable model species are, for example, the yeast *Saccharomyces cerevisiae*, the bacterium *Escherichia coli,* and the nematode worm *Caenorhabditis elegans*. Among insects, the yellow mealworm beetle *Tenebrio molitor*, the common honeybee *Apis mellifera*, and the fruit fly *Drosophila melanogaster* give some examples in this order. Additionally, in vertebrates, rodents such as the mouse *Mus musculus*, the domestic rooster, *Gallus gallus*, and the zebrafish *Danio rerio*. Many of these models have significantly increased our scientific knowledge in various fields, such as systematics, evolution, physiology, genetics, and behavior (Ankeny & Leonelli, 2011; Zuk et al., 2014), in addition to the use of some in biomedical research. Model species present characteristics that make them advantageous for their choice and use. For example, they are easy to collect, maintain, and manipulate in the laboratory and are generally small in size (both physical and genomic), with relatively short generation times and high fertility and mating rates, among other traits (Alber, 2009; Ankeny & Leonelli, 2011; Zuk et al., 2014).

Indeed, much of our current understanding of cellular processes, such as growth, division, and gene expression, has been made possible thanks to model species (Fields & Johnston, 2005). The use of model species is beneficial when the research goes beyond the characteristics that involve the natural history of a particular species, as in the case of the use of *Drosophila melanogaster* to identify and understand in detail mechanisms of inheritance and development (Fields & Johnston, 2005). Similarly, the use of rats (*Rattus norvegicus*) and common pigeons (*Columba livia*) was vital to develop theories of how the mind works, later extrapolated to other animals (Zuk et al., 2014). Model species play a crucial role in advancing knowledge across various research areas. However, a challenge arises when extrapolating data from these species to other organisms without considering their specific biological characteristics and phylogenetic distances. This requires careful consideration to ensure reliability and applicability to different species.

A Witness Field: Sexual Selection

The use of model species in the field of sexual selection is a clear example where the results of studies oriented to test hypotheses only in them were later generalized to the rest of the species, thus incurring significant errors in many cases (Zuk et al., 2014). Let us examine specific examples. Bateman's principle, which suggests that males benefit from a higher number of matings, while females gain significant reproductive benefits from only a few matings, was initially proposed based on data from *D. melanogaster* (Bateman, 1948; Zuk et al., 2014). However, if we set aside the model and consider particular characteristics of the species we are analyzing, such as mating system, mate guarding, and offspring care, perhaps this principle is not the one that best applies (Eberhard, 1996; Edvardsson & Canal, 2006; Tang-Martínez, 2016).

The fly *D. melanogaster* has also been a model species for studies on sexual conflict (Chapman et al., 1995; Holland & Rice, 1999). In this species, males transfer proteins in the ejaculate that reduce longevity and receptivity to future mating in females. This behavior is beneficial to the male as, for example, they decrease the likelihood of sperm competition, but not beneficial to the female because it reduces her level of polyandry (Chapman et al., 1995; Harshman & Prout, 1994; Manier et al., 2010; Wigby & Chapman, 2005; Zuk et al., 2014). This finding led to studies in numerous species that focused on sexual conflict, assuming in advance that matings are detrimental to females (Zuk et al., 2014). However, what if, in most cases, the male does not transfer some toxic substance along with the ejaculate in mating? What about thinking that females may benefit from the substances transferred along with the male ejaculate and by receiving stimulation or nurturing behaviors? What if we think that not everything is negative for females in sexual interactions? (Eberhard, 1985, 1996). In this sense, there are studies on the benefits that females gain with the transferred fluids and behaviors of males during mating (Simmons, 2001; Zuk et al., 2014). For example, in some insect species, seminal fluid transferred by males has been found to increase not only fecundity but also female survival (Edvardsson & Canal, 2006; García-Gonzalez & Simmons, 2010).

Another example of a model species in the field of sexual selection is the yellow dung fly *Scathophaga stercoraria*, where male behaviors in sexual interactions have been deeply analyzed and gave rise to the idea of sperm competition (Parker, 1970a, 1970b, 1970c, 1978). Here, particular interest was placed on the avoidance of sperm competition in a framework of postcopulatory sexual selection, with associated behaviors, such as prolonged copulation and postcopulatory mate care, among other behaviors associated with males. Clearly, the focus was on the males, their behavior, and their interests. This was the norm at the time, relegating the role of the female in mating (Ah-King, 2023). In this sense, the theory of sperm competition was considered a generalized process that occurred in different taxa and through which explanations were sought regarding reproductive behavior, physiology, and morphology (Smith, 1984; Zuk et al., 2014).

After several years, research began to take females into account. Indeed, with the development of studies on cryptic female choice, in the framework of postcopulatory sexual selection, females were no longer considered odd in terms of the active role they play in intersexual interactions (Eberhard, 1985, 1996, 2015). The idea of the female possibility of choosing which male to copulate with and the sperm used to fertilize eggs went from first being accepted as a rare phenomenon to becoming a rule in many taxa (Eberhard, 1996, 2015).

In recent years, sexual selection studies have witnessed a rise in the diversity of taxa employed for research. However, there remains a noticeable bias toward using model species of insects and birds and, to a lesser extent, arachnids, amphibians, and other groups (Zuk et al., 2014). It is worth acknowledging that model species have undeniably contributed significantly to advancing scientific knowledge across various disciplines. However, the big question is to what extent we can generalize knowledge between different taxa without making mistakes. Using only a limited number of species to explain a wide range of phenomena can lead to a problem: It prevents us from truly understanding the actual diversity present in nature. This limitation arises because there may be a tendency to overlook the importance of replicating studies in different groups. Researchers might assume that they already possess "generalizable" results derived from a few model species, leading many to view studying other groups as unnecessary or even a "waste of time." Consequently, such attitudes can hinder originality in research and potentially impede a comprehensive understanding of the natural world.

However, as we said above, the use of model species can lead to what is known as "confirmation bias," i.e., expecting to find what we want to find and not being open to the fact that our findings may differ from the prevailing patterns. This practice goes hand in hand with what we said about forcing generalizations from a few groups and thus potentially running the risk of continuing with scientific "inertia" when providing explanations for our results without exercising open and critical thinking.

In conclusion, it is vital to increase studies on other species beyond those taken as models (useful only in the first stage), for which knowledge of diversity is an essential point, an aspect that we will analyze below from a historical and evolutionary perspective.

7.4.1.2 Knowledge of Biodiversity and Evolutionary Framework

Biodiversity is understood as the different forms of life present on Earth at all levels, from genes to ecosystems, including ecological and evolutionary processes that take place within them (Risser, 1995). The scientific community now widely accepts the biological evolution of species over time. However, this was not always the case. For example, in the seventeenth and eighteenth centuries, the Creationist view of nature was deeply rooted. It is not surprising, then, that biological explanations of that time were mainly restricted to descriptions of plant and animal species, magnificently detailed in some cases, but without explaining how they came to be that way (in form and function). In this sense, Karl von Linné, in the middle of the eighteenth century, wrote numerous books with exhaustive descriptions of species based on their morphology and reproduction methods, which, according to the prevailing dogma of the time, were species created by God. However, von Linné's most significant contribution was his classification system for living beings, which he presented in his book published in 1735, Systema Naturae. He used a binomial nomenclature to name each organism, assigned to a genus and a species. This classification system is still used today.

At the end of the eighteenth century, some scientists, perhaps the most challenging to the established dogmas, began to debate the idea that the forms of life as we see them are not fixed but are changing over time. The current of thought known as Lamarckism, a theory of evolution proposed by the French naturalist Jean-Baptiste Lamarck, stood out at that time. In 1809, he published his book "*Zoological Philosophy*," in which he proposed that the different forms of life did not remain unchanged over time but evolved from simpler forms of life. The large number of species and their adaptations to the environment led Lamarck to think that species had evolved to adapt to changes in their home environments. The theory that Lamarck proposed was that individuals evolve by adapting to the changing conditions of habitats, where fossils were intermediate forms between present and past forms of life. Lamarck formulated two laws in his theory, which today is known as the theory of the "inheritance of acquired characters." In the first law, he states that it is the sustained and constant use of an organ or structure of an organism that maintains it. The disuse of such an organ or structure leads to its eventual disappearance. In the second law, he proposes that changes in the structures of organisms, acquired or lost, as a result of having been subjected to changes in their habitat for a prolonged period of time, are maintained in the following generations (Lamarck, 1809). In other words, what has been acquired can be inherited. Therefore, it would be the capacity of individuals to adapt to changing environments that would have produced evolution and the great diversity of species.

Let us think that Lamarck developed his theory of evolution in a context where what was accepted was religious dogma, with the spontaneous generation of species and where fossils, for example, were considered as different forms of life, without any connection with current species. Already here we could have an example where thanks to a greater knowledge of diversity, of its biology, of its nature, there is a progressive pressure to think differently from the predominant dogma; even more so in this case where the preconceived idea had no scientific basis. However, these new ideas were not immediately accepted by other scientists. Lamarck's postulates were not discussed in depth until approximately 50 years later, when Charles Darwin published his famous book "*The Origin of Species*" in 1859. In it, Darwin again takes up several of the concepts that had been treated by the French naturalist (e.g., use-disuse of traits, adaptation, inheritance) to explain the great variability on which natural selection would act. However, he does not consider the mechanism of inheritance of acquired characters proposed by Lamarck to be the driving force for the evolution of organisms. Darwin proposed, without the knowledge of genetics yet available, that during the process of reproduction of living beings, "errors" occurred, and this led to the offspring not being equal to their parents. This is what leads to the creation of new species, which, as the generations go by, develop different characteristics so that those that show more suitable characteristics are the ones that will be better able to adapt to the habitat where they live, the widely known, accepted, and spread the theory of natural selection. It is worth noting here how the voyages to numerous places where nature had been little or not explored until then, such as those made by Darwin on the Beagle, were key to a better understanding of the existing biodiversity. Indeed, they were key to the formulation of theories linked to ecology and evolution (Andersson & Iwasa, 1996; Kuijper et al., 2012).

During the twentieth century, in the 1930s, what is known as neo-Darwinism emerged. Dobzhansky, Mayr, and Simpson were among its initiators. This renewed vision of classical Darwinism proposes that the evolution of organisms is the product of a combination of processes such as those proposed by Darwin's theory of evolution through natural selection, Mendel's theory of genetic inheritance, and random DNA mutations as the basis of variation (Mayr & Provine, 1980). This combined theory is the one that has been the most widely accepted, in general, to explain biological evolution. However, it has been expanded and extended more recently due to advances in molecular biology, phenotypic plasticity, and epigenetic inheritance. Furthermore, it left some room for other alternatives, such as those proposed by the theory of symbiogenesis and a specific enduring framework of neolamarkism (see also Chap. 6). In all these frameworks, evolutionary theories changed as knowledge and interpretation of biodiversity increased from an evolutionary and phylogenetic point of view. It is a process of constant feedback.

We see that it is clear that to establish accurate patterns or rules, it is vital to have extensive knowledge of biodiversity. However, the knowledge of new information does not necessarily mean that the patterns or rules already stated are incorrect. Perhaps they should be revised and discussed according to specific organisms, regions, or both (Peretti, 2014). Moreover, knowledge of biodiversity does not end with its evolution. However, it is also helpful in many aspects of our lives, such as those associated with obtaining food, fuel, shelter, and medicines (Risser, 1995), aspects mentioned in various parts of this book.

In conclusion, regarding the passage of specific preestablished patterns or norms to become rarities or exceptions and the reverse in various fields, we could say that they are products of at least three aspects: (1) a strong theoretical bias prevailing at the time of interpretation of the data obtained, such as that which framed the predominant evolutionary concepts of each epoch; (2) rules or patterns established from partial data or information, not sufficiently representative, as, for example, what was already mentioned for model species (and see below geographical insufficiency and phylogenetic insufficiency); and (3) the availability of limited technological resources at the time when such pattern or rule was proposed (e.g., see item 7.4.1.3) (Peretti, 2014). For the rules established based on partial data, studies that include a great diversity of organisms, preferably from different regions of the planet, are essential. As mentioned above, the change from rule to exception due to insufficient data could be linked to two nonmutually exclusive processes, geographic inadequacies and phylogenetic inadequacies, concepts that we will review below.

Geographic (Intraspecific) Inadequacy

Inadequacy occurs when we extend the study area to unexplored regions for a particular species and find that a traditional rule is an exception in new places. For example, when we find that populations inhabiting distinct elevational or climatic zones exhibit eco-behavioral patterns that are very different from those previously described for "the species." In this sense, the case of the Neotropics is a clear example of how a region with different climatic conditions and vegetation types can differentially influence, for example, the sexual traits of species. If we only know a part of the distribution range, we may say that a given character has a particular morphology/behavior for a species, but this may not be the case (Macías-Ordóñez et al., 2013). In this sense, Blanckenhorn et al. (2006) and Macías-Ordóñez et al. (2013) make a fascinating analysis of how different ecological patterns could influence sexual traits. For example, patterns associated with Bergmann's rule, mentioned before, and Rensch's rule, which proposes that the size of males varies more than that of females, could influence traits associated with sexual dimorphism in both vertebrate and invertebrate species.

Not only abiotic factors to which individuals are exposed, but also a combination of biotic factors such as parasitism and predation can influence the differential development of morphological and behavioral characters of a species inhabiting different areas (Chown & Nicolson, 2004; Macías-Ordóñez et al., 2013; Schemske, 2009). An interesting concept to include is that of plasticity that individuals may have, which is linked to the extent of an ecological niche that different species populations may inhabit (Donohue et al., 2001). Effective plastic behavioral responses, for example, allow individuals of different populations to adapt to environmental variations (Donohue et al., 2001), which inevitably generates variations among populations of the same species. In this same framework, stating that certain behavioral or morphological traits are of a particular type without knowing the true geographic distribution of the species is tricky. This may lead us to mistakes similar to those we discussed regarding using model species to describe certain characters (see the development of the use of model species in item 7.4.1.1).

There are many examples of species with distribution ranges that cover areas with different altitudes and latitudes and that show interpopulation variation in certain characteristics. These cases remind us that if we do not know the species' real distribution, we can establish specific rules that are truly not generalizable. Among plants, some studies show an association between morphological variations according to the geographic regions where the species are located. In the case of *Quercus rugosa*, variations in leaf size have been reported. These show smaller sizes in the range of its distribution from south to north, a fact associated with precipitation levels (Rubio-Licona et al., 2011; Uribe Salas, 2009).

Another example is the significant variations among populations of *Arabidopsis thaliana* along climatic and elevation gradients in variables such as their ability to survive and recover from heat stress (Wolfe & Tonsor, 2014; Zhang et al., 2015). Thus, different populations of this species inhabiting areas with different temperatures and altitudes show differences in traits such as biomass allocation, fecundity, dormancy, and seed germination (Montesinos-Navarro et al., 2012; Zhang et al., 2015). This is one more example of how ignoring the distribution range of a particular species can lead us to say that some traits are the rule when, in fact, they are not.

Among animals, an example of a behavioral variation of different species populations with different distribution ranges is shown by the Mediterranean fruit fly, *Ceratitis capitata.* Here, males form aggregations and emit pheromones to attract females during the reproductive season (Diamantidis et al., 2008; Whittier et al., 1992). Although the male behavior of this species has been extensively analyzed in recent years (Eberhard, 1999; Yuval & Hendrichs, 1999; Diamantidis et al., 2008), comparative studies between populations with different geographic distributions remain limited (Diamantidis et al., 2008). In particular, this species has considerably increased its distribution range, encompassing a wide variety of environmental conditions (Diamantidis et al., 2008). Thus, a combination of different abiotic and biotic factors to which different populations are exposed is likely to cause different selective pressures. In turn, these selective pressures lead to a differentiation in the evolution of mating systems of individuals from different populations of the species with different distribution ranges (Briceño et al., 2007; Lux et al., 2003; Diamantidis et al., 2008). In this chapter, we discussed an example involving crickets, where natural circumstances led to a shift from a commonly observed rule to an exception. Specifically, the songs produced by male crickets to attract females for reproduction were found to have an unintended consequence—they also attracted predators and parasitoids (see 7.2.1.2). This illustrates how the complexities of nature can result in unexpected outcomes, altering the conventional understanding of certain behaviors or patterns in the animal kingdom. Thus, certain areas of a species' distribution can lead to differential selection pressure on this trait in populations with a higher risk of predation or parasitism (Zuk & Kolluru, 1998; Lehmann, 2003; Hedrick & Kortet, 2006). Thus, among populations of a species with different distribution ranges, there may be variations in male behavior, leading to male cricket singing in some populations and not in others, as well as variations in song duration to avoid predation (Hedrick, 2000; Hedrick & Kortet, 2006). For example, in populations of *Teleogryllus oceanicus* that inhabit ranges with intense predation pressure from bats, males emit their song from their roosts to avoid being attacked (Bailey & Haythornthwaite, 1998).

In vertebrates, we have the example of anurans, where phenotypic differences between populations of the same species inhabiting isolated geographic areas are common (Baraquet et al., 2012). For example, in the species *Hypsiboas cordobae*, it has been reported that certain morphometric variables are positively associated with the levels of altitude or latitude of the geographic area that the population has (Baraquet et al., 2012; Laugen et al., 2002).

A valid question arises: How many species and what distribution ranges are necessary to establish specific rules or patterns, ensuring that the discovery of new species with distinct characteristics does not invalidate them? This presents a biological paradox, as discussed in Chap. 1, regarding the challenge of finding rules applicable to a broad range of organisms at an intermediate or lower level. Exceptions will inevitably arise, except for laws or rules at a higher level (I or II). Nonetheless, caution should prevail when extrapolating concepts studied solely in specific organisms or environmental conditions to other groups or conditions.

Phylogenetic (Interspecific) Inadequacy

Phylogenetic inadequacy is the most common cause that determines the passage from rule to exception. This phenomenon occurs when, as the diversity of lineages included in our analysis increases, a rule initially applied to certain groups is now only an exception for the majority. In other words, in some examples, the discovery of new species and missing links, either living or fossil, changed the concept of their evolutionary relationships. In this sense, many preestablished patterns were proposed using model species from the Nearctic region, with distributions closer to the research centers. This highlights a significant risk of unintentional bias when applying a rule to organisms with different distributions (Peretti, 2014). For instance, numerous studies utilizing species from the Neotropics encounter the challenge of overcoming preconceived notions. This is largely influenced not only by the inherent partial knowledge of biodiversity, which is inevitable but also by insufficient taxonomic and phylogenetic studies that hinder a more comprehensive understanding of biodiversity in certain regions of the planet (Arnold, 2003; Dayton & Sala, 2001). In this sense, access to complete and updated biotic inventories that provide systematic, ecological, and biogeographic information on the species inhabiting a given area (Dennis & Ruggiero, 1996; Stork et al., 1996) is fundamental when trying to determine any kind of patterns or norms in the different species (Cotterill & Foissner, 2010).

In addition to establishing that natural selection is the mechanism by which species evolve, Darwin emphasized that species descend from common ancestors and that the figure of a tree could represent the phylogenetic relationships between them, the tree of life (Salvetti, 2019; Zardoya, 2010). Understanding the phylogenetic relationships between species is essential not only for their proper classification but also for establishing reliable and far-reaching rules or patterns. The naturalists of the time accepted the idea of the "tree of life" proposed by Darwin. Dedicated to observing, comparing, and classifying, according to morphological characteristics, they knew that each new species should be linked to other species (Salvetti, 2019). Currently, scientists no longer use morphological descriptions of species as a method of classification, but to reconstruct phylogenetic relationships, they rely on nucleotide or amino acid sequence data and statistical inference methods (Salvetti, 2019; Zardoya, 2010).

7.4.1.3 Linked to Current Equipment and Methodologies

The lack of availability of technological resources and adequate methodology has been and will continue to be a problem from which we will not be able to "escape" when identifying patterns to establish rules in biology. In other words, when they are proposed based on insufficient technology, it is to be expected that changes in conceptions and approaches will occur later. Over the years, these rules will cease to be valid and become, perhaps, just rarities or even disappear from knowledge. In the following, we will visualize this through some interesting examples to illustrate this point.

Monogamy in Sensu Stricto in Birds

A classic example of how a preestablished rule ceases to be accurate, mainly due to advances in available equipment and techniques, is the case of monogamy in birds (Bennett & Owens, 2002; Petrie & Kempenaers, 1998). Until a few decades ago, a monogamous mating system was the rule for 90% of birds (Lack, 1968). With the advancement of studies in the field of sexual selection, monogamy gradually became restricted to specific groups, such as gulls and other seabirds (Birkhead, 2000; Ismar et al., 2010). Progressively, with the development of novel molecular techniques, it was discovered that monogamy in birds was not the rule at all. Therefore, assigning 100% paternity to a single male was gradually taken as a mere rarity. Indeed, new studies based on genetic analyses of paternity clearly show that only 25% of bird species are monogamous, so this mating system has gone from being considered the norm to becoming the exception (Griffith et al., 2002). We can emphasize that this idea has changed with advances in genetic techniques, such as molecular fingerprinting. Today, monogamy in birds is considered an exception instead of a rule. In fact, when it occurs, it is more of a "social" type. That is, it does not prevent cases of infidelity and the litter of multiple parents.

However, the initial perception of monogamy as a widespread mating system in birds may have been indirectly influenced by human society's perspective on this matter. Essentially, the societal tendency to view monogamy as acceptable and promiscuity as punishable might have contributed to the unquestioned acceptance of a rule regarding avian monogamy. This bias by the prevailing human social paradigm in each historical epoch could undoubtedly be permeating other similar cases in biology. It is important to acknowledge that even the most serious and dedicated scientists may be influenced by prevailing human moral norms (e.g., concepts of "right and wrong") during foundational research on a particular subject. Therefore, we cannot dismiss the possibility of such influences shaping their perspectives.

Fortunately, technological, and methodological advances generally challenge dogmatic ideas and preconceived canons.

Thermoregulation in Ectothermic Animals

Thermoregulation refers to the capacity of organisms to actively control their temperature even when the temperature of the environment may be different. Homeothermic organisms, such as bird and mammal species, can control their body temperatures. However, ectothermic organisms are those that do not have such ability, and their temperature varies with the temperature of the environment, such as reptiles, amphibians, fish, and invertebrates (Cowles, 1962; Jarošík et al., 2004; Cowles & Bogert, 1944; Lara-Resendiz et al., 2013). This definition follows what was traditionally considered a rule. However, thanks to advances in temperature measurement methodologies, it has been possible to determine that animals traditionally considered ectotherms can maintain their temperature stability through not only behavioral but also physiological mechanisms (Hitchcock & McBrayer, 2006; Kearney & Predavec, 2000).

The behavioral mechanisms of what are considered ectothermic organisms, such as changes in body position, daily and seasonal activity patterns, and microclimate selection, arguably still "fit" somehow with the traditional definition. We can also name what happens in several species of insects and reptiles where their behavior is of great importance to regulate their temperature efficiently and obtain optimal metabolic performance (Casey, 1981; Huey, 1982; Kiefer et al., 2007). For example, butterflies regulate the opening angle between the wings and the body to regulate heat absorption (Kemp & Krockenberger, 2002; Barton et al., 2014), and among reptiles, the wood turtle *Glyptemys insculpta* moves to forest clearings to bask in the sun and raise its temperature (Dubois et al., 2009). Examples such as these show that some ectothermic animals would not be at the mercy of the climate but can actually modify their temperature to some extent.

However, in some cases, several species no longer fit the traditional definition of ectothermic organisms. Among the physiological mechanisms related to thermoregulation, the dark-light color change in reptile species regulates the absorption of solar radiation (Clusella-Trullas et al., 2009). For example, it has been seen in the desert iguana *Dipsosaurus dorsalis* that they can absorb more than 70% of the radiation when they have a darker coloration. In comparison, when individuals have a lighter coloration, this absorption drops to 60% (Norris, 1967). Another case we can name is the regulation of peripheral blood flow. Various animal species can control their temperature by controlling blood flow through vasodilation or vasoconstriction, modifying the diameter of blood vessels. For example, the marine iguana *Amblyrhynchus cristatus* is a species accustomed to living in warm areas but foraging in cold water currents. During this search, it can reduce its heart rate and blood flow to surface tissues to reduce heat loss (Willmer et al., 2009).

The analysis of thermal biology in reptiles is essential for studying their physiology and ecology (Lara-Resendiz et al., 2013), made possible by developing and using new technologies such as precision digital thermometers (Kearney & Predavec, 2000; Lara-Resendiz et al., 2013). However, endothermy has evolved in some species of teleost fishes and sharks of the family Lamnidae. Here, a countercurrent exchange of heat occurs, where heat passes from the venous blood, which is heated by muscle metabolism, to the cold blood in the arteries. Because heat passes between blood vessels carrying blood in different directions, it is retained in the muscles, allowing the organisms to have a higher temperature than the temperature of the water. The evolution of such mechanisms is attributed to the evolution of heat retention to have the ability to explore colder areas (Block & Finnerty, 1994). Among insects, species of various orders (e.g., Odonata, Diptera, Hymenoptera, Coleoptera, and Lepidoptera) can raise their body temperature by contracting thoracic flight muscles prior to flight (Heinrich, 1974; Heinrich, 1987). These species have a countercurrent hemolymph heat exchange system, where cold hemolymph enters the thorax from the abdomen, and warm hemolymph moves from the thorax to the abdomen. For example, during flight, moths of the genus *Eupsilia* reach their thoracic temperature of approximately 30°C even when the ambient temperature is much lower (Heinrich, 1987). This type of thermoregulation has also been detected in bees during foraging behaviors, so the higher the thoracic temperature they can reach, the greater the number of flowers they can visit per unit of time (Heinrich, 1987).

Although the examples found thus far differ from the traditional definition of ectothermic organisms, it is perhaps too early to say that we are facing a change from rule to exception. However, we believe that, with the study of a more significant number of species and greater technological advances, we will be able to find numerous examples showing that ectothermic organisms can regulate their body temperature in some way. Thus, this rule will no longer be valid.

Sexual Cannibalism: From Conjecture to Reality in Scorpions

Let us start by defining female sexual cannibalism behavior. This behavior refers to when, in general, females kill and consume the male before, during, or after mating (Elgar, 1992; Schneider & Elgar, 2001; Elgar & Schneider, 2004; Prenter et al., 2006). Thus, depending on when sexual cannibalism occurs, we can classify it as precopulatory and postcopulatory, depending on whether it occurs before or after insemination. Clearly, the costs and benefits for both females and males differ depending on whether sexual cannibalism occurs precopulatory or postcopulatory (Elgar, 1992; Elgar & Schneider, 2004). Therefore, the timing of sexual cannibalism is essential in determining whether it can be considered a reproductive strategy. If it occurs before insemination, i.e., precopulatory sexual cannibalism, it is definitely not beneficial for males and may be beneficial for females. However, if it occurs after insemination, postcopulatory sexual cannibalism, we can say that this behavior can be beneficial for one or both sexes (Andrade, 1996; Elgar & Schneider, 2004; Vahed, 1998).

In addition, some hypotheses have been proposed to explain the occurrence and evolution of this behavior. For example, the adaptive foraging hypothesis proposes that in a precopulatory instance, females evaluate the nutritional value of a male against his value as a mate (Andrade, 1998; Barry et al., 2008a; Blamires, 2011). Another hypothesis is the aggressive spillover hypothesis, which suggests that females that are more aggressive toward prey will also be more aggressive toward males in sexual interactions (Arnqvist, 1992; Barry et al., 2008b; Johnson, 2001; Polis, 1981). There is also the mate choice hypothesis, in which if a female rejects a male as a mate, then she cannibalizes him (Gatz, 1981; Persons & Uetz, 2005). Additionally, there is the mistaken identity hypothesis, which suggests that sexual cannibalism occurs when females fail to identify males as potential mates (Aisenberg & Peretti, 2011).

Sexual cannibalism is widely distributed among invertebrates, e.g., in gastropod, copepod, insect, and arachnid species (Bilde et al., 2006; Elgar, 1992). Among insects, there are numerous examples among the mantids. In *Pseudomantis albofimbriata*, although sexual cannibalism begins before sexual interaction, males can initiate copulation and sperm transfer, while females consume them. Here, sexual cannibalism is linked, e.g., with a better fecundity rate in males (Birkhead & Møller, 1998; Barry, 2004; Barry et al., 2008a; Barry et al., 2008b; Lelito & Brown, 2006). In *Tenodera sinensis*, the sexual cannibalism behavior of females to males was also recorded. In this case, egg analyses showed that eggs contained more amino acids derived from males when cannibalism occurred. Thus, males provide resources directly to the eggs, which offsets the cost of losing future matings (Brown & Barry, 2016). In the mantid *T. aridifolia*, there are also frequent records of sexual cannibalism (Watanabe & Yano, 2012). Another insect group where sexual cannibalism behaviors are widespread is water striders, e.g., *Gerris odontogaster* (Arnqvist, 1994; Spence & Cárcamo, 1991).

In spiders, there are also several cases of sexual cannibalism. In the spider *Araneus diadematus*, sexual cannibalism can occur before or after insemination (Elgar, 1989; Elgar & Schneider, 2004; Schneider & Elgar, 2001). In other spider species, the cannibalism observed is postcopulatory, as in the case of *Argiope bruennichi,* as females attack males after insemination (Elgar et al., 2000; Elgar & Schneider, 2004; Fromhage et al., 2003). In addition, an iconic example is what happens in the species *Latrodectus hasselti*, where males, during copulation, can turn their abdomen 180°, being very close to the mouthparts of females and can be consumed by them (Andrade, 1996, 1998; Elgar & Schneider, 2004). Possibly in this species, cannibalism occurs as there is a low probability of recopulation for males, and they increase their fertilization success when they "offer" their bodies for females to consume them (Andrade, 1996; Andrade & Banta, 2002; Peretti, 2014).

While numerous examples of sexual cannibalism occur among spiders, in another order of arachnids, scorpions, females were long considered to eat males immediately after mating. Indeed, this was considered the generalized pattern for the order (Polis & Sissom, 1990), particularly in books referring to sexual cannibalism (Elgar, 1992). In this sense, species such as *Bothriurus bonariensis*, *Urophonius jheringii*,

U. brachycentrus, and *Leiurus quinquestriatus* were included by Polis and Sissom (1990) among the species where postcopulatory sexual cannibalism was present. However, thanks to the advance of technology and a design of experiments more in line with the nature of this type of species, it was possible to prove that this behavior among scorpion species was far from being a rule (Elgar, 1992; Elgar & Schneider, 2004; Peretti et al., 1999). Several factors contribute to this misconception, including the oversight of the minimal space needed for a couple in the laboratory terrarium, insufficient field observations, and limited understanding of the ethological basis of sexual behavior (Peretti, 2014). Moreover, what distinguished these new analyses was the seemingly simple yet crucial aspect of ensuring adequate minimum distances between individuals. Indeed, the observations reporting the occurrence of postcopulatory sexual cannibalism were predominantly conducted in captivity, where males and females were confined in small habitats not only for mating but also for an extended period after copulation (Elgar & Schneider, 2004; Peretti et al., 1999). Thus, new observations have shown that sexual cannibalism, if it exists, could be more of a precopulatory than a postcopulatory type (Peretti, 2014; Peretti et al., 1999). It is not currently denied that postcopulatory sexual cannibalism may exist in scorpions. However, it is no longer considered the rule for the whole order as previously assumed, but in any case, an exception.

7.4.2 With "Potential" of R-E by the Change in Interpretations

7.4.2.1 Is Life Confined to Earth?

How can life be interpreted? Are we talking about a rule or an exception? The answer may lie in the level of interpretation we are assuming. For example, when we discussed life in the previous chapter, we did so from the perspective of the Earth. On Earth, life clearly went from being something rare in its beginnings (millions of years ago) to a common phenomenon in almost every corner, even in very hostile environments (Rothschild & Mancinelli, 2001). However, what happens if we take the universe as the focus? The view here changes. At this moment (the year 2022), life is confined only to the Earth, and we can safely say that life in the universe is rare. In fact, a theory called the rare Earth hypothesis postulates that while Earth-like planets exist, conditions such as those necessary to generate life are rare or perhaps even unique (Forgan & Rice, 2010; Ward & Brownlee, 2000). Earth-like planets or the conditions to harbor life are not rare; what is rare is life itself and the conditions that aided its emergence. Some researchers soften this last postulate by suggesting that in the universe, life in the form of bacteria or microorganisms would not be rare (Chela-Flores, 1998; Cockell, 2016; Rampelotto, 2010; Westall, 1999). However, uncertainty increases when talking about complex life, such as plants or animals, and even worse, intelligent life (Barnes, 2012; Chyba & Hand, 2005; Hoyle, 1983). Thinking a bit, the conditions shown by the Earth to harbor life are unique and rare. A combination of distance from the sun, type of sun star, size of the Earth, eccentricity of the orbit, presence and proximity of the Moon, atmosphere, plate tectonics, and volcanic activity make the Earth a complex and rare system in the universe. However, we must keep in mind that the universe is monstrously enormous (Levin et al., 1998), and some suggest it is even infinite (Teerikorpi et al., 2019). There are an estimated 8–10% of G-type stars similar to our sun (yellow dwarf star) in our galaxy alone (Stardate, Universe Guide; Ledrew, 2001). These stars are included in systems with several planets orbiting them (Gurumath, 2019). Thus, statistically, there should be a chance that at least one situation exactly like that of the Earth has occurred somewhere in the vast universe, something rare but not impossible. Calculations estimate that 6 trillion Earth-like planets exist in the Milky Way (Gough, 2020). We know, for example, that there are extrasolar Earth-like planets that are similar in mass, size, and temperature to the Earth. In addition, there are also "terrestrial twins" that are similar to the Earth, with liquid water in oceans and continental water masses (Seager, 2003). Between 3000 and 5000 potential exoplanets (Han et al., 2014) with interesting characteristics to resemble Earth or harbor minimal conditions for the emergence and maintenance of life have been studied (Han et al., 2014).

Among the many candidates with different interesting characteristics, we can discuss Kepler—452B and Trappist 1e as examples. The first presents a relationship with the star it orbits, similar to that of the Earth with the sun. The star is of the same type as the sun (G-type star) with a larger diameter. Similarly, the planet is rocky and 60% larger than the Earth. It is within the habitable area (zone where water remains liquid on the surface). It takes 385 days to complete orbit (Davachi, 2017; Jenkins et al., 2015). However, much attention has recently been given to the Trappist-1 system. Trappist-1 is a dwarf star that has seven orbiting rocky planets similar in size to Earth, and three of them would be in the habitability zone (Van Grootel et al., 2018) and could harbor life (Lingam & Loeb, 2017). Studies suggest that the planet Trappist-1e could harbor more water than terrestrial oceans (Bourrier et al., 2017) and possibly a planetary atmosphere (Acuña et al., 2021; Turbet et al., 2020), with the potential for the development of life.

Similarly, but closer to the solar system, we see potential candidates for harboring life (Chyba & Hand, 2005; Shapiro & Schulze-Makuch, 2009), but perhaps primitive and very different from what we are used to seeing here on Earth. This is the example of Europa, one of Jupiter's moons. Jupiter is a gas giant planet with a very powerful electromagnetic field (Khurana et al., 2002). That gravitational force is thought to cause ice to melt beneath Europa's permanent surface ice sheet, generating oceans of liquid water in the interior (Kivelson et al., 2009). It is speculated, still without concrete evidence, that these waters could contain even a primitive ecosystem under the ice sheet (Hiscox, 2000). There are also studies suggesting that in the not-so-distant past, both Venus and Mars may have harbored some form of life (Chyba & Hand, 2005). In fact, it is believed that Mars may have had liquid water running on the surface (Nazari-Sharabian et al., 2020). Similar cases of changes in primitive atmospheres have been evaluated on Earth-like exoplanets

(Seager, 2003). However, imagine if we have litimations analyzing nearby planets, anything beyond those has even more difficulties. It is important to recognize that even if conditions for life are present, the emergence of life is not a guaranteed outcome. While favorable conditions increase the likelihood of life, various factors and complexities may still prevent its appearance. Apart from all the planetary characteristics for harboring life, we also have the triggers for the appearance of life or its entry into the system, perhaps from space (see Chap. 6 on the origin of life for more details).

Indeed, at present, life in the universe appears to be an exceptional occurrence limited to Earth. However, it is essential to acknowledge that our exploration of the cosmos is still in its early stages, and despite significant advancements in technology (Seager, 2003), our current methods for detecting life might still be relatively primitive. When delving into the scientific papers concerning these discoveries, one can appreciate the significant advancements in indirectly detecting variables and constructing models to assess the suitability of planets. However, it is evident that we are still far from determining specific critical criteria for identifying pathways to life on these exoplanets.

Here, we see how solar system-like systems are not rare. In addition, planets with Earth-like characteristics are common. In fact, their presence in the universe would be a rule or a statistical consequence. However, we see that the presence of life would be the rare variable. Nevertheless, let us remember that the lack of life on them currently falls at an equal distance between the total absence and the technological inability to detect it. Thus, the dilemma that is very common in biology appears: What we are looking for does not exist, or we cannot find it?

The rule, at least on Earth, is that carbon is the ideal basis for life, just as water is the ideal solvent. However, again, is this true in the universe, or are we biased by what we know on Earth? We have heard much about the possibility of other life-based molecules based on other atoms and with other solvents. The most cited examples of alternative chemistry for life are silicon as a backbone and as an alternative to the solvents ammonium or benzene (Cockell, 2020). Silicon does not have the advantages of carbon bonding with other molecules, chirality, and abundance, so lately, the idea has been discarded (Jacob, 2016). Moreover, imagining organisms that respond or have adapted physically, chemically, and physiologically to these compounds becomes difficult. Carbon is more stable, does not produce such inert structures in its interaction with oxygen, and is found ten times more in the universe than silicon (Cockell, 2020). However, water is the most abundant solvent in the universe, meets the requirements to be a suitable solvent perfectly, and is perhaps the only solvent with this characteristic (Pohorille & Pratt, 2012). We see that carbon and water are, thus far, the rule of life because their characteristics are not matched by any other element or compound. However, we need more information on alternatives, including other atoms and solvents or a mixture of compounds, to dismiss other possibilities (Pohorille & Pratt, 2012).

As we move up the biological scale, thinking that life is common in the universe becomes much more complicated. We suggest that it is easy to talk about life in the universe in the form of microorganisms. However, when we move to more complex, multicellular organisms, the chances of their existence in the universe diminish. Chances go down even more when we talk about intelligent life. Indeed, the question persists: whether the rule is the existence of life on other planets beyond Earth (within the same or different galaxies), while intelligent life remains confined solely to our planet. In simpler terms, this idea allows for the possibility that while life could be widespread in the universe, intelligent beings might be unique to Earth, making it an exception in that aspect. There are many hypotheses and interpretations regarding intelligent life in the galaxy, and many say that we are very likely alone (Smith, 2011). Others say that while there are potentially millions of civilizations (Kerr, 2005), we have physical constraints that do not allow us to contact, which, by the way, may be quite dangerous (Michael, 2011). Thus, other hypotheses suggest the presence of hidden, undetected life in the universe. The dark forest hypothesis would solve the Fermi paradox and states that each civilization would be like a hunter who hides in the jungle to avoid being detected and hunted. Then, none of the civilizations give signals to go unnoticed (Yu, 2015).

However, perhaps their technological level is similar to ours, just enabling us to explore neighboring planets. We do not yet have the technological level to explore the universe in greater detail (Kurzgesagt, 2015; Webb, 2002). It is also thought that the duration of civilizations may be short, so we never get to explore space in detail before we perish as a civilization. Finally, a hypothesis speaks of a large filter that eliminates the possibility of civilizations reaching an advanced level (Peacock, 2018).

With these hypotheses in mind, we can understand that it would become quite rare and exceptional contact between possible galactic civilizations, assuming they existed. In this context, it is helpful to look at "local universes" that have been explored over the years on our own planet that generate ideas of what life might be like in other star systems and where to look for it. For example, exploring the depths of the oceans is an excellent starting point for analysis. The discovery of the first animals inhabiting these depths, represented by abyssal fish (Hoyt, 2014), was a remarkable event. However, this discovery did not imply that the idea of finding life in such extreme environments was exceptional. Instead, it opened up the possibility that life might thrive in diverse and challenging conditions, including the deep ocean, which was previously considered exceptional. That may be so in several places, but in others, the rule of the "absence or scarcity" of life, or animals punctually, has changed toward a rule that life is present in diverse forms, many microscopic. As Dr. Ian Malcom would say in the movie Jurassic Park from 1993, "*Because the history of evolution is that life escapes all barriers. Life breaks free. Life expands to new territories. Painfully, perhaps even dangerously. But life finds a way.*" Here, in the case concerning the exploration of ocean depths, we observe how technological advancements greatly facilitated this paradigm shift regarding life in such environments (Danovaro et al., 2014).

The reader may question why these examples appear in this item. They seem more like changes due to modifications in our knowledge or understanding of biology, Earth, and the universe than changes in the universe itself. It is a mixed situation. Certainly, our evolving understanding of ideas, access to data, and improved technology have significantly enhanced our knowledge of the universe and the ocean floor. As a result, we have gained valuable insights into the colonization of these spaces over time by various groups of living beings. This understanding has shed light on how certain organisms came to populate regions where they were previously absent. Exploring these aspects has led to fascinating discoveries about the dynamic processes of life dispersal and adaptation to new environments, providing a deeper understanding of the complexities of life's distribution and evolution.

In the ocean floor example, it could be asked whether the expansion of animals began from the depths to the surface or the other way around. Paleontological data based on traces (marks left by organisms in the sediment) suggest that organisms evolved in the late Precambrian, mainly in shallow water. Then, in the Cambrian, these forms would have moved to depth (Crimes & Fedonkin, 1994). Of course, this is the story the fossils tell us, but it may not be the whole story, or we may be missing an essential part of it, the part that was not left in the record. In any case, the compromise solution could come from the in-between. From there, the bottom of the oceans, sea surface, coast, and even the terrestrial environment were colonized. So yes, we have a situation strictly of the behavior of biology from being nothing at the bottom of the ocean or very rare to at least not so rare.

In the case of exobiology, the situation is, as of today, unknown. First, we discuss a conjecture. However, if we consider the possibility that life does exist beyond Earth, it opens up intriguing questions about its origin and distribution. We would need to investigate where it began, whether it emerged simultaneously in different locations (either strictly or over extended periods), or if it appeared sequentially throughout the universe. Additionally, we must consider the direction of its expansion, whether it originated from Earth and spread outward or originated elsewhere and potentially reached our planet, as suggested in some hypotheses (e.g., Lingam & Loeb, 2017). The exploration of these questions is fundamental to understanding the broader context of life's existence in the cosmos and how it may have evolved and dispersed throughout the universe. Again, beyond directionality, the example also applies to this item in that it alludes to biology. Thus, like others, we have "mixed" situations in this type of example.

7.5 Key Concepts and Ideas in this Chapter

- We have seen multiple examples of transitions from rules to exceptions in biology. Evolutionary causes such as natural selection, competition between species and between organisms within a species, and biotic and abiotic factors can cause certain traits or structures to have an obsolete function and eventually be lost, to be considered exceptions when they occur.
- Some factors, such as the decline or disappearance of a species, can produce changes in the interactions with other species, leaving a pattern as a rarity and sometimes having a drastic effect on the life of one of the interacting species.

- Species distribution patterns evolve along with relief, soil, climate, and relationships between organisms or human activity. Thus, some species are common in some areas but rare in others. Because of their more significant genetic variation, species with a broad global distribution can adapt better to environmental gradients.
- Cities profoundly modify natural habitats and create their microclimate, increase the average temperature, and offer shelter in their buildings and varied feeding possibilities. However, they are an inhospitable habitat for most species. Less adaptable species will become rare as environments become increasingly anthropized.
- Sometimes climate change causes specific populations to become isolated in climatic relicts, which can be very useful for studying the interactions between species and how these interactions can affect communities and the ecosystem. There are relictual species that are representatives of groups that have partially disappeared. These species have remained exceptions and give us information on the diversity of the past and existing groups in terms of ancestral species.
- On Earth, there were six great mass extinctions, in which more than half of the species that existed at that time disappeared without leaving descendants. These extinctions marked a turning point in species community composition that inhabited the earth, either by disappearance of some or by evolution of others.
- Some species survived the mass extinctions and evolved very slowly, resembling their ancestors in their general phenotype. Living fossils are unique lineages characterized by stable morphology, low taxonomic diversity, and a particular rarity. They can therefore be used as evidence of long-term evolution.
- The human being, through many factors such as artificial selection, the use of genetics, manipulation of races, or activities such as hunting or fishing, limits the possibility of evolution to continue its course in living beings. Thus, the original species that were the rule at the beginning of nature declined in abundance or ceased to exist and sometimes even gave rise to new species or varieties.
- Many times, the alterations that have occurred in the climate due to global warming favor species that have been introduced voluntarily or involuntarily by human action and can expand into new areas. This causes extinctions of native species, making them exceptions.
- The availability of technological resources and adequate methodology is essential to identify patterns and establish rules in biology. With insufficient technology, it is expected that changes in conceptions and approaches will occur, and often the established rules may cease to be valid, become rarities, or disappear from knowledge.
- Life can be a rule or an exception depending on the level of analysis. It is a rule on Earth but seems to be an exception throughout the universe. Life can be absent from everywhere else or just difficult to detect.
- Carbon and water are the rule for life as we know it. Other atoms and solvents theoretically could work, but it seems that carbon and water are truly a rule for life.

References

Abdala, F., & Ribeiro, A. M. (2010). Distribution and diversity patterns of *Triassic cynodonts* (Therapsida, Cynodontia) in Gondwana. *Palaeogeography, Palaeoclimatology, Palaeoecology, 286*(3–4), 202–217.

Acevedo, K., Bohórquez-Herrera, J., Moreno, F., Moreno, C., Molina, E., Grijalba-Bendeck, M., & Gomez Canchong, P. (2007). Tiburones y rayas (subclase Elasmobranchii) descartados por la flota de arrastre camaronero en el Caribe de Colombia. *Acta Biológica Colombiana, 12*(2), 69–80.

Acuña, L., Deleuil, M., Mousis, O., Marcq, E., Levesque, M., & Aguichine, A. (2021). Characterisation of the hydrospheres of TRAPPIST-1 planets. *Astronomy & Astrophysics, 647*, A53.

Agnolin, F. L., Agnolin, A. M., & Guerrero, E. L. (2001). *Tras las huellas del megaterio: plantas y animales que la última gran extinción olvidó*. Fundación de Historia Natural Félix de Azara-Universidad Maimónides.

Ah-King, M. (2023). *The female turn: How evolutionary science shifted perceptions about females*. Springer Nature.

Aisenberg, A., & Peretti, A. V. (2011). Male burrow digging in a sex role-reversed spider inhabiting water-margin environments. *Arachnology, 15*(6), 201–204.

Akcakaya, H. R., Burgman, M. A., Kindvall, O., Wood, C. C., Sjogren-Gulve, P., Hatfield, J. S., & McCarthy, M. A. (Eds.). (2004). *Species conservation and management: Case studies*. Oxford University Press.

Alber, C. (2009). Biology's next top model. *Nature, 458*(9), 695–698.

Alcamo, J., Döll, P., Henrichs, T., Kaspar, F., Lehner, B., Rösch, T., & Siebert, S. (2003). Global estimates of water withdrawals and availability under current and future "business-as-usual" conditions. *Hydrological Sciences Journal, 48*(3), 339–348.

Allison, S. D., & Vitousek, P. M. (2004). Rapid nutrient cycling in leaf litter from invasive plants in Hawai'i. *Oecologia, 141*, 612–619.

Almeida, J. P., Montúfar, R., & Anthelme, F. (2013). Patterns and origin of intraspecific functional variability in a tropical alpine species along an altitudinal gradient. *Plant Ecology & Diversity, 6*, 423–433.

Alroy, J. (2001). A multispecies overkill simulation of the end-Pleistocene megafaunal mass extinction. *Science, 292*(5523), 1893–1896.

Amemiya, C. T., Alföldi, J., Lee, A. P., Fan, S., Philippe, H., Mac Callum, I., et al. (2013). The African coelacanth genome provides insights into tetrapod evolution. *Nature, 496*(7445), 311–316.

Andersson, M., & Iwasa, Y. (1996). Sexual selection. *Trends in Ecology & Evolution, 11*(2), 53–58.

Andrade, M. C. (1996). Sexual selection for male sacrifice in the Australian redback spider. *Science, 271*(5245), 70–72.

Andrade, M. C. (1998). Female hunger can explain variation in cannibalistic behavior despite male sacrifice in redback spiders. *Behavioral Ecology, 9*(1), 33–42.

Andrade, M. C., & Banta, E. M. (2002). Value of male remating and functional sterility in redback spiders. *Animal Behaviour, 63*(5), 857–870.

Ankeny, R. A., & Leonelli, S. (2011). What's so special about model organisms? *Studies in History and Philosophy of Science Part A, 42*(2), 313–323.

Araujo, B. B., Oliveira-Santos, L. G. R., Lima-Ribeiro, M. S., Diniz-Filho, J. A. F., & Fernandez, F. A. (2015). Bigger kill than chill: The uneven roles of humans and climate on late quaternary megafaunal extinctions. *Quaternary International, 431*, 216–222.

Arija, C. M. (2010). Biología y Conservación del Lobo Ibérico: crónica de un conflicto. *REDVET. Revista electrónica de Veterinaria, 11*(6), 1–18.

Arita, H. T., & Santos-del-Prado, K. (1999). Conservation biology of nectar-feeding bats in Mexico. *Journal of Mammalogy, 80*, 31–41.

Arnell, N. W. (1999). Climate change and global water resources. *Global Environmental Change, 9*(S1), S31–S49.
Arnold, S. J. (2003). Too much natural history, or too little? *Animal Behaviour, 65*, 1065–1068.
Arnqvist, G. (1992). Courtship behavior and sexual cannibalism in the semi-aquatic fishing spider, *Dolomedes fimbriatus* (Clerck) (Araneae: Pisauridae). *Journal of Arachnology, 20*(3), 222–226.
Arnqvist, G. (1994). The cost of male secondary sexual traits: Developmental constraints during ontogeny in a sexually dimorphic water strider. *The American Naturalist, 144*(1), 119–132.
Ashley, M. V., Melnick, D. J., & Western, D. (1990). Conservation genetics of the black rhinoceros (*Diceros bicornis*), I: Evidence from the mitochondrial DNA of three populations. *Conservation Biology, 4*(1), 71–77.
Assmann, T. (1999). The ground beetle fauna of ancient and recent woodlands in the lowlands of North-West Germany (Coleoptera, Carabidae). *Biodiversity and Conservation, 8*, 1499–1517.
Assmann, T., Casale, A., Drees, C., Habel, J. C., Matern, A., & Schuldt, A. (2010). The dark side of relict species biology: Cave animals as ancient lineages. In *Relict species: Phylogeography and conservation biology* (pp. 91–103). Springer.
Ayup, M. M., Montti, L., Aragón, R., & Grau, H. R. (2014). Invasion of *Ligustrum lucidum* (Oleaceae) in the southern Yungas: Changes in habitat properties and decline in bird diversity. *Acta Oecologica, 54*, 72–81.
Bailey, W. J., & Haythornthwaite, S. (1998). Risks of calling by the field cricket *Teleogryllus oceanicus*: Potential predation by Australian long-eared bats. *Journal of the Zoological Society of London, 244*, 505e513.
Baillie, J. E. M., & Cokeliss, Z. (2004). Extinctions in recent time. In *2004 IUCN Red List of Threatened Species: A Global Species Assessment* (pp. 33–50).
Banks, P. B., & Bryant, J. V. (2007). Four-legged friend or foe? Dog walking displaces native birds from natural areas. *Biology Letters, 3*(6), 611–613.
Baraquet, M., Grenat, P. R., Salas, N. E., & Martino, A. L. (2012). Variación morfométrica y geográfica entre poblaciones de *Hypsiboas cordobae* (Anura: Hylidae) en Argentina. *UNED Research Journal, 4*(2), 147–155.
Barlow, C. (2001). Anachronistic fruits and the ghosts who haunt them. *Arnoldia, 61*, 14–21.
Barnes, L. A. (2012). The fine-tuning of the universe for intelligent life. *Publications of the Astronomical Society of Australia, 29*(4), 529–564.
Barnosky, A. D., Matzke, N., Tomiya, S., Wogan, G. O. U., Swartz, B., Quental, T. B., et al. (2011). Has the Earth's sixth mass extinction already arrived? *Nature, 471*(7336), 51–57.
Barry, K. L. (2004). *Mating behaviour of the sexually cannibalistic praying mantid, Pseudomantis albofimbriata*. Doctoral thesis,. Department of Biological Sciences, Macquarie University.
Barry, K. L., Holwell, G. I., & Herberstein, M. E. (2008a). Female praying mantids use sexual cannibalism as a foraging strategy to increase fecundity. *Behavioral Ecology, 19*(4), 710–715.
Barry, K. L., Holwell, G. I., & Herberstein, M. E. (2008b). Male mating behavior reduces the risk of sexual cannibalism in an Australian praying mantid. *Journal of Ethology, 27*, 377–383.
Barton, M., Porter, W., & Kearney, M. (2014). Behavioural thermoregulation and the relative roles of convection and radiation in a basking butterfly. *Journal of Thermal Biology, 41*, 65–71.
Barstow, M., Baldwin, H., & Rivers, M. C. (2017). *Nothofagus dombeyi*. The IUCN Red List of Threatened Species.
Bartomeus, I., Park, M. G., Gibbs, J., Danforth, B. N., Lakso, A. N., & Winfree, R. (2013). Biodiversity ensures plant–pollinator phenological synchrony against climate change. *Ecology Letters, 16*(11), 1331–1338.
Bateman, A. J. (1948). Intrasexual selection in *drosophila*. *Heredity, 2*(3), 349–368.
Beierkuhnlein, C. (2007). *Biogeographie - Die räumliche Organisation des Lebens in einer sich verändernden Welt* (p. 397). UTB Ulmer.
Bellis, L. M., Astudillo, A., Gavier-Pizarro, G., Dardanelli, S., Landi, M., & Hoyos, L. (2021). Glossy privet (*Ligustrum lucidum*) invasion decreases Chaco Serrano Forest bird diversity but favors its seed dispersers. *Biological Invasions, 23*, 723–739.
Bennett, P. M., & Owens, I. P. F. (2002). *Evolutionary ecology of birds: Life history*. Oxford University Press, Oxford.

Benton, M. J. (2003). *When life nearly died: The greatest mass extinction of all time*. Thames & Hudson.

Benton, M. J., & Twitchett, R. J. (2003). How to kill (almost) all life: The end-Permian extinction event. *Trends in Ecology & Evolution, 18*(7), 358–365.

Bhatnagar-Mathur, P., Vadez, V., & Sharma, K. K. (2008). Transgenic approaches for abiotic stress tolerance in plants: Retrospect and prospects. *Plant Cell Reports, 27*, 411–424.

Bhutiyani, M. R., Kale, V. S., & Pawar, N. J. (2008). Changing streamflow patterns in the rivers of northwestern Himalaya: Implications of global warming in the 20th century. *Current Science, 95*(5), 618–626.

Bilde, T., Tuni, C., Elsayed, R., Pekár, S., & Toft, S. (2006). Death feigning in the face of sexual cannibalism. *Biology Letters, 2*, 23–25.

BirdLife International (2024a) Species factsheet: Pygoscelis papua. Retrieved April 04, 2024, from https://datazone.birdlife.org/species/factsheet/gentoo-penguin-pygoscelis-papua

BirdLife International (2024b) Species factsheet: Vultur gryphus. Retrieved April 04, 2024, from https://datazone.birdlife.org/species/factsheet/andean-condor-vultur-gryphus.

Birkhead, T. (2000). *Promiscuity: An evolutionary history of sperm competition*. Faber and Faber.

Birkhead, T. R., & Møller, A. P. (1998). *Sperm competition and sexual selection*. Academic Press.

Bizzarro, J. J., Smith, W. D., Márquez-Farías, J. F., & Hueter, R. E. (2007). Artisanal fisheries and reproductive biology of the golden cownose ray, *Rhinoptera steindachneri* Evermann and Jenkins, 1891, in the northern Mexican Pacific. *Fisheries Research., 84*(2), 137–146.

Blamires, S. J. (2011). Nutritional implications for sexual cannibalism in a sexually dimorphic orb web spider. *Austral Ecology, 36*, 389–394.

Blanckenhorn, W. U., Stillwell, R. C., Young, K. A., Fox, C. W., & Ashton, K. G. (2006). When Rensch meets Bergmann: Does sexual size dimorphism change systematically with latitude? *Evolution, 60*, 2004–2011.

Blanco, J. C., & Cortés, Y. (2003). *El Lobo Ibérico, el misterio del monte*. Edilesa.

Block, B. A., & Finnerty, J. R. (1994). Endothermy in fishes: A phylogenetic analysis of constraints, predispositions, and selection pressures. *Environmental Biology of Fishes, 40*(3), 283–302.

Bollinger, R. R., Barbas, A. S., Bush, E. L., Lin, S. S., & Parker, W. (2007). Biofilms in the large bowel suggest an apparent function of the human vermiform appendix. *Journal of Theoretical Biology, 249*(4), 826–831.

Bond, D. P., & Grasby, S. E. (2020). Late Ordovician mass extinction caused by volcanism, warming, and anoxia, not cooling and glaciation. *Geology, 48*(8), 777–781.

Bond, W. J., & Silander, J. A., Jr. (2007). Springs and wire plants: Anachronistic defences against Madagascar's extinct elephant birds. *Proceedings of the Royal Society of London Series B: Biological Sciences, 274*, 1985–1992.

Borowsky, R., & Wilkens, H. (2002). Mapping a cave fish genome: Polygenic systems and regressive evolution. *Journal of Heredity, 93*, 19–21.

Bourrier, V., De Wit, J., Bolmont, E., Stamenković, V., Wheatley, P. J., Burgasser, A. J., et al. (2017). Temporal evolution of the high-energy irradiation and water content of TRAPPIST-1 exoplanets. *The Astronomical Journal, 154*(3), 121.

Boyce, D. G., Lewis, M. R., & Worm, B. (2010). Global phytoplankton decline over the past century. *Nature, 466*(7306), 591–596.

Bravo, D. N. (2003). El estudio de la distribución espacial en la biodiversidad: concepto y métodos. *Geographical Research Letters, 29*, 67–82.

Briceño, R. D., Eberhard, W. G., & Robinson, A. (2007). Copulation behavior of *Glossina pallidipes* (Diptera: Muscidae) outside and inside the female, and genitalic evolution. *Bulletin of Entomological Research, 97*, 1–18.

Brodie, J. F., Aslan, C. E., Rogers, H. S., Redford, K. H., Maron, J. L., Bronstein, J. L., & Groves, C. R. (2014). Tipping points and landscape traps: Conservation in a post-normal world. *Trends in Ecology & Evolution, 29*(12), 664–672.

Brook, B. W., & Bowman, D. M. (2004). The uncertain blitzkrieg of Pleistocene megafauna. *Journal of Biogeography, 31*(4), 517–523.

Brown, W. D., & Barry, K. L. (2016). Sexual cannibalism increases male material investment in offspring: Quantifying terminal reproductive effort in a praying mantis. *Proceedings of the Royal Society B: Biological Sciences, 283*(1833), 20160656.

Bulte, E., Horan, R. D., & Shogren, J. F. (2006). Megafauna extinction: A paleoeconomic theory of human overkill in the Pleistocene. *Journal of Economic Behavior & Organization, 59*(3), 297–323.

Burgos-Vázquez, M. I., Chávez-García, V. E., Cruz-Escalona, V. H., Navia, A. F., & Mejía-Falla, P. A. (2019). Reproductive strategy of the Pacific cownose ray *Rhinoptera steindachneri* in the southern gulf of California. *Marine and Freshwater Research., 70*(1), 93–106.

Burkhead, N. M. (2012). Extinction rates in North American freshwater fishes, 1900–2010. *Bioscience, 62*(9), 798–808.

Burkle, L. A., & Alarcón, R. (2011). The future of plant-pollinator diversity: Understanding interaction networks across time, space and global change. *American Journal of Botany, 98*, 528–538.

Busby, P. E., Newcombe, G., Dirzo, R., & Whitham, T. G. (2013). Genetic basis of pathogen community structure for foundation tree species in a common garden and in the wild. *Journal of Ecology, 101*(4), 867–877.

Butler, S. J., Vickery, J. A., & Norris, K. (2007). Farmland biodiversity and the footprint of agriculture. *Science, 315*(5810), 381–384.

Cane, J. H., & Tepedino, V. J. (2001). Causes and extent of declines among native North American invertebrate pollinators: Detection, evidence, and consequences. *Conservation Ecology, 5*(1).

Cardoso, P., Barton, P. S., Birkhofer, K., Chichorro, F., Deacon, C., Fartmann, T., et al. (2020). Scientists' warning to humanity on insect extinctions. *Biological Conservation, 242*, 108426.

Carleton, R. N. (2016). Fear of the unknown: One fear to rule them all? *Journal of Anxiety Disorders, 41*, 5–21.

Carpenter, K. E., Abrar, M., Aeby, G., Aronson, R. B., Banks, S., Bruckner, A., et al. (2008). One-third of reef-building corals face elevated extinction risk from climate change and local impacts. *Science, 321*(5888), 560–563.

Carroll, S. B. (1995). Homeotic genes and the evolution of arthropods and chordates. *Nature, 376*(6540), 479–485.

Casal, J. L., & Prenski, L. B. (Eds.). (2000). *Diagnóstico de los Recursos Pesqueros de la República Argentina*. INIDEP.

Casey, T. M. (1981). A comparison of mechanical and energetic estimates of flight cost for hovering sphinx moths. *Journal of Experimental Biology, 91*(1), 117–129.

Castro, J., et al. (1999). Seed predation and dispersal in relict scots pine forests in southern Spain. *Plant Ecology, 145*, 115–123.

Cavin, L., & Guinot, G. (2014). Coelacanths as "almost living fossils". *Frontiers in Ecology and Evolution, 2*, 49.

Ceballos, G., & Ehrlich, P. R. (2002). Mammal population losses and the extinction crisis. *Science, 296*(5569), 904–907.

Ceballos, G., & Ehrlich, P. R. (2009). Mammal population losses and the extinction crisis. *Proceedings of the National Academy of Sciences of the United States of America, 106*, 3841–3846.

Ceballos, G., Ehrlich, P. R., Barnosky, A. D., García, A., Pringle, R. M., & Palmer, T. M. (2015). Accelerated modern human–induced species losses: Entering the sixth mass extinction. *Science Advances, 1*(5), e1400253.

Chapman, T., Liddle, L. F., Kalb, J. M., Wolfner, M. F., & Partridge, L. (1995). Cost of mating in *Drosophila melanogaster* females is mediated by male accessory gland products. *Nature, 373*, 241–244.

Chela-Flores, J. (1998). Possible degree of evolution of solar-system microorganisms. In J. Chela-Flores & F. Raulin (Eds.), *Exobiology: Matter, energy, and information in the origin and evolution of life in the universe* (pp. 31–41). Springer.

Chimento, N. R., Agnolin, F. L., & Novas, F. E. (2013). Necrolestes: Un mamífero patagónico que sobrevivió a la extinción de los dinosaurios. *Revista del Museo Argentino de Ciencias Naturales Bernardino Rivadavia, 23*(134), 12–18.

Chown, S. L., & Nicolson, S. (2004). *Insect physiological ecology: Mechanisms and patterns*. Oxford University Press.

Chyba, C. F., & Hand, K. P. (2005). Astrobiology: The study of the living universe. *Annual Review of Astronomy and Astrophysics, 43*, 31–74.

Clusella-Trullas, S., Wyk, J. H., & Spotila, J. R. (2009). Thermal benefits of melanism in cordylid lizards: A theoretical and field test. *Ecology, 90*(8), 2297–2312.

Cockell, C. S. (2016). The similarity of life across the universe. *Molecular Biology of the Cell, 27*(10), 1553–1555.

Cockell, C. S. (2020). *Astrobiology: Understanding life in the universe*. John Wiley & Sons.

Colléony, A., Clayton, S., Couvet, D., Saint Jalme, M., & Prévot, A. C. (2017). Human preferences for species conservation: Animal charisma trumps endangered status. *Biological Conservation, 206*, 263–269.

Colunga-García, M. P., Larqué-Saavedra, A., Eguiarte, L. E., & Zizumbo-Villarreal, D. (2007). En lo ancestral hay futuro: del tequila, los mezcales y otros agaves. In *Centro de Investigación Científica de Yucatán, A.C.* CICY.

Colwell, R. K., Dunn, R. R., & Harris, N. C. (2012). Coextinction and persistence of dependent species in a changing world. *Annual Review of Ecology, Evolution, and Systematics, 43*, 183–203.

Contreras-Balderas, S. (1999). Annotated checklist of introduced invasive fishes in Mexico, with examples of some recent introductions. In R. Claudi & J. H. Leach (Eds.), *Non-indigenous freshwater fishes: Vectors, biology, and impacts* (pp. 33–54). Lewis Publ.

Cooke, B. D. (2012). Rabbits: Manageable environmental pests or participants in new Australian ecosystems? *Wildlife Research, 39*(4), 279–289.

Copeland, M. (2003). *Cockroach*. Reaktion Books.

Corlett, R. T. (2007). The impact of hunting on the mammalian fauna of tropical Asian forests. *Biotropica, 39*(3), 292–303.

Costa, A. C. G., & Machado, I. C. (2012). Flowering dynamics and pollination system of the sedge *Rhynchospora ciliata* (Vahl) Kükenth (Cyperaceae): Does ambophily enhance its reproductive success? *Plant Biology, 14*(6), 881–887.

Cotterill, F. P., & Foissner, W. (2010). A pervasive denigration of natural history misconstrues how biodiversity inventories and taxonomy underpin scientific knowledge. *Biodiversity and Conservation, 19*(1), 291–303.

Courchamp, F., Jaric, I., Albert, C., Meinard, Y., Ripple, W. J., & Chapron, G. (2018). The paradoxical extinction of the most charismatic animals. *PLoS Biology, 16*(4), e2003997.

Cowles, R. B. (1962). Semantics in biothermal studies. *Science, 135*(3504), 670–670.

Cowles, R. B., & Bogert, C. M. (1944). A preliminary study of the thermal requirements of desert reptiles. *Bulletin of the American Museum of Natural History, 83*, 263–296.

Cox, P. A. (1983). Extinction of the Hawaiian avifauna resulted in a change of pollinators for the ieie, Freycinetia arborea. *Oikos, 41*(2), 195–199.

Coyne, J. A., & Orr, H. A. (1989). Patterns of speciation in drosophila. *Evolution, 43*(2), 362–381.

Coyne, J. A., & Orr, H. A. (1997). "Patterns of speciation in drosophila" revisited. *Evolution, 51*(1), 295–303.

Coyne, J. A., & Orr, H. A. (2004). *Speciation* (p. 545). Sinauer Associates. Ss.

Cracraft, J. (2001). Avian evolution, Gondwana biogeography and the cretaceous–tertiary mass extinction event. *Proceedings of the Royal Society of London. Series B: Biological Sciences, 268*(1466), 459–469.

Craske, M. G. (1999). *Anxiety disorders: Psychological approaches to theory and treatment*. Westview Press.

Crawley, M. J. (1997). Biodiversity. In M. J. Crawley (Ed.), *Plant ecology* (pp. 595–632). Blackwell Science Ltd..

Crimes, T. P., & Fedonkin, M. A. (1994). Evolution and dispersal of deep-sea traces. *PALAIOS, 9*(1), 74–83.
Crisp, M. D., & Cook, L. G. (2009). Explosive radiation or cryptic mass extinction? Interpreting signatures in molecular phylogenies. *Evolution, 63*, 2257–2226.
Crowley, T. J., & North, G. R. (1988). Abrupt climate change and extinction events in earth history. *Science, 240*(4855), 996–1002.
Danovaro, R., Snelgrove, P. V., & Tyler, P. (2014). Challenging the paradigms of deep-sea ecology. *Trends in Ecology & Evolution, 29*(8), 465–475.
Dao, A. H., & Netsky, M. G. (1984). Human tails and pseudotails. *Human Pathology, 15*(5), 449–453.
Darwin, C. (1859). *On the origin of species by means of natural selection, or the preservation of favoured races in the struggle for life*. John Murray.
Davachi, N. (2017). Stable orbits for exomoons in Earth's cousin (Kepler-452b) orbiting a sun-like star. In *APS April meeting abstracts 2017* (p. Y6-002).
Dayton, P., & Sala, E. (2001). Natural history: The sense of wonder, creativity and progress in ecology. *Scientia Marina, 65*(Suppl 2), 199–206.
de Angelis, P. (2012). Assessing the impact of international trade on chambered nautilus. *Geobios, 45*(1), 5–11.
De Greef, W. D., Delon, R., Block, M. D., Leemans, J., & Botterman, J. (1989). Evaluation of herbicide resistance in transgenic crops under field conditions. *Bio-technology, 7*(1), 61–64.
de Roodt, A. R. (2014). Comments on environmental and sanitary aspects of the scorpionism by *Tityus trivittatus* in Buenos Aires City, Argentina. *Toxins, 6*(4), 1434–1452.
de Villalobos, R. (2002). *La valuación de recursos naturales extinguibles: el caso de la merluza en el mar continental argentino. Ecología política* (pp. 319–350). *Naturaleza, sociedad y utopía.*
Delucchi, G., & Torres Robles, S. (2009). Plantas exóticas en el Parque Costero del Sur: una categorización. In J. Athor (Ed.), *Parque Costero del Sur. Naturaleza, conservación y patrimonio cultural* (pp. 408–415). Fundación de Historia natural Félix de Azara.
Dennis, J. G., & Ruggiero, M. A. (1996). Biodiversity inventory: Building an inventory at scales from local to global. In R. C. Szaro & D. W. Johnston (Eds.), *Biodiversity in managed landscapes* (pp. 149–156). Oxford University Press.
Diamantidis, A. D., Papadopoulos, N. T., & Carey, J. R. (2008). Medfly populations differ in diel and age patterns of sexual signalling. *Entomologia Experimentalis et Applicata, 128*(3), 389–397.
Diamond, J. M. (1989). The present, past and future of human-caused extinctions. *Philosophical Transactions of the Royal Society of London. B, Biological Sciences, 325*(1228), 469–477.
Didham, R. K., Tylianakis, J. M., Gemmell, N. J., Rand, T. A., & Ewers, R. M. (2007). Interactive effects of habitat modification and species invasion on native species decline. *Trends in Ecology & Evolution, 22*(9), 489–496.
Dirzo, R., Young, H. S., Galetti, M., Ceballos, G., Isaac, N. J., & Collen, B. (2014). Defaunation in the Anthropocene. *Science, 345*(6195), 401–406.
Doadrio, I. (2001). La ictiofauna continental en el Plan Hidrológico Nacional. In P. Arrojo (Ed.), *El Plan Hidrológico Nacional a debate* (p. 487). Fundación Nueva Cultura del Agua.
Doadrio, I., & Elvira, B. (2002). Peces continentales. In J. M. Reyero (Ed.), *La naturaleza de España* (pp. 216–225). Organismo Autónomo Parques Nacionales.
Donald, P. F. (2004). Biodiversity impacts of some agricultural commodity production systems. *Conservation Biology, 18*(1), 17–37.
Doney, S. C. (2006). Plankton in a warmer world. *Nature, 444*(7120), 695–696.
Donohue, K., Pyle, E. H., Messiqua, D., Heschel, M. S., & Schmitt, J. (2001). Adaptive divergence in plasticity in natural populations of *Impatiens capensis* and its consequences for performance in novel habitats. *Evolution, 55*(4), 692–702.
Donovan, S. E., Jones, D. T., Sands, W. A., & Eggleton, P. (2000). Morphological phylogenetics of termites (Isoptera). *Biological Journal of the Linnean Society, 70*(3), 467–513.
Dorken, M. E., & Eckert, C. G. (2001). Severely reduced sexual reproduction in northern populations of a clonal plant, *Decodon verticillatus* (Lythraceae). *Journal of Ecology, 89*, 339–350.

Drees, C., Matern, A., Rasplus, J. Y., Terlutter, H., Assmann, T., & Weber, F. (2008). Microsatellites and allozymes as the genetic memory of habitat fragmentation and defragmentation in populations of the ground beetle *Carabus auronitens* (Col., Carabidae). *Journal of Biogeography, 35*, 1937–1949.
Drouin, G., Godin, J. R., & Page, B. (2011). The genetics of vitamin C loss in vertebrates. *Current Genomics, 12*, 371–378.
Dubois, Y., Blouin-Demers, G., Shipley, B., & Thomas, D. (2009). Thermoregulation and habitat selection in wood turtles *Glyptemys insculpta*: Chasing the sun slowly. *Journal of Animal Ecology, 78*(5), 1023–1032.
Dubrow, T. J., Wackym, P. A., & Lesavoy, M. A. (1998). Detailing the human tail. *Annals of Plastic Surgery, 20*(4), 340.
Ducarme, F., Luque, G. M., & Courchamp, F. (2013). What are "charismatic species" for conservation biologists. *BioSciences Master Reviews, 10*(2013), 1–8.
Dugdale, R. C. (1985). The effects of varying nutrient concentration on biological production in upwelling regions. *Sta, 14*, 8.
Duke, S. O. (1998). Herbicide-resistant crops-their impact on weed science. *Journal of Weed Science and Technology, 43*, 94–100.
Durant, J. M., Hjermann, D. Ø., Ottersen, G., & Stenseth, N. C. (2007). Climate and the match or mismatch between predator requirements and resource availability. *Climate Research, 33*(3), 271–283.
Eakin, C. M., Lough, J. M., & Heron, S. F. (2009). Climate variability and change: Monitoring data and evidence for increased coral bleaching stress. In *Coral bleaching: Patterns, processes, causes and consequences* (pp. 41–67).
Eberhard, W. (2015). Cryptic female choice and other types of Post-copulatory sexual selection. In A. Peretti & A. Aisenberg (Eds.), *Cryptic female choice in arthropods: Patterns, mechanisms and prospects* (pp. 2–23). Springer International Publishing.
Eberhard, W. G. (1985). *Sexual selection and animal genitalia*. Harvard University Press.
Eberhard, W. G. (1996). *Female control: Sexual selection by cryptic female choice*. Princeton University Press.
Eberhard, W. G. (1999). Sexual behavior and sexual selection in the Mediterranean fruit fly, *Ceratitis capitata* (Dacinae: Ceratitidini). In *Fruit Flies (Tephritidae)* (pp. 477–508). CRC Press.
Eckert, C. G. (2002). The loss of sex in clonal plants. *Evolutionary Ecology, 15*, 501–520.
Edgerton, M. D., Fridgen, J., Anderson, J. R., Jr., Ahlgrim, J., Criswell, M., Dhungana, P., et al. (2012). Transgenic insect resistance traits increase corn yield and yield stability. *Nature Biotechnology, 30*(6), 493–496.
Edvardsson, M., & Canal, D. (2006). The effects of copulation duration in the bruchid beetle *Callosobruchus maculatus*. *Behavioral Ecology, 17*, 430–434.
Ejankowski, W. (2010). Demographic variation of dwarf birch (*Betula nana*) in communities dominated by *Ledum palustre* and *Vaccinium uliginosum*. *Biologia, 65*, 248–253.
Elgar, M. A. (1989). Kleptoparasitism: A cost of aggregating for the orb-weaving spider *Nephila edulis*. *Animal Behaviour, 37*, 1052–1055.
Elgar, M. A. (1992). Sexual cannibalism in spiders and other invertebrates. In M. A. Elgar & B. J. Crespi (Eds.), *Cannibalism: Ecology and evolution among diverse taxa* (pp. 128–155). Oxford Science Publications.
Elgar, M. A., & Schneider, J. M. (2004). Evolutionary significance of sexual cannibalism. *Advances in the Study of Behavior, 34*, 135–163.
Elgar, M. A., Schneider, J. M., & Herberstein, M. E. (2000). Female control of paternity in the sexually cannibalistic spider *Argiope keyserlingi*. *Proceedings of the Royal Society of London. Series B: Biological Sciences, 267*(1460), 2439–2443.
Eller, F., Skálová, H., Caplan, J. S., Bhattarai, G. P., Burger, M. K., Cronin, J. T., et al. (2017). Cosmopolitan species as models for ecophysiological responses to global change: The common reed *Phragmites australis*. *Frontiers in Plant Science, 8*, 1833.
Ellers, J., Toby Kiers, E., Currie, C. R., McDonald, B. R., & Visser, B. (2012). Ecological interactions drive evolutionary loss of traits. *Ecology Letters, 15*(10), 1071–1082.

Emer, C., Galetti, M., Pizo, M. A., Jordano, P., & Verdú, M. (2019). Defaunation precipitates the extinction of evolutionarily distinct interactions in the Anthropocene. *Science. Advances, 5*(6), eaav6699.

Enserink, M. (1999). Biological invaders sweep. *Science, 285*(5435), 1834–1836.

Fasola, L., Muzio, J., & Chehébar, C. (2010). Range expansion and prey use of American mink in Argentinean Patagonia: Dilemmas for conservation. *European Journal of Wildlife Research, 57*, 283–294.

Fenster, C. B., Armbruster, W. S., Wilson, P., Dudash, M. R., & Thomson, J. D. (2004). Pollination syndromes and floral specialization. *Annual Review of Ecology, Evolution, and Systematics, 35*, 375–403.

Fernandez, F. A. (2015). Human dispersal and late quaternary Megafaunal extinctions: The role of the Americas in the global puzzle. In *Human origin sites and the world heritage convention in the Americas* (Vol. II, pp. 42–56). UNESCO Publishing.

Fernandez, R., Ceballos, S., Aragón, R., Malizia, A., Montti, L., Whitworth-Hulse, J., Castro-Díez, P., & Grau, R. (2020). A global review of *Ligustrum lucidum* (Oleaceae) invasion. *The Botanical Review, 86*, 93–118.

Fields, S., & Johnston, M. (2005). Whither model organism research? *Science, 307*(5717), 1885–1886.

Fjeldså, J., & Bowie, R. C. K. (2008). New perspectives on the origin and diversification of Africa's forest avifauna. *African Journal of Ecology, 46*, 235–247.

Foley, J. A., DeFries, R., Asner, G. P., Barford, C., Bonan, G., Carpenter, S. R., et al. (2005). Global consequences of land use. *Science, 309*(5734), 570–574.

Foltán, R., & Šedý, J. (2009). Behavioral changes of patients after orthognathic surgery develop on the basis of the loss of vomeronasal organ: A hypothesis. *Head & Face Medicine, 5*(1), 1–5.

Forgan, D. H., & Rice, K. (2010). Numerical testing of the rare earth hypothesis using Monte Carlo realization techniques. *International Journal of Astrobiology, 9*(2), 73–80.

Forrest, J. R. (2016). Complex responses of insect phenology to climate change. *Current Opinion in Insect Science, 17*, 49–54.

Franks, S. J., Weber, J. J., & Aitken, S. N. (2014). Evolutionary and plastic responses to climate change in terrestrial plant populations. *Evolutionary Applications, 7*, 123–139.

Fromhage, L., Uhl, G., & Schneider, J. M. (2003). Fitness consequences of sexual cannibalism in female *Argiope bruennichi*. *Behavioral Ecology and Sociobiology, 55*, 60–64.

Futagi, Y., Toribe, Y., & Suzuki, Y. (2012). The grasp reflex and Moro reflex in infants: Hierarchy of primitive reflex responses. *International Journal Of Pediatrics, 2012*, 191562.

Galetti, M., Moleón, M., Jordano, P., Pires, M. M., Guimarães Jr., P. R., Pape, T., & Svenning, J. C. (2018). Ecological and evolutionary legacy of megafauna extinctions. *Biological Reviews*, 93(2), 845–862.

Gallego, A. R. M. (2019). La cotorra Argentina: una exótica invasora dispuesta a quedarse. *Boletín de la Academia Malagueña de Ciencias, 21*, 107–114.

García, A. I. (1972). El ambiente natural de Tucumán en el siglo XVI a través de los cronistas y documentos de la época. *Anales de la Sociedad Argentina de Estudios Geográficos, 15*, 174–192.

García Fernández, J. J., Ojeda, R. A., Fraga, R. M., Díaz, G. B., & Baigún, R. J. (1997). *Mamíferos y aves amenazados de la Argentina*. FUCEMA, SAREM, AOP y APN.

García, J. T., García, F. J., Alda, F., González, J. L., Aramburu, M. J., Cortés, Y., et al. (2012). Recent invasion and status of the raccoon (*Procyon lotor*) in Spain. *Biological Invasions, 14*(7), 1305–1310.

Garcia-Castellanos, D., Estrada, F., Jiménez-Munt, I., Gorini, C., Fernández, M., Vergés, J., & De Vicente, R. (2009). Catastrophic flood of the Mediterranean after the Messinian salinity crisis. *Nature, 462*, 778–781.

García-Gonzalez, F., & Simmons, L. W. (2010). Male-induced costs of mating for females compensated by offspring viability benefits in an insect. *Journal of Evolutionary Biology, 23*, 2066–2075.

Gatz, A. J. (1981). Non-random mating by size in American toad, Bufo americanus. *Animal Behaviour, 29*(4), 1004–1012.
Gentry, H. S. (1982). *Agaves of continental North America*. University of Arizona Press.
Gérard, M., Vanderplanck, M., Wood, T., & Michez, D. (2020). Global warming and plant–pollinator mismatches. *Emerging Topics in Life Sciences, 4*(1), 77–86.
Ghazoul, J. (2005). Buzziness as usual? Questioning the global pollination crisis. *Trends in Ecology & Evolution, 20*(7), 367–373.
Gilman, S. E., Urban, M. C., Tewksbury, J., Gilchrist, G. W., & Holt, R. D. (2010). A framework for community interactions under climate change. *Trends in Ecology & Evolution, 25*(6), 325–331.
Giménez, R. R. (2009). *La ciudad y la biodiversidad urbana*. Observatorio de Medio Ambiente Urbano (OAMU).
Goin, F. J., Gelfo, J. N., Chornogubsky, L., Woodburne, M. O., & Martin, T. (2012). Origins, radiations, and distributions of south American mammals. In *Bones, clones and biomes. The history and geography of recent Neotropical mammals* (pp. 20–50).
Gomes, A., & Gomes, A. (2015). Scorpion venom research around the world: *Heterometrus* species. In *Scorpion Venoms* (pp. 351–367). Springer.
Gómez-Ruiz, E. P., & Lacher, T. E., Jr. (2019). Climate change, range shifts, and the disruption of a pollinator-plant complex. *Scientific Reports, 9*(1), 1–10.
Goudie, A. S. (2006). Global warming and fluvial geomorphology. *Geomorphology, 79*(3–4), 384–394.
Gough, E. (2020. Astronomers estimate there are 6 billion earth-like planets in the milky way.
Gould, S. J. (1977). Ontogeny and phylogeny. .
Graham, C. H., Moritz, C., & Williams, S. E. (2006). Habitat history improves prediction of biodiversity in rainforest fauna. *Proceedings of the National Academy of Sciences, 103*(3), 632–636.
Grandcolas, P., Nattier, R., Legendre, F., & Pellens, R. (2011). Mapping extrinsic traits such as extinction risks or modelled bioclimatic niches on phylogenies: Does it make sense at all? *Cladistics, 27*(2), 181–185.
Grandcolas, P., Nattier, R., & Trewick, S. (2014). Relict species: A relict concept? *Trends in Ecology & Evolution, 29*(12), 655–663.
Grebmeier, J. M., Overland, J. E., Moore, S. E., Farley, E. V., Carmack, E. C., Cooper, L. W., et al. (2006). A major ecosystem shift in the northern Bering Sea. *Science, 311*(5766), 1461–1464.
Greenwood, R., & Atkinson, I. (1977). Evolution of divaricating plants in New Zealand in relation to moa browsing. *Proceedings of the New Zealand Ecological Society, 24*, 21–33.
Griffith, S. C., Owens, I. P., & Thuman, K. A. (2002). Extra pair paternity in birds: A review of interspecific variation and adaptive function. *Molecular Ecology, 11*(11), 2195–2212.
Grottoli, A. G., Warner, M. E., Levas, S. J., Aschaffenburg, M. D., Schoepf, V., McGinley, M., et al. (2014). The cumulative impact of annual coral bleaching can turn some coral species winners into losers. *Global Change Biology, 20*(12), 3823–3833.
Guimarães Jr., P. R., Galetti, M., & Jordano, P. (2008). Seed dispersal anachronisms: Rethinking the fruits extinct megafauna ate. *PLoS One*, 3(3), 1745.
Guinder, V. A., Popovich, C. A., Molinero, J. C., & Perillo, G. M. (2010). Long-term changes in phytoplankton phenology and community structure in the Bahía Blanca estuary. *Argentina. Marine Biology, 157*(12), 2703–2716.
Gurumath, S. R. (2019). *Study of sun-like G stars and their exoplanets.*, Doctoral dissertation,. Vellore Institute of Technology and Indian Institute of Astrophysics.
Ha, M., & Nehm, R. H. (2014). Darwin's difficulties and students' struggles with trait loss: Cognitive-historical parallelisms in evolutionary explanation. *Science & Education, 23*(5), 1051–1074.
Habel, J. C., Assmann, T., Schmitt, T., & Avise, J. C. (2009a). Relict species: From past to future. In *Relict species* (pp. 1–5). Springer.
Habel, J. C., Augenstein, B., Nève, G., Rödder, D., & Assmann, T. (2009b). Population genetics and ecological niche modelling reveal high fragmentation and potential future extinction of the endangered relict butterfly Lycaena helle. In J. C. Habel & T. Assmann (Eds.), *Relict species, phylogeography and conservation biology*. Springer.

Hallam, A., & Wignall, P. B. (1997). *Mass extinctions and their aftermath*. Oxford University Press.
Hampe, A., & Bairlein, F. (2000). Modified dispersal-related traits in disjunct populations of bird-dispersed *Frangula alnus* (Rhamnaceae): A result of its quaternary distribution shifts? *Ecography, 23*, 603–613.
Hampe, A., & Jump, A. S. (2011). Climate relicts: past, present, future. *Annual Review of Ecology, Evolution, and Systematics, 42*, 313–333.
Han, E., Wang, S. X., Wright, J. T., Feng, Y. K., Zhao, M., Fakhouri, O., Brown, J. I., & Hancock, C. (2014). Exoplanet orbit database. II. Updates to exoplanets.org. *Publications of the Astronomical Society of the Pacific, 126*(943), 827.
Harrington, T. J., & Mitchell, D. T. (2002). Characterization of *Dryas octopetala* ectomycorrhizas from limestone karst vegetation, western Ireland. *Canadian Journal of Botany, 80*, 970–982.
Harrison, I. J., & Stiassny, M. J. (1999). The quiet crisis: A preliminary listing of the freshwater fishes of the world that are extinct or "missing in action". In R. D. E. MacPhee (Ed.), *Extinctions in near time* (pp. 271–331). Kluwer Academic–Plenum Publishers.
Harshman, L. G., & Prout, T. (1994). Sperm displacement without sperm transfer in *Drosophila melanogaster*. *Evolution, 48*, 758–766.
Harvey, A. W., Mooi, R., & Gosliner, T. M. (1999). Phylogenetic taxonomy and the status of *Allonautilus* Ward and Saunders, 1997. *Journal of Paleontology, 73*, 1214–1217.
Hautmann, M. (2012). *Extinction: End-Triassic mass extinction*. eLS. John Wiley & Sons, Ltd.
Hay, J. M., Subramanian, S., Millar, C. D., Mohandesan, E., & Lambert, D. M. (2008). Rapid molecular evolution in a living fossil. *Trends in Genetics, 24*(3), 106–109.
Hayward, M. W., Kamler, J. F., Montgomery, R. A., Newlove, A., Rostro-Garcia, S., Sales, L. P., et al. (2016). Prey preferences of the jaguar *Panthera onca* reflect the Post-Pleistocene demise of large prey. *Frontiers in Ecology and Evolution, 3*, 1–19.
Hedrick, A. V. (2000). Crickets with extravagant mating songs compensate for predation risk with extra caution. *Proceedings of the Royal Society of London, Series B, 267*, 671e675.
Hedrick, A. V., & Kortet, R. (2006). Hiding behaviour in two cricket populations that differ in predation pressure. *Animal Behaviour, 72*(5), 1111–1118.
Hegland, S. J., Nielsen, A., Lázaro, A., Bjerknes, A. L., & Totland, Ø. (2009). How does climate warming affect plant-pollinator interactions? *Ecology Letters, 12*, 184–195.
Heinrich, B. (1974). Thermoregulation in endothermic insects: Body temperature is closely attuned to activity and energy supplies. *Science, 185*(4153), 747–756.
Heinrich, B. (1987). Thermoregulation by winter-flying endothermic moths. *Journal of Experimental Biology, 127*(1), 313–332.
Hiller, M., Schaar, B. T., Indjeian, V. B., Kingsley, D. M., Hagey, L. R., & Bejerano, G. (2012). A "forward genomics" approach links genotype to phenotype using independent phenotypic losses among related species. *Cell Reports, 2*(4), 817–823.
Hinman, A. (1999). Eradication of vaccine-preventable diseases. *Annual Review of Public Health, 20*(1), 211–229.
Hiscox, J. A. (2000). Outer solar system, Europa, titan and the possibility of life. *Astronomy & Geophysics, 41*(5), 5–23.
Hitchcock, M. A., & McBrayer, L. D. (2006). Thermoregulation in nocturnal ectotherms: Seasonal and intraspecific variation in the Mediterranean gecko *Hemidactylus turcicus*. *Journal of Herpetology, 40*, 185–195.
Hoegh-Guldberg, O. (1999). Climate change, coral bleaching and the future of the world's coral reefs. *Marine and Freshwater Research, 50*(8), 839–866.
Hoegh-Guldberg, O. (2011). Coral reef ecosystems and anthropogenic climate change. *Regional Environmental Change, 11*, 215–227.
Hoegh-Guldberg, O., Mumby, P. J., Hooten, A. J., Steneck, R. S., Greenfield, P., Gomez, E., et al. (2007). Coral reefs under rapid climate change and ocean acidification. *Science, 318*(5857), 1737–1742.
Holland, B., & Rice, W. R. (1999). Experimental removal of sexual selection reverses intersexual antagonistic coevolution and removes a reproductive load. *Proceedings of the National Academy of Sciences, 96*, 5083–5088.

Holm, L., Doll, L., Holm, E., Pacheco, J., & Herberger, J. (1997). *World weeds: Natural histories and distributions*. Wiley.

Honig, M. A., Cowling, R. M., & Richardson, D. M. (1992). The invasive potential of Australian banksias in south African fynbos: A comparison of the reproductive potential of *Banksia ericifolia* and *Leucadendron laureolum*. *Australian Journal of Ecology, 17*(3), 305–314.

Hopkins, C. C. E. (2001). *Actual and potential effects of introduced marine organisms in Norwegian waters, including Svalbard.*, Research report 2001-1,. Directorate for Nature Management.

Houdebine, L. M. (2000). Transgenic animal bioreactors. *Transgenic Research, 9*, 305–320.

Howells, E. J., Ketchum, R. N., Bauman, A. G., Mustafa, Y., Watkins, K. D., & Burt, J. A. (2016). Species-specific trends in the reproductive output of corals across environmental gradients and bleaching histories. *Marine Pollution Bulletin, 105*(2), 532–539.

Hoyle, F. (1983). *The intelligent universe*. Michael Joseph.

Hoyt, E. (2014). Creatures of the deep. In *Search of the Sea's monsters and the world they live in*. Firefly Books.

Hsü, K. J., Montadert, L., Bernoulli, D., Cita, M. B., Erickson, A., Garrison, R. E., et al. (1977). History of the Mediterranean salinity crisis. *Nature, 267*(5610), 399–403.

Huey, R. B. (1982). Temperature, physiology, and the ecology of reptiles. In C. Gans & F. H. (Eds.), *Biology of the Reptilia*. Academic Press.

Hug, K. (2008). Genetically modified organisms: Do the benefits outweigh the risks? *Medicina, 44*(2), 87.

Hughes, T. P. (1994). Catastrophes, phase shifts, and large-scale degradation of a Caribbean coral reef. *Science, 265*(5178), 1547–1551.

Hughes, T. P., & Connell, J. H. (1999). Multiple stressors on coral reefs: A long-term perspective. *Limnology and Oceanography, 44*(3part2), 932–940.

Husby, J. (2007). Definitions of GMO/LMO and modern biotechnology. In *Biosafety first: Holistic approaches to risk and uncertainty in genetic engineering and genetically modified organisms* (pp. 569–581).

Ibarbalz, F. M., Henry, N., Brandão, M. C., Martini, S., Busseni, G., Byrne, H., et al. (2019). Global trends in marine plankton diversity across kingdoms of life. *Cell, 179*(5), 1084–1097.

Ismar, S. M., Daniel, C., Stephenson, B. M., & Hauber, M. E. (2010). Mate replacement entails a fitness cost for a socially monogamous seabird. *Naturwissenschaften, 97*, 109–113.

IUCN. (2023). The IUCN red list of threatened species. Version 2023-1. https://www.iucnredlist.org. Accessed 12 Dec 2023.

Jablonski, D., & Chaloner, W. G. (1994). Extinctions in the fossil record (and discussion). *Philosophical Transactions of the Royal Society of London B, 344*(1307), 11–17.

Jacob, D. T. (2016). There is no silicon-based life in the solar system. *SILICON, 8*(1), 175–176.

Jacobs, B. P., & Browner, W. S. (2000). *Ginkgo biloba*: A living fossil. *The American Journal of Medicine, 108*(4), 341–342.

Jaksic, F. M., Iriarte, J. A., Jiménez, J. E., & Martínez, D. R. (2002). Invaders without frontiers: Cross-border invasions of exotic mammals. *Biological Invasions, 4*, 157–173.

Jänne, J., Alhonen, L., Hyttinen, J. M., Peura, T., Tolvanen, M., & Korhonen, V. P. (1998). Transgenic bioreactors. *Biotechnology Annual Review, 4*, 55–74.

Janzen, D. H., & Martin, P. S. (1982). Neotropical anachronisms: The fruits the gomphotheres ate. *Science, 215*, 19–27.

Jarošík, V., Kratochvíl, L., Honěk, A., & Dixon, A. F. (2004). A general rule for the dependence of developmental rate on temperature in ectothermic animals. *Proceedings of the Royal Society of London. Series B: Biological Sciences, 271*(suppl_4), S219–S221.

Jeffery, W. R., & Martasian, D. P. (1998). Evolution of eye regression in the cavefish *Astyanax*: Apoptosis and the Pax-6 gene. *American Zoologist, 38*, 685–696.

Jenkins, J. M., Twicken, J. D., Batalha, N. M., Caldwell, D. A., Cochran, W. D., Endl, M., & Borucki, W. J. (2015). Discovery and validation of Kepler-452b: A 1.6 $R_\oplus$ super earth exoplanet in the habitable zone of a G2 star. *The Astronomical Journal, 150*(2), 56.

Johnson, C. N. (2009). Ecological consequences of late quaternary extinctions of megafauna. *Proceedings of the Royal Society B: Biological Sciences, 276*(1667), 2509–2519.

Johnson, J. C. (2001). Sexual cannibalism in fishing spiders (*Dolomedes triton*): An evaluation of two explanations for female aggression towards potential mates. *Animal Behaviour, 61*(5), 905–914.

Johnson, N. A., Lahti, D. C., & Blumstein, D. T. (2012). Combating the assumption of evolutionary progress: Lessons from the decay and loss of traits. *Evolution: Education and Outreach, 5*(1), 128–138.

Jones, M. E., Tennyson, A. J., Worthy, J. P., Evans, S. E., & Worthy, T. H. (2009). A sphenodontine (Rhynchocephalia) from the Miocene of New Zealand and palaeobiogeography of the tuatara (*Sphenodon*). *Proceedings of the Royal Society B: Biological Sciences, 276*(1660), 1385–1390.

Jordano, P., García, C., Godoy, J. A., & García-Castaño, J. L. (2007). Differential contribution of frugivores to complex seed dispersal patterns. *Proceedings of the National Academy of Sciences USA, 104*, 3278–3282.

Jouanin, L., Bonadé-Bottino, M., Girard, C., Morrot, G., & Giband, M. (1998). Transgenic plants for insect resistance. *Plant Science, 131*(1), 1–1.

Jump, A. S., & Peñuelas, J. (2005). Running to stand still: Adaptation and the response of plants to rapid climate change. *Ecology Letters, 8*, 1010–1020.

Kaebnick, G. E. (2013). *Humans in nature: The world as we find it and the world as we create it.* Oxford University Press.

Kalisz, S., Vogler, D. W., & Hanley, K. M. (2004). Context-dependent autonomous self-fertilization yields reproductive assurance and mixed mating. *Nature, 430*(7002), 884–887.

Kämpf, J., & Chapman, P. (2016). *Upwelling systems of the world.* Springer International Publishing Switzerland.

Kandori, I., Hirao, T., Matsunaga, S., & Kurosaki, T. (2009). An invasive dandelion unilaterally reduces the reproduction of a native congener through competition for pollination. *Oecologia, 159*, 559–569.

Kauhala, K., Talvitie, K., & Vuorisalo, T. (2015). Free-ranging house cats in urban and rural areas in the north: Useful rodent killers or harmful bird predators? *Folia Zoologica, 64*(1), 45–55.

Kayser, V., & Ramzan, I. (2021). Vaccines and vaccination: History and emerging issues. *Human Vaccines & Immunotherapeutics, 17*(12), 5255–5268.

Kearney, M., & Predavec, M. (2000). Do nocturnal ectotherms thermoregulate? A study of the temperate gecko *Christinus marmoratus*. *Ecology, 81*, 2984–2996.

Keller, G., & Kerr, A. C. (2014). *Volcanism, impacts, and mass extinctions: Causes and effects* (Vol. 505). Geological Society of America.

Kemp, D. J., & Krockenberger, A. K. (2002). A novel method of behavioural thermoregulation in butterflies. *Journal of Evolutionary Biology, 15*(6), 922–929.

Kerr, R. A. (2005). Are we alone in the universe? *Science, 309*(5731), 88.

Khurana, K., Kivelson, M., & Volwerk, M. (2002). The interactions of Europa and Callisto with the magnetosphere of Jupiter. In *34th COSPAR scientific assembly* (Vol. 34, p. 2316).

Kiefer, M. C., Van Sluys, M., & Rocha, C. F. (2007). Thermoregulatory behaviour in *Tropidurus torquatus* (Squamata, Tropiduridae) from Brazilian coastal populations: An estimate of passive and active thermoregulation in lizards. *Acta Zoologica, 88*(1), 81–87.

Kin, A., & Błażejowski, B. (2014). The horseshoe crab of the genus *limulus*: Living fossil or stabilomorph? *PLoS One, 9*, e108036.

King, M. C., & Wilson, A. C. (1975). Evolution at two levels in humans and chimpanzees. *Science, 188*(4184), 107–116.

Kivelson, M. G., Khurana, K. K., & Volwerk, M. (2009). Europa's interaction with the Jovian magnetosphere. In R. T. Pappalardo, W. B. McKinnon, & K. K. Khurana (Eds.), *Europa* (pp. 545–570). University of Arizona Press.

Klein, A. M., Vaissière, B. E., Cane, J. H., Steffan-Dewenter, I., Cunningham, S. A., Kremen, C., et al. (2007). Importance of pollinators in changing landscapes for world crops. *Proceedings of the Royal Society B: Biological Sciences, 274*(1608), 303–313.

Kloppenburg, J. (2014). Re-purposing the master's tools: The open-source seed initiative and the struggle for seed sovereignty. *Journal of Peasant Studies, 41*(6), 1225–1246.

Knobloch, I. W. (1951). Are there vestigial structures in plants? *Science, 113*(2938), 465.
Koh, L. P., & Wilcove, D. S. (2007). Cashing in palm oil for conservation. *Nature, 448*(7157), 993–994.
Komblihtt, A. (2009). El genoma del ornitorrinco y la evolución. *Ciencia Hoy, 19*(113), 74–75.
Kraaijeveld, K., Anvar, S. Y., Frank, J., Schmitz, A., Bast, J., Wilbrandt, J., & Ellers, J. (2016). Decay of sexual trait genes in an asexual parasitoid wasp. *Genome Biology and Evolution, 8*(12), 3685–3695.
Kuijper, B., Pen, I., & Weissing, F. J. (2012). A guide to sexual selection theory. *Annual Review of Ecology, Evolution, and Systematics, 43*, 287–311.
Kunkel, J. (2006). What Makes a Good Model System?. Retrieved from http://www.bio.umass.edu/biology/kunkel/modelsys.html.
Kurzgesagt. (2015). The fermi paradox—Where are all the aliens? (1/2). Kurzgesagt—In a nutshell. Retrieved from https://www.youtube.com/watch?v=sNhhvQGsMEc.
Lack, D. (1968). *Ecological adaptations for breeding in birds*. Methuen.
Lahti, D. C., Johnson, N. A., Ajie, B. C., Otto, S. P., Hendry, A. P., Blumstein, D. T., & Foster, S. A. (2009). Relaxed selection in the wild. *Trends in Ecology & Evolution, 24*(9), 487–496.
Lamarck, J. B. (1809). In H. Elliot (Ed.), *Philosophie zoologique* (p. 1963). Hafner Publishing Company.
Lambrinos, J. G. (2004). How interactions between ecology and evolution influence contemporary invasion dynamics. *Ecology, 85*, 2061–2070.
Lara-Resendiz, R. A., Arenas-Moreno, D. M., & Méndez-de la Cruz, F. R. (2013). Termorregulación diurna y nocturna de la lagartija *Phyllodactylus bordai* (Gekkota: Phyllodactylidae) en una región semiárida del centro de México. *Revista Chilena de Historia Natural, 86*(2), 127–135.
Lassuy, D. R. (1995). Introduced species as a factor in extinction and endangerment of native fish species. *American Fisheries Society Symposium, 15*, 391–396.
Laugen, A. T., Laurila, A., & Merilä, J. (2002). Maternal and genetic contributions to geographical variation in *Rana temporaria* larval life-history traits. *Biological Journal of the Linnean Society, 76*(1), 61–70.
Lavergne, S., & Molofsky, J. (2007). Increased genetic variation and evolutionary potential drive the success of an invasive grass. *Proceedings of the National Academy of Sciences USA, 104*, 3883–3888.
Leader-Williams, N. (1988). Patterns of depletion in a black rhinoceros population in Luangwa Valley. *Zambia. African Journal of Ecology, 26*(3), 181–187.
Ledrew, G. (2001). The real starry sky. *Journal of the Royal Astronomical Society of Canada, 95*, 32.
Legendre, F., Whiting, M. F., Bordereau, C., Cancello, E. M., Evans, T. A., & Grandcolas, P. (2008). The phylogeny of termites (Dictyoptera: Isoptera) based on mitochondrial and nuclear markers: Implications for the evolution of the worker and pseudergate castes, and foraging behaviors. *Molecular Phylogenetics and Evolution, 48*(2), 615–627.
Lehmann, G. U. C. (2003). Review of biogeography, host range and evolution of acoustic hunting in Ormiini (Insecta, Diptera, Tachinidae), parasitoids of night-calling bushcrickets and crickets (Insecta, Orthoptera, Ensifera). *Zoologischer Anzeiger, 242*(107e), 120.
Lehmann, G. U. C., Strauss, J., & Lakes-Harlan, R. (2007). Listening when there is no sexual signalling? Maintenance of hearing in the asexual bushcricket *Poecilimon intermedius*. *Journal of Comparative Physiology A, 193*, 537–545.
Lelito, J. P., & Brown, W. D. (2006). Complicity or conflict over sexual cannibalism? Male risk taking in the praying mantis *Tenodera aridifolia sinensis*. *The American Naturalist, 168*(2), 263–269.
Lettenmaier, D. P., Wood, E. F., & Wallis, J. R. (1994). Hydro-climatological trends in the continental United States, 1948–88. *Journal of Climate, 7*(4), 586–607.
Levin, J., Scannapieco, E., & Silk, J. (1998). Is the universe infinite or is it just really big? *Physical Review D, 58*(10), 103516.
Linares, J. C., Camarero, J. J., Bowker, M. A., Ochoa, V., & Carreira, J. A. (2010). Stand-structural effects on *Heterobasidion abietinum*-related mortality following drought events in *Abies pinsapo*. *Oecologia, 164*, 1107–1119.

Lingam, M., & Loeb, A. (2017). Enhanced interplanetary panspermia in the TRAPPIST-1 system. *Proceedings of the National Academy of Sciences, 114*(26), 6689–6693.
Liston, M., & Harrington, T. (2012). Mycorrhizal associations of *Arctostaphylos uvaursi* and *Helianthemum oelandicum* spp *piloselloides* in the Burren Dryas heaths. *Biological and Environmental Sciences, 112B*, 311–313.
Lizarralde, M. S., & Escobar, J. M. (2000). Mamíferos exóticos en la Tierra del Fuego. *Ciencia Hoy, 10*, 52–63.
Loiselle, B. A., & Blake, J. G. (2002). Potential consequences of extinction of frugivorous birds for shrubs of a tropical wet forest. In D. J. Levey & W. R. Silva (Eds.), *Seed dispersal and Frugivory: Ecology, evolution, and conservation* (pp. 209–224). CABI Publishing.
Lomolino, M. V., Riddle, B. R., & Brown, J. H. (2006). *Biogeography*. Sinauer.
Longrich, N. R., Vinther, J., Pyron, R. A., Pisani, D., & Gauthier, J. A. (2015). Biogeography of worm lizards (Amphisbaenia) driven by end-cretaceous mass extinction. *Proceedings of the Royal Society B: Biological Sciences, 282*(1806), 20143034.
Lovejoy, T. E., & Hannah, L. J. (Eds.). (2019). *Biodiversity and climate change*. Yale University Press.
Lowry, D. B., Modliszewski, J. L., Wright, K. M., Wu, C. A., & Willis, J. H. (2008). The strength and genetic basis of reproductive isolating barriers in flowering plants. *Philosophical Transactions of the Royal Society B: Biological Sciences, 363*(1506), 3009–3021.
Lux, S. A., Copeland, R. S., White, I. M., Manrakhan, A., & Billah, M. K. (2003). A new invasive fruit fly species from the Bactrocera dorsalis (Hendel) group detected in East Africa. *International Journal of Tropical Insect Science, 23*(4), 355–361.
Macalister, A. (1871). Additional observations on muscular anomalies in human anatomy (3rd series), with a catalogue of the principal muscular variations hitherto published. *Transactions of the Royal Irish Academy, 25*, 1–134.
Macías-Ordóñez, R., Machado, G., & Macedo, R. H. (2013). Macroecology of sexual selection: Large-scale influence of climate on sexually selected traits. In G. Machado & R. H. Macedo (Eds.), *Sexual selection perspectives and models from the Neotropics* (pp. 1–32). Elsevier.
Madrigal, L., & González-José, R. (2016). *Introducción a la Antropología Biológica*. Asociación Latinoamericana de Antropología Biológica.
Malhi, Y., Doughty, C. E., Galetti, M., Smith, F. A., Svenning, J. C., & Terborgh, J. W. (2016). Megafauna and ecosystem function from the Pleistocene to the Anthropocene. *Proceedings of the National Academy of Sciences, 113*, 838–846.
Manier, M. K., Belote, J. M., Berden, K. S., Novikov, D., Stuart, W. T., & Pitnick, S. (2010). Resolving mechanisms of competitive fertilization success in *Drosophila melanogaster*. *Science, 328*, 354–357.
Marshall, L. G., Webb, S. D., Sepkoski, J. J., Jr., & Raup, D. M. (1982). Mammalian evolution and the great American interchange. *Science, 215*(4538), 1351–1357.
Martin, E. B. (1983). *Rhino Exploitation*. WWF.
Maughan, H., Callicotte, V., Hancock, A., Birky, C. W., Jr., Nicholson, W. L., & Masel, J. (2006). The population genetics of phenotypic deterioration in experimental populations of *Bacillus subtilis*. *Evolution, 60*(4), 686–695.
Mayr, E., & Provine, W. B. (1980). *The evolutionary synthesis* (Vol. 231). Harvard University Press.
McDonnell, M. J., Pickett, S. T. A., & Pouyat, R. V. (1993). The application of the ecological gradient paradigm to the study of urban effects. In M. J. McDonnell & S. T. A. Pickett (Eds.), *Humans as components of ecosystems: Subtle human effects and the ecology of populated areas* (pp. 175–189). Springer-Verlag.
McKinney, M. L. (2002). Urbanization, biodiversity, and conservation. *Bioscience, 52*, 883–890.
McNeely, J., & Scherr, S. (2003). *Estrategias de ecoagricultura para ayudar a alimentar al mundo y salvar la biodiversidad silvestre*. Future Harvest Foundation y World Conservation Union (IUCN).
Medina, F. M., Bonnaud, E., Vidal, E., Tershy, B. R., Zavaleta, E. S., Josh Donlan, C., et al. (2011). A global review of the impacts of invasive cats on Island endangered vertebrates. *Global Change Biology, 17*(11), 3503–3510.

Memmott, J., Craze, P. G., Waser, N. M., & Price, M. V. (2007). Global warming and the disruption of plant–pollinator interactions. *Ecology Letters, 10*(8), 710–717.
Meredith, M. (1991). Sensory processing in the main and accessory olfactory systems: Comparisons and contrasts. *The Journal of Steroid Biochemistry and Molecular Biology, 39*(4), 601–614.
Michael, G. (2011). Extraterrestrial aliens: Friends, foes, or just curious? *Skeptic (Altadena, CA), 16*(3), 46–54.
Miller, B., Ceballos, G., & Reading, R. (1994). The prairie dog and biotic diversity. *Conservation Biology, 8*(3), 677–681.
Miller, B., Wemmer, C., Biggins, D., & Reading, Z. (1990). A proposal to conserve black-footed ferrets and the prairie dog ecosystem. *Environmental Management, 14*, 763–769.
Molina-Montenegro, M. A., Atala, C., & Gianoli, E. (2010). Phenotypic plasticity and performance of *Taraxacum officinale* (dandelion) in habitats of contrasting environmental heterogeneity. *Biological Invasions, 12*(7), 2277–2284.
Montesinos-Navarro, A., Pico, F. X., & Tonsor, S. J. (2012). Clinal variation in seed traits influencing life cycle timing in *Arabidopsis thaliana*. *Evolution, 66*, 3417–3431.
Morello, J., Matteucci, S., Rodriguez, A., & Silva, M. (2012). *Ecorregiones y complejos ecosistémicos Argentinos*. Buenos Aires.
Moros, M. (2004). *Seres extraordinarios: Anomalías, deformidades y rarezas humanas* (Vol. 1).
Mourenza, Á., Lorente-Torres, B., Durante, E., Llano-Verdeja, J., Aparicio, J. F., Fernández-López, A., Gil, J. A., Mateos, L. M., & Letek, M. (2022). Understanding microRNAs in the context of infection to find new treatments against human bacterial pathogens. *Antibiotics, 11*(3), 356.
Muench, D., Muench, M., & Gilders, M. A. (2000). *Primal forces*. Graphic Arts Center Publishing.
Mukhopadhyay, B., Shukla, R. M., Mukhopadhyay, M., Mandal, K. C., Haldar, P., & Benare, A. (2012). Spectrum of human tails: A report of six cases. *Journal of Indian Association of Pediatric Surgeons, 17*, 23.
Muller, G. B. (2002). Vestigial organs and structures. In *Encyclopedia of evolution* (Vol. 2, pp. 1131–1133).
Muñoz, A. A., & Cavieres, L. A. (2008). The presence of a showy invasive plant disrupts pollinator service and reproductive output in native alpine species only at high densities. *Journal of Ecology, 96*, 459–467.
Muñoz-Jiménez, J. L., & Alcántara-Carbajal, J. L. (2017). La cotorra Argentina (*Myiopsitta monachus*) en el Colegio de Postgraduados: ¿una especie invasiva? *Huitzil, 18*(1), 38–52.
Murphy, G. E., Romanuk, T. N., & Worm, B. (2020). Cascading effects of climate change on plankton community structure. *Ecology and Evolution, 10*(4), 2170–2181.
Naretto, S., Blengini, C. S., Cardozo, G., & Chiaraviglio, M. (2016). Pre and postcopulatory traits of *Salvator* male lizards in allopatry and sympatry. *Scientifica, 2016*, 1–9.
Navarro, F. S. (2011). *¿Un futuro sostenible?: El cambio global visto por un científico preocupado*. Universitat de València.
Naveh, Z., & Vernet, J. L. (1991). The palaeohistory of the Mediterranean biota. In *Biogeography of Mediterranean invasions* (pp. 19–32).
Nazari-Sharabian, M., Aghababaei, M., Karakouzian, M., & Karami, M. (2020). Water on Mars—A literature review. *Galaxies, 8*(2), 40.
Nee, S., May, R. M., & Harvey, P. H. (1994). Extinction rates can be estimated from molecular phylogenies. *Philosophical Transactions of the Royal Society of London. Series B: Biological Sciences, 344*(1307), 77–82.
Nei, M. (1972). Genetic distance between populations. *The American Naturalist, 106*(949), 283–292.
Nicola, G. G., Larraz, B., & Cano, A. (2013). Ictiofauna del Tajo. Los peces como indicadores biológicos. In I. B. Larraz & A. Saavedra (Eds.), *El río Tajo, Lecciones del pasado para un futuro mejor* (pp. 157–172). Editorial Ledoria.
Nikaido, M., Noguchi, H., Nishihara, H., Toyoda, A., Suzuki, Y., Kajitani, R., et al. (2013). Coelacanth genomes reveal signatures for evolutionary transition from water to land. *Genome Research, 23*(10), 1740–1748.

Norris, K. (2008). Agriculture and biodiversity conservation: Opportunity knocks. *Conservation Letters, 1*(1), 2–1.

Norris, K. S. (1967). Color adaptation in desert reptiles and its thermal relationships. In *Lizard ecology: A symposium* (pp. 162–229).

Ogg, J. G., Chen, Z. Q., Orchard, M. J., & Jiang, H. S. (2004). The triassic period. In F. M. Gradstein, J. G. Ogg, & A. G. Smith (Eds.), *A geologic time scale 2004* (pp. 271–306). Cambridge University Press.

Ohta, Y., & Nishikimi, M. (1999). Random nucleotide substitutions in primate nonfunctional gene for L-gulono-gamma-lactone oxidase, the missing enzyme in L-ascorbic acid biosynthesis. *Biochimica et Biophysica Acta (BBA)-General Subjects, 1472*(1–2), 408–411.

Okada, H., & Harada, H. (2007). Effects of tillage and fertilizer on nematode communities in Japanese soybean field. *Applied Soil Ecology, 35*, 582–598.

O'Leary, M. A., Bloch, J. I., Flynn, J. J., Gaudin, T. J., Giallombardo, A., Giannini, N. P., et al. (2013). The placental mammal ancestor and the post–K-Pg radiation of placentals. *Science, 339*(6120), 662–667.

Palmer, M. A., Lettenmaier, D. P., Poff, N. L., Postel, S. L., Richter, B., & Warner, R. (2009). Climate change and river ecosystems: Protection and adaptation options. *Environmental Management, 44*(6), 1053–1068.

Parker, G. A. (1970a). Sperm competition and its evolutionary consequences in the insects. *Biological Reviews, 45*, 525–567.

Parker, G. A. (1970b). Sperm competition and its evolutionary effect on copula duration in the fly *Scatophaga stercoraria. Journal of Insect Physiology, 16*, 1301–1328.

Parker, G. A. (1970c). The reproductive behaviour and the nature of sexual selection in *Scatophaga stercoraria* L. (Diptera: Scatophagidae) VII. The origin and evolution of the passive phase. *Evolution, 24*, 774–788.

Parker, G. A. (1978). The evolution of competitive mate searching. *Annual Review of Entomology, 23*, 173–196.

Pavés. (2021). *Las especies autóctonas víctimas del cambio climático*. Verde y azul. Retrieved from https://verdeyazul.diarioinformacion.com/las-especies-autoctonas-victimas-del-cambio-climatico.html.

Paviolo, A. J. (2010). Densidad de Yaguareté (*Panthera onca*) en la selva paranaense: su relación con la disponibilidad de presas, presión de caza y coexistencia con el puma (*Puma concolor*). *Mastozoología Neotropical, 17*(2), 397–398.

Peacock, K. A. (2018). Fermi and Lotka: The odds of survival in a dangerous universe. *Journal of the British Interplanetary Society, 71*, 207–211.

Pellens, R., & Grandcolas, P. (Eds.). (2015). *Biodiversity conservation and phylogenetic systematics: Protecting species in a major extinction crisis*. Springer-Verlag.

Penny, M. (1988). *Rhinos endangered species*. Facts on File Publications.

Peñuelas, J., & Filella, I. (2009). Phenology feedbacks on climate change. *Science, 324*(5929), 887–888.

Peres, C. A., & Palacios, E. (2007). Basin-wide effects of game harvest on vertebrate population densities in Amazonian forests: Implications for animal-mediated seed dispersal. *Biotropica, 39*(3), 304–315.

Peretti, A. V. (2014). Sexual selection in Neotropical species: Rules and exceptions. In R. Macedo & G. Machado (Eds.), *Sexual selection in Neotropical species: Rules and exceptions* (pp. 33–52). Elsevier.

Peretti, A. V., Acosta, L. E., & Benton, T. G. (1999). Sexual cannibalism in scorpions: Fact or fiction? *Biological Journal of the Linnean Society, 68*(4), 485–496.

Peristeraki, P., & Megalofonou, P. (2007). Sharks and rays. In *State of Hellenic fisheries* (Vol. 466, pp. 433–439). HCMR Publications.

Persons, M. H., & Uetz, G. W. (2005). Sexual cannibalism and mate choice decisions in wolf spiders: Influence of male size and secondary sexual characters. *Animal Behaviour, 69*(1), 83–94.

Petrie, M., & Kempenaers, B. (1998). Extra-pair paternity in birds: Explaining variation between species and populations. *Trends in Ecology & Evolution, 13*(2), 52–58.

Piaggio, A. J., Segelbacher, G., Seddon, P. J., Alphey, L., Bennett, E. L., Carlson, R. H., et al. (2017). Is it time for synthetic biodiversity conservation? *Trends in Ecology & Evolution, 32*(2), 97–107.

Pimentel, D., McNair, S., Janecka, J., Wightman, J., Simmonds, C., et al. (2001). Economic and environmental threats of alien plant, animal, and microbe invasions. *Agriculture, Ecosystems and Environment, 84*, 1–20.

Pimm, S., Raven, P., Peterson, A., Şekercioğlu, Ç., & Ehrlich, P. R. (2006). Human impacts on the rates of recent, present, and future bird extinctions. *Proceedings of the National Academy of Sciences USA, 103*, 10941–10946.

Pires, M. M., & Guimarães Jr., P. R. (2013). Interaction intimacy organizes networks of antagonistic interactions in different ways. *Journal of the Royal Society Interface*, 10, 20120649.

Pohorille, A., & Pratt, L. R. (2012). Is water the universal solvent for life? *Origins of Life and Evolution of Biospheres, 42*(5), 405–409.

Polis, G. A. (1981). The evolution and dynamics of intraspecific predation. *Annual Review of Ecology and Systematics, 12*, 225–251.

Polis, G. A., & Sissom, W. D. (1990). Life history. In G. A. Polis (Ed.), *The biology of scorpions* (pp. 161–223). Stanford University Press.

Porté, A. J., Lamarque, L. J., Lortie, C. J., Michalet, R., & Delzon, S. (2011). Invasive *Acer negundo* outperforms native species in non-limiting resource environments due to its higher phenotypic plasticity. *BMC Ecology, 11*(1), 1–13.

Porter, M. L., & Crandall, K. A. (2003). Lost along the way: The significance of evolution in reverse. *Trends in Ecology & Evolution, 18*(10), 541–547.

Post, E. (2013). *Ecology of climate change: The importance of biotic interactions*. Princeton University Press.

Potts, S. G., Biesmeijer, J. C., Kremen, C., Neumann, P., Schweiger, O., & Kunin, W. E. (2010). Global pollinator declines: Trends, impacts and drivers. *Trends in Ecology & Evolution, 25*(6), 345–353.

Premoli, A., Quiroga, P., & Gardner, M. (2013). Araucaria araucana. *Lista Roja de especies amenazadas de la UICN, 2013*, 2.

Prenter, J., MacNeil, C., & Elwood, R. W. (2006). Sexual cannibalism and mate choice. *Animal Behaviour, 71*(3), 481–490.

Presgraves, D. C. (2002). Patterns of postzygotic isolation in Lepidoptera. *Evolution, 56*(6), 1168–1183.

Prud'homme, B., Gompel, N., & Carroll, S. B. (2007). Emerging principles of regulatory evolution. *Proceedings of the National Academy of Sciences, 104*(Suppl 1), 8605–8612.

Radosevich, S. R., Holt, J. S., & Ghersa, C. M. (2007). Evolution of weeds and invasive plants. In S. R. Radosevich, J. S. Holt, & C. M. Ghersa (Eds.), *Ecology of weeds and invasive plants* (pp. 103–128). John Wiley & Sons.

Rampelotto, P. H. (2010). Resistance of microorganisms to extreme environmental conditions and its contribution to astrobiology. *Sustainability, 2*(6), 1602–1623.

Rastogi-Verma, S. (2013). Genetically modified plants: Public and scientific perceptions. *International scholarly research notices, 2013*, 820671.

Rauque Perez, C. A., Waicheim, M. A., Cordero, P., & Blasetti, G. (2017). ¿Qué se esconde tras la introducción de especie?: La invasión de la carpa común en Patagonia. *Desde La Patagonia Difundiendo Saberes, 14*, 23.

Renne, P. R., Deino, A. L., Hilgen, F. J., Kuiper, K. F., Mark, D. F., Mitchell, W. S., et al. (2013). Time scales of critical events around the cretaceous–Paleogene boundary. *Science, 339*(6120), 684–687.

Rezende, E. L., Lavabre, J. E., Guimarães Jr., P. R., Jordano, P., & Bascompte, J. (2007). Non-random coextinctions in phylogenetically structured mutualistic networks. *Nature, 448*(7156), 925–928.

Rhymer, J. M., & Simberloff, D. (1996). Extinction by hybridization and introgression. *Annual Review of Ecology and Systematics, 27*(1), 83–109.

Richard, E., & Juliá, J. P. (2000). El tapir (*Tapirus terrestris*): dieta y manejo en un bosque secundario de la ecoregión de selvas pedemontanas. In *Estatus en Argentina. Manejo de Fauna Silvestre en Amazonía y Latinoamérica. CITES Paraguay, Fund.* Moisés Bertoni y University of Florida, Asunción.

Risser, P. G. (1995). Biodiversity and ecosystem function. *Conservation Biology, 9*(4), 742–746.

Roche, J. R., Mitchell, F. J., & Waldren, S. (2009). Plant community ecology of *Pinus sylvestris*, an extirpated species reintroduced to Ireland. *Biodiversity and Conservation, 18*, 2185–2203.

Rodríguez, J., Cevallos, E., Rodríguez, R., Navarrete, Y. T., & González, M. A. (2020). Estudio comparativo de supervivencia de seis especies de peces nativos continentales en su medio natural. *Alternativas, 21*(3), 55–59.

Roels, S. A., & Kelly, J. K. (2011). Rapid evolution caused by pollinator loss in *Mimulus guttatus*. *Evolution, 65*(9), 2541–2552.

Roesler, I., Imberti, S., Casanas, H., & Volpe, N. (2012). A new threat for the globally endangered hooded grebe *Podiceps gallardoi*: The American mink *Neovison vison*. *Bird Conservation International, 22*(4), 383–388.

Rognini, P. (2018). Vestigial drifting drives in *Homo sapiens*. *Biological Theory, 13*(3), 199–211.

Root, T. L., Price, J. T., Hall, K. R., Schneider, S. H., Rosenzweig, C., & Pounds, J. A. (2003). Fingerprints of global warming on wild animals and plants. *Nature, 421*(6918), 57–60.

Ropper, A. H., & Brown, R. H. (2005). Normal development and deviations in development of the nervous system. In *Adams and Victor's Principles of Neurology* (pp. 493–518).

Rothschild, L. J., & Mancinelli, R. L. (2001). Life in extreme environments. *Nature, 409*(6823), 1092–1101.

Royer, A. M., Kremer, C., George, K., Pérez, S. G., Schemske, D. W., & Conner, J. K. (2016). Incomplete loss of a conserved trait: Function, latitudinal cline, and genetic constraints. *Evolution, 70*(12), 2853–2864.

Rozkovcová, E., Markova, M., & Dolejsi, J. (1999). Studies on agenesis of third molars amongst populations of different origin. *Sbornik lekarsky, 100*(2), 71–84.

Rubio-Licona, L. E., Romero-Rangel, S., Rojas-Zenteno, E. C., Durán-Díaz, Á., & Gutiérrez-Guzmán, J. C. (2011). Variación del tamaño de frutos y semillas en siete especies de encino (*Quercus*, Fagaceae). *Polibotánica, 32*, 135–151.

Rudel, T. K., Schneider, L., Uriarte, M., Turner, B. L., DeFries, R., Lawrence, D., et al. (2009). Agricultural intensification and changes in cultivated areas, 1970–2005. *Proceedings of the National Academy of Sciences, 106*(49), 20675–20680.

Ruppert, E. E., Fox, R. S., & Barnes, R. D. (2003). *Invertebrate zoology: A functional evolutionary approach* (7th ed., p. 1008). Cengage Learning.

Sahney, S., & Benton, M. J. (2008). Recovery from the most profound mass extinction of all time. *Proceedings of the Royal Society B: Biological Sciences, 275*(1636), 759–765.

Salgado, I. (2015). Mapache–*Procyon lotor* (Linnaeus, 1758). In A. Salvador & I. Barja (Eds.), *Enciclopedia Virtual de los Vertebrados Españoles*. Museo Nacional de Ciencias Naturales. http://www.vertebradosibericos.org/.

Salvetti, V. P. (2019). Ideas que cambian la historia: el boceto tree of life. *ALMA Cultura & Medicina, 5*(4), 24–42.

Sánchez-Guillén, R. A., Córdoba-Aguilar, A., Hansson, B., Ott, J., & Wellenreuther, M. (2016). Evolutionary consequences of climate-induced range shifts in insects. *Biological Reviews, 91*(4), 1050–1064.

Saraga-Babić, M., Lehtonen, E., Švajger, A., & Wartiovaara, J. (1994). Morphological and immunohistochemical characteristics of axial structures in the transitory human tail. *Annals of Anatomy-Anatomischer Anzeiger, 176*(3), 277–286.

Sasa, M. M., Chippindale, P. T., & Johnson, N. A. (1998). Patterns of postzygotic isolation in frogs. *Evolution, 52*(6), 1811–1820.

Saunders, W. B., & Landman, N. H. (2010). *Nautilus*. Springer Science & Business Media.

Schechter, S. P., & Bruns, T. D. (2008). Serpentine and non-serpentine ecotypes of *Collinsia sparsiflora* associate with distinct arbuscular mycorrhizal fungal assemblages. *Molecular Ecology, 17*, 3198–3210.

Schemske, D. W. (2009). Biotic interactions and speciation in the tropics. In R. K. Butlin, J. R. Bridle, & D. Schluter (Eds.), *Speciation and patterns of diversity* (pp. 219–239). Cambridge Univ. Press.

Schmitt, T. (2007). Molecular biogeography of Europe: Pleistocene cycles and postglacial trends. *Frontiers in Zoology, 4*, 11.

Schneider, J. M., & Elgar, M. A. (2001). Sexual cannibalism and sperm competition in the golden orb-web spider *Nephila plumipes* (Araneoidea): Female and male perspectives. *Behavioral Ecology, 12*(5), 547–552.

Schneider, W. T., Rutz, C., Hedwig, B., & Bailey, N. W. (2018). Vestigial singing behaviour persists after the evolutionary loss of song in crickets. *Biology Letters, 14*(2), 20170654.

Schweiger, O., Heikkinen, R. K., Harpke, A., Hickler, T., Klotz, S., Kudrna, O., Kühn, I., Pöyry, J., & Settele, J. (2012). Increasing range mismatching of interacting species under global change is related to their ecological characteristics. *Global Ecology and Biogeography, 21*(1), 88–99.

Scribailo, R. W., & Barrett, S. C. (1994). Effects of prior self-pollination on outcrossed seed set in tristylous *Pontederia sagittata* (Pontederiaceae). *Sexual Plant Reproduction, 7*, 273–281.

Seager, S. (2003). The search for extrasolar earth-like planets. *Earth and Planetary Science Letters, 208*(3–4), 113–124.

Seburn, D., Birch, M., Burns, M., Hounsell, R., Neville, S., & Nguyen, C. (2023). Do exotic red-eared sliders (*Trachemys scripta elegans*) bask with native Canadian turtles? *Herpetology Notes, 16*, 501–506.

Shapiro, R., & Schulze-Makuch, D. (2009). The search for alien life in our solar system: Strategies and priorities. *Astrobiology, 9*(4), 335–343.

Sheehan, P. M. (2001). The late Ordovician mass extinction. *Annual Review of Earth and Planetary Sciences, 29*(1), 331–364.

Simmons, L. G. (2001). *Sperm competition and its evolutionary consequences in insects*. Princeton University Press.

Smith, D. S., Bailey, J. K., Shuster, S. M., & Whitham, T. G. (2011). A geographic mosaic of trophic interactions and selection: Trees, aphids and birds. *Journal of Evolutionary Biology, 24*(2), 422–429.

Smith, F. A., Doughty, C. E., Malhi, Y., Svenning, J. C., & Terborgh, J. (2016). Megafauna in the earth system. *Ecography, 39*, 99–108.

Smith, H. A. (2011). Alone in the universe: Despite the growing catalog of extrasolar planets, data so far do not alter estimates that we are effectively on our own. *American Scientist, 99*(4), 320–327.

Smith, H. F., & Wright, W. (2018). Vestigial organ. In *Encyclopedia of animal cognition and behavior*. Springer, Cham.

Smith, R. L. (1984). *Sperm competition and the evolution of animal mating systems*. Academic Press.

Solomon, S., Qin, D., Manning, M., Averyt, K., Marquis, M., & (Eds.). (2007). *Climate change 2007—The physical science basis: Working group I contribution to the fourth assessment report of the IPCC* (Vol. 4). Cambridge University Press.

Song, H., Wignall, P. B., Tong, J., & Yin, H. (2013). Two pulses of extinction during the Permian–Triassic crisis. *Nature Geoscience, 6*(1), 52–56.

Spence, J. R., & Cárcamo, H. A. (1991). Effects of cannibalism and intraguild predation on pondskaters (Gerridae). *Oikos, 62*(3), 333–341.

Spiegelmann, R., Schinder, E., Mintz, M., & Blakstein, A. (1985). The human tail: A benign stigma: Case report. *Journal of Neurosurgery, 63*(3), 461–462.

Springate, D. A., & Kover, P. X. (2014). Plant responses to elevated temperatures: A field study on phenological sensitivity and fitness responses to simulated climate warming. *Global Change Biology, 20*, 456–465.

Steadman, D. W. (2006). *Extinction and biogeography of tropical Pacific birds*. Chicago University Press.

Steffan-Dewenter, I., Potts, S. G., & Packer, L. (2005). Pollinator diversity and crop pollination services are at risk. *Trends in Ecology & Evolution, 20*(12), 651–652.

Sternai, P., Caricchi, L., Garcia-Castellanos, D., Jolivet, L., Sheldrake, T. E., & Castelltort, S. (2017). Magmatic pulse driven by sea-level changes associated with the Messinian salinity crisis. *Nature Geoscience, 10*(10), 783–787.

Stork, N. E., Samways, M. J., & Eeley, H. A. (1996). Inventorying and monitoring biodiversity. *Trends in Ecology and Evolution, 11*, 39–40.

Straus, C., Vasilakos, K., Wilson, R. J. A., Oshima, T., Zelter, M., Derenne, J. P., & Whitelaw, W. A. (2003). A phylogenetic hypothesis for the origin of hiccough. *BioEssays, 25*(2), 182–188.

Sue, H. D. (2019). *The rise of reptiles: 320 million years of evolution*. John Hopkins University Press.

Sue, H. D., & Fraser, N. C. (2010). *Triassic life on land: The great transition*. Columbia University Press.

Tamatsu, Y., Tsukahara, K., Hotta, M., & Shimada, K. (2007). Vestiges of vibrissal capsular muscles exist in the human upper lip. *Clinical Anatomy: The Official Journal of the American Association of Clinical Anatomists and the British Association of Clinical Anatomists, 20*(6), 628–631.

Tang-Martínez, Z. (2016). Rethinking Bateman's principles: Challenging persistent myths of sexually reluctant females and promiscuous males. *The Journal of Sex Research, 53*(4–5), 532–559.

Teerikorpi, P., Valtonen, M., Lehto, K., Lehto, H., Byrd, G., Chernin, A., et al. (2019). Finite or infinite universe: Cosmological models. In *The evolving universe and the origin of life* (pp. 329–350). Springer.

Teichert, C., & Matsumoto, T. (1987). The ancestry of the genus *nautilus*. *Topics in Geobiology, 6*, 25–32.

Thompson, J. N. (2005). *The geographic mosaic of coevolution*. University of Chicago Press.

Thompson, J. R., Gosling, S. N., Zaherpour, J., & Laizé, C. L. R. (2021). Increasing risk of ecological change to major rivers of the world with global warming. *Earth's. Futures, 9*(11), e2021EF002048.

Tilley, S. G., Verrell, P. A., & Arnold, S. J. (1990). Correspondence between sexual isolation and allozyme differentiation: A test in the salamander *Desmognathus ochrophaeus*. *Proceedings of the National Academy of Sciences, 87*(7), 2715–2719.

Tinghitella, R. M., & Zuk, M. (2009). Asymmetric mating preferences accommodated the rapid evolutionary loss of a sexual signal. *Evolution, 63*, 2087–2098.

Tscharntke, T. (2002, August). Vertebrate effects on plant-invertebrate food webs. In *Multitrophic interactions in terrestrial systems: 36th symposium of the British Ecological Society* (Vol. 36, p. 277). Cambridge University Press.

Turbet, M., Bolmont, E., Bourrier, V., Demory, B. O., Leconte, J., Owen, J., & Wolf, E. T. (2020). A review of possible planetary atmospheres in the TRAPPIST-1 system. *Space Science Reviews, 216*(5), 1–48.

Turvey, S. T. (Ed.). (2009). *Holocene extinctions*. Oxford University Press.

Uribe Salas, M. D. (2009). *Variación morfológica y estructura genética en poblaciones naturales de Quercus rugosa Née (Fagaceae) en México*. Universidad Michoacana de San Nicolás de Hidalgo, México.

Vahed, K. (1998). The function of nuptial feeding in insects: Review of empirical studies. *Ethology, 113*, 105–127.

Valladares, F., Matesanz, S., Guilhaumon, F., Araujo, M. B., Balaguer, L., Benito-Garzon, M., et al. (2014). The effects of phenotypic plasticity and local adaptation on forecasts of species range shifts under climate change. *Ecology Letters, 17*(11), 1351–1364.

Valverde, J. A. (1971). El lobo español. *Montes, 159*, 229–241.

van der Kooi, C. J., & Schwander, T. (2014). On the fate of sexual traits under asexuality. *Biological Reviews, 89*(4), 805–819.

Van Grootel, V., Fernandes, C. S., Gillon, M., Jehin, E., Manfroid, J., Scuflaire, R., et al. (2018). Stellar parameters for Trappist-1. *The Astrophysical Journal, 853*(1), 30.
Vanbergen, A. J., & Initiative, T. I. (2013). Threats to an ecosystem service: Pressures on pollinators. *Frontiers in Ecology and the Environment, 11*(5), 251–259.
Van't Woudt, B. D. (1990). Roaming, stray, and feral domestic cats and dogs as wildlife problems. In *Proceedings of the vertebrate Pest conference* (Vol. 14, p. 14).
Vieilledent, G., Grinand, C., Rakotomalala, F. A., Ranaivosoa, R., Rakotoarijaona, J. R., Allnutt, T. F., & Achard, F. (2018). Combining global tree cover loss data with historical national forest cover maps to look at six decades of deforestation and forest fragmentation in Madagascar. *Biological Conservation, 222*, 189–197.
Vila, C., Castroriejo, J., & Urios, V. (1993). The Iberian wolf in Spain. (proceeding). In *Wolves in Europe-current status and prospects, Oberammergau (Germany), 1992*. Wildbiologische Gesellschaft.
Villenave, C., Bongers, T., Ekschmitt, K., Djigal, D., & Chotte, J. L. (2001). Changes in nematode communities following cultivation of soil after fallow periods of different length. *Applied Soil Ecology, 17*, 43–52.
Visser, M. E. (2016). Phenology: Interactions of climate change and species. *Nature, 535*, 236–237.
Visser, M. E., & Both, C. (2005). Shifts in phenology due to global climate change: The need for a yardstick. *Proceedings of the Royal Society B: Biological Sciences, 272*(1581), 2561–2569.
Vogel, J. (2020). An RNA biology perspective on species-specific programmable RNA antibiotics. *Molecular Microbiology, 113*(3), 550–559.
Vörösmarty, C. J., Green, P., Salisbury, J., & Lammers, R. B. (2000). Global water resources: Vulnerability from climate change and population growth. *Science, 289*(5477), 284–288.
Vrech, D. E., Olivero, P. A., Mattoni, C. I., & Peretti, A. V. (2018). From storage to delivery: Sperm volume and number of spermatozoa inside storage organs and ejaculates in males of *Timogenes elegans* (Scorpiones: Bothriuridae). *Journal of Arachnology, 46*(2), 231–239.
Walker-Larsen, J., & Harder, L. D. (2001). Vestigial organs as opportunities for functional innovation: The example of the *Penstemon* staminode. *Evolution, 55*(3), 477–487.
Ward, P. (1984). Is *nautilus* a living fossil? In N. Eldredge & S. M. Stanley (Eds.), *Living fossils* (pp. 247–256). Springer Verlag.
Ward, P. D., & Brownlee, D. (2000). Testing the rare earth hypothesis. In P. D. Ward & D. Brownlee (Eds.), *Rare earth: Why complex life is uncommon in the universe* (pp. 243–256). Springer.
Warner, M. E., Fitt, W. K., & Schmidt, G. W. (1999). Damage to photosystem II in symbiotic dinoflagellates: A determinant of coral bleaching. *Proceedings of the National Academy of Sciences, 96*(14), 8007–8012.
Waser, N. M. (1978). Interspecific pollen transfer and competition between co-occurring plant species. *Oecologia, 36*(1), 223–236.
Watanabe, H., & Yano, E. (2012). Behavioral response of male mantid *Tenodera aridifolia* (Mantodea: Mantidae) to windy conditions as a female approach strategy. *Entomological Science, 15*(4), 384–391.
Waters, C. N., Zalasiewicz, J., Summerhayes, C., Barnosky, A. D., Poirier, C., Gałuszka, A., et al. (2016). The Anthropocene is functionally and stratigraphically distinct from the Holocene. *Science, 351*(6269), aad2622.
Webb, S. (2002). *If the universe is teeming with aliens… Where is everybody?: Fifty solutions to the Fermi paradox and the problem of extraterrestrial life*. Copernicus Books.
Wellenreuther, M., Tynkkynen, K., & Svensson, E. I. (2010). Simulating range expansion: Male species recognition and loss of premating isolation in damselflies. *Evolution, 64*(1), 242–252.
Werth, A. J., & Shear, W. A. (2014). The evolutionary truth about living fossils. *American Scientist, 102*, 434–443.
Wessel, L. (2017). More than 700 species are already feeling the strain from climate change. *Science*. www.science.org/content/article/more-700-species-are-already-feeling-strain-climate-change.
Westall, F. (1999). The nature of fossil bacteria: A guide to the search for extraterrestrial life. *Journal of Geophysical Research: Planets, 104*(E7), 16437–16451.

Whittier, T. S., Kaneshiro, K. Y., & Prescott, L. D. (1992). Mating behavior of Mediterranean fruit flies (Diptera: Tephritidae) in a natural environment. *Annals of the Entomological Society of America, 85*(2), 214–218.
Wigby, S., & Chapman, T. (2005). Sex peptide causes mating costs in female *drosophila melanogaster*. *Current Biology, 15*, 316–321.
Willmer, P. (2014). Climate change: Bees and orchids lose touch. *Current Biology, 24*(23), R1133–R1135.
Willmer, P., Stone, G., & Johnston, I. (2009). *Environmental physiology of animals* (p. 644). John Wiley & Sons.
Wilson, G. P., Evans, A. R., Corfe, I. J., Smits, P. D., Fortelius, M., & Jernvall, J. (2012). Adaptive radiation of multituberculate mammals before the extinction of dinosaurs. *Nature, 483*(7390), 457–460.
Woinarski, J. C., Burbidge, A. A., & Harrison, P. L. (2015). Ongoing unraveling of a continental fauna: Decline and extinction of Australian mammals since European settlement. *Proceedings of the National Academy of Sciences, 112*(15), 4531–4540.
Wolfe, M. D., & Tonsor, S. J. (2014). Adaptation to spring heat and drought in northeastern Spanish *Arabidopsis thaliana*. *New Phytologist, 201*, 323–334.
Woolbright, S. A., Whitham, T. G., Gehring, C. A., Allan, G. J., & Bailey, J. (2014). Climate relicts and their associated communities as natural ecology and evolution laboratories. *Trends in Ecology & Evolution, 29*(7), 406–416.
Xia, B., Zhang, W., Wudzinska, A., Huang, E., Brosh, R., Pour, M., et al. (2021). The genetic basis of tail-loss evolution in humans and apes. *BioRxiv*. 2021–09.
Xiong, Y. E., Brandley, M. C., Xu, S., Zhou, K., & Yang, G. (2009). Seven new dolphin mitochondrial genomes and a time-calibrated phylogeny of whales. *BMC Evolutionary Biology, 9*(1), 1–13.
Yamamoto, Y., & Jeffery, W. R. (2000). Central role for the lens in cave fish eye degeneration. *Science, 289*(5479), 631–633.
Yamasaki, E., & Sakai, S. (2013). Wind and insect pollination (ambophily) of *Mallotus* spp. (Euphorbiaceae) in tropical and temperate forests. *Australian Journal of Botany, 61*(1), 60–66.
Young, H. S., McCauley, D. J., Galetti, M., & Dirzo, R. (2016). Patterns, causes, and consequences of anthropocene defaunation. *Annual Review of Ecology, Evolution, and Systematics, 47*, 333–358.
Yu, C. (2015). The dark Forest rule: One solution to the Fermi paradox. *Journal of the British Interplanetary Society, 68*, 142–144.
Yuval, B., & Hendrichs, J. (1999). Behavior of flies in the genus Ceratitis (Dacinae: Ceratitidini). In *Fruit Flies (Tephritidae)* (pp. 447–476). CRC Press.
Zapata-Ríos, G., & Branch, L. C. (2016). Altered activity patterns and reduced abundance of native mammals in sites with feral dogs in the high Andes. *Biological Conservation, 193*, 9–16.
Zardoya, R. (2010). El árbol de la vida: La filogenia de la biodiversidad.
Zaya, D. N., & Howe, H. F. (2009). The anomalous Kentucky coffeetree: Megafaunal fruit sinking to extinction? *Oecologia, 161*(2), 221–226.
Zayed, A., & Packer, L. (2005). Complementary sex determination substantially increases extinction proneness of haplodiploid populations. *Proceedings of the National Academy of Sciences, 102*(30), 10742–10746.
Zhang, N., Belsterling, B., Raszewski, J., & Tonsor, S. J. (2015). Natural populations of *Arabidopsis thaliana* differ in seedling responses to high-temperature stress. *AoB Plants, 7*, plv101.
Zuk, M., Bastiaans, E., Langkilde, T., & Swanger, E. (2014). The role of behaviour in the establishment of novel traits. *Animal Behaviour, 92*, 333–344.
Zuk, M., & Kolluru, G. R. (1998). Exploitation of sexual signals by predators and parasitoids. *The Quarterly Review of Biology, 73*(4), 415–438.
Zuk, M., Rotenberry, J. T., & Tinghitella, R. M. (2006). Silent night: Adaptive disappearance of a sexual signal in a parasitized population of field crickets. *Biology Letters, 2*, 521–524.

Chapter 8
Beyond Biology

There are no laws, no traditions, and no rules that can be universally applied, including this one

— Anonymous

Contents

A. V. Peretti et al., *Rules and Exceptions in Biology: from Fundamental Concepts to Applications*, https://doi.org/10.1007/978-3-031-55382-0_8

Abstract In chapter 8, we focus on the dynamics of changes between rules and exceptions in sciences linked to biology and in other sciences further removed from it. Within the sciences related to biology, we can name changes in medicine such as the interpretations of rare diseases and changes in the conception of diseases such as autism and obesity. Changes also occur in the fields of physics, chemistry, geology, and astronomy. In sciences further removed from biology, we also find examples of a dynamic between rules and exceptions. Within the social sciences, changes in interpretations linked, for example, to original culturfes and the concept of family; in the field linked to art with changes about painting, music, literature, and calligraphy, for example, and changes in the interpretations of themes associated with sports and religion where knowledge and interpretations are also changing throughout the time.

8.1 Introduction

In this chapter, we will move beyond the exploration of rules and exceptions solely in biology and extend the scope to other branches of science that are interconnected with biology. By doing so, we will uncover that the perception and analysis of rules and exceptions extend far beyond the realm of biology. In contrast, it is a more globalizing phenomenon, as it was exposed at the beginning of this book. Moreover, this chapter also includes other sciences and disciplines less related to biology, which are present in our daily lives. We intend to mention some cases to illustrate that there are many options and situations beyond the focus of this book that is centered on biology.

8.2 R&E in Sciences More Closely Related to Biology

8.2.1 Medicine and Related Sciences

Indeed, the evolution of medicine over time has been remarkable, leading to significant changes in its norms and rules. When we look back to ancient Greece, we can imagine how primitive and limited the facilities for caring for the sick were in comparison to modern healthcare settings. Despite the challenges of the past, ancient civilizations contributed valuable knowledge to medicine that has endured through the ages. For example, in the Temple of Kom Ombo in Egypt, carved hieroglyphs show medical instruments very similar to those still used in hospitals today, which are thought to have been used for cataract surgery (Butrous et al., 2020; Dobanovacki et al., 2012). They used honey as a natural antiseptic (Kuropatnicki et al., 2018), prepared medicines, and even practiced various surgeries (Raudvere, 2020). Thus, like these stories, we could cite many throughout history and civilizations.

Over centuries, medicine has experienced incredible advancements in understanding the human body's functions, diagnostics, and treatment techniques. Today, we benefit from a deeper comprehension of diseases, cutting-edge technologies, and refined medical practices, allowing for more precise and effective healthcare interventions. Currently, the already massive presence of specialized technology in health centers is an implicit rule in diagnosing and treating diseases. We now have monitoring devices, lasers, magnetic resonance devices with unimaginable resolutions, increasingly precise ultrasound scanners, and machines that aid rehabilitation. Today, it is a rule that medicine is a therapeutic and prophylactic practice. In daily medical practice, the physician examines the patient, orders tests, diagnoses, and medicates, if necessary, in general with synthetic drugs.

Western medicine is the most widespread and is the worldwide rule even in countries where alternative medicine is widely practiced. The so-called alternative treatments come from the oldest treatments based mainly on plant properties, meditation, and other ancient techniques. They are treatments that were perhaps the norm in other times and never disappeared, remaining as exceptions. Due to fashion or the failure of synthetic drugs, they often come back and become more common. In fact, in Latin America, multiculturalism and pluri-ethnicity show diverse medical systems (Duarte-Gómez, 2003). Examples include indigenous medicine, home medicine, yoga for stress relief, herbal teas for various conditions, acupuncture, and cannabis for pain relief.

Beyond the fact that almost half of the world's population does not have access to essential health services, the exceptions in medicine can be clearly seen in diseases. First, let us talk about a group collectively referred to as "rare" diseases.

8.2.1.1 Rare Diseases: A Way to Better Understand Common Patterns?

When thinking about exceptions in medicine, one of the most concrete examples is "Rare Diseases," which involve uncommon conditions (Table 8.1). These diseases affect the health to a greater or lesser extent of those who suffer from them. Different

Table 8.1 A list of examples of rare diseases. This table aims to understand the cause, and impact of rare diseases on the population and the individual

Rare disease	Cause	Effect	Population impact	Diagnosis	Consequences	References
Marfan syndrome	Genetic mutation	Weakness in connective tissue, affecting the eyes, heart, lungs, bones, and skin	Affects 1 in 5000 to 10,000 people worldwide	Clinical evaluation, genetic testing	Affects quality of life and increases the risk of potentially fatal complications such as aortic rupture	Dietz, (1996), Pyeritz, (2000, Du et al., (2021)
Huntington's disease	Genetic mutation	Degeneration of nerve cells in the brain, affecting movement, cognition, and behavior	Affects 1 in 10,000 people worldwide		Gradual loss of the ability to control movement and cognition, which can lead to disability and death	Vonsattel & DiFiglia, (1998), Walker, (2007), Ross & Tabrizi, (2011)
Cystic fibrosis	Genetic mutation	Accumulation of thick, sticky mucus in the lungs and other organs, causing recurrent respiratory infections, digestive problems, and other symptoms	Affects 1 in 2500 to 3500 people worldwide	Clinical evaluation, sweat test, genetic testing	Decreased life expectancy and increased risk of serious complications such as respiratory failure	Kock & Hoiby, (1993), Hodson et al., (2012)
Williams syndrome	Genetic deletion	Cognitive and learning difficulties, distinctive facial features, and cardiac and blood circulation problems	Affects 1 in 7500 people worldwide	Clinical evaluation, genetic testing	Can affect learning ability and social adaptation, and increase the risk of cardiac and blood circulation problems	Morris & Mervis, (2000, 2021)

Table 8.1 (continued)

Rare disease	Cause	Effect	Population impact	Diagnosis	Consequences	References
Gaucher disease	Genetic mutation	Accumulation of a fatty substance in the organs and tissues of the body, causing liver and spleen problems, and bone and joint pain	Affects 1 in 40,000 people worldwide	Clinical evaluation, genetic testing, blood analysis	Affects quality of life and increases the risk of serious complications such as organ failure	Lee, (1982), Sidransky, (2004), Pastores & Hughes, (2018)
Tuberous sclerosis	Genetic mutation	Formation of benign tumors in various organs of the body, including the brain, kidneys, and lungs, causing seizures, cognitive problems, and other symptoms	Affects 1 in 6000 people worldwide	Clinical evaluation, genetic testing, imaging analysis	Can affect learning ability and social adaptation, and increase the risk of serious complications such as organ failure	Gomez et al., (1999), Crino et al., (2006), Curatolo et al., (2008)
Niemann-pick disease	Genetic mutation	Accumulation of lipids in cells, causing neurological and hepatic problems, and pulmonary dysfunction	Affects 1 in 150,000 people worldwide	Clinical evaluation, genetic testing, blood analysis	Can affect quality of life and increase the risk of serious complications such as respiratory failure	Kolodny, (2000), Vanier, (2013)
Pompe disease	Genetic mutation	Inability to fully breakdown glycogen, causing progressive muscle weakness, respiratory and cardiac problems	Affects 1 in 40,000 people worldwide			Mellies & Lofaso, (2009), Kroos et al., (2012), Leslie & Bailey, (1993)

(continued)

Table 8.1 (continued)

Rare disease	Cause	Effect	Population impact	Diagnosis	Consequences	References
Tay–Sachs disease	Genetic mutation	Accumulation of a fatty substance in the brain, causing progressive neurological problems and pulmonary dysfunction	It mainly affects people of Ashkenazi Jewish descent and other rare ethnic groups	Clinical evaluation, genetic testing	It can affect the quality of life and increase the risk of serious complications such as respiratory failure	Gravel et al., (1991), Fernandes-Filho & Shapiro, (2004)
Rett syndrome	Genetic mutation	Progressive neurological problems, including motor, cognitive, and communication issues	Affects mainly girls and is caused by a mutation in the MECP2 gene		It can significantly impact quality of life and increase the risk of serious complications such as cognitive and physical disability	Hagberg et al., (1985), Smeets et al., (2012)

countries and organizations have approximations for considering a disease rare (see Richter et al., 2015). However, in general, it ranges from 5 to 86 per 100,000 people (Nguengang Wakap et al., 2020; Richter et al., 2015). The approximate number of rare diseases is not well defined, especially for their rarity, but ranges between 5000–8000 rare diseases. New rare diseases are discovered every year, and the truth is that being exceptions as such, they do not have the entire research interest of pharmaceutical companies or governments. This lack of interest arises mainly because there is no business behind them due to the low sales of potential drugs (Drummond et al., 2007). New drugs are being added to treat some of them (e.g., 560 new drugs in 2015) (Phrma.org, 2021).

Rare diseases include Mendelian, infectious, or rare cancers (Haendel et al., 2020). We can mention numerous possibilities resulting from viral or bacterial infections, allergies, degenerative diseases, and chromosomal disorders, among others (Haendel et al., 2020; National Center for Advancing Translational Sciences, 2022; Zhu et al., 2020). Mendelian diseases are somewhat easier to detect, as they appear in the family in predictable proportions, do not appear in in-laws, and manifest at approximately the same age (Buchanan et al., 2006). In contrast, complex diseases are the most common diseases affecting the population. The cause is multifactorial and depends on a combination of genetics, environment, and lifestyle factors, among other factors (Craig, 2008). Examples of the latter are complex chronic diseases such as cardiovascular conditions, asthma, diabetes, Alzheimer's disease, cancer, and psychiatric diseases that have become the focus of attention of epidemiologists and geneticists as they are more prevalent in health importance (Hunter, 2005).

Despite investment in research, the causes of these diseases remain poorly understood (Buchanan et al., 2006). Here, it is evident from the book's point of view that rare diseases are the exception in medicine, and complex diseases are the rule. Nevertheless, we should focus on the fact that the method of diagnosis and the feasibility of obtaining economic benefits for pharmaceutical companies are closely linked to labeling diseases as one or the other. We should not be surprised to find in the future that the casuistry of rare diseases is not as low as we think (McMaster, 2021), similar to what we will discuss with autism spectrum diseases later on.

From a scientific point of view, these diseases can often be the gateway to understanding poorly understood biological and physiological pathways. This process is similar to the study of teratologies to better understand some ontogenetic developmental processes, as discussed in Chap. 5. Here, we highlight what has been commented on in some journals and blogs (Craig, 2008; Peltonen & Uusitalo, 1997; Reid, 2016). The study of the molecular biology underlying "Rare Diseases" helps detect dysfunctions and novel biochemical pathways, which indirectly contributes to a better understanding of biological processes in general (Berman, 2014; Griggs et al., 2009). Many important discoveries arising from the study of rare diseases often contribute to a better understanding of more common conditions (Griggs et al., 2009). In his popular science blog, J. Reid (2016) highlights the benefits of these studies, particularly in clarifying the normal functioning of biochemical patterns.

For example, the study of mutations on the ob gene encoding leptin. This dysfunction produces insulin resistance and endocrine abnormalities. The study of congenital leptin deficiency and other Mendelian obesity disorders has taught much about the processes that generate overweight and obesity, such as those associated with insulin (Blüher & Mantzoros, 2009; Fatima et al., 2011; Reid, 2016). Thanks to advances in technology and molecular analyses, studies of rare diseases have been favored by advances in genetic labeling techniques, such as labeling with microRNAs (miRs). Recently, the expression of miR patterns in body fluids demonstrated their importance as noninvasive biomarkers for different diseases (Salvatore et al., 2011). The human genome project demonstrated that RNA transcripts are more numerous than protein-coding genes. This fact has generated a paradigm shift where the rule was that one RNA produced one protein (Hanna et al., 2019). The miRs are regulated and transcribed in the same manner as protein-coding genes. miR biogenesis involves separate transduction and processing steps in which the active site of 20–22 nucleotides is extracted from a longer hairpin RNA precursor. These 20–22 nucleotides are fused into partial duplexes between the short RNAs and the untranslated regions (UTRs) of targeted mRNAs, leading to silencing mRNA translation in mammals (Salvatore et al., 2011). miRs regulate gene expression and are related to many biological activities, such as tumorigenesis, immune response, neurotransmitter synthesis, insulin secretion, and even circadian rhythm (Zhou & Yang, 2012). Identifying these disease-linked mRNAs is crucial to understanding disease pathogenesis at the molecular level and helps design molecular tools for prevention, diagnosis, and treatment (Salvatore et al., 2011; Zhou & Yang, 2012).

Outside the technicalities, it is interesting to see that some diseases are rare in some regions but not others. For example, Tay–Sachs is rare in the general population but very common in the Ashkenazi Jewish group (Charrow, 2004; Myerowitz & Costigan, 1988). Tuberculosis is rare in developed societies such as the United States but is in the top 10 most frequent causes of death worldwide (Harding, 2020). Another example is Zika or Malaria, which are rare in Europe or the USA but very common in South America, Africa, or Southeast Asia (Beier et al., 1999; CDC, 2022a; Kazmi et al., 2020; Kumar et al., 2012; Yakob & Walker, 2016). Something similar is true for thalassemia. Thalassemia is an inherited blood disorder characterized by insufficient production of hemoglobin in red blood cells, which is responsible for transporting oxygen to all cells. The severity of the condition can vary, ranging from no symptoms or mild anemia to cases requiring blood transfusions (CDC, 2022b). Between its two forms, the beta form prevails punctually among the Mediterranean population (Greer et al., 2013), being relatively rare in the rest of the world, except in the Maldives, where the highest concentration of affected individuals is found (Waheed et al., 2016). With these examples, we see that the rarity of a disease can also be contextual (Haendel et al., 2020).

8.2.1.2 Exceptions that are Becoming More Common in Medicine

Next, let us discuss some examples of exceptions that become progressively more common in medicine. This shift includes certain diseases, medical conditions, and the use of specific biomedical tools.

A Case in Point: Autism, Real Increase or Revision of Diagnostic Parameters?

In recent decades, there has been a noticeable rise in the diagnosis and prevalence of disorders associated with autism spectrum disorders (Fombonne, 2009; Zeidan et al., 2022). It is estimated that the worldwide prevalence escalated from 1/5000 cases (0.02%) in the 1970s to the current data, which is approximately 1/100 children (1%) (Isaksen et al., 2013; Zeidan et al., 2022), a 50-fold increase. Even so, the prevalence is still believed to be quite underestimated (Fombonne, 2003). However, the question that quickly arises is why this increase occurred. Is it real? It is a subject of heated debate, and although there are different hypotheses, most of them largely point to methodological factors (Fombonne, 2009). The identification of cases has been facilitated by the expanded concept of the autism spectrum, which now includes individuals who do not exhibit apparent intellectual impairments. In addition, diagnostic criteria were expanded, and specialized services were developed to better understand the pathology (Fombonne, 2001, 2003; Isaksen et al., 2013; Infobae, 2015).

Nevertheless, it is not ruled out that the condition has indeed escalated and that environmental factors are causing this increase (Isaksen et al., 2013). Here, we see that the disease is perhaps not more common; it did not go from being the exception to the rule, but it did change the diagnostic paradigm and is now detected much more frequently. The same can happen with many rare diseases that can change their status from rare to more common as the diagnostic criteria are expanded, or there are ways to identify the disease better.

Immune Boosters: Who Says Vaccines Do Not Work?

Since the twentieth century, we have become accustomed to vaccines as an immune booster that has saved countless lives (Kayser & Ramzan, 2021). This practice has helped life expectancy grow and move beyond critical moments in early life where death from infectious diseases was very high (Okwo-Bele, 2015; WHO, 2019). The immunization schedule is estimated to save approximately three million lives worldwide annually (WHO, 2019). We have seen that some diseases, such as Smallpox caused by the *Variola virus* (Ray & Ryan, 2004), have been totally eradicated, and others are almost totally controlled thanks to the herd immunity that vaccines provide (Orenstein & Ahmed, 2017) (Fig. 8.1a–d). Methods of immunization by inoculation are older than we think. In ancient times, different people around the world used techniques to boost immunity (Kayser & Ramzan, 2021). Thus, the attempts of humans to boost immunity are not a rare action but rather a rule, and currently, vaccines are an essential part of our lives.Unlike the artificial processes of immunization that have accompanied us in our history, gene therapy is writing the history of medicine into the future (Sayed et al., 2022). We had already mentioned an example of gene therapy in the section on rare diseases when we discussed mRNA. This type of therapy is considered to be a paradigm shift in treating diseases because of its precision in targeting treatment. Although there is much talk about it, it is still in experimental stages, and there is still a long way to go. There are at least two perspectives linked to our book on this subject. The first is that gene therapies

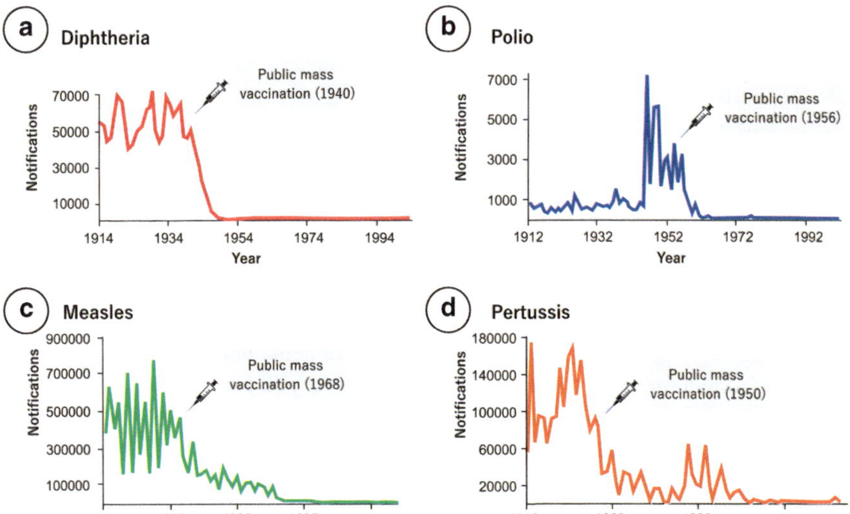

Fig. 8.1 Disease behavior after the development and introduction of vaccines in the UK. After the introduction of vaccines, the diseases decrease their incidence and may have some breakouts but not at the level seen before the development of the treatment. (**a**). Diphtheria. (**b**) Polio. (**c**) Measles. (**d**) Pertussis. Credits: Creazilla (NA). (License: Public domain. https://creazilla.com/nodes/19975-how-vaccines-work-vector)

will move from being rare or disease-oriented treatments to becoming a regular part of medical treatment and perhaps with room to expand to treating other diseases. The other perspective is to see viral vectors as something valuable and good that does not generate a disease (the rule until now) but, on the contrary, treats it and sometimes even cures it. This change that leads to the use of this type of therapy responds, of course, to the change in the use of technology and the advance of the same that makes the identification, replacement, or arrangement of genetic sequences a daily work and not confined to a "gloomy macabre" laboratory. We also must think that there will be a radical change in disease treatment and precision medicine (Kosorok & Laber, 2019; Mirnezami et al., 2012). In the current paradigm, the rule is a kind of "one size fits all approach" where the disease is considered, and standard patient treatment is applied. In the future, medicine aims to generate a health system where conditions are treated considering individual differences shown as genes, environment, and lifestyles that the patient has and thus adjust the therapeutic precision (Ashley, 2016). As precision therapy proves to be more effective, broad-spectrum medication may become increasingly rare. Alternatively, at the beginning, the cost of precision therapy will be high, remaining a rarity for most people.

Human Enhancement: "Pimp My Body"

Human enhancement is a discipline of medicine and bioengineering that seeks to develop techniques and technologies to enhance and surpass normal or diminished physical and intellectual capacities due to some conditions (Brey, 2009). It combines disciplines such as biotechnology, nanotechnology, genetic engineering, and pharmacology. This possibility of altering human capacities deserves a rethinking of the objectives of medicine. Many transhumanists believe this would generate a new race of posthumans with above-average characteristics. However, if we think about it, is a human being with an artificial organ or a robotic prosthesis much different from a person with an immune system boosted by a vaccine or someone who boosts their eyesight with glasses? Everything indicates that clinical advances since the first half of the last century produced accelerated progress that has led us to the formation of "superpeople," with improvements that take us back to science fiction (Bostrom & Roache, 2008). From the point of view of the book, these "superhumans" can be considered exceptions, but in the future, perhaps they will become the new rule. We can venture that in the long term, they could end up outnumbering and even displacing "normal" humans (which they would no longer be). However, for something like "superhumans" to become the rule, human societies must overcome many legal and ethical issues, such as identity, dignity, equality, and the psychology of people (Brey, 2009).

8.2.1.3 On Changing Conceptions in Medicine

As in the previous case, the reverse situation also occurs in medicine, where specific patterns (or medical principles) become progressively less common and end up being rarities or disappear. To illustrate this point, we will focus on the change in diagnostic criteria and medical recommendations in three common aspects of daily life: the problem of obesity, smoking, and how childbirth is treated.

Changing Perceptions of Fatness/Obesity

The perception of overweight or obesity has a strong cultural component, but it is thought to have varied over time in different cultures (Sobal, 2001). It is known that in many ancient cultures, there was a preference for women with a fat or moderately overweight phenotype (Bray, 1992). There are ancient representations of female icons associated with fertility, such as the Venus of Willendorf, which is approximately 25,000 years old. Similarly, today, some cultures still worship figures with a specific pattern of fatness over thin or skinny bodies, which they find even derogatory (Renzaho, 2004). For example, in sub-Saharan Africa, being overweight and obese are signs of social status. Hispanic families and families of European descent in America tended to see a certain amount of overweight as a sign of bodily well-being.

Obesity and being overweight have started to be seen as health problems in the last three decades (Renzaho, 2004). However, Hippocrates, the great physician of antiquity, already saw obesity or being overweight as a problem and identified good nutrition, exercise, and sun exposure as solutions to that problem (Alzate-Yepes, 2019; Renzaho, 2004). However, at that time, it was not a widely contemplated disease. Toward the beginning of the twentieth century, obesity started to attract the attention of the medical community. Nevertheless, it was still a powerful sign of social status in areas such as North America (Grivetti, 2001; Renzaho, 2004). The two world wars brought much poverty, hunger, and disease, which emphasized the relationship between fatness and health, deeply rooted in European immigrants who escaped the war.

Suddenly, there was a marked change in conception. Fatness went from being a common rule to being an exception and was stigmatized by society far beyond the health problems it can cause. After the Second World War, beauty criteria changed radically, strongly driven by the stereotypes that appeared in advertising, especially on television. Dancers, artists, models, and sports athletes were seen there, reinforcing the stereotype of the slim body (Renzaho, 2004). These stereotypes have not changed and continue to be a source of disconformity among people (Murnen & College, 2019). However, stereotypes have not changed in areas such as those inhabited by sub-Saharan tribes in Africa and areas of Polynesia, among others. These ethnicities continue to value the image of being overweight and relate it to well-being and power. Furthermore, these stereotypes migrate with the people. This is how in neighborhoods such as the Bronx in New York, we can see admiration for bodies that in other cities such as Buenos Aires would be cataloged as overweight. This is a heritage that Africans brought to America and is still reflected in their African–American descendants. Although currently, the rule seems to be admiration and reflection in thin and athletic bodies, there are many exceptions in groups that continue to see this type of physiognomy as derogatory.

From the Prescription of Cigarettes to the Use of Medical Cannabis

Many drugs, such as cocaine, nonmedical marijuana, ecstasy, and opium, are illegal, and their eradication is an issue on the agenda of the governments of almost all countries. Nevertheless, we have "legal drugs" such as cigarettes and alcohol. History shows a substantial shift from rule to exceptions in the use of cigarettes and other compounds such as cannabis. Today, smoking has a rather negative connotation in the general public. However, there was a time when doctors recommended smoking (Little, 2019). As much as this may seem like negligence, it was a move by the skillful marketing minds of the tobacco companies. In the first half of the twentieth century, smoking was seen as a sign of status and elegance and was often recommended for health (Martí, 2016). Between 1920 and 1950, it was common for doctors to prescribe smoking to reduce stress, throat conditions, cancer, and halitosis and even to improve the appearance of teeth (Martí, 2016; Seoane, 2017). Tobacco companies drove the business and used these practices to their advantage by using health professionals in advertisements, studies, and totally manipulated surveys (Seoane, 2017) (Fig. 8.2a–d).

By 1950, studies were accumulating, suggesting tobacco's harmful effects on health. In response, tobacco companies started fierce campaigns where not only health professionals appeared but also showbiz stars, children, and even babies talking about the benefits of cigarettes smoked by their parents. In 1939, Franz Hermann Muller warned that most lung cancer patients were smokers. This observation was joined by epidemiologists Richard Doll and Bradford Hill, who in 1951 published a study showing that 99.5% of lung cancer patients were smokers. Although tobacco is still consumed, the perception has changed substantially. Currently, apart from changing the consumption of cigarettes to electronic cigarettes, patches, or chewing gum, smoking has decreased thanks to restrictions in different countries, but tobacco is still consumed, and many are addicted.

Fig. 8.2 Some old cigarette advertisements in which doctors or athletes encouraged smoking. (**a**) Advertisement for Piccadilly cigarettes featuring a doctor, a figure skater, and two trapeze artists (Credits: Wellcome Images). (**b**) Image of a cigarettes ad, showing the many "benefits of smoking." (**c**) Another advertisement suggesting that athletes and doctors prefer smoking a particular brand of cigarettes (Credits: Don O'Brien). (**d**) The advertisement claims that the doctor's choice is the American choice (Credits: Joe Haupt)

Cannabis has been consumed by humans for 4000 years. It has been a therapeutic agent for different conditions, as with many other natural products used in traditional medicine. In the nineteenth century, cannabis extracts were used in Europe and the United States for myalgias and nausea (Quiñones-Laveriano & Catacora-Villasante, 2019). Cannabis was vetoed by the United Nations Single Convention on Narcotic Drugs in 1961 (Lande, 1962). Relatively recently, there has been a resurgence of general interest in cannabinoids to treat neurological conditions and especially chronic pain (Hill et al., 2017; Leonard, 2022). Thus, we see how cannabis has been a rule of use in the East (Brand & Zhao, 2017) but a rarity in the West. It then moved from more widespread use to a new exception by being vetoed and is now gaining momentum globally to perhaps become normal or a rule in pain management. While its reported efficacy is mixed, many patients claim it has been a major therapeutic game changer that has improved their lives.

Maternity and Childbirth

Some examples linked to motherhood show a change from rule to exception. First, new generations in developed countries tend not to have many children or not to have them at all. This idea arises because raising a child is too costly, and people are encouraged to reach projected goals efficiently and not to meet the standards needed by society (Hoffower, 2022; Jagannathan, 2021). This conception is responsible for the birth rate drop in many countries (Nargund, 2009). It is clear that in the last century, the idea reigned that to comply with the precepts of society, a person had to get a partner, get married, buy a house, have a child, and maybe buy a dog. This conception has been changing considerably, and today singlehood or the desire not to be a parent is much more common or not so negatively viewed than it was a few years ago.

The way of being born has also undergone changes with the new perceptions of parenthood. Before, births were natural except in situations where the child or the mother was in danger. Today, there is a tendency for deliveries to be by cesarean section and already scheduled in advance by the doctor. In fact, cesarean delivery rates have increased from 12% of total deliveries to 21% by 2015, reaching 44% in the Latin American region (Sandall et al., 2018). This change has advantages and disadvantages. On the one hand, if natural childbirth is opted for, the mother has labor that can extend for long hours and often ends in the same way in a cesarean section. If a cesarean section is chosen, at the moment, there is no pain, and the delivery is fast. However, the postoperative period is usually long and slightly painful. It has been seen that babies born from cesarean sections may have subtle health alterations resulting from abnormal exposures to hormones and bacteria during delivery (Sandall et al., 2018). Regardless of the method of birth, the perception of how to feed them has also changed. The rule used to be that you could feed profusely without too much of a set schedule. Today, specialists suggest a progression of foods to avoid problems in the digestive system, allergies, or intolerance to certain foods. These are everyday examples of changes from rule to exception in the health field.

8.2.2 Physics Miscellaneous

As we saw at the beginning of the book, we are generally accustomed to believing that over time, there are no exceptions to the theories and rules that are proposed in exact sciences. Although we know that theories can be modified or discoveries can be made, we tend to think these rules are a pattern that will not change and remain static. In biology, we saw countless exceptions to the patterns we thought were rules throughout this book. Although exceptions may be less likely in harder sciences, such as mathematics and physics, this does not mean that there are none, and this is what we will try to explain below concerning physics.

8.2.2.1 Triple Point in Thermodynamics

In the 1870s, physicist Josiah Gibbs found that a substance in thermodynamic equilibrium can be found simultaneously in three phases. This phenomenon can be observed in water, for example, which, upon reaching its "triple point" at a given combination of pressure and temperature, can exist as a solid, liquid, and gas simultaneously, a kind of "physical Schrödinger's Cat." These so-called triple points also occur in other substances, such as solids with different stable crystalline structures. The premise is that they always exhibit at most three phases simultaneously. Even in liquid crystals, for which it is known that they can adopt several phases depending on the arrangement of their molecules, never more than three phases were observed simultaneously (Cemič, 2005; Dehoff, 2005; Preston-Thomas, 1990). However, this rule is now being considered based on new findings. Published results show a system consisting of a mixture of elongated molecules and spheres of a polymer that are dissolved in a liquid. Because of this combination, the elongated molecules behave like liquid crystals. Furthermore, when there are sufficient elongated molecules in the solution, they are arranged parallel. However, since there are also spheres in the polymer, when the shrinkage of elongated molecules increases, they constrict the spheres. Therefore, two new liquid phases associated with the elongated molecules appear, and two solid phases with different crystalline structures are added. This dynamic results in a "quintuple point." Although it exists only as a simulated system, scientists believe it should be possible to reproduce it in the laboratory (Peters et al., 2020). This discovery calls into question one of the laws of thermodynamics, which could be transformed from a rule to an approximation.

8.2.2.2 The Direction of Time

Continuing with the laws of thermodynamics, the second law states that entropy can only increase and never decrease (Wark & Richards, 2001). Moreover, from this statement, it is understood that time moves in only one direction, the direction in which entropy increases. However, it should be noted that the laws of thermodynamics were established during the industrial revolution when work was being done on manufacturing steam engines that could be more efficient in energy use. Currently, the British scientist John Barbour questions this premise of time moving in one direction and suggests a new conception of time in his book "*The Janus Point: A New Theory of Time*," where he explains how time could move in two directions (Barbour, 2020).

For Barbour, the second law of thermodynamics was thought for machines where energy and heat were confined in a bounded space. However, in an unbounded space, particles can continue to travel and join others to form more complex structures, which will grow in all directions of space and time. That is, the passage of time would not be an increase in entropy but an increase in complexity, without limits. Confirming this hypothesis would represent an exception in the traditional vision of physics. It would show an increasingly complex, structured universe that is growing without borders.

8.2.2.3 Newton's Laws and the Motion of Bodies

At present, we consider that the laws formulated by Newton provide answers to most of the problems related to moving bodies, but there are also exceptions. On the one hand, they allow us to explain the motion of stars, for example, or of projectiles created by man or the mechanics of machine operation. However, classical dynamics is only valid for inertial systems, i.e., systems that move at a constant velocity and whose velocity is far from the speed of light (Serway & Jewett, 2009). An example of this type of system could be the earth, which, despite being under translational and rotational accelerations, can be considered for an observer as an inertial reference system due to the magnitude of these forces (Sears et al., 2013). The truth is that, in reality, there is always some force acting on bodies. Therefore, finding a system that can be treated as an inertial system is challenging. These facts show that although Newton's laws are rules, they are based on models where certain variables are taken into account, as in this case of motion. If those variables are not fulfilled, the motion of the bodies will not be governed by that rule either. Therefore, they could be seen as exceptions to it. Once again, we can see that every general rule has its exception and is not fulfilled on certain occasions since the given conditions are outside those necessary for this to occur. Although they are universal rules postulated long ago and that cannot be disproved, there will always be something that they cannot explain: the exceptions to the rule.

8.2.3 Chemical Miscellanea

8.2.3.1 A "Macropattern" of Diversity: The Periodic Table

The history of the discovery of chemical elements and the periodic table represents the quest for a universal language, enabling us to comprehend the composition of everything that exists and is known today. It is the search for the creation of a rule that made it possible to order so we can understand new discoveries. The story begins with the Greeks, who proposed the existence of the essential components of everything, proposing the four elements as the primary builders of everything: water, earth, fire, and air. From this first theory, which at that time could be considered a rule, in the Middle Ages, alchemists became interested in the mixture and transformation of elements to obtain the most precious metal: gold. In the search for gold, the alchemists identified up to 13 elements known today, such as tin, arsenic, and silver. Moreover, most known chemical elements were discovered during the Industrial Revolution due to the need for metal extraction methods (Santos, 2010).

The periodic table we have today comes from the table devised by the chemist Dimitri Mendeleev, based on Dalton's atomic theory, who published the atomic masses of certain elements. Mendeleev's periodic table orders the chemical elements according to their atomic weight and valence, which defines the combining power of the atoms. We now use a more extended version of the periodic table, where the elements are arranged according to atomic number, electron configuration, and chemical properties. The distribution of the elements horizontally in periods responds to the energy level of the atoms. Groups of elements of equal valence (corresponding to the number of outermost electrons) are arranged vertically, as proposed by Mendeleev, which has been a rule for the arrangement of elements according to their properties (Garbayo & Fernández, 2012).

Exceptions in the Chemical Elements

The Bohr model of the atom describes the atom as having a nucleus composed of protons and neutrons, with circular electron shells, similar to the orbits of planets. Each electron shell's energy level increases as they move away from the nucleus. To have the most stable and lowest energy configuration, the electrons fill the shells inside out. The number of electrons in the outermost shell is known as the valence shell. As a general rule, atoms are most stable when their outermost face is complete, with the most essential elements needing eight electrons in this shell to be stable (Lewis, 1916). This is known as the octet rule, which we will mention in the next item. However, there are exceptions for some elements in the table where this order of filling orbitals with increasing energy is not fulfilled. This exception is because the attractive forces of protons and electrons are added to the interference of the lower electron shells that cause energy deviation. The most stable configuration differs from the expected configuration (Ule, 2019). These elements are different from the rest, and we can observe an exception to the configuration described in general (the "rule") that is needed for the stability of the elements. Therefore, although the model described by Bohr and the ordering in the periodic table follow a specific pattern, sometimes these general patterns also have exceptions or oddities that are out of the ordinary. This example makes us wonder whether there truly is a general pattern that encompasses a totality or whether these patterns are always generalities that group the majority but from which something different always escapes.

8.2.3.2 Witness Case: Octet Rule

This rule mentioned above was proposed by Gilbert Newton Lewis in 1916 and explains how the atoms of chemical elements combine. The atoms of the different elements tend to maintain a stable electronic configuration by locating 8 electrons in their last energy levels. Therefore, elements with high electronegativity tend to gain electrons until they reach the octet, while those with low electronegativity tend to lose them for the same reason. Due to this phenomenon, molecules can acquire a stability similar to that of noble gases (elements located at the right end of the periodic table). This rule makes it possible to understand how atoms form their bonds and the behavior and chemical properties of the resulting molecules (Lewis, 1916).

Although it predicts the behavior of many substances, the octet rule has several exceptions in compounds that can reach stability without the need to have eight electrons in their last energy level. Phosphorus (P), sulfur (S), selenium (Se), silicon (Si), or helium (He) are hypervalent elements, i.e., they can accommodate more electrons. On the other hand, hydrogen (H) can accept a maximum of two electrons in a chemical bond. Beryllium (Be) is stable with four electrons, and boron (B) is stable with six electrons (Myers, 2003). As mentioned above, these exceptions demonstrate once again that the rules try to group or encompass, in this case, most of the chemical elements. However, it cannot capture the totality of the cases because most of the chemical elements respond to that pattern or rule. There will always appear exceptions that make us go out of the common and that are not part of the majority. In this chapter, we see that these exceptions occur not only in biology but also in other sciences, such as chemistry.

8.2.3.3 An Example of a "Duality" Rule and Exception

Hydrogen is the simplest of the atoms, since it has only one proton, electron, and its most common isotope, protium, unlike other isotopes, does not have any neutrons in the nucleus. It cannot be framed in any periodic table group, but it is often placed in group 1 (or family 1A) because it has only one electron in the valence shell. It is interesting to note that six elements (C, H, O, N, P, and S) make up 99% of all living matter. The atoms of these elements are tiny and form strong, stable covalent bonds. Except for hydrogen, they can all form covalent bonds with two or more atoms, giving rise to the complex molecules that characterize living systems (Curtis et al., 2008). However, hydrogen is the most abundant chemical element in the universe, appearing in more than 75% of matter mass and representing more than 90% of the atoms. Although it is not very reactive under normal conditions, it forms a multitude of compounds with most chemical elements. Because it has a relatively simple atomic structure, with a single proton and a single electron in its most abundant isotope (protium), the absorption spectrum of this atom could be explained quantitatively. This fact was the central point of Bohr's atomic model, a fact of great importance in developing the theory of atomic structure (Thomas Jefferson National Accelerator Facility — Office of Science Education).

What we want to call attention here is that we have a unique chemical element, not classifiable like the others in the table. Due to its notorious simplicity, it does not behave like other elements essential for life but is the most abundant and can form multiple compounds. In other words, from chemistry, we have in this element the "duality" of something unique, different from all the rest (everything is unique, but this element stands out) but at the same time common: "Rarity from the qualitative (i.e., simple), but pattern from the quantitative (i.e., abundance)." This example shows us again that something can be pattern and rarity simultaneously, depending on the variables or the focus we are concentrating on.

8.2.4 Geological and Related Notes

8.2.4.1 The First Law of Stratigraphy

This law was first postulated in the seventeenth century by the father of geology, Nicholas Steno, who stated that every rock stratum is older than that which overlies it and younger than that which underlies it. It responds to the process by which the eroded materials of already existing rocks will be deposited on each other over time. It is a law or rule that allows us to think and reconstruct the landscape and its history from the occurrence of changes and deformations of the rock structure. However, many situations showed that the strata did not respect a consecutive order according to what would correspond to their relative ages. These situations were exceptions to the general rule, which had not been considered previously. In some circumstances, tectonic deformations after the deposition of the substrate may cause the position of the observed layers to be somewhat reversed. Therefore, they could not be explained by this phenomenon in the same way as in other cases. Because of this fact, the first law of stratigraphy had to be modified to include exceptions not considered before. Thus, the law now postulates that "*every stratum is older than the one overlying it and younger than the one below it, as long as no tectonic deformations have taken place after the deposition of the strata*" (Press & Siever, 2000). Here, the general rule has exceptions. Although they were somehow considered in the statement of the law, they are situations that are far from what could be the pattern in an ideal condition or model.

8.2.4.2 The Age of Fossils and Strata

Biostratigraphy is the science that studies the distribution of fossils in rocks, allowing us to assign an age to them according to the evolutionary degree of the fossiliferous assemblages. The oldest living communities are arranged in lower strata and usually represent a lower diversity than the most modern ones, showing correspondence with biotic succession (Cabrera et al., 2018). Currently, geologists and paleontologists are guided by the international chronostratigraphic table, a permanently updated resource that contains all the necessary geochronological information and rock succession worldwide (Cohen et al., 2013). As living beings evolve, the ages of the successive layers of the Earth's crust can be perceived according to the associations with the fossils found there. However, in this rule, we also find exceptions. In this case, they are given by organisms that have persisted for a long time without changes, and some still exist today (Beierkuhnlein, 2007; Lomolino et al., 2006). These are the so-called "living fossils" we have already mentioned in previous chapters. Let us recall the ginkgo tree among plants and the nautilus in the group of invertebrates dating from the Paleozoic. Among the vertebrates, coelacanths (with much diversity between the Paleozoic and Mesozoic) and tuatara (recorded from the Jurassic and now confined to New Zealand), among others, remain (Ward, 1984; Werth & Shear, 2014). These "living fossils" would represent an exception to the rule of biostratigraphy and a problem since, by remaining unchanged, they give no possibility of knowing the age of the stratum in which they are found. Here, we have one more example that every rule can have its exception and that the rules are never 100% accurate in regard to covering all known cases in the different sciences.

8.2.5 Astronomy Miscellany

The influence of celestial bodies on the earth is undeniable. The sun gives us light and warmth that allows plants to grow and gives us a sense of security about the night, just as the moon influences the tides and eclipses that generate changes in the behavior of animals. With these basic premises, our ancestors assumed that the position and relationship between the sun, the moon, and the other planets (Luminaries sensu Kunth & Zarka, 2005) would influence various terrestrial events, human psychology, and the destiny of people (Zarka, 2009). The position of these luminaries at a particular point in time and space could be calculated, and their graphical representation produces the horoscope (Greek hora+skopein = hour+examine). Both astrology (the study of the heavens) and astronomy (the writing of the heavens) have a shared history from antiquity to the Renaissance (Zarka, 2009). The Renaissance saw the appearance of telescopes and the adoption of more scientific study parameters over preconceived esoteric ideas. Thus, astrology was left aside, and astronomy became much more robust. We see how astrology was the rule in ancient times, with the lack of 100% scientific studies. In fact, many courts and kings had their own astrologers who advised them in decisions on social, economic, harvest, and war matters. Such is the case of the German astronomer and mathematician Johannes Kepler, who, even without being totally convinced of the discipline, was an astrologer to the court of Emperor Rudolf II in Prague (Field, 1984; Zarka, 2009).

Moreover, physicians of antiquity diagnosed and prescribed remedies for illnesses based exclusively on the person's horoscope. At that time, astronomy, medicine, and astrology went hand in hand. Today, those rules have changed, and astronomy and medicine are strong scientific disciplines that are totally separate from astrology, which has been relegated to pseudoscience. However, for ordinary people, reading the horoscope in newspapers is still helpful, and many people still trust the predictions of an astrologer to make essential decisions in their lives.

The universe is not exempt from laws, rules, exceptions, and changes from one to another due to scientific advances. Physics in the universe was, is, and probably will always be the same. Many phenomena can be described with Newton's equations. Other more complex phenomena escape the analysis of Newton's laws but are addressed by Einstein's theory of relativity, which in simpler processes is equated with Newton's laws. If we look for rules toward the beginning of the creation of the universe, we will see that at the beginning, it was a chaotic furnace of space, mass, and energy condensed in an infinitesimal point. Obviously, there were no galaxies, stars, or planets (Ratra & Vogeley, 2008). After the Big Bang, the universe began to expand and cool. That is still the rule, and it is thought that the universe could expand infinitely at a slow speed. Another rule in the universe is the presence of hydrogen clouds, which have been noted to exist everywhere. As we said, hydrogen is ubiquitous in the universe, and helium and lithium are the main atoms formed in the Big Bang (Turner, 2009). Let us remember that hydrogen is a very reactive gas. However, the hydrogen nucleus of Jupiter behaves like a semiliquid metal, creating a substantial electromagnetic field (Nellis, 2000). This phenomenon, among other things, is the cause of very particular features in some of its moons where we see that this influence generates the melting of ice sheets forming subterranean seas (Chap. 7) or volcanic activity by melting deep rocks.

What about the other atoms that we know of the periodic table? They form in the heart of stars and, as a rule, are scattered across the universe when they explode (Cox, 1989). Another singular rule that often goes unnoticed is that in the universe, spheres are the natural rule (Cain, 2014; Tyson, 2019). Gravity is responsible for molding objects into this shape. Low gravity causes the shapes to deviate from a sphere for objects with more irregular shapes, such as the moons of Mars, Phobos, and Deimos (Tyson, 2019). Seems that flat planets should be an exception to this premise. Masses of gas can also coalesce into spheres of gas that form stars.

If we think of exceptions or oddities in the history of the Universe, perhaps the most obvious one is that at the beginning, after the Big Bang, all the cosmic structures that we study today did not exist. We know that it was because it started from the subatomic particles, a path that led to galaxies, stars, planets, and everything that exists today in the Universe, whether we can measure, identify, or not. We can also cite as an oddity the double star systems where one sphere exerts a significant gravitational force and the other sphere is distorted by it (Tohline, 2002).

Another rule that appears in the Universe is the gravitational influence of dark matter (Tucker & Wang, 1998). We do not know what it is, but we know its influence from gravitational effects. It does not influence at the planetary level, for example, but it influences the stars' motion at the galaxy's center (Pitjev & Pitjeva, 2013a, 2013b; Tyson, 2019). On the other hand, it is known that the force of gravity decreases its effect with distance. As an exception, we see that stars at the edges of galaxies move fast despite the greater distance to the galaxy's center (Tyson, 2019). Here, the dark matter would be involved, which is not identified, but its influence is measured.

We all know or think we know that the speed of light is constant. At least, it is strongly suggested. However, there are less common theoretical circumstances where the speed of light varies and therefore is not a constant (Farrell & Dunning-Davies, 2004) and is proposed as an alternative solution to some theoretical equations (Ellis & Uzan, 2005). A rare coincidence is that the Sun, Moon, and Earth have a very particular combination of size and distance that allows total eclipses to be seen. This combination, at the moment, is almost unique in the Universe (Tyson, 2019). These were scattered examples that show rules and exceptions at the astronomical level. These examples show how we can find rules and exceptions everywhere.

8.2.5.1 Some Dynamics between Rules and Exceptions in Astronomy

It is possible to appreciate many changes from rule to exception and vice versa regarding our knowledge of the Universe. Perhaps not because of a radical change in the natural behavior of the Universe but rather because of changes due to scientific and technological progress from antiquity to the present.

To cite an example of a change from rule to rarity or exception of nature, we can comment on the collisions of asteroids or other celestial bodies. Five hundred million years ago, after the formation of the solar system, the Earth was continuously bombarded by space debris still wandering in space. In fact, the Earth raised its temperature by friction as chunks of material entering the atmosphere generated heat and then fused with the crust, forming complex molecules (Tyson, 2019). The presence of an atmosphere counteracts to some extent the impact of a meteorite. However, if this is at least 1 km in diameter, the result can be very significant. Indeed, the history of the Earth confirms this. These phenomena, called, as we saw in the first chapters, contingency phenomena, have had a radical influence on evolutionary processes. For example, the tremendous collision of an asteroid toward the end of the Mesozoic era (Chicxulub event 65 mya) in the area comprised today by the Gulf of Mexico and its drastic effect on the atmosphere, solar irradiation, and the direct impact of the explosion annihilating 70% of all species, including dinosaurs (Chiarenza et al., 2020; Pope et al., 1997). The extinction of groups such as the major dinosaurs was accelerated by this event, perhaps already affected by other causes such as climatic cooling and the decline in herbivore diversity (Condamine et al., 2021).

We can see then that this topic of asteroid collisions in magnitude can be considered from the perspective of our book as (1) a phenomenon that was possibly the rule at the beginning of the Solar System, to become not an exception, but at least a relatively rare type of event. (2) From the perspective of what concerns the planet Earth and the evolutionary history of its life forms, it has been an exception to the times it has occurred. However, when it happened, the result was fundamental, as a contingency, for what would happen later with many animal and plant groups. This is an indirect but crucial extraplanetary abiotic artificer in the direction of evolution (Alvarez et al., 1980; Martin & Livio, 2022; Sleep et al., 1989). Concerning the latter, the idea that Jupiter had played a significant role in the development of life by acting as a sizable gravitational shield that would deflect Earth-bound objects was well established (Ward & Brownlee, 2000). However, in recent years, this idea has been rigorously evaluated. While popular science still suggests that idea, there is evidence that this is not the case with most bodies that can collide with the Earth

(Grazier, 2016; Horner et al., 2010; Horner & Jones, 2008, 2009, 2010). An exception is long-orbiting comets (Grazier, 2016).

In the latter context, currently, with the advancement of orbit detection techniques, extraplanetary body directions, and mathematical models, a collision of the asteroid named 99,942 Apophis with our planet is predicted approximately 2029 (Giorgini et al., 2005; Hirabayashi et al., 2021; Yoshida et al., 2021). It is speculated that, depending on the collision angle, whether it disintegrates in part upon entering the atmosphere, the results could be devastating. Perhaps a new contingency for new directions in biological evolution? Much of this is not discussed or worried about, but alarms sound on the effect on the human species (Morais et al., 2021). The efforts to detect with precision aspects referred to this future impact and, eventually, to destroy it beforehand with weapons at the service of extraplanetary science are the object of current research (Cheng, 2021; Maltagliati, 2022). Time is pressing. There are other potential collision cases cataloged as potentially hazardous asteroids and comets (Petrov et al., 2018; Trigo-Rodríguez, 2022), some of which may be in line with any of the inner planets of the solar system (Devyatkin et al., 2022).

Returning to our topic of interest, there are many cases in astronomy of transitions from rule to exception and vice versa due to changes in scientific conception and technological progress. For example, we know of paradigm shifts concerning the position of where we live (e.g., Kragh, 2006). Therefore, our planet was unique at the beginning of our history as human beings. Then, in antiquity, other planets were discovered, so the Earth went from being an exception (or something extraordinary) to being a rule in the then-known universe. Thus, the geocentric theory was created where it was postulated that although the Earth was not unique, it was the center of the known universe, and the planets and the sun revolved around it. Then, this changed once again, and it became known that the Earth and the other planets revolved around the sun. The sun was considered unique, and then it was realized that it was just another star in the universe. Thus far, the known universe was perhaps the Milky Way, and then it was known that the Milky Way was a galaxy like many others scattered throughout the universe. Finally, it is believed that there is one universe, but there are theories that speak of multiple universes, which would again take us out of our comfort zone. Although this was a crude and general account that seeks to quickly show how changes in conceptions of nature cause a paradigm shift of what is rule or exception, there are also more concrete examples.

In Einstein's time, the rule was that the universe was static (Hoyle et al., 2000; Sen, 1957). In fact, he developed a constant to fit within the statics paradigm (Harvey & Schucking, 2000; Straumann, 2002). Hubble realized that the universe was dynamic and expanding (Freedman, 2000), and Einstein regretted having generated the constant to fit the paradigm. In fact, he called it "*his biggest mistake.*" However, in recent years, Einstein's cosmological constant has been used again in cosmic acceleration studies (Barrow & Shaw, 2011a, 2011b; Peebles & Ratra, 2003).

Another rule is that galaxies are generally large clusters of stars. However, there are galaxies called dwarf galaxies (Hodge, 1971; Mateo, 1998). They are natural laboratories where galactic evolution models and paradigms can be evaluated

(Haynes, 2018). Tyson (2019) suggests that the rule is the presence of dwarf galaxies in the universe due to the amount that exists, which is clearly much larger than the so-called "normal" galaxies.

Astrophysics and cosmology are exciting fields that pose a significant challenge to understanding nature's workings. As we said at the beginning, the magnitudes are titanic and challenge the mind. There is much room for the emergence of exceptions and paradigm shifts, both associated with the progress of science and technology for measuring phenomena. The rules are strong, perhaps because phenomena are governed by well-established laws such as those of physics. These laws, at least, have been seen to work well in this universe. Interpretations could change a lot in the future with the unification of the fields into a theory of everything and the discovery of other universes where the laws we now use and know may vary (Halpern, 2004; Tegmark, 1998).

8.3 R&E in Other Knowledge Disciplines

8.3.1 Social and Related Sciences

In the vast field of social sciences, we will focus on two examples, native cultures and family conceptions, to show some cases of how social changes over time have modified what was once common to new conceptions. Sometimes this dynamic leaves the dominant patterns of the past as mere exceptions.

8.3.1.1 On Native Cultures and Concepts of Family

Native cultures are a clear example of how knowledge, including the ancestral wisdom of many peoples, has been lost over time and with them, at the cultural level, patterns of tradition. Let us consider how the original cultures native to each region have been modified over time. We can mention, for example, what we said about ancestral medicine based on the use of plants or the modification of each tribe's own beliefs about nature. Additionally, the habits of life as the way of obtaining food, the accepted clothing, beliefs, and festivities, to name some traditional practices of each social group that have been lost over time. The arrival of new beliefs and customs has caused many original cultures to be lost. Added to this are the racial and social prejudices that native cultures have suffered and which, even today, with all the progress in accepting the "different," still continue to suffer, perhaps transforming what were once rules into only exceptions. Native cultures have also been lost, not only by the mixture of the new cultures that were able to arrive but also, for example, by factors of migration of the youngest to large cities, loss of territories, and modes of exploitation of natural resources (Porcel Moscoso, 2018). The point is that traditions of different kinds that used to be the most frequent are no longer so or are only maintained by the older generations. Of course, this loss of native culture eventually leads to the loss of the cultural heritage of each place (Porcel Moscoso, 2018).

On the other hand, the traditional family conformation pattern considered the formation of monogamous marriage as what was accepted as valid in Western society. It was considered the only moral and legal way to establish that type of relationship. Marriage had to be socially recognized as such, and there was a sexual division of labor (Manrique, 1996). The "high society" families married their children based on economic and social interests, and it was typical that the first male-born child received the family inheritance (Freijo et al., 2004; Manrique, 1996). However, over time, the conception of the family has undergone a period of change. The two-parent, monogamous marriage model, considered the standard, has been adapting to a new social structure. It is now coexisting with other forms of familiar assemblages, such as single-parent homes, extended families, families resulting from assisted reproduction, families with adopted children, and homoparental families (Meler, 2008).

What is different today is that family models are beginning to be decided by individuals, and different behaviors are becoming increasingly frequent. For example, some couples do not get married. Some couples remarry, and some couples are open or exchange partners. There are homosexual couples and couples who live apart. Many of these behaviors in the past were absolute rarities and even morally condemned by society. Today, they are becoming increasingly frequent and accepted.

The change in the conception of family is closely linked to the advance and development of technology in several aspects (Manrique, 1996). Let us consider prenatal diagnosis, which has transformed pregnancy into a family project, changing it from a natural occurrence to a planned event in many cases (Manrique, 1996). Initially, this practice was almost inconceivable and remained exceptional. However, over time, advancements in more precise and commonly used techniques have led to its widespread adoption. Another significant change that took place, which is involved in family development, is that women began to work outside the home in paid jobs, something previously almost exclusive to men. This exception has become a pattern in much of the world. Women no longer only take care of the children and the housework but also "go out" to work, which granted them economic independence and influenced their role in society.

8.3.1.2 Something that is an "Exception" Among Human Societies

What is considered acceptable among human behaviors varies among different cultures, as norms are a social construct, and the meaning of what is right or wrong differs from one society to another (Vera et al., 2012). Each culture establishes different patterns of behaviors that may vary among members of society, considering, for example, children and adults or men and women (Maldonado, 2017). For instance, certain behaviors that deviate from societal norms may be tolerated when displayed by a child. However, the same behaviors would be deemed unacceptable if exhibited by an adult with fully developed psychological capacities. Although in the world numerous cultures vary in the behavioral patterns accepted as valid, behaviors such as rape, incest, theft, and murder are not considered acceptable for any member of that society. These are conducts that threaten the social, legal and political balance (Maldonado, 2017; Vera et al., 2012). Consequently, these behaviors are subject to direct penalties, such as legal repercussions and fines, or indirect consequences through education, such as the case of familial or community taboos.

Let us set aside extreme and punishable behaviors. The truth is that nearly all people, if not everyone, regardless of their honesty, have at some point in their lives told a lie or crossed some personal boundary. For example, we put ourselves at the supermarket express checkout with more products than allowed, cut in line, miss work without a justified reason, get on a plane without being called, cross a red light, to name a few cases of minor transgressions. In this sense, studies have shown that these small transgressions to preestablished rules are not associated with intelligence but with some conditions, such as certain emotional states or the degree of people's creativity. Individuals who are more creative can (and usually do) break the rules more easily. Another factor that can influence transgressions is the social group to which we belong. If our group tends to engage in rule-breaking behavior, there is a higher likelihood that we may do the same to fit in and be accepted, even if we would not act that way in different circumstances. Indeed, there is a positive aspect to consider. When presented with a challenge or a "dare to push further," we may break some established rules, but without causing harm to others. For instance, this could involve going beyond the usual expectations and seeking greater personal growth, satisfaction, and benefit.

Staying with the *status quo* is suitable and safe, but breaking the rules and going against the expected, allows one's talent to emerge and develop (Buckingham & Coffman, 2012). Just not to be misinterpreted, think about many brand advertisements where the slogan invites us directly or indirectly to break the rules. Let us see some examples: Nike: "*Just do it.*", Apple: "*Think different.*", Adidas: "*Impossible is nothing.* ", Virgin: "*Screw it, let's do it.*", Netflix: "*Watch what you want, when you want.*", Spotify: "*Listen to what you love.*", Airbnb: "*Belong anywhere.*", Uber: "*The way you move.*", Levi's: "*Live outside the lines.*", Nike Golf: "*Never follow.*", Under Armour: "*Rule yourself.*", Oakley: "*Eye protection for those who refuse to conform.*". These brands encourage people to think out of the box, do what they feel, be unique (outstanding), and never conform to the *status quo*.

Additionally, technological development is linked to essential changes in human behavior. Access to rapid communication (television, internet, social media) and travel around the world has enabled wide dissemination of new fashion standards emerging in the "more developed" cultures. It drives others to follow the same canons. Clothing, food, music, issues associated with art, and lifestyles in general are gradually changing. What was once common or seemed strange to us perhaps now no longer (Maldonado, 2017).

8.4 R&E Search on Subjective/Controversial Topics

8.4.1 Art and Culture

8.4.1.1 Image and Painting

Art can be seen as the most subjective form of human expression, as its purpose is to evoke emotions and convey a unique perspective of the world. This is why terms such as "exceptional," "monstrous," "ugliness," "genius," or "beauty" are frequently found in artistic works. These concepts can be interpreted from various angles and hold different values based on the historical context in which the artistic representations are analyzed (Kant, 1977; Rosenkranz, 1853; Tatarkiewicz, 1970). According to Kant (1977), artistic genius is a talent that produces original creations surpassing nature through its expressive potential, creating, expressing, communicating, and contemplating beauty (Flores, 2001).

However, the strange, the monstrous, the obscene, or "immoral" has always had a marginal place in art, being the works with these characteristics repudiated by academic or religious institutions, although widespread and in many cases claimed in later times. From Antiquity to the Middle Ages, ugliness is related to a lack of harmony that breaks the rules of proportion on which beauty is based (Rosenkranz, 1853). However, it is admitted that, despite the existence of beings and things considered ugly, art can represent them beautifully, and the beauty of this imitation makes the ugly acceptable (Aristotle, fourth century B.C.; Eco, 2007). However, to what extent does a representation of the ugly and monstrous make it somehow more attractive? While beauty does not produce a shock or wonder in the preconceived aesthetic pattern in the viewer, the ugly is configured as an original and shocking experience, which can result in extreme sensitivity and intensity in the receiver of an artistic work (Muñoz, 2017).

From ancient texts of the first centuries of the Christian Era, representations are known where monstrous men and animals appear, which are then reproduced in medieval encyclopedias (Bovey, 2002), giving rise to the fascination for the supernatural and an unknown world ruled by unique and monstrous figures: "*in the great symphonic concert of cosmic harmony, the monsters themselves constitute, even if only by contrast (…) to the beauty of the whole*" (Eco, 2015). In this way, monsters become increasingly present in literature and painting, from Dantesque descriptions of hell to the aberrant and fascinating paintings of Bosch (Fig. 8.3a–d). In the passage to the Modern Age, the monster lost its symbolic charge. It is only considered a mere "natural curiosity" to be studied in its anatomy more scientifically with a naturalistic interest.

Artistic movements are extraordinary examples of the dynamism between rules and exceptions, at the level of individual artists as exceptional geniuses admired or wholly rejected, or of movements and conceptions of the "ugly" and "beautiful" that were totally disruptive for their time, "revolutionary" artistic proposals or broke with the prevailing pattern or rule at a particular historical moment. During the

Fig. 8.3 Examples of illustrations of monstrosities. (**a**) Giovanni da Modena, The Inferno, 1410, Basilica di San Petronio, Bologna, Italy. (**b**) Coppo di Malcovaldo, Inferno-part. Giudizio Universale, c. (1260–70), mosaico. Firenze, Battistero, Italy. (**c**) Limbourg Brothers, Hell from the Tres Riches Heures du Duc de Berry, (1411–1416), Musée Condé, Chantilly, France. (**d**) Anonimo, Christus in de Limbus, (1575), Indianapolis Museum of Art, USA

Renaissance, artists such as Michelangelo, Raphael, and Leonardo da Vinci were inspired by classical Roman and Greek art, taking as rules balance, naturalism, perspective, symmetry, and "perfect" proportions (Summers, 1990). Humanist portraits with proportioned and anatomically correct sculptures and harmonious and symmetrical architecture are clear examples of the spirit of the Renaissance. The strange for this time became the rule during the Baroque, where vivid, extravagant colors prevailed. It became necessary to represent reality crudely, with distortion of forms with violent and contrasting effects, a certain tendency to imbalance and exaggeration, abandoning the corresponding "perfect" representations of the previous period. For example, Caravaggio illustrated the supernatural with a very human appearance, bringing man closer to the mystery of the transcendent through vulgarity and ugliness. However, this transition would be reversed with the appearance of Classicism during the Modern Age. Aesthetic and philosophical patterns of Classical Antiquity would re-emerge, such as the "*aurea mediocritas,*" a Latin term that alludes to the desire to reach a middle ground between extremes. This state is far from excess through the just measure of opposites (*concordia oppositorum*): the virtue is in the middle ground (Aristotle, fourth century B.C.). Classicism assumes the ideals of romanticism and humanism, aspiring to achieve absolute perfection through art. Nicolas Poussin, a great exponent of this movement, would say that Baroque painters such as Caravaggio, who departed from these rules, had come to destroy painting, qualifying them as "painters of ugliness."

An iconic example of painters rejected and considered rare in their time can be seen in the well-known "hall of the rejected" (Salon des Refusés) in 1863 in Paris. This collective show brought together great painters such as Manet, McNeill Whistler, Courbet, and Cézanne. They had been rejected from the Paris Salon exhibition of the Académie des Beaux-Arts (the most important annual artistic event or biennial in the world at the time) (Bozal, 1996; Hauptman, 1985). This fact coincides with the birth of Impressionism in painting as a disruptive artistic expression initiating the path toward avant-garde movements. Many of these artists broke academic rules by painting outdoors instead of indoors in their studios. The Salon of the Rejected was gaining popularity in subsequent years. Those artists first marginalized and exceptional from a negative assessment gave way to what would become the rule or norm in the later artistic period.

However, suppose we are discussing disruptive artistic movements. In that case, we cannot fail to mention Surrealism, where the extraordinary, fantastic, and strange were common currency in the leitmotiv of both pictorial and literary works (Nadeau, 1965). Surrealists often challenge perceptions and reality in their works of art. The limits of beauty were expanded; the spectrum of combinations grew, provoking a true creative explosion (Bigsby, 1972). In many cases, the images of pictorial space are represented with unnatural and disproportionate shapes and sizes, with what could be considered "technical errors," but to create empty and suggestive spaces for the observers to give definitive meaning to the representations. One of the precursors of Surrealism, Giorgio de Chirico proclaimed: "*To become immortal, a work of art must escape all human limits: logic and common sense will only interfere (…)*".

Later, artistic movements and artists would aim to intentionally provoke rejection or strangeness, presenting unique and provocative proposals that would stand out in some way or impose debates on the public (Gomes, 2021; Jennstål & Öberg, 2019) (Fig. 8.4a–f). "La Fontaine" (literally a signed urinal) by Marcel Duchamp inaugurated the postmodern era, and when presented at the Society of Independent Artists in 1917, it was not considered a work of art and was rejected (Tomkins, 1977). Although initially a joke by Duchamp to mock the artistic circle of the time and the prevailing notion of art, this work imposed a new style known as "ready-made" and was voted the most influential work of art of the twentieth century (Turner prize, Gordon's, 2004). Duchamp's only laws were those of exception, those that break and question the norms, and he was the pioneer of conceptual art (Paz, 1990). Something similar happened with Andy Warhol's "Campbell's Soup Cans" (1962), a pop art icon. When it was exhibited, it caused outrage, and the excessive questioning and criticism distinguished Warhol from other emerging pop artists (Schroeder, 1997). Warhol's reputation grew to become the most renowned and quoted American pop artist (Galenson, 2021). More recently, something similar happened with Maurizio Cattelan's "Comedian" (2019), a banana stuck to the wall

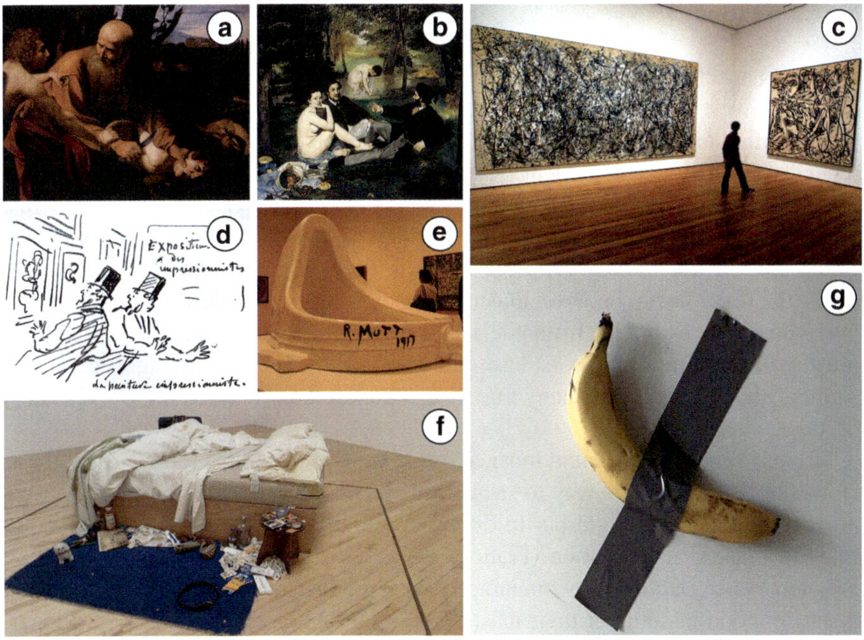

Fig. 8.4 Exceptional artistic manifestations that received the rejection of the public or the Academy. (**a**) Michelangelo Merisi da Caravaggio, Sacrifice of Isaac, 1595, Uffizi Firenze, Italy. (**b**) Edouard Manet, Luncheon on the Grass (1863), Musée d'Orsay. (**c**) Jackson Pollock, Blue Poles, (Number 11), 1952, National Gallery of Australia. (**d**) Cham, Caricature on Impressionism, on the occasion of their first exhibit, (1874), Bibliothèque nationale de France, Paris. (**e**) Marcel Duchamp, The Fountain, (1917), Original missing. (**f**) Tracey Emin, My Bed, 1998, Saatchi Gallery (**g**) Maurizio Cattelan, Comedian (2019), Private collection, presented for the first time in Miami Art Basel (2019), USA

with packing tape (Bryzgel, 2019) (Figure 8.4g) or "America" (2016), a fully functional golden toilet (Bown, 2019). These works hide a powerful critical component on an economic and political level but are still questioned as true "works of art" (Celik, 2021).

This same phenomenon can be transferred to other artistic disciplines, such as architecture and poetry, where disruptive artists have been rejected for their exceptional creations over time. The Eiffel Tower (1889) was considered "useless and monstrous" and was even ordered to be demolished, although it finally survived and today is considered the most emblematic symbol of Paris. In poetry, the French "damned poets" stand out (such as Corbière, Rimbaud, and Verlaine) who were talented but misunderstood by their contemporaries, who rejected the established artistic and social norms, developing a free and provocative art (Verlaine, 1884). This literary style gave rise to "malditism" for those poets who transgressed the established norms, and this nickname would end up being applied to many writers (Del Moral, 1985). The rare or exceptional will always be judged subjectively in art. Therefore, the dynamism between rules and exceptions transcends disciplines and temporality.

8.4.1.2 Music

Another aspect undoubtedly linked to art and culture is music, which has also changed over time. Although musical notes are invariable, what has been created and modified are musical styles. In this sense, musicology studies the phenomena related to music, its theoretical bases, its notational systems, its history, and its relationship with societies. It has been responsible for analyzing the musical structures consolidated throughout history (Franco, 2019; Palisca & Grout, 1996; Reich, 2002). The history of music refers to the study of manuscripts, iconographies, the relationship between music and society, and the different musical genres over time. For example, from the history of European classical music to rock music, different musical styles have had their "glory" period, and their great performers are considered by many true idols who left their mark on the world history of music over time.

In early times, the different musical forms were linked to the sacred use of these. However, they then gave way to secular music, i.e., music that is not linked to any practice of religious origin but to different cultures and traditions with their own rules in musical composition (Franco, 2019). Great composers such as Brahms, Wagner, Verdi, Debussy, Rachmaninoff, and Stravinsky, among many others, were misunderstood in some of their masterpieces for being original and "exceptional" (Slonimsky, 1965).

Music was significantly affected at the end of the nineteenth century by the industrial revolution, which allowed new and significant advances in various aspects of society. For example, the mechanics of instruments mainly provide new musical resources, giving greater creative flexibility to the compositions in their melodies (Franco, 2019). The transformation of music from technological innovation is undeniable. We can consider the sophisticated "tuning" of certain instruments that require a specific sound.

8.4.1.3 Literature

On the other hand, we can speak of the literacy process that has been gaining an essential space within societies and where in recent decades, more attention has been given to it (Tapia, 1994). The social phenomenon of literacy was an exclusive skill taught only to specific people or groups. For example, in ancient times, scribes and royal advisors, and people belonging to high society, only to the noblest families of the place, where literate people were responsible for teaching their children (Tapia, 1994). Even exclusive to sons, female daughters were only instructed in Christian morals and housework and only had access to literacy if they joined the cloistered nuns' convents. It was not until the middle of the seventeenth century that the literacy phenomenon became more massive with the implementation of schools. This strategy sought, among other things, to ensure that the children of the poorest families were not left without assigned responsibilities so that they could acquire education and professional training (Tapia, 1994).

Let us consider another point closely linked to literacy: access to books. Access to books was also for a tiny part of society. It was not as common to have a library at home as it is today, not only because of the cost of books but also because they were not printed on a massive scale. With the passing of time, having a "library" in every house went from being an exception to a rule. It was not strange then that there was a minor or extensive library in every house, which also reflects the increasingly preponderant role that education had. Physical libraries, then, went from being an exception to a rule, and with the advance of technology, they went from a rule to an exception again. Currently, how common is it to have libraries similar at home to those of our parents or grandparents? Most likely exceptional, as with internet access, for example, our library is already "virtual." The form of the library has changed from physical to virtual, but the "size" or the number of books or writings we have access to is undoubtedly much more significant.

8.4.1.4 Calligraphy

Another subject to which we can refer, associated with culture, to see how some patterns became exceptions with time, at least in their use, is calligraphy. According to a classic definition, calligraphy is "the art of writing with beautiful, artistic and well-formed handwriting according to different styles." In this sense, the morphology of the written letter was mainly associated with aesthetic aspects, legibility, and economy of the substrate. Essentially, where handwriting was most used was in religious and mercantile activities, leaving the orthographic and grammatical aspects of writing in the background (Bello, 2016). Although in many cultures, there has been a development of writing, and in its absence, one of its visual forms, calligraphy, the importance attributed to it has depended on the historical and cultural contexts (Gálvez Pizarro, 2004).

Calligraphy can be classified, for example, in the Latin writing system in Carolingian, Gothic, Chancellery, English commercial, foundational, and others. All of them are governed by different and demanding formal and aesthetic rules (Gálvez Pizarro, 2004). However, handwriting and calligraphy are being replaced due to technological advances and may be lost over time. Think, for example, of Johannes Gutenberg's creation of the printing press in the mid-fifteenth century, which significantly reduced handwriting for copying books, with movable type gaining ground. The printing press provided the technical conditions to ensure equality in the reproduction of copies, which was impossible with the manual writing of texts (Bello, 2016).

On the other hand, the forms of writing were simplified. As a result, the teaching of writing in schools was limited to letters with more cursive features, such as the Spanish Bastarda, similar to those used today (Bello, 2016). The innovations brought about by the development of typewriters first and computers years later further reduced the use of calligraphy. Word processing programs speed up, facilitate writing, and make the correction process much more manageable. In addition, when printing, the text will appear "clean" without amendments, as in handwritten or typewritten texts (Ferreiro, 2006). In addition, the use of personal computers and the Internet has produced changes in the way young people write, mainly when used in chats, e-mails, and cell phones, for example, so that changes in the forms of writing are associated with changes in social practices (Ferreiro, 2006).

Although formal calligraphy has largely lost its original function as a daily notation tool, it has been maintained over time in search of other types of utilities more linked to artistic expressions, with the design of advertising, logos, and labels, among others. At the beginning of the twentieth century, modern calligraphy resurfaced, where letters may or may not be legible (Claude et al., 1996; Mediavilla, 2005). In fact, the artistic movement called "calligraffiti" merges calligraphy with graffiti, amalgamating the traditional with the modern.

8.4.2 Sports

Within sports activities, we also find examples of rule changes. As previously mentioned, sports-related activities have become increasingly important in people's daily lives. Many people spend many hours playing sports, assigning them positive characteristics in terms of health and as a social, educational, moral, and cultural factor (Garoz Puerta, 2005). In addition to the vital role in economic activities already at the level of professional sports, sports such as Football, American Football, Baseball, and Basketball, for example, move large amounts of money worldwide (Triviño, 2011). An important role that sports has also had and still has is at the sociopolitical level. A clear example is the place taken by the practice of rugby in the unification between black and white people in South Africa (Triviño, 2011). With sufficient evidence, sport contributes positively to the development of our lives.

Numerous sports originated in ritual, medieval, and popular games that changed toward a sporting environment in the nineteenth century with the creation of competitive sports (Garoz Puerta, 2005; Triviño, 2011). In this sense, part of the sports study has focused on the different rules, their functions, and how they have evolved through history (Garoz Puerta, 2005; Triviño, 2011). If we refer to particular sports, we can say that in all of them, there are rules and that the rules serve to regulate the competition by granting equal opportunities to all competitors and sometimes bring safety to competitors, as in the evolution of F1 cars. Within the rules, we can differentiate between rules that refer to equipment, players, substitutes, coaches, and judges. Timing rules, rules that determine infractions, penalties and points, or goals, and rules of conduct in general of the participants (Arias et al., 2011; Garoz Puerta, 2005; Sánchez, 1987; Triviño, 2011).

Knowing and following the rules is fundamental in the development of the game. Therefore, if the aim is the modification of the conditions of the game, it is necessary to modify its rules. We should always keep the basic rules to not significantly alter the sport's spirit in question and maintain interest in the original game. It is crucial to analyze the impact that these modifications in its rules have before its definitive implementation (Arias et al., 2011; Nevill et al., 2008). In addition, many sports have modifications and adaptations in their rules when, for example, they are played by children or by people with disabilities, such as sports adapted for people in wheelchairs or blind people, where the rules are not the same, but the sport is the same.

Changes in the regulations of different sports are also influenced by changes in lifestyles. In this sense, they are added to sociocultural evolution (Garoz Puerta, 2005). Let us also think about sports video games that, with technological advances, are becoming increasingly frequent, not only those played by children or in an amateur way but also large competitions worldwide (e-sports). Here, the rules can be considered limitations on the actions of the players, whether the players can be injured or not, or the scenarios available for the game to take place, rules that do not exist in the "real" game (del Castillo, 2012; Gee, 2004; Juul, 2005; Prensky, 2002). We can then understand sports rules as changing and influenced, for example, by sociocultural evolution and by the evolution that technological advances bring with them (Garoz Puerta, 2005).

8.4.3 Religions and Beliefs

In religion, the concept of rules, principles, norms, and above all, the idea of "dogma" emerges strongly. According to the Dictionary of the RAE (Real Academia Española, 2022), dogmas are concepts taken as absolute truth that cannot be doubted. It is an undeniable principle. Religious dogmas are beliefs that define a religion, differentiate it from other religions, and are considered truths associated with people's faith. In the Catholic Church, dogmas are understood as absolute truths taught to the faithful. They are categorized into divine dogmas taught by Jesus, apostolic dogmas taught by the Apostles, and ecclesiastical dogmas taught by Popes. The Catholic Church takes dogmas as absolute truths, and if any doubt arises, they reaffirm it with another dogma. For example, Christianity takes as dogma Jesus Christ as the son of God. However, when doubt or inconsistency arose in this dogma, they created another article of faith to reaffirm it. In Judaism, Saadia Gaon was the one who systematized the dogmas of religion, the belief in one God, and Israel as the guide of the world. In addition, practices such as the use of the Talit, a fringed cloak worn by the faithful, and Tefillin, small leather boxes placed on the head, are common among the faithful. Islam establishes for its faithful that "*There is no God, but Allah and Mohammed is his prophet*." Islam preaches that God is merciful and all-powerful and has guided his faithful through his prophets. Its dogmas are embodied in the Quran, which is the textual word of God (Goldman, 1995; Watt, 2003), and it is religion that has seen the most significant increase in the number of followers in recent times.

Although religious dogmas have remained almost unchanged over time, we can say that the practice in many religions has changed. For example, in the Catholic Church, the mass was said in Latin, backward to the people, and facing God at the main altar. This practice ended with the Second Vatican Council between 1962 and 1965, with a mass celebrated in vulgar language. Pope John XXIII said… "*I want to open the windows of the church so that we can see outside and the faithful can see inside,*" which initially generated a rejection by other members of the Church who insisted on the conservation of Latin as the universal language of religion (Díaz Patri, 1994). This is a clear example of change from rule to exception in the Christian church.

On the other hand, common practices years ago among the faithful are now almost a rarity, such as attending Church on Sundays, which the whole family did as a ritual. It is increasingly common to hear from the mouths of the faithful, "*I believe in God, but not in the Church*"; in this sense, numerous corruptions linked to the Church have undoubtedly caused many of the faithful to move away from it. In addition, in recent years, the question has also arisen: are religions necessary for the development of a society? This question was not even asked before (Giner, 1993). In the past, a world without religion was not imaginable.

Currently, many of these concepts have changed. In fact, in today's societies, the institutional and political order can be maintained without the need for religion; for example, secular state countries do not take any religion as a state religion (Giner, 1993). The truth is that religions have lost their essential place in societies, occupied in the preindustrial world. The secular process, the passage from a religious to a civil sphere, of modern society led one to think that religion's days were numbered. Clearly, this was not the case, but it has confronted religions with significant crises and created new religious movements (Velasco, 2006).

8.5 Key Concepts and Ideas in this Chapter

- We reviewed rules and exceptions in other branches of science related to biology and other sciences and disciplines less related to biology.
- Advances in knowledge and technology associated with medicine lead to changes in its rules over time. Exceptions that are now more common include diseases such as autism, medical conditions, and biomedical tools such as vaccines, gene therapy, and precision medicine. There have been changes in the concepts and approaches in medicine for obesity, smoking, and parenthood.
- Although less likely to occur, exceptions are also observed in the "hard sciences," such as physics, chemistry, and geology. Transitions in rules and exceptions also occur in social sciences, such as changes in conceptions about native and family cultures as simple examples.
- Artistic movements are also clear examples of the dynamism between rules and exceptions both at the level of individual artists and subjective conceptions of "ugly" and "beautiful" or worthy of being called art. We review artistic proposals that break with the current pattern or rule in certain historical moments. Changes in rules and exceptions in music, literacy, calligraphy, sports, and religion are linked to sociocultural evolution and technological advances.
- Everyday activities such as sports and religion may also show a clear dynamic between Rule and Exceptions that boost its evolution.

References

Alvarez, L. W., Alvarez, W., Asaro, F., & Michel, H. V. (1980). Extraterrestrial cause for the Cretaceous–Tertiary extinction. *Science, 208*(4448), 1095–1108.

Alzate-Yepes, T. (2019). *Perspectivas en Nutrición humana* (Vol. 21, p. 21).

Arias, J. L., Argudo, F. M., & Alonso, J. I. (2011). Las reglas como variables didácticas. Ejemplo en baloncesto de formación. *Revista Internacional de Medicina y Ciencias de la Actividad Física y del Deporte/International Journal of Medicine and Science of Physical Activity and Sport, 11*(43), 491–512.

Ashley, E. A. (2016). Towards precision medicine. *Nature Reviews Genetics, 17*(9), 507–522.

Barbour, J. (2020). *The Janus point: A new theory of time*. Random House.

Barrow, J. D., & Shaw, D. J. (2011a). The value of the cosmological constant. *International Journal of Modern Physics D, 20*(14), 2875–2880.

Barrow, J. D., & Shaw, D. J. (2011b). New solution of the cosmological constant problems. *Physical Review Letters, 106*(10), 101302.

Beier, J. C., Killeen, G. F., & Githure, J. I. (1999). Entomologic inoculation rates and plasmodium falciparum malaria prevalence in Africa. *The American Journal of Tropical Medicine and Hygiene, 61*(1), 109–113.

Beierkuhnlein, C. (2007). *Biogeographie—Die räumliche Organisation des Lebens in einer sich verändernden Welt* (p. 397). UTB Ulmer.

Bello, K. (2016). El arte de la caligrafía en el siglo XVIII. Aproximaciones a la historia social de la escritura en el mundo hispánico (España y Nueva España). *Trashumante. Revista Americana de Historia Social, 7*, 8–27.

Berman, J. J. (2014). *Rare diseases and orphan drugs: Keys to understanding and treating the common diseases*. Academic Press.

Bigsby, C. W. E. (1972). *Dada and surrealism*. Methuen.

Blüher, S., & Mantzoros, C. S. (2009). Leptin in humans: Lessons from translational research. *The American Journal of Clinical Nutrition, 89*(3), 991S–997S.

Bostrom, N., & Roache, R. (2008). Ethical issues in human enhancement. In *New waves in applied ethics* (pp. 120–152).

Bovey, A. (2002). *Monsters and grotesques in medieval manuscripts*. University of Toronto Press.

Bown, M. (2019). Toilets of our time: Where art meets defecation. *Times Literary Supplement, 6073–6074*, 18–20.

Bozal, V. (1996). *Historia de las ideas estéticas y de las teorías artísticas contemporáneas*. Visor, La balsa de la Medusa.

Brand, E. J., & Zhao, Z. (2017). Cannabis in Chinese medicine: Are some traditional indications referenced in ancient literature related to cannabinoids? *Frontiers in Pharmacology, 8*, 108.

Bray, G. A. (1992). Obesity: Historical development of scientific and cultural ideas. In P. Björntorp & B. N. Brodoff (Eds.), *Obesity* (pp. 281–293). JB Lippincott.

Brey, P. (2009). Human enhancement and personal identity. In J. K. B. Olsen, E. Selinger, & S. Riis (Eds.), *New waves in philosophy of technology* (pp. 169–185). Palgrave Macmillan.

Bryzgel, A. (2019). *The $120,000 banana: How to have your art and eat it*. The Conversation.. https://theconversation.com/the-120-000-banana-how-to-have-your-art-and-eat-it-128571.

Buchanan, A. V., Weiss, K. M., & Fullerton, S. M. (2006). Dissecting complex disease: The quest for the Philosopher's Stone? *International Journal of Epidemiology, 35*(3), 562–571.

Buckingham, M., & Coffman, C. (2012). *Primero, rompa todas las reglas*. Editorial Buena Semilla, Bogotá—Colombia.

Butrous, G., Maron, B., & Yacoub, M. (2020). The lamp of medicine of ancient Egypt is still burning. *Global Cardiology Science & Practice, 2020*(1), e202016.

Cabrera, F., Corona, A., Daners, G., Soto, M., Ubilla, M., Verde, M., et al. (2018). *Fundamentos de paleontología*. Manuales didácticos.

Cain, F. (2014). *Why is everything spherical?* [Video file]. Retrieved from https://www.youtube.com/watch?v=c1zoV_oXMvE

CDC. (2022a). Health topics.. Recuperado de https://www.cdc.gov/health-topics.html
CDC. (2022b). *Thalassemia*. Centro para el control y la prevención de enfermedades. https://www.cdc.gov/ncbddd/spanish/thalassemia/
Celik, N. (2021). From artwork to advertisement: Reflections from social media ads of art Basel. *Glimpse, 22*(2), 31–42.
Cemič, N. (2005). *Thermodynamics in mineral sciences*. Springer.
Charrow, J. (2004). Ashkenazi Jewish genetic disorders. *Familial Cancer, 3*(3), 201–206.
Cheng, A. (2021). Apophis 2029 planetary defense Mission options. *The Astrophysical Journal Supplement Series, 253*(4).
Chiarenza, A. A., Farnsworth, A., Mannion, P. D., Lunt, D. J., Valdes, P. J., Morgan, J. V., & Allison, P. A. (2020). Asteroid impact, not volcanism, caused the end cretaceous dinosaur extinction. *Proceedings of the National Academy of Sciences, 117*(29), 17084–17093.
Claude, M., Marshall, A., Van Stone, M., Xuriguera, G., & Jackson, D. (1996). *Calligraphy: From calligraphy to abstract painting*. Scirpus Publications.
Cohen, S. C., Finney, P. L., & Gibbard, P. L. (2013). *Última versión de la Tabla Cronoestratigráfica Internacional en español. Comisión Internacional de Estratigrafía (IUGS).*
Condamine, F. L., Guinot, G., Benton, M. J., & Currie, P. J. (2021). Dinosaur biodiversity declined well before the asteroid impact, influenced by ecological and environmental pressures. *Nature Communications, 12*(1), 1–16.
Cox, P. A. (1989). The elements: Their origin, abundance, and distribution (pp. 8+207). Oxford University Press.
Craig, J. (2008). Complex diseases: Research and applications. *Nature Education, 1*(1), 184.
Crino, P. B., Nathanson, K. L., & Henske, E. P. (2006). The tuberous sclerosis complex. *New England Journal of Medicine, 355*(13), 1345–1356.
Curatolo, P., Bombardieri, R., & Jozwiak, S. (2008). Tuberous sclerosis. *The Lancet, 372*(9639), 657–668.
Curtis, H., Schnek, A., Barnes, S., & Massarini, A. (2008). *Biología de Curtis*. Editorial Médica Panamericana (7ma Ed.).
DeHoff, R. T. (2005). *Thermodynamics in material sciences*. Wiley.
del Castillo, H. (2012). Desarrollo de competencias a través de los videojuegos deportivos: alfabetización digital e identidad. *Revista de Educación a Distancia (RED), 33*.
Del Moral, J. R. (1985). Los "poetas malditos", de lo subjetivo a lo correcto. *Anales de Filología Francesa, 1*, 49–68.
Devyatkin, A. V., Lvov, V. N., & Tsekmeister, S. D. (2022). Special groups of potentially hazardous asteroids. *Solar System Research, 56*, 62–65.
Díaz Patri, G. S. (1994). El latín y la cultura clásica en la Iglesia de hoy. *Stylos, 3*, 135–161.
Dobanovacki, D., Milovanovic, L., Slavkovic, A., Milanka, T., Miskovic, S., Skoric Jokic, S., & Pecanac, M. (2012). Surgery before common era (B.C.E.). *Archive of Oncology, 20*, 28–35.
Drummond, M. F., Wilson, D. A., Kanavos, P., Ubel, P., & Rovira, J. (2007). Assessing the economic challenges posed by orphan drugs. *International Journal of Technology Assessment in Health Care, 23*(1), 36–42.
Du, Q., Zhang, D., Zhuang, Y., Xia, Q., Wen, T., & Jia, H. (2021). The molecular genetics of Marfan syndrome. *International Journal of Medical Sciences, 18*(13), 2752.
Duarte-Gómez, M. B. (2003). Medicina occidental y otras alternativas: ¿es posible su complementariedad? Reflexiones conceptuales. *Cadernos de Saúde Pública, 19*, 635–643.
Eco, U. (2007). *Historia de la belleza*. Lumen.
Eco, U. (2015). *Historia de la fealdad*. Lumen.
Ellis, G. F., & Uzan, J. P. (2005). *C* is the speed of light, isn't it? *American Journal of Physics, 73*(3), 240–247.
Farrell, D. J., & Dunning-Davies, J. (2004). The constancy, or otherwise, of the speed of light. arXiv preprint physics/0406104.
Fatima, W., Shahid, A., Imran, M., Manzoor, J., Hasnain, S., Rana, S., & Mahmood, S. (2011). Leptin deficiency and leptin gene mutations in obese children from Pakistan. *International Journal of Pediatric Obesity, 6*(5–6), 419–427.

Ferreiro, E. (2006). Nuevas tecnologías y escritura. *Revista Docencia del Colegio de Profesores de Chile, 11*(30), 46–53.

Field, J. V. (1984). A Lutheran astrologer: Johannes Kepler. *Archive for History of Exact Sciences, 31*, 189–272.

Flores, A. M. (2001). *Doble teoría del genio: sujeto y creación de Kant a Schopenhauer*. Universidad de Sevilla.

Fombonne, E. (2001). Is there an epidemic of autism? *Pediatrics, 107*(2), 411–412.

Fombonne, E. (2003). *The prevalence of autism. Jama, 289*(1), 87–89.

Fombonne, E. (2009). Epidemiology of pervasive developmental disorders. *Pediatric Research, 65*(6), 591–598.

Franco, D. (2019). La configuración de espacios fractales en la música minimalista (Configuration of Fractal Spaces in Minimal Music). *Razón Crítica, 6*, 133–154.

Freedman, W. L. (2000). The Hubble constant and the expansion age of the universe. *Physics Reports, 333*, 13–31.

Freijo, E. A., Delgado, A. O., & Jiménez, Á. P. (2004). *Familia y desarrollo psicológico*. Pearson educación.

Galenson, D. W. (2021). *Revising the canon: How Andy Warhol became the Most important American modern artist*. University of Chicago, Becker Friedman Institute for Economics Working Paper.

Gálvez Pizarro, F. (2004). *Educación tipográfica. Una introducción a la tipografía*. Universidad Diego Portales.

Garbayo, E. J. F., & Fernández, J. (2012). El icono de los químicos: la tabla periódica de los elementos. *Anales de la Real Sociedad Española de Química, 4*, 314–321.

Garoz Puerta, I. (2005). El desarrollo de la conciencia de regla en los juegos y deportes. *Revista Internacional de Medicina y Ciencias de la Actividad física y el Deporte, 5*(19), 238–269.

Gee, J. P. (2004). *Lo que nos enseñan los videojuegos sobre el aprendizaje y el alfabetismo*. Aljibe.

Giner, S. (1993). Religión civil. *Reis, 61*, 23–55.

Giorgini, J. D., Benner, L. A. M., Ostro, S. J., Nolan, M. C., & Busch, M. W. (2005). (99942) Apophis. *International Astronomical Union Circular, 8593*, 1.

Goldman, E. (1995). *Believers: Spiritual leaders of the world*. Oxford University Press.

Gomes, C. (2021). Morally provocative art. Contemporary ethical-aesthetic discourse and its limits. *Everydayness, 3*, 222.

Gomez, M. R., Sampson, J. R., & Whittemore, V. H. (Eds.). (1999). *Tuberous sclerosis complex*. Oxford University Press.

Gravel, R. A., Triggs-Raine, B. L., & Mahuran, D. J. (1991). Biochemistry and genetics of Tay-Sachs disease. *Canadian journal of neurological sciences, 18*(S3), 419–423.

Grazier, K. R. (2016). Jupiter: Cosmic Jekyll and Hyde. *Astrobiology, 16*(1), 23–38.

Greer, J. P., Arber, D. A., Glader, B., et al. (2013). Wintrobe's clinical hematology. *ISBN*. 9781451172683.

Griggs, R. C., Batshaw, M., Dunkle, M., Gopal-Srivastava, R., Kaye, E., Krischer, J., et al. (2009). Clinical research for rare disease: Opportunities, challenges, and solutions. *Molecular Genetics and Metabolism, 96*(1), 20–26.

Grivetti, L. (2001). Psychology and cultural aspects of energy. *Nutrition Reviews, 59*(1), S5–S12.

Haendel, M., Vasilevsky, N., Unni, D., Bologa, C., Harris, N., Rehm, H., et al. (2020). How many rare diseases are there? *Nature Reviews Drug Discovery, 19*(2), 77–78.

Hagberg, B. (1985). Rett's syndrome: Prevalence and impact on progressive severe mental retardation in girls. *Acta Paediatrica, 74*(3), 405–408.

Halpern, P. (2004). The great beyond: Higher dimensions, parallel universes and the extraordinary search for a theory of everything. *The Physicist, 26*, 3.

Hanna, J., Hossain, G. S., & Kocerha, J. (2019). The potential for microRNA therapeutics and clinical research. *Frontiers in Genetics, 10*, 478.

Harding, E. (2020). WHO global progress report on tuberculosis elimination. *The Lancet Respiratory Medicine, 8*(1), 19.

Harvey, A., & Schucking, E. (2000). Einstein's mistake and the cosmological constant. *American Journal of Physics, 68*(8), 723–727.

Hauptman, W. (1985). Juries, protests, and counter-exhibitions before 1850. *The Art Bulletin, 67*(1), 95–109.

Haynes, M. P. (2018). The future of dwarf galaxy research: What telescopes will discover. *Proceedings of the International Astronomical Union, 14*(S344), 3–16.

Hill, K. P., Palastro, M. D., Johnson, B., & Ditre, J. W. (2017). *Cannabis* and pain: A clinical review. *Cannabis and Cannabinoid Research, 2*(1), 96–104.

Hirabayashi, M., Kim, Y., & Brozović, M. (2021). Finite element modeling to characterize the stress evolution in asteroid (99942) Apophis during the 2029 earth encounter. *Icarus, 365*, 114493.

Hodge, P. W. (1971). Dwarf galaxies. *Annual Review of Astronomy and Astrophysics, 9*(1), 35–66.

Hodson, M., Bush, A., & Geddes, D. (2012). *Cystic fibrosis*. CRC Press.

Hoffower, H. (2022). *5 reasons more millennials are choosing not to have children*. Insider.

Horner, J., & Jones, B. W. (2008). Jupiter: friend or foe? *Astronomy & Geophysics, 49*(1), 1–22.

Horner, J., & Jones, B. W. (2009). Jupiter–friend or foe? II: The centaurs. *International Journal of Astrobiology, 8*(2), 75–80.

Horner, J., & Jones, B. W. (2010). Jupiter: friend or foe? An answer. *Astronomy & Geophysics, 51*(6), 6–16.

Horner, J., Jones, B. W., & Chambers, J. (2010). Jupiter–friend or foe? III: The Oort cloud comets. *International Journal of Astrobiology, 9*(1), 1–10.

Hoyle, F., Burbidge, G., & Narlikar, J. V. (2000). *A different approach to cosmology: From a static universe through the big bang towards reality* (Vol. 53, p. 71). Cambridge University Press.

Hunter, D. J. (2005). Gene–environment interactions in human diseases. *Nature Reviews Genetics, 6*, 287–298.

Infobae. (2015). *Una mirada profunda sobre el autismo ¿Por qué está en aumento? Infobae*. Retrieved May 3, 2022, from https://www.infobae.com/2015/04/18/1722286-una-mirada-profunda-el-autismo-por-que-esta aumento/#:~text=Cada%20vez%20son%20más%20los,1%25%.20de%20la%20población%20infantil.

Isaksen, J., Diseth, T. H., Schjølberg, S., & Skjeldal, O. H. (2013). Autism spectrum disorders–are they really epidemic? *European Journal of Paediatric Neurology, 17*(4), 327–333.

Jagannathan, M. (2021). *Americans increasingly say they don't plan to have kids—This is the no. 1 reason why*. MarketWatch. https://www.marketwatch.com/story/americans-increasingly-say-they-dont-plan-to-have-kids-this-is-the-no-1-reason-why-11637620277.

Jennstål, J., & Öberg, P. O. (2019). The ethics of deliberative activism: In search of reasonableness and dialogic responsiveness in provocative art exhibitions. *Policy Studies, 40*, 648–661.

Juul, J. (2005). Games telling stories. In *Handbook of computer game studies* (pp. 219–226).

Kant, M. (1977). *Crítica del juicio*. Espasa Calpe.

Kayser, V., & Ramzan, I. (2021). Vaccines and vaccination: History and emerging issues. *Human Vaccines & Immunotherapeutics, 17*(12), 5255–5268.

Kazmi, S. S., Ali, W., Bibi, N., & Nouroz, F. (2020). A review on Zika virus outbreak, epidemiology, transmission and infection dynamics. *Journal of Biological Research-Thessaloniki, 27*, 1–1.

Kosorok, M. R., & Laber, E. B. (2019). Precision medicine. *Annual Review of Statistics and Its Application, 6*, 263–286.

Kragh, H. S. (2006). *Conceptions of cosmos: From myths to the accelerating universe: A history of cosmology*. OUP Oxford.

Kroos, M., Hoogeveen-Westerveld, M., van der Ploeg, A., & Reuser, A. J. (2012). The genotype–phenotype correlation in Pompe disease. *American Journal of Medical Genetics Part C: Seminars in Medical Genetics, 160*(1), 59–68.

Kumar, A., Chery, L., Biswas, C., Dubhashi, N., Dutta, P., Dua, V. K., et al. (2012). Malaria in South Asia: Prevalence and control. *Acta Tropica, 121*(3), 246–255.

Kunth, D., & Zarka, P. (2005). L'astrologie aujourd'hui, un phénomène de société. *Que Sais-Je?, 2481*, 5–12.

Kuropatnicki, A. K., Kłósek, M., & Kucharzewski, M. (2018). Honey as medicine: Historical perspectives. *Journal of Apicultural Research, 57*(1), 113–118.

Lande, A. (1962). The single convention on narcotic drugs, 1961. *International Organization, 16*(4), 776–797.

Leonard, J. (2022). *What are the best cannabis strains for chronic pain?* Medical News Today.. https://www.medicalnewstoday.com/articles/322051.

Leslie, N., & Bailey, L. (1993). Pompe Disease. In: GeneReviews®. University of Washington, Seattle, Seattle (WA). PMID: 20301438.

Lewis, G. N. (1916). The atom and the molecule. *Journal of the American Chemical Society, 38*(4), 762–785.

Little, B. (2019). When cigarette companies used doctors to push smoking before studies showed that cigarettes caused cancer, tobacco companies recruited the medical community for their ads. History Channel Online. https://www.history.com/news/cigarette-ads-doctors-smoking-endorsement.

Lomolino, M. V., Riddle, B. R., & Brown, J. H. (2006). *Biogeography*. Sinauer.

Maldonado, J. (2017). *Comportamiento, desarrollo y cambio organizacional.* Universidad Nacional Autónoma de Honduras.

Maltagliati, L. (2022). A DART to deflect an asteroid. *Nature Astronomy, 6*, 184.

Manrique, R. (1996). *Del deseo a la familia: La construcción de lo familiar.* Biblioteca digital de la Universidad Nacional de Colombia. Retrieved from http://bdigital.unal.edu.co/1363/21/20CAPI19.pdf.

Martí, P. (2016). *Cuando el tabaco era bueno para la salud.* La Vanguardia. https://www.lavanguardia.com/vida/20160210/302057569916/tabaco-bueno-salud.html.

Martin, R. G., & Livio, M. (2022). Asteroids and life: How special is the solar system? *The Astrophysical Journal Letters, 926*(2), L20.

Mateo, M. (1998). Dwarf galaxies of the local group. *Annual Review of Astronomy and Astrophysics, 36*(1), 435–506.

McMaster, C. (2021). *Rare diseases: Not so rare*. News Wise.. Retrieved from https://www.newswise.com/articles/rare-diseases-not-so-rare.

Mediavilla, C. (2005). *Caligrafía: del signo caligráfico a la pintura abstracta.* Campgráfic.

Meler, I. (2008). Las familias. *Subjetividad y procesos cognitivos, 12*, 158–188.

Mellies, U., & Lofaso, F. (2009). Pompe disease: A neuromuscular disease with respiratory muscle involvement. *Respiratory Medicine, 103*(4), 477–484.

Mirnezami, R., Nicholson, J., & Darzi, A. (2012). Preparing for precision medicine. *New England Journal of Medicine, 366*(6), 489–491.

Morais, R. H., Santos, L. F., Silva, A. R., & Melicio, R. (2021). Hypothetical Apophis deep ocean impact—Energy analysis. *Acta Astronautica, 188*, 438–450.

Muñoz, D. (2017). *La Estética de lo Feo en las obras de arte*. Recuperado de. https://dustinmunoz.net/2017/02/16/la-estetica-de-lo-feo-en-las-obras-de-arte/.

Murnen, S. K., & College, K. (2019). Gender differences in responses to thin, athletic, and hypermuscular idealized bodies. *Body Image, 30*, 1–9.

Myerowitz, R., & Costigan, F. C. (1988). The major defect in Ashkenazi Jews with Tay-Sachs disease is an insertion in the gene for the alpha-chain of beta-hexosaminidase. *Journal of Biological Chemistry, 263*(35), 18587–18589.

Myers, R. (2003). *The basics of chemistry*. Greenwood Press.

Nadeau, M. (1965). *The history of surrealism*. Macmillan.

Nargund, G. (2009). Declining birth rate in developed countries: A radical policy re-think is required. *Facts, Views & Vision in ObGyn, 1*(3), 191–193.

National Center for Advancing Translational Sciences. (2022). *Genetic and rare diseases information center*. US National Institutes of Health.. https://ncats.nih.gov/gard/.

Nellis, W. J. (2000). Metallization of fluid hydrogen at 140 GPa (1.4 mbar): Implications for Jupiter. *Planetary and Space Science, 48*(7–8), 671–677.

Nevill, A., Atkinson, G., & Hughes, M. (2008). Twenty-five years of sport performance research in the journal of sports sciences. *Journal of Sports Sciences, 26*(4), 413–426.

Nguengang Wakap, S., Lambert, D. M., Olry, A., Rodwell, C., Gueydan, C., Lanneau, V., Murphy, D., Le Cam, Y., & Rath, A. (2020). Estimating cumulative point prevalence of rare diseases: Analysis of the Orphanet database. *European Journal of Human Genetics, 28*(2), 165–173.

Okwo-Bele, J. M. (2015). *Together we can close the immunization gap*. World Health Organization. Retrieved from https://apps.who.int/mediacentre/commentaries/vaccine-preventable-diseases/en/index.html.

Orenstein, W. A., & Ahmed, R. (2017). Simply put: Vaccination saves lives. *Proceedings of the National Academy of Sciences, 114*(16), 4031.

Palisca, C., & Grout, D. (1996). *A history of Western music*. Norton.

Paz, O. (1990). *Marcel Duchamp, appearance stripped bare*. Arcade Publishing.

Peebles, P. J. E., & Ratra, B. (2003). The cosmological constant and dark energy. *Reviews of Modern Physics, 75*(2), 559.

Peltonen, L., & Uusitalo, A. (1997). Rare disease genes—Lessons and challenges. *Genome Research, 7*(8), 765–767.

Peters, V. F. D., Vis, M., García, Á. G., Wensink, H. H., & Tuinier, R. (2020). Defying the Gibbs phase rule: Evidence for an entropy-driven quintuple point in colloid-polymer mixtures. *Physical Review Letters, 125*(12), 127803.

Petrov, N., Sokolov, L., Polyakhova, E., & Oskina, K. (2018). Predictions of asteroid hazard to the earth for the 21st century. *AIP Conference Proceedings, 1959*(1), 040012.

Phrma.org. (2021). Rare diseases. A report on orphan medicines in the pipeline. Retrieved from https://phrma.org/-/media/Project/PhRMA/PhRMA-Org/PhRMA-Org/PDF/MID-Reports/MID-Rare-Diseases-2.pdf.

Pitjev, N. P., & Pitjeva, E. V. (2013a). Constraints on dark matter in the solar system. *Astronomy Letters, 39*(3), 141–149.

Pitjev, N. P., & Pitjeva, E. V. (2013b). Relativistic effects and dark matter in the solar system from observations of planets and spacecraft. *Monthly Notices of the Royal Astronomical Society, 432*(4), 3431–3437.

Pope, K. O., Baines, K. H., Ocampo, A. C., & Ivanov, B. A. (1997). Energy, volatile production, and climatic effects of the Chicxulub cretaceous/tertiary impact. *Journal of Geophysical Research: Planets, 102*(E9), 21645–21664.

Porcel Moscoso, T. (2018). *Historia y conocimientos culturales de la etnia indígena amazónica Arasaeri y su revaloración de su identidad cultural en la región de Madre de Dios*. Universidad Nacional Amazónica de Madre de Dios, Perú.

Prensky, M. (2002). The motivation of gameplay: The real twenty-first century learning revolution. *On the horizon, 10*(1), 5–11.

Press, F., & Siever, R. (2000). *Understanding earth*. Freeman and Company.

Preston-Thomas, H. (1990). The international temperature scale of 1990 ITS90. *Metrologia, 27*, 3–10.

Quiñones-Laveriano, D. M., & Catacora-Villasante, M. (2019). Cannabis Medicinal. *Boletín Institucional. Instituto Nacional de Salud, 25*(9–10), 114–118.

Ratra, B., & Vogeley, M. S. (2008). The beginning and evolution of the universe. *Publications of the Astronomical Society of the Pacific, 120*(865), 235.

Raudvere, C. (2020). *Medical theory and practice in ancient Egypt*. University of Copenhagen.

Ray, C. G., & Ryan, K. J. (2004). *Sherris medical microbiology: An introduction to infectious diseases*. McGraw-Hill.

Real Academia Española. (2022). Diccionario de la lengua española, 23.ª ed. Retrieved April 09, 2022, from https://dle.rae.es/dogma.

Reich, S. (2002). *Writings on music*. Oxford University Press.

Reid, J. (2016). *Rare disease research helps us understand medicine for all diseases*. On Biology BMC blog network. Retrieved from https://blogs.biomedcentral.com/on-biology/2016/02/26/rare-disease-research-helps-understand-medicine-diseases/.

Renzaho, A. M. (2004). Fat, rich and beautiful: Changing socio-cultural paradigms associated with obesity risk, nutritional status and refugee children from sub-Saharan Africa. *Health & Place, 10*(1), 105–113.

Richter, T., Nestler-Parr, S., Babela, R., Khan, Z. M., Tesoro, T., Molsen, E., & Hughes, D. A. (2015). Rare disease terminology and definitions—a systematic global review: Report of the ISPOR rare disease special interest group. *Value in Health, 18*(6), 906–914.

Rosenkranz, K. (1853). *Estética de lo feo*. Verlag der Gebruder Bomtrager.

Salvatore, M., Magrelli, A., & Taruscio, D. (2011). The role of microRNAs in the biology of rare diseases. *International Journal of Molecular Sciences, 12*(10), 6733–6742.

Sánchez, R. M. (1987). Análisis estructural y funcional. *Apunts. Educación física y deportes, 1*(07–08), 21–23.

Sandall, J., Tribe, R. M., Avery, L., Mola, G., Visser, G. H., Homer, C. S., et al. (2018). Short-term and long-term effects of caesarean section on the health of women and children. *The Lancet, 392*(10155), 1349–1357.

Santos, S. E. (2010). *La historia del sistema periódico*. Editorial UNED.

Sayed, N., Allawadhi, P., Khurana, A., Singh, V., Navik, U., Pasumarthi, S. K., et al. (2022). Gene therapy: Comprehensive overview and therapeutic applications. *Life Sciences, 120375*, 120375.

Schroeder, J. E. (1997). Andy Warhol: Consumer researcher. In D. MacInnis & M. Brucks (Eds.), *Advances in consumer research* (pp. 476–482). Association for Consumer Research.

Sears, F. W., Zemansky, M. W., Young, H. D., & Freedman, R. A. (2013). *Física Universitaria: Volumen 1*. Pearson Educación de México.

Sen, D. K. (1957). A static cosmological model. *Zeitschrift für Physik, 149*(3), 311–323.

Seoane, P. (2017). Cuando fumar no era perjudicial para la salud. . https://www.lanacion.com.ar/sociedad/cuando-fumar-no-era-perjudicial-para-la-salud-nid2076884/.

Serway, R. A., & Jewett, J. W. (2009). *Física para ciencias e ingeniería con física moderna*. Cengage Learning Editores.

Sleep, N. H., Zahnle, K. J., Kasting, J. F., & Morowitz, H. J. (1989). Annihilation of ecosystems by large asteroid impacts on the early earth. *Nature, 342*(6246), 139–142.

Slonimsky, N. (1965). *Lexicon of musical invective*. Coleman-Ross.

Smeets, E. E. J., Pelc, K., & Dan, B. (2012). Rett syndrome. *Molecular Syndromology, 2*(3-5), 113–127.

Sobal, J. (2001). Social and cultural influences on obesity. In *International textbook of obesity* (pp. 305–322). John Wiley & Sons.

Straumann, N. (2002). The history of the cosmological constant problem. *arXiv preprint* gr-qc/0208027.

Summers, D. (1990). *The judgment of sense: Renaissance naturalism and the rise of aesthetics*. Cambridge University Press.

Tapia, S. D. (1994). La alfabetización de la población urbana castellana en el Siglo de Oro. *Historia de la educación: revista*. interuniversitaria.

Tatarkiewicz, W. (1970). *History of aesthetics*. Mouton, The Hague.

Tegmark, M. (1998). Is "the theory of everything" merely the ultimate ensemble theory? *Annals of Physics, 270*(1), 1–51.

Tohline, J. E. (2002). The origin of binary stars. *Annual Review of Astronomy and Astrophysics, 40*(1), 349–385.

Tomkins, C. (1977). *The world of Marcel Duchamp, 1887–1968*. Time-Life Books.

Trigo-Rodríguez, J. M. (2022). *Asteroid impact risk: Impact Hazard from asteroids and comets*. Springer Nature.

Triviño, J. L. P. (2011). La filosofía del deporte: temas y debates. *Dilemata, 5*, 73–98.

Tucker, R. W., & Wang, C. (1998). Dark matter gravitational interactions. *Classical and Quantum Gravity, 15*(4), 933.

Turner, M. S. (2009). Origin of the universe. *Scientific American, 301*(3), 36–43.

Tyson, N. D. (2019). *Astrophysics for people in a hurry*. Norton & Company.

Ule, G. A. C. (2019). El electrón diferencial (o diferenciador) ¿Para qué sirve? *Anales de Química de la RSEQ, 115*(3), 241.

Velasco, J. M. (2006). *Introducción a la fenomenología de la religión*. Editorial Trotta.

Vera, J. Á., Bautista, G., Ramírez, M., & Yáñez, A. I. (2012). Relación entre anomia social, alienación y conducta antisocial en jóvenes infractores mexicanos. *Revista Latinoamericana de Ciencias Sociales, Niñez y Juventud, 10*(2), 943–955.

Verlaine, P. (1884). Les Poètes maudits. .

Waheed, F., Fisher, C., Awofeso, A., & Stanley, D. (2016). Carrier screening for beta-thalassemia in the Maldives: Perceptions of parents of affected children who did not take part in screening and its consequences. *Journal of Community Genetics, 7*(3), 243–253.

Ward, P. (1984). Is *nautilus* a living fossil? In N. Eldredge & S. M. Stanley (Eds.), *Living fossils* (pp. 247–256). Springer Verlag.

Ward, P. D., & Brownlee, D. (2000). Testing the rare earth hypothesis. In P. D. Ward & D. Brownlee (Eds.), *Rare earth: Why complex life is uncommon in the universe* (pp. 243–256). Springer.

Wark, K., & Richards, D. E. (2001). *Termodinámica*. McGraw-Hill.

Watt, W. M. (2003). *Islam and the integration of society*. Psychology Press.

Werth, A. J., & Shear, W. A. (2014). The evolutionary truth about living fossils. *American Scientist, 102*, 434–443.

WHO. (2019). Immunization. Retrieved August 09, 2023, from https://www.who.int/news-room/facts-inpictures/detail/immunization.

Yakob, L., & Walker, T. (2016). Zika virus outbreak in the Americas: The need for novel mosquito control methods. *The Lancet Global Health, 4*(3), e148–e149.

Yoshida, F., Ishimaru, R., Okudaira, O., Ishibashi, K., Hong, P. K., Matsui, T., & Kim, M. J. (2021). Photometric observations of the potentially hazardous asteroid (99942) Apophis from Kawabe Cosmic Park. *Publications of the Astronomical Society of Japan, 73*(4), L13–L17.

Zarka, P. (2009). Astronomy and astrology. *Proceedings of the International Astronomical Union, 5*(S260), 420–425.

Zeidan, J., Fombonne, E., Scorah, J., Ibrahim, A., Durkin, M. S., Saxena, S., et al. (2022). Global prevalence of autism: A systematic review update. *Autism Research, 15*(5), 778–790.

Zhou, X., & Yang, P. C. (2012). MicroRNA: A small molecule with a big biological impact. *MicroRNA (Shariqah, United Arab Emirates), 1*(1), 1–1.

Zhu, Q., Nguyen, D. T., Grishagin, I., Southall, N., Sid, E., & Pariser, A. (2020). An integrative knowledge graph for rare diseases, derived from the genetic and rare diseases information center (GARD). *Journal of Biomedical Semantics, 11*(1), 1–13.

Chapter 9
Overview and Final Prospects

Contents

Abstract Finally, chapter 9 is intended to be a closing section of this book, summarizing in some way what has been said in the previous chapters. In this sense, we have discussed the dynamics between rules and exceptions, taking into account different factors involved in such dynamics. Some changes are generated by nature itself, and others are promoted, directly or indirectly, by human action. In addition to the dynamics of change produced by intellectual development, changes in dominant scientific paradigms, discoveries, and technological advances. We show the potential applications and benefits of considering rules and exceptions at two main levels: first, from a theoretical-scientific level, more directed to the scientific community, where publications that highlight different types of rarities are valued and the dynamism between rules and exceptions is appreciated; and second, from a level related to improving man's relationship with nature.

A. V. Peretti et al., *Rules and Exceptions in Biology: from Fundamental Concepts to Applications*, https://doi.org/10.1007/978-3-031-55382-0_9

9.1 Perspectives on Potential Applications

We conclude this journey with a summary of what has been presented in the various chapters of this book. Throughout these pages, we have reviewed the theoretical aspects specific to the life sciences, their knowledge and theories, their extensions to the daily lives of people, and their connection to biology. When analyzing rules, patterns, norms, exceptions, oddities, and anomalies, our research not only covers purely general or "basic" aspects but also extends to areas of significant practical relevance. Many cases are directly or indirectly related to important applied fields, such as agricultural engineering, biomedicine, and the environment. For example, the impact of climate change on rule-exception dynamics, biodiversity conservation, and the introduction of exotic species are clear examples of this intersection.

When talking about "applications" in science, several visions emerge. For example, these applications can be more direct within the framework of the specific discipline or more indirect, with important theoretical connotations. The most direct application is the one that is mainly related to a benefit for man, such as those oriented to technologies, agriculture, livestock, industrial or service productivity, among many others. On the other hand, more indirect applications (from an anthropocentric perspective) come from work in the social, philosophical, and epistemological sciences. In other words, the term "application" goes beyond a narrow anthropocentric utilitarianism, with synonyms such as "employ, use, manage, utilize, allocate, exploit" that should be more flexible. Whatever the type of application, it is possible that the concepts we have seen about the dynamics between rules and exceptions underlie and are important in these different scientific aspects.

Throughout this book, we have explored and described the rule-to-exception and exception-to-rule dynamics. In doing so, we have considered various influencing variables that shape these dynamics. Some cases illustrate entirely natural changes, while others reveal the active role of humans, directly or indirectly, in driving these shifts throughout history. In addition, we have emphasized the role of intellectual development in generating shifts between rules and exceptions that can result from shifts in scientific paradigms, new discoveries, technological advances, and new ideas. A key message of this book is that current situations may change in the future: the static succumbs to change. To illustrate the potential applications and benefits of rules and exceptions, we will organize this summary into two main levels. It is important to note that these levels are not mutually exclusive and may not always have clear boundaries. First, we will address a theoretical perspective aimed at scientists and the scientific community itself. Second, we will explore a more practical aspect, focusing on improving the connection between humanity and nature in various domains.

9.2 On a more Theoretically Scientific Level

9.2.1 *Strengthening Publications on Rarities*

We have emphasized the importance of exceptions, aberrations, teratologies, and abnormalities in biology and other disciplines. These phenomena are of great importance in studying ecological and evolutionary implications from various perspectives. Often, aberrant events have been overlooked in scientific reporting, but they have proved to be valuable contributions to certain fields, such as plant physiology, as exemplified by the study of homeotic genes and the ABC model. In addition, exceptional cases, such as cephalopods, offer excellent potential for understanding the evolution of intelligence in different animal groups. Taxonomic curiosities, such as carnivorous sponges or carnivorous plants, are particularly valuable from a comparative standpoint, as they provide essential information for exploring potential phylogenetic hypotheses.

However, evaluating, understanding, and comparing these examples is often challenging, mainly due to the lack of publications or in-depth studies of these rare phenomena. The discussion of aspects of rules and exceptions in many groups requires more emphasis on the publication of rare cases. Without adequate representation of these rare cases or events in the literature, it becomes more difficult and limited to share data, opinions, or engage in discussions at any level. Several factors contribute to these omissions, including confirmation bias, underfunding of research on rare phenomena (e.g., rare diseases), difficulties in publishing results from non-model species, and reluctance to publish papers with rejected hypotheses. These multifactorial reasons underscore the need for a concerted effort to promote and include information about these rare cases or events in the scientific literature.

For example, in animal behavior studies, it is common to observe some individuals exhibiting behaviors that deviate from the general pattern observed in other individuals. Unfortunately, researchers often ignore these outliers when analyzing the data, with no apparent justification. It would be interesting, however, if these data, even if not included in the formal analyses, were presented in the study. This would illustrate the natural variability within the population and potentially serve as a basis for exploring new questions in the future. In morphological studies, as mentioned earlier, teratologies or aberrations can be incredibly valuable from a comparative standpoint, providing essential insight into potential phylogenetic and evolutionary hypotheses. In addition, the study of endemic or locally rare organisms, especially in underrepresented regions such as the Neotropics, can provide critical information to address important questions. Such research can also help confirm or refute paradigms extrapolated from primary research sites, such as the Nearctic region.

In this context, fostering a constructive habit to strengthen the dynamic relationship between rules and exceptions involves incorporating more data on rare cases in our publications. One effective approach could be to include supplementary material annexes containing information on these cases. While these data might not be directly relevant to the specific study, they can be immensely valuable for other researchers. Presently, there is a growing emphasis on uploading supplementary data or original spreadsheets from publications on the web (Whitlock, 2011). With the digital and online publishing tools available today, it becomes feasible to share and access this type of information in a manner similar to image databases in taxonomic collections or morphological studies across various groups. Researchers could easily compare "rare" data with other similar information found in these appendices.

By collecting and openly sharing rare data, authors can potentially shed light on connections between ideas, casual observations, potentially innovative strategies, or variations in adaptive and evo-devo contexts. As responsible authors, we should commit to providing this information, and journals should actively include such data in the open-access databases linked to publications. This practice can be applied to experimental, descriptive, and comparative studies and will benefit the scientific community as a whole.

9.2.2 Appreciating the Dynamism Between R and E

This perspective is of utmost importance as it enriches us as students of life, enabling us to advance more rapidly in scientific knowledge while avoiding dogmas, even smaller ones. Embracing this approach entails developing a heightened critical capacity to question concepts based on nonstatic paradigms, not only from an epistemological standpoint but also at a strictly biological level, encompassing evolution, ontogeny, and generational aspects.

The cultivation of critical thinking starts early, beginning in the family and elementary school, where we encourage the development of reasoning skills and objectivity, along with a healthy dose of skepticism toward unsupported claims. This mode of thinking involves asking questions, formulating hypotheses, considering alternative explanations, and assessing the credibility of information sources (Halpern, 1998; Solbes-Matarredona, 2013; Vieira et al., 2010).

Indeed, there exists a close link between scientific thinking and critical thinking. Given the ongoing social and cultural changes, it is imperative to nurture citizens with a critical and flexible mentality capable of questioning accepted norms if they lack sufficient evidence. Importantly, this does not imply blindly doubting or criticizing established knowledge without grounds; rather, it emphasizes the importance of not accepting something as "true" if the supporting evidence is lacking or if there is evidence to disprove it. A well-informed, evidence-based approach is what we strive to cultivate.

In this sense, science education extends far beyond teaching theories and laws or conducting experiments to test hypotheses. It requires building a society that recognizes and embraces the dynamic nature of scientific activity. Such a society should be capable of evaluating, with reason and argumentation, what is currently considered true or a standard practice (Eslava, 2014). Science is not static and timeless; rather, it is an ongoing process of knowledge construction that heavily relies on the context in which it operates.

As we previously mentioned, scientific knowledge evolves and undergoes gradual changes based on evidence and discussions. As scientists, we should not become disappointed or frustrated when exceptions challenge established rules. Instead, we must take on the challenge of refining or reformulating those rules and continually striving for improvement. It is essential to remember that scientific growth and development often arise from breaking down dogmas that were once thought insurmountable. Aspiring to challenge scientific laws in a constructive manner with responsibility and professionalism is vital for the advancement of science.

9.2.3 Avoiding Bias in Scientific Education

Indeed, a crucial aspect of science education is to adopt a more open and less dogmatic approach. Encouraging openness and creativity in teachers can lead to the development of critical thinking in students. This perspective should extend beyond undergraduate education to graduate education, including Ph.D. students and researchers in their daily scientific work. This new vision should also influence the content of textbooks, with an emphasis on the presentation and appreciation of exceptional cases. As we have emphasized, the study of these exceptional cases is essential to capture the full complexity of biological systems. Consider, for example, the negative impact of a dogmatic professor on a group of students. If a professor rigidly adheres to a single perspective that he takes to be the absolute truth and fails to present alternative viewpoints or variants, he hinders the students' ability to construct knowledge independently. It also limits their ability to develop critical thinking skills and to evaluate and choose among different explanations or hypotheses.

To promote a more effective and enriching learning experience, we should encourage teachers to adopt a more inclusive and open-minded approach to teaching. By fostering an environment that allows for exploration, questioning, and consideration of diverse viewpoints, we can empower students to develop their critical thinking skills, enhance their ability to think independently, and prepare them to become versatile researchers in the scientific community. In science, bias refers to the difference between what we think we value and what we actually value (Cuéllar Rodríguez, 2019; Rodríguez, 2020). Bias is prevalent in all areas of scientific research. At each stage of the scientific process, from setting objectives to selecting methodologies, study variables, and statistical analyses, researchers may introduce bias. To ensure the integrity of research findings, it is essential to recognize and identify these biases to minimize and correct their effects (Manterola & Otzen, 2015; Rodríguez, 2020).

While objectivity is the ideal state in science, where personal goals and motivations should not influence research activities, it is difficult to completely avoid epistemic biases (Biddle & Kukla, 2017; Cippitani et al., 2021; Fraedrich, 2001; Reiss & Sprenger, 2014). The key is to acknowledge the existence of biases and work to minimize them. One effective way to address bias is to discuss it with other researchers and subject the work to peer review (Cippitani et al., 2021). Another type of bias in science relates to the publication of the scientific results obtained by researchers. Here, we can distinguish two important biases.

The first bias, publication bias, refers to the selective publication of research results, where studies with positive and statistically significant results are more likely to be published, while nonsignificant or negative results may be overlooked. This can result in an incomplete and biased representation of the true evidence in a given area of research. The root of this bias may lie in editorial decisions or societal pressure to publish "exciting" results, leading to underreporting of less favorable findings (Rodríguez, 2020). The second bias relates to scientists' adherence to established explanations or data from studies in model organisms. This tendency can inhibit the development of novel ideas or discoveries that challenge current rules or paradigms. According to this idea, many studies do not come to light because "*it did not work the way it was supposed to.*" Being more critical and constantly challenging the norms is essential for scientific progress and growth. It encourages researchers to explore alternative hypotheses and innovative approaches, potentially leading to breakthroughs and a deeper understanding of biological systems.

As scientists, it is crucial to recognize and actively address these biases. By promoting transparency, open-mindedness, and rigorous evaluation of all research results, we can improve the reliability and validity of scientific findings. Fostering a culture that embraces diverse perspectives and welcomes constructive criticism is fundamental to advancing science and achieving its ultimate goal of expanding knowledge and improving our understanding of the world.

9.3 On a Human–Nature Relationship Level

9.3.1 The Vindication of Singularity and Otherness

As we have seen, in systems theory and holism, there is a singularity where small changes can have significant effects due to the interconnectedness within the integrated "whole." Furthermore, exceptional cases can be linked to the concept of uniqueness, which refers to the intrinsic value of each living being and is closely related to the idea of "identity." At the social level, the notion of individuality plays a crucial role in shaping intra and interspecific relationships, social codification, belonging, and the hierarchization of human groups. Rose (2001) underscores the value of individuality by emphasizing that each person has his or her own life story. In addition, identifying someone as "other" (not belonging to our group) has significant implications for understanding social identity. The concept of "otherness" emerges from a philosophical, psychological, cognitive, and social process through which a group defines and distinguishes itself from other groups (Türkkan, 2010).

Gradually, concepts have emerged that emphasize the value of diversity and otherness by recognizing and embracing differences and ensuring their equitable representation in political and social agendas. It is crucial to emphasize that the emphasis on what makes individuals different and exceptional should never lead to discriminatory practices or associate negative connotations with those who are different (whether those differences are real or perceived). Such negative associations can potentially stigmatize members of other groups.

This emphasis is particularly important at a time when issues such as racial, religious, and gender discrimination, the marginalization of minorities and school bullying are among the most critical social problems in the world. Not only are they detrimental to society, but they also contribute significantly to mental health problems and suicidal behavior (Ahuja et al., 2015; Larraín, 2019). As a society, it is crucial to increase social awareness of the uniqueness of each individual to tackle discrimination. Achieving this goal can be challenging, especially in organizations where negative biases are deeply ingrained from an early age. An effective approach involves collaboration between interdisciplinary groups of scientists and educators to ensure that topics or examples with negative societal connotations are presented without bias from the beginning of education.

For instance, when teaching children about organisms often labeled "bad" in biology, it is essential to provide a comprehensive understanding of their biology, the precautions necessary when dealing with them, the benefits they offer, and how to care for them. By doing so, people would be less likely to develop fear and aversion toward creatures such as cockroaches, arachnids, snakes, or bats (among others) if they are introduced to their value and how to coexist with them from an early age.

To reach a wider audience and influence family perspectives, additional resources such as television, the Internet, and social networks can be used. These platforms can help dispel misconceptions and fears about various topics. In addition, addressing not only biological but also social and cultural issues in a professional manner can foster constructive discussions. However, we must remain vigilant about the biases mentioned in the previous section to ensure that the information presented is objective, accurate, and unbiased. This perspective, similar to the consideration of humans, is prominent in certain areas of research within animal behavior and biodiversity conservation practices. One such area of focus in animal behavior research is the study of animal personality and the incorporation of individuality into decision-making rules. This highlights the importance of recognizing the uniqueness of each organism as a critical variable (Carter et al., 2013; Schuett et al., 2010; Stamps & Groothuis, 2010a, 2010b).

Science can benefit from exploring what rare and exceptional species can offer to address current challenges. A notable example is the discovery of rare or relict species that can provide valuable benefits. Historically, such species have been instrumental in the development of biomedical compounds to combat diseases, including cancer. However, the focus should not be solely on finding species that offer cures for specific diseases. As we have seen, these new species have also shown promise in addressing environmental issues, such as the treatment of wastewater and pollutants. It is equally important to recognize and value the importance of common species that often go unnoticed because of their familiarity while providing critical ecosystem services. A pressing concern in this regard is the pollinator crisis, particularly the alarming decline in bee populations. Tackling this crisis and protecting pollinators are essential steps in preventing potential large-scale extinctions.

In contrast, related to uniqueness but extending to other organisms, we encounter the concept of "value" of animal and plant species that goes beyond purely economic considerations. It is crucial to recognize and appreciate the values inherent in species, which extend to various aspects, such as aesthetics, culture, social significance, recreation, and beliefs (Chan et al., 2012; Díaz et al., 2018). Species have intrinsic or inherent value independent of their usefulness to humans, often referred to as "existence value" (Vucetich et al., 2015). In addition, species may possess "nonuse value" simply because they contribute to the continuity of nature, allowing future generations to benefit from them (Pearson, 2016). In essence, species often have an intangible value that may or may not be linked to their utilitarian benefits. Acknowledging and appreciating these diverse values can lead to more comprehensive and inclusive conservation practices (Pearson, 2016).

Such practices recognize the complex relationships between humans and biodiversity and embrace different perspectives on valuing nature. In conclusion, each species is a unique product of thousands or millions of years of evolution, and each plays an important role in enabling ecosystems to function properly (Costanza et al., 1997; Kenter et al., 2015). In this way, we also become more aware of the value of each species, even rare ones, in ecosystems and native communities. Their disappearance will have serious short- or long-term consequences in the imbalance of biological and, ultimately, ecological interactions. We must make responsible and strategic use of umbrella species to design conservation strategies that benefit many groups (Roberge & Angelstam, 2004). On the other hand, in recent decades, there has been a greater social awareness, beyond solid environmental groups, fighting against the unscrupulous actions of humans that are detrimental to the environment, with consequences that ultimately lead to a decrease in biodiversity.

9.3.2 Direct Applications of Exceptions and R<->E Dynamism

In addition to their contributions to theoretical, phylogenetic, and evolutionary knowledge in biology, exceptions in nature can be of immense benefit to humans at various levels. Specific teratologies (abnormalities) observed in fungi, pollen, or embryos of different species and exceptional phenomena such as bioluminescence in fungi can be valuable bioindicators of environmental pollution. Moreover, the exceptional characteristics of certain organisms can inspire new industrial applications. For example, the remarkable abilities of tardigrades, such as anhydrobiosis (the ability to survive extreme desiccation) and radiotolerance, hold promise for advances in radiation protection, cryopreservation of biological material, and other industrial applications. Furthermore, some extraordinary organisms have the potential to make significant contributions to medicine and biomedicine. For example, amoeboid fungi have provided valuable insights into the study of human cancer and other diseases, aiding in drug mode of action studies and toxicological assessments. The "rare biosphere" of microbial communities can have a disproportionate impact on ecosystem functions, including biogeochemical cycles. Certain microbes possess essential enzymes relevant to industrial processes, such as thermophilic cellulases used in biofuel applications. Conversely, the removal of rare species from soil ecosystems can lead to an increase in undesirable pathogenic species, underscoring the critical role of rare species in biotechnological applications such as crop protection.

In various contexts, humans have the ability to modify or manipulate other species for various purposes, such as increasing productivity, disease resistance, docility, or purely aesthetic reasons. Today, even genetic engineering techniques are used to create transgenic organisms. While artificial selection can have positive results in terms of improving food productivity, there are instances where it leads to the creation of teratologies for human benefit.

From our perspective, what humans find beautiful can sometimes promote traits that natural selection would otherwise have eliminated, such as those associated with difficulties in feeding, locomotion, or reproduction. Artificial selection can also result in reduced genetic diversity within modified species, making them more susceptible to disease and morphological problems, leading to shorter life spans compared to their wild ancestors. Therefore, human intervention in the creation of genetically modified individuals should be undertaken with greater control and caution. While there are potential benefits to be gained, it is essential to minimize the adverse effects on the species and organisms created. A balance must be struck between achieving human goals and ensuring the long-term well-being of these modified organisms.

The interplay between rules and exceptions can extend beyond the strictly biological realm and have significant applied importance in various fields. For example, certain rare diseases often serve as gateways to understanding poorly understood biological and physiological pathways, proving valuable for the study and prevention of more common diseases. In addition, some medical phenomena previously perceived as rare, such as autism, have gained prominence on research and funding agendas due to increasing prevalence. Advances in technology and changes in symptomatology or diagnostic perceptions have led to the abandonment of certain concepts and practices in medicine. On the other hand, preventive medicine has witnessed the development of numerous vaccines to eradicate diseases, from a rarity a few decades ago to the norm today, with the vast majority of individuals receiving vaccinations. It is now rare to find individuals who have never been vaccinated. Of course, the acceptance of vaccines may vary depending on the sociocultural context. An alarming concern, however, is the increasing resistance of bacteria to antibiotics. The emergence of mutant pathogens can potentially cause some diseases that were once exceptions, to become the rule, especially if care is not taken in the use of these products.

The field of biotechnology is pushing the boundaries of what is possible in terms of modifying DNA configurations and the embryology of hybrid organisms.

Cross-species treatments, including those aimed at curing diseases, creating organs, and enhancing resistance, are becoming increasingly feasible. Evolutionary concepts serve as the foundation for advances in synthetic biology, robotics, and bioengineering. Synthetic biology continues to be the subject of intense global debate. While it promises many benefits, such as the creation of artificial cells, molecular-based tissue repair, cell reprogramming, and organ bioprinting to address organ shortages, it also raises concerns about potential drawbacks and biases. Despite these concerns, its application in medicine has the potential to yield significant positive outcomes. As technology advances, it is becoming increasingly rare to find individuals who are completely "natural," with no grafts, surgeries, prostheses, or foreign bodies inside them. Technology is advancing rapidly, allowing for procedures and manipulations that were unthinkable just a few years ago.

9.3.3 *Modifications and Environmental Risks of R<->E Dynamism*

Indeed, as we have observed, rarities can have multiple impacts, both positive and negative, on the environment and human activities. Some rare events may lead to direct applications in industry or inspire new scientific breakthroughs, while others may have adverse effects. One discussed example of a negative impact is the "witches' broom" case, which is causing significant economic damage to cocoa crops in Brazil and other regions of Latin America. However, such damaging cases also serve as catalysts for numerous research efforts aimed at mitigating their potential economic consequences.

On the other hand, rare species in specific ecosystems may be highly vulnerable compared to more widespread or abundant species. Despite their rarity, they may have significant value and play a critical role in providing various ecosystem services. For example, certain soil microorganisms act as nitrogen fixers, rare species stabilize food webs, keystone species maintain ecosystem structure, and predatory species have a disproportionate influence on community structure. The reckless exploitation of species through activities such as deforestation, hunting, fishing, and the degradation of natural environments has resulted in the transformation of once abundant species into exceptions, or worse, their complete disappearance from various regions of the planet. The primary cause of this mismanagement of natural resources and the transformation of species from common to exceptional is the lack of effective management policies and the reliance on government officials who may not be adequately qualified to make informed decisions in this regard.

Implementing appropriate measures to address these uses is critical to slowing the disappearance of species already threatened with extinction and possibly recovering some populations. While these measures may only be partially effective in the face of irreversible extinction, they can help reverse the damage to species that have experienced significant declines in abundance. As a result, these currently exceptional species may have a chance to avoid total extinction. With all that has been discussed about human activity on the dynamics of rule changes and exceptions, we need to become more aware of the implications of human action in modifying species through artificial selection and synthetic biology. Both to see its pros and cons at the economic level and to reintroduce nearly or completely extinct species.

Similarly, we must take better care to control the introduction of exotic organisms, whether for fishing, agriculture, forestry, poaching, or pleasure. Their impact on native organisms is severe, with cascading effects that are rarely predictable. Exotic organisms can cause the extinction or severe reduction in abundance of several native plant and animal species, in addition to other effects such as hybridization, parasitism, competition for resources, altered behavior, and reproduction. These negative effects spread, changing the structure and functioning of the environment and altering the population dynamics of certain regions. Thus, the introduction of exotic species is one of the greatest threats to biodiversity in the world. The destruction of forests and ecosystems also has cascading effects, often leading to catastrophes that were once exceptional but are now the norm in many parts of the world. Often, it is not measured, but the economic gain of any activity that favors the destruction of ecosystems is less than the material and human loss of the catastrophes caused directly or indirectly.

9.3.4 Applications of R<->E Dynamism on a more Holistic Level

Linked to the conceptual frameworks of holism and singularity, we have discussed the Gaia hypothesis (Lovelock & Margulis, 1974), in which everything is intimately connected despite distances, and the Earth functions as a unit with living beings interacting synergistically with their inorganic environment in a complex self-regulating system. In these cases, we recover the notion of the importance or value of each species. This uniqueness is linked to the exceptional characteristics of each living being or species, which can be translated into particularities in time and space and critical components of the biosphere. Lovelock was struck by the radical differences between Earth and the nearest planets when NASA invited him to the first attempt to discover the existence of life on Mars; this uniqueness of Earth's conditions led him to formulate this hypothesis (Lovelock & Margulis, 1974). The "unique" conditions of the Earth, with a self-regulating biosphere that makes its physical environment hospitable (complete homeostasis), allow life on our planet.

The Gaia hypothesis states that the initial conditions that made life possible have been modified by life itself. Therefore, the resulting conditions are the consequence and responsibility of the life that inhabits them. This hypothesis emphasizes the Earth's ability to "recover" to a state of equilibrium after events such as mass extinctions or global climatic changes. It can help predict the history and evolution of biota in the face of future scenarios, taking into account phenomena such as climate change. The question of whether life on Earth is an exception or whether life as a phenomenon can occur elsewhere in the universe is a fascinating and ongoing debate. Some scientists believe that life may be possible on other planets or galaxies, although it may not necessarily resemble life as we know it on Earth, a concept consistent with the mediocrity principle extended from the Copernican principle. Consequently, the search for life beyond Earth is one of the most important and challenging scientific endeavors of our time (Des Marais, 2000; Des Marais et al., 2008; Kiang et al., 2018; Race & Randolph, 2002).

In the search for potential life on other celestial bodies, such as planets, moons, or exoplanets, scientists look for evidence of current or past life through the detection of biosignatures (Kiang et al., 2018). For example, research has suggested that microbial life could exist in the subsurface of Mars, the atmosphere of Venus, and the oceans of some moons of giant planets (Schulze-Makuch et al., 2005). To help assess the evidence for life beyond Earth, a "Confidence of Life Detection" (CoLD) scale has been proposed (Green et al., 2021). Searching for planets that closely resemble Earth's characteristics and may be more conducive to life is undoubtedly a challenging task, despite the discovery of over 300 million "potentially habitable" planets in our galaxy. Despite these discoveries, no terrestrial planets with useful photon flux, energy, and energy efficiency comparable to Earth have been observed (Covone et al., 2021).

The "rare earth hypothesis" proposes that multicellular life forms such as those found on Earth are exceptional in the universe. This suggests that a large number of specific requirements, a large number of coincidences, and improbable astrophysical and geological circumstances are necessary for complex life to emerge. This hypothesis attempts to explain the "Fermi paradox," which asks why, given the potential for life to expand and occupy different niches, there is no evidence of intelligent extraterrestrial life on other planets or any indication of their visits to our own. The term "rare earth" comes from the book "Rare Earth: Why Complex Life Is Uncommon in the Universe" (2000) by Peter Ward, a geologist and paleontologist, and Donald E. Brownlee, an astronomer and astrobiologist. In their work, they argue that planets, planetary systems, and galactic regions capable of supporting complex life, such as Earth, the solar system, and our galaxy, are exceptionally rare.

Concluding Words In this book, we have presented a basic theoretical framework and provided numerous examples to stimulate reflection on the evolving concepts and facts that underlie our understanding of rules and exceptions. It is important to recognize that there are several viable alternatives and complementary approaches to the issues discussed here. While our current framework may be perceived as relatively simple, focusing on only two components (rules and exceptions) and direct associations along with the sources of dynamics (nature, humanity, epistemology, and technology), we welcome new hypotheses and perspectives to enrich and extend this theoretical foundation, making it more complex and comprehensive.

The three-context approach (time, space, and group) that we have used, while having its limitations, provides a necessary starting point for our discussion of rules and exceptions. It serves as a practical basis for future research, where a wider range of variables and a richer multidimensional analysis can be considered. In this regard, mathematical studies involving models that explore the prediction of exceptions based on natural and anthropic factors will prove invaluable in advancing our understanding of the dynamics between rules and exceptions. We leave a roadmap for further exploration in this area, trusting that it will inspire scientific research that combines epistemological and empirical aspects, drawing from the limitless examples provided by the magnificent nature that surrounds us.

References

Ahuja, A., Webster, C., Gibson, N., Brewer, A., Toledo, S., & Russell, S. (2015). Bullying and suicide: The mental health crisis of LGBTQ youth and how you can help. *Journal of Gay & Lesbian Mental Health, 19*(2), 125–144.

Biddle, J. B., & Kukla, R. (2017). The geography of epistemic risk. In *Exploring inductive risk: Case studies of values in science* (pp. 215–237).

Carter, A. J., Feeney, W. E., Marshall, H. H., Cowlishaw, G., & Heinsohn, R. (2013). Animal personality: What are behavioural ecologists measuring? *Biological Reviews, 88*(2), 465–475.

Chan, K. M., Satterfield, T., & Goldstein, J. (2012). Rethinking ecosystem services to better address and navigate cultural values. *Ecological Economics, 74*, 8–18.

Cippitani, R., Mirabile, A., & Onofri, M. (2021). "Objetividad científica" y sesgos en la toma de decisiones jurídicas: los casos de genética forense y de algoritmos. *Revista Justicia & Derecho, 4*(2), 1–22.

Costanza, R., d'Arge, R., De Groot, R., Farber, S., Grasso, M., Hannon, B., et al. (1997). The value of the world's ecosystem services and natural capital. *Nature, 387*(6630), 253–260.

Covone, G., Ienco, R. M., Cacciapuoti, L., & Inno, L. (2021). Efficiency of the oxygenic photosynthesis on earth-like planets in the habitable zone. *Monthly Notices of the Royal Astronomical Society, 505*(3), 3329–3335.

Cuéllar Rodríguez, S. (2019). 12 de noviembre. In *Sesgos en investigación clínica y edición científica [Conferencia]*. Real Academia Nacional de. Retrieved from https://www.ranf.tv/index.php/video/592/sesgos-en-investigacion-clinica-y-edicion-cientifica-12-de-noviembre-de-2019.

Des Marais, D. J. (2000). When did photosynthesis emerge on earth? *Science, 289*(5485), 1703–1705.

Des Marais, D. J., Nuth, J. A., III, Allamandola, L. J., Boss, A. P., Farmer, J. D., Hoehler, T. M., et al. (2008). The NASA astrobiology roadmap. *Astrobiology, 8*(4), 715–730.

Díaz, S., Pascual, U., Stenseke, M., Martín-López, B., Watson, R. T., Molnár, Z., et al. (2018). Assessing nature's contributions to people. *Science, 359*(6373), 270–272.

Eslava, E. (2014). Conceptos, pertinencia y acccceso: una mirada filosófica a la educación en ciencias naturales. *Revista Colombiana de Filosofía de la Ciencia, 14*(28), 67–82.

Fraedrich, D. (2001). Revival of objectivity in scientific method. *The Journal of Ayn Rand Studies, 3*(1), 29–46.

Green, J., Hoehler, T., Neveu, M., Domagal-Goldman, S., Scalice, D., & Voytek, M. (2021). Call for a framework for reporting evidence for life beyond earth. *Nature, 598*(7882), 575–579.

Kiang, N. Y., Domagal-Goldman, S., Parenteau, M. N., Catling, D. C., Fujii, Y., Meadows, V. S., et al. (2018). Exoplanet biosignatures: At the dawn of a new era of planetary observations. *Astrobiology, 18*(6), 619–629.

Halpern, D. (1998). Teaching critical thinking for transfer across domains. *American Psychologist, 53*(4), 449–455.

Kenter, J. O., O'Brien, L., Hockley, N., Ravenscroft, N., Fazey, I., Irvine, K. N., et al. (2015). What are shared and social values of ecosystems? *Ecological Economics, 111*, 86–99.

Larraín, J. (2019). Otredad, identidad e inmigración. *IARPP CHILE, 1*(1), 34.

Lovelock, J. E., & Margulis, L. (1974). Atmospheric homeostasis by and for the biosphere: The Gaia hypothesis. *Tellus, 26*(1–2), 2–10.
Manterola, C., & Otzen, T. (2015). Los sesgos en investigación clínica. *International Journal of Morphology, 33*(3), 1156–1164.
Pearson, R. G. (2016). Reasons to conserve nature. *Trends in Ecology & Evolution, 31*(5), 366–371.
Race, M. S., & Randolph, R. O. (2002). The need for operating guidelines and a decision-making framework applicable to the discovery of non-intelligent extraterrestrial life. *Advances in Space Research, 30*(6), 1583–1591.
Reiss, J., & Sprenger, J. (2014). Scientific objectivity. In E. Zalta (Ed.), *Stanford encyclopedia of philosophy*. Retrieved from http://plato.stanford.edu/entries/scientific-objectivity/.
Roberge, J. M., & Angelstam, P. E. R. (2004). Usefulness of the umbrella species concept as a conservation tool. *Conservation Biology, 18*(1), 76–85.
Rodríguez, S. C. (2020). Sesgos de publicación y edición científica: la metaciencia al rescate. *Panorama actual del medicamento, 44*(430), 37–47.
Rose, S. (2001). *Trayectorias de vida: biología, libertad, determinismo*. Ediciones Granica SA.
Schuett, W., Tregenza, T., & Dall, S. R. (2010). Sexual selection and animal personality. *Biological Reviews, 85*, 217–224.
Schulze-Makuch, D., Dohm, J. M., Fairén, A. G., Baker, V. R., Fink, W., & Strom, R. G. (2005). Venus, Mars, and the ices on Mercury and the Moon: Astrobiological implications and proposed mission designs. *Astrobiology*, *5*(6), 778–795.
Solbes-Matarredona, J. (2013). Contribución de las cuestiones sociocientíficas al desarrollo del pensamiento crítico (II): Ejemplos. *Revista Eureka sobre Enseñanza y Divulgación de las Ciencias, 10*, 171–181.
Stamps, J., & Groothuis, T. G. (2010a). The development of animal personality: Relevance, concepts and perspectives. *Biological Reviews, 85*(2), 301–325.
Stamps, J. A., & Groothuis, T. G. (2010b). Developmental perspectives on personality: Implications for ecological and evolutionary studies of individual differences. *Philosophical Transactions of the Royal Society B: Biological Sciences, 365*(1560), 4029–4041.
Türkkan, S. (2010). Other/alterity. The encyclopedia of literary and cultural theory.
Vieira, M. R., Tenreiro-Vieira, C., & Martins, E. (2010). Pensamiento crítico y literacia científica. *Alambique, 65*, 96–104.
Vucetich, J. A., Bruskotter, J. T., & Nelson, M. P. (2015). Evaluating whether nature's intrinsic value is an axiom of or anathema to conservation. *Conservation Biology, 29*(2), 321–332.
Whitlock, M. C. (2011). Data archiving in ecology and evolution: Best practices. *Trends in Ecology & Evolution, 26*(2), 61–65.

Index

A. V. Peretti et al., *Rules and Exceptions in Biology: from Fundamental Concepts to Applications*, https://doi.org/10.1007/978-3-031-55382-0

U

V

W

X

Y

Z

GPSR Compliance

The European Union's (EU) General Product Safety Regulation (GPSR) is a set of rules that requires consumer products to be safe and our obligations to ensure this.

If you have any concerns about our products, you can contact us on ProductSafety@springernature.com

In case Publisher is established outside the EU, the EU authorized representative is:

Springer Nature Customer Service Center GmbH
Europaplatz 3
69115 Heidelberg, Germany

Batch number: 10682282

Printed by Printforce, the Netherlands